Advances in Computational Methods in Sciences and Engineering 2005

Lecture Series on Computer and Computational Sciences
Editor-in-Chief and Founder: Theodore E. Simos

Volume 4B

Advances in Computational Methods in Sciences and Engineering 2005

Selected Papers from the International Conference of Computational Methods in Sciences and Engineering 2005 **(ICCMSE 2005)**

Recognised Conference by the European Society of Computational Methods in Sciences and Engineering (ESCMSE)

Editors:

Theodore Simos and George Maroulis

forms of interaction was central to the ICCSME 2003 held at Kastoria, in the north of Greece and to the ICCMSE 2004 held at Hotel Armonia, Athens, Greece.

In addition to the general programme the Conference offers an impressive number of Symposia. The purpose of this move is to define more sharply new directions of expansion and progress for Computational Science & Engineering.

We note that for ICCMSE there is a co-sponsorship by American Chemical Society. ICCMSE also is endorsed by American Physical Society.

More than 800 extended abstracts have been submitted for consideration for presentation in ICCMSE 2005. From these extended abstracts we have selected 500 extended abstracts after international peer review by at least two independent reviewers. These accepted papers will be presented at ICCMSE 2005.

After ICCMSE 2005 the participants can send their full papers for consideration for publication in one of the journals that have accepted to publish selected Proceedings of ICCMSE 2005. We would like to thank the Editors-in-Chief and the Publishers of these journals. The full papers will be considered for publication based on the international peer review by at least two independent reviewers.

We would like also to thank:

- The Scientific Committee of ICCMSE 2005 (see in page iv for the Conference Details) for their help and their important support. We must note here that it is a great honour for us that leaders on Computational Sciences and Engineering have accepted to participate the Scientific Committee of ICCMSE 2005.

- The Symposiums' Organisers for their excellent editorial work and their efforts for the success of ICCMSE 2005.

- The invited speakers for their acceptance to give keynote lectures on Computational Sciences and Engineering.

- The Organising Committee for their help and activities for the success of ICCMSE 2005.

- Special thanks for the Secretary of ICCMSE 2005, Mrs Eleni Ralli-Simou (which is also the Administrative Secretary of the European Society of Computational Methods in Sciences and Engineering (ESCMSE)) for her excellent job.

- Special thanks for the student of Professor Simos, Mr. Zacharias Anastassi for his excellent activities for typesetting this document.

Prof. Theodore Simos
President of ESCMSE
Department of Computer Science
and Technology
University of the Peloponnese
Tripolis
Greece

Prof. George Maroulis
Department of Chemistry
University of Patras
Patras
Greece

August 2005

Brill Academic Publishers
P.O. Box 9000, 2300 PA Leiden
The Netherlands

Lecture Series on Computer
and Computational Sciences
Volume 4, 2005, pp. iv-v

Conference Details

International Conference of Computational Methods in Sciences and Engineering 2005 (ICCMSE 2005), Resort Hotel Poseidin, Loutraki, Corinth, Greece, 21-26 October, 2005.

Recognised Conference by the European Society of Computational Methods in Sciences and Engineering (ESCMSE)

Chairman and Organiser

Professor T.E. Simos, President of the European Society of Computational. Methods in Sciences and Engineering (ESCMSE). Active Member of the European Academy of Sciences and Arts, Corresponding Member of the European Academy of Sciences and Corresponding Member of European Academy of Arts, Sciences and Humanities, Department of Computer Science and Technology, Faculty of Sciences and Technology, University of Peloponnese, GR-221 00 Tripolis, Greece.

Co-Chairman

Professor George Maroulis, Department of Chemistry, University of Patras, GR-26500 Patras, Greece.

Scientific Committee

Dr. Hamid R. Arabnia, USA
Dr. B. Champagne, Belgium
Prof. S. Farantos, Greece
Prof. I. Gutman, Serbia & Montenegro
Prof. P. Mezey, Canada
Prof. C. Pouchan, France.
Dr. G. Psihoyios, Vice-President ESCMSE
Prof. B.M. Rode, Austria.
Prof. A. J. Thakkar, Canada

Invited Speakers

Prof. Xavier Assfeld, France
Prof. B. Champagne, Belgique.
Prof. K. Balasubramanian, USA.
Prof. R.J. Bartlett, USA.
Prof. D. Clary, UK.

Prof. E.R. Davidson, USA.
Prof. S. Farantos, Greece.
Prof. W. Goddard, USA
Prof. E. Ghysels, USA
Prof. J.Jellinek, USA
Prof. B. Kirtman, USA.
Prof. M.Kosmas, Greece
Prof. J. Leszczynski, USA.
Prof. M. Meuwly, Switzerland.
Prof. A. Painelli, Italy.
Prof. M.G. Papadopoulos, Greece.
Prof. M. Pedersen, USA.
Prof. C. Pouchan, France.
Prof. V. Renugopalakrishnan, USA
Prof. B.M. Rode, Austria.
Prof. P. Schwerdtfeger, New Zealand.
Prof. G. Scuseria, USA.
Prof. P. Senet, France.
Prof. A.J. Thakkar, Canada.
Prof. A. van der Avoird, The Netherlands.
Prof. P. Weinberger, Austria.
Prof. S. S. Xantheas, USA

Brill Academic Publishers
P.O. Box 9000, 2300 PA Leiden
The Netherlands

*Lecture Series on Computer
and Computational Sciences*
Volume 4, 2005, pp. vi-vii

European Society of Computational Methods in Sciences and Engineering (ESCMSE)

Aims and Scope

The *European Society of Computational Methods in Sciences and Engineering (ESCMSE)* is a non-profit organization. The URL address is: http://www.uop.gr/escmse/

The aims and scopes of *ESCMSE* is the construction, development and analysis of computational, numerical and mathematical methods and their application in the sciences and engineering.

In order to achieve this, the *ESCMSE* pursues the following activities:

• Research cooperation between scientists in the above subject.
• Foundation, development and organization of national and international conferences, workshops, seminars, schools, symposiums.
• Special issues of scientific journals.
• Dissemination of the research results.
• Participation and possible representation of Greece and the European Union at the events and activities of international scientific organizations on the same or similar subject.
• Collection of reference material relative to the aims and scope of *ESCMSE*.

Based on the above activities, *ESCMSE* has already developed an international scientific journal called **Applied Numerical Analysis and Computational Mathematics (ANACM)**. This is in cooperation with the international leading publisher, **Wiley-VCH**.

ANACM is the official journal of *ESCMSE*. As such, each member of *ESCMSE* will receive the volumes of **ANACM** free of charge.

Categories of Membership

European Society of Computational Methods in Sciences and Engineering (ESCMSE)

Initially the categories of membership will be:

• **Full Member (MESCMSE):** PhD graduates (or equivalent) in computational or numerical or mathematical methods with applications in sciences and engineering, or others who have contributed to the advancement of computational or numerical or mathematical methods with applications in sciences and engineering through research or education. Full Members may use the title MESCMSE.

• **Associate Member (AMESCMSE):** Educators, or others, such as distinguished amateur scientists, who have demonstrated dedication to the advancement of computational or numerical or mathematical methods with applications in sciences and engineering may be elected as Associate Members. Associate Members may use the title AMESCMSE.

• **Student Member (SMESCMSE):** Undergraduate or graduate students working towards a degree in computational or numerical or mathematical methods with applications in sciences and engineering or a related subject may be elected as Student Members as long as they remain students. The Student Members may use the title SMESCMSE

• **Corporate Member:** Any registered company, institution, association or other organization may apply to become a Corporate Member of the Society.

Remarks:

1. After three years of full membership of the European Society of Computational Methods in Sciences and Engineering, members can request promotion to Fellow of the European Society of Computational Methods in Sciences and Engineering. The election is based on international peer-review. After the election of the initial Fellows of the European Society of Computational Methods in Sciences and Engineering, another requirement for the election to the Category of Fellow will be the nomination of the applicant by at least two (2) Fellows of the European Society of Computational Methods in Sciences and Engineering.

2. All grades of members other than Students are entitled to vote in Society ballots.

3. All grades of membership other than Student Members receive the official journal of the ESCMSE Applied Numerical Analysis and Computational Mathematics (ANACM) as part of their membership. Student Members may purchase a subscription to ANACM at a reduced rate.

We invite you to become part of this exciting new international project and participate in the promotion and exchange of ideas in your field.

VSP International
Science Publishers
P.O. Box 346, 3700 AH Zeist
The Netherlands

*Lecture Series on Computer
and Computational Sciences*
Volume 4, 2004, pp. 1057-1057

Preface of the Symposium : Stochastic Computational Techniques in Engineering and Sciences

T.E. Simos
Chairman of ICCMSE 2005

Department of Computer Science and Technology,
Faculty of Sciences and Technology,
University of Peloponnese,
GR-221 00 Tripolis, Greece

This symposium has been created after a proposal of Marcin Kaminski, Chair of Mechanics of Materials, Technical University of Łódź. After a successful review, we have accepted his proposal.

The organizer of the symposium, Marcin Kaminski, has selected, after international peer review four papers:

- Generalized Stochastic Perturbation Technique in Engineering Computations by M.M. Kamiński
- Deterministic and Probabilistic Sensitivity Analysis of Fatigue Fracture Model Parameters for a Curved Two Layer Composite by Ł. Figie and M. Kamiński
- Simulation of the Carnot Engine in the Frame of Statistical Mechanics by M. Gall1 and R. Kutner
- Research on Fault Tolerance Schemes Based on Energy-Aware by Li Zhong-wen, Yu Shui

I want to thank the symposium organizer for his activities and excellent editorial work.

Brill Academic Publishers
P.O. Box 9000, 2300 PA Leiden
The Netherlands

Lecture Series on Computer
and Computational Sciences
Volume 4, 2005, pp. 1058-1061

Generalized Stochastic Perturbation Technique in Engineering Computations

M.M. Kamiński[1]

Chair of Mechanics of Materials,
Faculty of Civil Engineering, Architecture and Environmental Engineering,
Technical University of Łódź, Al. Politechniki 6
PL 93-590 Łódź, Poland

Received 5 August, 2005; accepted in revised form 12 August, 2005

Abstract: Stochastic perturbation technique is, together with the simulation and spectral methods, one of the most powerful tools to model scientific and engineering problems with random parameters. Perturbation approach was implemented until now according to the second order and second moment in various Finite or Boundary Element and even Finite Difference or meshless methods. Now we propose a generalized n*th* order and n*th* moment perturbation method to be applied in conjunction with these computer techniques to effectively approximate any probabilistic moments and coefficients of the both transient as well as steady-state response of various engineering systems.

Keywords: stochastic perturbation technique, discrete numerical methods, symbolic computing

Mathematics Subject Classification: 34E10, 35J45

PACS: 02.50.Fz, 46.15.Ff

1. Introduction

The main interest is focused here on the solution of engineering problems represented by some boundary value problems, where their coefficients including material and/or geometrical design parameters are treated as random variables or fields. Engineering textbooks recommend in this case a variety of both analytical and numerical techniques to include this randomness into the model and decisively most popular, concerning computer applications, are simulation or spectral methods as well as perturbation techniques. As far as simulation methods are time consuming, spectral techniques need a lot of components in polynomial representation (slow convergence), then the perturbation technique seems to be promising considering a similarity of various orders equations and their solutions. As it was detected for lower order approaches, like second order second moment for instance [5], a precision of perturbation methods is strongly limited to the distributions with smaller coefficients of variation. This limitation is very serious considering most of engineering applications, where an uncertainty level can be of comparable order of the expectation for the verified random quantity. Concerning this limitation it would be necessary to improve lower order perturbation approach to reach the desired accuracy in computations using the optimal order of Taylor series expansions. It must be remembered that the length of this expansion strongly depends also on the order of probabilistic moment being determined, so that taking into account most of non-trivial applications, at least fourth order would be recommended. As it is demonstrated below, an application of symbolic mathematical programs like MAPLE for example gives the opportunity to automatically derive most of necessary equations – to compute various order solutions and, separately, for probabilistic moments of the structural response.

---

[1] Corresponding author. Active member of the Society of Industrial and Applied Mathematics.
E-mails: marcin@kmm-lx.p.lodz.pl, marcinka@p.lodz.pl; web: http://kmm.p.lodz.pl/pracownicy/Marcin_Kaminski/index.html

2. Stochastic perturbation technique in nth order - nth moment version

Similarly to the stochastic spectral techniques we use the following expansion of the random input using however Taylor series representation in case of random quantity $e=e(\omega)$ as

$$e = e^0 + \sum_{n=1}^{\infty} \frac{1}{n!} \varepsilon^n \frac{\partial^n e}{\partial b^n} (\Delta b)^n \tag{1}$$

and, then, using classical integral definition of it [1]

$$E[b] \equiv b^0 = \int_{-\infty}^{+\infty} b \, p(b) \, db \tag{2}$$

the expected values of the function $f(b(\omega))$ in Nth order approximation can be recovered for general distribution as [2]

$$E[f(b); b] = f^0 + \int_{-\infty}^{+\infty} \left(\sum_{n=1}^{N} \frac{1}{(n)!} \varepsilon^n \frac{\partial^n f}{\partial b^n} \Delta b^n \right) p(b) \, db \tag{3}$$

where $\varepsilon > 0$ is some small perturbation parameter and $p(b)$ is probability density function for the random input parameter b. Assuming that the random input has Gaussian distribution we can derive the closed form formulas for probabilistic moments like expectation

$$E[f(b); b] = f^0(b) + \frac{1}{2} \varepsilon^2 f^{,bb}(b) Var(b) + \frac{1}{4!} \varepsilon^4 f^{,bbbb}(b) \mu_4(b) + \frac{1}{6!} \varepsilon^6 f^{,bbbbbb}(b) \mu_6(b) + ... \tag{4}$$

and the variance

$$Var(f(b)) = \mu_2(b) f^{,b} f^{,b} + \mu_4(b) \left(\frac{1}{4} f^{,bb} f^{,bb} + \frac{2}{3!} f^{,b} f^{,bbb} \right) \\ + \mu_6(b) \left(\left(\frac{1}{3!} \right)^2 f^{,bbb} f^{,bbb} + \frac{2}{4!} f^{,bbbb} f^{,bb} + \frac{2}{5!} f^{,bbbbb} f^{,b} \right) + ... \tag{5}$$

Analogously we can derive any order probabilistic moments and the relevant coefficients and implement them into the computer package performing engineering computations. The main advantage of this method is that the solution for any next order approximation is in fact deterministic and similar to the classical deterministic problem, so that randomization according to this methodology does not lead to any essential changes in the computer software.

3. Computational implementation aspects

Let us review the generalized stochastic perturbation technique on the example of variational equation for the isotropic linear elasticity problem, which could be written for a displacement function **u** as [5]

$$\int_{\Omega} C_{ijkl} u_{k,l} \delta u_{i,j} d\Omega = \int_{\Omega} \rho f_i \delta u_i d\Omega + \int_{\partial\Omega} \tilde{t}_i \delta u_i d(\partial\Omega) \tag{6}$$

Expanding all random parameters according to eqn (1) and equating the terms of the same order we obtain up to nth order equations with the last one having the form

$$\int_{\Omega} \delta u_{i,j} C_{ijkl}^0 \frac{\partial^n u_{k,l}}{\partial b^n} d\Omega = \\ \int_{\partial\Omega_t} \delta u_i \frac{\partial^n \tilde{t}_i}{\partial b^n} d(\partial\Omega) - \int_{\Omega} \delta u_{i,j} \sum_{k=1}^{n} \binom{n}{n-k} C_{ijkl}^{,k} u_{k,l}^{,n-k} d\Omega + \int_{\Omega} \sum_{k=0}^{n} \binom{n}{n-k} \rho^{,k} f_i^{,n-k} \delta u_i d\Omega \tag{7}$$

Expanding this equation into (n+1) sequential order variational statements and using the Finite Element Method discretization we can obtain up to nth order structural response functions (displacements or stresses), where final probabilistic moments can be subtracted from. The following fundamental linear equations lead to this purpose [2]

$$\sum_{k=0}^{n} \binom{n}{k} K_{\alpha\beta}^{(k)} q_{\beta}^{(n-k)} = Q_{\alpha}^{(n)} \tag{8}$$

where $K_{\alpha\beta}$, q_β and Q_α denote the stiffness matrix, structural displacement and forcing vector. Upper indices for this equation components denote the relevant partial derivatives with respect to the input random variable. As it is known, initial integral equation in the Boundary Element Method is slightly different [3]

$$c(x)u(x) + \int_\Gamma P(x,y)u(y)d\Gamma(y) = \int_\Gamma U(x,y)p(y)d\Gamma(y) + \int_\Omega U(x,z)b(z)d\Omega(z) \tag{9}$$

with P being the stress vector operator. It also results in the system of algebraic equations

$$H_{\alpha\beta}u_\beta = G_{\alpha\beta}p_\beta + B_\alpha \; ; \; \alpha,\beta = 1,...,W \tag{10}$$

represented similarly to the Finite Element Method as

$$A_{\alpha\beta}X_\beta = F_\alpha \; ; \quad \alpha,\beta = 1,...,W; \tag{11}$$

but contrary to the FEM this system has non-symmetric L.H.S. matrix. Nevertheless, it leads in the case of stochastic perturbation technique to similar up to nth order algebraic equations, however numerical method for deterministic problem solution (and for each of higher than the zeroth order in perturbation-based solution) must include this lack of symmetry [3]

$$\sum_{k=0}^{n} \binom{n}{k} A_{\alpha\beta}^{(k)} X_\beta^{(n-k)} = F_\alpha^{(n)} \tag{12}$$

Contrary to these two techniques, the Finite Difference Method starts from partial (or ordinary) differential equation(s) of static (or dynamic) equilibrium, where partial derivatives of the structural response are approximated by the additional differences. The problem is displayed on the Love's equation for the shell equilibrium, where for zero initial curvature [4] its displacement **u** under the load q are computed from the following equation:

$$D\nabla^4 u + \rho h\ddot{u} = q \tag{13}$$

where D is the shell stiffness parameter and ρ its mass density. After standard transformation $u = U\exp(j\omega t)$ one can get its modified version as

$$D\left(U_{,xxxx} + 2U_{,xxyy} + U_{,yyyy}\right) + \rho h\omega^2 U = 0 \tag{14}$$

Further solution is based on a discrete representation of partial derivatives in the following form (assuming the same density of finite difference net in orthogonal directions given by the parameter δ). We can rewrite those derivatives for x_i and for the first two orders of perturbation solution as

$$\frac{\partial\left(U^0\right)}{\partial x_i} \cong \frac{U_1^0 - U_0^0}{\delta}, \; \frac{\partial\left(U^{,b}\right)}{\partial x_i} = \frac{\partial^2 U}{\partial x_i \partial b} \cong \frac{U_1^{,b} - U_0^{,b}}{\delta} = \frac{1}{\delta}\frac{\partial}{\partial b}\left(U_1 - U_0\right) \tag{15}$$

Thanks to these (and also higher order) expansions one can arrive at up to nth order equations for initial problem (13) as

$$\sum_{k=0}^{n}\binom{n}{k}D^{(k)}\left(\nabla^4 u\right)^{(n-k)} + \sum_{k=0}^{n}\binom{n}{k}\rho^{(k)}\left(\sum_{m=0}^{n-k}\binom{n-k}{m}h^{(m)}\left(\ddot{u}\right)^{(n-k-m)}\right) = q^{(n)} \tag{16}$$

4. Numerical illustration

Computational experiments concerns heat conduction in a prismatic beam with temperature equal to 0 at the left end and heated at the opposite edge. The analysis is performed in the system MAPLE using 10 stochastic finite elements (1D linear elements) using implementation of Eqn (8); the following data have been adopted: k=0.1 (heat conductivity coefficient), A=0.01 (cross-sectional area), E[L]=0.01, q=1.0 (heat flux), where E[L] denotes the finite element length. Perturbation parameter (marked as 'eps') and coefficient of variation of the random input ('alfa') are treated as the parameters of this study.

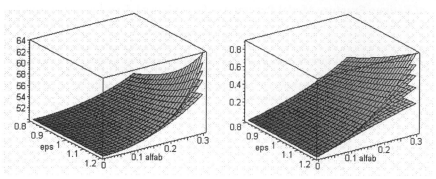

Figure 1: Expectations (at right) and coefficients of variation (at left) for the heated end temperature.

As it is apparent from these results, the coefficient of variation of input random variable is definitely more decisive for the probabilistic moments and characteristics of the solution vector component than the perturbation parameter. The extended 10th order perturbation technique displayed here appears to be effective even with 30% dispersion of the random input, which is quite unacceptable for the accuracy of previously applied 2nd order Stochastic Finite Element Method. It is documented here that the expected values and standard deviations are almost insensitive to the perturbation parameter for smaller coefficients of variation. The significance of this parameter increases with both order of perturbation method and the coefficient of variation for random input.

5. Concluding remarks

Stochastic perturbation technique presented here in its general nth order and nth moment version can find its application in efficient computations of the probabilistic moments in various engineering and scientific problems, where some input parameters are random variables or fields. In this second case for the multiple different uncertainty sources we need to insert the cross-correlations between them having in mind theoretical or practical background for non-zero covariances for various types of the parameters. The method can be used for analytical solutions as well as in conjunction with the well-established discrete numerical approaches like FEM, BEM, FDM or meshless techniques. Numerical analysis included here shows that the 10th order perturbation approach is efficient even for higher level of random dispersion of input random variables around its expectations. Symbolic mathematical packages interoperating with engineering modeling programs are recommended for the effective computational implementations of the methodology and for some practical numerical studies.

References

[1] W. Feller, *An Introduction to Probability Theory and Its Applications*. John Wiley & Sons, New York-London-Sydney, 1966.

[2] M. Kamiński, On generalized stochastic perturbation-based finite element method, *Communications in Numerical Methods in Engineering*, (in press).

[3] M. Kamiński, Stochastic second order perturbation BEM formulation. *Engineering Analysis with Boundary Elements* **2** 123-130 (1999).

[4] M. Kamiński, Stochastic perturbation approach to engineering structures vibrations by the Finite Difference Method, *Journal of Sound & Vibration* **251**(4) 651-670 (2002).

[5] M. Kleiber and T.D. Hien, *The Stochastic Finite Element Method*. Wiley, 1992.

Brill Academic Publishers
P.O. Box 9000, 2300 PA Leiden
The Netherlands

*Lecture Series on Computer
and Computational Sciences*
Volume 4, 2005, pp. 1062-1064

Deterministic and Probabilistic Sensitivity Analysis
of Fatigue Fracture Model Parameters for a Curved Two Layer Composite

Ł. Figiel[1] and M. Kamiński[2]

[1]Institute of Materials Research, German Aerospace Center,
Linder Höhe, 51147 Cologne, Germany. E-mail: Lukasz.Figiel@dlr.de
[2]Chair of Mechanics of Materials, Technical University of Łódź,
Al. Politechniki 6, 93-590 Łódź, Poland

Received 2 August, 2005; accepted in revised form 16 August, 2005

Abstract: Two sensitivity analysis concepts are proposed in this paper to elucidate effects of variations of fatigue fracture model parameters on the computational model output. Particularly, the deterministic and probabilistic sensitivities of the fatigue cycles number with respect to two parameters of a modified Paris fatigue fracture model are evaluated for a two-layer curved composite under cyclic shear. The deterministic sensitivity coefficients are obtained analytically. The probabilistic sensitivities are obtained in the form of the Pearson and Spearman correlation coefficients as well as from the trend lines. All approaches are combined with the finite element method to yield numerical values of sensitivities.

Keywords: sensitivity analysis, finite element method, composite materials, fatigue delamination model

1. Deterministic and probabilistic sensitivities of the modified Paris model

The fatigue life of a composite laminate with an initial defect of length a_i is determined from the modified Paris model as the cumulative fatigue cycles number

$$N_f = \sum_{i=1}^{I} N_i = \sum_{i=1}^{I} \left\{ \frac{(G_{Ti})^{-m_1}}{C_1} \int_{a_i}^{a_{i+1}} da \right\}, \tag{1}$$

where N_i denotes the fatigue cycles number required to move the crack from a_i to a_{i+1}; I is the total number of crack length increments; G_T denotes the total energy release rate (TERR); C_1 and m_1 are the empirical crack growth constant and the exponent.

Let us investigate a change of model parameters C_1 and m_1, that enter the nominal deterministic variable vector $b_o = \{C_1, m_1\}$, from b_o to $b = b_o + \Delta b$, to determine the sensitivity gradient of N_i with respect to b, $\partial N_i / \partial b$. This gradient is obtained analytically by partial differentiation of Eq (1), for a single crack growth increment from a_i to a_{i+1} and G_{Ti} being constant during the differentiation, subsequently with respect to m_1 and C_1

$$\frac{\partial N_i}{\partial m_1} = -\frac{1}{C_1}(G_{Ti})^{-m_1} \log G_{Ti} \int_{a_i}^{a_{i+1}} da \; ; \quad \frac{\partial N_i}{\partial C_1} = -\frac{1}{(C_1)^2}(G_{Ti})^{-m_1} \int_{a_i}^{a_{i+1}} da \; . \tag{2}$$

Eq (2) does not account for any deterministic correlation between C_1 and m_1. If this correlation exists between C_1 and m_1, e.g. Cortie and Garrett [1], then the analytical forms of sensitivity gradients of the fatigue cycles number with respect to m_1 and C_1 are consecutively given by

$$\frac{\partial N_i}{\partial m_1} = \left\{ e^{Bm_1-A}(G_{Ti})^{-m_1}(B - \log G_{Ti}) \right\} \int_{a_i}^{a_{i+1}} da \; ; \quad \frac{\partial N_i}{\partial C_1} = \left\{ \frac{(G_{Ti})^{\left(\frac{\log C_1-A}{B}\right)}}{(C_1)^2} \left(\frac{\log G_{Ti}}{B} - 1 \right) \right\} \int_{a_i}^{a_{i+1}} da \; . \tag{3}$$

The numerical evaluation of the gradients given by Eqs (2)-(3) requires the knowledge of the TERR, G_{Ti}, at subsequent crack lengths a_i. This information is obtained through the numerical solution of the relevant boundary value problem using the finite element method (FEM), Figiel and Kamiński [2]. The analytical approach enables to determine the sensitivity gradients only in a single analysis step.

Now $b=\{m_1,C_1\}$ is considered as the vector of random parameters with given probability distributions and any associated conditions (e.g. correlations). An output $N[b_k]=[N_{i1}[b_k],N_{i2}[b_k],...,N_{ij}[b]]$ corresponds to a sample $b_k=[b_{k1},b_{k2},...,b_{kj}]$ and $j=1,2,...,J$. The correlation coefficient between input b_k and output parameters N can be a qualitative measure of model parameters sensitivity through the Pearson product-moment correlation coefficient or the Spearman rank order correlation coefficient, respectively [3].

$$r_p = \frac{\sum_{j=1}^{J}\left\{\left(b_{kj}-\langle b_{kj}\rangle\right)\left(N_{ij}-\langle N_{ij}\rangle\right)\right\}}{\sqrt{\sum_{j=1}^{J}\left(b_{kj}-\langle b_{kj}\rangle\right)^2}\sqrt{\sum_{J=1}^{n}\left(N_{ij}-\langle N_{ij}\rangle\right)^2}}; \quad r_s = \frac{\sum_{j=1}^{J}\left\{\left(P_j-\langle P_j\rangle\right)\left(S_j-\langle S_j\rangle\right)\right\}}{\sqrt{\sum_{j}^{J}\left(P_j-\langle P_j\rangle\right)^2}\sqrt{\sum_{j}^{J}\left(S_j-\langle S_j\rangle\right)^2}}, \quad (4)$$

where $\langle.\rangle$ denotes the mean value; P_j is the rank of b_{kj}; S_j is the rank of N_{ij}. These coefficients take their values within a range $\langle-1,+1\rangle$, and the sign of the correlation coefficient points out a tendency (increasing or decreasing) of the output parameter due to a variation of the input.

The quantitative measure of probabilistic sensitivities is obtained using scatter plots. These plots are characterised by the so-called trendlines fitted by appropriate functions. The absolute probabilistic sensitivities are obtained by a simple differentiation of these functions with respect to model parameters, b_k, as follows:

$$S_{abs}^{pr}\left(E[b_k]\right)= d_1 + 2d_2 b_k +...+ \varpi d_p \left(b_k\right)^{\varpi-1}, \quad (5)$$

where d_o, d_1, d_2, ..., d_p are coefficients of the fitting functions. The sensitivities are then determined numerically at the nominal value of the random input parameter b_k.

All the aforementioned sensitivities (deterministic and probabilistic) are absolute – for comparative purposes they must be scaled to yield the relative sensitivity gradients, e.g. Kamiński [4].

2. Results

The two component boron/epoxy-aluminium (B/Ep-Al) curved composite laminate is a subject of computational studies. Details on geometrical and material properties of the composite as well as on the FEM discretisation and solution of the nonlinear boundary value problem are reported in Figiel and Kamiński [2].

The deterministic relative sensitivity gradients are numerically evaluated using analytical formulas given by Eqs (2)-(3). In both cases, uncorrelated and correlated, values of the sensitivity gradients indicate that the fatigue cycles number is much more sensitive to the exponent m_1, than to the coefficient C_1 of the fatigue law, as shown in Table 1. However, the very important observation is that the sensitivity gradients values obtained in the correlated case are much smaller, approximately one order, than those obtained in the uncorrelated one.

Table 1: Deterministic sensitivity coefficients

a/a_o	Uncorrelated		Correlated	
	C_1	m_1	C_1 (m_1)	m_1 (C_1)
1.167	-1.000	-47.185	0.024	-1.130
1.833	-1.000	-46.874	0.018	-0.818
2.500	-1.000	-47.846	0.038	-1.790
3.167	-1.000	-57.596	0.250	-11.540

For the purpose of probabilistic sensitivity analysis, the Latin Hypercube Sampling (LHS) method implemented in the system ANSYS was used to generate a sampled population. The exponent and the coefficient of the fatigue law were assumed to be random variables that satisfy the normal and lognormal probabilistic distributions, respectively. The mean values of both model parameters were set to $E[m_1]=10$ and $E[C_1]=1\times10^{-29}$, while their standard deviations $\sigma_{x,iv}$ were equal to 1, 5, 10 and 15% of the corresponding expected values. Initially, the generated sample consisted of $J=100$ sampling points. It is noted that the fitting of scatter plots was carried out by appropriate functions in the log-log scale, outside of ANSYS, using the algebra package Mathematica. The relative sensitivities of the fatigue cycles number with respect to m_1 are collected in Table 2 for four different crack lengths and $\sigma_{x,iv}=1\%\times E[m_1]$, when m_1 is uncorrelated with C_1, and if m_1 is correlated with C_1.

Table 2: Probabilistic sensitivity coefficients

a/a$_o$	Uncorrelated	Correlated
	m$_1$	m$_1$ (C$_1$)
1.167	-47.244	-1.130
1.833	-45.392	-0.818
2.500	-47.426	-1.790
3.167	-55.807	-11.536

Qualitatively, a similar trend is observed for both cases, namely, N_i decreases along with increasing m_1. Results of the relative probabilistic sensitivities are in a very good, either qualitative or quantitative, agreement with those obtained from analytical expressions. Then, the Pearson and Spearman correlation coefficients values were determined by ANSYS according to Eq (4) and the relevant results are collected in Table 3, for uncorrelated and correlated cases and $\sigma_{x,iv}=1\% \times E[m_1]$.

Table 3: Correlation coefficients

a/a$_o$	Pearson correlation coefficient, r$_p$				Spearman correlation coefficient, r$_s$			
	Uncorrelated		Correlated		Uncorrelated		Correlated	
	C$_1$	m$_1$	C$_1$(m$_1$)	m$_1$(C$_1$)	C$_1$	m$_1$	C$_1$(m$_1$)	m$_1$(C$_1$)
1.167	-0.999	-0.946	0.999	-0.999	-1.0	-1.0	1.0	-1.0
1.833	-0.999	-0.940	0.999	-0.999	-1.0	-1.0	1.0	-1.0
2.500	-0.999	-0.952	0.999	-0.999	-1.0	-1.0	1.0	-1.0
3.167	-0.998	-0.935	0.999	-0.998	-1.0	-1.0	1.0	-1.0

The outcome of computations shows that both model parameters are strongly correlated with the output parameter. Therefore, the correlation coefficients can be used to characterise qualitatively the model sensitivity to its parameters.

3. Conclusions

(i) Sensitivity studies demonstrated that the fatigue cycles number is much more sensitive to the exponent, than to the coefficient of the fatigue fracture model. (ii) Results of the sensitivity analysis results showed that the effect of model parameters correlation may considerably influence (decrease) effects of the model parameter variation on the predicted fatigue cycles number. (iii) The relative sensitivities determined by two approaches, analytical differentiation and trendline fitting, were in a very good agreement. (iv) The Pearson and the Spearman coefficients provided sufficient information about the model sensitivity, which was in a qualitative agreement with the outcome of either deterministic or probabilistic sensitivities. Unfortunately, these coefficients did not allow to detect quantitatively the most crucial model parameters.

Acknowledgments

The author wishes to thank the anonymous referees for their careful reading of the manuscript and their fruitful comments and suggestions.

References

[1] M.B. Cortie and G.G. Garrett, On the correlation between the C and m in the Paris equation for fatigue crack propagation. *Engineering Fracture Mechanics* **30** 49-58 (1988).

[2] Ł. Figiel and M. Kamiński, Mechanical and thermal fatigue delamination in curved layered composites. *Computers and Structures* **81** 1865-1873 (2003).

[3] D.J. Sheskin: *Handbook of Parametric and Nonparametric Statistical Procedures*. CRC Press, 1997.

[4] M. Kamiński, Sensitivity analysis of homogenised characteristics for some elastic composites. *Computational Methods in Applied Mechanics and Engineering* **192** 1973-2005 (2003).

Brill Academic Publishers
P.O. Box 9000, 2300 PA Leiden,
The Netherlands

*Lecture Series on Computer
and Computational Sciences*
Volume 4, 2005, pp. 1065-1068

Simulation of the Carnot Engine
in the Frame of Statistical Mechanics

M. Gall[1] and R. Kutner

Division of Physics Education, Institute of Experimental Physics,
Department of Physics, Warsaw University,
Smyczkowa Str. 5/7, PL-02678 Warsaw, Poland

Received 5 August, 2005; accepted in revised form 12 August, 2005

Abstract: We developed a highly interactive, multi-windows Java applet which made it possible to simulate and visualize within any platform and internet the Carnot engine in a real-time computer experiment. As usually the considerations of phenomenological thermodynamics began with a study of the basic properties of heat engines hence our approach, beside intrinsic physical significance, is also important from the educational, technological and even environmental points of view.

Keywords: Carnot engine; many-body problem; adiabatic and isothermal expansions and compressions; second law of thermodynamics

PACS: 89.90.+n; 05.40.+j

1   The main problem

We are interested in the relation between quasi-static thermodynamic processes and their numerical realization by the corresponding microscopical mechanism (considered within the statistical mechanics) [1]. We extend our previous microscopic model [2, 3] and algorithm to simulate not only heat flow but also the macroscopic movement of the piston which is the principal problem for thermodynamic processes [4].

As usually the considerations of phenomenological thermodynamics began with a study of the basic properties of heat engines - the subject which beside intrinsic physical significance has educational, technological and even environmental importance. For example, we consider the Carnot engine which is the principal one of the classical formulation of the second law of thermodynamics.

Since in reality it is impossible to construct a reversible Carnot engine **the main problem consists in that whether is it at least possible to simulate (with controlled accuracy) the Carnot cycle within a numerical experiment?** The positive answer to this question is connected with our microscopic model and algorithm.

2   Foundations of the model and algorithm

We recommend (thanks to the default option contained in the applet menu) a two-dimensional Boltzmann gas as the working medium. The energy of the gas particles is solely kinetic and their radius can be changed (again by using the menu of the applet). The reference option (chosen from

---

[1] Corresponding author. E-mail: mgall@poczta.onet.pl or erka@fuw.edu.pl

the menu) which neglects collisions between hard cores (or particles) has also been defined. The particles are enclosed in the container (cylinder), i.e. the total number of particles N is fixed (in a given numerical experiment but can be changed also by making use of the menu at the beginning of the experiment). The volume V of the gas is determined by the position of the mobile piston (cf. the snapshot of the monitor's screen shown in Fig.1); the gas pressure p is, of course, additionally dependent on the gas temperature T. We explain below how these thermodynamic parameters are calculated.

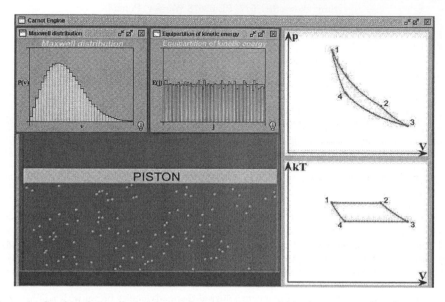

Figure 1: Combined snapshot pictures of monitor screen which show our results of a numerical experiment obtained directly after a single Carnot cycle (i.e. at the beginning of the next isothermal expansion process described by the curves $1-2$ shown in both small figures placed in the right-hand side of the figure). The container (closed by the piston) containing gas particles is presented in the left-hand side at the bottom of the figure. Above the piston we placed two small windows which show a comparison between experimental (pistils) and theoretical (solid curves) predictions of (i) Maxwell distributions (the first window) and (ii) mean energies of particles (the second window).

The necessary discretization of time t is, as usual, the basic element of computer simulation. We set $t_n = n\delta t$, where $\delta t (= t_{n+1} - t_n)$ is a discrete, elementary time-step and $n = 0, 1, 2, \ldots$, is the number of successive steps. We assume this time-step to be much shorter than the minimal linear size of the container, L_{min}, divided by the maximal velocity, v_{max}, of the gas particle $(\tau_{min} \overset{\text{def.}}{=} L_{min}/v_{max})$. Note that during the simulation the computer records temporal velocity of each gas particle (or momentum as the mass of each particle is set to $m = 1$) and hence yields its maximal value. Moreover, the computer also calculates the collision time τ_c (i.e. the mean time between successive particle collisions) as there is no reason to assume that the time-step δt is much shorter than the collision time. We expect the convenient hierarchy of the above defined characteristic times to be obeyed: $\delta t \leq \tau_c \ll \tau_{min} < \tau$, where relaxation time τ is the time needed for the system to reach thermal equilibrium.

The temporal temperature of the gas T_n in the nth time-step is defined within this discrete-time representation as proportional to the temporal average kinetic energy of all N particles $kT_n = \frac{1}{N} \sum_{j=1}^{N} \varepsilon_n^j$, where k is the Boltzmann constant and ε_n^j is the kinetic energy of the jth particle in the nth time-step. In fact, we use the equilibrium temperature necessary to perform the Carnot cycle as a quasi-static thermodynamic process.

2.1   The adiabatic thermodynamic process

The piston stroke. The single step of the piston, i.e. the piston stroke, consists of two stages: (1) a small displacement, δx, of the piston and (2) the waiting of the piston until thermal equilibrium of the system is established. Note that the second stage requires isolation of the system from the surroundigs. The movement of the piston, which consists of a sequence of such steps, simulates a quasi-static thermodynamic process of the gas.

To visually confirm the thermalization process we observe at the monitor screen the relaxation of the temporary velocity distribution, $P(v)$, to the Maxwell equilibrium one for the above mentioned particular state. Simultaneously, we observe the relaxation of the average (kinetic) energy of each particle (calculated within each piston stroke), $E(j)$, to their common equilibrium value. Thus the principle of energy equipartition of all particles is roughly confirmed (cf. the corresponding plot in Fig.1). However, the question arises when should we stop both our observations? The answer is straightforward: when no systematic changes are observed except only small fluctuations of the experimental quantities.

In each piston stroke (at the end of the time interval τ_w) the applet calculates the equilibrium temperature and the pressure of the gas (as well as its volume).

Rebounding of particles from a piston. We consider N_n particles which collided with the piston within a time interval $[t_n, t_{n+1}[$; **the problem of collisions of the particles with the mobile piston is one of the main questions** considered in the paper.

We can find the energy of every rebounded particle in the next $n^{th} + 1$ time-step (i.e. directly after the rebounding from the piston) as the arithmetic mean $\varepsilon_{n+1} = \frac{1}{N_n}(\sum_{j=1}^{N_n} \varepsilon_n^j + \delta W_n)$, where $\delta W_n = -p_n S \delta x$ is the work supplied to the system (if $\delta W_n > 0$) or performed by the system (if $\delta W_n < 0$) by means of the piston displacement δx, while the gas pressure p_n is a quantity averaged over time-intervals when the piston is temporarily immobilized and the system is in equilibrium. Since the pressure is a force exerted perpendicular to the surface S per its unit area, computer easily calculate the normal component of the molecular momentum transferred to the surface per unit time and area as a required quantity. This expression is an application of **principle of local energy equipartition** applied selectively in this model (i.e. only at the piston) which can be considered as a kind of a mean-field approach where fluctuations of particle energies directly after the rebounding are neglected. These considerations make it possible to calculate the gas temperature and its pressure at thermodynamic equilibrium as well as the kinetic energy of any particle rebounding from the mobile piston.

2.2   Isothermic thermodynamic process

Now, we are able to model the quasi-static isothermic thermodynamic process (expansion or compression) which is more complicated than the adiabatic one. Each successive single-step of the process is divided into the following (elementary) two stages: (1) heat transfer and (2) work performed by the piston. Of course, after each stage the system is isolated from the surroundings to make possible its spontaneous relaxation to thermal equilibrium.

Heat transfer. In our model we assumed for simplicity that heat, δQ, is exchanged between our system and the surroundings only across the bottom of the container. During this stage the piston is immobilized and plays the role of a top massive border wall (cf. section 2.1).

Like for the adiabatic process we can find the energy of a particle after its collision with the bottom during the heat transfer process by using the principle of local energy equipartition (applied selectively at the bottom of the container) $\varepsilon_{n+1} = \frac{1}{N_n}(\sum_{j=1}^{N_n} \varepsilon_n^j + \delta Q_n)$, where δQ_n is the heat exchanged between the system and the environment (which is positive when heat is supplied to the system and negative in the opposite case) within the time-step $\delta t(= t_{n+1} - t_n)$. This is how the first stage of the single-step of the process proceeds.

During the next, intermediate stage no heat is exchanged and no work is performed until the system reaches thermal equilibrium which is verified analogously as for the adiabatic process (see Sec.2.1).

Work performed by the piston. This part is analogous to the single-step of the adiabatic thermodynamic process. We put here $-\delta W_n = \delta Q_n$ to keep the temperature of the system fixed. Hence (and from definition of the work), we obtain the required displacement δx of the piston.

3   Concluding remarks

In Fig.1 we see a snapshot of the monitor's screen which shows some of our results. At first, the picture of the positions of gas particles in the container is shown. As it is seen, we consider here a small system consisting of only $N = 100$ particles (although the increase by two order of magnitude is also possible). Nevertheless, the Carnot cycles presented in (V, p) and (V, kT) thermodynamic variables were obtained in our numerical experiment (solid coloured curves) and they well agree with the theoretical predictions (dashed curves); although, gas pressure fluctuations are also seen. Of course, we obtained this result both for the gas with interacting and noninteracting particles and for rough as well mirror border walls and piston.

In the figure we also show the comparison of the experimental (pistils) with theoretical (solid curves) Maxwell distributions and mean energies of the colliding gas particles. This allows rough, visual control in real-time whether the system has reached the thermodynamic equilibrium in each piston stroke.

Concluding, in this work we simulated in a numerical experiment the quasi-static thermodynamic Carnot cycle thanks to the locally applied principle of the energy equipartition. This mean-field type principle made it possible to avoid the many-body problem and obtain macroscopic quantities from the microscopic model to simulate thermodynamic processes.

References

[1] J. Hurkała, M. Gall, R. Kutner, M. Maciejczyk: Real-Time Numerical Simulation of the Carnot Cycle, *European Journal of Physics* **26** 673-680 (2005).

[2] A. Galant, R. Kutner, A. Majerowski: Heat Transfer, Newton's Law of Cooling and the Law of Entropy Increase Simulated by the Real-Time Computer Experiments in Java, *Lecture Notes in Computer Science* **2657** 45-53 (2003).

[3] M. Gall, R. Kutner: Molecular mechanisms of heat transfer. Debye relaxation versus power-law, *Physica A* **352** 347-378 (2005).

[4] J.L. Lebowitz, J. Piasecki, Ya.G. Sinai: Scaling dynamics of a massive piston in an ideal gas. In: *Hard Ball Systems and the Lorentz Gas.* Encyclopaedia of Mathematical Sciences **101** 217-227. Springer, Berlin, 2000.

Brill Academic Publishers
P.O. Box 9000, 2300 PA Leiden
The Netherlands

Lecture Series on Computer
and Computational Sciences
Volume 4, 2005, pp. 1069-1072

Research on Fault Tolerance Schemes Based on Energy-Aware

Li Zhong-wen[1,2], Yu Shui[3]

[1]Information Science and Technology College, Xiamen University, China
[2]Zhongshan Institute of UESTC, China
[3]School of Computing and Mathematics, Deakin University, Australia

Received 25 July, 2005; accepted in revised form 8 August, 2005

Abstract: This paper shows that, by supporting processors' speed adjustment during task execution and inserting additional SCPs or CCPs according to characteristics of systems, a significant reduction in the execution time can be achieved. Average execution times to complete a task for the proposed approach are obtained, using the theory of probability. Numerical examples show that compared to previous method, the proposed approach significantly reduces the task execution time. Furthermore, an adaptive processors' speed algorithm, combined with DVS (dynamic voltage scaling) scheme, is put forwarded to reduce energy consumption.

Keywords: Fault Tolerance, DMR, DVS, SCP, CCP, Energy Consumption

1. Introduction

A combination of task duplication [1], DVS (Dynamic voltage scaling) [2] and checkpointing can be used to satisfy system's DVS requirement and improve the run-time reliability of embedded systems. In this paper, with theory of probability [3] and DMR scheme, we present an integrated approach that facilitates fault tolerance through additional CCPs or SCPs and power management through DVS. In addition, an adaptive processors' speed algorithm, combined with the DVS scheme, is put forwarded to achieve energy reduction. Assume that a task τ has a period T, a fixed quantity of computation cycles N in the fault-free condition. Like [4], we normalize the units of N such that the minimum processor speed is 1. To simplify analysis, we assume that a single processor with two speeds f_1 and f_2, and f_1 is the minimum processor speed, namely, $f_1=1$. Moreover, the processor can switch its speed in a negligible amount of time. In our scheme, if a failure occurs, we let the task execute at maximum speed after rolling back. Assume that faults arrive as a Poisson process with rate λ, and faults don't occur during checkpoints. Let t_s, t_{cp}, t_r note time to store the states of processors, time to compare processors' states and time to roll back the processors to a consistent state, respectively.

2. Checkpointing Schemes based on DVS

Assume that τ is divided equally into mn intervals of length $t_t = \lceil N/mn \rceil$, and at the end of each interval, a checkpoint is placed. A CSCP is placed every n intervals. Let ρ be the probability that no faults occurred in both processors, while executing a single interval. Processors' speed is f. We can write the following expression for ρ: $\rho(f) = e^{-2\lambda N \, mnf}$. The probability that no fault occurred in a CSCP interval is $\rho^n(f)$. Let δ denote last interval before the first fault occurred. Let φ be the number of CSCP with identical states before the fault is detected, that is, $\varphi = \lfloor \delta/n \rfloor$. δ and φ are geometric random variables with parameters $\rho(f)$ and $\rho^n(f)$, respectively. Let $T = \lceil N/m \rceil$.

2.1 Additional SCPs

The length of SCP interval T_1 is $T_1 = T/n$. SCPs are placed between CSCPs, the states of two processors are stored at $iT_1(i=1,2,\ldots, n\text{-}1)$. If two states do not agree at time $jT(j=1,2,\ldots, m)$, we need to

---
[1]Corresponding author. E-mail: lizw@xmu.edu.cn. This work is supported in part by Fujian young science & technology innovation foundation (2003J020), NCETXMU 2004 program and Xiamen University research foundation (0630-E23011).

find the most recent SCP with identical states and roll back to it. Let T_{SCP} note the average execution time of τ. It is consisted of two independent parts, we let $T_{SCP1}(f_1)$ and $T_{SCP2}(f_2)$ denote them, respectively. $T_{SCP1}(f_1)$ is the time from the beginning of the task τ execution to the first fault is detected and its rollback is completed. $T_{SCP2}(f_2)$ is the time to complete the remainder computation of τ, including the recovery for the first fault.

- Computing the average value of $T_{SCP1}(f_1)$

After the first fault is occurred in interval $\delta+1$, it will be detected at the $\varphi+1^{st}$ CSCP. Let r denote the research time for the rollback point. To calculate T_{SCP}, we need to find the progress, measured in intervals, and elapsed time from the beginning of the execution to the first fault is detected and its rollback is completed. Assume that the time to rollback processors is included in the time to find the most recent checkpoint with identical states. After the fault is detected, the task is rolled back to the last matching checkpoint. Therefore, the progress $A_{SCP1}(f_1)$ is equal to the last interval without error, namely, δ. Obviously, the average progress until the first fault is expressed by equation (1).

$$\overline{A}_{SCP1}(f_1) = \overline{\delta}(f_1) = \rho(f_1)[1-\rho(f_1)] + \cdots + k\rho^k(f_1)[1-\rho(f_1)] + \cdots = \rho(f_1)/[1-\rho(f_1)] \tag{1}$$

Let $D_{SCP1}(f_1)$ denote the time until the first fault is detected and the rollback to the last matching checkpoint is completed. The time to find the last matching checkpoint after the fault is detected is rt_{cp}. $D_{SCP1}(f_1)$ is expressed by (2). $\overline{r}$ nodes average value of r, using the Huffman tree, is approximately $\log_2 n$. For average value of $D_{SCP1}(f_1)$, please see (3). The average value of $T_{SCP1}(f_1)$ is showed in (4).

$$D_{SCP1}(f_1) = \frac{1}{f_1}[(\varphi(f_1)+1) \times (n(t_l+t_s)+t_{cp}) + rt_{cp}] \tag{2}$$

$$\overline{D}_{SCP1}(f_1) = \frac{1}{f_1}[(\overline{\varphi}(f_1)+1) \times (n(t_l+t_s)+t_{cp}) + \overline{r}t_{cp}] = \frac{1}{f_1}[\frac{n(t_l+t_s)+t_{cp}}{1-p^n(f_1)} + \overline{r}t_{cp}] \tag{3}$$

$$\overline{T}_{SCP1}(f_1) = \overline{A}_{SCP1}(f_1) \times \overline{D}_{SCP1}(f_1)/\overline{A}_{SCP1}(f_1) = \overline{D}_{SCP1}(f_1) \tag{4}$$

- Computing the average value of $T_{SCP2}(f_2)$

To calculate the average execution time of $T_{SCP2}(f_2)$, we need to find the progress $A_{SCP2}(f_2)$, measured in intervals, and elapsed time $D_{SCP2}(f_2)$ from the beginning of the execution under speed f_2 to the completion. The average value of $A_{SCP2}(f_2)$ and $D_{SCP2}(f_2)$ is expressed by (5) and (6), respectively.

$$\overline{A}_{SCP2}(f_2) = \overline{\delta}(f_2) = \rho(f_2)/[1-\rho(f_2)] \tag{5}$$

$$\overline{D}_{SCP2}(f_2) = \frac{1}{f_2}[(\overline{\varphi}(f_2)+1) \times (n(t_l+t_s)+t_{cp}) + rt_{cp}] = \frac{1}{f_2}[\frac{n(t_l+t_s)+t_{cp}}{1-p^n(f_2)} + \overline{r}t_{cp}] \tag{6}$$

We have the average value of $T_{SCP2}(f_2)$:

$$\overline{T}_{SCP2} = [nm - \overline{A}_{SCP1}(f_1)] \times [\overline{D}_{SCP2}(f_2)/\overline{A}_{SCP2}(f_2)] \tag{7}$$

Therefore, the average value of T_{SCP} is:

$$\overline{T}_{SCP} = \overline{T}_{SCP1}(f_1) + \overline{T}_{SCP2}(f_2) = \overline{D}_{SCP1}(f_1) + [mn - \overline{A}_{SCP1}(f_1)] \times [\overline{D}_{SCP2}(f_2)/\overline{A}_{SCP2}(f_2)] \tag{8}$$

2.2 Additional CCPs

In CCPs scheme, the length of CCP interval T_2 is $T_2 = \lceil T/n \rceil$. CCPs are placed between CSCPs, the states of two processors are compared at iT_2 ($i=1,2,\dots, n-1$) and $jT(j=1,2,\dots, m)$. If two states do not agree at iT_2 or jT, some errors have occurred during this interval, and two processors are rolled back to $(j-1)T$. T_{CCP} notes the average execution time of τ. Like T_{SCP}, T_{CCP} is consisted of two independent parts, let $T_{CCP1}(f_1)$ and $T_{CCP2}(f_2)$ denote them, respectively.

- Computing the average value of $T_{CCP1}(f_1)$

After the first fault is occurred at the end of interval $\delta+1$, a rollback to the end of interval $n\varphi$ is performed. Therefore, the progress between faults is $A_{CCP1}(f_1) = n\varphi$, and the average progress until the first fault is expressed by equation (9).

$$\overline{A}_{CCP1}(f_1) = n\overline{\varphi}(f_1) = np^n(f_1)/[1-p^n(f_1)] \tag{9}$$

The first fault is detected between CSCP φ and $\varphi+1$, after interval $\delta+1$. The elapsed time, $D_{CCP1}(f_1)$, until this fault is detected and the rollback is completed is expressed by (10), and (11) is its average value. The average time of $T_{CCP1}(f_1)$ is expressed by (12).

$$D_{CCP1}(f_1) = \frac{1}{f_1}[(\delta(f_1)+1)\times(t_l+t_{cp})+\varphi(f_1)\times t_s + t_r] \tag{10}$$

$$\overline{D}_{CCP2}(f_1) = \frac{1}{f_1}[(\overline{\delta}+1)(t_l+t_{cp})+t_s\overline{\varphi}(f_1)+t_r] = \frac{1}{f_1}[\frac{t_l+t_{cp}}{1-p(f_1)}+\frac{p^n \times t_s}{1-p^n(f_1)}+t_r] \tag{11}$$

$$\overline{T}_{CCP1}(f_1) = \overline{A}_{CCP1}(f_1)\times\overline{D}_{CCP1}(f_1)/\overline{A}_{CCP1}(f_1) = \overline{D}_{CCP1}(f_1) \tag{12}$$

• Computing the average value of $T_{CCP2}(f_2)$

Like the calculation of $A_{CCP1}(f_1)$, to calculate the average execution time of $T_{CCP2}(f_2)$, we need to find the progress $A_{CCP2}(f_2)$, measured in intervals, and elapsed time $D_{CCP2}(f_2)$ from the beginning of the execution under speed f_2 to the completion. The average value of $A_{CCP2}(f_2)$ and $D_{CCP2}(f_2)$ is expressed by (13) and (14), respectively. The average time to execute the whole task is in (15).

$$\overline{A}_{CCP2}(f_2) = n\times\overline{\varphi}(f_2) = np^n(f_2)/[1-p^n(f_2)] \tag{13}$$

$$\overline{D}_{CCP2}(f_2) = \frac{1}{f_2}[(\overline{\delta}+1)(t_l+t_{cp})+t_s\overline{\varphi}(f_2)+t_r] = \frac{1}{f_2}[\frac{t_l+t_{cp}}{1-p(f_2)}+\frac{p^n \times t_s}{1-p^n(f_2)}+t_r] \tag{14}$$

$$\overline{T}_{CCP2}(f_2) = [nm-\overline{A}_{CCP1}(f_1)]\times\overline{D}_{CCP2}(f_2)/\overline{A}_{CCP2}(f_2) \tag{15}$$

Therefore the average value of T_{CCP} is:

$$\overline{T}_{CCP} = \overline{T}_{CCP1}(f_1)+\overline{T}_{CCP2}(f_2) = \overline{D}_{CCP1}(f_1)+[mn-\overline{A}_{CCP1}(f_1)]\times\overline{D}_{CCP2}(f_2)/\overline{A}_{CCP2}(f_2) \tag{16}$$

3. Numerical examples

We carried out a set of experiments to evaluate our checkpointing schemes (namely, schemes with DVS) and to compare them with previous schemes (namely, schemes without DVS). Without loss of generality, we let $f_2 = 2f_1$ [2]. Assume $N=10000$. In figure 1, the performance of DMR scheme without DVS is compared to the performance of the proposed scheme with two or four SCPs between CSCPs.

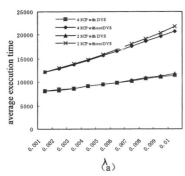
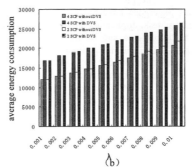

Figure 1. Comparison between schemes with and without DVS:
(a) average execution time, (b) average energy consumption

Figure 1.a shows that compared to previous schemes, our schemes significantly reduce the task execution time. Here $t_{cp} = 5\times10^{-4}$, $t_s = 5\times10^{-5}$ and $t_r = 10^{-5}$. The energy, E, consumed during the execution of task is proportional to the time it takes to execute the task and to the square of the speed during execution [4]. For the simplicity, let the proportionality constant equal 1. Figure 1.b describes E. Figure 1 shows us that schemes with DVS can consume more energy than schemes without DVS. In detail, E of 4SCP without DVS (2SCP without DVS) is 75.78 (76.73) percent of E consumed in scheme 4SCP with DVS (2SCP with DVS). We can get similar results when carried out simulation experiments of CCP schemes. Here, $t_{cp} = 5\times10^{-5}$, $t_s = 5\times10^{-4}$ and $t_r = 10^{-4}$.

4. Adaptive speed algorithm

In this section, we aim to reduce energy consumption based on the proposed schemes. It is done by letting processors possibly to execute task at low speed. Let T_{test} be an estimate of the time that the task has to execute in the presence of faults and with checkpointing. The expected number of faults for the duration T_{test} is λT_{test}. And let R_d denote the time left before the deadline. It is obtained by subtracting the time from D. R_t denotes the remaining execution time. Assume processors' speed is f, T_{test} of scheme with additional SCPs and scheme with additional CCPs are expressed by equation (17).

$$T_{test}(f, SCP) = \frac{1}{f}\{R_N + \lambda T_{test}(f)[n(t_I + t_s) + t_{cp}) + \overline{r}t_{cp}]\} \qquad T_{test}(f, CCP) = \frac{1}{f}\{R_N + \lambda T_{test}(f)[n(t_I + t_{cp}) + t_r)]\} \qquad (17)$$

The first term on the right-hand side of formulas denotes the time for forward execution, the second term denotes the recovery time for λT_{test} faults. Procedure *adapspeed*() is the adaptive speed algorithm.

```
Procedure adapspeed(f₁, f₂, D, N, λ, model)
1. model=CCP or SCP;
2. if model=SCP R_N=m[n(t_I+t_s)+t_cp]; else R_N=m[n(t_I+t_cp)+t_s];
3. R_d=D;
4. IF (T_test (f₁,model) <=R_d) f=f₁; else f=f₂;
5. While (R_N >0) do {
6.    If (R_N>R_d) break;
7.    Case 1: During normal execution, do {
8.            resume execution;
9.            if find fault then {
10.           roll back and restore status;
11.           Update R_N, R_d;
12.           go to 13}}
13.   Case 2: Upon fault occurrence, do {
14.   IF (T_test (f₁,model) <=R_d) f=f₁; else f=f₂;
15.   Resume execution;}}
```

5. Conclusion

In this paper, we have presented a unified method to improve the performance by unifying checkpointing, task duplication and dynamic voltage scaling for a real-time task. This approach provides fault tolerance and facilitates dynamic power management. Based on this, we design an adaptive processors' speed algorithm to reduce energy consumption. This algorithm uses a dynamic voltage scaling criterion that is based not only on the slack in task execution but also on the occurrences of faults during task execution.

Acknowledgments

The author wishes to thank Prof. Marcin Kaminski for his careful reading of the manuscript and his insightful comments on an earlier version of this paper.

References

[1] Ziv A, Bruck J. Analysis of Checkpointing Schemes with Task Duplication, IEEE Transactions on Computers , 1998, 47(2): 222-227

[2] Ying Z, Crishnendu C. Task feasibility analysis and dynamic voltage scaling in fault-tolerant real-time embedded systems, Proc. Of the design, automation and test in Europe conference and exhibition (DATE'04), 2004

[3] Ziv A, Bruck J. Performance Optimization of Checkpointing Schemes with Task Duplication. IEEE Transactions on Computers, 1997, 46(2): 1381-1386

[4] Melhem R, Mosse D, Elnozahy E. The interplay of power management and fault recovery in real-time systems, IEEE Tran. on computers, 2004, 53(2): 217-231

Brill Academic Publishers
P.O. Box 9000, 2300 PA Leiden
The Netherlands

*Lecture Series on Computer
and Computational Sciences*
Volume 4, 2005, pp. 1073-1073

2005 International Symposium of Computational Electronics: Physical Modeling, Mathematical Theory, and Numerical Algorithm

Yiming Li

The field of physical modeling, mathematical theory, and numerical algorithm plays a key role in scientific and engineering application domains and has obtained prominence through advances in technologies of electrical engineering, for example. Computational electronics then is a natural field that is ever ready to receive new efforts, and open forums in this area are always welcome. We therefore invite contributions to a symposium of International Conference of Computational Methods in Sciences and Engineering (ICCMSE 2005) for computational electronics. The purpose of this symposium is for scientists and engineers, computer scientists, applied mathematicians, physicists, and researchers to present their recent advances, ideas, results and to exchange experiences in the areas of modeling, simulation, optimization and other support for problems in electrical engineering.

Ten papers have carefully been reviewed and selected among fifteen very excellent submissions. Topics including quantum computing, device modeling, circuit simulation, signal control, optimization, chip implementation, and parallelization will be presented, from mathematical, statistical, and numerical points of view, in this symposium. Welcome to this symposium in International Conference of Computational Methods in Sciences and Engineering (ICCMSE 2005).

Look forward to seeing you in Korinthos, Greece!

Symposium Organiser
Prof. Dr. Yiming Li

Yiming Li

Yiming Li is an Associate Professor with the Department of Communication Engineering and the Microelectronics and Information Systems Research Center, the National Chiao Tung University at Hsinchu, Taiwan. His current research areas include computational electronics and physics, physics of semiconductor nanostructures, device modeling, parameter extraction, and circuit simulation, development of TCAD/ECAD tools and SOC applications, bioinformatics and computational biology, and advanced numerical method, parallel and scientific computation, and computational intelligence. Dr. Li has authored or coauthored over 120 research papers appearing in international book chapters, journals, and conferences.

Brill Academic Publishers
P.O. Box 9000, 2300 PA Leiden,
The Netherlands

*Lecture Series on Computer
and Computational Sciences*
Volume 4, 2005, pp. 1074-1077

Lyapunov Type Algorithm for Coupled Riccati Equations in MCV Problem

L.Cherfi [1]

Ecole Normale Supérieure de Cachan,
Laboratoire SATIE (UMR,CNRS, 8029)
61 Av. Président Wilson
94235 Cachan, France

Received 16 June, 2005; accepted in revised form 10 July, 2005

Abstract: In this study we propose a new proof for the convergence of the Lyapunov type algorithm [1]. This may be guaranteed by the existence of a positive definite solution of a standard algebraic Riccati equation. Moreover, we will show that this algorithm is well defined and converges to a unique pair of stabilizing solutions of the coupled algebraic Riccati equations under-study.

Keywords: Coupled Riccati equation, Riccati iteration, minimal cost variance, fixed point, numerical algorithm, control theory, algebraic Riccati equations,Risk sensitive problem.

Mathematics Subject Classification: 34A34, 65H20, 54H25, 32H50.

1   Introduction

This paper deals with the iterative stabilizing solutions of the coupled algebraic Riccati equations which arise in minimal cost variance [2, 3, 4]:

$$0 = A^T M + MA - MSM + \gamma^2 VSV + Q, \tag{1}$$
$$0 = A^T V + VA - 2\gamma VSV - MSV - VSM + 4MWM, \tag{2}$$

where

$$S = BR^{-1}B^T \ and \ W = E\tilde{W}E^T.$$

The goal of this work is to give a new proof for the convergence of the Lyapunov type algorithm which has been introduced in the paper of [1].

2   Lyapunov type algorithm

In the following let us assume that $Q > 0$, $S \geq 0$ and $W \geq 0$. Using the abbreviation $Z = M + \gamma V$, we get an elementary calculation that

$$0 = A^T Z + ZA - ZSZ + 4\gamma MWM + Q. \tag{3}$$

Here, we present the details of the above-mentioned algorithm for solving the coupled algebraic Riccati equations (1) and (2). Each iterative technique presented here will stop until the right-hand

---

[1]Corresponding author. PHD Student. E-mail:cherfi@satie.ens-cachan.fr

side of these equations is smaller than a chosen precision ϵ sufficiently small and fixed in advance.

Algorithm

• Start with $Z_0 \geq 0$ such that the matrix $A - SZ_0$ is Hurwitz. Put $l = 1$.

Define M_l, V_l, $N_1(M_l, V_l)$ and $N_2(M_l, V_l)$, $l \geq 1$, as follows

(a) determine M_l the unique solution of

$$(A - SZ_{l-1})^T M_l + M_l(A - SZ_{l-1}) + Z_{l-1}SZ_{l-1} + Q = 0, \tag{4}$$

and V_l the unique solution of

$$(A - SZ_{l-1})^T V_l + V_l(A - SZ_{l-1}) + 4M_{l-1}WM_{l-1} = 0. \tag{5}$$

(b) Put $A_{l-1} = A - SZ_{l-1}$, for $l \geq 1$, compute

$$N_1(M_l, V_l) \quad = \quad A_l^T M_l + M_l A_l + Z_l S Z_l + Q,$$

$$N_2(M_l, V_l) \quad = \quad A_l^T V_l + V_l A_l + 4M_l W M_l,$$

while

$$e_l = \max\{\|N_1(M_l, V_l)\|_\infty, \|N_2(M_l, V_l)\|_\infty\}, \tag{6}$$

is larger than ϵ, replace l by $l+1$ and go to (a), else stop.

Note that we obtain with the two equations (4) and (5), the iterative equation for $Z_l = M_l + \gamma V_l$, for $l \geq 1$

$$0 \quad = \quad (A - SZ_{l-1})^T Z_l + Z_l(A - SZ_{l-1}) + Z_{l-1}SZ_{l-1} + M_{l-1}WM_{l-1} + Q. \tag{7}$$

We consider the following standard Riccati equation

$$0 = A^T P + PA + Q - PSP. \tag{8}$$

Observe that (8) can be rewritten, $\forall l \geq 1$ as

$$0 = (A - SZ_l)^T P + P(A - SZ_l) + Z_l S Z_l - (P - Z_l)S(P - Z_l) + Q. \tag{9}$$

For a better understanding of the following developments we recall the following results from [?]:
Let X be a solution of the Lyapunov algebraic equation

$$0 = A^T X + XA + Q. \tag{10}$$

Lemma 1 *Suppose that $Q \geq 0$ and that the Lyapunov algebraic equation (10) has a solution $X \geq 0$. Then, the following assertions hold*

• *(i) If $Q > 0$, then A is Hurwitz and $X > 0$.*

Lemma 2 *If A is Hurwitz and $Q > 0$ (or $Q \geq 0$), then the unique solution X of (10) is positive definite (or positive semi-definite, respectively).*

Theorem 1 *Assume that the following conditions are satisfied*

1. *There exists a positive definite solution P of the algebraic Riccati equation (8).*

2. *There exists a positive semi-definite matrix Z_0 such that $(A - SZ_0)$ is Hurwitz.*

3. $\Delta S + S\Delta + \Delta S\Delta \geq 0$, *if* $\Delta \geq 0$.

Then

- *(i) The matrices* $(A - SZ_l)$ *are Hurwitz,* $\forall l \geq 0$.

- *(ii) The sequence* $(Z_l)_{l\geq 1}$ *is uniquely defined and is bounded below by* P.

- *(iii) The sequence* $(M_l)_{l\geq 1}$ *is uniquely defined and is bounded below by* P.

Remark 1 *The condition 3 of Theorem 1, ensures that* $X_1 S X_1 \leq X_2 S X_2$, *if* $X_1 \leq X_2$ *for any* X_1 *and* X_2 *symmetric positive semidefinite matrices.*

Proof of Theorem 1

First we prove (i) and (ii) simultaneously. For $l = 0$ in (7) and (9), we obtain in the first step the Lyapunov equations

$$0 = (A - SZ_0)^T Z_1 + Z_1(A - SZ_0) + Z_0 SZ_0 + 4\gamma M_0 W M_0 + Q, \tag{11}$$

$$0 = (A - SZ_0)^T P + P(A - SZ_0) + Z_0 SZ_0 + (P - Z_0)S(P - Z_0) + Q. \tag{12}$$

Subtracting (12) from (11) we obtain

$$0 = (A - SZ_0)^T(Z_1 - P) + (Z_1 - P)(A - SZ_0) + (P - Z_0)S(P - Z_0) + 4\gamma M_0 W M_0. \tag{13}$$

Since $A - SZ_0$ is Hurwitz, by lemma (1), together with $\gamma \geq 0$, $Q > 0$, $S \geq 0$ and $W \geq 0$, we infer that $0 < P \leq Z_1$. In addition for l=1, we obtain in (9)

$$0 = (A - SZ_1)^T P + P(A - SZ_1) + Z_1 SZ_1 - (P - Z_1)S(P - Z_1) + Q \tag{14}$$

Together with $0 < P \leq Z_1$, we infer from condition 3. that the matrix $Z_1 SZ_1 + (P - Z_1)S(P - Z_1) + Q > 0$ and by Lemma (1), (i), the matrix $(A - SZ_1)$ is Hurwitz. Therefore the statement $0 < P \leq Z_1$ and $(A - SZ_1)$ is Hurwitz is true for $l = 1$.

Let us assume that is true for $k = 0, \ldots, l$. The next steps are to prove that under these assumptions, the statement is also true for $l + 1$. Then by mathematical induction we conclude that is holds for all indices l. Consider the matrices P and Z_{l+1}, by subtracting (9) from (7) we obtain

$$0 = (A - SZ_l)^T Z_{l+1} + Z_{l+1}(A - SZ_l) + Z_l SZ_l + \gamma M_l W M_l + Q. \tag{15}$$

Since $(A - SZ_l)$ is Hurwitz and $0 < P \leq Z_l$, $\gamma \geq 0$, $Q > 0$, $S \geq 0$ and $W \geq 0$, by Lemma 2, we infer that $P \leq Z_{l+1}$.

At the step $l + 1$, the algebraic Riccati equations (8) can be written as

$$0 = (A - SZ_{l+1})^T P + P(A - SZ_{l+1}) + Z_{l+1} SZ_{l+1} - (P - Z_{l+1})S(P - Z_{l+1}) + Q. \tag{16}$$

Since $P > 0$ and $(P - Z_{l+1})S(P - Z_{l+1}) + Q > 0$, by Lemma 1 and by condition 3. we infer that the matrix $(A - SZ_{l+1})$ is Hurwitz.

To prove the assertion (iii) we need the two equations (4) and (9). By subtracting (9) from (4) we obtain

$$0 = (A - SZ_{l-1})^T(M_l - P) + (M_l - P)(A - SZ_{l-1}) + (P - Z_{l-1})S(P - Z_{l-1}) \tag{17}$$

Since $(A - SZ_{l-1})$ is Hurwitz and $(P - Z_{l-1})S(P - Z_{l-1}) \geq 0$, by Lemma 2, we infer that $P \leq M_l$.

Theorem 2 *Assume that the sequence* $M_{l-1} W M_{l-1} - M_l W M_l \geq 0$, $\forall l \geq 1$, *then the iterative sequence* $(Z_l)_{l\geq 1}$ *defined by (7) converges to a positive semi definite matrix* Z *such that* $(A - SZ)$ *is Hurwitz.*

Proof of Theorem 2 Consider the algebraic Lyapunov equation (7), we obtain after a short calculation

$$
\begin{aligned}
0 = & \ (A - SZ_l)^T(Z_l - Z_{l+1}) + (Z_l - Z_{l+1})(A - SZ_l) \\
& + 4\gamma[M_{l-1}WM_{l-1} - M_lWM_l] + (Z_{l-1} - Z_l)S(Z_{l-1} - Z_l).
\end{aligned} \tag{18}
$$

Tacking in count that the matrix $(A - SZ_l)$ is Hurwitz, $M_{l-1}WM_{l-1} - M_lWM_l \geq 0$ and $S \geq 0$, by Lemma 2, we infer that $Z_l - Z_{l+1} \geq 0$, $\forall l \geq 1$. Consequently $Z = lim_{l \to \infty} Z_l$ exists. Since $(Z_l)_{l \geq 1}$ is positive definite, so it converges to a positive semi-definite matrix.

Let $(e_l)_{l \geq 1}$ be an eigenvector of $A - SZ_l$ associated to the eigenvalue $(\lambda_l)_{l \geq 1}$ such that $\|e_l\| = 1$. Since $A - SZ_l$ is Hurwitz, let us put $\lambda_l = a_l + ib_l$ with $a_l < 0$.
Let M_l be the positive definite solution of the Lyapunov equation (4), i.e

$$
0 = \ (A - SZ_{l-1})^T M_l + M_l(A - SZ_{l-1}) + Z_{l-1}SZ_{l-1} + Q. \tag{19}
$$

Then by multiplication of (19) by e_{l-1}^* and e_{l-1} we obtain with $Q > 0$ and $S \geq 0$, that

$$
e_{l-1}^*[Q + Z_{l-1}SZ_{l-1}]e_{l-1} \geq e_{l-1}^*Qe_{l-1} \geq \lambda_{min}(Q) > 0. \tag{20}
$$

Together with $M_l > 0$, we infer from (20) that $\exists \alpha, \forall l \geq 0, Re(\lambda_l) \leq -\alpha < 0$. So, $lim_{l \to \infty} Re(\lambda_l) < 0$, then the matrix $(A - SZ)$ is Hurwitz.

Corollary 1 *Assume that the sequence $(Z_l)_{l \geq 1}$ is convergent, then the sequences $(M_l)_{l \geq 1}$ and $(V_l)_{l \geq 1}$ are also convergent to a pair of positif semi-definite matrices M and V respectively.*

3 Conclusion

The solution of the infinite time horizon MCV control problem is found in this paper. This has been done by applying an iterative linear method like any Lyapunov type ones to solve a coupled algebraic Riccati equations. The main result obtained here is that we establish a new proof for the convergence of this method by using an upper bound of the solution and with out any detectability assumption. Moreover the obtained solution is a unique and a stabilizing one. At the last of this paper we have discussed and compared ours results to other ones which are established in the literature.

Acknowledgment

The author wishes to thank the anonymous referees for their careful reading of the manuscript and their fruitful comments and suggestions.

References

[1] G.Freiling and S.R.Lee and G.Jank, Coupled matrix Riccati equations in minimal cost variance control problems : IEEE Automat. Control, 44(1999), 556-560.

[2] M.K.Sain, C.H.Won and B.F.Jr.Spencer, Cumulant in risk-sensitive control: The full-state feedback cost variance: Conference on Decision and Control, (1995), 1036-1041.

[3] Chang-Hee Won and M.K.Sain and S.R.Liberty, Infinite Time minimal cost variance control and coupled algebraic Riccati equations:Proceedings of the American Control Conference, (2003),5155-5160.

[4] M.K. Sain, On minimal Variance Control of linear systems with quadratic loss:Ph.D, Department of Electrical Engineering, University of Illinois, Urbana, January 1965.

Brill Academic Publishers
P.O. Box 9000, 2300 PA Leiden
The Netherlands

*Lecture Series on Computer
and Computational Sciences*
Volume 4, 2005, pp. 1078-1081

A Circuit Simulation Technique for Network Assignment Problem

Hsun-Jung Cho[1], Tsu-Tian Lee[2], and Heng Huang[3]

[1,3]Department of Transportation, National Chiao Tung University, Hsinchu, Taiwan

[2] Electronical and Control Department, National Chiao Tung University, Hsinchu, Taiwan

Received 24 July, 2005; accepted in revised form 12 August, 2005

Abstract: In this paper, under the network user equilibrium principle (Wardrop, 1952), a method to use the analogous electric circuit to simulate network traffic assignment, and to measure the current flow and voltage of the electric circuit to present the result of traffic assignment is proposed. The basic ideas are using the resistance and constant voltage source combines the analogous electric circuit, and using the current source presents the net traffic flow. Also, the nonlinear link cost function is approximated by the piecewise function. Finally, an example shows that an electric circuit simulating the network traffic assignment is workable.

Keywords: Traffic Assignment, User Equilibrium, Link Cost Function, Piecewise Linear Function, Electric Circuit

Mathematics Subject Classification: 90B10, 90B20, 90C30

1. Introduction

One of the major problems facing transportation engineers and urban planners is that of predicting the impact of given transportation scenarios. Travelers choose if and when to take a trip for some purpose, and which way to get there. These decisions depend, in part, on how congested the transportation system is and where the congested points are (Friesz, 1985).

Up to now, assignment of traffic flow using the analogous current flow in an electric circuit has been confined to the simple network in which there exist several alternative routes in parallel (Sasaki and Inouye, 1980).

In this paper, a novel concept is proposed to simulate the highway network and obtain the useful transportation information by specially designed electric circuit. The management of information and control systems can be systemized accordingly.

The key to the proposed approach is to find an appropriate design of electric circuit to describe the highway network. The mature and well-established electronic technology will then be applied to the system to make transportation management more effective.

From transportation planning point of view, for the network traffic assignment, there are several basic assumptions. First, the highway network has a single concentrating node and common destination. Second, current flow corresponds to traffic flow, and the difference of voltages corresponds to travel time. Third, the flow pattern obtained by the proposed method satisfies the equal travel time principle. Finally, the direction of every link is the same as that of the direction of the corresponding current flows to assure that constant voltage source will provide fixed voltage difference.

Sasaki and Inouye assume that the travel time is a linear function of traffic flow and that there exists a finite time even though the traffic flow approaches to zero. The link cost function is assumed to be nonlinear in this paper, so piecewise method would be used. Then network traffic assignment can be simulated by combining the concepts of the diode, resistance, constant voltage source, and current source and the theories of the analogous electric circuit. The approach to design an electric circuit similar to the feature of the cost function will be given in the following sections.

2. Mapping electric circuit to traffic

In physics, the electric current, I is the rate of change of charge flow through a cross section, that is

---

[1] Corresponding Author: Professor : hjcho@cc.cntu.edu.tw

$$I = \frac{\Delta Q}{\Delta t} \rightarrow I = \frac{dQ}{dt} \tag{1}$$

The definition of traffic flow is the number of vehicles passing a fixed point per hour. Under the relevant assumptions, traffic flow can be treated as electric current. The Boltzmann equation (Prigogine and Ferman, 1971) can be used in traffic flow theory by regarding vehicles as particles.

The resistance can be taken as the impedance of the road. High impedance means that it is hard to drive on the road. According to Ohm's law, voltage is equal to current multiplied by resistance, so for the same road, if the current increases, the voltage is increasing simultaneously. This phenomenon is similar to the fact that the travel time increases as the traffic flow increases. Therefore the voltage can be mapped to the travel time. As to the cost function, the impedance of the road is a nonlinear function with respect to traffic flow.

3. Link Cost Function

Cost function describes the functional relationship between traffic flow and traffic time. Let $c_a(f_a)$ be the average travel time experienced by each user of link a when f_a units of vehicles flow along the link. Here, $c_a(f_a)$ is an increasing function: it is taken to be nonlinear because of the effects of congestion on the travel time for each user of any link. The form used in this paper will be the one used in the U.S. Federal Highway Administration traffic assignment models.

Now define

$$c_a(f_a) = t_0 \left(1 + \beta \left(\frac{f_a}{C} \right)^4 \right) \tag{2}$$

These functions are shown in Figure 1: t_0 is link free flow travel time; C is link capacity and β is empirically determined parameter for each link which is computed from its length, speed limit, geometric design including number of lanes, and traffic lights (LeBlanc et al., 1975). As in the figure, the travel time per user increase very slowly at first; it remains almost constant for low levels of flow. However, as the flow begins to reach the level for which the link was designed, the travel time experienced by each user begins to increase rapidly.

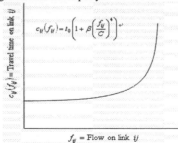

Figure 1: Travel time functions used in the U.S.F.H.W.A. traffic assignment model.

4. Calibration of Link Cost Function and Construct Link Model

In order to construct the piecewise model, here are some basic assumptions:
1. There is only an origin-destination pair in the network.
2. The voltage corresponds to the travel time; the current corresponds to the traffic flow.
3. The relation between travel time and traffic flow on the link is nonlinear and a fixed travel time to pass the link exists.
4. Neglecting the travel time in the node.
5. Every path has the same total travel time when the system is in user equilibrium condition.
6. Every link is directional and as the same way of current.
7. The traffic flow on the link can not be over its capacity.
8. The diode is the ideal diode.

In order to design an electric circuit that has the same characteristic curve as the cost function, the diode is used to control the piecewise curve as a switch. Because of the characteristic of diode, it controls the piecewise characteristic by turn-on or turn-off two states. When the diode D1 (Figure 2) is in turn-off state, the diode is equivalent to an open circuit, and only has the reversed bias (voltage), so the entire circuit represents the characteristic as R1. Conversely, when D1 is in turn-on state, the diode is equivalent short circuit (actually a very small resistance). Furthermore, R1 and D1 is similar to parallel, so the entire circuit has zero resistance, the increased voltage as the current increasing is very small, and I-V curve rises slightly. However, the resistance R1 calculated before is too small to neglect the diode self resistance, and the diode threshold voltage is also taken into consideration, so the adjustment has to be made. The travel time modifies to 1 minute correspond to 100 volt, and the resistance is proportionally increases 100 times with fixed current (flow). Because the voltage V2 determines whether the diode is in turn-on state or not, let V2 be the product of R1 and I1, the equivalent circuit should be represented as the characteristic of resistance R1 at current flow I1.

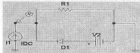

Figure 2: A part of piecewise model

After combination of each part (Figure 3), the beginning voltage and the characteristic value are not the same, so a constant voltage source (V4) is added in order to match the characteristic value. Because all parts are in series, the characteristic of entire circuit is the superposition of each part. Then the circuit is simulated with PSPICE, the simulation I-V curve is as below.

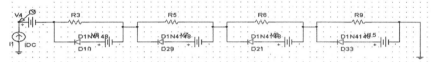

Figure 3: The entire circuit model of a link

If taking a 5 kilometers link for example, Figure 4 is the comparison between the cost function and the simulation of the electric circuit. It is found easily that the difference is can be tolerated.

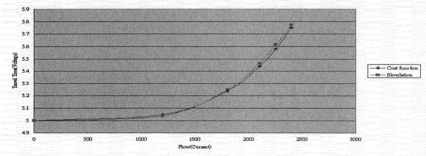

Figure 4: Graphic comparison between cost function and simulation

5. Construct Network Model

If there is a network with one Origin-Destination pair and four links like Figure 5 , based on the link model above, each link can be constructed an equivalent circuit model with its characteristic. Then, a network model can be obtained by connecting each link model according to the network structure. Furthermore, there is a current source at the origin to control the input traffic flow and the destination is ground. Hence, the entire network model is shown as Figure 6. If the input current source is controlled at a fixed value, then the equilibrium travel time of O-D pair and the flow on each link can be obtain by the result of electric circuit simulation because of the voltage difference between O-D is the same no matter which path the current passes by.

Figure 5: Example network

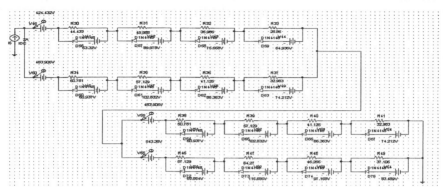

Figure 6: The model of example network

6. Conclusion

According to the simulated results of this paper proposed circuit model, it can be proved that using electric circuit to simulate traffic assignment is possible. If the result focuses on the total travel time when the system is in user equilibrium condition, it can be found that the difference of the equilibrium between real cost function and our model is very small. Besides, there are several advantages of using electric circuit to simulate traffic assignment. First, the velocity of the electron in circuit is very quick, so the result of user equilibrium can be obtained fast. Second, the range of electric current can be inputted variously as using the software of electric circuit simulation, so the equilibrium of traffic assignment resulting from different traffic flow is diverse. Finally, using electric circuit to simulate traffic assignment can systematize the information and control system, and it not only reduces data transmitting time, but also decreases the risk of data losing.

Acknowledgments

This research was partially supported by the National Science Council, R.O.C., under contract number NSC 93-2211-E-009-032, also partially supported by the Ministry of Education, R.O.C. under contract number EX-91-E-FA06-4-4.

References

[1] A.S. Sedra and K.C. Smith, *Microelectronic Circuit*, Fourth Edition, Oxford, New York, 2003.

[2] T. Friesz, Transportation Network Equilibrium, Design and Aggregation: Key Developments and Research Opportunities, *Transportation Research A* **19** 413-427(1985).

[3] L.J. LeBlanc, E.K. Morlok and W.P. Pierskalla, An Efficient Approach to Solving the Road Network Equilibrium Traffic Assignment Problem, *Transportation Research B* **9** 309-318(1975)

[4] E.K. Morlok, Types of Transportation Supply Functions and Their Applications, *Transportation Research B* **14** 9-27(1980).

[5] I. Prigogine and R. Herman, *Kinetic Theory of Vehicular Traffic*, American Elsevier Publishing Company, New York, 1971.

[6] T. Sasaki and H. Inouye, Traffic Assignment by Analogy to Electric Circuit, *Proceedings of the International Symposium on Transportation and Traffic Theory*, 1980.

Brill Academic Publishers
P.O. Box 9000, 2300 PA Leiden
The Netherlands

*Lecture Series on Computer
and Computational Sciences*
Volume 4, 2005, pp. 1082-1083

A Four Base Computational meThod for the Implementation of a Quantum Computer using Silicon Devices: Circuit and Simulation

D. Ntalaperas, K.Theodoropoulos ,N. Konofaos, A.N.Tsakalidis

Computer Engineering and Informatics Dept., University of Patras, GR-26500 Patras, Greece,
and
Research Academic Computer Technology Institute, Patras, Greece.

Received 11 July, 2005; accepted in revised form 12 August, 2005

Abstract: In this paper a model for the implementation of a quantum computer based on semiconductor devices is being presented. The mechanism in which a two level computational basis is represented, as well as a detailed description of the quantum CNOT gate is given. A simulation of a quantum circuit being implemented in the above model is also given among with some estimation of the errors introduced, when the computation is performed in actual physical systems.

Keywords: Quantum Computing, Semiconductors, CNOT

PACS: 03.67.-a, 03.67.Lx, 03.67.Pp, 71.55.-i , 72.20.Jv, 73.20.-r

1. Extended Abstract

Quantum mechanical systems can be in a superposition of states, allowing for a two level system to be in more than one basis states (eigenstates) at the same time. By using this property, quantum systems which can perform parallel computation following numerous computational paths simultaneously, can be built. These quantum computers are already known to outperform their classical counterparts in a variety of computational tasks [1,2]. In this work an implementation technique is presented which makes use of the properties of the electronic traps occurring in silicon based semiconductor devices.

The different trap levels are used in order to represent the computational basis for the quantum computer. The correspondence between physical states and logical ones is such that the number of electrons in two adjacent traps, say trap A and trap B, is used to denote the value of the control qubit of the basis. Specifically, if the occurrence of exactly one electron in the two traps is denotes a value of the control qubit equal to 1 and 0 otherwise, the following mapping from physical states to logical ones can be made:

$|00>$ → Both traps are unoccupied
$|01>$ → Both traps are occupied
$|10>$ → Trap A occupied. Trap B unoccupied
$|11>$ → Trap A unoccupied. Trap B occupied

Using suitable physical procedures, logical operations can be performed. One such procedure is the action of "electron flip" [3], in which the states change occupancy, hence the following table is obeyed:

$|00>$ → $|00>$
$|01>$ → $|01>$
$|10>$ → $|11>$
$|11>$ → $|10>$

This table is the same as the truth table for the quantum Controlled – NOT operation (CNOT Gate), one of the three gates requited for universal quantum computation [4].

By chargeing or decharging a trap (depending always on its current state), a superimposed state for one trap can be obtained. This operation can be mapped to the logical equivalent of the Hadamard Gate the second of the three universal quantum gates, which obeys the following:

$$|0> \quad \rightarrow \quad 1/\sqrt{2} \ (|0> + |1>) \qquad \text{(charging)}$$
$$|1> \quad \rightarrow \quad 1/\sqrt{2} \ (|0> - |1>) \qquad \text{(decharging)}$$

This operation can be performed in either qubit (control or target) of the computational basis. Therefore, if these two operations are performed in sequence, and assuming the existence of the third universal quantum gate (PHASE gate), quantum algorithms can be executed by applying a series of external signals onto the states system. .

For example, the implementation of the quantum Fourier Transform, needed for Shor's factorization algorithm, can be obtained if n different trap pairs are used, where n denotes the length of the number to be factored. In this case, the PHASE operation is also needed, denoted by R_n denotes and given by:

$$a|0> + b|1> \quad \rightarrow \quad a|0> + exp(2\pi j/2^n) \ |1>$$

Errors in the above procedure can be introduced by either decoherence or by electron spreading between trap levels which are not used for the computation.

For the errors of the first type, it should be noted that the coupling factors between the semiconductor electrons and the external signal are fairly small [4]. Even so, this error adds up during the computation. If l is the number of steps required for the entire computation, the total error would be of the order of $l\delta$, where δ is an upper limit of the error due to the coupling between the two systems.

The error due to spreading can be addressed by assuming a physical procedure that can manipulate the semiconductor properties accordingly (i.e. a controlled introduction of impurities into the semiconductor, such that the energy gap between traps is large enough for an electron to be located in more than two energy levels at the same time). The error due to overlapping between traps ε would remain constant through the computation.

Combining the above, the total error would be $l\delta + \varepsilon$, a term which can then be subtracted by the final outcome for given small values of δ and ε. Simulations of the above model have been carried out, in which errors have been introduced and various results for the accepted values of δ and ε have been obtained. Hence, by taking into account the obtained values and connecting them to the existing semiconductor technologies, the theoretical feasibility of a quantum computer can be decided.

References

[1] L.K. Grover, Phys. Rev. Lett. 79 , pp. 325-328, (1996).

[2] P.W Shor, in Proc. 35th Annu. Sump. Foundations of Computer Science (ed. Goldwasser, S.) pp.124-134 (IEEE Computer Society, Los Alamitos, CA 1994).

[3] R.Bube, *"Photoelectronic Properties of Solids",* John Wiley and Sons, New York, 1997.

[4] M A, Nielsen I L Chuang, *Quantum Computation and Quantum Information*, Cambridge University Press, 2000.

[5] K.Theodoropoulos, D.Ntalaperas, I.Petras, A.Tsakalidis, N.Konofaos, in Proc. ICPS2004, 27th Int. Conference on the Physics of Semiconductors, July 26-30, 2004, Flagstaff, Arizona, USA, published by AIP conference proceedings, vol. 772, pp. 1463-1465, 2005.

Brill Academic Publishers
P.O. Box 9000, 2300 PA Leiden
The Netherlands

*Lecture Series on Computer
and Computational Sciences*
Volume 4, 2005, pp. 1084-1087

Parallel Implementation of Dynamic Origin-Destination Calculation on a PC-based Linux Cluster

David Bernstein[1], Chien-Lun Lan [2]

[1] College of Integrated Science and Technology, James Madison University 800 S. Main St.
Harrisonburg, VA 22807
[2] Department of Transportation, National Chiao Tung University, Hsinchu, Taiwan

Received 19 July, 2005; accepted in revised form 14 August, 2005

Abstract: Origin-Destination (O-D) information is important in many transportation related domains. The conventional ways to obtain O-D data are costly and also impossible to acquire in real-time. To obtain real-time O-D information in a reasonable way, state space model with Gibbs sampler and Kalman filter is then introduced by researchers. The Gibbs sampler method mentioned above requires considerable quantities of iterations and takes massive computation time, thus parallel computing method has been introduced to increase the computing power. This paper implements parallel computation for the O-D estimation algorithm on a Linux cluster and gives a numerical example which shows a satisfying result.

Keywords: Dynamic Origin-Destination, State Space Model, Gibbs Sampler, Traffic, PC cluster

Mathematics Subject Classification: 60K30, 65Y05, 90B20

1. Introduction

Origin-destination (O-D) data is very important in many transportation related domains (Bernstein, 1996, 1997, 2001; Chang, 1995, 1999, 2005; Jou, 2001, 2002), such as transportation planning, urban and regional planning, and traffic assignment. In the Intelligent Transportation Systems (ITS) domain, real-time O-D information also plays an important role in providing real-time information for the Advanced Traffic Management System (ATMS) and the Advanced Traveler Information System (ATIS). With real-time information, numerous high-value ITS applications e.g., just-in-time delivery, time shortest path emergency vehicle routing and congestion avoidance would be feasible. The traditional way to acquire O-D information includes license plate recognizing, automatic vehicle identification and so on are costly. Due to this reason, researchers have been seeking estimation methods to derive valuable O-D flow information from less expensive traffic data, mainly link traffic counts of surveillance systems.

Jou [10] introduced state space model into dynamic O-D estimation which estimates both O-D matrices and transition matrix simultaneously without any prior information of state variables, while other studies assume that the transition matrix is known or at least approximately known, which is unrealistic for a real world network. Gibbs sampler, a particular type of Markov Chain Monte Carlo (MCMC) method, has been introduced in the solution algorithm to overcome the shortcoming of the assumption of known transition matrix. The algorithm requires considerable quantities of iterations and takes massive computation time, thus parallel computing technique is then been introduced to improve the performance.

The remainder of this paper is organized as follows, the dynamic origin-destination estimation (DODE) by state space model is introduced in section 2, and the parallel implementation of DODE is addressed in section 3.

---

[1] Corresponding author, Associate Professor, E-mail: bernstdh@jmu.edu

2. Dynamic Origin-Destination Estimation

State space model has been introduced to estimate O-D flow from link traffic counts. The standard state space model is coupled with two parts: transition equations and observation equations. First, the state equation which assumed that the O-D flows at time t can be related to the O-D flows at time $t-1$ by the following autoregressive form,

$$x_t = Fx_{t-1} + u_t, \quad t = 1,2,3,...,n \tag{1}$$

where x_t is the state vector which is unobservable, F is a random transition matrix, $u_t \sim N_p(0,\Sigma)$ is independently and identically distributed noise term, where N_p denotes the p-dimensional normal distribution, Σ is the corresponding covariance matrix. x_t, the state variable at time t, is defined to be the path flow belonging to an O-D pair.

Next, the observation equation,

$$y_t = Hx_t + v_t, \quad t = 1,2,3,...n \tag{2}$$

where y_t is the $q \times 1$ observation vector, representing the link traffic counts from the q detectors on the road network. The number of O-D pairs is denoted by p. H is a $q \times p$ zero-one matrix, which denotes routing matrix for a network. v_t is also a noise term that $v_t \sim N_q(0,\Gamma)$. All of x_t and F are unobservable, thus Kalman filter is not suitable to directly estimate and forecast the state vector. Hence, Gibbs sampler is used to tackle the problem of simultaneous estimation of F and x_t by latest available information.

There are two major elements to be incorporated in the solution method, i.e. filtering states by observations and sampling scheme of F and state variables. Since the observations y_t are not used in the conditional distribution, the Kalman filter and the Gibbs sampler must be combined. The Gibbs sampler is a technique for generating random variables from a distribution indirectly, without having to calculate the density. It is a Markovian updating scheme that proceeds as follows. Given an arbitrary starting set of values $Z_1^{(0)}, Z_2^{(0)}, Z_3^{(0)}, ..., Z_k^{(0)}$, and then draw $Z_1^{(1)} \sim \left[Z_1 | Z_2^{(0)}, Z_3^{(0)}, ... Z_k^{(0)}\right]$, $Z_2 \sim \left[Z_2 | Z_1^{(0)}, Z_3^{(0)}, ... Z_k^{(0)}\right]$, and so on. Each variable is visited in the natural order and a cycle requires k random variate generations. After i iterations we have $(Z_1^{(i)}, Z_2^{(i)}, Z_3^{(i)}, ..., Z_k^{(i)})$. Under mild conditions, Geman and Geman showed that $(Z_1^{(i)}, Z_2^{(i)}, Z_3^{(i)}, ..., Z_k^{(i)}) \to [Z_1, Z_2, Z_3, ..., Z_k]$ and hence for each s, $Z_s^{(i)} \to [Z_s]$ as $i \to \infty$.

The solution algorithm is shown as follows,

- Step 1 (Initialization)
    1. Use prior information to generate $F^{(0)}$
    2. Given Σ and Γ
    3. Given $x_0 \sim N(\mu_0, V_0)$
- Step 2 (Generate $x_t^{(g)}$, $t = 0,1,2,...,n$)
    1. Generate $x_0^{(g)}$ from $N(\mu_0, V_0)$
    2. Generate $x_1^{(g)}$ from $x_1 | x_0^{(g)}, F^{(g)} \sim N(F^{(g)} x_0^{(g)}, \Sigma)$
    3. Use the Kalman filter to filter $x_1^{(g)}$
    4. Repeat 2, 3 for t = 2,3,...,n
- Step 3 (Generate $F'^{(g)}$)
    1. Calculate $\quad A^{(g)} = \{a_{ij}^{(g)}\} \quad , \quad a_{ij}^{(g)} = \left(X_{n(i)}'^{(g)} - X_{n-1}'^{(g)} \hat{F}_i'^{(g)}\right)'\left(X_{n(i)}'^{(g)} - X_{n-1}'^{(g)} \hat{F}_j'^{(g)}\right) \quad$ and
       $\hat{F}_i'^{(g)} = \left(X_{n-1}^{(g)} X_{n-1}'^{(g)}\right)^{-1} X_{n-1}^{(g)} X_{n(i)}'^{(g)}$
    2. Calculate $X_{n-1}^{(g)} X_{n-1}'^{(g)}$
    3. Generate $w \sim Wishart\left(X_{n-1}^{(g)} X_{n-1}'^{(g)}, n-p\right)$
    4. Generate $Z = \left(z_1', z_2', z_3', ..., z_p'\right), z_k \overset{iid}{\sim} N_p\left(0, A^{(g)}\right)$

5.   Generate $F'^{(g)} = \left(\left(w^{\frac{1}{2}} \right)' \right)^{-1} Z$

- Step 4 (Iteration)
  Repeat Step 2 and Step 3 m times, and then we have $\{X^{(1)},...,X^{(m)}\}$.

- Step 5 (Estimate X and F')
  Repeat Step 1 to Step 4 k times, then we have $\{X^{(m)}{}_{(1)},...,X^{(m)}{}_{(k)}\}$. Finally, estimate X and F' by

$$\hat{X} = \frac{1}{k}\sum_{n=1}^{n=k} X^{(m)}{}_{(n)} \text{ and } \hat{F}' = \frac{1}{k}\sum_{n=1}^{n=k} \hat{F}'^{(m)}{}_{(n)} .$$

3.   The Implementation of Parallel Computation and its Results

Computation power is crucial to achieve real-time information requirement. Parallel computing is then been introduced to satisfy this requirement. The solution algorithm introduced in section 2 is been divided into several independent computation parts by dividing it at step 5. Given n computation nodes and k Gibbs iterations, each node will take care of k/n iterations. Each computation process stores its own $X^{(m)}{}_{(n)}$ and $\hat{F}'^{(m)}{}_{(n)}$, when the number of iterations is reached, all results are then gathered together to estimate X and F'. In this situation, communication between computing nodes is minimum, and computing power can be easily increased without communication bandwidth limitation. Figure 1 describes the parallel architecture; a similar architecture had been proposed by Li [11]. In the pre-processor section, parameters used in our algorithm are initialized, so does the necessary input data. When assigning jobs, these input data are sent to computing nodes in the cluster through TCP/IP base intranet with Message Passing Interface (MPI) Library. The computational procedure for the parallel process consists of:

Step 1.   Load input data and parameters. Initialize MPI environment.
Step 2.   Count the existing computing nodes in the cluster environment. Decide the count of samples should be generated by each computing nodes. Send data to each computing nodes.
Step 3.   Each computing nodes generate its own $X^{(m)}$ and $F'^{(m)}$ by given input data for given times. And then send the results to server after computation.
Step 4.   After all the data had been sent to server, the server estimate $\hat{X}$ and $\hat{F}'$ by $X^{(m)}$ and $F'^{(m)}$ samples from each computing nodes.
Step 5.   Stop MPI environment. Output data.

The parallel environment of this research consists of 16 computing nodes; each contains 2 Intel XEON 3.2GHz processors and 1 GB memory. Nodes are connected with gigabits Ethernet switch for MPI protocol and a fast Ethernet switch for Network File System (NFS) and Network Information System (NIS).

Our test network is part of a real road network in Taiwan, the Hsinchu Science-Based Industrial Park, which consists of 89 nodes and 244 links; and according to the segmentation of traffic zone there exists 99 O-D pairs with 246 paths. Without parallel computing, approximately 40 hours of computation time is required for a single CPU, but with the parallel computing, a quite good value of the speedup and efficiency of the parallel scheme is achieved. The speedups is the ratio of the code execution time on a single processor to that on multiple processors, and efficiency is defined as the speedup divided by the number of processors(Gropp, 1999; El-Rewini, 1998). Figure 2 shows the speedups and efficiencies of the parallel computing for 100 samples on a 32 CPU Linux-cluster with MPI library, which leads to a satisfying result.

4.   Conclusions

This paper provides a parallel implementation of estimating dynamic origin-destination matrices for general road network by using the state space model with Kalman filter and Gibbs sampler. The parallel implementation shows a good result of nearly 80% of computing power remains for each CPU under a 32 CPUs cluster environment. That leads to the conclusion: with this parallel scheme, real-time estimation of O-D matrices can be easily achieved by increasing a reasonable amount of CPUs.

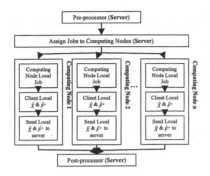

Figure 1. The flow chart of parallel algorithm

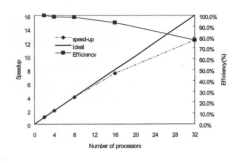

Figure 2. Speedups and efficiencies for the parallel computing of $k = 100$

Acknowledgments

This research was supported in part by the Ministry of Education of Taiwan, ROC under Grants EX-91-E-FA06-4-4 and in part by the National Science Council of Taiwan, ROC under Grants NSC-93-2218-E-009-042 and NSC-93-2218-E-009-043 respectively.

References

[1] D.A. Bader, J. JaJa and R. Chellappa, Scalable data parallel algorithms for texture synthesis using Gibbs random fields, *IEEE Transactions on Image Processing* **4** 1456-1460(1995).

[2] David Bernstein, T.L. Friesz, Z. Suo and R.L. Tobin, Dynamic Network User Equilibrium with State-Dependent Time Lags, *Networks and Spatial Economics* **1** 319-347(2001).

[3] David Bernstein, T.L. Friesz, Infinite Dimensional Formulations of Some Dynamic Traffic Assignment Models, *Network Infrastructure and the Urban Environment*, Springer, New York, 112-124, 1997.

[4] David Bernstein, T.L. Friesz and R. Stough, Variational Inequalities, Dynamical Systems and Control Theoretic Models for Predicting Time-Varying Urban Network Flows, *Transportation Science* **30** 14-31(1996).

[5] Gang-Len Chang, Xianding Tao, Advanced computing architecture for large-scale network O-D estimation, *Proceedings of the 6th 1995 Vehicle Navigation and Information System Conference* 317-327(1995).

[6] Gang-Len Chang, X.D. Tao, An integrated model for estimating time-varying network origin-destination distributions, *Transportation Research A* **33** 381-399(1999).

[7] Pei-Wei Lin, Gang-Len Chang, Saed Rahwanji, Robust Model for Estimating Freeway Dynamic Origin-Destination Matrix, *Transportation Research Board* (2005).

[8] Pierre Bremaud, *Markov chains: Gibbs fields, Monte Carlo simulation, and queues*, Springer, 1999.

[9] Yow-Jen Jou, Ming-Chorng Hwang, Rapid Transit System Origin-Destination Matrix Estimation, *2002 World Metro Symposium & Exhibition*, Taipei, April 2002.

[10] Yow-Jen Jou, Ming-Chorng Hwang and Jia-Ming Yang, A Traffic Simulation Interacted Approach for the Estimation of Dynamic Origin-Destination Matrix, *IEEE International Conference on Systems, Man & Cybernetics* **2** 868-873(2004).

[11] Y. Li, S.M. Sze and T.-S. Chao, A Practical Implementation of Parallel Dynamic Load Balancing for Adaptive Computing in VLSI Device Simulation, *Engineering with Computers* **18** 124-137(2002).

Brill Academic Publishers
P.O. Box 9000, 2300 PA Leiden,
The Netherlands

*Lecture Series on Computer
and Computational Sciences*
Volume 4, 2005, pp. 1088-1091

Hybrid Evolutionary Approach to Optimal Design of CMOS LNA Integrated Circuits

Yiming Li[1] and Hung-Mu Chou

Department of Communication Engineering and Microelectronics and Information Systems
Research Center, National Chaio Tung University, Hsinchu 300, Taiwan

Received 10 July, 2005; accepted in revised form 12 August 2005

Abstract: In this work we propose a hybrid intelligent circuit optimization technique for low noise amplifier (LNA) circuit. This method combines with the genetic algorithm (GA), Levenberg-Marquardt (LM) method, and circuit simulator to perform automatic LNA circuit optimization. For a given LNA circuit, the optimization method considers the electrical specification such as S parameters: $S_{11}, S_{12}, S_{21}, S_{22}$, K factor, the noise figure, and the input third-order intercept point, simultaneously. The optimization procedure starts with loading the necessary parameters for circuit simulation, and then calls the circuit simulator for circuit simulation and evaluation. Sixteen optimized parameters of the LNA circuit composed with 0.18 μm metal-oxide-silicon filed effect transistors (MOS-FETs) are acquired by our developed optimization prototype, where the aforementioned seven specifications are all matched. The proposed circuit optimization method shows its robustness and practicability on radio-frequency (RF) circuit and wireless system on chip (SoC) design.

Keywords: Low Noise Amplifier, Genetic Algorithm, Levenberg-Marquardt, Circuit Design, Optimization

Mathematics Subject Classification: Methods of successive approximation, Nonlinear stabilities, Other matrix algorithms, Necessary conditions and sufficient conditions for optimality.

PACS: 02.60.-x, 02.60.Cb, 02.70.-c, 02.60.Pn

1   Introduction

The low noise amplifier (LNA) plays an important role in radio frequency (RF) circuit design. In modern integrated circuit (IC) design flow and chip implementation, the designers must perform a series of functional examination and characteristic analysis. In order to achieve the specification, the designers must continuously tuning the design coefficients and perform the circuit simulation [1][2] to get optimized active device model parameters, passive device parameters, circuit layout, and width of wires. This task usually requires the expert designers to accomplish such complicated work. Therefore, an systematic global optimization method is required to improve this procedure.

To perform circuit optimization, many methods are adopted, such as numerical, statistical, and soft computing method [3]. However, due to the highly nonlinear property of circuit performance, the numerical method often trapped in local optima; on the other hand, the statistical and soft computing methods suffer the massive computation issue [4]. To solve this problem, we, in this

---

[1]Corresponding author. E-mail: ymli@faculty.nctu.edu.tw

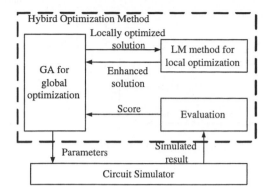

Figure 1: The architecture of the hybrid intelligent computational technique.

work, apply a hybrid intelligent optimization technique combined with both numerical and soft computing method to solve circuit optimization problem [5]. The optimization procedure starts with loading the necessary parameters for circuit simulation, and then calls the circuit simulator for circuit simulation and evaluation. Sixteen optimized parameters of the LNA circuit composed with 0.18 μm metal-oxide-silicon filed effect transistors (MOSFETs) are acquired by our developed optimization prototype, where the aforementioned seven specifications are all matched. The proposed circuit optimization method shows its robustness and practicability on RF circuit and wireless system on chip (SoC) design.

The paper is organized as follows. In Sec. 2 we presented a framework of the hybrid optimization methodology. The achieved results are discussed in the section 3. Finally, we draw the conclusions.

2   The Framework of the Hybrid Optimization Methodology

We construct a hybrid search algorithm which takes genetic algorithm (GA) to perform global search, and while the evolution seems to be saturated, the Levenberg - Marquardt (LM) method is then enhancing the searching behavior to perform the local search. The architecture of the proposed hybrid search algorithm is shown in Fig. 1. As shown in this figure, the GA searches the entire problem space first. During this period, the candidates GA searched are passed to certain simulator to retrieve results, the results are then passed to evaluator to measure the fitness score. After a rough solution is obtained, the LM method makes a local optima search and set the local optima as the initial value for the GA to perform further optimization.

3   Results and Discussion

We first perform the parameter optimization on the device model and then the circuit level for the tested LNA circuit, shown in Fig. 2. There are more than 15 parameters have to be extracted in the designed complementary metal-oxide-semiconductor (CMOS) LNA circuit. Three experiments are performed under different consideration. The First experiment considered only the optimization of passive devices. The second experiment optimize active devices along with passive devices. Because the low power consumption that is especially important for portable communications systems. Finally, the target of the third experiment contains passive, active devices, and power dissipation. Table 1 left side shows the optimized parameters of all experiments and the right side shows

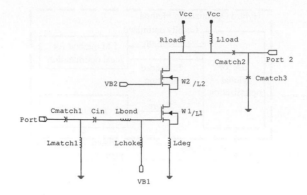

Figure 2: The explored LNA circuit in out numerical experiment.

Table 1: A list of optimized parameters (the left table) and the results comparison (the right one), where AGAP means as great as possible.

Element	Test1	Test2	Test3	Spec.	Target	Test1	Test2	Test3
Cmatch1	512.132f	579.524f	657.738f	S_{11}	<-10dB	-14.1dB	-35.3dB	-35.1dB
Cmatch2	4.6104f	5.092p	4.505p	S_{22}	<-10dB	-22.6dB	-30.6dB	-19.1dB
Lbond	1.0782n	1.21n	1.058n	S_{12}	<-25dB	-39.3dB	-37.9dB	-38.3dB
Lmatch1	6.202n	6.202n	5.257n	S_{21}	AGAP	12.7dB	15.4dB	11.3dB
Rload	3P5.1	3P5.1	3P5.1	K	>1	10.7	6.84	11.1
VB1	0.75V	0.75V	0.69V	NF	<2	0.979	0.657	1.17
L	–	0.18u	0.25u	*IIP3*	>-10	-1.3	-5.45	0.3

the optimized result of above experiments. The robustness of the proposed hybrid optimization technique is investigated. As shown in Fig. 3a, it reveals that the geometry parameters would make the most improvement, while the input and output categories makes little improvement after 120 generations. This phenomenon indicates that the geometry of the active devices are more difficult to optimized than passive device parameters. Figure 3b shows the score convergence behavior comparison of the standard GA and the hybrid optimization technique. As shown in this figure, the proposed methodology is superior to the pure GA after 60 generations. The proposed method shows no significant advantage at the beginning because the LM method has not triggered yet. Once the LM method is active, it based on the result of GA to perform local optimization, and the GA follows the local optima obtained by the LM method to keep evolving. Under this mechanism, our proposed methodology shows better trend of convergence and the robustness of our proposed methodology hence be held.

4 Conclusions

Many researches have been made for circuit optimization recently. However, these methodologies are mainly based on traditional optimization techniques, which are lack of efficiency or accuracy due to local solution limitations. We have proposed a hybrid optimization methodology to au-

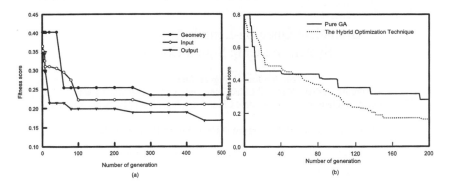

Figure 3: (a) The sensitivity analysis. (b) A score convergence behavior comparison of the GA and the hybrid optimization technique.

tomatically search a set of optimal solutions of a specified LNA problem. Numerical simulation shows that this approach is promising for CMOS RF circuit simulation and optimal SoC design.

Acknowledgment

This work is supported in part by the National Science Council (NSC) of Taiwan under Contract NSC-93-2215-E-429-008 and Contract NSC-94-2752-E-009-003-PAE, by the Ministry of Economic Affairs, Taiwan under Contract 93-EC-17-A-07-S1-0011, by the Taiwan Semiconductor Manufacturing Company under a 2004-2005 grant, and by the Toppoly Optoelectronics Corp. under a 2003-2005 grant.

References

[1] R.J. Pratap, S. Sarkar, S. Pinel, J. Laskar, and G.S. May, Modelling and optimization of multilayer RF passives using coupled neural networks and genetic algorithms, *Microwave Symposium Digest* **3** 1557 - 1560 (2004).

[2] K. Y. Huang, Y. Li, and C.-P. Lee, A Time Domain Approach to Simulation and Characterization of RF HBT Two-Tone Intermodulation Distortion, *IEEE Trans. on Microwave Theory and Tech.* **51** 2055-2062 (2003).

[3] Y. Li and Y.-Y. Cho, Intelligent BSIM4 Model Parameter Extraction for Sub-100 nm MOSFETs era, *Japanese Journal of Applied Physics* **43** 1717-1722 (2004).

[4] W. Daems, G. Gielen, G. W. Sansen, Simulation-based generation of posynomial performance models for the sizing of analog integrated circuits, *IEEE Trans. on CAD ICs and Sys.* **22** 517-534 (2003).

[5] Y. Li and S.-M. Yu, A Hybrid Intelligent Computational Methodology for Semiconductor Device Equivalent Circuit Model Parameter Extraction, *Book of Abs. Int. Workshop on Sci. Comput. Elec. Eng.* 79 (2004).

Brill Academic Publishers
P.O. Box 9000, 2300 PA Leiden
The Netherlands

*Lecture Series on Computer
and Computational Sciences*
Volume 4, 2005, pp. 1092-1095

Statistical Queuing-Length Approach to
System-on-Chip Signal Control

Yow-Jen Jou[1], Ming-Te Tseng [2]
[1] Institute of Statistic, National Chiao Tung University, Hsinchu, Taiwan
[2] Department of Transportation, National Chiao Tung University, Hsinchu, Taiwan

Received 18 July, 2005; accepted in revised form 15 August, 2005

Abstract: The chief goal of this paper is to describe a methodology for predicting the dynamic queue length of traffic signal control system on chip (SoC). There are two key steps. The first is to predict the dynamic traffic volumes for an oversaturated traffic intersection's link. This step is to fit a censored model for the traffic flow by using the historical relevant information of traffic flow, including past volumes and occupancy data. Then, the predicted traffic volumes will be cumulated to predict the queue length. Also, algorithm has been developed and a sample is given to verify the model.

Keywords: regression, censor model, time series, oversaturated, queue, traffic signal control

Mathematics Subject Classification: AMS-MOS, PACS

PACS: 62P25, 62P30, 02.70.Rr, 07.05.Kf

1. Introduction

Road capacity, planning and design are the essential for traffic performance. Urban traffic often congests in peak hours, so how to effectively plan and design roads and correctly predict car-flow to solve traffic awkward situations becomes the prime issues for the government.

The traffic control strategy, such as PASSER II [8] and TRANSYT-7F [10], are off line and static computational algorithms. The traffic information, which these strategy need, is collected by hand or some detector. The traffic control parameters, such as green time and cycle time, are computing off line, then are input to traffic controller. Full or semi actuated control strategy are on line processing the vehicle detector's data and generating the traffic control parameters. But it does not predict any traffic information. SCOOT [2] and SCATS [7] are online adaptive traffic control strategy. They process the traffic information from vehicle detector online, predict the near future's traffic information and output the traffic control parameters. The predict method of these online adaptive control strategy are focused on under-saturated traffic information. CHO and Tseng [3] provide an online oversaturated traffic control strategy which is executed on a SoC. But this algorithm doesn't have a completely predicting model for estimating the oversaturated queue length.

Hence, we provide a dynamic model to estimate the queue length of traffic signal control SoC. There are two key steps to do the estimation. The first is to predict the dynamic traffic volumes for an oversaturated traffic intersection's link. This step is to fit a censored model [6, 9, 11] for the traffic flow by past volumes and occupancy data. Then, the predicted traffic volumes will be cumulated to predict the queue length. Under normality, the maximum likelihood estimators of the parameters [4, 5] will be obtained and the appropriateness of the model will be checked. Also, algorithm has been developed and a sample is given to verify the model.

2. The data and Model

Before constructing the model, we need to know the characteristic of oversaturated traffic information which is collected by vehicle detectors. First, let's focus on the traffic volume of a vehicle detector, the

---

[1] Corresponding Author. E-mail: yjjou@stat.nctu.edu.tw

volume will be zero when there is no traffic or traffic queue is up to vehicle detector, the volume will have a maximal value when traffic flow reach to its maximal capacity. Then, we care about the relation between traffic volume and occupancy. As figure 1 show, the volume will tend to zero when occupancy is almost 0% or 100%. Occupancy is 0% if there are no traffic and 100% for oversaturated traffic.

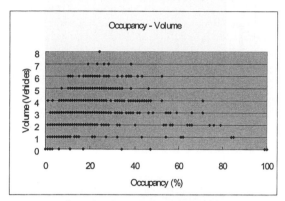

Figure 1 : Occupancy - Volume Plot

Now let's construct the censored dynamic data model. First, we need to consider the censor in our model. We have to find out the relationship between the occupancy and the volume. Observing the figure 1, we can find that there is a quadratic curve relation between occupancy and volume. Also we find that the maximum volume is eight. Moreover, the current volume is affected by the prophase volume. Therefore, there will be the square term and the AR (1) from time series in our model, and we will define the censor value to eight.

Let x_t represents the occupancy, and y_t represents the volume. Then we can write the model as follows.

$$y_t^* = \gamma y_{t-1}^* + \beta_1 x_t^2 + \beta_2 x_t + \alpha + \varepsilon_t, \quad t = 1,2,....1000, \quad \varepsilon_t \overset{iid}{\sim} N(0,\sigma^2)$$

$$y_t = \begin{cases} y_t^*, & if \ y_t^* < 8 \\ 8, & if \ y_t^* \geq 8 \end{cases}$$

By iteration of Newton, find the parametric estimator when the error < 0.00001.

3. Algorithm

The algorithm for the censored dynamic data model has three steps.

Step1 : $\hat{\theta}_0 = (\hat{\gamma}_0, \hat{\beta}_0', \hat{\alpha}_0, \hat{\sigma}_0^2)'$

Step2 : $\hat{\theta}_k = \hat{\theta}_{k-1} - \left[\dfrac{\partial^2 \ln L}{\partial \theta^2} \Big|_{\theta=\hat{\theta}_{k-1}} \right]^{-1} \left[\dfrac{\partial \ln L}{\partial \theta} \Big|_{\theta=\hat{\theta}_{k-1}} \right], \quad k = 1, 2, ...$

Step3 : If $\left| \hat{\theta}_k - \hat{\theta}_{k-1} \right| < 0.00001,$ stop and $\hat{\theta}_k$ is the answers

If $\left| \hat{\theta}_k - \hat{\theta}_{k-1} \right| \geq 0.00001,$ $\hat{\theta}_k = \hat{\theta}_{k-1}$ and repeat Step1 ~ Step3

4. Example

After applying figure 1 and figure 2's data by seven times Newton, we get the estimators of the model. The complete equation is:

$$y_t^* = 0.058536 y_{t-1}^* - 0.002 x_t^2 + 0.173517 x_t + 0.756959 + \varepsilon_t, \quad \varepsilon_t \overset{iid}{\sim} N(0,1.179615)$$

Figure 3 is the forecast result of the censored model. Compare the sample volume with the forecast volume by this model, we has the error of residual as figure 4.

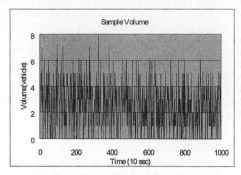

Figure 2 : Volume – Time Plot

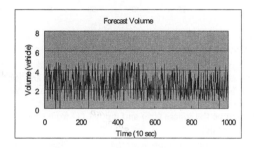

Figure 3 : Forecast Volume –Time Plot

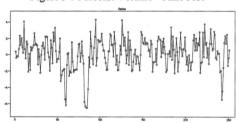

Figure 4: Residual--Time Plot

In order to get the queue length of the over-saturated network link, we need to sum the volume. The total vehicle volume here is the queue length of the over-saturated network link. Figure 5 show the cumulative volume for the sample data and its forecasting data. The two cumulative lines almost overlay each other. It means that the predicted queue length is almost the same as real queue length.

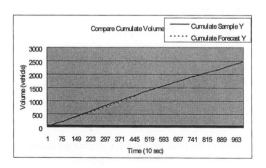

Figure 5 : Cumulate Sample Volume and Cumulate Forecast Volume Plot

5. Conclusion

We provide a censored dynamic data model to estimate the traffic information of traffic signal control SoC. The traffic information that we care about is the queue length. The traffic condition is oversaturated. We use traffic volume and occupancy to predict the queue length. First, we use the censored dynamic data model to estimate the traffic volume. The traffic volume and occupancy are the parameters of the model. The algorithm also been developed to estimate the parameters of the model. After predicting the volume, then we take summary of volume into the queue length. The example shows that the predicting queue length of the model is almost the same as real queue length.

Acknowledgments

This work is supported in part by the Ministry of Education of Taiwan, ROC, under grant EX-91- E-FA06-4-4 and in part by the National Science Council of Taiwan under contracts: NSC-92-2622- E-009-011-CC3 and NSC-93-2211-E-009-032.

References

[1] G. Abu-Lebdeh and R.F. Benekohal, Development of Traffic Control and Queue Management Procedures for Oversaturated Arterials, *Transportation Research Record* **1603** 119-127(1997).

[2] D. Bretherton, K. Wood and N. Raha, Traffic Monitoring and Congestion Management in the SCOOT Urban Traffic Control System, *Transportation Research Record* **1634** 118-124(1998).

[3] H.J. Cho and M.T. Tseng, *A system on chip approach to traffic signal control*, the international conference on computer design, 2005

[4] A.C. Cohen, Estimating the Mean and Variance of Normal Populations from Singly Truncated and Doubly Truncated Samples, *Annals of Mathematical Statistics* **21** 557-569(1950).

[5] A. Hald, Maximum Likelihood Estimation of the Parameters of a Normal Distribution Which is Truncated at a Known Point, *Skandinavisk Aktuarietidskrift* **32** 119-134(1949).

[6] L. Hu, Estimation of a Censored Dynamic Panel Data Model, *Department of Economics Northwestern University*, 2002.

[7] P.R. Lowrie, *SCATS-Sydney Coordinated Adaptive Traffic System-A Traffic Responsive Method of Controlling Urban Traffic*, Roads and Traffic Authority, Sydney, NSW, Australia, 1992.

[8] *Passor II-90 Users Guide Volume 3 in a Series: Methodology for Optimizing Signal Timing*, 1991.

[9] T. Amemiya, Regression Analysis When the Dependent Variable is Truncated Normal, *Econometrica* **41** 997-1016(1973).

[10] *TRANSYT-7F User Guide Volume 4 in a Series: Methodology for Optimizing Signal Timing*, 1998.

[11] J. Tobin, Estimation of Relationship for Limited Dependent Variables, *Econometrica* **26** 24-36(1958).

Brill Academic Publishers
P.O. Box 9000, 2300 PA Leiden,
The Netherlands

Lecture Series on Computer
and Computational Sciences
Volume 4, 2005, pp. 1096-1099

A Mixed Riccati-Lyapunov Algorithm for Coupled Algebraic Riccati Equations in MCV Problems

L.Cherfi [1]

Ecole Normale Supérieure de Cachan,
Laboratoire SATIE (UMR,CNRS, 8029)
61 Av. Président Wilson
94235 Cachan, France

Received 7 July, 2005; accepted in revised form 9 August, 2005

Abstract: A new algorithm for solving coupled algebraic Riccati equations appearing in minimal cost variance problems is introduced. It is shown that each sequence of this algorithm has an upper and a lower bound. As an application of the developed technique, some examples are given to illustrate the performance of the proposed method.

Keywords: Coupled Riccati equation, Riccati iteration, minimal cost variance, fixed point, numerical algorithm, control theory, algebraic Riccati equations,Risk sensitive problem.

Mathematics Subject Classification: 34A34, 65H20, 54H25, 32H50.

1 Introduction

We shall consider structures whose dynamic motion can be described in state variable form by the following stochastic equation

$$dx(t) = Ax(t) + Bu(t)dt + Edw(t), \qquad x(t_0) = x_0, \qquad (1)$$

where $x(t) \in R^n$ is the state vector, x_0 is a random action which is independent of w, $u(t) \in R^m$ is the control vector and $w(t)$ is a vector of Brownian motion of dimensional d defined on a probability space $(\Omega, \mathcal{F}, \mathcal{P})$ and such that $E\{dw(t)dw'(t)\} = \bar{W}dt$, with $\bar{W} \in R^{n \times n}$.
The matrices A, B and E are matrices of dimensions $n \times n$. The performance of system (1) is evaluated by the minimization of the cost variance of the following criterion

$$J(x, u) = \int_0^\infty \left\{ u^T Ru + x^T Qx \right\} ds. \qquad (2)$$

Given a positif scalar γ, the solutions of the infinite time horizon full-state feedback MCV problem (1) and (2) are found using the Hamiltonian Jacobi theory [3, 4, 5]. In the case of infinite time horizon MCV control problems a pair of coupled algebraic Riccati equations arises. For a process $x(t)$ from (1), having an admissible controller u which minimizes the variance $E\{J(x, u)\}$, the optimal control is given by

$$u^*(x) = -R^{-1}B^T[M + \gamma V]x, \qquad (3)$$

---

[1]Corresponding author. PHD Student. E-mail:cherfi@satie.ens-cachan.fr

where M and V are the stabilizing solutions of the coupled algebraic Riccati equations

$$0 \;=\; A^T M + MA - MSM + \gamma^2 VSV + Q, \tag{4}$$
$$0 \;=\; A^T V + VA - 2\gamma VSV - MSV - VSM + 4MWM, \tag{5}$$

where

$$S = BR^{-1}B^T \text{ and } W = E\tilde{W}E^T.$$

The goal of this work is to introduce a new algorithm (called mixed Riccati-Lyapunov type algorithm) to calculate a pair of stabilizing solutions of the coupled algebraic Riccati equations (4) and (5). This one is based on a decoupling of these equations, by using appropriately one step delay. At each step of this algorithm we have to solve one standard algebraic Riccati equation and one Lyapunov algebraic equation. To assume that this algorithm is well defined, we impose the assumption that $(A, B, \sqrt{Q})$ is stabilizable-detectable [1].

2   Mixed Riccati-Lyapunov algorithm

In the following let us assume that $Q \geq 0$, $S \geq 0$ and $W \geq 0$. Using the abbreviation $Z = M + \gamma V$, we get an elementary calculation that

$$0 \;=\; A^T Z + ZA - ZSZ + 4\gamma MWM + Q. \tag{6}$$

Here, we present the details of the above-mentioned algorithm for solving the coupled algebraic Riccati equations (4) and (5). Each iterative technique presented here will stop until the right-hand side of these equations is smaller than a chosen precision ϵ sufficiently small and fixed in advance.
Algorithm • Start with $M_0 = 0$.
• Define Z_l, M_{l+1}, $N_1(Z_l, M_l)$ and $N_2(Z_l, M_l)$, $l \geq 0$, as follows
(a) determine Z_l the unique semidefinite stabilizing of the algebraic Riccati equation

$$0 = A^T Z_l + Z_l A - Z_l S Z_l + 4\gamma M_l W M_l + Q \tag{7}$$

and M_{l+1} the unique semidefinite solution of the Lyapunov equation

$$(A \quad SZ_l)^T M_{l+1} + M_{l+1}(A - SZ_l) + Z_l SZ_l + Q = 0. \tag{8}$$

(b) Put $A_l = A - SZ_l$, then compute

$$N_1(Z_l, M_l) \;=\; A^T Z_l + Z_l A - Z_l S Z_l + 4\gamma M_l W M_l + Q,$$

$$N_2(Z_l, M_l) \;=\; A_l^T M_l + M_l A_l + Z_l S Z_l + Q,$$

while

$$e_l = \max\{\|N_1(Z_l, M_l)\|_\infty, \|N_2(Z_l, M_l)\|_\infty\}, \tag{9}$$

is larger than ϵ, replace l by $l+1$ and go to (a), else stop.

Assume that the following conditions are satisfied
A1) there exists a unique positive semidefinite stabilizing solution of the algebraic Riccati equation

$$0 = A^T P + PA - PSP + Q, \tag{10}$$

A2) there exists a positive semidefinite (stabilizing) solution of the algebraic Riccati equation

$$0 = A^T \bar{P} + \bar{P}A - \bar{P}(S - 4\gamma W)\bar{P} + Q, \tag{11}$$

A3)

$$\Delta W + W\Delta + \Delta W\Delta \geq 0, \quad if \Delta \geq 0. \tag{12}$$

Theorem 1 *Let the assumptions A1, A2 and A3 hold. Then the iteration schemes defined by (7) and (8) starting with $M_0 = 0$ defines two sequences $(M_l)_{l \geq 1}$ and $(Z_l)_{l \geq 0}$ with the following properties*

$$0 \leq P \leq M_{l+1} \leq Z_l \leq \bar{P}, \quad \forall l \geq 0. \tag{13}$$

Proof of Theorem 1 For $i = 0$, with $M_0 = 0$ we obtain in the first step the Riccati and the Lyapunov equations

$$0 = A^T Z_0 + Z_0 A - Z_0 S Z_0 + Q, \tag{14}$$
$$0 = (A - S Z_0)^T M_1 + M_1 (A - S Z_0) + Z_0 S Z_0 + Q. \tag{15}$$

By assumption A1 the required positive semidefinite stabilizing solutions P of the algebraic Riccati equation (10) exists. So, there exists a unique positive semidefinite stabilizing solution Z_0 of the algebraic Riccati equation (14). By Lemma 2, $Q \leq Q + 4\gamma \bar{P} W \bar{P} \Rightarrow Z_0 \leq \bar{P}$. But since $(A - S Z_0)$ is Hurwitz, we obtain by subtracting (15) from (14) the following equation

$$0 = (A - S Z_0)^T (Z_0 - M_1) + (Z_0 - M_1)(A - S Z_0), \tag{16}$$

so, $M_1 = Z_0 \leq \bar{P}$. Therefore the statement $M_{k+1} \leq Z_k \leq \bar{P}$, is true for $k = 0$.
Let us assume that this is true for $k = 0, \ldots, l - 1$. The next steps are to prove this assumption is also true for l. Then by induction we conclude that it holds for all indices l.
For $k = l$, we obtain the following equation

$$0 = A^T Z_l + Z_l A - Z_l S Z_l + 4\gamma M_l W M_l + Q. \tag{17}$$

As $M_l \leq \bar{P}$, we have $\bar{P} = M_l + \Delta_l$ with $\Delta_l \geq 0$ then by assumption A3, we obtain $Q + 4\gamma M_l W M_l \leq Q + 4\gamma \bar{P} W \bar{P}$. By using the comparison results (see Lemma 2) for the solutions of (11) and (17) it follows that $Z_l \leq \bar{P}$.
Consider now the Lyapunov equation for M_{l+1}

$$0 = (A - S Z_l)^T M_{l+1} + M_{l+1}(A - S Z_l) + Z_l S Z_l + Q, \tag{18}$$

we obtain by subtracting (18) from (17)

$$0 = (A - S Z_l)^T (Z_l - M_{l+1}) + (Z_l - M_{l+1})(A - S Z_l) + 4\gamma M_l W M_l. \tag{19}$$

Since $(A - S Z_l)$ is Hurwitz and since $W \geq 0$, we infer by Lemma 1 that $M_{l+1} \leq Z_l$.
In order to prove the upper bound of the sequences $(Z_l)_{l \geq 0}$ and $(M_{l+1})_{l \geq 0}$, let us rewrite (10) as

$$0 = (A - S Z_l)^T P + P(A - S Z_l) + Z_l S Z_l - (P - Z_l)S(P - Z_l) + Q. \tag{20}$$

By subtracting (20) from (17), we obtain

$$0 = (A - S Z_l)^T (Z_l - P) + (Z_l - P)(A - S Z_l) + 4\gamma M_l W M_l + (P - Z_l)S(P - Z_l). \tag{21}$$

Since $(A - S Z_l)$ is Hurwitz, together with $W > 0$ and $S \geq 0$, by Lemma 1 we obtain $P \leq Z_l$. In analogous way we obtain by subtracting (20) from (8)

$$0 = (A - S Z_l)^T (M_{l+1} - P) + (M_{l+1} - P)(A - S Z_l) + (P - Z_l)S(P - Z_l). \tag{22}$$

Since $A - S Z_l$ is Hurwitz and $S \geq 0$, by Lemma 1, we obtain $P \leq M_{l+1}, \forall l \geq 0$.

3 Conclusion

In this paper we have presented a new algorithm to calculate a pair of stabilizing solutions (when it exists) of the coupled algebraic Riccati equations. These equations are known as the Riccati feedback MCV problems. Moreover we used comparison results for standard algebraic Riccati equations to get a lower and an upper bound of the solutions.

4 Appendix

For convenience of the other reader we summarize here without proof some well known facts from literature [1], which have been used. First consider the Lyapunov equation

$$A^T X + XA + Q,\tag{23}$$

here $A, Q \in R^{n \times n}$ are given matrices $Q = Q^T$.

Lemma 1 *If A is Hurwitz (i.e, if all eigenvalues of A have negative real part) then (23) has the unique solution*

$$X = \int_0^\infty e^{\tau A} Q e^{\tau A^T},\tag{24}$$

therefore $Q \geq 0$ implies $X \geq 0$.

Second, we consider the algebraic Riccati equations for $i = \{1, 2\}$:

$$0 = A_i^T X_i + X_i A_i + Q_i - X_i S_i X_i,\tag{25}$$

where A_i, Q_i and S_i are $n \times n$ real matrices with $Q_i = Q_i^T$ and $S_i = S_i^T$. Define two matrices

$$K1 = \begin{bmatrix} Q_1 & A_1^T \\ A_1 & -S_1 \end{bmatrix} \quad and \quad K2 = \begin{bmatrix} Q_2 & A_2^T \\ A_2 & -S_2 \end{bmatrix}.$$

Lemma 2 *Let X_i, $(i = 1, 2)$ be the positive semidefinite (stabilizing) solutions of the algebraic Riccati equations (25). If $K_1 \leq K_2$ then $X_1 \leq X_2$.*

Acknowledgment

The author wishes to thank the anonymous referees for their careful reading of the manuscript and their fruitful comments and suggestions.

References

[1] H.Abou-Kandil and G.Freiling and V.Ionescu and G.Jank, *Matrix Riccati Equations in Control and Systems Theory*. Birkauser Verlag, Germany, 2003.

[2] G.Freiling and S.R.Lee and G.Jank, Coupled matrix Riccati equations in minimal cost variance control problems : IEEE Automat. Control, 44(1999), 556-560.

[3] M.K. Sain, C.H.Won and B.F.Jr.Spencer, Cumulant in risk-sensitive control: The full-state feedback cost variance: Conference on Decision and Control, (1995), 1036-1041.

[4] Chang-Hee Won and M.K.Sain and S.R.Liberty, Infinite Time minimal cost variance control and coupled algebraic Riccati equations:Proceedings of the American Control Conference, (2003),5155-5160.

[5] M.K. Sain, On minimal Variance Control of linear systems with quadratic loss:Ph.D, Department of Electrical Engineering, University of Illinois, Urbana, January 1965.

Brill Academic Publishers
P.O. Box 9000, 2300 PA Leiden,
The Netherlands

Lecture Series on Computer
and Computational Sciences
Volume 4, 2005, pp. 1100-1103

Numerical Algorithms for Solving Coupled Algebraic Riccati Equations

L.Cherfi [1]

Ecole Normale Supérieure de Cachan,
Laboratoire SATIE (UMR,CNRS, 8029)
61 Av. Président Wilson
94235 Cachan, France

Received 28 June, 2005; accepted in revised form 20 July, 2005

Abstract: In order to obtain the closed-loop strategies in Nash differential game with infinite horizon, one needs to solve a system of coupled algebraic Riccati equations. Under standard conditions it is not yet known if solutions for such equations exist. One way to achieve that goal is to consider discrete dynamical systems, whose fixed points (if they exist) are solutions of the problem under study. These discrete dynamical systems of coupled algebraic Riccati equations can also serve as numerical algorithms to compute possible solutions. In this paper, we propose a new discrete dynamical system. Through the study of pertinent examples, we show numerically that this algorithm behaves better than the existing ones, both in terms of convergence speed and detection of a stabilizable solution (when it exists).

Keywords: Coupled Riccati equation, Nash equilibrium, Lyapunov iteration, Riccati iteration, Fixed point, Numerical algorithm, Control theory, algebraic Riccati equations.

Mathematics Subject Classification: 34A34, 65H20, 54H25, 32H50,

1 Introduction

Consider a dynamical system whose state x evolves according to

$$\dot{x} = Ax + B_1 u_1 + B_2 u_2, \quad x(t_0) = x_0, \tag{1}$$

where $x \in R^n$, $u_i \in R^{m_i}$ is the control of the i-th player, $(i = 1, 2)$. The matrices A, B_1 and B_2 are constant and have the appropriate dimensions.

A quadratic cost functional is associated with each control agent $(i = 1, 2)$:

$$J_i(u_1, u_2) = \int_0^\infty \left\{ u_i^T(t) R_{ii} u_i(t) + x^T(t) Q_i x(t) \right\} dt, \tag{2}$$

where all weighting matrices are real symmetric with $Q_i \geq 0$ and $R_{ii} > 0$, $(i = 1, 2)$.

Assume that both players are required to formulate an optimal pair of strategies. In the case of non cooperative games, they can use Nash equilibrium with the closed-loop information structure

---

[1]Corresponding author. PHD Student. E-mail:cherfi@satie.ens-cachan.fr

i.e. feedback memoryless (see [2]) for details) .

It is well known (see [2]), that the closed-loop Nash equilibrium strategies is given by

$$u_1^*(x) \;=\; -R_{11}^{-1} B_1^T P x, \tag{3}$$
$$u_2^*(x) \;=\; -R_{22}^{-1} B_2^T R x, \tag{4}$$

where the real symmetric matrices P and R are the stabilizing solutions of the coupled algebraic Riccati equations:

$$0 \;=\; A^T P + PA + Q_1 - PS_{11}P - RS_{22}P - PS_{22}R, \tag{5}$$
$$0 \;=\; A^T R + RA + Q_2 - RS_{22}R - RS_{11}P - PS_{11}R, \tag{6}$$

with $S_{11} = B_1 R_{11}^{-1} B_1^T$ and $S_{22} = B_2 R_{22}^{-1} B_2^T$.

2   Riccati type algorithm

In this section, we give a new algorithm to calculate a pair of solutions (when it exists) (P, R) of the coupled algebraic Riccati equation (5) and (6). We present the details of the above-mentioned algorithm by decoupling theses equations using the Riccati type iterations. Each iterative technique presented here will stop until the right-hand side of these equations is smaller than a chosen precision ϵ sufficiently small and fixed in advance.

2.0.1   Algorithm C

- Start with the initial matrix $P_0 \geq 0$. Put $l = 1$.

- Define P_l, R_l, $N_1(P_l, R_l)$ and $N_2(P_l, R_l)$, $l \geq 1$ as follows:

- (a) Determine R_l the strong solution of

$$
\begin{aligned}
0 \;=\;\; & (A - S_{11} P_{l-1})^T R_l + R_l (A - S_{11} P_{l-1}) \\
& - R_l S_{22} R_l + Q_2.
\end{aligned}
\tag{7}
$$

- (b) determine P_l the unique strong solution of

$$0 = (A - S_{22} R_l)^T P_l + P_l (A - S_{22} R_l) - P_l S_{11} P_l + Q_1. \tag{8}$$

- (c) Put

$$A_l^r = A - S_{22} R_l, \;\; and \;\; A_{l-1}^p = A - S_{11} P_{l-1}, \tag{9}$$

for $l \geq 1$, compute

$$N_1(P_l, R_l) = (A_l^r)^T P_l + P_l A_l^r - P_l S_{11} P_l + Q_1, \tag{10}$$

and

$$N_2(P_l, R_l) = (A_l^p)^T R_l + R_l A_l^p - R_l S_{22} R_l + Q_2, \tag{11}$$

while

$$e_l = \max\{\|N_1(P_l, R_l)\|_\infty, \|N_2(P_l, R_l)\|_\infty\}, \tag{12}$$

is larger than ϵ replace l by $l + 1$ and go to (a), else stop.

3 Numerical examples

In order to demonstrate the efficiency of the introduced algorithm C, we have run for different initializations the interesting example which can be found in ([1], page 342) and called below "Example 1″. For comparison, we give simulations results with other existing algorithms [3] and [1]. We called the Lyapunov type algorithm [3] by A and the Riccati type algorithm from [1] by B.

Example 1 *Consider the coupled algebraic Riccati equations (5) and (6) associated with the following coefficients matrices:*

$$A = \begin{bmatrix} 20 & 50 \\ -25 & 15 \end{bmatrix}, S_{11} = \begin{bmatrix} 18 & 18 \\ 18 & 18 \end{bmatrix}, S_{22} = \begin{bmatrix} 4 & 8 \\ 8 & 16 \end{bmatrix}, Q_1 = \begin{bmatrix} 8 & 2 \\ 2 & 4 \end{bmatrix}, Q_2 = \begin{bmatrix} 8 & 2 \\ 2 & 14 \end{bmatrix}.$$

We obtain the following results:

- Algorithm A: after 40 iterations, algorithm A converges to a pair of solutions, which is approximated by

$$P_{40} = \begin{bmatrix} 2.0139 & 7.0416 \times 10^{-2} \\ 7.0416 \times 10^{-2} & 1.4959 \end{bmatrix}, R_{40} = \begin{bmatrix} 6.3421 \times 10^{-1} & -1.1875 \times 10^{-1} \\ -1.1875 \times 10^{-1} & 2.9936 \times 10^{-1} \end{bmatrix},$$

with

$$N_1(P_{40}, R_{40}) = 4.9641 \times 10^{-5}, \text{ and } N_2(P_{40}, R_{40}) = 1.3217 \times 10^{-5}.$$

- Algorithm B: after 40 iterations, algorithm B exhibits two cluster values, which are approximated by

$$P_{40} = \begin{bmatrix} 2.9894 \times 10^{-1} & 6.8680 \times 10^{-2} \\ 6.8680 \times 10^{-2} & 1.3624 \times 10^{-1} \end{bmatrix}, R_{40} = \begin{bmatrix} 3.1659 & 4.3560 \times 10^{-1} \\ 4.3560 \times 10^{-1} & 2.3835 \end{bmatrix}$$

and

$$P_{40} = \begin{bmatrix} 2.0139 & 7.0416 \times 10^{-2} \\ 7.0416 \times 10^{-2} & 1.4959 \end{bmatrix}, R_{40} = \begin{bmatrix} 6.3421 \times 10^{-1} & -1.1875 \times 10^{-1} \\ -1.1875 \times 10^{-1} & 2.9936 \times 10^{-1} \end{bmatrix},$$

with

$$N_1(P_{40}, R_{40}) = 3.400 \times 10^{-4} \text{ and } N_2(P_{40}, R_{40}) = 1.200 \times 10^{-4}.$$

- Algorithm C: after only 12 iterations, the algorithm C converges to a pair of solutions which is approximated by

$$R_{12} = \begin{bmatrix} 6.3420 \times 10^{-1} & -1.1875 \times 10^{-1} \\ -1.1875 \times 10^{-1} & 2.9936 \times 10^{-1} \end{bmatrix}, P_{12} = \begin{bmatrix} 2.0139 & 7.0415 \times 10^{-2} \\ 7.0415 \times 10^{-2} & 1.4959 \end{bmatrix},$$

with

$$N_1(P_{12}, R_{12}) = 7.6383 \times 10^{-14}, \text{ and } N_2(P_{12}, R_{12}) = 3.1963 \times 10^{-5}. \tag{13}$$

Remark 1 *These numerical examples presented here and other simulations (several other choices of initial conditions in particular) provide strong numerical evidence that algorithm C is the best of the three algorithms for the computation of a pair of stabilizing solutions of the coupled algebraic Riccati equations (5) and (6);*

4 Conclusion

In this paper we have presented a new algorithm to calculate a stabilizing solution (when it exists) of the coupled algebraic Riccati equations. These equations are known as the Riccati closed-loop Nash game. We use numerical examples to show the performance and the advantages of this algorithm. We also succeed to have better convergence than the existing ones.

5 Appendix

Simulations results for the different values of the coefficients of the coupled algebraic Riccati equations (5) and (6) which are not presented here correspond to two sequences P_l and R_l, $\forall l \geq 1$, which are both convergent but not monotone. Next, we give sufficient conditions for the monotonicity of the sequences P_l and R_l, $\forall l \geq 1$ appearing in algorithm C.

Lemma 1 *Assume that the following matrices*

$$\Gamma_l^p = (R_l - R_{l+1})^T S_{22} P_{l+1} + P_{l+1} S_{22} (R_l - R_{l+1}) - (P_l - P_{l+1})^T S_{11} (P_l - P_{l+1}),$$

and

$$\Gamma_l^r = (P_{l-1} - P_l)^T S_{11} R_{l+1} + R_{l+1} S_{11} (P_{l-1} - P_l) - (R_l - R_{l+1})^T S_{22} (R_l - R_{l+1}),$$

are positive and negative definite (respectively) $\forall l \geq 1$, then the sequence (P_l) is increasing and the sequence (R_l) is decreasing.

Acknowledgment

The author wishes to thank the anonymous referees for their careful reading of the manuscript and their fruitful comments and suggestions.

References

[1] H.Abou-Kandil and G.Freiling and V.Ionescu and G.Jank, Matrix Riccati Equations in Control and Systems Theory: Birkhauser Verlag, (2003), Germany.

[2] T.Başar and G.J. Olsder, Dynamic Noncooperative Game Theory: Academic Press, (1995), London.

[3] Z.Gajic and S.Shen,Parallel Algorithms for Optimal Control of Large Scale Linear Systems: Springer, (1993),Berlin,Heidelberg.

Brill Academic Publishers
P.O. Box 9000, 2300 PA Leiden
The Netherlands

*Lecture Series on Computer
and Computational Sciences*
Volume 4, 2005, pp. 1104-1107

Parallel Simulation of Deep Sub-Micron Double-Gate Metal-Oxide-Semiconductor Field Effect Transistors

Shao-Ming Yu [a,b], Hung-Mu Chou [b,c], and Shih-Ching Lo [d,1]

[a] Department of Computer and Information Science, National Chiao Tung Univeristy, Hsinchu 300, Taiwan
[b] Microelectronics and Information Systems Research Center, National Chiao Tung Univeristy, Hsinchu 300, Taiwan
[c] Department of Communication Engineering, National Chiao Tung Univeristy, Hsinchu 300, Taiwan
[d] National Center for High-Performance Computing, Hsinchu 300, Taiwan

Received 1 July, 2005; accepted in revised form 1 August, 2005

Abstract: Drift-Diffusion Density Gradient model (DD-DG) is the most popular model for simulating carrier transport phenomena in sub-micron semiconductor device, especially in two- or three-dimensional space. In deep sub-micron regime, the width effects cannot be neglected while simulating, i.e., three-dimensional simulation must be considered. However, three-dimensional computing is time-consuming. Fortunately, the dilemma of time consuming or rough approximation can be overcame by advanced computing technique. In this paper, we employ a parallel direct solving method to simulate double-gate metal-oxide-semiconductor field effect transistors (DG-MOSFET). The computational benchmarks of the parallel simulation, parallel speedup, load balance, and efficiency are studied in this work. Parallel numerical simulation of semiconductor devices is shown to be an indispensable tool for fast characterization and optimal design of semiconductor devices.

Keywords: Quantum effects, DG-MOSFET, Drift-Diffusion model, Density Gradient model, Numerical simulation, Parallel computing.

PACS: 02.60.Cb, 75.40.Mg, 85.35.Ds, 85.30.De.

1. Introduction

As the progress of semiconductor fabrication technology for the advanced metal oxide semiconductor field effect transistor (MOSFET) has been of great interests in recent years, computer- aided simulation for semiconductors, which provides a software-driven approach to explore new physics and device also acquires a crucial role in the development of semiconductor. Semiconductor device models, such as Drift-Diffusion (DD) and Hydrodynamic (HD) models [1], are the classical device simulation. In order to understand the characteristics of deep sub-micron devices, it is important to take quantum mechanical effects into account with the classical models. Therefore, quantum correction models, which produce a similar results to quantum mechanically calculated one but requires only about the same computation cost as that of the classical calculation, are developed. Among the quantum correction models, Density Gradient (DG) model is considered a good approximation of the quantum effect [2-5]. With the continuous decrease of device dimensions, one- and two-dimensional simulations are not accurate enough to describe and explore the physical transportation phenomena for electrons and holes in a semiconductor device. Therefore, a three-dimensional simulation is considered to examine characteristic fluctuations in deep sub-micron double-gate MOSFETs. Three-dimensional simulation is time-consuming. Fortunately, the dilemma of time-consuming or rough approximation can be overcome by parallel computing. Parallel numerical simulation of semiconductor devices has

---

[1] Corresponding author. E-mail: sclo@nchc.org.tw

been proven to be an indispensable tool for fast characterization and optimal design of semiconductor devices [6-8]. In this paper, we employ a parallel direct solving method to simulate a 40 nm DG-MOSFET and show the computational efficiency.

2. Drift-Diffusion Density-Gradient Model

The DD-DG model is employed to simulate drain current (I_{DS}) and threshold voltage (V_{TH}). The equations are given as follows.

$$\nabla \varepsilon \cdot \nabla \phi = -q(p - n + N_D - N_A),$$ (1)

$$q\frac{\partial n}{\partial t} - \nabla \cdot \mathbf{J_n} = -qR,$$ (2)

$$q\frac{\partial p}{\partial t} + \nabla \cdot \mathbf{J_p} = -qR,$$ (3)

$$n = N_C \exp((E_F - E_C - \Lambda)/k_B T),$$ (4)

$$\Lambda = \frac{\gamma \hbar^2 \beta}{12m}\left[\nabla^2(\phi + \Lambda) - \frac{\beta}{2}(\nabla\phi + \nabla\Lambda)^2\right],$$ (5)

where ε is the electrical permittivity, q is the elementary electronic charge, n and p are the electron and hole densities, and N_D and N_A are the number of ionized donors and acceptors, respectively. $\mathbf{J_n} = -qn\mu_n\nabla\phi_n$ and $\mathbf{J_p} = -qp\mu_p\nabla\phi_p$ are the electron and hole current densities. μ_n and μ_p are the electron and hole mobility, and ϕ_n and ϕ_p are the electron and hole quasi-Fermi potentials, respectively. $\phi_n = -\nabla\phi - \nabla n(kT/\mu_n)$ and $\phi_p = -\nabla\phi + \nabla p(kT/\mu_n)$. R is the generation-recombination term. According to DG method [2-5], Λ is an additional potential. Equation (5), which is proved that can apply to 3D simulation, is suggested by Wettstein, Schenk and Fichtner [5]. N_C is the conduction band density of states, E_C is the conduction band energy, and E_F is the electron Fermi energy. γ and β are fitting factors. According to box discretization, each PDE is discretized as

$$\sum_{j \neq i} \sigma_{ij} \cdot \varepsilon(\phi_i - \phi_j) + \Omega_i \cdot (p_i - n_i + N_{Ai} - N_{Di}) = 0,$$ (6)

$$\sum_{j \neq i} \sigma_{ij} \cdot \mu_n(n_i B(\phi_i - \phi_j) - n_j B(\phi_j - \phi_i)) + \Omega_i \cdot (R_i + \partial n_i/\partial t) = 0,$$ (7)

$$\sum_{j \neq i} \sigma_{ij} \cdot \mu_p(p_j B(\phi_j - \phi_i) - p_i B(\phi_i - \phi_j)) + \Omega_i \cdot (R_i + \partial p_i/\partial t) = 0,$$ (8)

$$\Omega_i \Lambda_i = \frac{\gamma \hbar^2}{6m_i}\sum_j \sigma_{ij}\left(1 - \exp\left[\frac{\phi_i + \Lambda_i}{2k_B T} - \frac{\phi_j + \Lambda_j}{2k_B T}\right]\right),$$ (9)

where $\sigma_{ij} = d_{ij}/l_{ij}$ in two-dimensional space and $\sigma_{ij} = D_{ij}/l_{ij}$ in three-dimensional space. l_{ij} is the distance between i and j. d_{ij} is the area of box ij and D_{ij} is the volume of box ij. In two-dimensional space, Ω_i is the area of the box face between i and j. Ω_i is the volume of the box face between i and j in three-dimensional space. After the systematic equation is discretized in spatial domain, the Backward Euler method is employed to time matching. The discrete scheme is solved by finite difference method self-consistently. After the drain current is obtained, the threshold voltage is determined. The definition of threshold voltage employed in this study is the Gm maximum method. The method firstly find out the gate voltage at the maximum of Gm, then make a tangent line of the drain current – gate voltage (I_{DS}-V_{GS}) curve at the gate voltage. Finally, extrapolated intercept of the tangent line to the V_{GS}-axis; the extrapolation is defined as V_{TH}.

3. Parallel Algorithm

Parallel algorithm employed in this work is base on the parallel direct method of linear system. There are four main steps. The first one is permuting and scaling the nonsymmetric matrix so as to maximize the elements on the diagonal of the matrix. Then, reordering the matrix to obtain the coefficient matrix **A** such that the factorization incurs low fill-in. The third step is the factorization step. Let the matrix **A** be factorized as $\mathbf{A} = \mathbf{LU}$. The main complication is due to the need for efficiently handling the fill-in in the factors $\mathbf{L}$ and $\mathbf{U}$. The solving strategy of the fill-in reduction is integrated by multilevel recursive or minimum-degree based approaches [7]. The numerical factorization algorithm utilizes the supernode structure of the numerical factors $\mathbf{L}$ and $\mathbf{U}$ to reduce the number of memory references. The result is a greatly increased sequential factorization performance. Furthermore, a left-right looking super node algorithm [7] for the parallel sparse numerical factorization on shared-memory multiprocessors is used. The last step is solving the triangular systems.

4. Numerical Results

In the numerical studies, a 40 nm DG-MOSFETs is simulated. Oxide thickness is 2 nm, channel width is 20 nm and channel thickness is 10 nm. Doping concentration of source and drain are 1×10^{20} cm^{-3}.

The uniform channel doping concentration (N_A) is 1×10^{18} cm^{-3}. Numerical results of MOSFETs are simulated by ISE-DESSIS ver. 10 on PC cluster with 8 processors. The CPU is Intel Xeon 2.8GHz 533MHz and RAM is 2GB. Figure 1 (a) illustrates the comparison of two- and three-dimensionally simulated I_{DS}-V_{GS} curves. In the subthreshold region, the difference of I_{DS}-V_{GS} curves between them is quite different and the two curves become closer while $V_{GS} > V_{TH}$. Therefore, the V_{TH} is quite different. In the case of the two-dimensional simulation, $V_{TH} = 0.22$ V. On the other hand, $V_{TH} = 0.32$ V in the three-dimensional case. The difference of V_{TH} might induce confusion of signal transmission in a circuit. The result of three-dimensional simulation is smaller than it of two-dimensional simulation because the scattering of electron and hole is considered in the three-dimensional simulation, i.e., the width effects cannot be ignored in the simulation of deep sub-micron semiconductor devices. For IC design, a lot of data of devices characteristics is necessary. Family curves with different applied biases should be simulated. In this study, five curves are computed for example. Figure 1 (b) illustrates the three-dimensional results with $V_{DS} = 0.05, 0.2, 0.5, 0.7,$ and 1.0 V. Simulating the five curves needs 10472 seconds for 2D computation, however, 3D computation takes 31662 seconds by 1 CPU.

To measure the efficiency of parallel computing, speedup (S) is employed to discuss the performance, which is defined as follows:

$$S(p) = T_1 / T(p) \tag{10}$$

where p is number of processors, T_1 is the runtime of the serial solution and $T(p)$ is the runtime of the parallel solution with p processors. $T(1) = 31662$, $T(2) = 16753$, $T(3) = 8789$ and $T(4) = 5287$ seconds. Therefore, $S(2) = 1.89$, $S(3) = 3.60$, $S(4) = 5.99$. The results are given in Fig. 2. According to Fig. 2, when the number of processors increases the speedup decreases because of the communication among processors. The parallel algorithm accelerates 3D computing and achieves a reliable accuracy successfully.

5. Conclusions

In this paper, the numerical result of DD-DG model is investigated so as to compare the difference of 2D and 3D simulation in deep sub-micron regime. Also, a parallel algorithm is employed to solve the 3D problem. With the simulation results, we conclude that the results show the difference between two- and three-dimensional simulations. For obtaining accurate results, three-dimensional simulation is suggested to simulate a deep sub-micron semiconductor device. Although 3D simulation is time-consuming, the parallel computing presented in this study can be applied to 3D simulation successfully. As mentioned above, the results are observed by simulation. To verify the results accuracy, comparisons between the numerical results and the measured data should be included in a future work.

Acknowledgments

This work is supported in part by the national science council (NSC) of Taiwan under Contract NSC-93-2215-E-429-008 and Contract NSC 94-2752-E-009-003-PAE, by the grant of the Ministry of Economic Affairs, TAIWAN Contract No. 93-EC-17-A-07-S1-0011, and by the TAIWAN Semiconductor Manufacturing Company (TSMC), Hsinchu, Taiwan under a 2004-2005 grant.

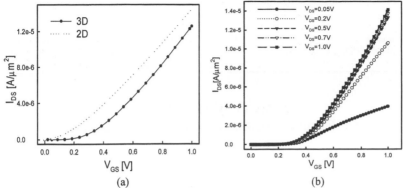

Fig. 1. Simulated I_{DS}-V_{GS} curves of (a) 2D and 3D simulation with V_{DS} = 1.0 V and (b) 3D simulation with V_{DS} = 0.05, 0.2, 0.5, 0.7, 1.0 V for 40 nm DG-MOSFET.

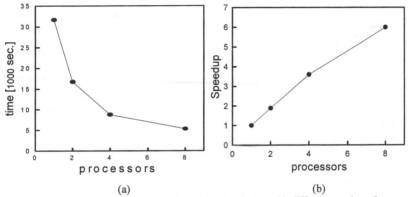

Fig. 2. (a) Computing time and (b) speedup of 3D simulation with different number of processor.

References

[1] H. K. Gummel, "A Self-Consistent Iterative Scheme for One-Dimensional Steady State Transistor Calculations," IEEE Transactions on Electron Devices, Vol. 11, pp.455, 1964.

[2] M. G. Ancona and H. F. Tierstein, "Macroscopic Physics of the Silicon Inversion Layer," Physical Review B, Vol. 35, No. 15, pp.7959-7965, 1987.

[3] Y. Li *et al.*, "Numerical Simulation of Quantum Effects in High-k Gate Dielectrics MOS Structures using Quantum Mechanical Models," Computer Physics Communications, Vol. 147, No. 1-2, pp. 214-217, 2002.

[4] T. W. Tang *et al.*, "Discretization Scheme for the Density-Gradient Equations and Effect of Boundary Conditions," Journal of Computational Electronics, Vol.1, No. 3, pp. 389-393, 2002.

[5] A. Wettstein, A. Schenk and W. Fichtner, "Quantum Device-Simulation with the Density-Gradient Model on Unstructured Grids," IEEE Transactions on Electron Devices, Vol. 48, No. 2, pp.279-284, 2001.

[6] Y. Li *et al.*, "A Novel Parallel Approach for Quantum Effect Simulation in Semiconductor Devices," International Journal of Modelling and Simulation, Vol. 23, No. 2, pp.1-8, 2003.

[7] O. Schenk, K. Garter and W. Fichtner, "Efficient Sparse LU Factorizaytion with Left-right Looking Strategy on Shared Multiprocessors," BIT, Vol. 40, No. 1, pp.158-176, 2000.

Brill Academic Publishers
P.O. Box 9000, 2300 PA Leiden
The Netherlands

*Lecture Series on Computer
and Computational Sciences*
Volume 4, 2005, pp. 1108-1108

Preface to Symposium:
Matter at Extreme Conditions: Theory and Application

Organizer: M. Riad Manaa
Lawrence Livermore National Laboratory
Livermore, California 94551

The subject of "Matter at Extreme Conditions" encompasses a wide range of phenomena the thrust of which is to address the physical and chemical behaviors of materials exposed to "abnormal" conditions of high pressures, temperature extremes, or external fields. Recent advances in theoretical methodologies and first principle computational studies have predicted unusual properties and unraveled a few surprises when matter is subjected to such strains: a reversed and anomalous Doppler effects in shocked periodic media, the possible existence of low temperature liquid metallic state of hydrogen, and a superionic phase of water at high temperature and pressure. A unified approach from quantum mechanical principles allows for exploring such diverse and disparate subjects as ultracold plasmas in a strong magnetic field, and the dynamic aspects of Bose-Einstein condensates. These topics, which are aptly presented in this symposium, are but a few examples of interesting discoveries and methodologies in this active and exciting area of research.

The development of reactive force fields from quantum mechanical principles for use in conjunction with molecular dynamics provide us with an invaluable tool for large-scale simulations to study the chemical transformations and decomposition products of complex organic systems at extreme conditions. Simulations implementing classical fields can also provide an unprecedented access to the short time scales of chemical events that occur in dense fluids at high-temperature, and for the study of atomic clusters under strong laser pulses. Such simulations have a wide aspect of applications: chemical reactions of hydrocarbon induced by shock impact of interest in planetary science, detonation of energetic materials, and laser-matter interactions, to name a few examples that will also be presented in this symposium.

This work was performed under the auspices of the U.S. Department of Energy by the Lawrence Livermore National Laboratory under contract number W-7405-Eng-48.

M. Riad Manaa

Email:manaa1@llnl.gov

Chemist, Chemistry and Materials Science Directorate
Lawrence Livermore National Laboratory

Ph.D. The Johns Hopkins University, Baltimore, MD, USA (1992)
National Research Council associate, 1993-1995.

RESEARCH:
Ab initio/MD simulations of physical and chemical properties of materials under extreme conditions of pressure and temperature. Structure, reactivity, and energetic of energy-rich molecular systems. Computational predictions of hetero-fullerenes and nanotubes properties. Electronic structure theory of nonadiabatic and spin-forbidden processes.

Brill Academic Publishers
P.O. Box 9000, 2300 PA Leiden
The Netherlands

*Lecture Series on Computer
and Computational Sciences*
Volume 4, 2005, pp. 1109-1113

Reactive Force Fields based on Quantum Mechanics for Applications to Materials at Extreme Conditions

Adri van Duin, Sergey Zybin, Kim Chenoweth, Si-Ping Han, William A. Goddard III,

Materials and Process Simulation Center,
California Institute of Technology,
Pasadena, CA, USA 91125
Received 12 July, 2005; accepted in revised form 11 August, 2005

Abstract: Understanding the response of energetic materials (EM) to thermal or shock loading at the atomistic level demands a highly accurate description of the reaction dynamics of multimillion-atom systems to capture the complex chemical and mechanical behavior involved: nonequilibrium energy/mass transfer, molecule excitation and decomposition under high strain/heat rates, formation of defects, plastic flow, and phase transitions.

To enable such simulations, we developed the ReaxFF reactive force fields based on quantum mechanics (QM) calculations of reactants, products, high-energy intermediates and transition states, but using functional forms suitable for large-scale molecular dynamics simulations of chemical reactions under extreme conditions.

The elements of ReaxFF are:
- charge distributions change instantaneously as atomic coordinates change,
- all valence interactions use bond orders derived uniquely from the bond distances which in turn describe uniquely the energies and forces,
- three body (angle) and four body (torsion and inversion) terms are allowed but not required,
- a general "van der Waals" term describes short range Pauli repulsion and long range dispersion interactions, which with Coulomb terms are included between all pairs of atoms (no bond or angle exclusions),
- no environmental distinctions are made of atoms involving the same element; thus every carbon has the same parameters whether in diamond, graphite, benzene, porphyrin, allyl radical, or TATB,
- ReaxFF uses the same functional form and parameters for reactive simulations in hydrocarbons, polymers, metal oxides, and metal alloys, allowing mixtures of all these systems into one simulation.

We will present an overview of recent progress in ReaxFF developments, including the extension of ReaxFF to nitramine-based (nitromethane, HMX, PETN, TATB) and peroxide-based (TATP) explosives. To demonstrate the versatility and transferability of ReaxFF, we will present applications to solid composite propellants such as Al/Al2O3-metal nanoparticles embedded into solid explosive matrices (RDX, PETN)

1. Introduction

The atomistic mechanisms of the fundamental chemistry and physics of condensed-phase materials under extreme conditions that experimentally may be challenging to investigate (e.g. ultrafast laser heating, propellant combustion, shock compression, and detonation) are still not well understood. Quantum mechanics (QM) can provide a reliable description of the above processes with a valuable insight into the energy barriers and reaction pathways. However, QM methods are still practical only for systems too small to describe coupling of chemical and mechanical behavior at the requisite level of spatiotemporal resolution (e.g. for simulation of voids, dislocations, grain boundaries, and interfaces).

Recently, we developed the ReaxFF force field in order to describe chemical reactions in classical molecular dynamics (MD) simulations that are many orders of magnitude faster than the QM methods. We have found that ReaxFF can reproduce QM-energies for reactive systems, including reactants, transition states and products, for a wide range of materials, such as hydrocarbons,[1] nitramines,[2,3] silicon/silicon oxides,[4,5] aluminum/ aluminum oxides,[6] transition metal interactions with first-row elements,[7] and magnesium/ magnesium hydrides.[8] This allows dynamic simulations of materials under a wide range of heat or shock loading conditions, including initiation and detonation of solid EM and EM composites. The ReaxFF functional form is suitable for highly scalable parallel MD codes, making possible MD simulations of chemical reactions in systems with millions of atoms.[9]

Furthermore, we found that at increased pressures the rate of PDMS decomposition drops considerably, leading to the formation of fewer methyl radicals and methane molecules. Finally, we studied the influence of various chemical environments on the stability of PDMS. We found that the addition of water, nitrogen monoxide, or a SiO_2-slab has no direct effect on the short-term stability of PDMS but addition of reactive species such as ozone leads to significantly lower PDMS decomposition temperature. The addition of nitrogen monoxide does retard the initial production of methane and C_2 hydrocarbons until the nitrogen monoxide is depleted. These results, and their good agreement with available experimental data, demonstrate that ReaxFF provides a useful computational tool for studying the details of the rapid decomposition mechanism and the chemical stability of polymers.

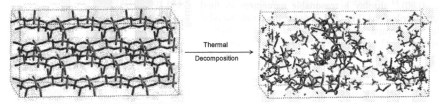

Figure 4. The rapid decomposition process of PDMS by NVT-MD simulations with the ReaxFF.

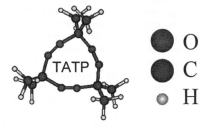

To study the initial chemical events related to the detonation of triacetonetriperoxide (TATP) commonly using in Improvised Explosive Devices (IED). we carried out molecular dynamics (MD) simulations using the ReaxFF reactive force field.[20] Our simulations demonstrate that thermal initiation of condensed phase TATP-is entropy-driven (rather than enthalpy driven), since the initial reaction (leading mainly to the formation of acetone, O_2, and several unstable $C_3H_6O_2$-isomers) is almost energy-neutral. The O_2 generated in the initiation steps is subsequently utilized in exothermic secondary reactions, leading finally to formation of water and a wide range of small hydrocarbons, acids, aldehydes, ketones, ethers and alcohols.

All simulations and force field optimizations reported here used the ReaxFF-program,[17] which carries out both MD-simulations and force field optimization using the ReaxFF-system energy description. A Berendsen-thermostat and barostat[18] was used to control the temperature and pressure in the NVT and NPT-simulations.

Acknowledgments

Funding was provided by DARPA-PROM, ONR, and ARO-MURI.

References

[1] A.C.T. van Duin, S. Dasgupta, F. Lorant and W.A. Goddard III, *J. Phys. Chem. A* **2001**, *105*, 9396.

[2] A. Strachan, A.C.T. van Duin, D. Chakraborty, S. Dasgupta and W.A. Goddard III, *Phys. Rev. Letters* **2003**, *91*, 098301.

[3] A. Strachan, E. Kober, A.C.T. van Duin, J. Oxgaard and W.A. Goddard III, *J. Chem. Phys.* **2005**, *122*, 054502.

[4] A.C.T. van Duin, A. Strachan, S. Stewman, Q. Zhang, X. Xu and W.A. Goddard III, *J. Phys. Chem. A* **2003**, *107*, 3803.

[5] K. Chenoweth, S. Cheung, A.C.T. van Duin, W.A. Goddard III and E.M. Kober, *J. Am. Chem. Soc.* **2005**, *127*, 7192.

[6] Q. Zhang, T. Cagin, A.C.T. van Duin, W.A. Goddard III, Y. Qi and L. Hector, *Phys. Rev. B* **2004**, *69*, 045423.

[7] K. Nielson, A.C.T. van Duin, J. Oxgaard, W. Deng and W.A. Goddard III, *J. Phys. Chem. A.* **2005**, *109*, 493.

[8] S. Cheung, W. Deng, A.C.T. van Duin and W.A. Goddard III, *J. Phys. Chem. A* **2005**, *109*, 851.

[9] Vashishta et al – Multimillion simulations using the ReaxFF reactive force field. To be published

[10] H.H.Cady and A.C.Larson, Acta Crystallogr. B, **1975**, *B31*, 1864.

[11] The SeqQuest DFT/LCAO program is available at

[12] A. Strachan, A.C.T. van Duin, and W.A. Goddard III, unpublished.

[13] M.R. Manaa, L.E. Fried, and E.J. Reed, J. Comp.–Aid. Design, **2003**, *10*, 75

[14] S. Han, A.C.T. van Duin, W.A. Goddard III, and A. Strachan, in preparation.

[15] M.R. Manaa, E.J. Reed, L.E. Fried, G. Galli, and F.Gygi, J. Chem. Phys. **120**, 10146, (2004).

[16] M.T.Nguyen, H.T.Le, B.Hajgato, T.Veszpremi, and M.C.Lin, J.Phys.Chem. **107**, 4286 (2003).

[17] The ReaxFF-program is available for distribution to academic users. Please contact ACTvD (duin@wag.caltech.edu) or WAG (wag@wag.caltech.edu).

[18] H.J.C. Berendsen, J.P.M. Postma, W.F. van Gunsteren, A. DiNola and J.R. Haak *J. Chem. Phys.* **1984**, *81*, 3684.

[19] Manaa et al., *J. Chem. Phys.,* v 120, 10146, (2004)

[20] Adri C.T. van Duin, Yehuda Zeiri, Faina Dubnikova, Ronnie Kosloff. and William A. Goddard III J. Am. Chem. Soc. In press

shock propagation) with frequency within the 1st bandgap of the post-shock crystal as shown in Figure 1. The frequency of this radiation is far from the 1st bandgap edge in the pre-shock crystal. The incident light is reflected and acquires a *reversed* Doppler shift, i.e. lowered frequency in this case. We have observed this reversed Doppler effect in finite-difference simulations.

The physical origin of the effect is an unusual nature of the reflecting surface that the shock front represents. Rather than endowing reflected light with a fixed phase shift (like a metal mirror, for example) the shock front endows reflected light with a time-dependent phase shift of magnitude greater than the usual Doppler effect phase shift.

We find that that nature of the reflected light depends on the thickness of the shock front. For shock front thicknesses greater than about 1 lattice constant of the crystal, a single frequency is reflected from the shock front. In the limit of thin shock front thicknesses, multiple equally-spaced frequencies are reflected from the shock front rather than one frequency. This behavior can be interpreted as a periodic modulation of the reflected light by the discrete nature of the shocked lattice.

We have developed a simple analytical theory describing the frequencies observed in light reflected from a shocked periodic medium. This theory agrees well with our finite-difference simulations.

3 Experimental observation

Seddon and Bearpark (SB) have recently observed a reversed Doppler effect in an electrical transmission line. [1] Rather than a photonic crystal, a 1D radio-frequency electrical transmission line comprised of coupled inductor-capacitor resonators was utilized. Our theoretical description of this experiment is closely related to the photonic crystal theory and considers the time-dependent phase shift endowed upon the reflected light. The experimentally observed frequency shift is in quantitative agreement with our theory and only one frequency is reflected, consistent with our theory in the thick shock front limit.

A different proposed theoretical explanation involves the reflection of unusual electromagnetic waves (backward waves with $v_{phase}v_{group} < 0$) from a reflecting shock front with fixed reflection phase shift. [1] In this scenario the reversed Doppler shift results from the unusual nature of the electromagnetic waves within this description rather than an unusual reflecting surface. This picture is closely related to the effect predicted to occur in negative-index materials (rather than transmission lines) made by Veselago. [4] However, we find that the waves in a model of this system are not backward waves and therefore cannot be the origin of the effect.

Another theory proposes that spatial harmonics in the Bloch state of this system give rise to the anomalous Doppler effect. [9] Like the backward wave picture, this theory is based on the assumption that the shock front acts as a usual reflecting surface, with no time-dependent phase shift. Within this picture, a phase-matching condition on the radiation at the shock front predicts that the usual Doppler effect should also be observed, in addition to an inverse Doppler shift. However, we find that the usual Doppler shift is not observed in the experiments [1] or any numerical simulations. [1, 9]

Acknowledgments

The authors are particularly grateful to Nigel Seddon for many helpful discussions. This work was supported in part by the Materials Research Science and Engineering Center program of the National Science Foundation under Grant No. DMR-9400334. This work was performed in part under the auspices of the U.S. Department of Energy by University of California, Lawrence Livermore National Laboratory under Contract W-7405-Eng-48.

References

[1] N. Seddon and T. Bearpark *Science* **302**, p. 1537, 2003.

[2] L. M. Barker and R. E. Hollenbach *J. App. Phys.* **43**, p. 4669, 1972.

[3] E. J. Reed, M. Soljačić, and J. D. Joannopoulos *Phys. Rev. Lett.* **91**, p. 133901, 2003.

[4] V. G. Veselago *Sov. Phys. USPEKHI* **10**, p. 509, 1968.

[5] A. M. Belyantsev and A. B. Kozyrev *Tech. Phys.* **47**, p. 1477, 2002.

[6] E. Yablonovich *Phys. Rev. Lett.* **58**, p. 1059, 1987.

[7] S. John *Phys. Rev. Lett.* **58**, p. 2486, 1987.

[8] J. D. Joannopoulos, R. D. Meade, and J. N. Winn, *Photonic Crystals*, Princeton University Press, Princeton, NJ, 1995.

[9] A. B. Kozyrev and D. W. van der Weide *Phys. Rev. Lett.* **94**, p. 203902, 2005.

[10] E. J. Reed, M. Soljačić, M. Ibanescu, and J. D. Joannopoulos *Science* **304**, p. 778, 2004.

DVR merges aspects of the grid and basis methods, drawing on the close association between a spectral basis of n functions and an underlying quadrature rule with n points (x_i) and weights (w_i), for example Legendre polynomials and the Gauss-Legendre quadrature. The methods provides a high degree of accuracy with a minimum of spatial points. The FE aspect limits connections between elements to functions on the boundary, thus rendering the kinetic energy matrix elements highly structured and sparse. In fact, the structure permits dividing the matrix into two parts, each being block diagonal.

We present a diverse set of examples from matter under extreme conditions in which these methods have proved crucial in obtaining viable solutions. These include:

1. Interaction of ultrashort laser radiation with molecules and the generation of attosecond pulses[e.g.[7]];

2. Ultracold plasmas of anti-hydrogen in a strong magnetic field [e.g. [8]]; and

3. Dynamic aspects of Bose-Einstein condensates [e.g. [9]].

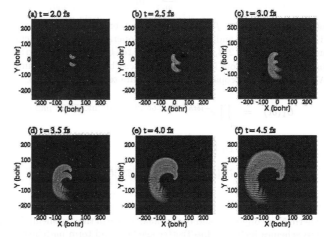

Figure 1: Ionization of highly-dissociated H_2 by a femtosecond circularly polarized laser pulse. Interaction of the freed wavepackets with the nuclei generate attosecond pulse.

Acknowledgment

Work performed under the auspices of the U.S. Department of Energy, contract W-7405-ENG-36.

References

[1] M. Suzuki, J. Math. Phys. **32**, 400 (1991).

[2] H. De Raedt, Comp. Phys. Rep. **7**, 1 (1987).

[3] L. A. Collins, J. D. Kress, and R. B. Walker, Comp. Phys. Comm. **114**, 15 (1998).

[4] B. Schneider and L. Collins, J. Non-Crystal. Solids **351**, 1551 (2005).

[5] M. Tuckerman, B. Berne, and G. Martyna, J. Chem. Phys. **97**, 1990 (1992).

[6] T. N. Rescigno and C. W. McCurdy, Phys. Rev. A **62**, 032706 (2000); B. I. Schneider and D. L. Feder, Phys. Rev. A **59**, 2232 (1999); B. Schneider and N. Nygaard, Phys. Rev. E **70**, 056706 (2004).

[7] S. Hu and L. Collins, Phys. Rev. Lett. **94**, 073004 (2005).

[8] S. Hu, D. Vrinceanu, S. Mazevet, and L. Collins, Phys. Rev. Lett. (submitted); D. Vrinceanu *et. al.* **92**, 133402 (2004).

[9] L. Collins *et. al.*, Phys. Rev. A **71**, 033628 (2005).

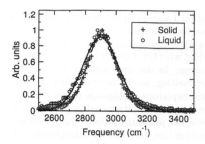

Figure 1: Vibron frequency distributions in solid and liquid hydrogen at 50 GPa and 800 K from velocity autocorrelation analysis of *ab initio* MD trajectories.

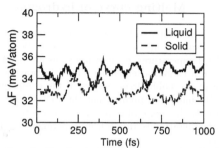

Figure 2: Quantum corrections to the classical ionic free energies of solid and liquid hydrogen at 130 GPa and 700 K.

of first-order quantum corrections to the ionic free energies using the Wigner-Kirkwood expansion (Fig. 2). Similar cancellation is expected for errors originating from the GGA.

3   Dissociation in fluid hydrogen

Dissociation in fluid hydrogen is studied by carrying out a series constant-volume and constant-pressure *ab initio* MD simulations over a range of pressures, from 15 to 200 GPa. For a given set of conditions, thermal equilibrium can be attained within several ps of simulation time. Upon sufficient heating, a transition to a non-molecular fluid takes place over $\sim 10^2$ fs simulation time. This liquid-liquid (LL) phase transition has negligible hysteresis effects of super-heating or super-cooling and, therefore, can be studied with single-phase simulations alone.

At pressures above ~ 35 GPa, the LL transition is sharp and is accompanied by a drop in the pressure, which indicates a first-order phase transition. To confirm this conclusion, correlation lengths have to be considered. Simulations with progressively lager supercells (up to 800 atoms) have been carried out, until satisfactory convergence has been achieved. It is important to note that the finding of a critical point for the LL phase transition around 30 GPa resolves previous controversies between theory and experiment [10].

The LL phase line approaches the melting curve as pressure increases, however, at 200 GPa, the two lines still remain ~ 400 K apart. Further extrapolations of the DFT-MD results indicate a triple point at around 300 GPa and 400 K, beyond which the hydrogen solid melts into a metallic liquid. It must be emphasized that the accuracy of the computed phase lines is crucial for making reliable projections for the low-temperature phases of hydrogen. Indeed, a low-temperature liquid metal can exist only in a limited range of densities; eventually, upon further compression, it will give way to a metallic solid.

Unlike the case of melting, quantum ion corrections and GGA exchange-correlation errors are significant in the DFT-MD simulations of the LL phase transitions. These can be estimated from a Wigner-Kirkwood expansion and quantum Monte Carlo calculations, respectively. Neglecting the zero-point motion favors the molecular fluid phase. The GGA, on the other hand, tends to favor the atomic fluid. Preliminary results indicate that that the combined effect of the two types of errors is small at the highest pressures considered, but closer to the critical pressure, it leads to lowering of the transition temperatures (reliable quantitative results are not available yet).

4 Conclusion

We have used *ab initio* MD simulations based on DFT to compute the solid-liquid and liquid-liquid phase boundaries of hydrogen over a large pressure and temperature range. The accuracy of the computed LL phase transition can be further improved by including corrections for zero-point motion and exchange-correlation errors. The results indicate the existence of a low-temperature liquid metallic hydrogen around 400 GPa, however, the simulation of such a phase requires full consideration of ion quantum motion.

Acknowledgment

S.B. was supported by the National Sciences and Engineering Research Council of Canada. Part of this work was performed under the auspices of the U.S. Dept. of Energy at the University of California/Lawrence Livermore National Laboratory under contract no. W-7405-Eng-48.

References

[1] See, for example, I.F. Silvera, *Rev. Mod. Phys.* **52**, 393 (1980); H.K. Mao and R.J. Hemley, *Rev. Mod. Phys.* **66**, 671 (1994); and references therein.

[2] E. Wigner and H.B. Huntington, *J. Chem. Phys.* **3**, 764 (1935).

[3] N.W. Ashcroft, *Phys. Rev. Lett.* **21**, 1748 (1968).

[4] N.W. Ashcroft, *J. Phys.: Condens. Matter* **12**, A129 (2000).

[5] E. Babaev, A.Sudbe and N.W. Ashcroft, *Nature* **431**, 666 (2004).

[6] S.T. Weir, A.C. Mitchell and W.J. Nellis, *Phys. Rev. Lett.* **76**, 1860 (1996).

[7] D.M. Straus and N.W. Ashcroft, *Phys. Rev. Lett.* **38**, 415 (1977).

[8] K. Nagao, S.A. Bonev, A. Bergara and N.W. Ashcroft *Phys. Rev. Lett.* **90**, 035501 (2003).

[9] S.A. Bonev, E. Schwegler, T. Ogitsu and G. Galli, *Nature* **431**, 669 (2004).

[10] See, for example, B. Militzer and E.L. Pollock, *Phys. Rev. E* **61**, 3470 (2000); S. Scandolo, *PNAS* **100**, 3051 (2003); S.A. Bonev, B. Militzer and G. Galli, *Phys. Rev. B* **69**, 014101 (2004); and references therein.

[11] E. Gregoryanz, A.F. Goncharov, K. Matsuishi, H.-k. Mao and R.J. Hemley, *Phys. Rev. Lett.* **90**, 175701 (2003).

[12] R. Car and M. Parrinello, *Phys. Rev. Lett.* **55**, 2471 (1985).

Figure 1: Snapshot of the simulations at 115 GPa. A well defined network has been formed, and the protons dissociate very rapidly, continually breaking and reforming bonds.

Maximally localized Wannier centers [11] of several trajectories were calculated, and a distribution function was determined. The outer peak at 0.46–0.50 Å corresponds to electrons participating in a covalent bond. Based on the above distribution, one can define the minimum at roughly 0.42 Å as a dividing surface wherein a maximally localized Wannier center located at distances shorter than this, relative to its parent oxygen atom, represents a lone pair, and those found at greater distances represent covalent bonds [4]. We used this definition to compute the percentage of O–H bonds with a Wannier center along the bond axis. Surprisingly, the results for pressures of 34 – 75 GPa consistently showed that 85-95% of the O–H bonds are covalent. For 95 and 115 GPa, we find about 50 – 55% of the bonds are covalent. This is consistent with symmetric hydrogen bonding, for which the split between ionic and covalent bonds would be 50/50.

In conclusion, we have performed first principles simulations of water at pressures up to 115 GPa (3.0 g/cc) and 2000K. Along this isotherm we can define three different phases. First, from 34 GPa to 58 GPa (2.0-2.4 g/cc), we observe a molecular fluid phase with superionic diffusion of the hydrogens. Second, at 75 GPa (2.6 g/cc), we find a stable bcc oxygen lattice with superionic proton conduction. O–H bonds within this "solid" phase are found to be mostly covalent, despite their exceedingly short lifetimes of ca. 10 fs. Third, at 95 – 115 GPa (2.8 – 3.0 g/cc) we find a transformation to a phase dominated by transient networks of symmetric O-H hydrogen bonds. Given the smooth nature of the calculated P-V isotherm, the transition to the network phase does not appear to be first order. The network can be attributed to the symmetrization of the hydrogen bond, similar to the ice VII to ice X transition

Acknowledgment

The authors wish to thank the anonymous referees for their careful reading of the manuscript and their fruitful comments and suggestions. This work was performed under the auspices of the U. S. Department of Energy by the University of California Lawrence Livermore National Laboratory under contract No. W-7405-Eng-48.

References

[1] W. B. Hubbard, *Science*, **214**, 145 (1981).

[2] P. Demontis, R. LeSar, and M. L. Klein, *Phys. Rev. Lett.*, **60**, 2284 (1988).

[3] C. Cavazzoni, G. L. Chiarotti, S. Scandolo, E. Tosatti, M. Bernasconi and M. Parrinello, *Science*, **283**, 44 (1999).

[4] E. Schwegler, G. Galli, F. Gygi and R. Q. Hood, *Phys. Rev. Lett.*, **87**, 265501 (2001).

[5] W. J. Nellis, N. C. Holmes, A. C. Mitchell, D. C. Hamilton and M. Nicol, *J. Chem. Phys*, **107**, 9096 (1997).

[6] R. Chau, A. C. Mitchell, R. W. Minich and W. J. Nellis, *J. Chem. Phys*, **114**, 1361 (2001).

[7] E. Katoh, H. Yamawaki, H. Fujihisa, M. Sakashita and K. Aoki, *Science*, **295**, 1264 (2004).

[8] , N. Goldman, L. E. Fried, I.-F. W. Kuo and C. J. Mundy, *Phys. Rev. Lett.*, **94**, 217801 (2005).

[9] , A. F. Goncharov, N. Goldman, L. E. Fried, J. C. Crowhurst, I.-F. W. Kuo, C. J. Mundy and J. M. Zaug, *Phys. Rev. Lett.*, **94**, 125508 (2005).

[10] , D. Chandler, *J. Chem. Phys.*, **68**, 2959 (1978).

[11] , P. L. Silvestrelli and M. Parrinello, *Phys. Rev. Lett.*, **82**, 3308 (1999).

3. Tests of ZND Theory at the Nanoscale

To use molecular dynamics to test ZND theory we employ the long studied AB model of a detonating diatomic molecular solid [4,5]. The reactive empirical bond order potentials characterizing this model are capable of simultaneously following the dynamics of millions of atoms in a rapidly changing environment while including the possibility of exothermic chemical reactions that proceed along chemically reasonable reaction paths from the cold solid-state reactants to the hot gas-phase molecular products. They also incorporated the strong intramolecular forces binding atoms into diatomic molecules with reasonable bond strengths, lengths, and vibrational frequencies and the weak intermolecular forces that bind these molecules into molecular solids with reasonable van der Waals binding energies, intermolecular separations, and sound speeds. This model is known to detonate with a short reaction zone length [4,5].

In the simulations the detonation is initiated using a high velocity flier plate with periodic boundary conditions employed in the direction perpendicular to the detonation propagation. The system then evolves for several ps until a well-defined detonation front is established with a characteristic von Neumann spike. A piston with a given velocity in the direction of the shock propagation is then inserted into the system where the velocity distribution of the particles matches the piston velocity. Particles behind the newly inserted piston are discarded and the equations of motion of the system are integrated for at least 20 ps. This results in a large region in front of the piston with the same average values of P_1 and V_1. This process was repeated for several piston velocities and the results compiled in a simulated detonation Hugoniot shown in Fig. 1 for a 2D simulation.

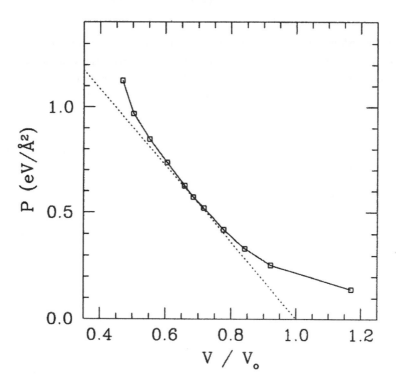

Figure 1. Simulated Hugoniot for the 2D AB system (lines drawn as guide to eye) along with dotted Rayleigh line. The curve at pressures below that of the CJ point is not defined as a Hugoniot.
Calculated points on this curve are represented by squares connected by solid lines that are guides to the eye. Also shown is the Rayleigh line (dotted) determined by the CJ hypothesis. This hypothesis selects the Rayleigh line with minimum D, that is the minimum slope line that originates at the initial

state. As can be seen from Fig. 1 the requirement of minimum slope makes this line tangent to the Hugoniot curve. The point of tangency is the CJ point. We find that D determined from the slope of this Rayleigh line agrees with D measured directly from the simulations within less than a percent. We have also found that the non-equilibrium flow from the CJ point to the detonation front is steady within our numerical error and the flow behind the CJ point although non-steady is self-similar. These results taken together provide strong confirmation of ZND theory in general and of the CJ hypothesis in particular for this model system.

Acknowledgments

This work was supported by the U.S. Office of Naval Research both directly and through the Naval Research Laboratory. One of the authors (MLE) received additional funding from the Naval Academy Research Council.

References

[1] W. Fickett, *Introduction to Detonation Theory*, U. Calf. Press, Berkeley, 1985.

[2] W. C. Davis, The Detonation of Explosives, *Scientific American* **256**, 106 (1987).

[3] Ya. B. Zel'dovich and Yu. P. Raizer, *Physics of Shock Waves and High-Temperature Hydrodynamic Phenomena*, Vols. 1 and 2, (Academic Press, New York, 1966, 1967).

[4] J. J. Erpenbeck, Molecular dynamics of detonation. I. Equation of state and Hugoniot curve for a simple reactive fluid, *Phys. Rev. A* **46**, 6406-6416 (1992).

[5] D. W. Brenner, D. H. Robertson, M. L. Elert, and C. T. White, Detonations at nanometer resolution using molecular dynamics, *Phys. Rev. Lett.* **70** 2174-2177 (1993).

[6] C. T. White, D. R. Swanson, and D. H. Robertson, Molecular Dynamics Simulations of Detonations, *Chemical Dynamics in Extreme Environments*, (Editor: Rainer A. Dressler) World Scientific, London, 547-592 (2001).

Figure 1: Initial configuration of naphthalene crystal for shock simulation. The two crystals are propelled toward each other at a fixed initial velocity, generating shock waves which propagate back through the crystals.

2.2 Acetylene and Ethylene

Acetylene (C_2H_2) and ethylene (C_2H_4) are both pi-bonded C_2 species capable of undergoing addition polymerization reactions. Acetylene has been observed in comets [14] and in the atmospheres of Jupiter and Titan, and its polymerization has been proposed [13] as a mechanism for the formation of PAHs there. MD simulations show the onset of significant polymerization reactions for acetylene at an impact speed of 10 km/s, with more oligomers and longer chains being formed at higher impact velocities. As expected due to energetic and steric effects, the reaction threshold for ethylene was found to be significantly higher at 15 km/s.

2.3 Methane

Reactions of methane at high temperature and pressure are important for the study of the composition of Neptune and Uranus, whose interiors are assumed to contain large quantities of methane ice at pressures of hundreds of GPa. Molecular dynamics simulations of shocked solid methane can provide insight into these reactions. Large-scale AIREBO simulations indicate that the threshold for hydrogen abstraction reactions occurs at about 20 km/s impact speed. Significant carbon chemistry, including the formation of ethylene and other C_2 species, occurs at 25 km/s. This corresponds to a pressure of approximately 100 GPa and transient temperatures of 6000 to 9000 K.

Acknowledgments

This work was supported by the Office of Naval Research and the Naval Academy Research Council. MLE would like to thank ENS Shannon M. Revell, USN for performing the naphthalene calculations.

References

[1] H. Eyring and M. Polanyi, Simple gas reactions, *Z. physik. Chem.* **12** 279-311 (1931); J. Tully, Dynamics of gas-surface interactions: reaction of atomic oxygen with adsorbed carbon on platinum, *J. Chem. Phys.* **73** 6333-6342 (1980).

[2] J. Tersoff, New empirical approach for the structure and energy of covalent systems, *Phys. Rev. B* **37** 6991-7000 (1980); J. Tersoff, Empirical interatomic potential for carbon, with applications to amorphous carbon, *Phys. Rev. Lett.* **61** 2879-2882 (1988).

[3] M. L. Elert, D. W. Brenner, and C. T. White, Some one-dimensional molecular dynamics simulations of detonation, *Shock Compression of Condensed Matter – 1989* (Editors: S. C. Schmidt, J. N. Johnson, and L. W. Davison) Elsevier, 275-278 (1990).

[4] D. W. Brenner, M. L. Elert, and C. T. White, Incorporation of reactive dynamics in simulations of chemically-sustained shock waves, *Shock Compression of Condensed Matter – 1989* (Editors: S. C. Schmidt, J. N. Johnson, and L. W. Davison), Elsevier, 263-266 (1990).

[5] C. T. White, D. H. Robertson, M. L. Elert, and D. W. Brenner, Molecular dynamics simulations of shock-induced chemistry: application to chemically sustained shock waves, *Microscopic Simulations of Complex Hydrodynamic Phenomena*, (Editors: M. Mareschal and B. L. Holian), Plenum Press, 111-123 (1992).

[6] D. W. Brenner, D. H. Robertson, M. L. Elert, and C. T. White, Detonations at nanometer resolution using molecular dynamics, *Phys. Rev. Lett.* **70** 2174-2177 (1993).

[7] D. W. Brenner, Empirical potential for hydrocarbons for use in simulating the chemical vapor deposition of diamond films, *Phys. Rev. B* **42** 9458-9471 (1990).

[8] D. W. Brenner, O. A. Shenderova, J. A. Harrison, S. J. Stuart, B. Ni, and S. B. Sinnott, A second-generation reactive empirical bond order (REBO) potential energy expression for hydrocarbons, *J. Phys.: Condens. Matter* **14** 783-802 (2002).

[9] S. J. Stuart, A. B. Tutein, and J. A. Harrison, A reactive potential for hydrocarbons with intermolecular interactions, *J. Chem. Phys.* **112** 6472-6486 (2000).

[10] A. Strachan, A.C.T. van Duin, D. Chakraborty, S. Dasgupta and W.A. Goddard III, Shock waves in high-energy materials: the initial chemical events in nitramine RDX, *Phys. Rev. Lett.* **91** 098301:1-4 (2003).

[11] M. L. Elert, S. V. Zybin, and C. T. White, Molecular dynamics study of shock-induced chemistry in small condensed-phase hydrocarbons, *J. Chem. Phys.* **118** 9795-9801 (2003).

[12] M. L. Elert, S. V. Zybin, and C. T. White, Shock-induced chemistry in hydrocarbon molecular solids, *Chemistry at Extreme Conditions,* (Editor: M. R. Manaa), Elsevier Press, 351-368 (2005).

[13] C. Sagan, B. N. Khare, W. R. Thompson, G. D. McDonald, M. R. Wing, J. L. Bada, T. Vo-Dinh, and E. T. Arakawa, Polycyclic aromatic hydrocarbons in the atmospheres of Titan and Jupiter, *Astrophys. J.* **414** 399-405 (1993).

[14] T. Y. Brooke, A. T. Tokunaga, H. A. Weaver, J. Crovisier, D. Bockelée-Morvan, and D. Crisp, Detection of acetylene in the infrared spectrum of comet Hyakutake, *Nature,* **383** 606-609 (1996).

Brill Academic Publishers
P.O. Box 9000, 2300 PA Leiden
The Netherlands

*Lecture Series on Computer
and Computational Sciences*
Volume 4, 2005, pp. 1138-1141

Multibillion-Atom Molecular Dynamics Simulations of Shockwave Phenomena on BlueGene/L

Timothy C. Germann[1], Brad Lee Holian, Kai Kadau, and Peter S. Lomdahl

X-7 (Materials Science)
Applied Physics Division,
Los Alamos National Laboratory,
Los Alamos, NM 87545 USA

Received 19 August, 2005; accepted in revised form 21 August, 2005

Abstract: The IBM BlueGene/L supercomputer at Lawrence Livermore National Laboratory, with 131,072 processors connected by multiple high-performance networks, enables an exciting new range of physical problems to be investigated. Using either pairwise interactions such as the Lennard-Jones potential, or the embedded atom method (EAM) potential for simple metals, system sizes of more than 300 billion atoms (or a cube of copper more than 1.5 microns on each side) can be modeled. In order to obtain any new physical insights, however, it is equally important that the analysis of such systems be tractable. This is in fact possible, in large part due to our highly efficient parallel visualization code, which enables the rendering of atomic spheres, Eulerian cells, and other geometric objects in a matter of minutes, even for tens of thousands of processors and billions of atoms. We will describe the performance scaling and results obtained for shock compression and release of a defective EAM Cu sample, illustrating the plastic deformation accompanying void collapse as well as the subsequent void growth and linkup upon release.

Keywords: molecular dynamics, parallel computing, shock waves, visualization

PACS: 02.70.Ns, 07.05.Bx, 07.05.Tp, 61.82.Bg, 62.20.Fe, 62.50.+p

1. BlueGene/L Architecture

The IBM/LLNL BlueGene/L supercomputer consists of 65,536 nodes, each with two IBM PowerPC 440 processors (at 700 MHz clock speeds) and 512 MB of memory [1][2]. Each processor contains a superscalar double floating point unit (FPU), allowing up to 4 floating-point operations (a fused multiply-add instruction on each FPU) to be issued by each processor each clock cycle [2]. The theoretical peak performance is thus 5.6 Gflop/s (8 flops × 700 MHz) per node, or 360 Tflop/s. Although the default operation mode assigns one processor on each node to computation and the other to communication (in so-called "coprocessor" mode), it is also possible to run independent MPI tasks on each processor ("virtual node" mode). This is of course most effective for CPU-bound applications that do not demand an excessive amount of memory (under 256 MB per CPU) or communication; fortunately, this class includes classical molecular dynamics codes, as we show below.

There are 3 independent internal networks on BlueGene/L (plus two other system control-related networks not of interest to the general user). The principal interconnect is a 3D torus connecting each node to its 6 nearest neighbors, which is particularly amenable to grid-based applications such as our cell-based spatial decomposition algorithm, described below. A binary-tree network enables efficient global reduction and broadcast operations (only 16 hops are required to access the entire $2^{16} = 65,536$ node system), and a special single-bit binary tree barrier network enables global synchronization of all nodes in less than 10 µs.

---

[1] Corresponding author. E-mail: tcg@lanl.gov

2. $\underline{S}$calable $\underline{P}$arallel $\underline{S}$hort-range $\underline{M}$olecular Dynamics (SPaSM)

David Beazley and Peter Lomdahl at Los Alamos originally developed the SPaSM code for the Thinking Machines CM-5 in the early 1990s [3]. Although initially restricted to short-range pair potentials (such as the frequently-studied Lennard-Jones (LJ) interaction), SPaSM has since been extended to include several more complex (and hopefully more realistic) potentials. These include the semiempirical embedded atom method (EAM) potentials for simple metals, as well as the modified EAM (MEAM) potentials with angular forces to describe the partial covalent character of more complex metals, and the purely covalent Stillinger-Weber potential for silicon and several of its compounds. An empirical model description of the chemistry in simple energetic materials is also implemented, in particular the reactive empirical bond order (REBO) potential of Don Brenner, Carter White, and colleagues.

In an extensive series of benchmark timings made in April 2005 on half of the complete BlueGene/L system (32,768 nodes, i.e. 65,536 processors), we found [4] excellent scalability of SPaSM for analytic LJ pair potentials, as well as an EAM potential for copper [5]. EAM potentials are typically twice as expensive as pair potentials (for similar interaction cutoff distances), since they involve two pairwise interaction steps: the first to compute the electron density ρ at each atom i due to all neighboring atoms j, and the second to accumulate the total energy as a sum of a pair potential and an energy $F(\rho)$ to embed each nucleus into a background electron density:

$$E_i = \frac{1}{2} \sum_j \varphi(r_{ij}) + F\left[\sum_j \rho_j(r_{ij}) \right].$$

Timing measurements [4] were made for 1-160 billion atoms on 2,048 to 32,768 nodes, using the analytic LJ 6-12 potential truncated at two different cutoff distances, $r_{cut} = 2.5\sigma$ or 5.0σ, where σ is the bond distance. For comparison, we also benchmarked the EAM Cu potential used in the present work; although it has a very short interaction range (corresponding to $r_{cut} \approx 2.0\sigma$ in LJ units), it is in a nonanalytic form which requires table lookups for each of the functions $\varphi(r)$, $\rho(r)$, and $F(\rho)$. The time required for each timestep (including force calculation, position and velocity updates, and message-passing of any particles crossing processor boundaries) is roughly 8-9 µs / (particle / processor) for the short-range LJ potential, or 8-9 seconds for 64 billion atoms on 65,536 processors; 49-51 µs / (particle / processor) for the longer-range LJ potential, and 40-42 µs / (particle / processor) for EAM Cu.

To eliminate the need for enormous I/O sizes and times required to store the positions (and for restarting or certain analyses, velocities) of billions of atoms, simulation analysis is done on-the-fly. This included the conversion of a recently developed high-resolution parallel visualization code [6] into a callable function, as described elsewhere [4].

3. Shock Loading and Unloading of Single-Crystal Copper

Since molecular dynamics simulations integrate the detailed trajectories of each and every atom in the computational cell, the timestep dt must be small enough (typically 1-2 fs) to resolve the fastest individual vibrational motions (with typical periods of 200 fs). This leads to a balance between system size and simulation time, since the computational cost is proportional to the product of these two. Diffusive or nucleation-dominated processes are the most challenging ones, as they require a long and often unpredictable length of time, exponentially dependent on the ratio between the activation energy E_a and system temperature kT.

For large-scale simulations, one would thus like to focus on fast processes, ideally with predictable simulation time requirements. Examples can be found in a variety of high strain-rate processes, including fracture (tensile and/or shear loading), sliding friction (shear loading, often along with compression), and the propagation of shock compression waves. In recent years, we have shown that nonequilibrium molecular dynamics simulations can provide unprecedented insight into shock-induced plasticity [7] and solid-solid phase transitions [8]. A typical simulation geometry used to study both the shock compression as well as release processes is shown in Fig. 1, which can be used to reliably estimate the required simulation time. One such simulation, involving a perfect crystal flyer plate and target with "only" 205 million atoms, is shown in Figs. 2 and 3, illustrating the spall process as well as the visualization capability. In this case, using a "centrosymmetry" order parameter [9] to distinguish fcc from non-fcc atoms provides a clear view of the plastic deformation and spall failure which occurs.

Acknowledgments

We thank Steve Louis, Michel McCoy, and James Peery for enabling early access to BlueGene/L. Los Alamos National Laboratory is operated by the University of California for the U.S. Department of Energy under contract no. W-7405-ENG-36.

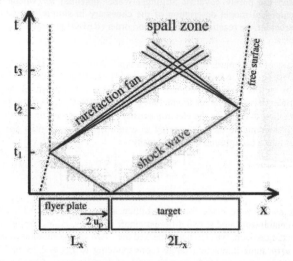

Figure 1: Schematic *x-t* diagram showing the loading (red shock fronts) and unloading (blue expanding rarefaction fans) geometry, and relationship to required simulation times: $t_1 = L_x/u_s$ is the shock transit time through the flyer plate, $t_2 = 2t_1$ the transit time through the target, and $t_3 \approx 3t_1$ the time at which the expanding rarefaction fans collide and induce a tensile region.

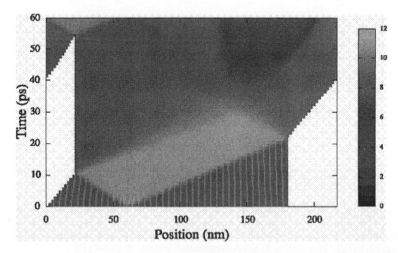

Figure 2: Density (in g/cc) evolution for a single crystal of copper shock-compressed in the $\langle 100 \rangle$ direction (both flyer plate and target), to particle velocity $u_p = 1.0$ km/s, resulting in the onset of spallation around 50 ps. The initial striations result from an incommensurability between the perfect crystal lattice and the computational bin width used to calculate densities. The target crosses the periodic boundary around 40 ps, resulting in a subsequent re-shock which is not of interest here.

Figure 3: Non-fcc atoms, for the simulation shown in Fig. 2 at 50 ps. Hcp stacking fault atoms are grey, and all others red. From left to right, one can see the following features: (a) stacking faults intersecting the free surface of the target, which is just about to impact the flyer plate free surface; (b) residual damage (primarily in the form of point defects such as Frenkel pairs) due to shock loading and unloading in the flyer plate; and (c) the nucleation of a number of voids, which will grow and coalesce to form a spall plane at later times.

References

[1] N.R. Adiga et al., An Overview of the BlueGene/L Supercomputer, in *SC2002 – High Performance Networking and Computing*, Baltimore, MD, November 2002.

[2] G. Almasi, S. Chatterjee, A. Gara, J. Gunnels, M. Gupta, A. Henning, J. Moreira, and B. Walkup, Unlocking the Performance of the BlueGene/L Supercomputer, in *SC2004 – High Performance Computing, Networking and Storage Conference*, Pittsburgh, PA, November 2004.

[3] D.M. Beazley and P.S. Lomdahl, Message-passing multi-cell molecular dynamics on the Connection Machine 5, *Parallel Computing* **20**, 173-195 (1994).

[4] T.C. Germann, K. Kadau, and P.S. Lomdahl, 25 Tflop/s Multibillion-Atom Molecular Dynamics Simulations and Visualization/Analysis on BlueGene/L, to appear in *Supercomputing 2005*.

[5] A.F. Voter, Parallel replica method for dynamics of infrequent events, *Phys. Rev. B* **57**, 13985-13988 (1998).

[6] K. Kadau, T.C. Germann, and P.S. Lomdahl, Large-Scale Molecular-Dynamics Simulation of 19 Billion Particles, *Int. J. Modern Phys. C* **15**, 193-201 (2004).

[7] B.L. Holian and P.S. Lomdahl, Plasticity Induced by Shock Waves in Nonequilibrium Molecular-Dynamics Simulations, *Science* **280**, 2085-2088 (1998).

[8] K. Kadau, T.C. Germann, P.S. Lomdahl, and B.L. Holian, Microscopic View of Structural Phase Transitions Induced by Shock Waves, *Science* **296**, 1681-1684 (2002).

[9] C.L. Kelchner, S.J. Plimpton, and J.C. Hamilton, Dislocation nucleation and defect structure during surface indentation, *Phys. Rev. B* **58**, 11085-11088 (1998).

Brill Academic Publishers
P.O. Box 9000, 2300 PA Leiden,
The Netherlands

*Lecture Series on Computer
and Computational Sciences*
Volume 4, 2005, pp. 1142-1145

Microscopic Dynamics of Atomic Clusters in Strong Laser Pulses

Md. Ranaul Islam and Ulf Saalmann

Max Planck Institute for the Physics of Complex Systems
Nöthnitzer Straße 38 · 01187 Dresden · Germany

Received 3 July, 2005; accepted in revised form 18 July, 2005

Abstract: We analyze microscopically the ionization dynamics of large Xenon clusters with up to 10^4 atoms subjected to strong femtosecond laser pulses. By means of classical molecular dynamics, with the force calculation based on a hierarchical tree code, we are able to follow the laser-induced motion of ions and electrons. We will discuss the mechanisms responsible for the very effective coupling of laser light into atomic clusters by studying the time evolution of parameters characterizing the strongly excited state of the cluster.

Keywords: Molecular dynamics and other numerical methods, Plasma and collective effects in clusters

PACS: 36.40.Gk, 31.15.Qg, 36.40.Wa, 33.80.Wz

1 Atomic clusters subjected to strong laser pulses

Over the last ten years a large number of experiments have been performed in order to reveal the mechanism of light absorption by atomic clusters from short intense laser pulses [1, 2]. The reason for this interest is two-fold: On one hand, clusters are prototypes for studying laser impact on finite atomic systems, since the various cluster types (metallic, rare-gas, hetero-nuclear or molecular clusters) can be easily produced with a desired size. On the other hand, they are compact sources of X-ray photons and highly-charged, fast fragments. Here as well, the outcome can be easily controlled by variation of cluster type and size in the target.

As has been observed in many of the experiments [3, 4], gases of clusters can provide very effective conditions for light absorption. Being the bridge between solids and atomic gases, they combine the advantages of both: the locally solid-like atomic density enables intense energy absorption and the finite size prevents fast distribution of this energy. Furthermore, clusters may rapidly expand on a femtosecond time scale, i. e. typically during the laser pulse.

A number of phenomenological models have been developed to understand the laser-induced cluster dynamics [2], the most widely used being the so-called "nano-plasma model" [5]. Despite their successes in providing a qualitative picture of the process, microscopic approaches are necessary to check their applicability and explore details of the complex dynamics of electrons and ions driven by the laser field.

2 Microscopic description of the ionization dynamics

We present a microscopic approach [6, 7] which accounts for both steps in the ionization of atomic clusters: (i) excitation of electrons from bound states, also called inner ionization, which then form a plasma confined by the cluster ions and (ii) further heating of these plasma electrons which then eventually leave the cluster, also called outer ionization.

Numerically, bound electrons are not treated explicitly and inner ionization corresponds to the "creation" of new electrons. The rates for inner ionization may be determined using a tunnelling formula and electron-impact cross sections. An even simpler approach [7] describes the inner ionization classically: An electron is "created" at a particular ion with the correct binding energy, if there is no other electron classically bound to that ion. This method naturally includes effects due to Coulomb interactions of neighbouring ions and electrons, and thus accounts for barrier-suppression and electron-impact ionization without additional assumptions.

Using classical molecular dynamics, the description of the plasma heating or outer ionization is the computationally most expensive part. For cluster sizes n of interest ($n > 100$ atoms) traditional force calculation schemes fail because of their unfavourable n^2 scaling due to the long-range nature of the Coulomb interaction. Hierarchical tree codes [8], however, scale typically as $n \cdot \log n$ and can be easily parallelized. We started from a gravitational n-body code [9] and adapted it to the case of Coulomb interacting particles with positive and negative charges. Additionally, it takes account of the C_{5v} symmetry of icosahedral clusters, which is conserved for linear polarized laser impact. The code allows us to follow the dynamics of all charged particles over a few hundred femtoseconds with typical time steps of attoseconds [10].

3   Time-dependent picture of laser-induced cluster ionization

The microscopic description gives a detailed and time-dependent insight into the ionization processes. Here, we will present the time evolution of quantities calculated from our microscopic approach, which can be compared to those obtained using phenomenological models, e. g. [5]. Figure 1 shows the time evolution for clusters of two sizes, Xe_{1151} and Xe_{9093}, subjected to a laser pulse of $4 \times 10^{14} \, W/cm^2$, with 20 fs rise and fall time, respectively, and a 180 fs plateau. The absorbed energy per atom (*1st row*) amounts to about 10 and 15 keV, respectively, resulting in complete disintegration of the clusters with keV-fragments. Whereas the larger cluster absorbs more energy per atom, its average ionic charge (*2nd row*) is lower. This is due to the higher space charge of Xe_{9093} compared to Xe_{1151}. It is this space charge which provides the confining potential for the plasma of inner-ionized electrons. Since this charge increases in time, it can bind a plasma with higher and higher temperatures (*3rd row*). The temperature increase is due to laser heating.

Another consequence of the positive net charge is the expansion of the clusters because of Coulomb repulsion. The corresponding pressure, also called Coulomb pressure [5], is dominating (*4th row*) over the pressure due to the hot electron plasma, also called hydrodynamic pressure [5]. The Coulomb pressures decreases towards the end of the pulse since the clusters expand.

Although the intensity of the laser is constant over the whole time shown in Fig. 1, the rates of energy absorption (*1st row*) and cluster charging (*2nd row*) change over time. This can be understood in terms of a simple harmonic oscillator model [7]

$$\ddot{X}(t) + 2\Gamma_t \dot{X}(t) + \Omega_t^2 X(t) = F_0 \cos(\omega t). \tag{1}$$

for the motion of the centre-of-mass X of the electron plasma, bound in a harmonic potential with eigenfrequency Ω_t and damping Γ_t, the latter being due to inner and outer ionization. Because it is driven by a periodic laser field (of strength F_0 and frequency ω), the induced dynamics is explicitly given by

$$X(t) = A_t \cos(\omega t - \phi_t) \quad \text{with} \quad \begin{aligned} A_t &= F_0 / \sqrt{\left(\Omega_t^2 - \omega^2\right)^2 + (2\Gamma_t \omega)^2}, \\ \phi_t &= \arctan\left(2\Gamma_t \omega / (\Omega_t^2 - \omega^2)\right). \end{aligned} \tag{2}$$

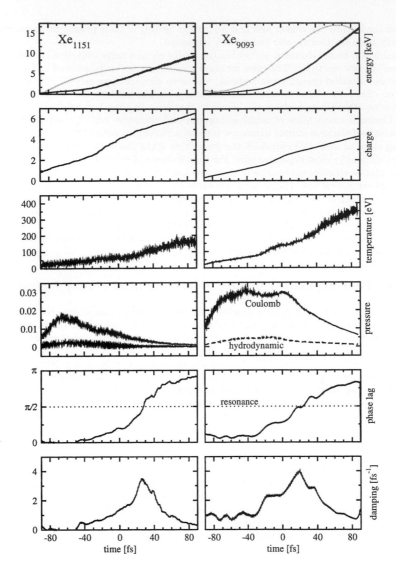

Figure 1: Time evolution for clusters of two sizes, Xe_{1151} (*left column*) and Xe_{9093} (*right*), subjected to a laser pulse of $4 \times 10^{14}\,W/cm^2$. We show only the plateau region of the laser pulse, $t = -90 \ldots + 90\,fs$. From *top* to *bottom*: total absorbed energy per atom, cluster charge divided by number of atoms, temperature of the confined electron plasma, Coulomb vs. hydrodynamic pressure, phase lag ϕ_t of the plasma oscillation w.r.t. to the laser field as defined in Eq. (2), and damping Γ_t of this oscillation estimated from a simple harmonic oscillator model, cf. Eq. (1).

Since the model parameters Ω_t and Γ_t depend adiabatically on time, as indicated by the index t in Eq. (1), the amplitude A_t and the phase lag ϕ_t of the oscillatory motion in Eq. (2) do as well. The phase lag ϕ_t (*5th row*) passes for both cluster sizes at $t \approx 20\,\text{fs}$ the value of $\pi/2$, which is a clear indication of a resonance, i. e. $\Omega_t = \omega$ [7]. Such a resonance is connected with optimal energy absorption [10]. Around this time also the damping Γ_t is maximal (*6th row*). Because it is difficult to extract this damping term directly, we calculated A_t and ϕ_t from the model by fitting the centre-of-mass motion of the microscopic calculation. Knowing A_t and ϕ_t we have calculated the damping Γ_t by inverting Eqs. (2). Obviously, inner and outer ionization, which causes the damping, happens in particular at resonance. This leads to a characteristic dependence of the cluster ionization on pulse length [3] or delay in pump-probe experiments [4]. Our microscopic results demonstrate resonant energy absorption to be the responsible mechanism for this behaviour.

References

[1] J. Posthumus (ed.) *Molecules and clusters in intense laser fields*. Cambrige Univ. Press 2001.

[2] V. P. Krainov and M. B. Smirnov, Phys. Rep. **370**, 237 (2002).

[3] L. Köller, M. Schumacher, J. Köhn, S. Teuber, J. Tiggesbäumker, and K. H. Meiwes-Broer, Phys. Rev. Lett. **82**, 3786 (1999).

[4] T. Ditmire, J. Zweiback, V. P. Yanovsky, T. E. Cowan, G. Hays, and K. B. Wharton, Nature **398**, 489 (1999).

[5] T. Ditmire, T. Donnelly, A. M. Rubenchik, R. W. Falcone, and M. D. Perry, Phys. Rev. A **53**, 3379 (1996).

[6] C. Rose-Petruck, K. J. Schafer, K. R. Wilson, and C. P. J. Barty, Phys. Rev. A **55**, 1182 (1997).

[7] U. Saalmann and J. M. Rost, Phys. Rev. Lett. **91**, 223401 (2003).

[8] S. Pfalzner and P. Gibbon, *Many-body tree methods in physics*. Cambridge University Press 1996.

[9] J. E. Barnes and P. Hut, Nature **324**, 446 (1986).

[10] U. Saalmann, J. Mod. Opt. (2005) in press.

Brill Academic Publishers
P.O. Box 9000, 2300 PA Leiden
The Netherlands

Lecture Series on Computer
and Computational Sciences
Volume 4, 2005, pp. 1146-1147

Open-shell organic molecules
Electric, optical, and magnetic properties

Benoît Champagne and Masayoshi Nakano

Recently, open-shell organic molecules have attracted increased attention from the viewpoint of the rational design of functional materials exhibiting outstanding electric, optical, and magnetic properties as well as because they are unique intermediate species of reactions in solution. This session focuses on recent progresses made in theoretical and computational studies on the physical and chemical phenomena where open-shell systems play an essential role.

In the field of molecular magnets, studies started with the discovery of high-spin carbene (Ito) and a proposition of ferromagnetic polymers (Mataga) in 1967, and have boosted intensive experimental and theoretical works since the preparation of the first organic ferromagnet by Kinoshita and Awaga in 1991. The molecular magnetism area now addresses the topics of photo-induced and photo-switched molecular magnets, spin-spin clusters, single molecular magnets, and biospin systems. Since organic radicals are reaction intermediates, these studies are also closely related to the unraveling of molecular functions and reaction mechanisms of radical species in biological systems. In this area, theoretical tools have been elaborated for interpreting electron spin resonance (ESR) spectra and providing structural information on the reactive species, intermediate, and products as well as their environment. In nonlinear optics (NLO), most organic □-conjugated compounds exhibiting large nonlinear optical and electro-optic effects have so far been closed-shell systems, though pioneering studies have been carried out on open-shell systems in view of designing multi-functional materials. More recently, theoretical investigations have addressed the relationships between spin multiplicity, diradical character, and the potential of open-shell systems for applications as third-order NLO materials.

Benoît Champagne

Born on October 14, 1967, Benoît CHAMPAGNE received his first degree in 1989 and his Ph.D. in 1992, both in Chemistry at the University of Namur. The subject of his Ph.D. thesis was the elaboration of polymer band structure methods for evaluating the polarizabilities of polymers. He accomplished post-doctoral stays at the Quantum Theory Project (Gainesville, Florida) with Prof. Yngve ÖHRN and with Prof. Bernard KIRTMAN at the University of California in Santa Barbara. Since 1995, he holds a permanent position at the Belgian National Fund for Scientific Research (FNRS). He presented in 2001 his habilitation thesis on the development of methods for evaluating and interpreting the vibrational hyperpolarizabilities. He is presently Research Director of the FNRS. His research interests are the development and applications of theoretical schemes for determining the (nonlinear) optical and electric properties of molecular and polymeric systems in gas and condensed phases.

Masayoshi Nakano

Masayoshi NAKANO was born on January 16, 1964 in Osaka, Japan and has obtained an M. Eng. in 1988 and Ph.D. in 1991 in Chemical Engineering at Osaka University. The subject of his Ph.D. thesis was "Theoretical Studies on the Nonlinear-Optical Properties of Organic Substances". During 1991-1992, he stayed with Prof. Keiichiro Nasu at the Institute for Molecular Science (IMS) (as JSPS Research Fellowships for Young Scientists). In 1992, he served as a research associate, in 2002 an associate professor, under the direction of Prof. Kizashi Yamaguchi in Department of Chemistry, Osaka University. In 1998, he received the 47th Young Scientist Award from the Chemical Society of Japan for "Development of nonperturbative calculation method and hyperpolarizability density analysis, and molecular design of novel nonlinear optical compounds". Since 2005, he is Professor of Department of Materials Engineering Science, Osaka University. His research interests are quantum chemical study on the nonlinear optical molecules and molecular aggregates, energy migration dynamics of dendritic systems using quantum master equation approach, and interaction dynamics between atomic/molecular systems and quantized field using dissipative quantum mechanics.

Brill Academic Publishers
P.O. Box 9000, 2300 PA Leiden
The Netherlands

*Lecture Series on Computer
and Computational Sciences*
Volume 4, 2005, pp. 1148-1151

Hyper-Rayleigh Scattering of Neutral and Charged Helicenes

Edith Botek[1], Milena Spassova, and Benoît Champagne

Laboratoire de Chimie Théorique Appliquée, Facultés Universitaires Notre-Dame de la Paix,
rue de Bruxelles, 61, B-5000 Namur (Belgium).

Received 8 July, 2005; accepted in revised form 2 August, 2005

Abstract: Oxidation effects on the hyper-Rayleigh scattering (HRS) second-order NLO responses of homo- and hetero-helicenes have been investigated using the time-dependent Hartree-Fock approach and the Austin Model 1 semiempirical Hamiltonian.
Keywords: Hyper-Rayleigh Scattering, Non-linear Optics, Oxidation, Helicenes.

PACS: 33.15.Kr, 42.65.An

1. Introduction

Second-order nonlinear optical (NLO) phenomena and materials exhibiting them are important in view of applications in photonics and electrooptics. When designing these materials, the first quantity under scrutiny is the first hyperpolarizability tensor, β, the microscopic second-order NLO response. While the coherent second harmonic generation (SHG) response is cancelled in isotropic media due to symmetry [1], incoherent second harmonic scattering, called hyper-Rayleigh scattering (HRS), is (weakly) allowed [2] and produces frequency-doubled light. HRS, which depends on the instantaneous fluctuations in the second order susceptibility, enables the measurement of β without electric field-induced orientation to build a poled structure. This is why the HRS technique is also suitable for determining the first hyperpolarizability of non-dipolar molecules and of ionic species.

In NLO, chiral molecules are particularly interesting due to their intrinsic non-centrosymmetry. Consequently, their bulk assemblies display non-zero second-order NLO effects while these assemblies are thermodynamically more stable than poled structures. Among these chiral systems, helicenes, ortho-fused rings, present a large potential due to their helical π-electron network. Since the helix itself is the conjugated pathway, suitable chemical substitutions by donor (D) and/or acceptor (A) groups can lead to enhancement of the NLO responses. So far, charging effects on second-order NLO responses have not yet been considered for helicenes. These effects are addressed here from a theoretical point of view while synthetic efforts are continuing [3].

2. Theory and method of calculation

The (total) hyper-Rayleigh scattering intensity for plane-polarized incident light and observation made perpendicular to the propagation plane reads:

$$\beta_{HRS} = \sqrt{\left\langle \beta_{ZZZ}^2 \right\rangle + \left\langle \beta_{XZZ}^2 \right\rangle} \tag{1}$$

while the associated depolarization ratio is given by:

$$R = \left\langle \beta_{ZZZ}^2 \right\rangle / \left\langle \beta_{XZZ}^2 \right\rangle \tag{2}$$

The different β tensor components entering into Eqs. (1) and (2) are evaluated at the time-dependent Hartree-Fock (TDHF) level [4] using the Austin Model 1 (AM1) [5] semi-empirical Hamiltonian. They are given by

$$\beta_{ijk}(-2\omega; \omega, \omega) = -\text{Tr}\left[\mu_i D^{jk}(\omega, \omega)\right] \tag{3}$$

---

[1] Corresponding author. E-mail: edith.botek@fundp.ac.be

where ω (2ω) is the angular frequency of the incident (scattered) oscillating field, μ_i is the ith component of the electric dipole moment matrix and $D^{jk}(\omega,\omega)$ is the second-order derivative of the density matrix with respect to the electric field oscillating at frequency ω applied along the j and k directions. For open-shell species the perturbed density matrix is evaluated as the sum of its α and β spin components [6, 7]:

$$D^{jk}(\omega,\omega) = D^{jk(\uparrow)}(\omega,\omega) + D^{jk(\downarrow)}(\omega,\omega) \qquad (4)$$

The superscripts ($\uparrow$) and ($\downarrow$) refer to α and β spin, respectively, denoting the use of the UHF wavefunction. All calculations were performed using a modified version of MOPAC 2000 [8] program.

Aiming at analyzing the first hyperpolarizability values, the first (dipole-allowed) excitation energies are estimated at the same level of approximation from the frequency dispersion of the polarizability, which is described by the relation,

$$[\alpha(-\omega;\omega)/\alpha(0;0)] - 1 = A\omega^2 + B\omega^4 + C\omega^6 + ... \qquad (5)$$

Indeed, following Ref. [9], an upper bound to the first excitation energy is given by $1/\sqrt{A}$, its [1,0] Padé approximant. The A values are obtained by fitting Eq. (5) to a set of frequency-dependent polarizabilities obtained for photon energies of 0.05, 0.10, ... 0.5 eV.

3. Results and discussion

HRS first hyperpolarizabilities, depolarization ratios, and first dipole-allowed excitation energies have been calculated for the selected helicenes, with and without heterocycles in the helix, with and without substituents, in both neutral and mono-oxidized forms. Among these, the "classical" [6]-helicene, its tetramethoxy-bisquinone derivative, [6]H-OCH$_3$, modeling tetra(dodecyloxy) analogs of Katz and coworkers [10], dithia-[7]-helicene and its tetramethoxy-bisquinone derivative, tetrathia-[7]-helicene [19] and its bis(trimethylsilyl)-substituted form, heptathiophene [11], and pentathia-[9]-helicene [12].

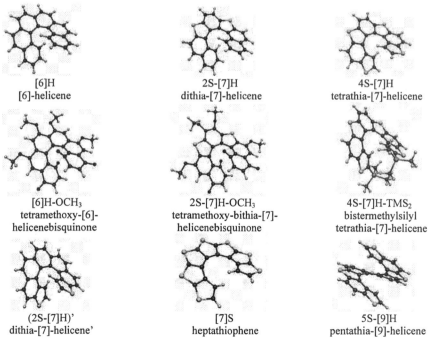

[6]H	2S-[7]H	4S-[7]H
[6]-helicene	dithia-[7]-helicene	tetrathia-[7]-helicene
[6]H-OCH$_3$	2S-[7]H-OCH$_3$	4S-[7]H-TMS$_2$
tetramethoxy-[6]-helicenebisquinone	tetramethoxy-bithia-[7]-helicenebisquinone	bistermethylsilyl tetrathia-[7]-helicene
(2S-[7]H)'	[7]S	5S-[9]H
dithia-[7]-helicene'	heptathiophene	pentathia-[9]-helicene

β_{HRS} increases with the frequency and this increase is stronger for compounds having smaller excitation energies. When the incident light frequency gets closer to the half of the excitation energy, resonance effects occur; the TDHF method is not adequate and often does not converge towards a self-consistent solution. Among the neutral species, the tetramethoxy-bisquinone derivatives ([6]H-OCH$_3$ and 2S-[7]H-OCH$_3$) present the largest static responses. Then, the dithia-[7]-helicenes have larger β_{HRS}

Brill Academic Publishers
P.O. Box 9000, 2300 PA Leiden
The Netherlands

Lecture Series on Computer
and Computational Sciences
Volume 4, 2005, pp. 1152-1154

Theory and Spectroscopy of Organic High-spin Molecules

Paul M. Lahti[1]

Department of Chemistry,
University of Massachusetts,
Amherst, MA 01003 USA

Received 8 August 2005; accepted 8 August 2005

Abstract: Computational estimates can provide critical insight concerning the identity and interpretation of spectral signatures for organic open-shell molecules, especially electronic absorption and electron spin resonance spectra. Hybrid density functional methods are particularly convenient for studying organic dinitrenes and heterospin nitrenoradicals, despite the known drawbacks of such methods for some aspects of open-shell systems such as multiconfigurational states.

Keywords: nitrene, nitroxide, open-shell molecules, open-shell electronic spectra, high spin EPR spectra

Subject Classification: Excited states and pairing interactions in model systems, Exchange and superexchange interactions
PACS: 31.15.Ew, 33.20.Lg, 33.35.+r, 70.10.Li, 71.15.Mb, 75.30.Et, 82.50.Hp

1. Synergy of Spectroscopy and Computation for Investigating Open-shell Molecules

Two of the most important experimental tools used to investigate open-shell organic molecules are electron spin resonance (ESR) and absorption spectroscopy. ESR probes electron distributions and inter-spin interactions in detectable paramagnetic molecules, and can give some information about electronic state ordering and energy gaps, so long as the gaps are 30-40 J/mol $\Delta E < 4$-5 kJ/mol. But, many open-shell molecules have such novel structures that their ESR spectroscopic signatures are not a priori obvious. Absorption spectroscopy gives qualitative insight about the nature of the molecular structures, often by comparison to similar structures. Comparisons are not always obvious for open-shell molecules, given the strong influence of their unpaired electrons upon electronic transition.

Although SCF-MO-CI and CASSCF-based methods have been much used to obtain high quality predictions about the electronic ground states, state energy spacings, and bonding in open-shell molecules,[1] density functional theoretical (DFT) methods have become a convenient alternative for basic comparison to experimental results. Hybrid UB3LYP and similar DFT methods give good comparisons to experimental spin density distributions for many organic radicals,[2] and should be applicable to single-determinant open-shell states with multiple unpaired electrons to estimate distribution of unpaired electron spin density. Also time dependent DFT (TD-DFT) methods have proven useful for predicting absorption spectra of many conjugated molecules, including open-shell species.[3]

This presentation will describe ESR and absorption spectroscopy of unusual open-shell organic molecules that incorporate triplet arylnitrene (**NPh**) units in combination with organic radicals to allow up to three unpaired electrons to interact in the system. The spectroscopic results will be compared to computational predictions of the electronic properties of the same systems, in order to understand better the structure natures of these unusual molecules.

2. Structure and Generation of Nitreno-Radical Molecules.

We wished to understand the degree of electron sharing and delocalization between an NPh unit and a radical unit that are connected by a π-electron network. The radical units that were studied were nitronylnitroxide, iminoylnitroxide, and tert-butylnitroxide: **NN**, **IN**, and **NIT**, respectively. The intramolecular communication from one spin unit to the other will

---

[1]Corresponding author. E-mail: lahti@chem.umass.edu

NPh   NN   IN   NIT   NPhRad

Radical = Rad

depend greatly upon the molecular connectivity, but the influence of non-carbon atoms in the network will perturb molecular orbital (MO) levels and may limit delocalization. For example, in the para-connected **pNPhNN** system, redistribution of the electrons in the NPh π-system allows the molecule to be formulated either as **pNPhNN** or as **pNPhNN-Q**; the latter has more unpaired electron spin density on its N-O groups, and less on the nitrene nitrogen

NPh π-system
and nitrene
valence electrons

pNPhNN   pNPhNN-Q

atom. Such differences would greatly affect ESR spectroscopy (depends upon density distributions) and UV-vis spectroscopy (influenced by π-electron delocalizability).

    The **NPhRad** systems **pNPhNN, pNPhIN, pNPhNIT, mNPhNN, mNPhIN,** and **mNPhNIT** were all generated[4-5] by photochemical cleavage of their corresponding azidophenyl precursors (N replaced by N$_3$ in the structures shown) in frozen solutions of deoxygenated 2-methyltetrahydrofuran (77 K and below). The final samples containing the reactive

pNPhIN   pNPhNIT   mNPhNN   mNPhIN   mNPhNIT

intermediates were studied by absorption (200-900 nm) and ESR (9.6 GHz) spectroscopy while being kept frozen. All spectral features associated with the **NPhRad** intermediates were lost upon brief thawing of their solutions to a fluid state.

3.   Spectral Features of the NPhRad Systems.

In addition to non-reacted starting material peaks, the ESR spectra of the **NPhRad** systems all showed peaks characteristic of quartet state systems, except for the **pNPhNIT** sample, which showed only S = 1/2 radical peaks. The zero field splitting (zfs) parameters and quartet natures of the spectra were obtained by simulation,[6] and are listed in Table 1. Major UV-vis spectral peaks attributed to the **NPhRad** molecules are also listed in the table.

    Curie analysis was carried out for the intensities of main ESR peaks as a function of temperature. Linear plots are consistent with the ESR peak belonging to a ground state, although it is formally possible that the state is degenerate with another state to within a few tens of J/mol. Curved plots show that the ESR peak belongs to a thermally excited state that is depopulated at lower temperature. For three unpaired electrons, quartet and doublet states are possible. Table 1 lists the experimental Curie-based ground state findings.

Table 1: Experimental and computed properties of **NPhRad** systems (see citations).

b_1 Compound	b_2 zfs, D(E) in cm^{-1}	UV-vis (nm)	Quartet Ground State?	Computed Ground State
pNPhNN [4]	0.277 (<0.002)	580	yes	quartet
pNPhIN [5]	0.300 (<0.002)	483	maybe	quartet
pNPhNIT	(no quartet seen)	--	no	doublet

mNPhNN [4]	0.300 (<0.002)	580	no	doublet
mNPhIN [5]	0.352 (0.006)	300 (wk)	no	doublet
mNPhNIT [4]	0.336 (0.004)	380 (wk)	yes	quartet

4. Computational Natures of the NPhRad Systems.

UB3LYP/6-31G* computations were carried out on models for the **NPhRad** systems using the Gaussian[7] program; all CH_3 groups were replaced by hydrogen atoms in the computations. Geometries were optimized for quartet and doublet states separately. Mulliken spin populations were obtained directly from these computations. TD-DFT computations were carried out using the 6-31G* basis set, using the optimized quartet and doublet geometries. Details have been given elsewhere, or will be reported in this talk. Table 1 lists the computed ground state spin multiplicities for the **NPhRad** systems.

5. Brief Summary of Conclusions

The computations show only small changes in spin population on the nitrene nitrogens of the **NPhRad** systems by comparison to one another, and even by comparison to **NPh**. Even para-linked systems are reasonably well described as nitreno radicals with only limited spin delocalization by comparison to separate spin units. The exception is **pNPhNIT**, which gives a doublet ground state with strong spin pairing of the NIT electron to the nitrene π-electron. The asymmetric spin distribution of IN makes significant differences quartet-doublet splittings for systems that incorporate it, by comparison to the NN analogues. The NIT systems exhibit stronger exchange interactions than either IN or NN, because they have a large spin density site directly attached to the **NPh**. The experimental UV-vis results are in fairly good accord with TD-DFT/6-31G* spectral transitions predicted for the low spin states of the meta NN and IN linked systems, and for the high spin state of the para NN system. The UB3LYP/6-31G* predicted ground states are in good accord with the ground spin state multiplicities assigned by the ESR zfs and Curie analyses, save possibly for **pNPhIN** (small computed favoring of the quartet, small apparent experimental favoring of the doublet).

Acknowledgments

This work was supported by the U.S. National Science Foundation. The author gratefully acknowledges the work of his coauthors in references 4-5.

References

[1] Cf. T. Bally and W. T. Borden, "Calculations on Open-Shell Molecules, A Beginner's Guide", Reviews in Computational Chemistry (Editors K. B. Lipkowitz, R. Larter, and T. R. Cundari; Wiley-VCH, Hoboken, NY :1999), Volume 13, 1-97.

[2] Cf. D. Goldfarb and D. Ari, *Ann. Rev. Biophys. Biomol. Struct.*, **33**, 441 (2004).

[3] E. K. U. Gross, J. F. Dobson, and M. Petersilka, in *Density Functional Theory, Topics in Current Chemistry* (Editor R.F. Nalewajski, Springer, Heidelberg, Germany:1996).

[4] P. M. Lahti, P. R. Serwinski, B. Esat, Y. Liao, R. Walton, and J. Lan. *J. Org. Chem.* **69**, 5247 (2004).

[5] P. Taylor, P. R. Serwinski, and P. M. Lahti, *Org. Lett.*, 7, 0000 (2005).

[6] Y. Teki, Ph. D. Dissertation, Osaka City University, Osaka, Japan (1985). Y. Teki, T. Takui, H. Yagi, K. Itoh, and H. Iwamura, *J. Chem. Phys.* **1985**, *83*, 539. K. Sato, Ph. D. Dissertation, Osaka City University, Osaka, Japan (1994).

[7] M. J. Frisch, G. W. Trucks, H. B. Schlegel, et al., Gaussian 03, Revision B.03; Gaussian Inc.: Pittsburgh, PA, 2003.

Brill Academic Publishers
P.O. Box 9000, 2300 PA Leiden
The Netherlands

Lecture Series on Computer
and Computational Sciences
Volume 4, 2005, pp. 1155-1557

CASSCF-DFT based on an Interacting Reference System

K. Nakata[1], S. Yamanaka[1], K. Kusakabe[2], T Takada[3], H. Nakamura[4], R. Takeda[1], K. Yamaguchi[1]

1 Department of Chemistry, Faculty of Science, University of Osaka,
2 Department of Chemistry, Graduate School of Engineering Science, University of Osaka,
3 NEC Corporation,
4 Institute for Protein Research, Osaka University.

Received 4 July, 2005; accepted in revised form 27 July, 2005

1. CASSCF-DFT

The Kohn-Sham (KS) density functional theory (DFT) [1,2] is now a powerful tool in computational chemistry. But it is known that it does not work well for near degeneracy of several configurations. In particular, the density that is not noninteracting v-representable is beyond the scope of KS-DFT [3]. Since this defect is due to not only the approximation of exchange-correlation functional but also the single determinant feature of KS-DFT in the case of a modest finite basis set, one possible remedy is to combine a multireference (MR) wavefucntion (WF) and a DFT treatment. The first generation to merge MR-WF with DFT remains a double counting problem of electron correlation [4]. The prescription to cover the electron correlation by KS-DFT is quite different to that of the MR-WF theory, so this is a very difficult issue. Recent developers to resolve this problem by various ways [5,6]. The issue of how to define the spin-polarizability of a singlet MR-WF solution is also settled [7].

Another direction of recent developments of MR-DFT is obviously towards the multireference version of density functional theory. Savin and his co-workers have developed such a MR-DFT based on adiabatic connection by dividing the Coulomb interaction term in Hamiltonian into the short-range and long-range interaction [6]. We recently developed an iterative CASCI-DFT (ICASCI-DFT) method, [8]. This is one practical theory combining MR-DFT with the extended KS-DFT proposed by Kusakabe, who also introduced MR-WF in order to define the universal functional of the partially interacting system [9]. The mathematical framework of ICASCI-DFT is given by a CASCI equation with a residual correlation potential, so its variational space of ICASCI-DFT is same to that of the CASCI approach. In addition, from the computational point of view, the orbital transformation procedure is skipped except that in the first cycle. Nevertheless, we can obtain the effective CAS wavefunction that handles both nondynamical and dynamical correlation effects.

However, when an appropriate set of molecular orbitals is not available, the orbital relaxation procedure becomes essential to obtain the good density. So, in this study, we present the general formulation of the MCSCF-DFT. Very recently, Roos and his co-workers have developed a precursive approach of CASSCF-DFT [10]. They intend to add a total correction of DFT to CASSCF wavefunction approach that is pioneered by Roos et al., in order to obtain the any physical properties. This is a straightforward (and probably powerful) way to realize the CASSCF-DFT approach. The different point of our formualism compared with theirs is that the emphasis of our formalism is on the multireference implementation of DFT, in other words, the equivalent formalism of the Euler equation of DFT. This leads us to the relation between correlation functional and its potential given below.

We start from the division of the energy into three parts,

$$E[\rho(\mathbf{r})] = Min_{\rho(\mathbf{r}) \to N} \left[F^P[\rho(\mathbf{r})] + E_{RC}[\rho(\mathbf{r})] + \int d\mathbf{r} \rho(\mathbf{r}) V_{ext}(\mathbf{r}) \right], \qquad (1a)$$

where first, second, and third terms in parentheses at the right side are the modified universal functional , residual correlation, and external potential terms, respectively. The modified universal functional is defined for the variational space of MCSCF wavefunction:

---

[1] Corresponding author. Shusuke Yamanaka, E-mail: syama@chem.sci.osaka-u.ac.jp

$$F^p[\rho(\mathbf{r})] = Min^p_{\Psi \to \rho(\mathbf{r})} \langle \Psi | \hat{T} + \hat{V}_{ee} | \Psi \rangle. \tag{1b}$$

The variation of CI coefficients lead to the equation

$$\delta E[\rho(\mathbf{r})] = \sum_{ij} H^{core}_{ji} \delta P_{ij} + \sum_{ijkl} \langle kl | ij \rangle \delta \Pi_{ijkl} + \sum_{ij} \langle i | \frac{\delta E_{RC}}{\delta \rho} | j \rangle \delta P_{ji} \tag{1c}$$

where H^{core}_{ji}, P_{ji}, Π_{ijkl} are the core Hamiltonian, a spinless-one particle density matrix (1-DM), a spinless two-particle density matrix (2-DM), respectively, and the physicists' notation is used for electron-repulsion integrals. Here it is arbitrary whether the basis is of atomic orbitals or of molecular orbitals. We here assume the latter, according to the convention of the MR-WF formalism [11]. The effective MR-DFT equation is given by

$$\left(\hat{H}^{core} + \hat{V}^{eff}_1 + \hat{V}^{eff}_2 \right) \Psi \rangle = E^{eff} | \Psi \rangle. \tag{2a}$$

The deviation of the effective energy due to the variation of CI coefficients is given by

$$\delta E^{eff} = \sum_{ij} \left\{ \left(H^{core}_{ji} + V^{eff}_{1\,ji} \right) \delta P_{ij} + \delta V^{eff}_{1\,ji} P_{ij} \right\} + \sum_{ijkl} \left(V^{eff}_{2\,klij} \delta \Pi_{ijkl} + \delta V^{eff}_{2\,klij} \Pi_{ijkl} \right). \tag{2b}$$

Thus, a density of the solution of the effective CASCI equation satisfied the original Euler euqation, if we set

$$V^{eff}_{2\,klij} = \langle kl | ij \rangle \tag{3a}$$

$$V^{eff}_{1\,ji} = \langle j | \hat{V}^{RC} | i \rangle \tag{3b}$$

where the $V^{RC}(\mathbf{r})$ is yielded from the residual correlation term , of which the relation is same to ICASCI-DFT [8].

The remaining issue is the variational procedure for molecular orbitals. The mathematical (not computational) formulation can be archived by a straightforward manner as follows. The first order deviation of the real system is given by

$$\delta E[\rho(\mathbf{r})] = \sum_{ij} \langle \delta i | \hat{H}^{core} + \frac{\delta E_{RC}}{\delta \rho} | j \rangle P_{ji} + \sum_{ijkl} \langle \delta ij | kl \rangle \Pi_{klij} + c.c. \tag{4}$$

The deviation of effective energy is

$$\delta E^{eff} = \sum_{ij} \left\{ \delta i | \hat{H}^{core} + \hat{V}^{eff}_1 | j \rangle + \langle i | \delta \hat{V}^{eff}_1 | j \rangle \right\} P_{ji} + \sum_{ijkl} \langle \delta j | kl \rangle \Pi_{klij} + c.c. \tag{5}$$

which is obviously equivalent to eq. (4) by noting the relations given by eq. (3). Introducing the Lagrange multiplier for orthogonal conditions, the usual treatment of this equation yields the effective one-electron problem:

$$\sum_j \left(F_{ij} - \varepsilon_{ij} \right) \phi_j = 0, \tag{6}$$

$$F_{ij} = \sum_{ij} P_{ji} \left(H^{core}_{ij} + \langle i | \frac{\delta E_{RC}}{\delta \rho} | j \rangle \right) + \sum_{jk,il} \Pi_{jlik} \langle i | V^{pc} | j \rangle, \tag{7}$$

$$\langle i | V^{pc} | j \rangle \equiv \int d\mathbf{r} d\mathbf{r}' \frac{\phi^*_i(\mathbf{r}) \phi^*_l(\mathbf{r}') \phi_k(\mathbf{r}') \phi_j(\mathbf{r})}{|\mathbf{r} - \mathbf{r}'|}. \tag{8}$$

Those are usual equations of MCSCF [11], but with including the correlation term of DFT.

The practical approach we developed is the CASSCF-DFT method. The computational details of our CASSCF-DFT are follows: first, we use the Newton-Rapson method to solve CASSCF procedure. Second, the original parallelization scheme [12] for integral transformations for CASSCF is employed, which is crucial for reducing the computational time. The numerical results are presented in the session in the International Conference of Computational Methods in Sciences and Engineering 2005. The extension of numerical implementation to any MCSCF-DFT is straightforward: All we need is the replacement of a corresponding residual correlation term.

Acknowledgments

This research has been partially supported by a Grant-in-aid for NAREGI Nano-science Project, from the Ministry of Education, Culture, Sports, Science and Technology of Japan,

References

[1] W. Kohn and L. J. Sham, Phys. Rev. 140, A1131 (1964).

[2] R. G. Parr, W. Yang, *Density-Functional Theory of Atoms and Molecules*, Oxford University Press, New York, 1989.

[3] M. Levy and J. P. Perdew, *Density functional methods in physics* (Editor: R. M. Dreilzler and J. Pprovidencia), 11-30 (1985).

[4] G. C. Lie, E. Clementi, J. Chem. Phys. 60 (1974) 1275; ibitd, 1288;F. Moscardò, E. San-Fabiän, Phys. Rev. A 44 (1991) 1549.

[5] B. Miehlich, H. Stoll, A. Savin, Mol. Phys. 91, 527(1997);J. Gräfenstein, D. Cremer, Chem. Phys. Lett., 316, 569 (2000); J. Gräfenstein, D. Cremer, Mol. Phys. 103, 279 (2005).

[6] R. Takeda, S. Yamanaka, K. Yamaguchi, Chem. Phys. Lett. 366, 321 (2002); Int. J. Quantum Chem. 96, 463 (2004).

[7] R. Pollet, A. Savin, T. Leininger, H. Stoll, J. Chem. Phys. 116, 1250 (2002); A. Savin, F. Colonna, R. Pollet, Int. J. Quantum Chem. 93, 166 (2003); J. Toulouse, F. Colonna, and A. Savin, Phys. Rev. A, 70, 062505 (2004).

[8] S. Yamanaka, et al, to be publishcd.

[9] K. Kusakabe, J. Phys. Soc. Jpn. 70, 2038, (2001).

[10] S. Gusarov, P-A. Malmqvist, R. Lindh, B. O. Roos, Theor. Chem. Acc. 122, 84, (2004).

[11] R. Mcweeny: *Methods of Molecular Quantum Mechanics. Sedond Edition.*, Academic Press, San Diego, 1992.

[12] K. Nakata, T. Murase, T. Sakuma, T. Takada, J. Comp. Appl. Math., 149, 351 (2002).

Brill Academic Publishers
P.O. Box 9000, 2300 PA Leiden
The Netherlands

*Lecture Series on Computer
and Computational Sciences*
Volume 4, 2005, pp. 1158-1161

Quantum Chemical Calculations Of Third-Order Nonlinear Optical Properties For Organic Open-Shell Systems: Nitronyl Nitroxide Radicals And Several π-Conjugated Systems Having Unique Structures.

S. Yamada[*1], M. Nakano[2], R. Kishi[2], S. Ohta[2], S. Furukawa[2], H. Takahashi[2] and K. Yamaguchi[1]

Department of Chemistry,
Graduate School of Sciences,
Osaka University, Toyonaka, Osaka 560-0043, Japan
And
Division of Materials Engineering,
Department of Materials Engineering Science,
Graduate School of Engineering Science,
Osaka University, Toyonaka, Osaka 560-8531, Japan

Received 5 August, 2005; accepted in revised form 5 August, 2005

Abstract: The static second hyperpolarizability (γ) for several open-shell systems, which are predicted to exhibit a negative γ by our classification rule of γ based on symmetric resonance structures with invertible polarization (SRIP), has been studied. The remarkable electron correlation dependence has been observed in the magnitude and sign of γ. We also investigate the applicability of the density functional theory (DFT) methods for the γ values of these systems. By tuning the mixing parameter of DFT/HF exchange term, a DFT method can reproduce the γ values at the higher-order electron correlation method.
Keywords: open-shell system, second hyperpolarizability, π conjugated system

1. Introduction

The interaction between material and light is one of the most basic and important phenomena in nature. Therefore, a large number of studies on the interaction have been done experimentally and theoretically. Development of laser brings the powerful light source, and has made it possible to investigate nonlinear optical (NLO) response properties. In recent years, NLO properties have been studied actively both experimentally and theoretically for several materials. Above all, organic materials have attracted a great deal of attention because of their high-speed response originating in fluctuation of π electrons. Although examples of investigation on NLO properties of organic molecules increase, there are only a few investigations for organic radical species. One of the reasons is that the most organic radicals are unstable and then experimental measurement of NLO properties for radical species is difficult. However, we are convinced that an importance of the study on NLO properties for open-shell systems increase in near future. The radical species have some instability in their electronic states, so that the electronic states are changed easily by the slight perturbations. Therefore, it is expected that response properties of radical species are changed remarkably by slight physical and chemical perturbations. Such sensitivity of response properties for radical species would be useful for controlling the NLO properties. In order to predict the useful open-shell NLO systems, detailed quantum calculations is indispensable. We introduce some theoretical study on NLO properties of open-shell systems having unique structures by using quantum chemical calculations.

---

[*] Corresponding author. E-mail: yamada@chem.sci.osaka-u.ac.jp

2. Theory

We here briefly explain our structure-property relation in the static third-order NLO properties for molecular systems.[1] The perturbative formula for static γ can be partitioned into three types of contributions as follows:

$$\gamma^{(I)+(II)+(III)} = \gamma^{(I)} + \gamma^{(II)} + \gamma^{(III)} = \sum_{n=1} \frac{(\mu_{n0})^2 (\Delta\mu_n)^2}{E_n^3} - \sum_{n=1} \frac{(\mu_{n0})^4}{E_n^3} + \sum_{\substack{m,n=1 \\ (m \neq n)}} \frac{(\mu_{n0})^2 (\mu_{mn})^2}{E_n^2 E_m} \tag{1}$$

Here, μ_{n0} is the transition moment between the ground and the nth excited states, μ_{nn} is the transition moment between the mth and the nth excited states, $\Delta\mu_n$ is the difference of dipole moments between the ground and the nth excited states and E_n is the transition energy from ground to the nth excited state. From these equations, apparently, the contributions of $\gamma^{(I)}$ and $\gamma^{(III)}$ are positive in sign, whereas the contribution of $\gamma^{(II)}$ is negative. On the basis of this partitioning, we have proposed a structure-property relation in γ. We here particularly consider a case, *i.e.*, $|\gamma^{(I)}|=0$, $|\gamma^{(II)}|>|\gamma^{(III)}|$ ($\gamma<0$), in which the compounds are symmetric ($\Delta\mu_n=0$) and exhibit strong virtual excitation between the ground and the first excited states ($|\mu_{n0}|>|\mu_{nn}|$). From this condition, the symmetric systems with large polarization are predicted to exhibit negative γ, which is rare in general for conventional π-conjugated molecules, but important for the application in NLO.   Namely, a system with large contribution of symmetric resonance structures with invertible polarization, *i.e.*, SRIP, satisfies our criteria for the system to have a negative γ.

3. Results and discussions

3.1 γ for nitronyl nitroxide radical and $\chi^{(3)}$ for crystal of para-phenyl nitronyl nitroxide

The γ of a neutral radical, *i.e.*, nitronyl nitroxide (N^2-oxidoformamidin-N^1-yloxyl; NN) radical is investigate. Figure 1(a) shows the molecular geometry and a coordinate system of NN radical. The purpose of our study is to obtain a qualitative electron correlation dependence and the sign of γ for NN radical, which satisfies our structure-property correlation rule. Figure 1(b) shows the corresponding resonance structures for the NN radical. As shown in Figure 1(b), this molecule involves two nitroxide groups (N-O·) arranged symmetrically. In the resonance structures, this characteristic structure leads to a transfer of the radical spin and charge from one side to the other side of the NO group. Such resonance structure (SRIP) implies an important contribution to the stability of the ground state of this system. Namely, the sign of γ_{xxxx}, the component of the γ in the direction of the O-N-C-N-O unit, is expected to be negative.

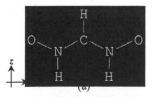 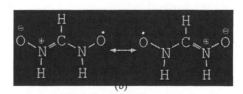

Figure 1: (a) The geometry and coordinates of NN radical (b) SRIP for NN radical

 As shown in Figure 2, a remarkable electron correlation dependence of the γ_{xxxx} of NN radical is observed.[2] The HF method provides a large negative γ, while double (D) excitation effects by the MP2 method the sign and enhance the magnitude of γ. The singlet (S) and triplet (T) excitation effects involved in the MP4 method decrease γ. Particularly, the S effects involved in the CCSD method remarkably decrease γ value and then reverse its sign. This feature implies that the contribution from the higher-order S effects is indispensable for a reliable description of the γ value for this system. The γ_{xxxx} by the CCSD(T) method is shown to be nearly equal to that by the CCSD method. Therefore, γ_{xxxx} by the CCSD method seems to be sufficiently converged.

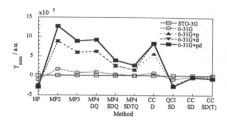

Figure 2: Variations in the γ_{xxxx} of nitronyl nitroxide radical for various electron-correlation methods.

Figures 3(a) and 3(b) show the structures obtained from X-ray spectra for para-nitrophenyl nitronyl nitroxide (*p*-NPNN), which is one of stable radical, and its β-phase crystal. The *x* direction of the *p*-NPNN monomer is along the direction of the largest polarization in the NN radical group. The γ_{xxxx} of the NN radical is expected to be negative based on our classification rule of γ. Further, the directions of the polarizations in the NN units nearly coincide with the direction of the *X* axis of the crystal of *p*-NPNN as shown in Figure 3(b). This structure suggests an emergence of negative macroscopic third-order susceptibility ($\chi^{(3)}$) in the direction of the *X* axis.

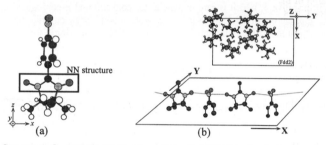

Figure 3: (a) Structure of *p*-NPNN, (b) Configuration of *p*-NPNN in the β-phase crystal.

For the calculation of γ for *p*-NPNN monomer and its cluster models, the INDO CHF method is employed.[3] From a comparison between the experimental β_z (-3.57x10^{-30}esu) and the calculated β_z (-3.22x10^{-30}esu) by the INDO CHF method, this method is considered to be adequate for a qualitative study. Each component of γ for a monomer of *p*-NPNN is shown in Table 1. The largest component, γ_{xxxx}, is found to be negative. This feature is in good agreement with our prediction based on our classification rule. Namely, a feasibility of electron fluctuation on the NN radical unit is shown to enhance the ground state polarizability and hence tends to provide negative γ. Each component of $\chi^{(3)}$ for the β-phase crystal of *p*-NPNN calculated by the INDO CHF method are given in Table 2. We apply the oriented gas model to estimate the $\chi^{(3)}$. As expected from the characteristics of γ_{xxxx} for the NN radical and alignment structure of *p*-NPNN in the β-phase crystal, the largest component, $\chi^{(3)}_{XXXX}$, is found to be negative. From a comparison of relative magnitudes of these components between for *p*-NPNN monomer and for its crystal, the difference in magnitudes is more enhanced in the crystal than in the monomer.

Table 1: Microscopic γ components of *p*-NPNN calculated using INDO CHF method [a.u.].

γ_{xxxx}	γ_{yyyy}	γ_{zzzz}	γ_{xxyy}	γ_{xxzz}	γ_{yyzz}
-33900	-284	14700	-2350	41	4137

Table 2: Macroscopic $\chi^{(3)}$ values of of *p*-NPNN crystal in β-phase [10^{-14} esu/cm^3].

$\chi^{(3)}_{XXXX}$	$\chi^{(3)}_{YYYY}$	$\chi^{(3)}_{ZZZZ}$	$\chi^{(3)}_{XXYY}$	$\chi^{(3)}_{XXZZ}$	$\chi^{(3)}_{YYZZ}$
-15.27	-0.08	5.90	-0.77	0.01	1.59

3.2 second hyperpolarizability of tetrathiapentalene (TTP) using highly correlated ab initio MO and the density functional theory methods

We have proposed a new class of third-order nonlinear optical molecules, which are cation radical condensed-ring and alternant conjugated systems involving sulfur (S) atoms.[4] One of the condensed-ring systems is tetrathiapentalene (TTP), which is also expected to be donor (D) molecular elements constructing high electrical conductive aggregates (See Figure 4(a)). As shown in Figure 4(b), cationic radical state of TTP (TTP$^+$)is expected to possess large contributions of SRIP and is known to be donor (D) molecular elements constructing high electrical conductive aggregates, so that we have proposed a multi-functional aggregate systems combining unique optical nonlinearity (negative γ) and high electrical conductivity.

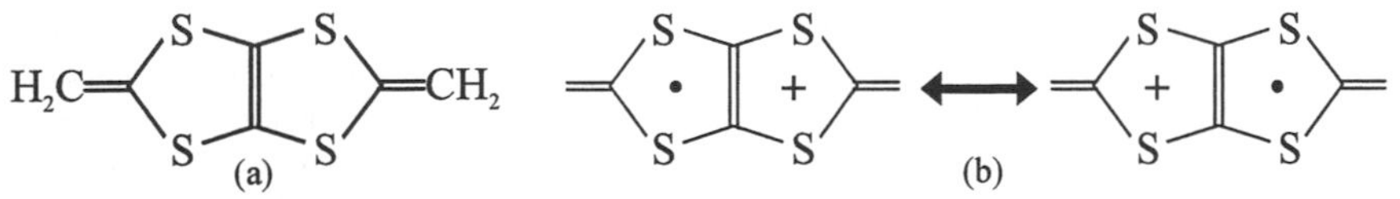

Figure 4: (a) Structure of TTP, (b) SRIP of TTP in cationic radical state.

Since the application of higher-order electron correlation methods to the calculation of real molecules of interest is difficult, it is indispensable to improve the DFT methods to reproduce reliable high-order polarizabilities for systems with charged defects. We here attempt to tune the mixing parameter of DFT/HF exchange term of the hybrid DFT method to reproduce the γ values calculated by the higher-order electron correlation method. As shown in Table 3, the γ values obtained by the BLYP and BHandHLYP methods are smaller and larger than that of CCSD(T) method, respectively. This shows that pure DFT exchange term included in the BLYP method gives excessive positive contributions, while the HF exchange term included in BHandHLYP method gives excessive negative contributions with respect to that at the CCSD(T) level. Therefore, the γ value of TTP$^+$ at the CCSD(T) level is expected to be reproduced by tuning the mixing of HF/DFT exchange parameter in the hybrid DFT method. This hybrid DFT method tuned for γ values for TTP$^+$ at the CCSD(T) level is also required to reproduce the γ value of neutral TTP at the CCSD(T) level. As a tuning parameter to satisfy these conditions, we find the mixing exchange parameter, DFT/HF = 0.7/0.3. The new DFT method is shown to semiquantitatively well reproduce γ values at the CCSD(T) level for both TTP and TTP$^+$. These results suggest that the new hybrid DFT method is expected to be applicable to the calculation of hyperpolarizabilities for the larger-size TTP clusters with several charged defects.

Table 3: γ_{xxxx} values of TTP and TTP$^+$ calculated by CCSD(T) and DFT methods.

Method	TTP	TTP$^+$
CCSD(T)	14100	-289520
BLYP	19160	-53750
BHandHLYP	9100	-408000
new hybrid DFT	**12040**	**-309350**

References

[1] (a) M. Nakano and K. Yamaguchi, *Chemical Physics Letters* **206**, 285-288(1993). (b) M.Nakano, I. Shigemoto, S. Yamada and K. Yamaguchi, *J. Chem. Phys.* **103** 4175-4191 (1995).

[2] M. Nakano, S. Yamada and K. Yamaguchi, *Bulletin the Chemical Society of Japan.* **71**, 845-850 (1998).

[3] S. Yamada, M. Nakano, I. Shigemoto, S. Kiribayashi and K. Yamaguchi, *Chemical Physics Letters* **267**, 438-444 (1997).

[4] M. Nakano, S. Yamada and K. Yamaguchi, *Chemical Physics Letters* **311**, 221-230 (1999).

Brill Academic Publishers
P.O. Box 9000, 2300 PA Leiden
The Netherlands

*Lecture Series on Computer
and Computational Sciences*
Volume 4, 2005, pp. 1162-1164

Pairing Mechanism Induced by Exchange Interaction

D. Yamaki [1 a)], K. Yasuda [a)], H. Nagao [b)] and K. Yamaguchi [c)]

a) Graduate School of information science,
Nagoya University, Chikusa-ku,
Nagoya 464-8601, Japan

b) Faculty of Science, Kanazawa University,
Kakuma, Kanazawa 920-1192, Japan

c) Graduate School of Science, Osaka University,
Toyonaka, Osaka 560-0043, Japan

Received 4 August 2005; accepted in revised form 21 August 2005

Abstract: A modified Hubbard (MH) model is derived by adding exchange type two-electron operators to the ordinary Hubbard model, in order to consider the exchange effect on the electronic states. Relationships between formation of Cooper pairs and the exchange type interactions are discussed through the maximum eigenvalues of the second-order reduced density matrices of ground states of the 4-electron 4-site model. The results suggest a cooperative pairing mechanism involving both the exchange interactions and repulsion effects.

Keywords: Hubbard model, t-J model, Cooper pair, high-T_c superconductivity, second-order reduced density matrix, exchange interaction

PACS: 31., 74.20.Mn

1. Introduction

The Hubbard model has been used for discussion of mechanism of high-T_c superconductivity. Effective models like this can be used for interpretations and energy-corrections of ab initio calculations [1,2]. This Hamiltonian is characterized by parameter t and U for one-electron hopping and on-site two-electron-repulsion, respectively:

$$\hat{H}_{\text{Hubbard}} = -t\sum_{i,\sigma}(a_{i\sigma}^+ a_{j\sigma} + a_{j\sigma}^+ a_{i\sigma}) + U\sum_{i,\sigma\neq\tau} a_{i\sigma}^+ a_{i\sigma} a_{i\tau}^+ a_{i\tau} \cdot \tag{1}$$

From the Hubbard Hamiltonian with large U/t, the t-J Hamiltonian is derived:

$$\hat{H}_{\text{t-J}} = -t\sum_{i,\sigma}(a_{i\sigma}^+ a_{j\sigma} + a_{j\sigma}^+ a_{i\sigma}) + J\sum_{i,j}\hat{S}_i \cdot \hat{S}_j \cdot \tag{2}$$

The t-J Hamiltonian is characterized by parameter t and J for one-electron hopping and effective exchange interactions, respectively. In this case the parameter J depends on the Hubbard parameter U.

The wavefunctions of the Hubbard model are more flexible than those of the t-J model, however the Hubbard model cannot represent some states that the t-J model can. For example, the ground state of the 2-site 2-electron Hubbard model is always singlet, while that of t-J model can be triplet. This is because the effective exchange interactions of t-J model are directly controlled by parameter J. The Hubbard model is less flexible than the t-J model for the exchange interaction [3,4]. Therefore it is interesting to investigate Hubbard models with additional terms for the exchange interactions.

In this paper we derive a Hubbard model with the exchange-type interactions by adding two-electron interactions. We discuss the effect on the electronic states in order to elucidate relationships between

---

[1] Corresponding author. Research Fellow of the Japan Society for the Promotion of Science. E-mail: yamaki@info.human.nagoya-u.ac.jp

formation of Cooper pairs and the additional change interactions within small cluster model, i.e., 4-site model.

2. Modified Hubbard Model

Let us consider symmetric two-site model with ab initio type two-electron integrals. This model has three types of integrals in the chemist notation: $U = (11|11) = (22|22)$, $x = (12|21) = (12|12) =...$ and $V=(11|22)=(22|11)$. We remove the integral V because it is not an independent parameter in the case of the two-electron system. Then we have a modified Hubbard (MH) model which is a Hubbard model with additional term including the parameter x:

$$\hat{H}_{MH} = -t\sum_{i,\sigma}(a_{i\sigma}^+a_{j\sigma} + a_{j\sigma}^+a_{i\sigma}) + U\sum_{i,\sigma\neq\tau}a_{i\sigma}^+a_{i\sigma}a_{i\tau}^+a_{i\tau}$$
$$+ x\sum_{i,j}\left(a_{i\sigma}^+a_{j\sigma}a_{i\tau}^+a_{j\tau} + a_{i\sigma}^+a_{j\sigma}a_{j\tau}^+a_{i\tau}\right)$$

(3)

The additional term of the MH model consists of two types of operators. One is exchange of two electrons. The other is double hopping type. We give the same coefficient x to both types of operators assuming the symmetries of ab initio two-electron integrals.

The energy difference between lowest triplet and lowest singlet state of 2-site 2-electron MH model is given by

$$\Delta E_{T-S}^{MH} = \Delta E_{T-S}^{Hubbard} - 2x = \frac{1}{2}\left(-U + \sqrt{U^2 + 16t^2}\right) - 2x \cdot$$

(4)

The first term, which depends on U, is the energy difference of the ordinary Hubbard model. This is always positive. This term vanishes at the limit $U/t \rightarrow \infty$. The second term, which depends on x, is the effect of the additional interactions. The integral x becomes the effective exchange integral itself at the limit $U/t \rightarrow \infty$.

3. Calculation

To investigate x- and U- dependency of electronic states, numerical diagonalizations were performed for the 4-site 4-electron MH model as shown in Figure 1.

The maximum eigenvalues of second-order density matrices were calculated. We chose the maximum eigenvalue as indicator of the formation of Cooper pairs. The maximum eigenvalues of ordinary Hartree-Fock wavefunctions are 1. Existence of larger eigenvalues than 1 means that formation of the Cooper pairs and their condensation.

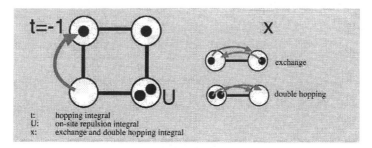

Figure 1: Four-site modified Hubbard (MH) model

4. Result and Discussion

Figure 2A shows parameters x- and U- dependency of the maximum eigenvalues of second-order density matrices of the ground state. The eigenvalues on the line x = 0 in Figure 2A are those of the ordinary Hubbard model. These are smaller than 1 in the region with positive U. The ordinary Hubbard model (x = 0) does not exhibit the pairing without attractive interaction U<0. On the line U = 2x in the region with positive U and positive x, large eigenvalues are found despite the repulsive interactions. It is found that the pairing occurs under existence of the repulsion and exchange-type interactions.

Figure 2B shows parameters x- and U- dependency of main configurations of the ground states. We divided the x-U-parameter-space into four regions: I, II, III and IV. In Region I, the two electron pairs occupy two adjacent sites in the main configurations. In Region II, the two electron pair occupy two adjacent sites as same as region I but the phase is different. In Region III, the two electron pairs occupy two separate sites. In Region IV, all the 4 sites are singly occupied. The large eigenvalues are found in Region II and on the border between Region III and IV.

On the border between Regions I and II, some configurations are vanished because of interference. On the other hand, on the border between Region III and IV, the main configurations of two regions are mixed and their coefficients are degenerate.

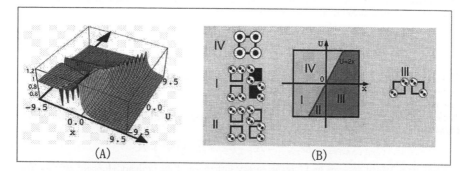

Figure 2: Parameters x- and U-dependency of (A) Maximum Eigenvalues of Second-order Density Matrices and (B) Main Configurations of the Ground State of the MH Model

5. Conclusions

We have derived a modified Hubbard model with exchange-type interactions. The energy difference between the lowest triplet and the singlet states consists of the ordinary Hubbard term depending on U and the exchange-type term depending on x for two-electron two-site system.

In order to use an efficient scheme to determine model parameter from ab initio results, the same approximation need to apply to the model and ab initio Hamiltonian [2]. This model is suited for this scheme because it contains only ordinary one- and two- electron operators and does not include spin operators. This is one of advantages to t-U-J model [3,4].

The 4-site 4-electron model exhibits formations of Cooper pairs even in the regions of repulsive interrelations. This suggests a cooperative pairing mechanism involving both exchange interactions and repulsion effects.

Acknowledgments

D. Y. was supported by Research Fellowships of the Japan Society for the Promotion of Science for Young Scientists.

References

[1] D. Yamaki et al. Int. J. Quant. Chem. 96, 10 (2004).

[2] D. Yamaki et al. Int. J. Quant. Chem. 103, 73 (2005).

[3] L. Arrachea. D. Zanchi, Phys. Rev. B 71, 064519 (2005).

[4] S. Daul, D. J. Scalapino, and S. R. White, Phys. Rev. Lett. 84, 4188 (2000).

Brill Academic Publishers
P.O. Box 9000, 2300 PA Leiden
The Netherlands

*Lecture Series on Computer
and Computational Sciences*
Volume 4, 2005, pp. 1165-1166

Iterative CASCI-DFT for excited states

S. Yamanaka[1], K. Nakata[1], T Takada[3], M. Nakano[2], and K. Yamaguchi[1].

1 Department of Chemistry, Faculty of Science, University of Osaka,
2 Department of Chemistry, Graduate School of Engineering Science, University of Osaka,
3 NEC Corporation,

Received 8 August, 2005; accepted in revised form 16 August, 2005

As in the time-independent (TI) Kohn-Sham (KS) density functional theory (DFT) [1,2], the time-dependent (TD) version of KS-DFT with the adiabatic approximation has successfully applied for the excitation spectra.   Since the prescriptions for electronic correlation are similar among time-independent and   time-dependent DFT,   the TD-DFT description using an usual adiabatic approximation also breaks down for the states and the systems where KS-DFT breaks down.   The electron correlations that both of TI and TD-DFT do not work for are explicit two-(and higher order) excitations in the context of wavefunction theory (WFT).   In contrast, the so-called dynamical correlation effects can be covered well by both of them in implicit ways.

As for the ground state, the new class of the electronic structure theory combining multireference (MR) WFT has attracted much attention [3-8].   The developers intend to make full use of the capability of KS-DFT for the electron correlation effects besides a MR characters.   Most of the pioneering redevelopers of the MR-DFT devoted their efforts to construct a DFT correction scheme for a MR-wavefunction energy [3-5].   In contrast, some of them go towards an effective MR method with involving the DFT correction, where we can obtain not only the energy-related properties, but also any properties at a high-qualitative level including both nondynamical and dynamical correlation corrections.   We also proposed the iterative complete-active-space configuration interaction-DFT (ICASCI-DFT) approach [8], of which the system of equations is given by the effective CASCI equation with involving the DFT correlation potential.   In this study, we present an application of ICASCI-DFT approach for low-lying excited states of simple molecules.

The ICASCI-DFT is based on the minimization principle of the DFT,

$$E[\rho(\mathbf{r})] = Min_{\rho(\mathbf{r}) \to N}\left[F^P[\rho(\mathbf{r})] + E_{RC}[\rho(\mathbf{r})] + \int d\mathbf{r}\, \rho(\mathbf{r}) V_{ext}(\mathbf{r})\right]. \qquad (1a)$$

Here, first, second, and third terms in parenthesis at the right side are the universal functional of CASCI-DFT , residual correlation, and external potential terms, respectively.   The modified universal functional is defined for the variational space of CASCI-DFT wavefunction:

$$F^P[\rho(\mathbf{r})] = Min_{\Psi \to \rho(\mathbf{r})}^{p}\langle \Psi | \hat{T} + \hat{V}_{ee} | \Psi \rangle. \qquad (1b)$$

The corresponding effective CASCI-DFT equation is given by

$$\left(\hat{H} + \hat{V}^{RC}\right)|\Psi\rangle = E^{eff}|\Psi\rangle. \qquad (2)$$

Here, $\hat{H}$ is the usual ab initio Hamiltonian and $\hat{V}^{RC}$ is the residual correlation potential of DFT [8]. A little care is needed for the fact that the eq. (2) is diagonalized by the basis within a specific CASCI space. Eq. (2) becomes a nonlinear equation since the RC potential term depends on CI coefficients, so should be solved in a self-consistent field manner.

---

[1] Corresponding author. E-mail: syama@chem.sci.osaka-u.ac.jp
[2] Corresponding author. E-mail: yama@chem.sci.osaka-u.ac.jp

TS is in the singlet biradical state (see Fig. 1). Hence, the electronic coupling between the solute and solvent is a matter of great significance to investigate the reaction mechanism.

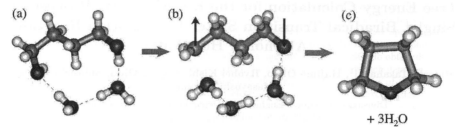

Figure 1: Dehydration reaction of 1,4-butanediol assisted by proton transfers; (a) reactant, (b) TS, (c) product

Free energy change associated with a chemical reaction, of course, plays an essential role in determining the reaction pathway. The quantum chemical approach based on the first principles is essential for the study of chemical reactions, however, it generally requires much computational cost even at low levels of theory. In addition, a substantial amount of ensemble for molecular configurations is needed to attain the convergence of the free energy when the system consists of many particles. Therefore, it is a heavily demanding task to compute the free energy for a reaction in a condensed system. In this paper, we present a novel approach to the free energy calculation by the hybrid quantum mechanical/molecular mechanical method combined with the theory of energy representation (QM/MM-ER) recently developed by Takahashi and Matubayasi[2]. Within the theory of energy representation proposed by Matubayasi[3], the distribution of the solute-solvent interaction plays a fundamental role in determining the excess chemical potential of the solute. This is differed from the conventional theory of solution where the free energy is expressed in terms of the spatial distribution functions of the solvent around the solute. It is worthy of note that a set of interaction sites, which is an artificially simplified model of a real molecule, is no longer needed within the framework of the theory of energy representation. The internal degrees of freedom of the solute molecule can also be naturally incorporated into the free energy calculations. By virtue of these advantages, one can straightforwardly take into consideration the spatially diffuse nature of the electron density and its fluctuation, which are inherent in the quantum mechanical object, through the QM/MM-ER approach.

We apply the QM/MM-ER method to compute the free energy change associated with the dehydration reaction of alcohol assisted by the proton translocation along the water wire. The solvation free energies for the reactant and TS are decomposed into several contributions, such as electron density polarization or fluctuation. In the following, we review an outline of the methodology and the computational details. The results and discussion is presented in the subsequent section.

2   Methodology

The important features of our methodology (QM/MM-ER) are summarized in two points. One is that the real-space grid approach[4] is employed to express the one-electron wave functions in Kohn-Sham DFT and the other is that the solvation free energy of a solute is described in terms of the distribution functions of the solute-solvent interaction energy. The following subsections are devoted to the review of these method and as well as the description for the computational details.

(a) The real-space grid approach

Within the framework of the Kohn-Sham DFT, the one-electron Schrödinger equation for the QM/MM simulation can be described as,

$$\left(-\frac{1}{2}\nabla^2 + \int \frac{n(\mathbf{r}')}{|\mathbf{r} - \mathbf{r}'|} d\mathbf{r}' + \nu_{\mathrm{ps}} + \frac{\delta E_{\mathrm{xc}}(n(\mathbf{r}))}{\delta n(\mathbf{r})} + V_{\mathrm{pc}}(\mathbf{r}) \right) \varphi_i(\mathbf{r}) = \varepsilon_i \varphi_i(\mathbf{r}) \tag{1}$$

where $n(\mathbf{r})$ is the electron density, ν_{ps} represents the pseudopotentials of atoms, E_{xc} is the exchange correlation functional, and finally V_{pc} expresses the electrostatic field formed by the point charges on the solvent molecules described by molecular mechanics. The one-electron wave functions are expressed by the values on the real-space grids uniformly distributed over a cubic QM cell. Accordingly, the Laplacian in Eq. (1) is approximated by a finite-difference scheme as proposed by Chelikowsky et al[4], thus,

$$-\frac{1}{2}\nabla^2 \varphi_i(x_i, y_j, z_k) = -\frac{1}{2h^2} \left[\sum_{n=-N}^{N} C_n \varphi_i(x_i + nh, y_j, z_k) + \sum_{n=-N}^{N} C_n \varphi_i(x_i, y_j + nh, z_k) \right. \\ \left. + \sum_{n=-N}^{N} C_n \varphi_i(x_i, y_j, z_k + nh) \right], \tag{2}$$

where (x_i, y_j, z_k) is the coordinate of the grid, h is the grid-spacing, and C_n are the expansion coefficients. The advantage of the real-space approach in the DFT is that it is straightforward to compute the exchange-correlation energy E_{xc} of the system. Actually, even in the LCAO approach, the integration for the E_{xc} energy is performed over the real-space grids that are distributed radially around the nuclei. By virtue of the locality of the E_{xc} operator in DFT, most of the operators in Eq. (1) becomes local and the Hamiltonian matrix elements in the real-space representation are confined within a small region of space. Therefore, the parallel implementation to the real-space approach is amenable and the high performance computing can be realized.

(b) Theory of energy representation

Within the conventional theory of solution, the solvation free energy of a solute of interest is described in terms of the spatial distribution functions of solvent around the solute. In practice, the set of site-site radial distribution functions, which is a reduced form of the full coordinate description, is commonly used due to the computational convenience. In the method of energy representation, on the other hand, the molecular configuration is represented by the distribution function of the solute-solvent interaction energy. We introduce the energy distribution function $\rho(\varepsilon)$ below.

Let $\nu(\mathbf{x})$ be the solute-solvent interaction potential of interest, where $\mathbf{x}$ is the full coordinate (position and orientation) of the solvent molecule relative to the solute fixed at the origin with a fixed orientation. The instantaneous distribution $\hat{\rho}(\varepsilon)$ in the energy representation can be written as

$$\hat{\rho}(\varepsilon) = \sum_i \delta\left(\nu(\mathbf{x}_i) - \varepsilon\right) \tag{3}$$

where $\mathbf{x}_i$ is the full coordinate of the ith solvent molecule and the sum is taken over solvent molecules. Here, we define the energy distribution $\rho(\varepsilon)$ as the ensemble average of Eq. (3) in the solution and the distribution $\rho_0(\varepsilon)$ as the ensemble average for the pure solvent system where the solute molecule is placed in the neat solvent as a test particle. Then, the solvation free energy $\Delta\bar{\mu}$ of the solute can be expressed exactly as

$$\Delta\bar{\mu} = -k_{\mathrm{B}}T \int d\varepsilon \left[(\rho(\varepsilon) - \rho_0(\varepsilon)) + \beta\omega(\varepsilon)\rho(\varepsilon) - \beta \left(\int_0^1 d\lambda \omega(\varepsilon; \lambda) \right) (\rho(\varepsilon) - \rho_0(\varepsilon)) \right] \tag{4}$$

where $\omega(\varepsilon)$ is the indirect part of the solute-solvent potential of mean force and β is the inverse of the Boltzmann constant k_B multiplied by the temperature T. λ in Eq. (4) is the coupling parameter associated with the gradual insertion of the solute in the solvent. In the practical implementation, integration with respect to λ is performed approximately by adopting a hybrid functional of PY and HNC[3].

In the development of Eq. (3), it is assumed that the solute-solvent interaction is pairwise additive. However, the electron density of the QM solute is determined under the interaction with a number of solvent molecules. The solute-solvent interaction then involves many-body effects and is not pairwise. We consider the QM solute with the electron density $\tilde{n}(\mathbf{r})$ fixed at its average distribution in solution. Then, the total solvation free energy $\Delta\mu$ of the QM solute can be decomposed as

$$\Delta\mu = \Delta\bar{\mu} + \bar{E} + \delta\mu \tag{5}$$

where $\Delta\bar{\mu}$ is the solvation free energy of the solute with the density $\tilde{n}(\mathbf{r})$, $\bar{E}$ is the average distortion energy of the QM solute, and the remaining term $\delta\mu$ represents the contribution due to the many-body effect. $\Delta\bar{\mu}$ is the free energy originating from the two-body part in the QM/MM potential and can be computed directly from Eq. (4). The free energy $\delta\mu$ in Eq. (5) can be estimated separately by introducing another energy coordinate corresponding to the electron density deviation from $\tilde{n}(\mathbf{r})$. The major contribution to the total free energy can be given, of course, by the leading two terms in r.h.s of Eq. (5) in ordinary cases.

(c) Computational details

At first, we have optimized the molecular geometries for the reactant and transition state (TS) by means of KS-DFT with B3LYP functional and 6-31G* basis set. The geometry optimizations have been performed by Gaussian 98. Since the transition state has the singlet biradical electronic structure, spin-unrestricted Kohn-Sham wave functions have been used to construct the electron density. In the QM/MM simulation, the QM solute has been described by KS-DFT with BLYP functional, where the QM cell has been discretized by 64 grids in one dimension. The MM water solvent has been represented by 252 TIP4P water molecules of which Newton's equations of motion have been solved numerically by leap-frog algorithm with the time step set at 1.0 fs. The temperature is set at 575 or 690 K. The QM/MM simulations have been carried out for 50 ps to obtain average electron density $\tilde{n}(\mathbf{r})$, followed by the 100 ps and 200 ps simulations to construct the energy distribution functions $\rho(\varepsilon)$ and $\rho_0(\varepsilon)$, respectively.

3   Results and discussions

According to the spin-unrestricted B3LYP calculation, the activation energy for the reaction has been computed as 76.0 kcal/mol with the zero-point energy correction. While the BLYP functinal has estimated the energy barrier as 60.2 kcal/mol. The underestimation of the activation energy is a general trend in the BLYP functional. The Mulliken's spin density analysis has revealed that the TS is significantly spin-polarized in the gaseous phase and the up-spin localizes at the carbon (0.298) and the counter spin at the oxygen (0.303) due to the bond breakings accompanied by the proton transfers. As a consequence, the spin-contamination increases up to 0.251 at the TS. A dramatic change in the electronic structure takes place when the solute is immersed in the polar solvent. The QM/MM simulations have revealed that the spin-contamination becomes almost zero at the TS, on the other hand, the charge is notably polarized indicating that the biradical electronic structure of the TS changes to zwitterionic structure by the influence of the water solvent. Accordingly, the dipole moment of the TS has increased to 14.0 Debye in solution from 7.4 Debye at the isolation. The hydration effect of the polar solvent will more stabilize the TS as compared with the reactant, which will substantially lower the activation free energy in the solution.

We have performed the QM/MM-ER simulations to compute the solvation free energies for the reactant and TS. The free energy components at 575 K are summarized in Table1. The total solvation free energy for the TS is much smaller than that for the reactant as expected. The distortion energy E_{dist} for the TS is very large as a consequence that the electronic structure changes to zwitterionic form in the solution, however, the destabilization energy is compensated by the free energy gain $\Delta\mu_{pol}$ due to the electron polarization. It has been found that the interaction between the non-polarized solute and solvent gives major contribution to the total solvation free energy for both molecular species. It should also be noted that the free energies $\delta\mu$ due to the electron density fluctuations are non-negligible. The free energy barrier in the gaseous phase has been computed as 62.0 kcal/mol by adopting the result of BLYP functional. Then the free energy barrier in the solution can be obtained as 49.7 kcal/mol. according to the experimental observation, the free energy barrier was estimated to be 50 kcal/mol in excellent agreement with the present calculation. Thus, it has been suggested that the dehydration reaction is catalyzed by the proton transfer mechanism where the interaction between the spin-polarized solute and solvent plays a role.

Table 1: The solvation free energies and their components in unit of kcal/mol. The thermodynamic condition is $\rho = 0.6$ g/cm^3 and T = 575 K. $\Delta\mu$ in Eq. (5) is further decomposed into the contributions due to the electron density polarization and the non-polarized solute.

			$\Delta\mu_{sol}$			
	E	$\Delta\bar{\mu}_{pol}$	$\Delta\bar{\mu}_{nonpol}$	$\delta\mu$	$\Delta\mu_{onsager}$	$\Delta\mu_{sol}$
reactant	4.1	-3.3	-3.4	-3.9	-0.2	-6.7
TS	15.5	-18.3	-8.8	-5.6	-1.8	-19.0

References

[1] Y. Nagai, N. Matubayasi, and M. Nakahara, Bull. Chem. Soc. Jpn., **77**, 691 (2004).

[2] H. Takahashi, N. Matubayasi, M. Nakahara, and T. Nitta, J. Chem. Phys., **121**, 3989 (2004).

[3] N. Matubayasi and M. Nakahara, J. Chem. Phys., **113**, 6070 (2000).
    N. Matubayasi and M. Nakahara, J. Chem. Phys., **117**, 3605 (2002); **118**, 2446 (2003).

[4] J. R. Chelikowsky, N. Troullier, and Y. Saad, Phys. Rev. Lett., **72**, 1240 (1994).

Brill Academic Publishers
P.O. Box 9000, 2300 PA Leiden
The Netherlands

*Lecture Series on Computer
and Computational Sciences*
Volume 4, 2005, pp. 1172-1174

Reactivity descriptors: Conceptual and Computational Developments

The reactivity descriptors based on density functional theory, have played very important part in study of stability and site-and regio-selectivity in chemical reactions. These descriptors broadly fall into two categories, one global set of descriptors, and another local descriptors. Global descriptors, define, in general the stability and reactivity of the system as a whole. There have been several in this category. Similarly, several local descriptors have been proposed for the purpose of describing site and regio-selectivity of chemical systems. Of these, Fukui function and local softness have played significant role. Global and local versions of hard-soft-acid-base principle have emerged as important tools in study of stability and reactivity. In the early days, the conceptual development of these has helped in understanding in simple terms. However, in recent years attempts have been made to quantify the descriptors and their role in molecular recognition. This session will highlight the latest conceptual and computational developments in the area of reactivity descriptors.

Curriculum Vitae of Dr. Sourav Pal

1.	**Name and address**	Dr. Sourav Pal Physical Chemistry Division National Chemical Laboratory Poona 411 008 India
2.	**Date and place of Birth**	12th May 1955 Ranchi
3.	**Address with telephone/Fax/e-mail No., etc**	Dr. Sourav Pal CII-1 NCL Colony Pune 411 008 (020) 25893300, Extn. 2299 (020) 25890754 (direct) e-mail pal@ems.ncl.res.in
4.	**Area of specialization**	Theoretical Chemistry/ chemical physics

5. Academic Qualifications (Bachelor's degree onwards with University, year and subject)

S.No	Degree	Subject	Class/ CGPA	Year	University	Additional Particulars
1	M.Sc(Integrated-5yrs)	Chemistry	Ist Class	1977	Indian Institute of Technology (Kanpur)	
2	Ph.D	Chemistry		1985	IACS (Calcutta)	under the supervision of Prof. Debashis Mukherjee
3	Post-Doc Research Work	Quantum Chemistry		April '86 to Oct. '87	University of Floriday, Gainesville, FL,USA	with Prof. R.J.Bartlett

6. Field of specialization
     Theoretical Chemical Physics with specialization in quantum chemistry

7. Present Position: Scientist F and Head of the Physical Chemistry Division, at National
                         Chemical Laboratory, Pune, with basic pay of Rs. 19,550 per month

8. Special Fellowship/ visiting appointment:

National Science Talent Search Scholar, 1972
Alexander von Humboldt Fellowship in Germany from Nov '87 to March '88, Sept'91 to Dec'91 and April'94 to August '94, May, 2000 to July, 2000.
Visiting Professor at the Institute for Molecular Sciences, Okazaki, Japan from March, 1997 to September, 1997
Visiting Scientist at the University of Arizona, Tucson, May 1995
Selected as "Dai-Ichi Karkaria Ltd" Endowment Fellow for 2004-05 by UICT, Mumbai

9. Work done:

A: Frontier Theoretical Development on Molecular Electric Properties using
     extended coupled-cluster and Hilbert-space and Fock-space multi-reference
     coupled-cluster response
B: Theoretical Investigation of stability and reactivity using Hard-Soft Acid-Base
     relation
C:: Study of Interaction Potential in Electron - Molecule Scattering
D:: Ab initio Born-Oppenheimer molecular dynamics using localized Gaussian basis
     sets for reactions in zeolites .
E. Magnetic properties using coupled-cluster response approach:
F. Complex-scaled and complex absorbing potential coupled-cluster technique for electron-molecule response
G: Theoretical study of metallo-aromaticity and anti-aromaticity using electron localization function and ring current

10. Awards and Honours

Recepient of the Shanti Swarup Bhatnagar Prize in Chemical Sciences, 2000 Elected as a Fellow of the Indian National Science Academy, New Delhi, 2003
Elected as a Fellow of the National Academy of Sciences, India, Allahabad, 1998
Elected as a Fellow of the Indian Academy of Sciences, Bangalore, 1996
Recepient of the Chemical Research Society of India medal, 2000
Elected as a Fellow of the Maharashtra Academy of Sciences, 1994
Recepient of the NCL Research Foundation Scientist of the year (1999) award
Recepient of the P.B.Gupta Memorial lecture Award of the Indian Association for the Cultivation of Science, Calcutta for 1993
Received Council of Scientific and Industrial Research (CSIR) Young Scientist award in Chemical Sciences for 1989
Received Indian National Science Academy (INSA) medal for Young Scientist 1987
Received NCL Research Foundation Best Paper Award in Physical Sciences for the year 1995, 1996, 1997, 1999, 2000,2002

11. Membership of Editorial Boards of Journals / Societies

Chosen as a member of the Editorial Board of International Journal of Molecular Sciences from 2000
Member, Editorial Board, Journal of Chemical Sciences, published by the Indian Academy of Sciences, Bangalore from 2004
Member, Editorial Board, International Journal of Applied Chemistry from 2005.
Elected as a Life Member of the Society for Scientific Values
Member, American Physical Society, USA

12. Invitations to International Conferences

Delivered several (about 100) invited lectures at several International and Indian Conferences. The most notable ones being the plenary lecture at the Sanibel Symposium, St. Augustine, Florida, 2003, key note lecture at the International Conference on Computational Methods in Science and Engineering, Attica, Greece, invited lecture at the First Asian Pacific Conference on theoretical and computational Chemistry, Okazaki, Japan, 2004 and the invited lectures at two satellite conferences of the International Congress on Quantum Chemistry in 1994 and 1997.

Invited to chair a session at the coupled cluster theory and electron correlation workshop to be held at Cedar Key, Florida, USA in June 1997. This workshop is a satellite meeting of the 9th International Congress on Quantum Chemistry held every three years. Chaired several other sessions at International Conferences

13. Books Authored: Co-authored a book " Mathematics in Chemistry" with Dr. K
V. Raman, Vikas Publishing House Pvt Ltd., New Delhi, 2004

14. Publications: About 115 in reputed international journals. (Journal of American Chemical Society, The Journal of Chemical Physics, Journal of Physical Chemistry, Physical Review, Chemical Physics Letters etc.)

Brill Academic Publishers
P.O. Box 9000, 2300 PA Leiden
The Netherlands

*Lecture Series on Computer
and Computational Sciences*
Volume 4, 2005, pp. 1175-1177

Some Problems Related to Electronegativity Equalization

Paul.W. Ayers[1]

Department of Chemistry,
McMaster University,
Hamilton, Ontario, Canada L8S 4M1

Received 17 July, 2005; accepted in revised form 14 August, 2005

Abstract: The principle of electronegativity equalization, as originally proposed by Sanderson, indicates that when chemical reagents come into contact, they undergo electron transfer until each reagent has the same chemical potential. This powerful principle not only explains why electrons are transferred from less electronegative to more electronegative atoms in molecules, but also is important for explaining solvent effects on chemical reactions. (In reactions in solution, the electronegativity of the reagents are effectively "pinned" to the chemical potential of the solvent.) There is a paradox related to the electronegativity equalization principle, however: two atoms that are very far apart should have the same electronegativity, but this seems counterintuitive: how can the presence of a Cesium atom in Paris effect the chemical potential of a Fluorine atom in Loutraki? Berkowitz has called this "spooky action at a distance" the "EPR paradox of conceptual density-functional theory." By analyzing the structure of the exact density functional, however, one can resolve the paradox: in the limit of infinite separation, molecules have no well-defined chemical potential because the density functional variational principle is stationary not only for the ground state (Cs in Paris and F in Loutraki), but also for some excited states (e.g., Cs^+ in Paris and F^- in Loutraki; Cs^- in Paris and F^+ in Loutraki). Practical electronegativity equalization calculations that need to be able to reproduce not only the "dense" limit, but also the asymptotic interactions will need to address this matter. To this end, a practical method based on the finite-temperature grand-canonical ensemble (with a geometry-dependent temperature) is proposed.

Keywords: electronegativity, chemical potential, electronegativity equalization, conceptual density-functional theory

PACS: 82.90.+j, 31.10.+z

1. Extended Abstract

One of the most fundamental precepts of molecular chemistry is the fact that the electronegative elements tend to be negatively charged while electropositive elements tend to be positively charged. Sanderson interpreted this using the electronegativity equalization principle: electrons are transferred from electropositive elements to electronegative elements until the electronegativity of all the atoms in a molecule are equal.[1] The basic idea was then elaborated upon by Parr and coworkers, who identified the electronegativity with the negative of the electronic chemical potential,[2]

$$\mu \equiv \left(\frac{\partial E[v;N]}{\partial N} \right)_{v(r)} = -\chi , \qquad (1)$$

where the electronic chemical potential is defined as the partial derivative of the electronic energy, $E[v;N]$ with respect to the number of electrons at constant geometry. (In density-functional theory, information about the molecular geometry is encapsulated in the external potential which, for an M atom molecule, is

$$v(\mathbf{r}) = \sum_{\alpha=1}^{M} \frac{-Z_\alpha}{|\mathbf{r} - \mathbf{R}_\alpha|} \qquad (2)$$

where $\{Z_\alpha\}_{\alpha=1}^{M}$ and $\{\mathbf{R}_\alpha\}_{\alpha=1}^{M}$ denote the atomic numbers and positions of the molecule's constituent nuclei.)

---

[1] E-mail: ayers@mcmaster.ca

The electronegativity equalization principle is important in chemistry for a number of reasons. On a purely qualitative level, electronegativity equalization explains why the electronegative elements tend to be associated with negatively charged molecular sites: electropositive atoms donate electrons to electronegative elements until the amount the electronegative atoms "want" electrons is equal to the "desire" of the electropositive elements to forfeit them. The electronegativity equalization principle is also apposite to solution phase reactions. In dilute solution, the electronegativity of the reagents is effectively "pinned" to the electronegativity of the solvent, so that changes in the solvent's electronegativity (as commonly adjusted by changing the solvent's pH or ionic strength or mixing different solvents) will cause charge transfer to and from the reagents, and change their reactivity.

At a quantitative level, there has been much work in computational electronegativity equalization schemes which, because they include the effects of charge transfer between reagents, go beyond the usual electromagnetic interactions that are typically included in molecular force fields. It seems to be in this context that the "electronegativity equalization paradox" was first observed. Write the energy, as a function of the number of electrons, as

$$E_A(N_A) = E_A^0 + (N_A - N_A^0)\mu_A + \tfrac{1}{2}(N_A - N_A^0)^2 \eta_A + \dots \tag{3}$$

and note that, if you combine two atoms, A and B, with different chemical potentials, $\mu_A < \mu_B$, electron transfer from the base, B, to the acid, A, will occur until, eventually, the chemical potentials of the two atoms are equal. If we consider the case where the two atoms are very far apart, then this seems to contradict the well-known observation that diatomic molecules dissociate into neutral atoms because the smallest atomic ionization potential is greater than the largest atomic electron affinity. Instead, it predicts that either diatomic molecules will dissociate into fragments with fractional electron number or, perhaps, that the electronegativity of a dissociated molecule is ill-defined.

The fractional electron number possibility was resolved, in the negative, by Perdew, Parr, Levy, and Balduz, who noted that $\mu = \left(\frac{\partial E}{\partial N}\right)_{v(r)}$ does not have a unique definition for an isolated electronic system at zero temperature.[3] But this does not entirely resolve the paradox. Perdew et. al argued that the one-sided derivatives, pertaining to the addition

$$\mu^+ = \left(\frac{\partial E}{\partial N}\right)_{v(r)}^+ = -EA \tag{4}$$

or subtraction

$$\mu^- = \left(\frac{\partial E}{\partial N}\right)_{v(r)}^- = -IP \tag{5}$$

of electrons *are* uniquely defined, and are simply related to the electron affinity or the ionization potential of the molecule. For a molecule at a finite bond length, this is not problematic. However, when one considers a molecule in the limit of infinite separation, a problem arises. For concreteness, consider Cesium Fluoride. At dissociation, one has

$$\mu^+[CsF_{R\to\infty}] = -IP_{Cs}$$
$$\mu^-[CsF_{R\to\infty}] = -EA_F. \tag{6}$$

Next, from Parr, Donnelly, Levy, and Palke,[2] we know that the chemical potential is simply related to the functional derivative of the energy density functional functional with respect to the electron density,

$$\mu_{CsF}^{\pm} \equiv \left.\frac{\delta E_v[\rho]}{\delta\rho(r)}\right|_{\rho=\rho_{CsF}}^{\pm} \tag{7}$$

Very strangely, this suggests that the change in energy associated with adding a small fraction of an electron to the Cesium atom is related to the electron affinity of the *Fluorine* atom, even though the Fluorine atom is infinitely far away.

$$E_v\left[\rho_{CsF} + \varepsilon\left(\rho_{Cs^-} - \rho_{Cs}\right)\right] - E_v[\rho_{CsF}] = \varepsilon \int \left(\rho_{Cs^-}(r) - \rho_{Cs}(r)\right)\left.\frac{\delta E_v[\rho]}{\delta\rho(r)}\right|_{\rho=\rho_{CsF}}^{+} dr \tag{8}$$
$$= -\varepsilon(EA_F)$$

Berkowitz has referred to this "spooky action at a distance" as the "EPR paradox" of density functional theory. Indeed, it is patently ridiculous that the chemical potential of an atom (or molecule) should be affected by other atoms/molecules infinitely far away.

This presentation will seek to resolve the paradox. The key is to note that the energy functional is stationary not only for the ground state of the system, but also for certain select excited states.[4] In particular, at infinite separation, the energy functional is stationary for any way of distribution electrons between the atoms, and so the energy functional of the cation will be stationary not only for the canonical structure Cs^+F, but also for CsF^+, $Cs^{+2}F^-$, Cs^-F^{+2}, with other possibilities excluded only because one cannot bind more than one additional electron to Cesium or Fluorine. Because of this, the ionization potential of CsF is not uniquely defined, and Eq. (6) should be replaced by

$$\mu^+\left[CsF_{R\to\infty}\right] = \left(-IP_{Cs}\right) \text{ or } \left(-IP_F\right)$$

$$\text{or } \left(E_{Cs^{+2}} + E_{F^-} - E_{Cs} - E_F\right) \text{ or } \left(E_{Cs^-} + E_{F^{+2}} - E_{Cs} - E_F\right) \tag{9}$$

$$\mu^-\left[CsF_{R\to\infty}\right] = \left(-EA_{Cs}\right) \text{ or } \left(-EA_F\right)$$

The functional derivative, in turn, does not exist since different changes in electron density will be associated with different ionization potentials. (A similar situation arises in perturbation theory; when there are multiple energies for a single state, there is no "linear response" Hellmann-Feynman-type formula relating changes in Hamiltonian to changes in energy and, instead, the change in energy depends on the specific change in Hamiltonian under consideration. Similarly, the change in energy here depends on the specific way the density changes, and cannot be expressed as a simple "linear functional" (i.e., functional derivative) like Eq. (8).

The practical implication of this analysis is that computational electronegativity equalization schemes need to be revised if they are going to successfully address the nature of charge transfer between systems that are far apart or, more generally, systems of low density (e.g., in the gas phase). I'll conclude this presentation with a few brief speculations on how this problem might be addressed. In particular, one might consider introducing a temperature (which removes the discontinuity between the chemical potential for electron addition and subtraction) that approaches zero as molecules dissociate. An appropriate model of the temperature should be related to the strength of interactions (ergo, the ease of charge transfer) between the molecular fragments, and might be constructed by choosing a temperature that reproduces the increase in energy that accompanies the "deformation" of a molecular fragment's density by an approaching fragment.

References

[1] R. T. Sanderson, *Science* **114** 670-672 (1951).

[2] R. G. Parr, R. A. Donnelly, M. Levy, and W. E. Palke, *J. Chem. Phys.* **68** 3801-3807 (1978).

[3] J. P. Perdew, R. G. Parr, M. Levy, and J. L. Balduz, Jr., *Phys. Rev. Lett.* **49** 1691-1694 (1982).

[4] J. P. Perdew and M. Levy, *Phys. Rev. B*, **31** 6264-6272 (1985).

Brill Academic Publishers
P.O. Box 9000, 2300 PA Leiden
The Netherlands

*Lecture Series on Computer
and Computational Sciences*
Volume 4, 2005, pp. 1178-1182

The Electronegativity Equalization Method and its application in Computational Medicinal Chemistry

P. Bultinck

Department of Inorganic and Physical Chemistry,
Faculty of Sciences,
Ghent University,
Krijgslaan 281, B-9000 Gent, Belgium

Received 7 July, 2005; accepted in revised form 12 August, 2005

Abstract: This paper summarizes the essential features of the Electronegativity Equalization Method as introduced by Mortier et al. based on the Sanderson principle. EEM expressions are derived from Density Functional Theory and subsequently the algebraic equations are solved to obtain expressions for the very efficient calculation of atomic charges, Fukui functions, softness, … The role of the calibration is described and the integrability of the EEM method in a Computational Medicinal Chemistry environment.

Keywords: EEM, Electronegativity Equalization, Computational Medicinal Chemistry

1. Introduction

One of the most important quantities in so-called Conceptual Density Functional Theory [1-3] is the electronegativity of a chemical entity.

$$\chi = -\left(\frac{\partial E}{\partial N}\right)_v = \left(\frac{\delta E}{\delta \rho}\right)_v \qquad (1)$$

where ρ is the electron density, E the energy and N is the number of electrons. It is immediately clear that there is a very close relationship between the electronegativity and the chemical potential of the system.

An often used approximation for the calculation of the partial derivative $\left(\frac{\partial E}{\partial N}\right)_v$ shows the equivalence to the original Mulliken definition :

$$\chi = \frac{I + A}{2} \qquad (2)$$

The electronegativity equalization principle formulated by Sanderson [4-5], states that when molecules are formed, the electronegativities of the constituent atoms become equal, yielding the molecular, equalized (Sanderson) electronegativity. Several formalisms have evolved from this principle. The present lecture will build on the Electronegativity Equalization Method (EEM) of Mortier et al. [6]

2. EEM algebraic relationships

In EEM the electronegativity of an atom α in an N-atom molecule is shown to be :

$$\chi_{eq} = \chi_\alpha = \chi_\alpha^0 + \Delta\chi_\alpha + 2(\eta_\alpha^0 + \Delta\eta_\alpha)q_\alpha + \sum_{\beta \neq \alpha}^{N} \frac{q_\beta}{R_{\alpha\beta}} \qquad (1)$$

In this expression, χ_α^0 and η_α^0 represent the isolated atom electronegativity and hardness, respectively, q_α represents the atomic charge on atom α and $R_{\alpha\beta}$ the interatomic distance between atoms α and β. $\Delta\chi_\alpha$ and $\Delta\eta_\alpha$ are corrections to the respective isolated atom values due to the incorporation of the atom in a molecule (or crystal). The external potential represented by the last term accounts for the influence of the surrounding atoms (or molecules).

Using χ_α^* ($=\chi_\alpha^0 + \Delta\chi_\alpha$) and η_α^* ($=\eta_\alpha^0 + \Delta\eta_\alpha$) for the effective atomic electronegativity and hardness values respectively, the operational, simplified formula used throughout the study is given by :

$$\chi_{eq} = \chi_\alpha = \chi_\alpha^* + 2\eta_\alpha^* q_\alpha + \sum_{\beta \neq \alpha}^{N} \frac{q_\beta}{R_{\alpha\beta}} \tag{2}$$

Realizing that the χ_{eq} is equal for all atoms, one can construct an algebraic system of equations allowing us to compute atomic charges and χ_{eq} from simple matrix equations. N of the necessary equations are obtained by equilibrating the individual atomic electronegativities to the molecular electronegativity ($\chi_{eq} = \chi_\alpha = \chi_\beta = \dots$) while one supplementary equation is obtained by constraining the sum of the atomic charges to equal the total molecular charge ($Q = \sum_\alpha^N q_\alpha$). In matrix form this may be written as :

$$\begin{bmatrix} 2\eta_1^* & 1/R_{12} & \cdots & 1/R_{1N} & -1 \\ 1/R_{21} & 2\eta_2^* & \cdots & 1/R_{2N} & -1 \\ \vdots & \vdots & \cdots & \vdots & \vdots \\ 1/R_{N1} & 1/R_{N2} & \cdots & 2\eta_N^* & -1 \\ 1 & 1 & \cdots & 1 & 0 \end{bmatrix} \begin{bmatrix} q_1 \\ q_2 \\ \vdots \\ q_N \\ \chi_{eq} \end{bmatrix} = \begin{bmatrix} -\chi_1^* \\ -\chi_2^* \\ \vdots \\ -\chi_N^* \\ Q \end{bmatrix} \tag{3}$$

Finally, in addition to the molecular electronegativity and atomic charges, the EEM framework also allows a straightforward and transparent calculation of other fundamental properties such as the total electronic energy, hardness and reactivity indices, such as Fukui functions and local softness [1-3].

3. EEM Calibration and resulting quality of charge prediction

EEM holds the potential of generating, at a very modest computational cost, atomic charges that are both connectivity- and geometry-dependent [7-9]. The problem naturally is the fact that one needs to have the values for χ_α^* and η_α^*. χ_α^* and η_α^*, however, are unknown, and cannot be calculated directly. Hence, the need for calibration.

These calibrations are found to be very difficult if sufficient precision is sought. As an example, we report the calibration of parameters for C, H, N, O and F; although a much wider range of elements can be calibrated if needed. Based on a molecular set of several dozens of drug-size molecules, several types of atomic charges were computed and then parameters calibrated.

The actual calibration is a stepwise process. First, χ^* and η^* values for all elements are assigned randomly. These values are then used to calculate EEM charges on all atoms through standard matrix algebra. Equation (3) is used as fitness function to evaluate the quality of the fit between the DFT charges and the EEM charges.

$$\Delta q = \sum_{z=1}^{N_{el}} \frac{\sum_{i=1}^{M} \sum_{\alpha=1}^{N_{i,z}} (q_{\alpha iz}^{EEM} - q_{\alpha iz}^{DFT})^2}{N_z} \tag{3}$$

This fitness function is then minimized by updating the χ^* and η^* values by means of a combination of a local and global optimizer. First many randomly chosen sets of parameters were submitted to the simplex method in multidimensions. The 100 best unique sets (i.e. those sets that give the lowest value for the fitness function) were used as input in a genetic algorithm. The parameter set giving the best fit between the EEM and DFT charges is recorded and considered to represent the optimal set of calibrated parameters.

Figures 1 gives the EEM charges (using the optimal effective electronegativity and hardness values) versus the DFT charges. Also included are the parameters for the best fitting linear function between both types of charges. There is an obvious linear correlation between both sets of charges. This illustrates the ability of the principle of electronegativity equalization to yield quantitative atomic charges.

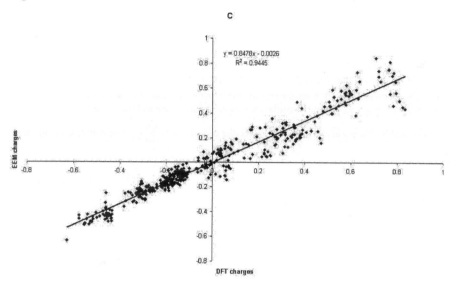

Figure 1: Agreement between B3LYP/6-31G* and EEM charges for carbon atoms in a 138 molecule test set.

The applicability of EEM to different types of atomic charges, however, is not always equally good [7-9].

4. EEM based molecular descriptors

Since the EEM equations hold the electronegativities as well as atomic charges, which are directly related to the atomic integrated electron density, new derivations can be made to obtain the molecular hardness and atomic Fukui functions. The EEM equations as given above can be differentiated with respect to N [10]. Similar derivations have also been reported by Baekelandt et al. [11]. To allow the calculation of the atomic Fukui functions and molecular hardness, the effective hardness and electronegativity values are assumed constant under the change of N. As an example, he following expression relating Fukui functions and the molecular hardness with the calibrated atomic hardness parameters is found :

$$0 = 2\eta_\alpha^* f_\alpha + \sum_{\beta \neq \alpha}^{N} \frac{f_\beta}{R_{\alpha\beta}} - 2\eta_{mol} \tag{4}$$

Such a result is quite similar to the EEM set of equations. Adding the constraint of normalization of the Fukui function, one finds the following matrix equation :

$$
\begin{bmatrix}
2\eta_1^* & 1/R_{12} & \cdots & 1/R_{1N} & 1 \\
1/R_{21} & 2\eta_2^* & \cdots & 1/R_{2N} & 1 \\
\vdots & \vdots & \cdots & \vdots & \vdots \\
1/R_{N1} & 1/R_{N2} & \cdots & 2\eta_N^* & 1 \\
1 & 1 & \cdots & 1 & 0
\end{bmatrix}
\begin{bmatrix}
f_1 \\
f_2 \\
\vdots \\
f_N \\
-2\eta_{mol}
\end{bmatrix}
=
\begin{bmatrix}
0 \\
0 \\
\vdots \\
0 \\
1
\end{bmatrix}
\tag{7}
$$

Another set of descriptors are the global molecular softness and the local atomic softness.

Several EEM equations can be derived to allow the calculation of the following molecular descriptors : atomic charges and equalized molecular electronegativity, atom condensed Fukui functions and molecular hardness, and finally atom condensed softness and the total softness.

Next to these basic descriptors, Karelson [12-13] has described many derived QSAR parameters like the globally most positive and negative atomic charge in the molecule, and similarly for each element present in the molecule, a polarity-like parameter as the difference between the most positive and most negative charge as well as various sums of absolute or squared values of partial charges. Other common charge based descriptors are averages of atomic partial charges absolute values. Fukui functions, softness and hardness and derived quantities have been found to be useful molecular descriptors by different authors.

EEM allows the calculation of many of these parameters in a Computational Medicinal Chemistry environment at very high speed, typically more than one million molecules/hour [14].

Acknowledgments

P.B. wishes to thank the Fund for Scientific Research in Flanders (FWO-Vlaanderen) for continuous support.

References

[1] R.G. Parr and W. Yang: *Density Functional Theory of Atoms and Molecules*. Oxford Science Publications, Oxford, 1989.

[2] P. Geerlings, F. De Proft and W. Langenaeker, Conceptual density functional theory, *Chemical Reviews* **103** 1793-1873 (2003).

[3] P.K. Chattaraj, S. Nath and B. Maiti, Reactivity Descriptors, *Computational Medicinal Chemistry for Drug Discovery* (Editors: P. Bultinck, H. De Winter, W. Langenaeker and J.P. Tollenaere), Dekker, New York (USA), 2004, pp. 295-322.

[4] R.T. Sanderson *Science* **114** 670 (1951).

[5] R.T. Sanderson: *Polar Covalence*. Academic Press, New York, 1983.

[6] W.J. Mortier, S.K. Ghosh, S. Shankar, Electronegativity equalization method for the calculation of atomic charges in molecules, *Journal of the American Chemical Society* **108** 4315-4320 (1986).

[7] P. Bultinck, W. Langenaeker, P. Lahorte, F. De Proft, P. Geerlings, M. Waroquier and J.P. Tollenaere, The electronegativity equalization method i : parametrization and validation for atomic charge calculations, *Journal of Physical Chemistry A*, **106** 7887-7894 (2002).

[8] P. Bultinck, W. Langenaeker, P. Lahorte, F. De Proft, P. Geerlings, C. Van Alsenoy and J.P. Tollenaere, The Electronegativity Equalization Method Ii : Applicability Of Different Atomic Charge Schemes, *Journal of Physical Chemistry A*, **106** 7895-7901 (2002).

[9] P. Bultinck, R. Vanholme, P. Popelier, F. De Proft and P. Geerlings, High-Speed Calculation Of Aim Charges Through The Electronegativity Equalization Method, *Journal of Physical Chemistry A* **108** 10359-10366 (2004).

[10] P. Bultinck and R. Carbó-Dorca, Algebraic relationships between Conceptual DFT quantities and the electronegativity equalization hardness matrix, Chemical Physics Letters, **364** 357-362 (2002).

[11] B.G. Baekelandt, G.O.A. Janssens, H. Toufar, W.J. Mortier, R.A. Schoonheydt,, Acidity and Basicity in Solids: Theory, Assessement and Utility, (Editors: J. Fraissard and L. Petrakis) NATO ASI Series C444; Kluwer Academic Publishers, Dordrecht, 1994, p. 95.

[12] M. Karelson, *Molecular Descriptors in QSAR/QSPR*, J. Wiley and Sons, New York, 2000.

[13] M. Karelson, Quantum-Chemical Descriptors in QSAR, *Computational Medicinal Chemistry for Drug Discovery* (Editors: P. Bultinck, H. De Winter, W. Langenaeker and J.P. Tollenaere), Dekker, New York (USA), 2004, pp. 641-668.

[14] P. Bultinck W. Langenaeker, R. Carbó-Dorca and J.P. Tollenaere, Fast calculation of quantum chemical molecular descriptors from the electronegativity equalization method, *Journal of Chemical Information and Computer Science* **43** 422-428 (2003).

Brill Academic Publishers
P.O. Box 9000, 2300 PA Leiden
The Netherlands

*Lecture Series on Computer
and Computational Sciences*
Volume 4, 2005, pp. 1183-1183

The Influence of Electric Field on the Global and Local Reactivity Descriptors: Reactivity and Stability of the weakly Bonded Complexes

K. R. S. Chandrakumar
Theoretical Chemistry Section, RC&CD Division, Bhabha Atomic Research Centre,
Mumbai - 400 085, India. Email: krsc@magnum.barc.ernet.in

Sourav Pal
Theoretical Chemistry Group, Physical Chemistry Division, National Chemical Laboratory, Pune - 400 008, India. Email: s.pal@ncl.res.in

Received 12 August, 2005; accepted 13 August, 2005

Abstract: The response of the global and local reactivity density based descriptors (chemical potential, hardness, softness, Fukui function and local softness) in the presence of external electric field has been studied for some of the simple prototype molecular systems. In addition to the analysis on the reactivity of these systems, the influence of the electric field on the interaction energy of the complexes formed by these systems has also been studied using the recently proposed semi-quantitative model based on the local Hard-Soft Acid-Base principle. Using the inverse relationship between the global hardness and softness parameters, a simple relationship is obtained for the variation of hardness in terms of the Fukui function under the external electric field. It is shown that the increase in the hardness values for a particular system in the presence of external field does not necessarily imply that the reactivity of the system would be deactivated or vice versa.

Brill Academic Publishers
P.O. Box 9000, 2300 PA Leiden
The Netherlands

*Lecture Series on Computer
and Computational Sciences*
Volume 4, 2005, pp. 1184-1185

Application of the Reactivity Index to Propose Intra and Intermolecular Reactivity in Catalytic Materials

Abhijit Chatterjee[1]

Material Sceince,
Accelrys,
Nishishinbashi TS Bldg. 11F, 3-3-1 Nishishinbashi, Minato-ku, Tokyo
105-0003, Japan,

Received 29 July, 2005; accepted in revised form 18 August, 2005

Abstract: The hard soft acid-base (HSAB) principles classify the interaction between acids and bases in terms of global softness. Pearson proposed the global HSAB principle[1]. The global hardness was defined as the second derivative of energy with respect to the number of electrons at constant temperature and external potential, which includes the nuclear field. The global softness is the inverse of this. Pearson also suggested a principle of maximum hardness (PMH)[2], which states that, for a constant external potential, the system with the maximum global hardness is most stable. In recent days, DFT has gained widespread use in quantum chemistry. Some DFT-based local properties, e.g. Fukui functions and local softness, have already been used for the reliable predictions in various types of electrophilic and nucleophilic reactions. This study aims to review the development and application of reactivity index in key catalytic process. The reactivity index finds its application in material designing. In our study[3] we proposed a reactivity index scale for heteroatomic interaction with zeolite framework. The scale holds well for unisite interaction or in other way with one active site preset in the molecule, the scale does not hold good for systems with two or more active sites. The activity of different representative templating molecules along with zeolite framework is investigated using a range of reactivity indexes using density functional theory (DFTWe investigated the local softness of the interacting templates to compare their affinity with the zeolite framework cluster models. The cluster models are chosen to mimic the secondary building units of zeolite crystals for both silicalite and silica aluminates. The conformational flexibility was brought out as common features of those representative organic templates. The influence of the nature of the functional group and alkyl group on the electronic interaction is studied systematically. An a priori rule is formulated to choose the best template for a particular zeolite (e.g., ZSM-5) synthesis. The role of water during nucleation process is monitored in terms of solvation energy to rationalize the fundamental mechanismof crystal growth[4].

The activity of nitrogen heterocyclics present in biomacromolecules and its suitable sorbent from the dioctahedral smectite family is investigated using a range of reactivity index using density functional theory (DFT)[5]. For the first time, a novel function λ has been defined for quantitative description of weak adsorption cases, which was so far qualitative inside the domain of DFT. From the values of the local softness, it is concluded that the local acidities of the inorganic material systems are dependent on several characteristics, which are of importance within the framework of the HSAB principle. We first rationalized an understanding of the electronic structures of a range of nitrogen heterocyclics ranging from indole, imidazole, pyrrole, and pyridine, followed by the local softness calculation to locate its active site. We compared its activity with that of the OH group of isomorphously substituted ($Fe3+$, $Mg2+$, $Fe2+$, and $Li+$) dioctahedral smectite family. Two types of interactions were identified between heterocyclics and smectite.The ordering for best sorption follows the order $Mg2+ > Fe2+ > Fe3+ > Li+$, whereas the order for best sorbent is imidazole > pyridine > pyrrole > indole. The results rationalize the experimental observation. We have as well performed calculation to derive group softness[6] for inter and intra molecular reactivity for nitro aroatics and its adsorption over clay matrices. We will ellaboarte the development of the methodology and its application in many different systems.[7-9].

---

[1] E-mail: achatterjee@accelrys.com, Phone: +81-3-3578-3861, Fax: +81-3-3578-3873

References

1. Pearson, R. G. J. Am. Chem. Soc., **105**, 7512 (1983).
2. Pearson, R. G.J. Chem. Educ., **64**, 561 (1987).
3. Chatterjee, A., Iwasaki, T. and Ebina, T. J. Phys. Chem. A 103 2489 (1999).
4. Chatterjee, A. and Iwasaki, T. J. Phys. Chem. A 105, 6187 (2001).
5. Chatterjee, A., Iwasaki, T. and Ebina, T. J. Phys. Chem. A 105, 10694(2001).
6. Chatterjee, A., Ebina, T., Iwasaki, T. and Mizukami, F. J. Chem. Phys. 118, (2003).
7. Chatterjee, A., Ebina, T., Onodera, Y. and Mizukami, F. J. Mol. Graphics & Modeling 22, 93(2003).

8. Chatterjee, A., Suzuki, T.; Takahashi, Y.; Tanaka, D.A.P. Chemistry – A European Journal 9, 3920 (2003).

9. Chatterjee, A., Ebina, T., Onodera, Y. and Mizukami, F. J. Chem. Phys. 120, 3414 (2004).

Brill Academic Publishers
P.O. Box 9000, 2300 PA Leiden,
The Netherlands

*Lecture Series on Computer
and Computational Sciences*
Volume 4, 2005, pp. 1190-1192

Symposium on Electron Densities and Density Functionals

Ajit J. Thakkar[1]

Department of Chemistry,
University of New Brunswick,
Fredericton, NB E3B 6E2, Canada

Received 20 July, 2005; accepted 20 July, 2005

Abstract: The symposium on "Electron Densities and Density Functionals" has attracted 20 papers on a wide variety of topics.

Keywords: Electron density, electron-pair density, momentum density, reduced density matrix, density-functional theory.

PACS: 31.15.Ew, 31.10.+z, 31.15.-e, 34.20.-b, 71.15.Mb, 71.20.-b

1 Introduction

The symposium on "Electron Densities and Density Functionals" has attracted 20 papers on using one-electron number (or charge) densities for interpretative and analytic purposes, on density-functional theory and methods, on one-electron momentum densities, and on reduced-density matrices. A terse outline of these papers by authors from eight countries is presented below.

2 Chemical insight from electron densities

Richard Bader gives an overview of the topology of the kinetic energy density and its connection with bonding and reactivity [1]. Russell Boyd and coworkers present a study of weak bonding in DNA based on electron densities [2]. Patrick Bultnick et al. focus on molecular aromaticity using density-based multi-center bond indices, and similarity indices based on density overlaps [3]. Pierre Becker and coworkers survey the use of experimental and/or theoretical number and momentum densities to reconstruct approximate one-electron reduced density matrices that can be used to analyze bonding and reactivity [4].

Axel Becke and Johnson describe a model of the dispersion interaction based upon the dipole-moment density of the exchange hole [5]. Paul Popelier et al. examine how topological partitions of molecules can be used for the construction of intermolecular potentials [6].

3 Density-functional theory

Gustavo Scuseria and Heyd discuss the performance of the meta-GGA TPSS, hybrid HSE and other density functionals for the calculation of lattice constants and band gaps [7]. Weitao Yang presents the potential functional perspective of density-functional theory [8]. David Tozer and Teale zero in on exchange methods in Kohn-Sham theory for the computation of NMR shielding constants [9].

---

[1] E-mail: ajit@unb.ca

Paul Ayers constructs multidimensional numerical integration formulas, with a weight function related to the electron density, for use in density-functional codes [10]. Frank Jensen examines polarization-consistent basis sets optimized for DFT computations [11].

Zahariev and Alexander Wang consider how unnormalized densities can be included in the domain of the universal density functional [12]. The nuclear Fukui function in spin-polarized density-functional theory is studied by Frank de Proft and coworkers [13]. Ignacio Porras describes a simple, semi-explicit, density-functional method for atoms [14].

4 Other work

Carmela Valdemoro and coworkers address variational calculations using second-order reduced density matrices [15]. Peter Gill introduces several novel two-electron distributions connected to the Wigner representation of density matrices [16]. Nicholas Besley also discusses Wigner and Husimi intracules [17]. Jesus Ugalde and coworkers examine electron-pair densities for simple models of H_2 treated as a four-body system [18].

Juan-Carlos Angulo discusses properties of the Fourier transform of the momentum density— the reciprocal form factor [19]. Tony Tanner describes how he stumbled into the creation of an on-line bibliography on Compton scattering and momentum densities [20].

References

[1] R. F. W. Bader. Topology of the positive definite kinetic energy density and its physical consequences, this volume.

[2] C. F. Matta, N. Castillo, and R. J. Boyd. An electron density study of the characterization of extended weak bonding in DNA: β-stacking (base-base), base-backbone, and backbone-backbone interactions, this volume.

[3] P. Bultinck, R. Carbó-Dorca, and R. Ponec. Generalized population analysis and molecular quantum similarity for molecular aromaticity, this volume.

[4] P. Becker, J. M. Gillet, and B. Courcot. Electron densities and reduced density matrix: a crucial information from steady states to reacting systems, this volume.

[5] A. D. Becke and E. R. Johnson. Dipole moment density of the exchange hole, this volume.

[6] P. Popelier, M. Rafat, M. Devereux, S. Liem, and M. Leslie. Towards a force field via quantum chemical topology, this volume.

[7] G. E. Scuseria and J. Heyd. New density functionals applied to old problems, this volume.

[8] W. Yang. Potential and orbital functionals: Analytic energy gradients for OEP and self-interaction-free exchange-correlation energy functional for thermochemistry and kinetics, this volume.

[9] A. M. Teale and D. J. Tozer. Exchange methods in Kohn-Sham theory, this volume.

[10] P. W. Ayers. Using the electron density as a weight function for multi-dimensional integration, this volume.

[11] F. Jensen. Polarization consistent basis sets, this volume.

[12] F. E. Zahariev and Y. A. Wang. Extension of the universal density functional to the domain of unnormalized densities, this volume.

[13] F. De Proft, P. Geerlings, and E. Chamorro. The nuclear Fukui function: Generalization within spin-polarized conceptual density functional theory, this volume.

[14] I. Porras. A simple semi-explicit density functional approach for electron systems, this volume.

[15] C. Valdemoro, L. M. Tel, and E. Pérez-Romero. A new G-matrix dependent energy expression for singlet states: a possible variational application, this volume.

[16] P. M. W. Gill. Two-electron reductions of many-electron wavefunctions, this volume.

[17] N. A. Besley. Intracules in phase space, this volume.

[18] E. V. Ludeña, X. Lopez, and J. M. Ugalde. Pair densities for the Hooke and Hooke-Calogero models of the non-Börn-Oppenheimer hydrogen molecule, this volume.

[19] J. C. Angulo. The reciprocal form factor of many-electron systems, this volume.

[20] A. C. Tanner. Compton scattering from 1897 to 1987: The accidental bibliography, this volume.

Ajit Thakkar

Ajit Thakkar, born in Poona, India in 1950, left home at 17 to explore the West. A circuitous route led him to Queen's University in Kingston, Ontario. A summer job programming calculations of transport cross sections using Fortran IV, dreadful JCL, and punched cards on an IBM 360/50 drew him to computational chemistry. In 1976, he completed a PhD in theoretical chemistry guided by Vedene Smith and influenced by Robert Parr. His faculty career began at the University of Waterloo and, since 1984, continued at the idyllic Fredericton campus of the University of New Brunswick. He has written more than 220 research articles on molecular properties, electron densities, and intermolecular forces.

Brill Academic Publishers
P.O. Box 9000, 2300 PA Leiden,
The Netherlands

*Lecture Series on Computer
and Computational Sciences*
Volume 4, 2005, pp. 1193-1196

Extension of the Universal Density Functional to the Domain of Unnormalized Densities

Federico E. Zahariev and Yan Alexander Wang[1]

Department of Chemistry
University of British Columbia
Vancouver, BC V6T 1Z1, Canada

Received 22 July, 2005; accepted in revised form 16 August, 2005

Abstract: Since the early 1980's, there have been numerous attempts to generalize the variational density domain of the Hohenberg-Kohn universal density functional to unnormalized densities. Recently, several papers by Lindgren and Salomonson [Phys. Rev. A **67**, 056501 (2003); **70**, 032509 (2004); and Adv. Quantum Chem. **43**, 95 (2003)] and by Gál [Phys. Rev. A **63**, 022506 (2001); **64**, 062503 (2001); J. Phys. A **35**, 5899 (2002)] appeared in the literature. We point out that all such efforts do not agree with some results of density functional theory.

Keywords: Universal density functional, Density functional theory, Density domain, Functional derivative

PACS: 31.15.Ew, 02.30.Sa, 02.30.Xx

1 Results of Lindgren and Salomonson

Many people have worked on the Hohenberg-Kohn (HK) universal density functional [1, 2, 3] in the domain of unnormalized densities [3, 4, 5, 6, 7, 8, 9, 10, 11, 12, 13, 14, 15, 16]. In particular, Lindgren and Salomonson recently published three papers on density-functional differentiability [4, 5, 6], whose results resemble those of Nguyen-Dang *et al.* [7], Bergmann and Hinze [8], Gál [9, 10, 11], and Parr and Liu [12].
Taking Eq. (10) in their first paper in *Phys. Rev. A* by Lindgren and Salomonson [4]

$$\left[\frac{\delta T[\rho]}{\rho(\mathbf{r})}\right]_{\rho=\rho_0} + v(\mathbf{r}) - \mu = 0 , \tag{1}$$

and Eq. (21) in the same paper [4]

$$\left[\frac{\delta T[\rho]}{\rho(\mathbf{r})}\right]_{\rho=\rho_0} = \frac{E_0}{N} - v(\mathbf{r}) , \tag{2}$$

we immediately have

$$\mu = \frac{E_0}{N} , \tag{3}$$

where $\rho_0(\mathbf{r})$ is the ground-state (GS) density, E_0 is the Kohn-Sham (KS) energy for N noninteracting electrons [2, 3], $v(\mathbf{r})$ is the total KS effective potential, and $T[\rho]$ is the KS kinetic energy or

---

[1]Corresponding author, email: yawang@chem.ubc.ca

$T_s[\rho]$ in the conventional notation. Eq. (3) foretold Eq. (34) of the latest paper in *Phys. Rev. A* by these two authors [6], which has been confuted recently [16].

Let us focus on Eq. (2), multiply both sides of this equation by $\rho_0(\mathbf{r})$, integrate over the entire space, and get the following result:

$$\left\langle \rho_0(\mathbf{r}) \left[\frac{\delta T[\rho]}{\rho(\mathbf{r})} \right]_{\rho=\rho_0} \right\rangle = \left\langle \left[\frac{E_0}{N} - v(\mathbf{r}) \right] \rho_0(\mathbf{r}) \right\rangle = E_0 - \langle v(\mathbf{r})\rho_0(\mathbf{r}) \rangle = T[\rho_0] \,, \qquad (4)$$

or simply put: the KS kinetic energy is homogeneous in density of order 1. Similar result [16] can be derived based upon Eqs. (59) and (69) of their second paper in *Adv. Quantum Chem.* [5]:

$$\left\langle \rho_0(\mathbf{r}) \left[\frac{\delta F[\rho]}{\rho(\mathbf{r})} \right]_{\rho=\rho_0} \right\rangle = F[\rho_0] \,, \qquad (5)$$

where $F[\rho]$ is the HK universal density functional [1, 2, 3].

In fact, Eq. (4) is the same result first advocated by Parr and Liu [12], and then refuted by Wang [15], Joubert [13], Chan and Handy [14], and Gál [9, 10, 11]. Interestingly, identical statements like Eqs. (3)-(5) have been proposed by Nguyen-Dang *et al.* [7], of Bergmann and Hinze [8], and of Parr and Liu [12] before, Lindgren and Salomonson just reached the same result in a different context.

In this talk, we are going to comment on such proposals by these authors [3, 4, 5, 6, 7, 8, 9, 10, 11, 12] and reveal some subtleties in their derivation.

2   Gál's Proposal

Following some earlier efforts [3], Gál worked out an explicit way to define a functional derivative in Hilbert space of N electrons [10], without an extension of the density domain of the universal functional followed by a subsequent inclusion of a Lagrangian multiplier term that enforces the normalization of the density of the stationary solution.

The idea is to separate the shape of the density from its normalization through the following generalization of the density domain [3, 17, 10]:

$$\rho^N(\mathbf{r}) = N \frac{g(\mathbf{r})}{\langle g(\mathbf{r}) \rangle} = N\sigma(\mathbf{r}) \,, \qquad (6)$$

where positive function $g(\mathbf{r})$ can be normalized to any positive real number and $\sigma(\mathbf{r})$ is the shape function with unit normalization $\langle \sigma(\mathbf{r}) \rangle = 1$. Consequently, any density functional $A[\rho^N(\mathbf{r})]$ can be defined in terms of N and $g(\mathbf{r})$:

$$A[\rho^N(\mathbf{r})] = A[N\sigma(\mathbf{r})] = A[g(\mathbf{r}), N] \,. \qquad (7)$$

It is straightforward to show that from the definition in Eq. (6) for a given $\rho^N(\mathbf{r})$, Gál's expression of the functional derivative with a fixed N is well defined without any arbitrary additive constant [10]:

$$\left. \frac{\delta A[\rho]}{\delta_N \rho(\mathbf{r})} \right|_{\rho=\rho^N} = \left. \frac{\delta A[\rho]}{\delta \rho(\mathbf{r})} \right|_{\rho=\rho^N} - \frac{1}{N} \left\langle \rho^N(\mathbf{r}) \left[\frac{\delta A[\rho]}{\delta \rho(\mathbf{r})} \right]_{\rho=\rho^N} \right\rangle \,. \qquad (8)$$

The symbol $\delta A[\rho]/\delta_N \rho(\mathbf{r})$ in Gál's notation means that the density variation is such that it always lies within the space of densities normalized to N (see Ref. [10] for more details).

Led by the early success [9, 10], Gál and coworker have applied this line of treatment and shown that a first-degree homogeneous KS-type kinetic-energy functional can be defined and should be utilized in place of the conventional KS kinetic-energy functional [11], echoing the similar results advocated by Nguyen-Dang *et al.* [7], Bergmann and Hinze [8], Parr and Liu [12], and Lindgren and Salomonson [4, 5, 6].

It seems that Gál's proposal should be the canonical version to be adopted by the entire DFT community, provided that the definition in Eq. (6) is unique and genuine. Unfortunately, such an assessment cannot be substantiated, not only because Gál's proposal is in direct conflict with the latest results by Zahariev and Wang [16], but also because the definition in Eq. (6) is, in fact, not unique and genuine. With a little bit further generalization of the definition in Eq. (6), we write the normalized density as

$$\rho^{N,m}(\mathbf{r}) = g(\mathbf{r}) \left(\frac{N}{\langle g(\mathbf{r}) \rangle} \right)^{1+(m-1)\delta_{N,\langle g(\mathbf{r}) \rangle}} = \begin{cases} g(\mathbf{r}) \left(\frac{N}{\langle g(\mathbf{r}) \rangle} \right)^m, & \text{if } \langle g(\mathbf{r}) \rangle = N \ ; \\ \rho^{N,1}(\mathbf{r}) \equiv \rho^N(\mathbf{r}), & \text{if } \langle g(\mathbf{r}) \rangle \neq N \ . \end{cases} \tag{9}$$

Here, $\delta_{N,\langle g(\mathbf{r}) \rangle}$ is the Kroneckner delta, and m can be any real number. It is obvious that Eq. (9) coincides with Eq. (6) numerically: $\rho^{N,m}(\mathbf{r}) = \rho^N(\mathbf{r})$, but $A[\rho^{N,m}(\mathbf{r})]$ and $A[\rho^N(\mathbf{r})]$ *do* have different Gâteaux functional derivatives at fixed N.

Because of the directional nature of the Gâteaux differential, the density variation, once chosen, will not change its shape during the entire variational path. Moreover, if the variation $\delta g(\mathbf{r})$ does not change the normalization of $g(\mathbf{r})$, say $\langle \delta g(\mathbf{r}) \rangle = 0$, the normalization of $g(\mathbf{r}, \varepsilon) = g(\mathbf{r}) + \varepsilon \delta g(\mathbf{r})$ will remain the same throughout the entire variational process. Thus, for the set $\{ g(\mathbf{r}, \varepsilon) | g(\mathbf{r}, \varepsilon) = g(\mathbf{r}) + \varepsilon \delta g(\mathbf{r}), \langle g(\mathbf{r}, \varepsilon) \rangle \neq N \}$, the Gâteaux functional derivative of a general functional $A[\rho^{N,m}(\mathbf{r})]$ will be identical to that of $A[\rho^N(\mathbf{r})]$. However, for the set $\{ g(\mathbf{r}, \varepsilon) | g(\mathbf{r}, \varepsilon) = g(\mathbf{r}) + \varepsilon \delta g(\mathbf{r}), \langle g(\mathbf{r}, \varepsilon) \rangle = N \}$, the Gâteaux functional derivative of a general functional $A[\rho^{N,m}(\mathbf{r})]$ can be shown to be

$$\left. \frac{\delta A[\rho]}{\delta_N \rho(\mathbf{r})} \right|_{\rho=\rho^N} = \left. \frac{\delta A[\rho]}{\delta \rho(\mathbf{r})} \right|_{\rho=\rho^N} - \frac{m}{N} \left\langle \rho^N(\mathbf{r}) \left[\frac{\delta A[\rho]}{\delta \rho(\mathbf{r})} \right]_{\rho=\rho^N} \right\rangle , \tag{10}$$

which is different from that of $A[\rho^N(\mathbf{r})]$ shown in Eq. (8). Just like that Eq. (9) is a generalization of Eq. (6), Eq. (8) becomes a special case of Eq. (10).

Moreover, because m is an arbitrary real number, Eq. (10) restores the ambiguity (of an arbitrary additive constant) associated with the Gâteaux functional derivative of a general functional in Hilbert space at a fixed N. It now becoms self-evident that the only physically meaningful way to resolve this ambiguity is through the extension based upon the statistical ensemble in Fock space and the key to unlock this ambiguity is by the careful inspection of the chemical potential at a fixed integral N [16].

3   Conclusions

In conclusion, among the numerous attempts to generalize the variational density domain of the HK universal density functional to unnormalized densities, the latest proposal by Lindgren and Salomonson [4, 5, 6] suffers from the same logical flaw of earlier works by Nguyen-Dang *et al.* [7], Bergmann and Hinze [8], and Parr and Liu [12]. Gál's proposal [9, 10, 11] is interesting in its own right, but the definition of the normalized density in terms of well-behaved positive functions [3, 17] is neither unique nor genuine, and hence should not be adopted in general.

Acknowledgment

The financial support from the Natural Sciences and Engineering Research Council (NSERC) of Canada is gratefully acknowledged.

References

[1] P. Hohenberg and W. Kohn, Phys. Rev. **136**, B864 (1964).

[2] W. Kohn and L. J. Sham, Phys. Rev. **140**, A1133 (1965); N. Hadjisavvas and A. Theophilou, Phys. Rev. A **30**, 2183 (1984).

[3] R. G. Parr and W. Yang, *Density-Functional Theory of Atoms and Molecules* (Oxford University, New York, 1989).

[4] I. Lindgren and S. Salomonson, Phys. Rev. A **67**, 056501 (2003).

[5] I. Lindgren and S. Salomonson, Adv. Quantum Chem. **43**, 95 (2003).

[6] I. Lindgren and S. Salomonson, Phys. Rev. A **70**, 032509 (2004).

[7] T. T. Nguyen-Dang, R. F. W. Bader, and H. Essen, Int. J. Quantum Chem. **22**, 1049 (1982).

[8] D. Bergmann and J. Hinze, in *Electronegativity, Structure and Bonding 66*, edited by K. D. Sen and C. K. Jorgensen (Springer-Verlag, Berlin, 1987), p. 145.

[9] T. Gál, Phys. Rev. A **62**, 044501 (2000).

[10] T. Gál, Phys. Rev. A **63**, 022506 (2001); Erratum, *ibid.* **63**, 049903 (2001); T. Gál, J. Phys. A **35**, 5899 (2002).

[11] T. Gál, Phys. Rev. A **64**, 062503 (2001); Erratum, *ibid.* **65**, 039906 (2002); T. Gál, N. H. March, and Á. Nagy, Phys. Lett. A **302**, 55 (2002); T. Gál and Á. Nagy, THEOCHEM **501-502**, 167 (2000).

[12] R. G. Parr and S. Liu, Chem. Phys. Lett. **276**, 164 (1997); S. Liu and R. G. Parr, *ibid.* **278**, 341 (1997); R. G. Parr and S. Liu, *ibid.* **280**, 159 (1997).

[13] D. P. Joubert, Chem. Phys. Lett. **288**, 338 (1998); D. Joubert, Phys. Rev. A **64**, 54501 (2001).

[14] G. K.-L. Chan and N. C. Handy, Phys. Rev. A **59**, 2670 (1999).

[15] Y. A. Wang, unpublished results (1997).

[16] F. E. Zahariev and Y. A. Wang, Phys. Rev. A **70**, 042503 (2004).

[17] E. H. Lieb, in *Density Functional Methods in Physics*, edited by R. M. Dreizler and J. da Providência (Plenum, New York, 1985), p. 31; E. H. Lieb, in *Physics as Nature Philosophy: Essays in Honor of Laszlo Tisza on His 75th Birthday*, edited by H. Feshbach and A. Shimony (MIT, Cambridge, Massachusetts, 1982), p. 111; E. H. Lieb, Int. J. Quantum Chem. **24**, 243 (1983).

Brill Academic Publishers
P.O. Box 9000, 2300 PA Leiden,
The Netherlands

*Lecture Series on Computer
and Computational Sciences*
Volume 4, 2005, pp. 1197-1200

The Reciprocal Form Factor of Many-Electron Systems

J.C. Angulo[1]

Departamento de Física Moderna and Instituto Carlos I de Física Teórica y Computacional,
Facultad de Ciencias,
Universidad de Granada,
E-18071 Granada, Spain

Received 13 July, 2005; accepted in revised form 12 August, 2005

Abstract:
Analytical expressions previously obtained for the reciprocal form factor $B(r)$ are numerically analyzed within a Hartree-Fock framework for ground state atomic systems throughout the Periodic Table and a study of monotonicity properties of $B(r)$ is carried out, also for singly-charged ions and Bare Coulomb Field systems.

Keywords: reciprocal form factor; momentum density; Compton profile; maximum entropy; minimum cross-entropy; bare Coulomb field.

PACS: 31.10.+z; 31.15.-p.

1 Introduction

The Fourier transform of the one-particle momentum density $\gamma(\mathbf{p})$

$$B(\mathbf{r}) \equiv \int e^{-i\mathbf{p}\cdot\mathbf{r}}\gamma(\mathbf{p})d\mathbf{p}$$

is used in the analysis of fundamental chemical concepts when studying many-electron systems (e.g. neutral atoms, singly-charged ions, molecules), such as hybridization and bonding, as well as in the interpretation of experimental Compton profiles [1, 2].

This function, called *reciprocal form factor* (RFF) or *internally folded density* constitutes a bridge between the complementary position and momentum spaces. In fact, there is an equivalence between the 'position space form factor' $B(\mathbf{r})$ and the momentum density on the one hand, and between the momentum space form factor $F(\mathbf{p})$ and the charge density on the other. Consequently, many properties and theoretical results concerning this quantity have been investigated [1].

For most purposes, it is sufficient to deal with the spherically averaged RFF

$$B(r) = \frac{1}{4\pi}\int B(\mathbf{r})d\Omega = 4\pi\int_0^\infty p^2\gamma(p)j_0(pr)dp$$

related to the spherically averaged momentum density $\gamma(p) = (1/4\pi)\int \gamma(\mathbf{p})d\Omega$ by means of a Fourier-Bessel transform.

Among the obtained results obtained in the study of $B(r)$, let us mention different kinds of tight model-independent estimations [7, 8] (e.g. maximum and minimum cross entropy approximations,

---

[1]E-mail: angulo@ugr.es

[4] A.J. Thakkar, A.L. Wonfor and W.A. Pedersen, *Journal of Chemical Physics* 87(1987)1212.

[5] E.S. Kryachko and T. Koga, *Journal of Mathematical Physics* 28(1987)8.

[6] T. Koga, M. Omura, H. Teruya and A.J. Thakkar, *Journal of Physics B* 28(1995)3113.

[7] J. Antolín, J.C. Cuchí, A. Zarzo and J.C. Angulo, *Journal of Physics B* 29(1996)5629.

[8] J. Antolín, J.C. Cuchí and J.C. Angulo, *International Journal of Quantum Chemistry* 87(2002)214.

[9] E. Romera and J.C. Angulo, *Journal of Chemical Physics* 120(2004)7369.

[10] J.C. Angulo and E. Romera, *International Journal of Quantum Chemistry* (in press).

Brill Academic Publishers
P.O. Box 9000, 2300 PA Leiden,
The Netherlands

*Lecture Series on Computer
and Computational Sciences*
Volume 4, 2005, pp. 1201-1204

Intracules in Phase Space

N. A. Besley[1]

School of Chemistry,
University of Nottingham,
University Park,
Nottingham, NG7 2RD, UK

Received 20 June, 2005; accepted in revised form 22 July, 2005

Abstract: The many dimensions of an electronic wave function make it difficult to analyze and interpret. It is often useful to extract low-dimensional functions that can provide chemical insight readily. The most familiar of these functions is the one electron density. However, often the interaction between electrons is important and so it is desirable to have reduced functions that retain explicit two electron information. Intracules are two-electron distribution functions that fulfill this requirement. In this contribution, we will describe our recent work on the computation and analysis of intracules derived from phase space quantum distributions.

Keywords: Intracules, phase space, Wigner, Husimi

PACS: 31.15Ar, 31.15Ew

1 Introduction

Much of modern electronic structure theory is based on the one-electron density $\rho(\mathbf{r})$. The relative simplicity of $\rho(\mathbf{r})$ compared to the many electron wavefunction $\Psi(\mathbf{r}_i)$, leads to computationally cost effective methods. Furthermore, chemically useful information can be more easily gleaned from $\rho(\mathbf{r})$. However, this simplicity comes at a price and when $\Psi(\mathbf{r}_i)$ is reduced to $\rho(\mathbf{r})$ much information is lost. In particular, it is often valuable to retain explicit two-electron information. Intracules are two-electron distribution functions and are an intermediate quantity between $\rho(\mathbf{r})$ and $\Psi(\mathbf{r}_i)$.

Intracules have been studied for many years. Traditionally, intracules have been computed in either position or momentum space. A position intracule, $P(u)$, is the spherically averaged intracule density in position space, and represents the probability that two electrons are separated by a distance $u = |\mathbf{r}_1 - \mathbf{r}_2|$.

$$P(u) = \int \rho(\mathbf{r}_1, \mathbf{r}_2)\delta(\mathbf{r}_{12} - \mathbf{u})\mathrm{d}\mathbf{r}_1\mathrm{d}\mathbf{r}_2\mathrm{d}\Omega_u \tag{1}$$

where $\rho(\mathbf{r}_1, \mathbf{r}_2)$ is the two-electron density and $\mathrm{d}\Omega_u$ represents integration over the angular parts of $\mathbf{u}$. Similarly, momentum intracules, $M(v)$, give the probability that two electrons have relative momentum $v = |\mathbf{p}_1 - \mathbf{p}_2|$.

$$M(v) = \int \pi(\mathbf{p}_1, \mathbf{p}_2)\delta(\mathbf{p}_{12} - \mathbf{v})\mathrm{d}\mathbf{p}_1\mathrm{d}\mathbf{p}_2\mathrm{d}\Omega_v \tag{2}$$

---

[1]Corresponding author. E-mail: nick.besley@nottingham.ac.uk

where $\pi(\mathbf{p}_1, \mathbf{p}_2)$ is the momentum two-electron density. The study of intracules has a rich history, with substantial contributions from Coulson, Smith, Thakkar, Boyd, Banyard, Cioslowski, Ugalde and Koga (for example, see references [1, 2, 3, 4] and references therein). Much of this work has focused on the computation of intracules for accurate correlated wavefunctions and studying phenomena such as the effects of correlation [5] and exchange [6] on the electron distribution.

$P(u)$ and $M(v)$ provide a representation of the electron distribution in either position *or* momentum space but neither alone provides a complete description. It is desirable to have a *combined* position and momentum description since this new function could be potentially important in the study of electron correlation. In this contribution, we summarise our efforts in the definition and computation of phase space intracules.

2 Wigner and Husimi Intracules

A well known consequence of Heisenberg's uncertainty principle is a joint position and momentum wavefunction cannot exist. However, phase space quantum distributions have been defined. The oldest of these distributions is the Wigner distribution [7]. The uncertainty principle manifests itself within the Wigner distribution through negative regions. Consequently, the Wigner distribution is often termed a "quasi-probability" distribution. A Wigner intracule, $W(u, v)$, is defined as [8]

$$
\begin{aligned}
W(u,v) \;=\; & \int W_2(\mathbf{r}_1, \mathbf{p}_1, \mathbf{r}_2, \mathbf{p}_2)\delta(\mathbf{r}_{12} - \mathbf{u})\delta(\mathbf{p}_{12} - \mathbf{v}) \\
& \mathrm{x} \quad \mathrm{d}\mathbf{r}_1 \mathrm{d}\mathbf{r}_2 \mathrm{d}\mathbf{p}_1 \mathrm{d}\mathbf{p}_2 \mathrm{d}\Omega_u \mathrm{d}\Omega_v
\end{aligned}
\tag{3}
$$

where W_2 is the second-order reduced Wigner function [9]

$$
\begin{aligned}
W_2(\mathbf{r}_1, \mathbf{p}_1, \mathbf{r}_2, \mathbf{p}_2) \;=\; & \frac{1}{\pi^6} \int \rho_2(\mathbf{r}_1 + \mathbf{q}_1, \mathbf{r}_1 - \mathbf{q}_1, \mathbf{r}_2 + \mathbf{q}_2, \mathbf{r}_2 - \mathbf{q}_2) \\
& \mathrm{x} \quad e^{-2i(\mathbf{p}_1 \cdot \mathbf{q}_1 + \mathbf{p}_2 \cdot \mathbf{q}_2)} \mathrm{d}\mathbf{q}_1 \mathrm{d}\mathbf{q}_2
\end{aligned}
\tag{4}
$$

and ρ_2 is the spinless reduced second order density matrix. $W(u, v)$ is a measure of the probability that electrons are separated by a distance u and have relative momentum v.

The Husimi function is an alternative phase space quantum distribution that contains the uncertainty principle within its definition [10]. Consequently, it can be interpreted more rigorously as a probability distribution. A Husimi intracule, $H(u, v)$, is defined as [11]

$$
\begin{aligned}
H(u,v) \;=\; & \int \eta_2(\mathbf{r}_1, \mathbf{p}_1, \mathbf{r}_2, \mathbf{p}_2)\delta(\mathbf{r}_{12} - \mathbf{u})\delta(\mathbf{p}_{12} - \mathbf{v}) \\
& \mathrm{x} \quad \mathrm{d}\mathbf{r}_1 \mathrm{d}\mathbf{r}_2 \mathrm{d}\mathbf{p}_1 \mathrm{d}\mathbf{p}_2 \mathrm{d}\Omega_u \mathrm{d}\Omega_v
\end{aligned}
\tag{5}
$$

where $\eta_2(\mathbf{r}_1, \mathbf{p}_1, \mathbf{r}_2, \mathbf{p}_2)$ is the second-order reduced Husimi function

$$
\begin{aligned}
\eta_2(\mathbf{r}_1, \mathbf{p}_1, \mathbf{r}_2, \mathbf{p}_2) \;=\; & \frac{1}{\pi^6} \int e^{-\kappa(\mathbf{r}_1' - \mathbf{r}_1)^2} e^{-\kappa(\mathbf{r}_2' - \mathbf{r}_2)^2} e^{-\frac{1}{\kappa}(\mathbf{p}_1' - \mathbf{p}_1)^2} e^{-\frac{1}{\kappa}(\mathbf{p}_2' - \mathbf{p}_2)^2} \\
& \mathrm{x} \quad W_2(\mathbf{r}_1', \mathbf{p}_1', \mathbf{r}_2', \mathbf{p}_2') \mathrm{d}\mathbf{r}_1' \mathrm{d}\mathbf{r}_2' \mathrm{d}\mathbf{p}_1' \mathrm{d}\mathbf{p}_2'
\end{aligned}
\tag{6}
$$

and κ is a parameter that controls the extent of localization in position or momentum space.

3   Computation of Intracules

The computation of $P(u)$ [12], $M(v)$ [13], $W(u,v)$ [14] and $H(u,v)$ [11] has been described in detail. If the molecular orbitals are expanded within a basis set

$$\psi_a(\mathbf{r}) = \sum_{\mu}^{N} c_{\mu a}\phi_\mu(\mathbf{r}) \tag{7}$$

the intracules can be expressed as

$$P(u) = \sum_{\mu\nu\lambda\sigma} \Gamma_{\mu\nu\lambda\sigma}(\mu\nu\lambda\sigma)_P \tag{8}$$

$$M(v) = \sum_{\mu\nu\lambda\sigma} \Gamma_{\mu\nu\lambda\sigma}(\mu\nu\lambda\sigma)_M \tag{9}$$

$$W(u,v) = \sum_{\mu\nu\lambda\sigma} \Gamma_{\mu\nu\lambda\sigma}(\mu\nu\lambda\sigma)_W \tag{10}$$

$$H(u,v) = \sum_{\mu\nu\lambda\sigma} \Gamma_{\mu\nu\lambda\sigma}(\mu\nu\lambda\sigma)_H \tag{11}$$

where $\Gamma_{\mu\nu\lambda\sigma}$ is the two-particle density matrix and $(\mu\nu\lambda\sigma)_P$, $(\mu\nu\lambda\sigma)_M$, $(\mu\nu\lambda\sigma)_W$ and $(\mu\nu\lambda\sigma)_H$ are the position, momentum, Wigner and Husimi integrals, respectively.

$$(\mu\nu\lambda\sigma)_P = \int \phi_\mu(\mathbf{r})\phi_\nu(\mathbf{r})\phi_\lambda(\mathbf{r}+\mathbf{u})\phi_\sigma(\mathbf{r}+\mathbf{u})d\mathbf{r}d\Omega_u \tag{12}$$

$$(\mu\nu\lambda\sigma)_M = \frac{v^2}{2\pi^2}\int \phi_\mu(\mathbf{r})\phi_\nu(\mathbf{r}+\mathbf{q})\phi_\lambda(\mathbf{q}+\mathbf{u})\phi_\sigma(\mathbf{u})$$
$$\text{x}\quad j_0(qv)d\mathbf{r}d\mathbf{q}d\mathbf{u} \tag{13}$$

$$(\mu\nu\lambda\sigma)_W = \frac{v^2}{2\pi^2}\int \phi_\mu(\mathbf{r})\phi_\nu(\mathbf{r}+\mathbf{q})\phi_\lambda(\mathbf{r}+\mathbf{q}+\mathbf{u})$$
$$\text{x}\quad \phi_\sigma(\mathbf{r}+\mathbf{u})j_0(qv)d\mathbf{r}d\mathbf{q}d\Omega_u \tag{14}$$

$$(\mu\nu\lambda\sigma)_H = \frac{\sqrt{2}}{\pi^5}(\pi\kappa)^{\frac{3}{2}}v^2\int e^{-2\kappa q^2}e^{-\frac{\kappa}{2}|\mathbf{r}_1-\mathbf{r}_2-\mathbf{u}|^2}j_0(2qv)$$
$$\text{x}\quad \phi_\mu(\mathbf{r}_1+\mathbf{q})\phi_\nu(\mathbf{r}_1-\mathbf{q})\phi_\lambda(\mathbf{r}_2-\mathbf{q})\phi_\sigma(\mathbf{r}_2+\mathbf{q})d\mathbf{r}_1 d\mathbf{r}_2 d\mathbf{q}d\Omega_u \tag{15}$$

The evaluation of these integrals is the difficult step when computing intracules. For gaussian basis functions, position and momentum integrals can be expressed in terms of relatively simple closed form expressions. For the Wigner and Husimi integrals no simple closed form expression could be found, consequently, the evaluation of these integrals is considerably more problematic. Two approaches were adopted, the first used quadrature, while the second expressed the integrals in terms of an infinite series [14].

Hartree-Fock Wigner and Husimi intracules have been reported for a number of atomic and molecular systems. The form of these intracules can be rationalised based on the electronic structure. In addition, differences between Wigner and Husimui representations can be understood. The small negative regions observed in Wigner intracules are not present in the corresponding Husimi intracules. The computation of intracules has been extended to excited states using the singles configuration interaction wavefunction [15]. These excited state intracules have provided fundamental new insight into the effects of exchange in the Be-like ions.

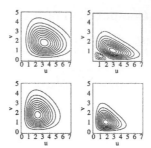

Figure 1: Wigner and Husimi intracules for the ground (bottom) and lowest triplet state (top) of H_2. Negative contours are represented by broken lines.

Acknowledgment

The author wishes to thank Prof. Peter M. W. Gill and Darragh P. O'Neill for their contributions to this work. Support from the Engineering and Physical Sciences Reseach Council (UK) is also acknowledged (GR/R77636).

References

[1] A.J. Thakkar, in: R.M. Erdahl and V.H. Smith Jr. (Eds.), *Density Matrices and Density Functionals*, Reidel, Dordrecht, 553 (1987).

[2] R.J. Boyd and J.M. Ugalde, in: S. Fraga (Eds.), *Computational Chemistry, Part A*, Elsevier, Amsterdam. 273 (1992).

[3] E. Valderamma, J.M. Ugalde and R.J. Boyd, in: J. Cioslowski (Eds.), *Many-Electron Densities and Reduced Density Matrices*, Kluwer Academic/Plenum, New York, 231 (2000).

[4] T. Koga, in: J. Cioslowski (Eds.), *Many-Electron Densities and Reduced Density Matrices*, Kluwer Academic/Plenum, New York, 267 (2000).

[5] C.A. Coulson, A.H. Neilson, *Proc. Phys. Soc. London* **78** 831 (1961).

[6] J. Katriel, *Phys. Rev. A* **5** 1990 (1972).

[7] E. Wigner, *Phys. Rev.* **40** 749 (1932).

[8] P.M.W. Gill, D.P. O'Neill, N.A. Besley, *Theor. Chem. Acc.* **109** 241 (2003).

[9] M. Hillery, R.F. O'Connell, M.O. Scully, E.P. Wigner, *Phys. Rep.* **106** 121 (1984).

[10] K. Husimi, *Proc. Phys. Math. Soc. Jpn* **22** 264 (1940).

[11] N.A. Besley, *Chem. Phys. Lett.* **409** 65 (2005).

[12] A.M. Lee, P.M.W. Gill, *Chem. Phys. Lett.* **313** 271 (1999).

[13] N.A. Besley, A.M. Lee, P.M.W. Gill, *Mol. Phys.* **100** 1763 (2002).

[14] N.A. Besley, D.P. O'Neill, P.M.W. Gill, *J. Chem. Phys.* **118** 2033 (2003).

[15] N.A. Besley, P.M.W. Gill, J. Chem. Phys. **120** 124 (2004).

Brill Academic Publishers
P.O. Box 9000, 2300 PA Leiden,
The Netherlands

*Lecture Series on Computer
and Computational Sciences*
Volume 4, 2005, pp. 1205-1208

Two-Electron Reductions of Many-Electron Wavefunctions

P.M.W. Gill[1]

Research School of Chemistry,
Australian National University,
Canberra ACT 0200, Australia

Received 20 June, 2005; accepted in revised form 14 July, 2005

Abstract: We discuss a number of two-electron distributions that can be extracted from a many-electron wavefunction, via the Wigner distribution. Two of these, the Position and Momentum intracules, have been known for many years and discussed by many authors. Two others, the Wigner and Action intracules, were introduced two years ago. Four new distributions, the Omega, Lambda, Dot and Angle intracules, are presented here for the first time. We argue that some of these intracules may be useful for understanding electron correlation.

Keywords: Two-electron distributions; intracules; Wigner distribution

PACS: 02.70.-c ; 31.25.-v

1   Eight Intracules

Although a simple Fourier transform allows us to interconvert the position-space wavefunction $\Psi(r_1,\ldots,r_n)$ and momentum-space wavefunction $\Phi(p_1,\ldots,p_n)$ of an n-electron system, quantum mechanics forbids the construction of a *joint* position-momentum wavefunction. Likewise, the position-space probability density function $|\Psi(r_1,\ldots,r_n)|^2$ and its momentum-space analogue $|\Phi(p_1,\ldots,p_n)|^2$ are unproblematic but there is no comparable *joint* probability density. Notwithstanding this, Wigner [1] managed to construct a function

$$W(r_1,\ldots,r_n,p_1,\ldots,p_n) = \pi^{-3n} \int \cdots \int \Psi(r_1 + q_1,\ldots,r_n + q_n)^*$$
$$\Psi(r_1 - q_1,\ldots,r_n - q_n)e^{2i(p_1 \cdot q_1 + \ldots + p_n \cdot q_n)} \, dq_1 \ldots dq_n \tag{1}$$

that possesses many of the properties that such a joint probability density should have.

Nonetheless, because the Wigner function depends on $6n$ coordinates, it is conceptually even more difficult than the wavefunction from which it was derived. Fortunately, however, it can be reduced [2] by integrating over the coordinates of all electrons but two, thereby yielding the second-order reduced Wigner distribution

$$W_2(r_1,p_1,r_2,p_2) = \int \cdots \int W(r_1,\ldots,r_n,p_1,\ldots,p_n) \, dr_3 \ldots dr_n \, dp_3 \ldots dp_n \tag{2}$$

This a much simpler object than the full Wigner distribution $W(r_1,\ldots,r_n,p_1,\ldots,p_n)$, but is nonetheless a function of 12 variables and remains conceptually formidable. Once again, we ask whether it is possible to effect another reduction without losing important information.

---

[1]E-mail: peter.gill@anu.edu.au

A key insight comes from the recognition that the physics of electron correlation depends less on the *absolute* positions and momenta of two electrons than on their relative position $r_{12} = r_1 - r_2$ and relative momentum $p_{12} = p_1 - p_2$. One also suspects that the absolute directions of the vectors r_{12} and p_{12} are less important than their magnitudes r_{12} and p_{12} but that the dynamical angle θ_{uv} between them may be significant. It is therefore plausible that most of the important information in $W_2(r_1, p_1, r_2, p_2)$ is captured by the three key variables, r_{12}, p_{12} and θ_{uv}, whose joint probability density is

$$
\begin{aligned}
\Omega(u, v, \omega) &= \int W_2(r_1, p_1, r_2, p_2)\delta(\theta_{uv} - \omega)\delta(r_{12} - u)\delta(p_{12} - v)\, dr_1\, dr_2\, dp_1\, dp_2 \\
&= \frac{1}{8\pi^3}\int \rho_2(r, r + q, r + u + q, r + u)e^{iq\cdot v}\delta(\theta_{uv} - \omega)\, dr\, dq\, d\Omega_u\, d\Omega_v \qquad (3)
\end{aligned}
$$

where Ω_u and Ω_v are the angular parts of u and v, respectively, and ρ_2 is the spinless reduced second-order density matrix [3]. We call this novel function the Omega intracule and it is easy to prove that $\Omega(u, v, \omega) = \Omega(u, v, \pi - \omega)$.

There are several ways in which the Omega intracule can be reduced further. The most obvious is simply to integrate over one of its three arguments. For example, integration over the dynamical angle yields a function

$$
W(u, v) = \int_0^\pi \Omega(u, v, \omega)\, d\omega = \frac{1}{8\pi^3}\int \rho_2(r, r + q, r + u + q, r + u)e^{iq\cdot v}\, dr\, dq\, d\Omega_u\, d\Omega_v \qquad (4)
$$

that we have previously called the Wigner intracule [4]. This can be interpreted as the joint probability density of r_{12} and p_{12}, without regard to ω. We have shown how $W(u, v)$ can be calculated from Hartree-Fock wavefunctions employing gaussian basis functions, for both ground and excited states of atoms and molecules.

The Omega intracule can also be reduced by combining two of its arguments. Because the product $s = uv$ appears to be important in the context of electron correlation [5], it is of interest to combine the u and v coordinates in this way to yield the function

$$
\Lambda(s, \omega) = \int_0^\infty \Omega(u, s/u, \omega)\, u^{-1}\, du \qquad (5)
$$

that gives the joint probability density of s and ω. This intracule has not previously been reported.

The two-dimensional $W(u, v)$ and $\Lambda(s, \omega)$ intracules are reduced forms of the $\Omega(u, v, \omega)$ intracule but, as Fig. 1 shows, each can itself be further reduced to yield various one-dimensional intracules. For example, integrating the Wigner intracule over v yields the function

$$
P(u) = \int_0^\infty W(u, v)\, dv = \int \rho_2(r, r, r + u, r + u)\, dr\, d\Omega_u \qquad (6)
$$

that gives the probability density of finding two electrons at a distance u. This was the original intracule and was introduced in the seminal paper by Coulson and Neilson [6]. Unlike most of the others discussed here, it is a rigorous probability density and has been widely studied.

If, instead, the Wigner intracule is integrated over u, one obtains the function

$$
M(v) = \int_0^\infty W(u, v)\, du \qquad (7)
$$

that gives the (rigorous) probability density of finding two electrons with relative momentum v. This intracule was introduced by Banyard and Reed [7] and has been studied subsequently in a number of groups.

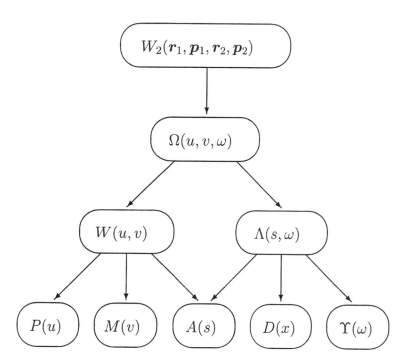

Figure 1: Hierarchical relationships between intracules

2   Theory

The present model, for an atom or ion of nuclear charge Z and number of electrons N, is based on the following semi-explicit density functional (atomic units used throughout):

$$E[\rho] = E_1[\{\phi_i(\rho)\}] + E_2[\rho] + U_{12}[\{\phi_i(\rho)\}, \rho], \tag{1}$$

where

$$E_1 = T_1[\{\phi_i(\rho)\}] + V_1[\{\phi_i(\rho)\}] + U_{11}[\{\phi_i(\rho)\}], \tag{2}$$

and

$$E_2 = T_2[\rho] + V_2[\rho] + U_{22}[\rho] + K_2[\rho], \tag{3}$$

obtained by the splitting of space in two regions:

$$R_1 = \{\vec{r}/|\vec{r}| < r_0\} \qquad \text{and} \qquad R_2 = \{\vec{r}/|\vec{r}| \geq r_0\}, \tag{4}$$

and the particular terms are:

$$T_1 = \sum_{i=1}^{N} \int_{R_1} d\vec{r} \, \phi_i^*(\vec{r}) \left(-\frac{1}{2}\nabla^2\right) \phi_i(\vec{r}), \tag{5}$$

$$V_1 = \sum_{i=1}^{N} \int_{R_1} d\vec{r} \, \phi_i^*(\vec{r}) v_{\text{ext}}(\vec{r}) \phi_i(\vec{r}), \tag{6}$$

$$U_{11} = \frac{1}{2} \sum_{i,j=1}^{N} \int_{R_1} d\vec{r} \int_{R_1} d\vec{r'} \, \phi_i^*(\vec{r})\phi_j^*(\vec{r'}) \frac{1}{|\vec{r}-\vec{r'}|} \phi_i(\vec{r})\phi_j(\vec{r'}), \tag{7}$$

$$U_{12} = \sum_{i=1}^{N} \int_{R_1} d\vec{r} \, \phi_i^*(\vec{r})\phi_i(\vec{r}) \int_{R_2} d\vec{r'} \, \frac{\rho(\vec{r'})}{|\vec{r}-\vec{r'}|}, \tag{8}$$

$$T_2 = \frac{3}{10} \left(3\pi^2\right)^{2/3} \int_{R_2} d\vec{r} \, [\rho(\vec{r})]^{5/3}, \tag{9}$$

$$V_2 = \int_{R_2} d\vec{r} \, v_{\text{ext}}(\vec{r})\rho(\vec{r}) \tag{10}$$

$$U_{22} = \frac{1}{2} \int_{R_2} d\vec{r} \int_{R_2} d\vec{r'} \, \frac{\rho(\vec{r})\rho(\vec{r'})}{|\vec{r}-\vec{r'}|}, \tag{11}$$

and

$$K_2 = -\frac{3}{4} \left(\frac{3}{\pi}\right)^{1/3} \int_{R_2} d\vec{r} \, [\rho(\vec{r})]^{4/3}. \tag{12}$$

This functional comes from the energy in the single-particle approximation, where the electron orbitals in region are approximated by local plane waves and thus the energy density is replaced by the well-known Thomas-Fermi-Dirac (TFD) form. Other approximation performed is that exchange between orbitals in region 1 is neglected.

The Euler-Lagrange equations obtained by the minimization of this functional with respect to the density (via the orbitals in region 1), with the restriction of proper normalization, are satisfied if:

1) The electron orbitals satisfy the single-particle equations, for $r < r_0$:

$$\left[-\frac{1}{2}\nabla^2 + V(\vec{r}) \right] \phi_i(\vec{r}) = \epsilon_i \phi_i(\vec{r}) \tag{13}$$

2) The density verifies the following relation, for $r \geq r_0$:

$$\frac{1}{2}\left(3\pi^2\right)^{2/3}[\rho(\vec{r})]^{2/3} - \left(\frac{3}{\pi}\right)^{1/3}[\rho(\vec{r})]^{1/3} + V(\vec{r}) + \lambda = 0 \tag{14}$$

where

$$V(\vec{r}) = v_{\text{ext}}(\vec{r}) + \int \frac{\rho(\vec{r'})}{|\vec{r} - \vec{r'}|}\, d\vec{r'} \tag{15}$$

Additional conditions are: continuity of the density and its derivative at r_0, and the cutoff of the density (usual condition of the TFD model), in a point r_l at which the pressure of the homogeneous electron gas is zero.

For an isolated atom or ion, where spherically symmetric solutions of the problem can be found, these two coupled equations are solved in the following way: assuming r_0 is small, we will solve analytically the single-particle equations via a expansion of the potential, in which the coefficients depend on global properties of the density.

In this work, the expansion is done analytically for small values of r (the atomic cusp-condition is used):

$$V(r) = -\frac{Z}{r} + \langle r^{-1}\rangle - \frac{2\pi}{3}\rho(0)r^2 - \frac{2\pi}{3}\rho(0)Zr^3 \tag{16}$$

The summation of the asymptotic solution of the single-particle equations leads to the expression up to third order in r:

$$\rho_1(r) = \sum_{i=1}^{N} \phi_i^*(r)\phi_i(r) = A\left[1 - 2Zr + (2C + Z^2)r^2 + \frac{1}{3}(Z^3 - 10ZC)r^3 + O(r^4)\right] \tag{17}$$

where two global parameters A and C have been defined. The particular values of both are fixed so that this density and its first derivative matches at $r = r_0$ to the solution of Eq.(14). This density, as in the standard TFD procedure, is written as:

$$\rho_2(r) = \frac{2^{3/2}}{3\pi}\left\{ a + \sqrt{\frac{Z}{r}}[\chi(r/b)]^{1/2} \right\}^3 \tag{18}$$

where $\chi(x)$ is found by solving numerically the differential equation:

$$\frac{d^2\chi}{dx^2} = x\left[\beta + \left(\frac{\chi}{x}\right)^{1/2}\right]^3 \tag{19}$$

where a, β and the scale factor b are the usual constants defined in the TFD model.

The two conditions for the resolution of this second-order differential equation between $x_0 = r_0/b$ and a point $x_l = r_l/b$ are the value of the potential at r_0 and the normalization of the total density. The value of x_l is usually chosen by the condition of null pressure of the electron cloud.

With these condition, there is only one parameter which remains free, i.e. the value of the matching radius r_0. In previous work it was fixed by the condition of continuity of the energy density. The values obtained were very close to $r_0 = 1/(2Z)$ when obtained consistently with the whole resolution procedure. However, this is a somewhat arbitrary condition because the energy density is not uniquely defined. In addition to this, fixing the value of r_0 from the start would drastically reduce the complexity of the resolution.

3   Test for a non-interacting electron system

For simplifying the approach, which is one of the main goals of this work, we have studied the role of r_0, as well as the limit of precision that can be expected from this method, by its application to a model system: a closed shell atom in a bare coulomb field.

This is a very interesting benchmark system, which has been used in the past for studying corrections due to strongly bound electrons, where the effects of the electron-electron repulsion can be expected not to play an important role.

The energy of a system of this type can be calculated exactly and expanded in terms of the number of electrons N in the form:

$$E = -\left(\frac{3}{2}\right)^{1/3} N^{1/3} Z^2 + \frac{1}{2} Z^2 - \frac{1}{18} \left(\frac{3}{2}\right)^{1/3} N^{-1/3} Z^2 + O(N^{-2/3}) \tag{20}$$

The application of our method can be performed analytically and leading to a density given by

$$\rho_2(r) = \frac{(2Z)^{3/2}}{3\pi^2 r^{3/2}} \left(1 - \frac{r}{r_l}\right)^{3/2} \tag{21}$$

for $r \geq r_0$, and by Eq.(17) for $r < r_0$. Using the conditions of continuity of ρ and ρ' at r_0 and the normalization to N, we finally find that the expansion of the energy of Eq.(20) can be exactly reproduced using the value:

$$r_0 = \frac{0.494375}{Z} \left(1 - 0.007637 N^{-2/3}\right) \tag{22}$$

very close to the values obtained by the continuity of the energy density.

4   Concluding remarks

The above described test suggests that the present method can be simplified further by holding for r_0 the optimal value found for the non-interacting electron system. The result of this is that a very simple subroutine can provide a zero-order estimation of the density and average properties of an atom or ion, and we plan to study also in the future the inclusion of external fields.

With respect to the accuracy about average properties, we can expect similar accuracy as for the energy can be found with the semiclassical formula with Scott and exchange corrections [5], that provide a shell-average estimation of the energy with an error below 0.5% for most atoms.

References

[1] R.G. Parr and W. Yang, *Density-Functional Theory of Atoms and Molecules*. Oxford University Press, N.Y., 1989.

[2] I. Porras and A. Moya, *Phys. Rev. A*, **59**, 1859-1864(1999).

[3] I. Porras and A. Moya, Perspectives in Relativistic Thomas-Fermi Calculations for Atomic Systems *Quantum Systems in Chemistry and Physics, vol 1, 195* (Editors: A. Hernández Laguna, J. Maruani, R. McWeeny and S. Wilson), Kluwer, Dordrecht, 2000.

[4] I. Porras and A. Moya, *Int. J. Quantum Chem.*, **99**, 288-296(2004).

[5] B.-G. Englert, *Semiclassical Theory of Atoms*, Springer-Verlag, Berlin, 1988.

Brill Academic Publishers
P.O. Box 9000, 2300 PA Leiden,
The Netherlands

*Lecture Series on Computer
and Computational Sciences*
Volume 4, 2005, pp. 1213-1216

New Density Functionals Applied to Old Problems

Gustavo E. Scuseria[1] and Jochen Heyd

Department of Chemistry, Rice University,
Houston, Texas 77005, USA

Received 30 June, 2005; accepted in revised form 8 August, 2005

Abstract: This presentation addresses our current efforts to develop more accurate exchange-correlation functionals for Density Functional Theory. The functionals discussed include the new Tao-Perdew-Staroverov-Scuseria (TPSS) meta-GGA and the Heyd-Scuseria-Ernzerhof screened Coulomb potential exchange hybrid denoted HSE, especially designed with solids in mind. The latter is not only much faster than regular hybrids but can also be used in metals and systems with negligible band gaps. Although HSE was initially developed to mimic thermochemical PBEh (hybrid) results on molecules, it has found an important niche in applications to solids. In particular, we discuss the significant improvement in band gap estimates that this functional yields compared to LSDA and GGAs.

Keywords: DFT, screened Coulomb potential, hybrid, HSE, TPSS

PACS: 31.15.Ew, 71.15.Mb, 71.20.-b

1 Introduction

For many years, the local spin-density approximation[1] (LSDA) has reigned supreme for applications in solid state physics. Structural properties such as lattice constants and bulk moduli are predicted with good accuracy. However, LSDA results for electronic properties like band gaps are of much lower quality[2]. Improvements in density functional theory (DFT) have first led to generalized-gradient approximations (like the Perdew-Burke-Ernzerhof[3] or PBE functional) and now meta-generalized-gradient approximations (such as the Tao-Perdew-Staroverov-Scuseria[4] or TPSS functional). Hybrid functionals, which include a certain amount of Hartree-Fock (HF) exchange, have been widely used for studying molecules. They yield significantly better agreement with experiment than pure DFT approaches. However, the application of hybrid functionals to solids is hampered by the long spatial range of the HF exchange interactions[5, 6]. This long range increases the computational effort of hybrid calculations in solids to a point where only few systems can be studied. The Heyd-Scuseria-Ernzerhof (HSE) hybrid functional[6] circumvents this problem by using a screened Coulomb potential to evaluate the HF exchange interactions.

The present study aims at assessing the performance of both the new TPSS meta-GGA functional and the HSE hybrid for the prediction of lattice constants and band gaps in solids. First, we will give a brief overview of the design of the HSE functional. Then, results for a test set of 40 solids with band gaps ranging from 0.2 eV to 7.2 eV will be shown. Details of both the test set and the computational methods can be found in Ref. [7].

---

[1] Corresponding author. E-mail: guscus@rice.edu

2   Screened Coulomb Potentials and the HSE Hybrid Functional

A screened Coulomb potential is based on a splitting of the Coulomb operator into short range (SR) and long range (LR) components. The HSE functional uses the error function to accomplish this split since it leads to computational advantages in evaluating the short range HF exchange integrals [6]. The following partitioning is used:

$$\frac{1}{r} = \underbrace{\frac{\mathrm{erfc}(\omega r)}{r}}_{\mathrm{SR}} + \underbrace{\frac{\mathrm{erf}(\omega r)}{r}}_{\mathrm{LR}} \tag{1}$$

where the complementary error function $\mathrm{erfc}(\omega r) = 1 - \mathrm{erf}(\omega r)$, and ω is an adjustable parameter. For $\omega = 0$, the long range term becomes zero and the short range term is equivalent to the full Coulomb operator. The opposite is the case for $\omega \to \infty$.

The HSE hybrid functional is based on the hybrid PBE (PBEh) exchange-correlation functional[8, 9]. However, the exchange energy term is split into SR and LR components and the HF long range is neglected but compensated by the PBE long range. This results in the following form for the exchange-correlation energy:

$$\begin{aligned} E_{\mathrm{xc}}^{\mathrm{HSE}} = {} & a E_{\mathrm{x}}^{\mathrm{HF,SR}}(\omega) + (1-a) E_{\mathrm{x}}^{\omega \mathrm{PBE,SR}}(\omega) \\ & + E_{\mathrm{x}}^{\omega \mathrm{PBE,LR}}(\omega) + E_{\mathrm{c}}^{\mathrm{PBE}} \end{aligned} \tag{2}$$

where $E_{\mathrm{x}}^{\mathrm{HF,SR}}$ is the SR HF exchange. $E_{\mathrm{x}}^{\omega \mathrm{PBE,SR}}$ and $E_{\mathrm{x}}^{\omega \mathrm{PBE,LR}}$ are the short and long range components of the PBE exchange functional. ω is the screening parameter and $a = 1/4$ is the HF mixing constant (determined analytically via perturbation theory[10, 11]). The effect of different ω values has been examined for a large number of enthalpies of formation[6, 12]. The ω-dependence of this property was only slight, with a range of possible values ($0.15\,a_0^{-1} < \omega < 0.30\,a_0^{-1}$). This fact, coupled with preliminary results for solids, have previously led us to choose $\omega = 0.15\,a_0^{-1}$. The value is system independent and is used in the present study.

Neglecting the long-range HF exchange leads to a drastic reduction in computational cost. As an example, Figure 1 shows a comparison of the CPU time required for a geometry optimization of bulk silicon. HSE, using only the SR component of HF exchange, requires only a small fraction of the CPU time needed by the established hybrid functionals PBEh and B3LYP[13]. In general, HSE calculations are only two to three times more time consuming than pure DFT calculations.

This significant speedup enabled us to test HSE on a wide range of solids. A previous assessment on a small set of 21 insulators, semiconductors, and metals yielded very encouraging results[14]. Both predicted lattice constants and bulk moduli were in good agreement with experiment. Especially noteworthy is the fact that HSE can be applied to three dimensional metallic systems for which traditional hybrid functions fail due to a derivative discontinuity[15].

3   Lattice Constants and Band Gaps in Solids

In this work, we focus on a set of 40 (mostly) semiconducting systems. Figure 2 shows the mean absolute errors (MAEs) for predicting lattice constants with both the established LSDA and PBE functionals, as well as the new TPSS and HSE functionals. In general, LSDA underestimates lattice constants in nearly all cases while the GGA functional PBE always overestimates the lattice constants. The meta-GGA TPSS improves upon the GGA results but still predicts lattice constants which are too long. The screened hybrid functional HSE reduces the overestimation of PBE (on which it is based) drastically, leading to the best predictions overall. The MAEs for the four functionals are 0.047 Å, 0.076 Å, 0.063 Å, and 0.037 Å for LSDA, PBE, TPSS, and HSE, respectively. All other error measures paint a similar picture.

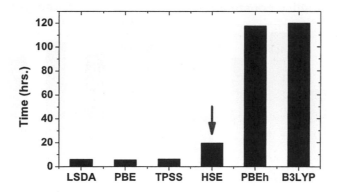

Figure 1: Timings for a complete geometry optimization of bulk silicon using a modified 6-311G* basis set. The calculations were performed on a four processor AMD Opteron 848 system.

Pure DFT functionals are know to always underestimate bandgaps[16]. All three pure functionals tested in this work predict bandgaps significantly below the experimental values. The MAEs for LSDA, PBE, and TPSS are 1.14 eV, 1.25 eV, and 1.12 eV, respectively. Bandgaps are always underestimated, in extreme cases by as much as 2.88 eV. In addition, several small-bandgap systems (Ge, GaSb, InN, InAs, and InSb) are predicted to be quasi-metallic. The HSE hybrid functional, on the other hand, yields a drastically reduced MAE of only 0.26 eV and predicts even small-bandgap systems correctly.

4   Conclusions

We assessed both the TPSS meta-GGA and the screened Coulomb hybrid functional HSE on a set of 40 solids with experimental bandgaps ranging from 0.2 to 7.2 eV. When compared to the GGA PBE, TPSS yields superior results. For lattice constants, however, it is still inferior to the LSDA. HSE offers both improved lattice constants compared to LSDA and significantly improved bandgaps. The computational effort involved in HSE calculations presents only a modest increase (a factor of two to three) over pure DFT calculations. Given the vast improvement in accuracy, such an extra effort is certainly justified.

Acknowledgment

This work was supported by the National Science Foundation (Grant No. CHE-0457030)

References

[1] W. Kohn and L. J. Sham, *Phys. Rev.* **140**, A1133 (1965).

[2] J. P. Perdew, *Int. J. Quantum Chem.* **30**, 451 (1986).

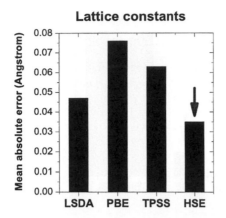
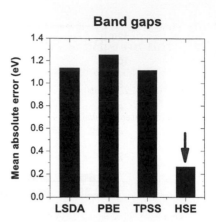

Figure 2: Mean absolute errors for lattice constants and bandgaps for a set of 40 bulk solids. Experimental bandgaps range from 0.2 eV to 7.2 eV.

[3] J. P. Perdew, K. Burke, and M. Ernzerhof, *Phys. Rev. Lett.* **77**, 3865 (1996).

[4] J. Tao, J. P. Perdew, V. N. Staroverov, and G. E. Scuseria, *Phys. Rev. Lett.* **91**, 146401 (2003).

[5] W. Kohn, *Int. J. Quantum Chem.* **56**, 229 (1995).

[6] J. Heyd, G. E. Scuseria, and Matthias Ernzerhof, *J. Chem. Phys.* **118**, 8207 (2003).

[7] J. Heyd, J. E. Peralta, G. E. Scuseria, and R. L. Martin, *in preparation* (2005).

[8] M. Ernzerhof and G. E. Scuseria, *J. Chem. Phys.* **110**, 5029 (1999).

[9] C. Adamo and V. Barone, *J. Chem. Phys.* **110**, 6158 (1999).

[10] D. C. Langreth and J. P. Perdew, *Solid State Commun.* **17**, 1425 (1975).

[11] J. P. Perdew, M. Ernzerhof, and K. Burke, *J. Chem. Phys.* **105**, 9982 (1996).

[12] J. Heyd and G. E. Scuseria, *J. Chem. Phys.* **120**, 7274 (2004).

[13] A. D. Becke, *J. Chem. Phys.* **98**, 1372 (1993).

[14] J. Heyd and G. E. Scuseria, *J. Chem. Phys.* **121**, 1187 (2004).

[15] N. W. Ashcroft and N. D. Mermin, *Solid State Physics* (Saunders College Publishing, Orlando, Florida, 1976), p. 335.

[16] J. P. Perdew and M. Levy, *Phys. Rev. Lett.* **51**, 1884 (1983).

Brill Academic Publishers
P.O. Box 9000, 2300 PA Leiden,
The Netherlands

*Lecture Series on Computer
and Computational Sciences*
Volume 4, 2005, pp. 1217-1221

Pair densities for the Hooke and Hooke–Calogero models of the non–Börn–Oppenheimer hydrogen molecule

Eduardo V. Ludeña*,†, **Xabier Lopez**†, **Jesus M. Ugalde**†

*Centro de Química, Instituto Venezolano de Investigaciones Científicas, IVIC, Apartado 21827,
Caracas 1020-A (Venezuela)

†Kimika Fakultatea; Euskal Herriko Unibertsitatea and Donostia International Physics Center
(DIPC); P. K. 1072; 20080 Donostia; Euskadi (Spain)

Received 23 June, 2005; accepted in revised form 22 July, 2005

Abstract: Pair densities for the Hooke and Hooke–Calogero models of the non–Börn–Oppenheimer hydrogen molecule are reported.

Keywords: Electron-pair densities, Hooke model, Hooke-Calogero model, hydrogen, non-Börn-Oppenheimer

PACS: 31.90.+s, 31.10.+z, 31.15.Ew

1   Introduction

Exact solutions of the Schrödinger equation for multiparticle systems with interparticle Coulombic interactions are unknown. This led to the famous remark by Dirac:

> "The fundamental laws necessary for the mathematical treatment of a *large part* of physics and the *whole* chemistry are thus completely known, and the difficulty lies only in the fact that application of these laws leads to equations that are too complex to be solved"

This fact has stimulated the ingenuity of many scientist since then, and as a result of the enormous effort made to model systems of interest as to make them tractable, a number of new concepts and tools have developed. Concepts like electronic configuration, valence orbitals, σ/π separation, electron correlation, Coulomb and Fermi holes, exchange–correlation functionals, etc., have been created in the coarse of electronic structure research effort and many of them have been pivotal to the development of the field.

The quest for approximate solutions to the Schrödinger equations has, nonetheless, followed two broad different strategies:

- Expressing the interparticle interaction potential as a Coulomb potential, the attention has been focused on devising approximate solutions to the resulting Schrödinger equation. Wave function theory and density functional theory fall into this category.

- Alternatively, one can model (make educated approximations) the interparticle interaction potential and then try to solve the resulting Schrödinger equation exactly.

Within the context of the latter approach, a much studied model is the Hookean two-electron atom[1, 2, 3, 4, 5, 6], a system possessing a nucleus with charge +2 interacting through a harmonic potential with the electrons which, in turn, repel each other through the usual Coulomb interaction. The replacement of the Coulombic central confining electron–nucleus potential occurring in real systems by a harmonic potential makes the problem separable in terms of center–of–mass and relative coordinates. The solution for the center of mass motion is a three-dimensional harmonic oscillator wavefunction. For the relative motion, as first determined by Emilio Santos[2], the ensuing equation has analytic solutions only for discrete values of the harmonic confinement strength parameter

Our approach for the non–Börn–Oppenheimer H_2 molecule falls also into the latter category. Thus, we have modeled the confining electron–nucleus Coulombic potential with a harmonic potential, while keeping the remaining interparticle interactions Coulombic.

2 The Hookean H_2

Let us change the coordinates of our four particle problem (two electrons at r_1 and at r_2, respectively, and two nuclei of mass M at R_A and at R_B, respectively) to a new set of coordinates (r, R, P, Q), given by:

$$r = r_1 - r_2; \quad R = R_A - R_B \tag{1}$$

$$\begin{bmatrix} P \\ Q \end{bmatrix} = R(\theta) \begin{bmatrix} (r_1 + r_2)/2 \\ \sqrt{M}(R_A + R_B)/2 \end{bmatrix} \tag{2}$$

being $R(\theta)$ a unitary rotation of angle $\theta = \tan^{-1}\sqrt{M}$. It can be shown[7] that in the new set of coordinates, the Schrödinger equation can be decoupled into the following integro–differential equations:

$$\left[-\frac{1}{2}\nabla_T^2 + \frac{k_T}{2T} + \frac{W_T^2}{2}T^2 \right] \Psi_T(T) = E_T' \Psi_T(T) \tag{3}$$

$$-\frac{1}{2}\nabla_P^2 \Psi_P(P) = E_P \Psi_P(P) \tag{4}$$

$$\left[-\frac{1}{2}\nabla_Q^2 + \frac{4(M+1)\omega^2}{M}Q^2 \right] \Psi_Q(Q) = E_Q \Psi_Q(Q) \tag{5}$$

where ω is the confinement strength parameter of the electron–nucleus potential, $(\omega^2/2)(r_i - R_\alpha)^2$, with $i=1, 2$ and $\alpha= A, B$. Notice that Eq. (3) gathers the equations corresponding to the relative motion of both nuclei and electrons, which can be recovered by setting,

$$\text{Nuclei} \quad \rightarrow \quad k_T = M, \qquad W_T^2 = M\frac{\omega^2}{2}, \qquad E_T' = \frac{M}{2}E_R \tag{6}$$

$$\text{Electrons} \quad \rightarrow \quad k_T = 1, \qquad W_T^2 = \frac{\omega^2}{2}, \qquad E_T' = \frac{1}{2}E_r \tag{7}$$

Analytical closed–form solutions for Eq. (3) exist only for a discrete set, $\{A\}$, of values of the confinement strength parameter W_T. Therefore in accordance with Eqs. (6) and (7) we obtain two separate sets of confinement parameters for which Eq. (3) has analytical closed–form solutions. Namely,

$$\{\omega^R\} = \{\omega \mid \omega = \sqrt{2/M}\,W_T \mid W_T \in \{A\}\} \tag{8}$$

$$\{\omega^r\} = \{\omega \mid \omega = \sqrt{2}\,W_T \mid W_T \in \{A\}\} \tag{9}$$

We have found, by explicitly solving Eq. (3), that these two sets are disjoint:

$$\{\omega^R\} \cap \{\omega^r\} = \emptyset \tag{10}$$

Consequently, given a confinement strength parameter ω, there not exist analytical closed–form solutions for both nuclei and electrons simultaneously. If we chose analytical closed–form solutions for either nuclei or electrons, then approximate solutions for the other set of particles must be sought.

However, at this point, we are again at a crossroad point similar to the one alluded to in the Introduction. Namely, we can chose an analytical closed–form solution for either nuclei or electrons and then search for approximate solutions for the other set of particles or model the internuclei interaction potential so that we have an exactly solvable model again.

We have followed both ways. Thus, we have found variational solutions of Eq. (3) for the nuclei (recall Eq. (6)) when $\omega \in \{\omega^r\}$. Also, we have modeled the internuclei Coulombic interaction potential with a squared potential, $\mid \mathbf{R}_A - \mathbf{R}_B \mid^{-2}$, following the earlier suggestion by Calogero[8]. In the latter case a complete analytical closed–shell solution has been developed[9].

3 The Pair Densities

Due to the exchange of coordinates $\Psi(\mathbf{R}_A, \mathbf{R}_B, \mathbf{r}_1, \mathbf{r}_2) = \Psi_r(\mathbf{r}) \ \Psi_R(\mathbf{R}) \ \Psi_Q(\mathbf{Q}) \ \Psi_P(\mathbf{P})$, the electronic ($e$) and nuclear ($n$) intracule and extracule densities are straightforwardly evaluated as:

$$I_e(\mathbf{r}) = |\Psi_r(\mathbf{r})|^2; \ I_n(\mathbf{R}) = |\Psi_R(R)|^2 \tag{11}$$

and the corresponding extracule densities by the following ansatz[10]:

$$E_e(\mathbf{s}) = \Psi_Q^2 \left[-\sqrt{\frac{M+1}{M}} \, \mathbf{s} \leftarrow \mathbf{Q} \right]; \ E_n(\mathbf{S}) = \Psi_Q^2 \left[\sqrt{M(M+1)} \, \mathbf{S} \leftarrow \mathbf{Q} \right] \tag{12}$$

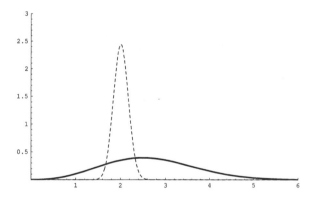

Figure 1: Intracular electronic (solid line) and nuclear (dashed line) probability distribution functions for a Hookean H_2 molecule with $\omega = \sqrt{2}/4$.

An interesting consequence of the coupling between the electronic and nuclear harmonic confinement strength parameters (see Eqs. (6) and (7)) is that to a weakly correlated electron motion there corresponds a very highly correlated nuclear motion. In fact, it is observed that while the electron coupling constant is in the high ω regime (weak correlation) the nuclear coupling constant is in the lower one (strong correlation). This allows us to conjecture that the onset of Wigner

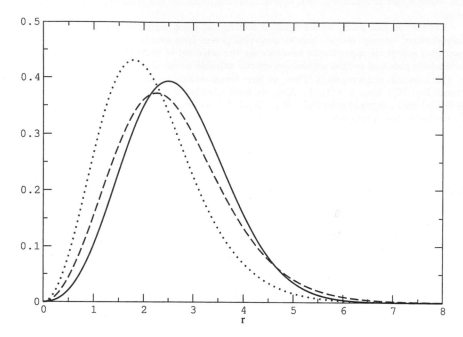

Figure 2: Intracular electronic probability distributions of the Hookean H_2 molecule of Figure 1 (solid line) and of the Coulombic H_2 molecule for internuclear distances R=1.4 a.u. (dotted line) and R=2.009 a.u (dashed line).

crystallization is a characteristic feature of nuclear motion. These aspects of nuclear and electron correlation can be made more clear by inspection the intracule densities.

Thus, Figure 1 shows the calculated intracular probability distribution functions for $\omega = \sqrt{2}/4 \in \{\omega^r\}$.

Observe that the nuclear intracule distribution function is sharply peaked at its maximum probability distance, $R = 2.009$ bohrs, which indicates that nuclei are quasi–clamped at this relative distance, while electrons are substantially more disperse. This is the essence of the Börn–Oppenheimer approximation, which comes up naturally within our model.

Additionally, when we compare, see Figure 2, the electron pair distribution for the Hookean model of H_2 with the corresponding full CI electronic intracule distribution for the Coulombic H_2 molecule [11] calculated at 2.00 bohrs, we see that they are remarkably similar.

This suggests that the inclusion of the proper Coulomb potential for the electron–electron interaction (a condition which is fulfilled in the present model) is the main relevant aspect affecting the dynamics of electronic motion. By extension, we may also conjecture that the shape of the nuclear pair distribution function closely approaches that of the real H_2 molecule, although due to the harmonic electron-nuclear interaction its peak is displaced to the larger equilibrium distance of 2.009 bohrs.

References

[1] Kestner, N. R.; Sinanöglu, O. *Phys. Rev.* **1962**, *128*, 2687.

[2] Santos, E. *Anal. R. Soc. Esp. Fis. y Quim.* **1968**, *64*, 177.

[3] White, R. J.; Byers Brown, W. *J. Chem. Phys.* **1970**, *53*, 3869.

[4] Taut, M. *Phys. Rev.* A **1993**, *48*, 3561.

[5] Cioslowski, J; Pernal, K. *J. Chem. Phys.* **2000**, *113*, 8434.

[6] O'Neil, D. P.; Gill, P. M. W. *Phys. Rev. A.*, **2003**, *68*, 022505.

[7] Ludeña, E. V.; Lopez, X.; Ugalde, J. M. *J. Chem. Phys.*, **2005**, *123*, xxx.

[8] Calogero, F. *J. Math. Phys.* **1969**, *10*, 2197.

[9] Ludeña, E. V.; Lopez, X.; Ugalde, J. M. *J. Phys. B.*, **2005**. *Submitted.*

[10] Lopez, X.; Ugalde, J. M.; Ludeña, E. V. *Chem. Phys. Lett.*, **2005**. *Submitted.*

[11] Mercero, J. M.; Valderrama, E. V.; Ugalde, J. M. In *Metal-Ligand Interactions*; Russo, N; Salahub, D. R.; Witko, M. (Eds.): Kluwer Academic Publishers, Dordrecht, 2003; 205–239.

Brill Academic Publishers
P.O. Box 9000, 2300 PA Leiden,
The Netherlands

*Lecture Series on Computer
and Computational Sciences*
Volume 4, 2005, pp. 1222-1225

A New G-matrix Dependent Energy Expression for Singlet States: a Possible Variational Application

C. Valdemoro, L. M. Tel*, E. Pérez-Romero*

Instituto de Matemáticas y Física Fundamental,
Consejo Superior de Investigaciones Científicas,
Serrano 123, 28006 Madrid, Madrid, Spain
* Departamento de Química Física, Universidad de Salamanca,
37008 Salamanca, Spain

Received 20 June, 2005; accepted in revised form 23 July, 2005

Keywords: Energy, Singlet, G-matrix, Correlation matrix, S-representability

PACS: 02.10-Yn, 02.10.-v, 31.10z, 31.25.-v

Abstract

It is well known that the knowledge of the second-order Reduced Density Matrix (2-RDM) suffices to calculate the electronic energy of an N-electron system. But the energy can also be expressed as a functional of the G-matrix. In their remarkable paper of 1969 [1] Mihailović and Rosina studied the excitations as ground state variational parameters by focusing on the G-matrix [2]. In this paper we also look at the electronic energy as a functional of the G-matrix, or equivalently of the Correlation Matrix, but our approach is different. Thus, in view of the recent results of Alcoba and Valdemoro [5], we focus our attention on the spin properties of the different spin-blocks of this matrix in singlet states as well as in the relations linking the first-order Reduced Density Matrix both to the 2-RDM and to the G-matrix in these spin-states. By taking into account explicitly these properties a new energy functional is obtained for singlet states. Apart from the physical insight provided by the structure of this functional, the number of the unknowns and, hence, the number of constraints to be introduced in a variational approach, in order to determine an appropriate set of independent variables, is drastically reduced. We will show here how, for singlet states, the spin-properties of the Correlation and G-matrices lead to an expression which depends linearly of a single spin-block of the G-matrix. This unique dependence, joint to the very restrictive N- and S-representability necessary conditions that this G-matrix spin-block must satisfy, suggests the possibility of using its elements, or functions of them, as variational variables without having to consider constraints involving other matrices explicitly.
The main idea is, therefore, to start from the usual expression:

$$E = \sum_{i,j,m,l}^{K} \sum_{\sigma,\sigma'} {}^{0}H_{ij;ml} \, D_{i_\sigma j_{\sigma'};m_\sigma l_{\sigma'}} \tag{1}$$

and to evaluate an expression of the electronic energy, **valid only for singlet states**, through an explicit use of all the spin-properties of the G-matrix.

In eq. (1), $^0\mathbf{H}$ is a reduced hamiltonian matrix formed by one- and two-electron integrals

$$^0H_{ij;ml} = \frac{1}{N-1}\left(h_{i;m}\,\delta_{jl} + h_{j;l}\,\delta_{im}\right) + g_{ij;ml}$$

and the 2-RDM elements are defined as

$$2!\,D_{i_\sigma j_{\sigma'};m_\sigma l_{\sigma'}} = <\Phi|\,b_{i_\sigma}^\dagger\,b_{j_{\sigma'}}^\dagger\,b_{l_{\sigma'}}\,b_{k_\sigma}\,|\Phi> \tag{2}$$

In this expression, the indices i, j, m, l represent the elements of a finite basis set of K orthonormal orbitals. The symbols σ, σ' denote the spin-functions (α or β) and Φ is the N-electron function being considered. In what follows, the 1-RDM, contraction of the 2-RDM, will be denoted by the letter d.

It is well known that for singlet states the following relations are satisfied:

$$D_{ij;ml} = D_{\bar{i}\bar{j};\bar{m}\bar{l}} \qquad\qquad D_{i\bar{j};m\bar{l}} = D_{\bar{i}j;\bar{m}l}$$

$$d_{i;m} = d_{\bar{i};\bar{m}}$$

where overbarred indices refer to β spin. The other spin-blocks vanish. Therefore, the energy can be written as

$$E = \sum_{i,j,m,l} {}^0H_{ij;ml}\,\{D_{ij;ml} + D_{i\bar{j};m\bar{l}}\} \tag{3}$$

It has been shown [6]-[9] that $\mathbf{D}$ can be decomposed as:

$$2!\,D_{ij;ml} = -\,d_{i;l}\,d_{j;m} + \delta_{jl}\,d_{i;m} - \mathcal{C}_{ij;lm} \tag{4a}$$

$$2!\,D_{i\bar{j};m\bar{l}} = \delta_{jl}\,d_{i;m} - \mathcal{C}_{i\bar{j};\bar{l}m} \tag{4b}$$

where $\mathcal{C}$ is the correlation matrix defined in [9]:

$$\mathcal{C}_{ij;lm} = \sum_{\Phi'\neq\Phi} \langle\Phi|\,b_i^\dagger\,b_l\,|\Phi'\rangle\,\langle\Phi'|\,b_j^\dagger\,b_m\,|\Phi\rangle \equiv G_{il;mj}$$

where G is the positive semidefinite matrix defined by Garrod and Percus in 1964 [2] and further studied, among others, by Garrod and Rosina [3] and Mihailović and Rosina [4].

From the results reported in [5], it can be shown that for SINGLET states

$$\mathcal{C}_{ij;lm} = d_{il}\,(\delta_{jm} - d_{\bar{j};\bar{m}}) + \mathcal{C}_{i\bar{j};\bar{l}m} - \mathcal{C}_{i\bar{j};\bar{m}l} \tag{5}$$

Recalling that $d_{\bar{j};\bar{m}} = d_{j;m}$ and taking into account the previous relation one may write:

$$2!\,D_{ij;ml} = -\,\delta_{jm}\,d_{i;l} + \delta_{jl}\,d_{i;m} - \mathcal{C}_{i\bar{j};\bar{l}m} + \mathcal{C}_{i\bar{j};\bar{m}l} \tag{6}$$

That is,

$$2!\,D_{ij;ml} = -\,\delta_{jm}\,d_{i;l} + \delta_{jl}\,d_{i;m} - G_{i\bar{l};m\bar{j}} + G_{i\bar{m};l\bar{j}} \tag{7}$$

When replacing this relation and eq. (4b) into eq. (3), one obtains:

$$\boxed{\begin{aligned} E = & \sum_{i,l}^{K}\{2\,h_{i;l}^c - h_{i,l}^x\}\,d_{i;l} \\ & - \sum_{i,j,l,m}^{K} {}^0H_{ij;ml}\left(2\,G_{i\bar{l};m\bar{j}} - G_{i\bar{m};l\bar{j}}\right) \end{aligned}} \tag{8}$$

where

$$h_{i;l}^c = \sum_j {}^0\mathrm{H}_{ij;lj} \qquad\qquad h_{i;l}^x = \sum_j {}^0\mathrm{H}_{ij;jl}$$

Relation (8) shows that for singlet states E depends **linearly** on $\mathbf{d}$ (α or β spin-block) and on the $(\alpha\beta; \alpha\beta)$ spin-block of the $\mathbf{G}$-matrix. (Note that all the previously known effective two-electron terms involved products of two 1-RDM elements.)

In order to obtain this energy relation, several important necessary N- and S-representability conditions have been used. Thus, it may not be any need to impose them as auxiliary conditions in a variational procedure. Indeed, the number of variables has been cut by 3 with respect to the initial expression of E.

Denoting $\mathbf{h} = 2\mathbf{h}^c - \mathbf{h}^x$ and ${}^0\widetilde{\mathrm{H}}_{lj;im} = 2\,{}^0\mathrm{H}_{ij;lm} - {}^0\mathrm{H}_{ij;ml}$ one may write

$$E = tr(\mathbf{h}\,\mathbf{d}^{(\alpha;\alpha)}) + tr\left({}^0\widetilde{\mathbf{H}}\,\mathbf{G}^{(\alpha\beta;\alpha\beta)}\right) \tag{9}$$

Clearly, the first term in this energy expression represents an **effective one-body energy** while the second term represents a **Correlation energy** (or effective two-body energy).

In order to evaluate the 1-RDM relation for singlet states, let us recall that

$$\sum_i G_{i\bar{i};m\bar{l}} = 0 = \sum_i G_{m\bar{l};i\bar{i}} \tag{10}$$

because it implies the action of the spin-operators S_+ or S_- upon a singlet state. Then, performing the two contractions of the 2-RDM, according to eq. (7), one obtains

$$d_{i;l} = \frac{1}{K}\left(\frac{N}{2}\,\delta_{i,l} + \sum_t \left(G_{i\bar{t};l\bar{t}} - G_{\bar{t}i;\bar{t}l}\right)\right) \tag{11}$$

This expression shows the dependence of a singlet 1-RDM on the elements of the spin-block $G^{(\alpha\beta;\alpha\beta)}$. Replacing this relation into relation (9) one obtains

$$\boxed{\begin{aligned} E = {} & \frac{N\,tr\,\mathbf{h}}{2\,K} + \frac{1}{K}\sum_{i,l,t}\left(G_{i\bar{t};l\bar{t}} - G_{\bar{t}i;\bar{t}l}\right)h_{il} \\ & - tr\left({}^0\widetilde{\mathbf{H}}\,\mathbf{G}^{(\alpha\beta;\alpha\beta)}\right) \end{aligned}} \tag{12}$$

As can be seen, the energy is given by a constant term and by G-dependent terms. Obviously, all the latter expressions can be rewritten in terms of the $\mathcal{C}$ matrix but for variational purposes it is more convenient to use explicitly the G-matrix which is a symmetric positive semidefinite matrix. A further discussion of the alternative approaches to the variational treatment of the energy according to eq. (12) will be given elsewhere [10]

Aknowledgements

The authors are thankful to D. A. Alcoba for many interesting discussions. The authors also aknowledge the financial support granted to this work by the Ministerio de Ciencia y Tecnologia under project BFM2003-05133.

References

[1] M. V. Mihailović and M. Rosina, *Nucl. Phys.*, **A130**, 386 (1969).

[2] C. Garrod and J. K. Percus, *J. Math. Phys.*, **5**, 1756-1776 (1964).

[3] C. Garrod and M. Rosina, *J. Math. Phys.*, **10**, 1855 (1969),

[4] C. M. V. Mihailović and M. Rosina, *Nucl. Phys. A*, **237**, 221 (1975).

[5] D. R. Alcoba. C. Valdemoro, *Int. J. Quantum Chem,*, **102**, 629-644 (2005).

[6] C. Valdemoro, L. M. Tel, D. R. Alcoba, E. Pérez-Romero and F. J. Casquero, *Int. J. Quantum Chem.*, **90**, 1555-1561 (2002).

[7] C. Valdemoro, D. R. Alcoba, L. M. Tel and E. Pérez-Romero, *Int. J. Quantum Chem.*, **85**, 214-224 (2001).

[8] C. Valdemoro, D. R. Alcoba and L. M. Tel, *Int. J. Quantum Chem.*, **93**, 212-222 (2003).

[9] C. Valdemoro, M. P. de Lara-Castells, E. Pérez-Romero, L. M. Tel, *Advances in Quantum Chemistry*, **31**, 37 (1999).

[10] To be published in a volume in memory of the late B. G. Wybourne.

where $r(\theta_i)$ is constructed from the inverse of the transformation in Eq. (5). We refer to the integration formula defined by Eq. (7) as a pseudo-Gaussian quadrature formula. It will be accurate whenever $f(r(\theta))$, viewed as a function of $\theta \in [0,1]^d$, is readily approximated by a low-degree polynomial. (Alternatively, one can use trigonometric functions on the unit hypercube, in which case we need for $f(r(\theta))$ to have a low-order Fourier series expansion). The situation for derivatives is similar. Using the identities

$$\nabla_r f(r) = \nabla_\theta f(r(\theta)) \cdot \nabla_r \theta$$

$$\nabla_r^2 f(r) = [\nabla_r \theta]^T \nabla_\theta \nabla_\theta f(r(\theta)) \cdot \nabla_r \theta + \nabla_\theta f(r(\theta)) \cdot \nabla_r^2 \theta \tag{8}$$

we can use differentiation formulae on the unit hypercube to differentiate functions in real space.

It remains to define appropriate weight functions and test the efficiency of the pseudo-Gaussian formulae. As a simple example, we can consider the interaction between closed-shell atoms using the Gordon-Kim approximations.[9] In that case, an obvious weighting function is the sum of the isolated atomic densities,

$$w(r) = \tilde{\rho}(r) = \rho_A(r) + \rho_B(r) . \tag{9}$$

For evaluating integrals associated with independent-electron models (cf. Eq. (2)), a suitable weight function is a product of approximate atomic densities,

$$w(r_1, r_2) = \tilde{\rho}(r_1)\tilde{\rho}(r_2) \approx |\gamma(r_1, r_2)|^2 . \tag{10}$$

Similarly, for expectation values of the wave function, one can consider

$$w(r_1, r_2, \ldots r_N) = \tilde{\rho}(r_1)\tilde{\rho}(r_2)\cdots\tilde{\rho}(r_N) \approx N^N |\Psi(r_1, \ldots r_N)|^2 . \tag{11}$$

This contribution will present preliminary results from our ongoing research along these lines.

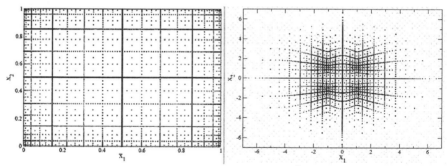

Figure 1: The abscissas of quadrature points on the unit square, $[0,1]^2$, and in real space, $\mathbb{R}^2$, where the weight function for the transformation is the density of four "hydrogen" atoms, located at the points $(\pm 1, \pm 1)$. Note how the points in the grid "concentrate" in the regions where the density is highest.

Acknowledgments

The author wishes to thank NSERC and the Canada Research Chairs for funding.

References

[1] W. Gautschi, Orthogonal polynomials: applications and computation. *Acta Numerica* 5 45-119(1996).

[2] W. Gautschi, Algorithm-726 - Orthpol - a package of routines for generating orthogonal polynomials and Gauss-type quadrature-rules. *ACM Trans. Math. Software*, **20** 21-62(1994).

[3] Y. Xu, *Common zeros of polynomials in several variables and higher dimensional quadrature.* Wiley, New York, 1994

[4] G. Monegato, Stieltjes Polynomials and Related Quadrature-Rules. *Siam Review*, **24** 137-158(1982).

[5] S. A. Smolyak, Quadrature and interpolation formulas for tensor products of certain classes of functions. *Dokl. Acad. Nauk SSSR*, **4** 240-243(1963).

[6] H. J. Bungartz and M. Griebel, A note on the complexity of solving Poisson's equation for spaces of bounded mixed derivatives. *Journal of Complexity*, **15** 167-199(1999).

[7] M. Griebel and S. Knapek, Optimized tenser-product approximation spaces. *Constructive Approximation*, **16** 525-540(2000).

[8] G. W. Wasilkowski and H. Wozniakowski, Explicit cost bounds of algorithms for multivariate tensor product problems. *Journal of Complexity*, **11** 1-56(1995).

[9] R. G. Gordon and Y. S. Kim, Theory for the forces between closed-shell atoms and molecules. *Journal of Chemical Physics*, **56** 3122-3133(1972)

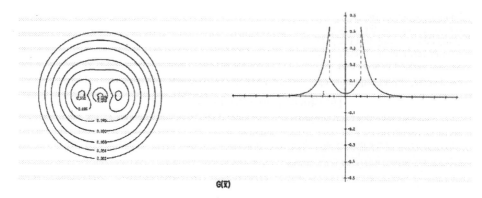

G(x̄)

Figure 1. Contour maps and profiles along the internuclear axis of the electron density $\rho(r)$ and the kinetic energy density $G(r)$ in atomic units (au) for H_2 at a separation of 1.4 au. The reader is asked to compare in particular the two profiles indicating that the accumulation of electron density in the internuclear region is accompanied by a great reduction in $G(r)$. The centre point in $\rho(r)$ is a (3,-1) critical point, the two curvatures of $\rho(r)$ perpendicular to the axis being negative, that along the axis being positive. The same point in $G(r)$ is a (3,+3) critical point, a local minimum and $G(r)$ increases in all direction away from the critical point. $G(r_m) = 0$ at the Hartree-Fock level, but has the value 0.0165 au for the correlated function. The reduction in $G(r)$ in the internuclear region is most pronounced for a hydrogen atom whose distributions are determined primarily by an 's-like' basis function. The discontinuous drop in $G(r)$ at the positions of the nuclei is a consequence of the nuclear-electron coalescence cusp condition on $(\rho(0)$.

References

[1] J. Lennard-Jones and J. A. Pople, Proc. Roy Soc. London. Series A **210**, 190 (1951). "There is only one source of attraction between two atoms, and that is the force between electrons and nuclei. But there are three counteracting influences: the nuclei repel each other, the electrons repel each other, and the kinetic energy of the electrons increases when a chemical bond is formed."

[2] J. C. Slater, *Quantum Theory of Molecules and Solids. I* (McGraw-Hill Book Co. Inc., New York, 1963).

[3] R. F. W. Bader, *Atoms in Molecules: a Quantum Theory* (Oxford University Press, Oxford UK, 1990).

[4] R. F. W. Bader and F. De-Cai, J. Chem. Theory and Comp. **1**, 403 (2005).

[5] K. Ruedenberg , Rev. Mod. Phys. **34**, 326 (1962). Ruedenberg specifies that bonding be referenced to a 'contractive promotion' state (rather than the proper states of the separated atoms) in which the density of each atom is contracted towards its own nucleus, an idea based on the increase in the orbital exponent found for the H atom in forming H_2. This choice has the obvious result of decreasing V and increasing T beyond the values demanded by the virial theorem, "the virial theorem is far from valid in the promotion state." The subsequent sharing of electrons between the two atoms to give the 'interference effect' and the final correct density, then results in an *increase* in V and a *decrease* in T, their values reverting to those required by the virial theorem from the physically unattainable extremes assigned to them by the contractive promotion. Bader and Preston[5] , as described in the text, studied the local behaviour of the kinetic energy density in conjunction with the virial theorem thereby obviating the need to invent a reference state with properties designed to give a desired result. The model of contractive promotion can be applied only to H and He. Atoms past He undergo an 'expansive promotion' on interacting with another atom, one that decreases the electron density at and in the immediate vicinity of the nuclei, as demonstrated using the near Hartree-Fock functions with optimized STO orbital exponents from LMSS, Chicago.

[6] R. F. W. Bader and H. J. T. Preston, Int. J. Quantum Chem. **3**, 327 (1969).

[7] G. Das and A. C. Wahl, J. Chem. Phys. **44**, 87 (1966).

[8] R. F. W. Bader, J. Chem. Phys. **73**, 2871 (1980).

[9] R. F. W. Bader and H. Essén, J. Chem. Phys. **80**, 1943 (1984).

[10] Y. Tal and R. F. W. Bader, Int J. Quantum Chem. Symp. **12**, 153 (1978).

[11] R. F. W. Bader and P. M. Beddall, J. Chem. Phys. **56**, 3320 (1972).

[12] W. A. Bingel , Naturforsch A **18**, 1249 (1963).

[13] P. E. Cade and W. M. Huo, At. Data Nucl. Data Tables **12** , 415 (1973).

[14] P. E. Cade and W. M. Huo, At. Data Nucl. Data Tables **15**, 1 (1975).

[15] P. E. Cade and A. C. Wahl, At. Data Nucl. Data Tables **13** , 339 (1974).

[16] R. F. W. Bader, P. J. MacDougall, and C. D. H. Lau, J. Am. Chem. Soc. **106**, 1594 (1984).

Brill Academic Publishers
P.O. Box 9000, 2300 PA Leiden
The Netherlands

*Lecture Series on Computer
and Computational Sciences*
Volume 4, 2005, pp. 1234-1234

Dipole Moment Density of the Exchange Hole

A.D. Becke and E.R. Johnson

Department of Chemistry
Queen's University
Kingston, ON K7L 3N6
Canada

Received 27 June, 2005; accepted in revised form 28 June, 2005

PACS: 34.20.-b, 31.15.Ew

We have recently proposed [1] a model of the dispersion interaction between nonoverlapping systems in which the reference-point dependent dipole moment of the *exchange hole* is used to generate dipole-induced-dipole attractions. The model requires no time dependence, virtual orbitals, or electronic correlation in the usual sense. Interatomic and intermolecular C6 dispersion coefficients of remarkable accuracy can be obtained by performing Hartree-Fock calculations on the monomers followed by some simple integrations involving the occupied orbitals.

Furthermore, our C6 expression can be analytically decomposed into interatomic pair terms. We introduce a novel energy-based damping of each pair term at small interatomic separation, and thereby obtain a post-Hartree-Fock model of intermolecular interactions [2]. The model has been tested on intermolecular energies and separations of a variety of complexes with good results. Errors in intermonomer separations are of order 0.1 A, and the mean absolute error in the intermolecular binding energies is of order 15 percent.

The fundamental quantity on which the model is based is the (density weighted) square of the exchange-hole dipole moment at each point in a system. This easy to compute function, which we will call the "dipole density", might play an important role in other realms of chemistry beyond the dispersion interaction. A density-functional version of the model is also possible [3], in which the dipole density depends only on local density, gradient and Laplacian of the density, and the kinetic energy density.

References

[1] A.D. Becke and E.R. Johnson, J. Chem. Phys. **122**, 154104 (2005).

[2] E.R. Johnson and A.D. Becke, J. Chem. Phys., in press.

[3] A.D. Becke and E.R. Johnson, J. Chem. Phys., submitted.

Brill Academic Publishers
P.O. Box 9000, 2300 PA Leiden
The Netherlands

*Lecture Series on Computer
and Computational Sciences*
Volume 4, 2005, pp. 1235-1235

Electron densities and reduced density matrix : a crucial information from steady states to reacting systems

Pierre BECKER, Jean Michel GILLET, Blandine COURCOT

Laboratoire SPMS
CNRS and Ecole Centrale Paris,
Grande Voie des Vignes, 92295 Chatenay Malabry, France

Received 24 June, 2005; accepted in revised form 25 June, 2005

Abstract: Charge, spin and momentum densities can be accurately measured. We have developed new procedures to analyze them through a unique model based on the reduced one particle density matrix (1RDM). Moreover, we have shown that 1RDM can be expanded as a sum of local contributions, that refer to atoms or molecules imbedded in their interacting environment (cluster partitioning method)

Besides applications to various inorganic materials, we successfully analyzed Compton scattering data on ice, taking proton disorder into account.

More recently, the method was used for pharmaceuticals and is presently extended towards modeling their bioactive interaction. Two examples will be discussed : FZ41 molecule, that is considered to be HIV-1 integrase inhibitor precursor, and busulfan. For those molecules, we have access to both experimental and theoretical data.

Finally, we started a study of evolution of electron density with geometry changes, in particular driven by a chemical reactive mechanism. We studied the evolution of charge density and 1RDM along the reaction path for the simple reaction $H + H_2 \rightarrow H_2 + H$ and tried to relate those with reaction dynamics, through Hellmann Feynman force.

Keywords: charge and momentum density, cluster partitioning of 1RDM, pharmaceuticals, joint refinement.

PACS: 31.15.-e, 31.90.+s

References

[1] Gillet, J.M., Becker, P. (2004). *J.Phys. Chem Solids* **65**, 2017.
[2] Ragot, S, Gillet, J.M., Becker, P. (2002). *Physical Review B* **65**, 235115.

Brill Academic Publishers
P.O. Box 9000, 2300 PA Leiden
The Netherlands

Lecture Series on Computer
and Computational Sciences
Volume 4, 2005, pp. 1236-1239

Generalized Population Analysis and Molecular Quantum Similarity for Molecular Aromaticity

P. Bultinck[1], R. Carbó-Dorca

Department of Inorganic and Physical Chemistry,
Faculty of Sciences,
Ghent University,
Krijgslaan 281, B-9000 Gent, Belgium

R. Ponec

Institute of Chemical Process Fundamentals,
Czech Academy of Sciences,
Prague 6, Suchdol, 165 02 Czech Republic

Received 15 June 2005; accepted in revised form 14 July, 2005

Abstract: This paper presents two novel approaches to quantify molecular aromaticity. The first uses Generalized Population Analysis to define multicenter indices. The values of these indices characterize the extent of delocalized (cyclic) bonding between several atoms. In the second approach the ideas of molecular quantum similarity are used to assess aromaticity of benzenoid rings in polyaromatic hydrocarbons. Both indices correlate very well and allow fast calculation of quantitative aromaticity measures.

Keywords: Generalized Population Analysis, Quantum Similarity, Aromaticity

1. Introduction

Aromaticity, although introduced long time ago, remains a challenging chemical concept. The basic reason is that there is no clear definition based on quantum chemical description, yet it persists as a very powerful idea, allowing the interpretation of many chemical phenomena.

As a consequence there exist many indices aimed at quantifying molecular aromaticity. These may be classified as structural, energetic, reactivity based, and magnetic criteria [1]. An especially intriguing finding is that some of these measures are mutually orthogonal or divergent [2], an unacceptable situation in front of the necessary aromaticity index uniqueness. Moreover, there exist many ambiguities about how to exactly compute many of these measures.

The present paper introduces two novel approaches which are tested for a set of polyaromatic hydrocarbons (PAH), solving many of the problems related to other ways to assess aromaticity.

2. Generalized Population Analysis

There exist many different approaches to population analysis. In this work, the two probably most popular ones will be considered, namely Mulliken's and Bader's atoms-in-molecules (AIM) approach. The essential difference consists into that the Mulliken technique divides the electron density in the

---
[1] Corresponding author. E-mail: Patrick.Bultinck@UGent.be

Hilbert space of basis functions, whereas in AIM the 3D Cartesian space is divided in so-called atomic basins.

It is very well known that molecular aromaticity is related to conjugation of electron density over different atoms. The obvious textbook example is benzene, where six centers take part in the delocalization. In order to quantify the amount of delocalization over n centers, we consider the n-th order electron density, which can in a single determinant approach be obtained from the one electron density matrix. The n-th order density matrix gives the simultaneous possibility of finding every electron in a different corresponding location in 3D space. In order to obtain atom condensed probabilities, meaning the probability of every electron to be attached to a corresponding different atom, one needs to introduce proper projection operators. In the Mulliken approach such a projector for atom A is given by:

$$\Pi_A = \sum_{v \in A} \sum_{\mu} S_{v\mu}^{(-1)} |v\rangle\langle\mu| \tag{1}$$

Where v and μ are basis functions centered respectively on atom A and all atoms in the molecule. In the AIM approach we first divide 3D space in atomic basins and then introduce the logical operator:

$$\Pi_A(\mathbf{r}) = \delta(\mathbf{r} \in \Omega_A) \tag{2}$$

Meaning that $\Pi_A(\mathbf{r}) = 1$ if point $\mathbf{r}$ belongs to the basin of atom A and otherwise $\Pi_A(\mathbf{r}) = 0$.

In order to obtain a multicenter index, or a generalized bond order, an adequate tensorial product of these operators is applied on the n-th order electron density. Considering the exchange terms in the higher order density, it can be shown after some manipulations that Mulliken n-center generalized bond indices are obtained as [3]:

$$E_{AB...Z} = \frac{1}{2^{n-1}} \sum_{v \in A} \sum_{\sigma \in B} \cdots \sum_{\lambda \in C} \sum_{i=1}^{(n-1)!} P_i \{\mathbf{PS}\}_{v\sigma} \{\mathbf{PS}\}_{\sigma\alpha} \cdots \{\mathbf{PS}\}_{\lambda v} \tag{3}$$

with $\mathbf{P}$ and $\mathbf{S}$ respectively the charge and bond order matrix and the overlap matrix. P_i is a permutation operator which creates all permutations over all basis functions, except that of the first and last basis function, which are kept constantly equal to v.

AIM based multicenter indices are obtained in the same manner but using the operator (2). It is easy to see that for one center only, Mulliken and AIM atomic populations are recovered, as well as Mulliken based Wiberg-Mayer or AIM Delocalization indices for two centers. With these formulae, however, they can be extended to an arbitrary number of atoms.

Practical applications [3-5] for hydrocarbons have shown that aromaticity indices via Generalized Population Analysis:

- Allow, being generally defined, aromaticity quantification in a very uniform way, irrespective of the ring size.
- Can be computed very fast, especially within the Mulliken algorithm. For the AIM approach, use is made of a specially adapted version of Morphy03.
- Are equally well suited for the study of organic, and inorganic aromaticity, homoaromaticity, ...
- Allow to differentiate in a very straightforward way between σ and π aromaticity.

3. Molecular Quantum Similarity

Beside the GPA approach, another novel way to measure aromaticity is based on molecular quantum similarity (MQS) theory [6]. This involves comparing two molecules A and B via the computation of the Molecular Quantum Similarity Measure (MQSM) involving their densities:

Brill Academic Publishers
P.O. Box 9000, 2300 PA Leiden
The Netherlands

Lecture Series on Computer
and Computational Sciences
Volume 4, 2005, pp. 1240-1243

The Nuclear Fukui Function: Generalization within Spin-polarized Conceptual Density Functional Theory

F. De Proft,[1] P. Geerlings, and E. Chamorro[1]

Eenheid Algemene Chemie (ALGC),
Faculteit Wetenschappen,
Vrije Universiteit Brussel (VUB), Pleinlaan 2,
1050 Brussels, Belgium

Received 27 June 2005; accepted in revised form 22 July, 2005

Abstract: An extension of Cohen's nuclear Fukui function is presented within the framework Spin-Polarized Density Functional Theory (SP-DFT). Next, calculations of these SP-DFT quantities are presented for a simple set of molecules. Results have been interpreted in terms of chemical bonding in the context of Berlin's theorem, which provides a separation of the molecular space into binding and antibinding regions.

Keywords: Conceptual Density Functional Theory, Nuclear reactivity, Spin-polarization

PACS: 31.15.Ew, 31.15.-e

1. Introduction

Density Functional Theory (DFT) provides a natural framework for the development of a conceptual theory of chemical reactivity.[1-3] Within this theory, chemical reactivity is described using response functions of the system's energy E with respect to changes either the number of electrons N, the external (i.e. due to the nuclei) potential $v(r)$ or both.[1-7] These derivatives have been identified with commonly used chemical concepts such as the electronegativity,[8] hardness and softness.[9] Moreover, theoretical justification could be provided for a series of principles, such as e.g. Sanderson's electronegativity equalization principle[10] and Pearson's Hard and Soft Acids and Bases and Maximum Hardness principles.[9, 11]

The reactivity indices introduced within conceptual DFT can be divided into electronic and nuclear categories. The electronic descriptors are intended to measure the response of the electron density $\rho(\mathbf{r})$ and/or related quantities to global or local external perturbations, without any explicit reference to the nuclear rearrangement. The corresponding response of the nuclei upon external perturbations constitutes the so-called *nuclear reactivities*.[12-14] From the Hohenberg-Kohn theorems,[15] it is clear that density changes are coupled to external potential changes, and a complicated response kernel would be needed to translate electron density changes into external potential variations. Variational principles for describing chemical reactions and reactivity indexes based both on the electronic quantities[16, 17] as well as on the external potential $v(\mathbf{r})$ itself[18] have been discussed very recently by Ayers and Parr.

2. Nuclear Reactivity Indices

In order to model the changes in the external potential $v(\mathbf{r})$, the changes of the forces on the different nuclei in the molecule can be examined. In this context, and within a Born-Oppenheimer framework, Cohen et al.[12-14] have introduced the nuclear Fukui function (NFF) Φ_α as the derivative of the force $\mathbf{F}_\alpha$ on nuclei α, , with respect to the number of electrons N at a fixed external potential $v(\mathbf{r})$:

---

[1] Corresponding authors : echamorro@unab.cl and fdeprof@vub.ac.be

$$\Phi_\alpha \equiv \left[\frac{\partial \mathbf{F}_\alpha}{\partial N}\right]_{v(\mathbf{r})} = \int \frac{f(\mathbf{r})Z_\alpha(\mathbf{r}-\mathbf{R}_a)}{|\mathbf{r}-\mathbf{R}_a|^3} d\mathbf{r} \tag{1}$$

where the Hellmann-Feynman force on nucleus α, $\mathbf{F}_\alpha$ is given by

$$\mathbf{F}_\alpha = Z_\alpha \left[\int \frac{\rho(\mathbf{r})(\mathbf{r}-\mathbf{R}_\alpha)}{|\mathbf{r}-\mathbf{R}_\alpha|^3} d\mathbf{r} - \sum_{\beta \neq \alpha} \frac{Z_\beta(\mathbf{R}_\alpha - \mathbf{R}_\beta)}{|\mathbf{R}_\alpha - \mathbf{R}_\beta|^3}\right] \tag{2}$$

This NFF quantity measures the initial response of the nuclear framework when the system is perturbed in its number of electrons N. This initial response will determine the nature of the change in the external potential $v(\mathbf{r})$. Indeed, the NFF Φ_α can be be interpreted as the sum of electrostatic force-like contributions due to electronic Fukui function $f(\mathbf{r})$, [19]

$$f(\mathbf{r}) = \left[\frac{\partial \rho(\mathbf{r})}{\partial N}\right]_{v(\mathbf{r})}, \tag{3}$$

as noted from the last term in Eq. (1). The nuclear Fukui function of Eq. (1) has been proven useful in the calculation and interpretation of geometrical changes resulting from changes in the number of electrons[20-23]

On the other hand, spin-polarized density functional theory (SP-DFT) provides a more general framework for the analysis of chemical reactivity, incorporating changes both in the total electron density as the spin density charge $\rho_S(\mathbf{r})$, after explicitly considering the spin-up and spin down components of the electron density:

$$\rho(\mathbf{r}) = \rho_\uparrow(\mathbf{r}) + \rho_\downarrow(\mathbf{r}) \text{ and } \rho_S(\mathbf{r}) = \rho_\uparrow(\mathbf{r}) - \rho_\downarrow(\mathbf{r}) \tag{4}$$

Hence, spin polarization will be also incorporated in a natural way in the theoretical treatment of chemical reactivity.[24] A general formalism for electronic chemical reactivity in a conceptual DFT context was first presented by Galvan and coworkers,[24] the number of applications being relatively small up to the present moment, and the nuclear Function was not yet explored within this theoretical framework.

3. Aim, Results and Discussion

In the present work and with the aim to get further insight into a more complete description of chemical reactivity, we introduce and derive relations for nuclear reactivity indexes within the SP-DFT framework.[25] Such a type of quantities should be important in the description of the response of nuclei in processes involving charge transfer and/or spin polarization upon excitation or de-excitation, as occurring both in chemical reactions and in spectroscopic experiments.

We will focus on a more complete SP-DFT description of electronic responses and introduce new nuclear Fukui functions that are the logical generalizations of the definition put forward by Cohen. In analogy to Eq. (1), we introduce the general spin-polarized nuclear Fukui functions $\Phi_{N,\alpha}$ and $\Phi_{S,\alpha}$ as,

$$\Phi_{N,\alpha} \equiv \left[\frac{\partial \mathbf{F}_\alpha}{\partial N}\right]_{N_S, v(\mathbf{r}), B(\mathbf{r})} \tag{5}$$

and

$$\Phi_{S,\alpha} \equiv \left[\frac{\partial \mathbf{F}_\alpha}{\partial N_S}\right]_{N, v(\mathbf{r}), B(\mathbf{r})} \tag{6}$$

It has to be emphasized however that the new descriptor $\Phi_{N,\alpha}$ of Eq. (5) is completely analogous to the nuclear Fukui function in Eq. (1). However, in the present case, the derivative is also performed at a fixed value for the spin number, N_S, at constant external potential and magnetic field. In the Cohen definition given in Eq. (1), only the external potential is kept fixed in the definition of the derivative. The quantity $\Phi_{S,\alpha}$ defined in Eq. (6) is a new descriptor and can properly be called a *spin nuclear Fukui function*, since it is measuring the change in the force on a nucleus α upon a spin polarization process (i.e., a change in the spin number N_S at constant N and external potential and magnetic field).

Thus, such a type of nuclear descriptor measures in effect the onset of the response of the external potential to a given perturbation in the spin number). It could thus e.g. be used to probe changes in the nuclear geometry when a molecule is excited from a singlet ground state to a triplet excited state.

We have also discussed the nuclear Fukui functions of Eq. (5) and (6) within the context of Berlin's Theorem for chemical binding within a Kohn-Sham scheme. It has been shown that a simple connection with the Berlin function F_B for molecules,[26-28]

$$F_B = -\sum_\alpha \mathbf{R}_\alpha \cdot \mathbf{F}_\alpha = \int_{f_v > 0} \rho(\mathbf{r}) f_v(\mathbf{r}) d\mathbf{r} + \int_{f_v < 0} \rho(\mathbf{r}) f_v(\mathbf{r}) d\mathbf{r} - \sum_\alpha \sum_{\beta \neq \alpha} \frac{Z_\alpha Z_\beta}{|\mathbf{R}_\alpha - \mathbf{R}_\beta|} \qquad (7)$$

can be established, where the local Berlin function $f_v(\mathbf{r})$ is defined as,

$$f_v(\mathbf{r}) \equiv -\sum_\alpha Z_\alpha \frac{\mathbf{R}_\alpha \cdot (\mathbf{r} - \mathbf{R}_\alpha)}{|\mathbf{r} - \mathbf{R}_\alpha|^3} \qquad (8)$$

The binding function F_B of Eq. (7) is the virial of the forces necessary to hold all the nuclei in the molecule fixed.[28] At the equilibrium geometry, it thus will have a value of zero. Note that the first integral in Eq. (7) describes a positive or "binding" (shrunken) global effect on the molecule geometry, whereas the two last terms are associated with a negative or "antibinding" (enlarging) effect on it. In this work, we have explored computationally the spin-polarized nuclear Fukui functions and their relation with binding and antibinding effects on the molecular geometry. Some numerical results for a small set of molecules, including H_2O, H_2CO, and some simple nitrenes (NX), and phosphinidenes (PX), X = H, Li, F, Cl, OH, SH, NH_2, and PH_2 will be presented In most of these cases considered, it is indeed found that nuclear Fukui function vectors properly describe the associated changes in the nuclear framework upon spin polarization processes in these systems.[25]

Acknowledgements

E. Chamorro is a Visiting Postdoctoral Fellow from the Fund for Scientific Research- Flanders (Belgium) "Fonds voor Wetenschappelijk Onderzoek - Vlaanderen" (F.W.O.) within the context of the Scientific Research Community of Quantum Chemistry (Density Functional Theory). He also wishes to acknowledge the Universidad Andres Bello through grant UNAB DI 16-04, to Fondecyt (Chile), grant 1030173, and the Millennium Nucleus for Applied Quantum Mechanics and Computational Chemistry (Mideplan-Conicyt, Chile), grant P02-004-F for continuous support.

References

[1] R. G. Parr and W. Yang, *Density Functional Theory of Atoms and Molecules.* (Oxford University Press, Oxford, 1989).

[2] H. Chermette, Journal of Computational Chemistry 20, 129 (1999).

[3] P. Geerlings, F. De Proft, and W. Langenaeker, Chemical Reviews 103, 1793 (2003).

[4] R. F. Nalewajski, Journal of Chemical Physics 78, 6112 (1983).

[5] A. Cedillo, International Journal of Quantum Chemistry, Quantum Chemistry Symposium 28 (Proceedings of the International Symposium on Atomic, Molecular, and Condensed Matter Theory and Computational Methods, 1994), 231 (1994).

[6] P. Senet, Journal of Chemical Physics 105, 6471 (1996).

[7] P. Senet, Journal of Chemical Physics 107, 2516 (1997).

[8] For an account on the different electronegativity scales, see e.g. J. Mullay, in Electronegativity (Structure and Bonding, Vol. 66) ; K. D. Sen and C. K. Jørgenson, Editors, Springer-Verlag: Berlin, Heidelberg, 1987, p. 1.

[9] R. G. Pearson, Journal of the American Chemical Society 85, 3533 (1963).

[10] R. G. Pearson, Chemical Hardness, John Wiley and sons, 1997.

[11] R. T. Sanderson, Polar Covalence, Academic Press, New York, 1983.

[12] M. H. Cohen, M. V. Ganduglia-Pirovano, and J. Kudrnovsky, Journal of Chemical Physics 103, 3543 (1995).

[14] M. H. Cohen and M. V. Ganduglia-Pirovano, Journal of Chemical Physics 101, 8988 (1994).

[15] M. H. Cohen, Topics in Current Chemistry 183 (Density Functional Theory IV), 143 (1996).

[12]   P. Hohenberg and W. Kohn. Physical Review B 136, 864 (1964).

[16] P. W. Ayers, R. C. Morrison, and R. K. Roy, Journal of Chemical Physics 116, 8731 (2002).

[17] P. W. Ayers and R. G. Parr, Journal of the American Chemical Society 122, 2010 (2000).

[18] P. W. Ayers and R. G. Parr, Journal of the American Chemical Society 123, 2007 (2001).

[19] R. G. Parr and W. Yang, Journal of the American Chemical Society 106 (14), 4049 (1984).

[20] F. De Proft, S. Liu, and P. Geerlings, Journal of Chemical Physics 108 (18), 7549 (1998).

[21] R. Balawender, F. De Proft, and P. Geerlings, Journal of Chemical Physics 114 (10), 4441 (2001).

[22] R. Balawender and P. Geerlings, Journal of Chemical Physics 114 (2), 682 (2001).

[23] P. Geerlings, F. De Proft, and R. Balawender, Reviews of Modern Quantum Chemistry 2, 1053 (2002).

[24] M. Galvan, A. Vela, and J. L. Gazquez, Journal of Physical Chemistry 92 (22), 6470 (1988).

[25] E. Chamorro, F. De Proft and P. Geerlings, Journal of Chemical Physics, accepted for publication, in press.

[26] T. Berlin, Journal of Chemical Physics 19, 208 (1951).

[27] T. Koga, H. Nakatsuji, and T. Yonezawa, Journal of the American Chemical Society 100, 7522 (1978).

[28] X. Y. Wang and Z. R. Peng, International Journal of Quantum Chemistry 47, 393 (1993).

Brill Academic Publishers
P.O. Box 9000, 2300 PA Leiden
The Netherlands

*Lecture Series on Computer
and Computational Sciences*
Volume 4, 2005, pp. 1244-1246

Polarization Consistent Basis Sets

F. Jensen[1]

Department of Chemistry,
University of Southern Denmark,
DK-5230 Odense, Denmark

Received 21 June 2005; accepted in revised form 15 July, 2005

Abstract: Polarization consistent basis sets are designed to provide a hierarchical sequence of basis set for systematically approaching the basis set limit with density functional methods.

Keywords: Basis set, density functional theory, convergence.

PACS: 31.10.+z, 31.15.Ew

1. Polarization Consistent Basis Sets

Ab initio electronic structure methods have the advantage that the quality of the results can be improved in a systematic fashion by extending the treatment of electron correlation and the size of the basis set. For a given method, the inherent error is obtained as the basis set limiting result. Establishing the basis set limit is non-trivial, as the convergence of the correlation energy is slow. Based on perturbation analysis, the convergence with respect to the highest angular momentum L included in the basis set has been shown to be an inverse power series, with the leading term being proportional to L^{-3}.[1]

$$\Delta E_{corr}(L) = \Delta E_{corr}(\infty) + \frac{A}{L^3} + \frac{B}{L^4} + \frac{C}{L^5} + \cdots \tag{1}$$

Retaining only the leading term and using the correlation consistent basis sets developed by Dunning and co-workers,[2] produces an extrapolation procedure which has been quite successful for estimating the basis set limit, as exemplified by the Wn methods developed by Martin and co-workers.[3]

Density functional methods have the same potential for approaching the exact result, although the improvements with respect to the exchange-correlation energy are much less systematical than for wave function based methods. The inherent error of a given functional is again obtained as the basis set limiting value. In contrast to the correlation energy, however, relatively little is known about the convergence rate with respect to the size of the basis set. Kutzelnigg and Klopper have shown that the Hartree-Fock energy for the hydrogen atom expanded in a Gaussian basis set converges exponentially with the square root of the number of functions n.[4]

$$E_{HF}(n) = E_{HF}(\infty) + Ae^{-B\sqrt{n}} \tag{2}$$

Numerical results for a few molecular systems also indicate an exponential convergence with respect to the highest angular momentum included in the basis set. Given the similarity of density functional and Hartree-Fock methods, the basis set convergence is expected to be very similar, as has indeed been verified numerically.

Given the fundamental different convergence behavior of the many-particle (correlation) and the independent-particle (Hartree-Fock or density functional) energies, the requirements for the optimum

---

[1] Corresponding author. E-mail: frj@dou.dk

basis set composition must also be different. In the last few years we have analyzed the basis set convergence for density functional methods, and proposed a hierarchy of basis sets denoted polarization consistent basis sets to systematically approach the basis set limit.[5] The overriding principle of construction is that functions which contribute the same amount of energy are included at the same stage. The basis set exponents are explicitly optimized at the density functional level, at the atomic level for s- and p-functions, and for a selection of molecules for the polarization functions. The resulting set of primitive functions is subsequently contracted, with the degree of contraction determined by the condition that the contraction error is significantly smaller than the inherent basis set error determined by the number of primitive functions. The basis sets are denoted pc-n ($n = 0,1,2,3,4$), where n indicates the polarization beyond the isolated atom. Table 1 shows a comparison of the polarization consistent and correlation consistent basis sets in terms of composition for a first row element.

Table 1. Basis set compositions.

L_{max}	Basis	Contracted	Primitive	Basis	Contracted	Primitive
1	pc-0	3s2p	5s3p			
2	pc-1	3s2p1d	7s4p	cc-pVDZ	3s2p1d	9s4p
3	pc-2	4s3p2d1f	10s6p	cc-pVTZ	4s3p2d1f	10s5p
4	pc-3	6s5p4d2f1g	14s9p	cc-pVQZ	5s4p3d2f1g	12s6p
5	pc-4	8s7p6d3f2g1h	18s11p	cc-pV5Z	6s5p4d3f2g1h	14s8p
6				cc-pV6Z	7s6p5d4f3g2h1i	16s10p

In agreement with the faster (exponential) convergence of density functional methods, the lower angular momentum functions become more important for the pc-basis sets relative to the cc-basis sets as the size of the basis sets increases.

Based on molecular properties, like atomization energies, equilibrium geometries and vibrational frequencies, we have shown that the pc-n basis sets provide a faster convergence than other basis sets,[6] and approach the basis set limit in an exponential fashion, as illustrated by the atomization energy in Figure 1.

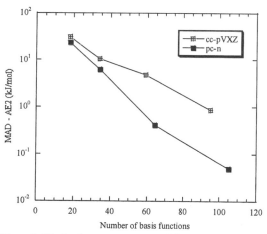

Figure 1: The basis set convergence of the atomization energy.

The pc-n basis sets have been derived using a general contraction scheme, since this allows full control over the contraction errors. Most electronic structure programs, however, are optimized for segmented contracted basis sets. We have thus investigated whether it is possible to replace the s- and p-functions with a segmented contracted set of functions.[7] These investigations have shown that:
- Duplication of functions is required for a segmented contraction, thereby increasing the size of the underlying primitive basis set.
- Many different local minima exist in the combined parameter space of the exponents and contraction coefficients.

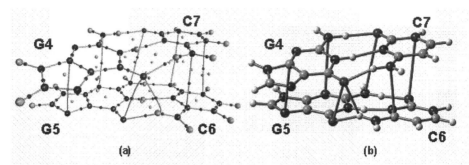

Figure 1. (a) Molecular graph of a GC/GC duplex of Watson-Crick dimers. The base numbering refers to the placement of the base in the experimental structure[9] from which the model has been taken. The bond properties corresponding to this molecular graph are listed in Table 1 below. (b) A simplified ball-and-stick representation of the molecular graph where bond paths located in the electron density are represented by sticks.

Table 1. Closed-shell bonding in the $\downarrow \left| \begin{array}{l} \text{G4 - C7} \\ \text{G5 - C6} \end{array} \right| \uparrow$ dimer duplex

Bases	Bond	bl	bpl	bpl-bl	Δ_b	$\rho^2\Delta_b$	81	82	83	,	G_b	V_b	H_b
Watson-Crick H-bonding													
G4-C7	H2a-O2	1.848	1.871	0.023	0.031	0.111	-0.047	-0.045	0.202	0.052	0.026	-0.025	0.001
G4-C7	H1-N3	1.831	1.852	0.021	0.040	0.103	-0.063	-0.059	0.225	0.069	0.029	-0.033	-0.003
G4-C7	O6-H4a	1.843	1.867	0.024	0.033	0.110	-0.049	-0.046	0.206	0.060	0.027	-0.026	0.001
G5-C6	H2a-O2	1.834	1.857	0.024	0.032	0.116	-0.049	-0.046	0.211	0.052	0.028	-0.026	0.001
G5-C6	H1-N3	1.832	1.853	0.021	0.039	0.103	-0.063	-0.059	0.225	0.070	0.029	-0.032	-0.003
G5-C6	O6-H4a	1.855	1.883	0.028	0.031	0.110	-0.046	-0.043	0.200	0.069	0.026	-0.025	0.001
Stacking (intrastrand)													
G4-G5	O6-O6	2.904	2.916	0.012	0.010	0.039	-0.008	-0.003	0.050	1.470	0.009	-0.008	0.001
G4-G5	N2-N3	3.742	3.747	0.006	0.003	0.009	-0.002	-0.001	0.012	0.139	0.002	-0.002	0.000
G4-G5	C4-N7	3.171	3.184	0.013	0.007	0.025	-0.004	-0.001	0.030	1.930	0.005	-0.004	0.001
C7-C6	N4-C4	3.107	3.146	0.039	0.008	0.027	-0.005	-0.002	0.034	1.461	0.006	-0.005	0.001
C7-C6	N1-O2	3.729	3.732	0.003	0.002	0.009	-0.001	-0.001	0.011	0.403	0.002	-0.001	0.000
C7-C6	C5-N1	3.618	3.648	0.030	0.004	0.011	-0.002	-0.002	0.015	0.195	0.002	-0.002	0.000
Stacking (inter-strand)													
G4-C6	O6-H4a	2.262	2.419	0.156	0.014	0.055	-0.014	-0.010	0.079	0.433	0.012	-0.010	0.002
G5-C7	N2-O2	3.877	3.879	0.002	0.002	0.006	-0.001	-0.001	0.009	0.104	0.001	-0.001	0.000

- Geometry obtained from Ref.[9] The electron density was calculated at the DFT-B3LYP/6-311++G(d,p) level at the experimental geometry.

- *Labeling convention:* The column labeled "Bond" lists the two atoms involved in the bonding, the first atom belonging to the first base listed under "Bases" and the second belonging to the second base. Thus, the first entry (G4-C7, H2a-O2) means that the bonding is between atom H2a belonging to the base G4 and atom O2 belonging to base C7. Standard IUPAC atomic numbering is adopted to label atoms.

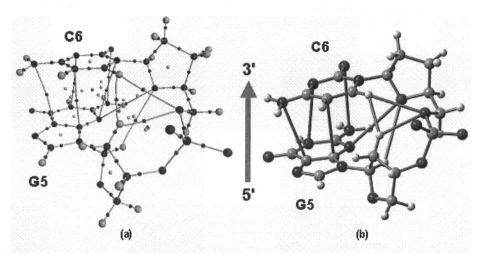

Figure 2. (a) Molecular graph of a G/C duplex of single nucleobases attached to their backbone. The base numbering refers to the placement of the base in the experimental structure[9] from which the model has been taken. The bond properties corresponding to this molecular graph are listed in Table 2 below. (b) A simplified ball-and-stick representation of the molecular graph where bond paths located in the electron density are represented by sticks.

Table 2. Closed-shell bonding in the dinucleotide $\begin{vmatrix} C6 \\ G5 \end{vmatrix} \uparrow$ with the backbone

Moiety	Bond	bl	bpl	bpl-bl	Δ_b	ρ $^2\Delta_b$	81	82	83	,	G_b	V_b	H_b
Stacking													
G5-C6	O2-N2	3.435	3.438	0.003	0.004	0.013	-0.003	-0.002	0.019	0.257	0.003	-0.003	0.000
G5-C6	N3-N1	3.453	3.462	0.009	0.005	0.015	-0.003	-0.002	0.019	0.773	0.003	-0.003	0.000
G5-C6	N4-C5	3.257	3.271	0.014	0.007	0.022	-0.004	-0.003	0.029	0.287	0.005	-0.004	0.001
G5-C6	C5-C4	3.381	3.397	0.016	0.006	0.018	-0.003	-0.001	0.022	4.234	0.004	-0.003	0.001
C-Backbone													
C6-BB(C6)	H6-O5'	2.965	2.999	0.034	0.004	0.013	-0.002	-0.001	0.016	1.137	0.003	-0.002	0.001
C6-BB(G5)	H6-H2'b	2.184	2.275	0.092	0.007	0.022	-0.007	-0.005	0.034	0.233	0.004	-0.003	0.001
G-Backbone													
G-BB(C6)	N3-O4'C	3.674	3.680	0.006	0.002	0.010	-0.002	-0.001	0.012	0.840	0.002	-0.001	0.001
Backbone-Backbone													
BB(G5)-BB(C6)	H2'b-O5'	2.488	2.524	0.036	0.010	0.035	-0.009	-0.006	0.050	0.494	0.008	-0.006	0.001
BB(C6)-BB(G5)	O4'-H1'	3.300	3.334	0.034	0.002	0.007	-0.001	-0.001	0.009	0.161	0.001	-0.001	0.000

- Geometry obtained from Ref.[9]. The electron density was calculated at the DFT-B3LYP/6-311++G(d,p) level at the experimental geometry.

- *Labeling convention:* The column labeled "Bond" lists the two atoms involved in the bonding, the first atom belonging to the first nucleotide listed under "Moiety" and the second belonging to the second nucleotide. When the base label is listed, the atom belongs to the base, but when BB(X) is listed the atom belongs to the backbone of the nucleotide in which the base X exists. For example, the first entry under "backbone-backbone" (BB(G5)-BB(C6), H2'b-O5') means that the bonding is between atom H2'b belonging to the backbone of the *nucleotide* G5 and atom O5' belonging to backbone of the *nucleotide* C. The labeling of stacking interactions follows the same conventions described in the footnote of Table 1. Standard IUPAC atomic numbering is adopted to label atoms.

References

[1] Matta C. F., Castillo N., Boyd R. J. "Closed-shell bonding in DNA: π-Stacking of nucleic acid bases (base-base), base-backbone, and backbone-backbone interactions". To be submitted.

[2] Saenger W. *Principles of Nucleic Acid Structure*. New York: Springer-Verlag, **1984**.

[3] Delcourt S. G., Blake R. D. "Stacking energies in DNA". *J. Biol. Chem.*; **1991**, *266*, 15160-15169.

[4] Sponer J., Leszczynski J., Hobza P. "Hydrogen bonding and stacking in DNA bases: A review of quantum-chemical *ab initio* studies". *J. Biomol. Struct. Dynam.*; **1996**, *14*, 117-135.

[5] Sponer J., Berger I., Spackova N., Leszczynski J., Hobza P. "Aromatic base stacking in DNA: From *ab initio* calculations to molecular dynamics simulations". *J. Biomol. Struct. Dynam.*; **2000**, *S2*, 383-407.

[6] Hobza P., Šponer J. "Toward true DNA base-stacking energies: MP2, CCSD(T), and complete basis set calculations". *J. Am. Chem. Soc.*; **2002**, *124*, 11802-11808.

[7] Mignon P., Loverix S., Steyaert J., Geerlings P. "Influence of the $\pi-\pi$ interaction on the hydrogen bonding capacity of stacked DNA/RNA bases". *Nucl. Acid Res.*; **2005**, *33*, 1779-1789.

[8] Zhikol O. A., Shishkin O., Lyssenko K. A., Leszczynski J. "Electron density distribution in stacked benzene dimers: A new approach towards the estimation of stacking energies". *J. Chem. Phys.*; **2005**, *122*, 144104-1-144104-8.

[9] Dornberger U. , Flemming J., Fritzsche H. "Structure determination and analysis of helix parameters in the DNA decamer d(CATGGCCATG)2 comparison of results from NMR and crystallography". *J. Mol. Biol.*; **1998**, *284*, 1453-1463.

[10] Shui X., Sines C. S., McFail-Isom L., VanDerveer D., Williams L. D. "Structure of the potassium form of CGCGAATTCGCG: DNA deformation by electrostatic collapse around inorganic cations". *Biochemistry*; **1998**, *37*, 16877-16887.

[11] Shui X., McFail-Isom L., Hu G. G., Williams L. D. "The *B*-DNA dodecamer at high resolution reveals a spine of water on sodium". *Biochemistry*; **1998**, *37*, 8341-8355.

[12] Frisch, M. J., et al.. *Gaussian 03, Revision B.03.* **2003**, Gaussian Inc.: Pittsburgh PA.

[13] Bader R. F. W. *Atoms in Molecules: A Quantum Theory*. Oxford, U.K.: Oxford University Press, **1990**.

[14] Biegler-König, F. W., Schönbohm, J., and Bayles, D. "AIM 2000 program can be downloaded from the Internet at http://gauss.fh-bielefeld.de/aim2000".

[15] Biegler-König F. W., Schönbohm J., Bayles D. "AIM 2000 - A program to analyze and visualize atoms in molecules". *J. Comput. Chem.*; **2001**, *22*, 545-559.

Brill Academic Publishers
P.O. Box 9000, 2300 PA Leiden
The Netherlands

Lecture Series on Computer
and Computational Sciences
Volume 4, 2005, pp. 1251-1255

Towards a Force Field via Quantum Chemical Topology

P.Popelier[1], M.Rafat[1], M.Devereux[1], S.Liem[1] and M.Leslie[2]

[1] School of Chemistry, University of Manchester, Manchester M60 1QD, Great Britain
[2] Daresbury Laboratory, Daresbury, Warrington, Cheshire WA4 4AD, Great Britain

Received 27 June 2005; accepted in revised form 18 July, 2005

Abstract: Here we focus on how topological atoms can be employed in the construction of intermolecular potentials, and in a further phase the provision of transferable units for the design of a novel force field, including intra-molecular terms. The intra-molecular part of the project is in the initial stages of construction and will not be explicitly discussed here. We survey the intermediate steps taken in the construction of a QCT force field, with emphasis on the convergence of the multipole expansion of the Coulomb/electrostatic interaction.

Keywords: Quantum Chemical Topology, DNA base pairs, peptides, atom types, force fields, intermolecular interactions, multipole moments, electron density, convergence

PACS: 31.15.-e, 34.20.-b

1. Introduction

The theory of Quantum Chemical Topology (QCT)[1,2] was introduced by Bader and co-workers with an eye on generalising quantum mechanics to subspaces. QCT uses so-called zero-flux surfaces to partition a molecule into (topological) atoms. As a corollary QCT also unambiguously partitions a cluster of molecules into molecules. In Figure 1 we show an example of the partitioning of the molecular density of a simple molecule into atoms.

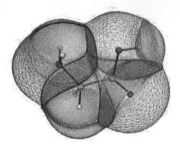

Figure 1: A three-dimensional semi-transparent view of the atomic basins in glycine. The molecular graph is also shown.

Progress in visualization of interatomic surfaces has recently been published[3]. QCT has been reviewed in numerous sources and hence the background will not be repeated here. It is worth mentioning that a recent book chapter appeared[4], critically reviewing QCT's contribution to chemical bonding, along with some speculative ideas on its impact on potentials.

Some time ago we published a detailed justification for the use of the acronym QCT as a footnote in reference[5]. Of course, in this work we are really using the theory of "Atoms in Molecules"(AIM), which is also referred to in the literature as the Quantum Theory of Atoms in Molecules(QTAIM). The use of the acronym QCT does not mean to downplay the physics behind AIM by referring to the topology as a central idea. Instead the name QCT seeks to capture better what this theory is delivering as work that uses and extends it evolves. The term AIM is widely used but is actually too narrow because, strictly speaking, it only makes sense as a term if one analyses the electron density topologically. Only then does one recover an atom in a molecule. A topological analysis of the Laplacian of the electron density (which is part of AIM) or the topology of the Electron Localization Function (ELF) does not generate atoms in molecules. However, they can both be put "under the umbrella of" QCT since they share the central topological idea. Also, returning to the electron density one could use the topological analysis to recover molecules in van der Waals complexes, an important idea in intermolecular forces. Again, as a name AIM would not describe this result, since "Molecules in Clusters" would be correct. The name QCT also embraces any future developments based on a topological analysis of other 3D or higher-dimensional scalar functions.

2. Convergence of the atomic electrostatic potential

First[6] we focused on the electrostatic component of intermolecular interaction given its importance in biological systems. An understanding of the *convergence* of the electrostatic multipole expansion was vital to make solid progress (Figure 2, left). The electrostatic potential was chosen as a starting point because it is a special case of electrostatic interaction, where one of the interacting partners is just a proton. We computed for the first time the exact electrostatic potential generated by a QCT atomic fragment, called the AEP. The statement[7] that the multipole expansion associated with bounded fragments in real space has poor convergence proves to be wrong. One does not need an excessively large number of multipoles to reproduce the exact *ab initio* molecular electrostatic potentials within the QCT theory. The atomic population (or rank-zero multipole moment) is just one term of the expansion of a physically observable quantity, namely the electrostatic potential. Hence the QCT populations cannot be judged on their reproduction of the electrostatic potential. Instead, they must be seen in the context of a multipole expansion of the AEP. Finally we computed the exact AEP and its value obtained via multipole expansion for molecules including molecular nitrogen, water, ammonia, imidazole, alanine (Figure 2, right) and valine.

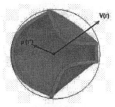 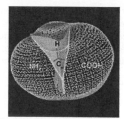

Figure 2. (left) Since topological atoms are finite it is possible to monitor their formal convergence. (right) The deviations in the exact AEP and the AEP obtained from the multipole moments up to the octupole ($\ell=3$) for the C_α atoms in alanine. The part of the picture in front of the plotting plane is deleted in order to show the interior of the object. The largest deviations occur near the cusp-like edges of the atom and the region of closest proximity. Color code (in kJ/mol):white<0.1<grey<0.2<blue<0.3< green < 0.4 < yellow< 0.5.

We then tried to understand[8] the cause of the excellent convergence. How can it be compatible with the admittedly highly non-spherical shape of the topological atoms? The answer lies in the exponentially decaying electron density. The convergence behaviour of the actual electron density inside an atomic basin is due to its decay rather than to the atom's shape. Indeed when the atom is filled with a uniform density the convergence worsens often by more than an order of magnitude. We confirm that finite atomic shapes have undesirable convergence properties but that this phenomenon is in practice not relevant due to the profile of the actual electron density inside.

3. Multipole moments and the partitioning of the Coulomb energy

Then we focused[9] on the Coulomb energy between atoms in *supermolecules*. We proposed an atom-atom partitioning of the Coulomb interaction. The atom-atom Coulomb interaction energy should *not* be confused with the electrostatic component of the intermolecular interaction, which is uniquely defined within the perturbation approach. Instead, the atom-atom Coulomb interaction energy uses the total molecular or supermolecular electron density as its input. Atom-atom contributions to the molecular intra-and intermolecular Coulomb energy are computed exactly, i.e. via a six-dimensional integration simultaneously over two atomic basins, and by means of the spherical tensor multipole expansion, up to rank $L=\ell_A+\ell_B+1=5$. The convergence of the multipole expansion is able to reproduce the exact interaction energy with an accuracy of 0.1-2.3 kJ/mol at L=5 for atom pairs, each atom belonging to a different molecule constituting a van der Waals complex, and for non-bonded atom-atom interactions in single molecules. The atom-atom contributions do not show a significant basis set dependence (3%) provided electron correlation and polarisation basis functions are included. The proposed atom-atom Coulomb interaction energy can be used both with post-Hartree-Fock wave functions, without recourse to the Hilbert space of basis functions, and experimental charge densities in principle. This method was applied to $(C_2H_2)_2$; $(HF)_2$; $(H_2O)_2$; butane ; 1,3,5-hexatriene ; acrolein and urocanic acid, thereby covering a cross section of hydrogen bonds, and covalent bonds with and without charge transfer. The Coulomb interaction energy between two molecules in a van der Waals complex can be computed by summing the additive atom-atom contributions between the molecules. Our method is able to extract from the supermolecule wavefunction an estimate of the molecular interaction energy in a complex, without invoking the reference state of free non-interacting molecules. Provided quadrupole-quadrupole interactions are included the convergence is adequate between atoms belonging to different interacting molecules. Within a single molecule the convergence is reasonable except for bonded neighbours.

In a subsequent paper[10] we searched for improved convergence of the QCT multipole expansion of electrostatic interaction. To assess the quality and the speed of this convergence we made systematic comparisons with an accurate and well-known anisotropic electrostatic model called Distributed Multipole Analysis (DMA)[11]. A set of small van der Waals systems was investigated as well as a set of much larger DNA base pairs. For the first time it was shown how topological multipole moments can be distributed to off-nuclear sites. The introduction of extra sites improved the convergence of QCT without detracting from the way it partitions molecular information among atoms. In the QCT context, the addition of extra sites is more beneficial to the convergence of the electrostatic interaction energy of *small* systems. However, for large systems excellent convergence is found with QCT *without* the introduction of extra sites. This advantage further encourages the development of a topological intermolecular force field.

Recent, as of yet unpublished work, showed that the Coulomb of 1,3 and 1,4 interactions can also be expressed *in terms of a converging multipole series*, by shifting the expansion sites by a given amount. We believe this is a very important result because it eliminates the need for bonded terms, in particular valence angle and torsion angle potential (and their cross terms). As a result an (intramolecular) force field can maximally benefit from available electrostatic information, in the form of precomputed multipole moments, without the introduction of parameters that need fitting.

4. Intermolecular forces: multipole moments in action

Here[12] we focused on a topological intermolecular potential within the (long-range) perturbation approach. We are interested in a *careful* test of this proposal in the context of the successful Buckingham-Fowler model, now using improved algorithms compared to the eighties. Particular attention was paid to the convergence of both the energy and the geometry of a set of van der Waals complexes, with respect to the rank L of the multipole expansion. For the first time this convergence behaviour has been contrasted with exact values, obtained without multipole expansion, via 6D integration over two atomic basins. Although the QCT results converge more slowly than the DMA results, excellent agreement is obtained between the two methods at high rank $(L=6)$, both for geometry as well as intermolecular electrostatic interaction energy. A direct, complete and explicit comparison between QCT and DMA was achieved. Contrary to views expressed before by Stone this work opens an avenue to introduce the topological approach in the construction of an accurate intermolecular force field. It is here that the high degree of transferability of the functional groups defined by QCT will be extremely useful.

Following the success of the topological electrostatic model on van der Waals complexes we assessed its performance on the important biological problem of DNA base-pairing[13]. Geometries and intermolecular interaction energies predicted by QCT supplemented with a hard-sphere or the Lennard-Jones potential have been compared with other methods in two stages. First with supermolecular HF, MP2 and B3LYP calculations at 6-31G(d,p) level and then with other potentials such as Merz-Kollman(MK), Natural Population Analysis (NPA) and DMA at 6-311+G(2d,p) level. The geometries for all 27 base pairs predicted by QCT and B3LYP differ by 0.08Å and 3.5^0 for 55 selected intermolecular geometrical parameters, while the energies show an average discrepancy of 6 kJ/mol. The B3LYP functional proves to be a reliable alternative to MP2 since their energies are in excellent agreement (~1 kJ/mol). Globally the QCT interaction energy curve follows the same pattern as that of MK, NPA and DMA. The MK model systematically underestimates the interaction energy and NPA shows undesirable fluctuations. Surprisingly, the convergence of the QCT multipole expansion is somewhat better than that of DMA, but both have similar basis set dependence. A test of QCT on a DNA tetrad suggests that it is able to predict geometries of more complex nucleic acid oligomers than base pairs.

Our work clearly demonstrates that the electrostatic description dominates DNA base pair patterns but more work is needed to predict the three most stable base pairs better. A current inadequacy of our QCT potential is that it is combined with empirical potentials, and hence not completely derived from *ab initio* calculations. Ultimately the QCT potential should draw all its information from *ab initio* calculations on monomers. Given the success of topological distributed polarisabilities induction energies could be included next, while work on the modeling of exchange-repulsion is warranted.

A similar analysis (to be submitted) was performed for water clusters and a selection of hydrated amino acids.

5. Simulation of liquids using a high rank QCT potential

For the first time a QCT potential based on high-rank atomic MMs was used in a molecular dynamics simulation[14]. Other than the parameter L, which keeps track of the rank of the electrostatic interaction, this potential contains only two adjustable parameters of the Lennard–Jones type. A system of 216 water molecules was simulated including long-range interactions represented by a high rank multipolar Ewald summation. High-order multipolar interactions are essential to recover the typical features of a liquid-like structure. Liquid simulations at five different temperatures showed a maximum in the density and a temperature profile that agrees fairly well with experiment. The density of simulated water at 300 K and 1 atm is about 0.1% off the experimental value, while the calculated potential energy of the liquid is within 3% of the experimental result. The experimental value of the self-diffusion coefficient is underestimated by 35%. The value of the heat capacity C_p is overestimated by 40% and the thermal expansion coefficient by 37%. The calculated correlation coefficients between the calculated QCT profile and the experimental profile of $g_{OO}(r)$, $g_{OH}(r)$, and $g_{HH}(r)$ are 0.976, 0.970, and 0.972, respectively.

6. The computation of atom types

Finally, in preparation of the peptide/protein fore field we *computed* atom types[5,15,16] via a cluster analysis on 760 atoms drawn from amino acids and derived molecules. Discrepancies between QCT atom types and some atom types featuring in AMBER emerged. Our technique can be used to define – by computation – atom types in the design of force fields for other materials.

7. Conclusion

We are systematically exploring the alternative route that Quantum Chemical Topology offers in force field design. With the benefit of being able to start afresh, we strive to proceed in a conceptually minimalist way but by making full use of the computer power presently available. A crucial feature of QCT is that, topologically, it makes no distinction between intra-molecular and intermolecular interactions. The full impact of this observation may greatly simplify the treatment of solvation and the general way in which current force field treat interactions. Work is in progress to introduce polarisation purely at the level of fluctuating multipole moments with the aid of neural networks. We hope that this approach will deliver first in a more accurate prediction of bulk properties of neat liquids.

Acknowledgments

The authors wishe to thank EPSRC and GSK for their support of this work.

References

(1)   Bader, R. F. W. *Atoms in Molecules. A Quantum Theory*; Oxford Univ. Press: Oxford, GB, 1990.
(2)   Popelier, P. L. A. *Atoms in Molecules. An Introduction.*; Pearson: London, Great Britain., 2000.
(3)   Rafat, M.; Devereux, M.; Popelier, P. L. A. Rendering of quantum topological atoms and bonds *J. of Mol. Graph. & Model.* **2005**, *in press, available on line 20 June 2005.*
(4)   Popelier, P. L. A. In *Structure and Bonding. Intermolecular Forces and Clusters*; Wales, D. J., Ed.; Springer: Heidelberg, Germany, 2005; Vol. 115; pp 1-56.
(5)   Popelier, P. L. A.; Aicken, F. M. Atomic Properties of Amino Acids: computed Atom Types as a Guide for future Force Field Design. *ChemPhysChem* **2003**, *4*, 824-829.

(6)   Kosov, D. S.; Popelier, P. L. A. Atomic partitioning of molecular electrostatic potentials. *J. Phys.Chem.A* **2000**, *104*, 7339-7345.

(7)   Stone, A. J. *The Theory of Intermolecular Forces*; Clarendon: Oxford, GB, 1996.

(8)   Kosov, D. S.; Popelier, P. L. A. Convergence of the multipole expansion for electrostatic potentials of finite topological atoms. *J.Chem.Phys.* **2000**, *113*, 3969-3974.

(9)   Popelier, P. L. A.; Kosov, D. S. Atom-atom partitioning of intramolecular and intermolecular Coulomb energy. *J.Chem.Phys.* **2001**, *114*, 6539-6547.

(10)   Joubert, L.; Popelier, P. L. A. Improved convergence of the Atoms in Molecules multipole expansion of electrostatic interaction *Molec.Phys.* **2002**, *100*, 3357-3365.

(11)   Stone, A. J. Distributed Multipole Analysis, or How to Describe a Molecular Charge-Distribution *Chem.Phys.Lett.* **1981**, *83*, 233-239.

(12)   Popelier, P. L. A.; Joubert, L.; Kosov, D. S. Convergence of the Electrostatic Interaction Based on Topological Atoms *J.Phys.Chem.A* **2001**, *105*, 8254-8261.

(13)   Joubert, L.; Popelier, P. L. A. The prediction of energies and geometries of hydrogen bonded DNA base-pairs via a topological electrostatic potential. *Phys.Chem.Chem.Phys.* **2002**, *4*, 4353-4359.

(14)   Liem, S.; Popelier, P. L. A.; Leslie, M. Simulation of liquid water using a high rank quantum topological electrostatic potential *Int.J.of Quant.Chem.* **2004**, *99*, 685-694.

(15)   Popelier, P. L. A.; Aicken, F. M. Atomic Properties of selected Biomolecules : Quantum Topological Atom Types of Carbon occurring in natural Amino Acids and derived Molecules. *J.Am.Chem.Soc.* **2003**, *125*, 1284-1292.

(16)   Popelier, P. L. A.; Aicken, F. M. Atomic Properties of selected Biomolecules : Quantum Topological Atom Types of Hydrogen, Oxygen, Nitrogen and Sulfur occurring in natural Amino Acids and derived Molecules. *Chemistry. A European Journal* **2003**, *9*, 1207-1216.

Brill Academic Publishers
P.O. Box 9000, 2300 PA Leiden
The Netherlands

*Lecture Series on Computer
and Computational Sciences*
Volume 4, 2005, pp. 1256-1257

Compton Scattering from 1897 to 1987: The Accidental Bibliography

A. C. Tanner

Department of Chemistry
Austin College
Sherman, Texas, USA

Received 18 July 2005; accepted in revised form 4 August, 2005

Abstract: A bibliography on Compton scattering and related subjects has been compiled and made available on the web. At present it consists of approximately 4,000 records from somewhat before 1900 to somewhat before 1990. References to topics related to Compton scattering, such as quantum mechanics in the momentum representation, are also included. This talk will characterize the bibliography, give a bit of personal history of how it came about and plans for development. Computers have played an obvious and expected role in its construction and presentation.

Keywords: Compton scattering, electron momentum distributions

Mathematics Subject Classification: Bibliography, Elastic and Compton scattering, Atomic scattering, cross sections, and form factors; Compton scattering, Interactions of particles and radiation with matter

PACS: 01.30.Tt, 13.60.Fz , 32.80.Cy , 78.70.-g

1. Introduction

A bibliography on Compton scattering has been compiled and made available on the web, http://trillian.austincollege.edu:8080/april/initial.jsp. At present it consists of approximately 4,000 records from somewhat before 1900 to somewhat before 1990. References to topics related to Compton scattering are also included; some examples are quantum mechanics in the momentum representation, applications of Compton scattering in medicine and geology, inverse Compton scattering, scattering as a probe of nuclear structure.

'Compton scattering' refers to the inelastic scattering of radiation (usually X- or gamma-radiation), i.e. the scattered radiation is less energetic than the initial. It is one of the basic interactions between matter and radiation.

Originally the bibliography represented my personal collection and was kept on index cards. For a time, it resided on a mainframe computer; then it migrated to a personal computer where it was, and still is, managed by the personal bibliography software, Pro-Cite. Finally, I was encouraged to make it available on the web, where it now resides in a form which can be searched in rudimentary ways.

2. History

The bibliography had its origins in the mid-1970s when I was a graduate student of I. R. Epstein, who encouraged my to write a first draft of a review article on the chemical viewpoint of electron momentum distributions and Compton profiles. I collected and read several dozens of references on the topic while preparing the first draft. It was pointed out to me, after my thesis defense, that I really should have had a broader knowledge of fields neighboring the subject of my thesis. After graduate school, I held a post-doctoral position with R. Fox and J. Felsteiner, who were interested in experimental and theoretical aspects of Compton scattering from solids. I was excited, and to an extent exasperated, to find another substantial body of literature which was not quite, but almost orthogonal, to the papers I had read in graduate school on molecular aspects of Compton scattering. The reprimand in the aftermath of my thesis defense echoed in my ears and I set myself to correcting my perceived deficiencies. During a second post-doc, which dealt with the theory of intermolecular forces, I tried to find time to collect additional references on Compton scattering and momentum distributions of atoms, molecules and solids. In 1978, DIALOG (the competitor to Chemical Abstracts) offered free computer searches, probably both to advertise itself and get feedback from users. I took the opportunity to make a thorough DIALOG-search on Compton scattering. In yet a third post-doctoral position with A. J. Thakkar, I returned to research on momentum-space topics and the further collection of references. Finally after finding a permanent teaching position, I made systematic searches in Chemical Abstracts, first paper, then on-line in the late 1980s.

At that point, John West, the director of the Austin College library, pointed out the fact, obvious to him, that I had a rather large bibliography that might be useful to others. In 1997, Professor Thakkar also encouraged me to make the bibliography available to others. The obvious mode was web-access rather than a less-useful paper version. I cast about for software packages that would make it easy for me to mount it on the web but was disappointed to find nothing that would serve the purpose. In 2004, I sought the assistance of one of our extremely able Computer Science faculty, Michael Higgs and one of his best students, Leo Lansford. Higgs decided it would be best to write a program rather than trying to find packaged software to do the job, a viewpoint that should always be considered when setting computers to perform a task. The result of Leo's work, with a heavy input of advice from Mike, is the web-version now available. There are numerous improvements planned, e.g. more complicated searching (nested searches for example) and the addition of keywords of my own construction. In addition, the bibliography is badly in need of updating. It will of course always be in need of updating as work on Compton scattering is almost certain to continue into the foreseeable future.

3. Focus

Actually the bibliography is somewhat unfocused, as well as unfinished, for two main reasons. First, I'm not a professional bibliographer (just an accidental one) and so work on the bibliography has been scattered over three decades. As a result, the function of my collection has changed. With changing and broadening interests, there have been changes and expansion in the topics covered. Secondly, because Compton scattering is a basic interaction between radiation and matter, it touches on an enormously broad range of topics, nearly all of them outside my initial narrow focus on molecular aspects. When I later decided to expand the bibliography I had to retrace my steps more than once. For example in the interior of stars, Compton scattering is a major mechanism for the degradation of gamma rays produced in stellar nuclear processes; in interstellar and intergalactic media, both electrons and high-energy radiation are present so that Compton scattering occurs over a grand scale in the universe. The phenomenon is also important in probing the sub-microscopic structure of matter, the momentum distributions of electrons, the composition of the nucleus and the nature of elementary particles. Compton scattering has been the subject of historical-philosophical studies. In addition, whenever X- or gamma-rays are used in medicine, in crystallographic studies and so forth, the contributions or interferences of Compton scattering must be accounted for. And finally Compton scattering is one way of analyzing geological formations and in well-logging in the petroleum industry.

Acknowledgments

I am grateful to numerous people: A. J. Thakkar for his encouragement of my work on the bibliography and to attend this conference; John West for his encouragement and assistance with computer searches; Mike Higgs and Leo Lansford for actually getting the bibliography on the web; and to the Austin College Richardson Fund to support work on the bibliography and to attend this conference

Brill Academic Publishers
P.O. Box 9000, 2300 PA Leiden
The Netherlands

*Lecture Series on Computer
and Computational Sciences*
Volume 4, 2005, pp. 1260-1261

Potential and Orbital Functionals: Analytic Energy Gradients for OEP and Self-Interaction-Free Exchange-Correlation Energy Functional for Thermochemistry and Kinetics

Weitao Yang

Department of Chemistry, Duke University
Durham, NC 27708, USA

Received 8 June, 2005; accepted in revised form 15 June, 2005

Abstract: The potential functional perspective of density functional theory will be presented with emphasis on two recent results, the development of analytic energy gradients for potential/orbital functional calculations and the construction of orbital functional that is self-interaction free and accurate both for thermochemistry and kinetics.

Keywords: Potential Functionals, Orbital Functionals, Self Interaction, Exchange-Correlation Energy Functional, Analytic Energy Gradients.

Theoretical Chemistry, Physics
PACS: 31.15.Ew

We have developed a variational principle for ground-state energy as a functional of external potentials. This potential functional formulation is dual to the density functional approach and provides a solution to the v-representability problem in the original Hohenberg-Kohn theory. A second potential functional for Kohn-Sham non-interacting systems establishes the theoretical foundation and also the direct optimization computational approach for the optimized effective potential (OEP) approach. The Yang-Wu direct optimization approach to OEP is an implementation of such a potential functional variational principle in finite basis sets.

Applications of the exact exchange functional to static linear and nonlinear response properties of polymers largely solve the problem that conventional DFT functionals dramatically overestimate these properties, although there remains a significant correlation contribution that cannot be accounted for with current correlation functionals.

The analytic energy gradients of the OEP method have also been developed. Their implementation in the Yang-Wu approach has been carried out and the validity is confirmed by comparison with corresponding gradients calculated via numerical finite difference. These gradients are then used to perform geometry optimizations on a test set of molecules. It is found that exchange-only OEP (EXX) molecular geometries are very close to Hartree-Fock results and that the difference between B3LYP and OEP-B3LYP results is negligible. When the energy is expressed in terms of a functional of Kohn-Sham orbitals, or in terms of Kohn-Sham potential, the OEP becomes the only way to perform density functional calculations and the present development in the OEP method should play an important role in the applications of orbital or potential functionals.

To go beyond the exact exchange model, we have developed a second-order perturbation theory (PT) energy functional within density-functional theory (DFT). Based on PT with the Kohn Sham (KS) determinant as a reference, this new {\it ab initio} exchange correlation functional includes an exact exchange (EXX) energy in the first order and a correlation energy including all single and double excitations from the KS reference in the second-order. For two-electron atoms and small molecules described with small basis sets, this new method provides excellent results, improving both MP2 and any conventional DFT results significantly. For larger systems, however, it performs poorly, converging to very low unphysical total energies. The failure of PT based energy functionals is analyzed, and its origin is traced back to near degeneracy problems due to the orbital- and eigenvalue-dependent algebraic structure of the correlation functional. Effort to overcome this challenge will be described.

To develop further the theory, we demonstrate that the Harris--Jones potential-based adiabatic connection constitutes a simple formulation for constructing functionals of the Kohn-Sham potential. We show the adiabatic connection curve monotonically decreases with increasing coupling constant parameter. Particularly the initial slope is related to perturbation theory on the Kohn-Sham determinant, providing a simple route to include many-body effects in density functional theory.

For practical applications, we have developed a self-interaction-free exchange correlation functional that is accurate both for thermochemistry and reaction energy barriers. The functional is constructed by using a Pade interpolation of the adiabatic connection that contains exact exchange, GGA and meta-GGA integrals. This functional has one empirical parameter chosen to optimize the performance over a large set of 407 molecules. Overall the functional improves significantly upon functionals such as BLYP, B3LYP and TPSS, while correctly describing one electron systems, having exchange-correlation potential with accurate long-range behavior, and being also size extensive.

Acknowledgments

The work has been supported by the National Science Foundation.

References

[1] Weitao Yang, Paul W. Ayers, and Qin Wu,. Potential functionals: Dual to density functionals and solution to the v-representability problem. *Phys. Rev. Lett.*, 92:146404, 2004.

[2] Weitao Yang and Qin Wu, Direct method for optimized effective potentials in density functional theory. *Phys. Rev. Lett.*, 89:143002, 2002.

[3] Paula Mori-Sanchez, Qin Wu, and Weitao Yang., Accurate polymer polarizabilities with exact exchange density functional theory. *J. Chem. Phys.*, 119(21):11001, 2003.

[4] Paula Mori-Sanchez, Qin Wu, and Weitao Yang, Orbital-dependent correlation energy in density-functional theory based on a second-order perturbation approach: success and failure. *J. Chem. Phys.*, in press, 2005.

[5] Felipe A. Bulata, Alejandro Toro-Labbe, Benoit Champagne, Bernard Kirtman and Weitao Yang, DFT (hyper)polarizabilities of push-pull π-conjugated systems. Treatment of exact exchange and role of correlation, *J. Chem. Phys.*, in press, 2005.

[6] Paula Mori-Sanchez, Aron J. Cohen, and Weitao Yang, "Self-interaction free exchange-correlation functional for thermochemistry and kinetics", in preparation, 2005.

[7] Qin Wu, Aron J. Cohen and Weitao Yang, Analytic energy gradients of the optimized effective potential method, *J. Chem. Phys.*, in press, 2005.

Brill Academic Publishers
P.O. Box 9000, 2300 PA Leiden,
The Netherlands

Lecture Series on Computer
and Computational Sciences
Volume 4, 2004, pp. 1268-1271

Reconciling Multifractal and Multifractional Processes in Financial Modeling

Sergio Bianchi[1] and Augusto Pianese

Dept. of Istituzioni, Metodi Quantitativi e Territorio
Faculty of Economics
University of Cassino (Italy)

Received 21 July, 2005; accepted in revised form 16 August, 2005

Keywords: Multifractals, MMAR, multifractionality, stock prices

Mathematics Subject Classification: 60G10, 60G15, 60G18.

1 Extended Abstract

In the last years the analysis of the scaling properties of financial time series has become the subject of an increasing number of works. Since the pioneering papers due to Mandelbrot in the Sixties, both empirical and theoretical issues have been addressed in the field. Concerning the former, Müller et al. (1990) find evidence of a scaling law behavior in four FX spot rates against the U.S. Dollar and observe that the distributions of the price changes strongly differ for different interval sizes. More recently, Schmitt et al. (2000) find nonlinearity in the log variations of the daily US Dollar/French Franc exchange rate and argue the multifractal nature of the FX data. Bonanno et al. (2000) report $1/f$ behavior[2] of the spectral density of the logarithm of stock price and $1/f$-like behavior of the spectral density of the daily number of trades for several stocks traded in the New York Stock Exchange. Turiel et al. (2002) show that the fluctuations of returns in stock market time series exhibit multifractal properties and exploit the geometry of the series. Analyzing the scaling properties of the DAX, Górski et al. (2002) conclude that even more complicated dynamics than the multifractal one should be used to model price fluctuations. Analyzing the qth order moments of the German index DAX, Ausloos and Ivanova (2002) find a hierarchy of power law exponents. Plotting the averaged absolute returns as a function of time intervals for different powers, Gençay and Xu (2003) find that the 5-min returns of the USD-DEM series show different slopes for these powers and argue that the nonlinearity of the scaling exponent indicates multifractality of returns. Similar results are found by Matia et al. (2003), who analyze the daily prices of 29 commodities and 2449 stocks and find that the former have a significantly broader multifractal spectrum than the latter. Fillol (2003) investigates the multifractal properties of the French Stock Market and provides Monte Carlo simulations showing that the Multifractal Model of Asset Returns (MMAR) is a better model to replicate the scaling properties observed in the CAC40 series than alternative specifications like GARCH or FIGARCH.

At the same pace of the empirical analyses, advances have been reached within the theoretical framework. Muzy et al. (2000) provide an interpretation of multifractal scaling laws in terms of

---

[1]Corresponding author.

[2]Shortly, it is called "$1/f$ noise" the type of noise whose power spectra $P(f)$ as a function of the frequency f behaves like $P(f) = f^{-\alpha}$, for some positive real α.

volatility correlations and show that in this context $1/f$ power spectra naturally emerge. Brachet et al. (2000) address the question of scaling transformation of the price process by establishing a new connection between non-linear group theoretical methods and multifractal methods developed in mathematical physics. Using two sets of financial time series, they show that the scaling transformation is a non-linear group action on the moments of the price increments, whose linear part has a spectral decomposition that puts in evidence a multifractal behavior in the price increments. Pochart and Bouchaud (2002) generalize the construction of the multifractal random walk due to Bacry et al. (2001) to take into account the asymmetric character of the financial returns. They show how one can include in this class of models the observed correlation between past returns and future volatilities, in such a way that the scale invariance properties of the motion are preserved. Through a market simulation model which displays multifractality, Yamasaki and Machin (2003) reproduce many stylized facts of speculative markets. From this model they analytically derive the MMAR for the macroscopic limit.

In addition to those above recalled, many other works seem to agree about the multifractal nature of financial time series. Anyway, as usual, things are more complicate, at least because the continuum of local scales characterizing multiscaling processes has been defined in two different ways. On the one hand, the *Multifractal Model of Asset Returns* (MMAR), introduced by Mandelbrot et al. (1997), is based on the concept of global scaling and qualifies as *multifractal* those stationary increments processes $X(t)$ whose absolute moments satisfy the scaling relation

$$\mathbb{E}\left(|X(t)|^q\right) = c(q)t^{\tau(q)+1} \text{ for } t \in \mathcal{T}, \, q \in \mathcal{Q} \tag{1}$$

with $\mathcal{T}$ and $\mathcal{Q}$ intervals on the real line such that $0 \in \mathcal{T}$, $[0,1] \in \mathcal{Q}$ and q such that $\mathbb{E}\left(|X(t)|^q\right) < \infty$. In (1) the deterministic functions $c(q)$ and $\tau(q)$ are respectively called *prefactor* and *scaling function*; the key quantity is $\tau(q)$, which takes into account the influence of time on the absolute moment of order q and synthesizes all the information about the rate of growth of the moments of $X(t)$ as t varies.

Given the above definition, the Multifractal Model of Asset Return (MMAR) assumes the log-prices $\{X(t) = \ln P(t) - \ln P(0); 0 \le t \le T\}$ to be a compound process satisfying the following assumptions [Mandelbrot et al. (1997); Calvet and Fisher (2002)]

1. $X(t) = B_H[\vartheta(t)]$, where B_H is an fBm of exponent H and $\vartheta(t)$ is a stochastic multifractal trading time, that is a multifractal transformation of chronological time into what can be thought as a "trading time" (see Assumption 2). As both the fBm and the multifractal trading time are self-similar, self-similarity is preserved when the two are compounded.

2. The multifractal trading time $\vartheta(t)$, which governs the istantaneous volatility of the log-price process, is the cumulative distribution function of a multifractal measure[3] μ defined on $[0, T]$.

3. B_H and $\vartheta(t)$ are independent.

Differently, Péltier and Lévy Véhel (1995) define the *multifractional Brownian motion* (mBm) as a generalization of the very well-known *fractional Brownian motion* (fBm), obtained by replacing the parameter H with the functional parameter $H(t)$. The mBm admits the following moving average representation

$$M(t) = V_{H(t)}^{1/2} \int_{\mathbb{R}} f_t(s)dB(s) \tag{2}$$

---

[3]The notion of multifractal measure substantially founds the whole literature related to multifractal analysis. For introduction and motivation see, e.g., Mandelbrot (1999) or Harte (2001).

[8] Morgenthaler, S., and Tukey, J. W. (2000), "Fitting Quantiles: Doubling, *HR*, *HQ*, and *HHH* Distributions," *J. Comp. Graph. Stat.*, 9, 180-195.

[9] Parzen, E. (1979), "Nonparametric Statistical Data Modeling," *Journal of the American Statistical Association*, 74, 105-121.

[10] Sheather, S. J., and Marron, J. S. (1990), "Kernel Quantile Estimators," *Journal of the American Statistical Association*, 85, 410-416.

Brill Academic Publishers
P.O. Box 9000, 2300 PA Leiden,
The Netherlands

Lecture Series on Computer
and Computational Sciences
Volume 4, 2005, pp. 1285-1288

Interval Forecasting of Daily Exchange Rate Returns using Realised Volatility

Michael P. Clements

Department of Economics, University of Warwick, Coventry, CV4 7AL, UK

Ana Beatriz C. Galvão

Ibmec São Paulo, Rua Maestro Cardim, 1170, São Paulo-SP, 01323-001, Brazil

Jae H. Kim[1]

Department of Econometrics and Business Statistics, Monash University,
Caulfield East, Victoria 3145, Australia

Received 12 July, 2005; accepted in revised form 16 August, 2005

1 Extended Abstract

The increasing availability of high-frequency intraday data for financial variables such as stock prices and exchange rates has fuelled a rapidly growing research area in the use of realized volatility estimates to forecast daily and lower frequency return volatilities and distributions. Andersen and Bollerslev (1998) suggested using realized volatility as the measure of unobserved volatility for the evaluation of daily volatility forecasts from ARCH/GARCH models[2], instead of the usual practice of proxing volatility using daily squared returns. Recent contributions have gone beyond the use of realized volatility as a measure of actual volatility for evaluation purposes, and consider the potential value of intraday returns data for forecasting volatility at lower frequencies (such as daily). Andersen, Bollerslev, Diebold and Labys (2003, henceforth ABDL03) is a key paper that sets out a general framework for modelling and forecasting with high-frequency, intraday return volatilities,[3] drawing on contributions that include Comte and Renault (1998) and Barndorff-Nielsen and Shephard (2001). The high-frequency intraday data is aggregated by summing squared intraday returns to obtain realized volatility (or realized log volatility) daily series. The realized volatility series is then modelled using vector autoregressions (VARs). Barndorff-Nielsen and Shephard (2003,2004) propose realized power variation - the sum of intraday absolute returns - as an alternative measure to realized volatility when there are jumps in the price process.

Ghysels, Santa-Clara and Valkanov (2004b) predict the future volatility of equity returns using the high-frequency returns directly: realized volatility is projected on to intraday squared and absolute returns using the MIDAS[4] (MIxed Data Sampling) approach of Ghysels, Santa-Clara and

---

[1]Corresponding and presenting author. E-mail jae.kim@buseco.monash.edu.au

[2]See Engle, 1982 and Bollerslev, Engle and Nelson, 1994.

[3]Related contributions include: Andersen, Bollerslev, Diebold, and P. Labys (2000, 2001), with applications to exchange rates; Barndorff-Nielsen and Shephard (2002, 2003) on asymptotic theory and inference, amongst others. See Poon and Granger (2003) for a recent review.

[4]A feature of this approach is that the response to the higher-frequency data variables is modelled using distributed lag polynomials. The information content of the higher-frequency returns data is thus exploited in tightly parameterised models, and the problem of selecting the appropriate lag orders is in part automatically taken care of: see the references for details.

Valkanov (2004a) and Ghysels, Santa-Clara, Sinko and Valkanov (2004). They also project realized volatility on to lagged daily realized volatility and realized power variation. Some authors such as Blair, Poon and Taylor (2001) have investigated adding daily realized volatility as an explanatory variable in the variance equation of GARCH models estimated on daily returns data.

In this and related work the future conditional variance (or volatility) is taken to be quadratic variation (or its log transformation), measured by realized volatility. ABDL03 justify the use of quadratic variation to measure volatility, and the use of realized volatility to measure quadratic variation. They show that, in the absence of microstructure effects, as the sampling frequency of the intraday returns increases, the realized volatility estimates converge (almost surely) to quadratic variation. But in the presence of microstructure effects, the appropriate intraday sampling frequency is unclear - sampling at the highest frequencies may introduce distortions.

Thus the various models and sets of explanatory variables are evaluated alongside GARCH models on the basis of mean-squared error calculations that compare the volatility predictions to future realized volatilities.

As an alternative to evaluating the volatility predictions against future realized volatility (or log thereof), we instead obtain estimates of the quantiles (i.e., interval forecasts and Value-at-Risk estimates) of the future distributions of returns from the volatility predictions. This is an indirect method of evaluating the volatility predictions. Whilst it circumvents the need to specify a 'true' future volatility (and the possible dependence of the realized volatility proxy on the selected intraday frequency), the downside is of course that the evaluation of the volatility predictions is now inextricably linked with the method used to generate the forecast quantiles. There are a number of ways in which conditional quantiles can be computed. For example, Bao, Lee and Saltoğlu (2005) consider a variety of methods.

We do not view the sole purpose of the evaluation of the interval forecasts as being to assess the quality of the volatility predictions. On the contrary, good quality interval forecasts are valued in their own right, and form the basis of the popularly employed Value-at-Risk (VaR) analysis used in risk management. One might suppose that one of the main uses of volatility predictions is as an input in VaR analysis, in which case there is a strong case for evaluating the quality of the interval forecasts directly. We consider the empirical coverage rates of intervals and VaR, as well as the tests of interval forecasts developed by Christoffersen (1998), Engle and Manganelli (1999) and Clements and Taylor (2003), amongst others.

We have used five spot exchange rates: the Australian dollar, Canadian dollar, Euro dollar, U.K. pound, and Japanese yen, all vis-à-vis the U.S. dollar, from 4 Jan. 1999 to 31 October 2003. Following ABDL (2001, 2003), 30-minute intraday returns are used to calculate the realized volatility estimates. This is on the basis of the observation made by ABDL (2000) that the realized volatility estimates remain stable as the sampling frequency increases up to approximately 30-minute interval. This sampling frequency, according to ABDL(2000, 2003), provides a satisfactory balance between the market microstructure frictions (or noise) from high frequency sampling and the accuracy of the continuous record asymptotics from low frequency sampling.

The intraday returns are calculated as the first difference of the logarithmic average of the bid-ask quotes over the 30-minute interval. Weekend days, public holidays, and other inactive trading days are excluded from the sample, following ABDL (2003). This gives a total of 1240 trading days, each with 48 intraday observations for a 24-hour trading day. All data are obtained from the Reuters screen.

References

Andersen, T. G. and Bollerslev, T., 1998, Answering the Skeptics: Yes, Standard Volatility Models do Provide Accuate Forecasts, *International Economic Review*, 39, 885–905.

Andersen, T. G., Bollerslev, T., Diebold, F. K. and P. Labys, 2000, Exchange Rate Returns Standardized by Realized Volatility Are (Nearly) Gaussian, *Multinational Finance Journal*, 4, 159-179.

Andersen, T. G., Bollerslev, T., Diebold, F. K. and P. Labys, 2001, The Distribution of Realized Exchange Rate Volatility, *Journal of the American Statistical Association*, 96, 42-55.

Andersen, T. G., Bollerslev, T., Diebold, F. K. and P. Labys, 2003, Modeling and Forecasting Realized Volatility, *Econometrica*, 71, 579-625.

Bao, Lee and Saltoğlu, 2005, Evaluating predictive performance of value-at-risk models; a reality check, mimeo, UT San Antonio.

Barndorff-Nielsen, O. E. and N. Shephard, 2001, Non-Gaussian Ornstein-Uhlenbeck-based models and some of their uses in financial economics (with discussion). *Journal of the Royal Statisitcal Society, Series B*, 63, 167–241.

Barndorff-Nielsen, O. E. and N. Shephard, 2002, Econometric analysis of realised volatility and its use in estimating stochastic volatility models, *Journal of the Royal Statisitcal Society, Series B*, 64, 252–280.

Barndorff-Nielsen, O. E. and N. Shephard, 2003, Realised power variation and stochastic volatility, Bernoulli, 9, 243–265.

Barndorff-Nielsen, O. E. and N. Shephard, 2004, Power and bipower variation with stochastic volatility and jumps. *Journal of Financial Econometrics*. Forthcoming.

Blair, B. J., Poon, S. H. and Taylor, S. J., 2001, Forecasting S&P 100 volatility: the incremental information content of implied volatilities and high frequency returns, *Journal of Econometrics*, 105, 5-26.

Bollerslev, T. G., Engle, R. F., and D. B. Nelson, 1994, ARCH models, in Engle, R. F., and D. McFadden (eds.), *The Handbook of Econometrics, Volume 4*, 2959–3038: North-Holland.

Christoffersen, P. F., 1998, Evaluating interval forecasts, *International Economic Review* 39, 841–862.

Clements, M. P., and N. Taylor, 2003, Evaluating Prediction Intervals for High-frequency data, *Journal of Applied Econometrics* 18, 445–456.

Comte, F. and E. Renault, 1998, Long memory in continuous-time stochastic volatility models, *Mathematical Finance*, 8, 291–323.

Engle, R. F., 1982, Autoregressive conditional heteroscedasticity, with estimates of the variance of United Kingdom inflation, *Econometrica* 50, 987–1007.

Engle, R. F., and S. Manganelli, 1999, CAViaR: Conditional autoregressive Value-at-Risk by regression quantiles, Department of Economics, UCSD Discussion Paper 99-20.

Ghysels, E., Santa-Clara, P. and R. Valkanov, 2004a, The MIDAS Touch: Mixed Data Sampling Regression Models, mimeo, Chapel Hill, N.C.

Ghysels, E., Santa-Clara, P. and R. Valkanov, 2004b, Predicting Volatility: Getting the most out of return data sampled at different frequencies, *Journal of Econometrics*, Forthcoming.

Ghysels, E., Santa-Clara, P., Sinko, A. and R. Valkanov, 2004, MIDAS Regressions: Further Results and New Directions, mimeo, Chapel Hill, N.C.

Poon, S. H. and Granger, C. W. J., 2003, Forecasting volatility in financial markets: A review, *Journal of Economic Literature*, 41, 478-539.

Thombs, L. A., and W. R. Schucany, 1990, Bootstrap prediction intervals for autoregression, *Journal of the American Statistical Association* 85, 486–492.

Brill Academic Publishers
P.O. Box 9000, 2300 PA Leiden,
The Netherlands

*Lecture Series on Computer
and Computational Sciences*
Volume 4, 2005, pp. 1289-1294

Realized Risk and Return: Relationships and Asymmetries in US Size and Value Portfolios

Michail Koubouros[1] and Dimitrios Thomakos

Department of Economics,
School of Management and Economics,
University of Peloponnese,
GR-221 00 Tripolis, GREECE

Received 16 August, 2005; accepted in revised form 16 August, 2005

Abstract: We study the temporal dependencies of monthly realized volatility and its relationship with returns, using a set of daily observations on size and book-to-market sorted portfolios from the US stock market from July 1963 to December 2004. Following the linear regression framework developed by Bollerslev and Zhou (*Journal of Econometrics*, 2005) we ask whether the documented (at an aggregate level) volatility feedback, leverage effects and asymmetric response of realized risk to past returns are present, and if different, across size and value portfolios. We find that small and growth portfolios exhibit stronger volatility feedback effects and large and growth portfolios stronger leverage effects with weak-form asymmetries. Our results are robust to various definitions of realized risk.

Keywords: Volatility Feedback Effect, Leverage Effect, Asymmetries, Size, Value

1 Introduction

The concept of realized volatility has been around for a number of years (see for example Merton (1980), Poterba and Summers (1986), French et al. (1987), Schwert (1990) and Campbell et al. (2001)). However, little was known about the properties of the realized volatility estimates until recently, with the advent of higher frequency datasets and the ease of computation of daily realized volatility. Andersen and Bollerslev (1998) and Barndorff-Nielsen and Shephard (2001a and b) have shown, using the theory of quadratic variation, that the realized volatility estimator is a consistent estimator of the actual volatility.

In this paper we study the properties or realized volatility at a portfolio level using data from the US. More specifically, using a large set of daily returns on size and value portfolios from the major US stock markets, an the methodology recently examined by Bollerslev and Zhou (2005), we first examine the stochastic behavior of monthly realized volatility, and second, we ask for the existence (and, if so, differences) on the so called volatility feedback and leverage effects (see, for example, Pindyck (1984), French et al. (1987), Campbell and Hentschel (1992), and Nelson (1991), Engle and Ng (1993), Bekaert and Wu (2000) and Andersen et al. (2001c)). To preview our results from the analysis of realized volatility, there appear a significant negative relationship between realized returns and contemporaneous measures of realized risk (the volatility feedback effect) found to be stronger in small-cap and low BE/ME (growth) stocks, but also, there is evidence of financial leverage effect and week-form asymmetries that are mostly profound in big-cap and low BE/ME (growth stocks).

---

[1]Corresponding author. E-mail: m.koubouros@uop.gr

Secondly, the volatility of high frequency data displays intraday seasonality. For example in the Japanese stock market there is a distinct lunch break, so the plot of its volatility is W-shaped. Precise estimation will not result from the use of such data without preprocessing. Therefore, we adjust the seasonality in the following way. Let $\varphi(t_i)$ be a deterministic function of calendar time and r_i be the non-adjusted return series. Then we define a new series r_i^* as $r_i^* = r_i/\varphi(t_i)$ where r_i^* is called seasonally adjusted return and $\varphi(t_i)$ is called the time of day function. The $\varphi(t_i)$ can be estimated by kernel or spline smoothing. Our particular choice here is Friedman's super smoother. Finally we obtain the seasonally adjusted rate of return. All the variants of GARCH models are estimated based on the seasonally adjusted returns, and the time of day function $\varphi(t_i)$ is used to overlay the intraday seasonality on the one step ahead forecast of value at risk.

2 Empirical Results on Multivariate GARCH Models

In this paper we analyze high-frequency data of Japanese worldwide leading companies in the automobile industry: NISSAN (code: 7201), TOYOTA (7203) and HONDA (7267). They are listed on the Tokyo Stock Exchange (First Section) from 2 July to 28 September 2001. Throughout this paper we consider the value at risk only for long position. Hence we are only interested in the left tail of the distributions of asset returns.

Before jumping at the empirical results, we briefly explain the structure of statistical devices used here. Let us consider univariate case at first. We do not assume any mean structure for the (seasonally adjusted) rate of return r_i^* but it has possibly the time varying conditional variance h_i. Hence models should be written as $r_i^* = \epsilon_i$, $\epsilon_i = \sqrt{h_i}\xi_i$ for some i.i.d. r.v. ξ_i. In case of Normal GARCH(1,1), for example, we assume $\xi_i \sim$ i.i.d.$N(0,1)$ and

$$h_i = \omega + \alpha\epsilon_{i-1}^2 + \beta h_{i-1}$$

where $\omega > 0$, $0 \le \alpha, \beta \le 1$nd $\alpha + \beta < 1$. Then VaR can be estimated as

$$\text{VaR}(i) = z_\alpha\sqrt{\hat{h}_i\varphi(t_i)}$$

where z_α is $100 \times \alpha$ percentile of Normal distribution. Similarly, we consider Normal model, Student t model, Student GARCH model and RiskMetrics^{TM}. Due to limited space, we omit the detail of the results on univariate GARCH. To put it shortly, the performance of models depends on the size of lower probability α. The models based on Normal distribution show good performance for 5% and 2.5% lower probability cases while the models with t-distribution outperform in forecasting 1% and 0.5% VaR.

When it comes to multivariate GARCH models, they take form of

$$R_i = E_i, \quad E_i = H_i^{1/2}\Xi_i,$$

where R_i is N-variate return process, H_i is an $N \times N$ positive definite conditional covariance matrix of R_i, Ξ_i is an $N \times 1$ i.i.d. random vector with $\text{E}(\Xi_i) = 0$ and $\text{Var}(\Xi_i) = I_N$ with I_N the N-dimensional identity matrix. Considered multivariate GARCH models are (a) VECH proposed by Bollerslev et al. (1998), (b) BEKK originally invented by Baba, Engle, Kraft, Kroner but published as Engle and Kroner (1995) afterwards, (c) Diagonal model which is the special case of VECH, (d) CC (Conditional Correlation) model by Bollerslev (1990) and finally (e) DCC (Dynamic Conditional Correlation) model recently proposed by Engle (2002), Christodoulakis and Satchell (2002), Tse and Tsui (2002). After calculation of exceedance rate, namely the ratio of the number of times the actual return exceeds VaR to the sample length, we carried out the likelihood ratio test employed in Kupiec (1995). Null hypothesis is that the observed exceedance ratio (or failure risk) p is equal to the expected tail probability p_0. Test statistic follows χ^2 distribution with degree of freedom 1, and the corresponding p-values are reported in Table 1. The head of each column describes the expected failure rate. A large p-value shows the coherency of the empirical and the expected exceedance rate, hence a larger p-value implies the better forecasting performance. From the table, BEKK and DCC models have better forecasting performance uniformly with respect to the assumed failure rate. In terms of p-values BEKK and DCC are almost on par with each other. However BEKK requires 24 parameters in this case while DCC needs only 11 unknown parameters. Hence we conclude that DCC showed the best performance for the settings of our empirical analysis. Figure 1 graphically shows how DCC-GARCH model behaves. The dotted line represents the actual data, and the solid line is the 95% VaR calculated by DCC-GARCH model.

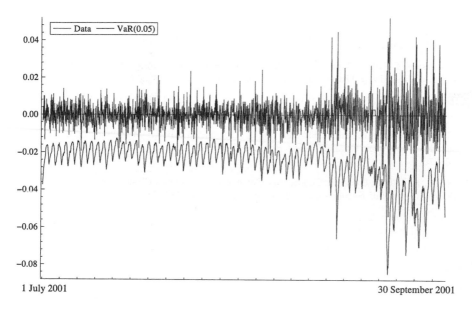

Figure 1: VaR and Actual Return (DCC-GARCH model)

Acknowledgment

This study was carried out under the ISM Cooperative Research Program No. 2005-ISMCRP-1006 and supported by a Grant-in-Aid for the 21st Century COE "Frontiers of Computational Science"

	0.050	0.025	0.010	0.005	0.001
VECH	0.5743	0.4917	0.2437	0.0216	0.0406
BEKK	0.9112	0.2003	0.1081	0.0838	0.1337
Diagonal	0.9112	0.3241	0.1652	0.0216	0.0105
CC	0.9112	0.2568	0.1652	0.0216	0.0406
DCC	0.4298	0.5915	0.2437	0.0838	0.1337

Table 1: P-values of LR Test Statistics (Multivariate GARCH Models)

(Nagoya University). The second author expresses his thanks to the support by Grant-in-Aid for Scientific Research sponsored by JSPS (17200020, 17500189).

References

[1] Bauwens, L., L. Sébastien, and J. Rombouts (2005). Multivariate GARCH models: a survey, *Journal of Applied Econometrics*, forthcoming.

[2] Bollerslev, T. (1990). Modeling the Coherence in Short-run Nominal Exchange Rates: A Multivariate Generalized ARCH model, *Review of Economics and Statistics*, **72**, 498-505.

[3] Bollerslev, T., R. Engle, and J. Wooldridge (1988). A Capital Asset Pricing Model with Time Varying Covariances, *Journal of Political Economy*, **96**, 116-131.

[4] Christodoulakis, G., and S. Satchell (2002). Correlated ARCH: Modelling the Time-varying Correlation between Financial Asset Returns, *European Journal of Operations Research*, **139**, 351-370.

[5] Engle, R. (2002). Dynamic Conditional Correlation - a Simple Class of Multivariate GARCH Models, *Journal of Business and Economic Statistics*, **20**, 339-350.

[6] Giot, P. (2000). Time transformations, intraday data and volatility models, *Journal of Computational Finance*, **4**, 31-62.

[7] Giot, P. (2002). Market risk models for intraday data, Discussion Paper 30, University of Namur.

[8] Giot, P., and S. Laurent (2003). Value-at-risk for long and short trading positions, *Journal of Applied Econometrics*, **18**, 641 - 663.

[9] Kroner, F., and V. Ng (1998). Modelling Asymmetric Comovements of Asset Returns, *The Review of Financial Studies*, **11**, 817-844.

[10] Kupiec P. (1995). Techniques for verifying the accuracy of risk measurement models, *Journal of Derivatives*, **2**, 173-184.

[11] Morimoto, T. (2005). Estimating and forecasting instantaneous volatility through a duration model: An assessment based on VaR, To appear in *Applied Financial Economics*.

[12] Tsay,R. (2002). Analysis of Financial Time Series, Wiley, New York.

[13] Tse, Y., and A. Tsui (2002). A Multivariate GARCH Model with Time-Varying Correlations, *Journal of Business and Economic Statistics*, **20**, 351-362.

Brill Academic Publishers
P.O. Box 9000, 2300 PA Leiden,
The Netherlands

Lecture Series on Computer
and Computational Sciences
Volume 4, 2005, pp. 1303-1306

Using Option Theory and Fundamentals to Assessing Default Risk of Listed Firms

George A. Papanastasopoulos[1]

Department of Economics,
School of Management and Economics,
University of Peloponnese,
GR-221 00 Tripolis Campus, Greece

Received 11 July, 2005; accepted in revised form 16 August, 2005

Abstract: The purpose of this study is to explore and extend the usefulness of the two major default risk modeling approaches, the fundamental approach and the classic option pricing approach. First, we use market information from option based measures of financial performance such as leverage, profitability and business risk to assessing default risk for listed firms. The above option based measures are used in a binary probit regression to examine their informational context and properties as leading indicators of corporate distress and to estimate default probabilities for listed firms. We find that the default probabilities estimated from the option based measures of financial performance have more explanatory power and predictive ability in assessing corporate distress than the distance to default rates generated from the same option pricing model. Then, we combine the two modeling approaches by enriching the above option based measures with accounting based measures of financial performance. The results suggest that by adding accounting information from financial statements to market information from option theory we can imporove both in sample and out of sample predictabilty of defaults. Our main conclusion, is that financial statement provide signficant and incremental information and thus, option theory does not generate sufficient statistics to assessing default risk of listed firms.

Keywords: Option Theory, Fundamentals, Default Risk

1 Extended Abstract

Default risk refers to the uncertainty associated with a firm's ability to meet its required or contractual obligations. Over the last 3 decades, default risk measurement has developed to a signficant body of research in Corporate Finance and Accounting. A large number of academic researchers form all over the world have been dedicated to develop default risk models. Two major categories of default risk models may be distnquished on the basis of the theoretical approach they adopt. The first category is traditional models that adopt fundamental analysis and the second category is structural models that adopt option theory.

The philosophy of fundamental approach that goes back to Beaver (1966), Altman (1967), Ohlson (1980) and Zmijewski(1984) is to pre-identify which characteristics of financial performance such as size, liquidity, leverage , profitability, efficiency and cash flow adequacy are important in assessing the default probability of a firm. Traditional models that adopt fundamental approach evaluate the significance importance of the above factors, mapping a reduced set of accounting

---

[1]Corresponding author. E-mail: papanast@uop.gr

based measures of the above characteristics, mainly financial ratios, accounting variables and other information from financial statements into a quantitative score. In some cases, this score can be literally interpreted as a default probability while in other cases can be used as a system to classify firms into a failing group or a solvent group of firms with a certain degree of accuracy or misclassification rate.

The philosophy of option theory that goes back to Black-Scholes (1973) and Merton (1974), is to consider equity as a call **option** on the assets of the firm : equityholders have the right but not the obligation to buy the" firm's assets" from their debtholders by repaying debt. In Merton's framework the default probability is simply the probability that the market value of firm's assets is less than the face value of its debt at debt's maturity :

$$EDP = N(-\frac{\ln(\frac{A_0}{D^T}) + (\mu - \delta - \frac{\sigma^2}{2})T}{\sigma_A \sqrt{T}}) \tag{1}$$

where A_0 notes the market value of assets , D^T the face value of debt with maturity T, μ is the expected return on asset value, σ_A is asset volatility and δ is the dividend rate of the firm. The term $\frac{\ln(\frac{A_0}{D^T}) + (\mu - \delta - \frac{\sigma^2}{2})T}{\sigma_A \sqrt{T}}$ is called distance to default rate (DD) and measures the number of standard deviations that the firm asset value is away from its default point D^T. To empirically implement the option pricing approach we need to estimate the unknowns parameters A_0, σ_A and μ. These parameters can be estimated by solving simultaneously the Black-Scholes european call option pricing formula and the optimal hedge equation from Ito's lemma with numerical recipies.

As we said, the purpose of this paper is to explore and extend the usefulness of the two major modeling approaches in default risk measurement, the fundamental approach and the classic option approach. For this purpose, we use a sample of 270 solvent and 40 defaulted listed firms from the U.S. and apply probit regressions that include one firm year observation for each firm. We use the loan-default as the definition of failure since it is more consistent with economic reality than the legal definition of bankruptcy. Moreover, we use the whole sample described above and not a matched pair sample of solvent and defaulted firms since matched samples may lead in estimation biases on the significance tests, cause unstable discriminant coefficients and affect the accuracy of the classification results.

We start our empirical analysis by estimating a model with accounting based maesures of financial performance. In this way, we eveluate their importance in default risk measurement. Moreover, we generate default probabilities that utilize accounting information from financial statements. Beacuse of the large number of accounting based measures found to be signficant in the literature, a set of 25 potentially helpful accounting based measures (mainly financial ratios) was complied for evaluation. It is obvious that an important aspect in estimating a traditional model is the selection of the final set of the independent variables from an intial set of accounting based measures. The procedure of reducing the intial set of accounting based measures to an acceptable number is an attempt to determine the relative importance within a given variable set. Several methods have been proposed in the literature to select from an intial set of distress indicators a final set. It has been stated that these methods focus solely on the statistical grounds of the variables and ignore their economic importance. In order to arrive at the final set of accounting based maesures we utilize the following procedures. Fist, we consider all possible combination of our accounting based measures when taken five at a time. From, the above combinations we select those combinations with accounting based measures that have statistical signficance at $p < 0,1$ level and no intercorrelation between them. This means, that the correlation among the independent variables in those combinations is less than $0,7$ in absolute value. Then we evaluate their economical signficance by neglecting all combinations that asingn a counter-intuitive sign for one or more coefficients. Finally, we arrive at the optimal set of accounting based measures by selecting the combination with the highest explanatory power (Mc- Fadden , R-squared ratio) and the lowest information criteria (Akaike, Schwarz, Hannan-Quinn). Therefore, with the above itterative procedure we select from our initial 25 accounting based measures those five that are doing the best overall job together in default risk measurement.

Then, we estimate a model with the three primary option variables that determine the default probability of a firm. In this way, we evaluate the informational context and the properties of these variables as leading indicators of corporate distress. Moroever, we generate default probabilities of listed firms that utilize market information from option theory. These option motivated variables are the market leverage ratio $\frac{D^T}{A_0}$, asset volatility σ_A and the expected return on asset value (or asset drif) μ that capture several characteristics of financial performance that are important in assessing the default probability of a firm. The market leverage ratio capture leverage effects, asset volatility capture business risk effects and the expected return on assets capture profitability effects. Recall, that default probaility is increasing with market leverage ratio and asset volatility and decreasing with the expected return on assets. Hence, using them as explanatory variables in a model we relate different default risk factors in an analytical way and allow non-linear effects among them. Furthermore, we explore the usefulness of option theory in default risk measurement by estimating a model with the distance to default rate DD as unique explanatory variable. Note, that the distance to default rate DD is generated from the same option pricing model as the three primary option based measures of financial performance.

The above option motivated variables do not capture factors such as size, liquidity, efficiency and cash flow adequacy. Moroever, structural models that adopt option approach rely on theories about market efficiency. Therefore, equity prices should reflect all relevant and available information about firm's fundamentals. However, Hillegeist , Keating, Cram and Lundstedt (2004) , Benos and Papanastasopoulos (2005) document that structural models do not generate sufficients statistics that capture all available information about the default probability of a firm. Thus, questionable is whether accounting based measures of financial performance reflect information about the default probability of a firm beoynd that contained in option based measures. In order to capture this possibility, we have enrich the above option motivated variables that utilize market information with accounting variables and financial ratios that utilize financial statement information into a hybrid model of default risk measurement.

In the final part of our empirical analysis, we compare the explanatory and the classification power of the above models. To compare the explanatory power of each of the above models we use relative information tests. Explanatory power is assessed by comparing the Mc-Fadden (R-squared) ratio and the information criteria (Akaike, Schwarz, Hannaan-Quinn), for each default risk model and the model with the highest Fc-Fadden ratio and the lowest information criteria is deemed the best. To compare their classification power we use prediction-oriented tests. These tests examine the prediction accuracy and error generated by each default probability estimate when discriminating firms as defaulters and non-defaulters. We divide firms by their estimated default probability into ten equal portfolios. The optimal cut off point, is then the first default probability in that portfolio that has the highest ratio of defaulters to the total number of firms in the portfolio (concetration ratio). Note that the above criterion recognizes that the cost of Type I error (default risk cost of lending a financially weak firm that defaults) is significantly larger than the cost of Type II error (opportunity cost of not lending a financial healthy that does not default). However, the in-sample clasiification accuracy does not come as a surprise as each model is evaluated using the same data that we use to estimate them. Thus, we also examine the ability of each model to rank defaulters and non defaulters accurately using the above optimal cut off thresholds and a different sample of 72 solvent and 28 defaulted firms.

We find that the default probabilities estimated from the option based measures of financial performance have more explanatory power and predictive ability in assessing corporate distress

is that the given portfolio SD dominates all other portfolios, thus, it is not dominated by any other portfolio. The alternative hypothesis is that we can construct a portfolio that stochastically dominates the market portfolio. The result provides a random variable that dominates the limiting random variable corresponding to the test statistic under the null hypothesis. The inequality yield a test that will never reject more often than a probability level, say 5% for any given portfolio satisfying the null hypothesis. The result also indicates that the test is capable of detecting any violation of the full set of restrictions of the null hypothesis. Of course, in order to make the result operational, we need to find an appropriate critical value. Since the distribution of the test statistic depends on the underlying distribution, this is not an easy task, and we decide hereafter to rely on a method to simulate p-values. The method is based on block bootstrapping. Block boostrap methods extend the nonparametric i.i.d. bootstrap to a time series context. They are based on "blocking" arguments, in which data are divided into blocks and those, rather than individual data, are resampled in order to mimic the time dependent structure of the original data.

In our paper, we also test for FSDE, although it gives necessary and not sufficient optimality conditions. The FSD criterion places on the form of the utility function beyond the usual requirement that it is nondecreasing, i.e., investors prefer more to less. Thus, this criterion is appropriate for both risk averters and risk lovers since the utility function may contain concave as well as convex segments. Owing to its generality, the FSD permits a preliminary screening of investment alternatives eliminating those which no rational investor will ever choose. The SSD criterion adds the assumption of global risk aversion. This criterion is based on a stronger assumption and therefore, it permits a more sensible selection of investments. The test statistic for SSD is formulated in terms of standard linear programming (LP). Note that numerical implementation of FSD is much more difficult since we need to develop mixed integer programs (MIP). Nevertheless, standard widely available algorithms can be used to compute both test statistics.

2 Empirical application

To illustrate the potential of the proposed test statistics, we test whether different SDE criteria (FSDE, and SSDE) rationalize the market portfolio. Although we focus the analysis on testing the SSD efficiency of the market portfolio, we additionally test FSD to examine the degree of the subjects' rationality (in the sense that they prefer more to less).

We use the 6 Fama and French benchmark portfolios constructed as the intersections of 2 ME portfolios and 3 BE/ME portfolios. The portfolios are constructed at the end of June. ME is market cap at the end of June. BE/ME is book equity at the last fiscal year end of the prior calendar year divided by ME at the end of December of the prior year. Firms with negative BE are not included in any portfolio. The annual returns are from January to December. We use data on monthly returns (month-end to month-end) from July 1963 to October 2001 (460 monthly observations) obtained from the data library on the homepage of Kenneth French[2]. Excess returns are computed from the raw return observations by subtracting the return on the one-month U.S. Treasury bill. We also use the Fama and French market portfolio, which is the value-weighted average of all non-financial common stocks listed on NYSE, AMEX, and Nasdaq, and covered by CRSP and COMPUSTAT. Further, we use the one-month U.S. Treasury bill as the riskless asset.

First we analyze the statistical characteristics of the data covering the period from July 1963 to October 2001 (460 monthly observations) that are used in the test statistics. Portfolio returns exhibit considerable variance in comparison to their mean. Moreover, the skewness and kurtosis indicate that normality cannot be accepted for the majority of them. These observations suggest to adopt the FSD and SSD tests which account for the full return distribution and not only the mean and the variance.

---

[2]http://mba.turc.dartmouth.edu/pages/faculty/ken.french

One interesting feature is the comparison of the behavior of the market portfolio with that of the individual portfolios. We observe that the test portfolio is Mean-Standard Deviation inefficient. This means that we can construct portfolios that achieve a higher expected return for the same level of standard deviation, and a lower standard deviation for the same level of expected return. If the investor utility function is not quadratic, then the risk profile of the benchmark portfolios cannot be totally captured by the variance of these portfolios. Generally, the variance is not a satisfactory measure. It is a symmetric measure that penalizes gains and losses in the same way. Moreover, the variance is inappropriate to describe the risk of low probability events. Finally, the mean-variance approach is not consistent with second-order stochastic dominance. This is well illustrated by the mean-variance Paradox and motivates us to test whether the market portfolio is FSD and SSD efficient.

2.1   The SDE test results

We use the block bootstrap method to compute the p-values for testing SDE. They are computed by comparing the test statistics with a number of recentered bootstrap samples. The p-values are approximated at any order j by

$$\tilde{p}_j \approx \frac{1}{R} \sum_{r=1}^{R} \mathbb{I}\{\tilde{S}_{j,r} > \hat{S}_j\}, \qquad j = 1, 2,$$

where $\hat{S}_j$ is the original test statistic, $\tilde{S}_{j,r}$ is the test statistic corresponding to each bootstrap sample, and the averaging is made on 300 replications. This number guarantees that the approximations are accurate enough, given time and computer constraints. We reject the null hypothesis if the p-value is lower than the significance level of 5%.

For the FSD efficiency, we found that the market portfolio is FSD highly and significantly efficient. The p-value of the test statistic is $\tilde{p}_1 \approx 0.55$. The market portfolio is found to be efficient in 167 out of 300 replications.

Our test results suggest also that the market portfolio is highly and significantly SSD efficient. The p-value of the test statistic is $\tilde{p}_2 \approx 0.59$. The market portfolio is found to be efficient in 179 out of 300 replications. These results indicate that the whole distribution rather than the mean and the variance plays an important role in comparing portfolios. This efficiency of the market portfolio is interesting for investors. If the market portfolio is not efficient, individual investors could diversify across diverse asset portfolios and outperform the market.

2.2   Rolling window analysis

We carried out an additional test to validate the SSD efficiency of the market portfolio and the stability of the model results. It is possible that the efficiency of the market portfolio could change over time, as the risk and preferences of investors change. Therefore, the market portfolio could be efficient in the total sample, but inefficient in some subsamples. Moreover, the degree of efficiency may change over time. To control for that, we perform a rolling window analysis, using a window width of 10 years. The test statistic is calculated separately for 340 overlapping 10-year periods, (July 1963-June 1973), (August 1963-July 1973),...,(November 1991-October 2001).

Interestingly, the market portfolio is SSD efficient in the total sample period. The SSD efficiency is not rejected for all subsamples. The p-values are always greater than 15%, and in some cases they reach the 80% − 90%. This result confirms the SSD efficiency that was found in the previous subsection, for the whole period. This means that we cannot form an optimal portfolio from the set of the 6 benchmark portfolios that dominates the market portfolio by SSD. The line exhibit large fluctuations, thus the degree of efficiency is changing over time, but remains always above the critical value.

This is a strong indication of the SSD efficiency of the market portfolio, as the resulting p-values remain always above the critical level of 5%.

To conclude, the empirical results indicate that the market portfolio is FSD and SSD efficient. This result is also confirmed in a rolling window analysis. In contrast, the market portfolio is Mean-Variance inefficient, indicating the weakness of the variance to capture the risk.

Brill Academic Publishers
P.O. Box 9000, 2300 PA Leiden,
The Netherlands

*Lecture Series on Computer
and Computational Sciences*
Volume 4, 2005, pp. 1311-1312

An Investigation of the Lead-Lag Relationship in Returns and Volatility between Cash and Stock Index Futures : the case of the CAC40 Index

Abdelwahed Trabelsi[1] and Wejda Ochi

BESTMod Laboratery, Institut Supérieur de Gestion de Tunis, Tunisia

Received 21 July, 2005; accepted in revised form 16 August, 2005

Abstract: This paper investigates the lead-lag relationship in daily returns and volatilities between price movements of stock index futures and the underlying cash index in the CAC40 markets. Previous empirical results show that there is a long-run contemporaneous relationship, whith an asymetric lead-lag short-run behavior between the cash and futures markets. It seems that cash index returns lead futures by responding to shocks in futures market. After examining whether daily volatility in futures prices systematically lead daily volatility in the cash index, we show that there is a bi-directional volatility spillover effects between the two markets.

Keywords: Stock index futures; Volatility spillovers

1 Extended Abstract

It is known that in perfectly efficient futures and cash markets,where all available information is fully and instantaneously utilized to determine market prices, informed investors are indifferent between trading in either market, and new information is reflected in both simultaneously. However, the existence of market frictions, such as transactions costs or capital market microstructure effects, makes one market reacts faster to new information, and the other market is slow to react. This phenomen is called a lead-lag relationship between the two markets.

This paper investigates the lead-lag relationship in daily returns and volatilities between price mouvements of the CAC40 index futures and the underlying cash index. Our empirical study has two main emphases. Firstly, we analyze this relationship between cash and futures returns. We try to determine which market is faster to react to new information and permits to reach the long-run equilibrium. Secondly, this same relationship is examined in term of volatility. The aim of this issue is to detect the existence of spillover volatility effects between the two markets.

We consider the close cash and futures prices of the CAC40 negociated respectively on the "Bourse de Paris" and the MATIF for the period January,4, 2000 to December, 31, 2004. The CAC40 futures prices are always those of the nearby contract.

We define S_t and F_t as the logarithm of the CAC40 cash and futures prices, respectively.

Different statistical tests are undertaken. It is shown that the two price series are first order integrated. Furthermore, there is a cointegration relationship between them. The Granger causality test indicates that both variables Granger cause each other, then it is said that there is a two-way feedback relationship between S_t and F_t.

---

[1]Corresponding author. Email: abdel.trabelsi@isg.rnu.tn.

According to these results, a VECM (3) model is used to analyze the relationship between them. The results indicate that here is a long-run bi-directional causal relationship between the cash and futures markets. However, the short-run relationship between the two markets is unidirectional. Indeed, the cross-market coefficients are significant only in the futures equation, which implies that cash market changes preceed those in the futures market, and therefore, cash prices lead futures prices.

In order to investigate for volatility spillovers between the cash and futures markets, an extended bi-variate VECM-GARCH model has been retained. The results of the variance equations estimation are in line with earlier results as they showed, in one hand, the bi-directional long-run relationship between the two markets and the unidirectional short-run relationship from cash to futures market in the other hand.

Estimation of the volatility equations indicates that this parameter is time-varying in both cash and futures markets.

Significance of the estimated positive coefficient of the squared lagged basis in the variance equation of the futures market indicates that the spread increases the volatility of cash and futures markets.

The coefficients of volatility spillover parameter from cash to futures and from futures to cash are significant at conventional significance level. This result implies that the volatility relationship between cash and futures markets is bi-directional. As a consequence, any piece of information that is released by one of the two markets has an effect on the volatility of the other one.

These results confirm the strong lead-lag relationship between cash and futures markets always demonstrated in the financial litterature. The caracteristics of this relationship (unidirectional or bi-directional) and of the volatility spillover effects, and thus information transmission, between the two markets can be used as a mean of decision making regarding hedging activities.

Brill Academic Publishers
P.O. Box 9000, 2300 PA Leiden,
The Netherlands

*Lecture Series on Computer
and Computational Sciences*
Volume 4, 2005, pp. 1313-1313

Nonlinear Dynamical Signatures in Volume-Price Dynamics

C. E. Vorlow[1]

Department of Economic Research and Forecasting,
EFG Eurobank Ergasias S.A., Othonos 6, 10557 Athens, Greece,
and
Durham Business School,
Durham University, Mill Hill Lane, Durham DH13LB, UK.

Received 16 August, 2005; accepted in revised form 16 August, 2005

Abstract: Research so far has indicated that stock price fluctuations may be governed in part by complex dynamics, often indistinguishable from noise, due to their weak signatures and the nature of financial time series. We employ Cross Recurrence Quantification Analysis (CRQA) and entropy based criteria to investigate the interrelationship between volume and price-return dynamics for US, UK and Greek stock market data. We examine daily and minute by minute data in order to determine whether the time-resolution of the observed fluctuations and price "nano-structure" are relevant to the classification of the dynamics. As a second stage, we apply a sliding window to determine the degree to which CRQA measurements are sensitive to the sampling process as well as the length of the sequences analyzed and whether this sensitivity could provide a useful input to a technical trading strategy.

Keywords: Nonlinear and complex dynamics; time-series analysis; recurrence plots;

PACS: 05.45.Tp; 43.60.Hj; 05.45.Ac; 89.65.Gh; 02.50-.r;

---

[1]Corresponding author. Dept. of Economic Research and Forecasting, EFG Eurobank Ergasias S.A. & Durham Business School. E-mail: cvorlow@eurobank.gr, costas@vorlow.org. Web: http://www.vorlow.org. The views expressed in this article are of the author and do not reflect those of EFG Eurobank Ergasias S.A., which is not responsible for, nor endorses, any opinion, recommendation, conclusion, solicitation, offer or agreement or any information contained in this written communication. Hence, EFG Eurobank Ergasias S.A. cannot accept any responsibility for the accuracy or completeness of this research submitted to the public domain.

Brill Academic Publishers
P.O. Box 9000, 2300 PA Leiden
The Netherlands

Lecture Series on Computer
and Computational Sciences
Volume 4, 2005, pp. 1314-1319

Idiosyncratic Risk Matters! A Switching Regime Approach. Evidence from US and Japan.

Timotheos Angelidis[*]

Athens Laboratory of Business Administration. Athinas Ave. & 2a Areos street,
Vouliagmeni GR-166 71, Greece

Nikolaos Tessaromatis[**]

Athens Laboratory of Business Administration. Athinas Ave. & 2a Areos street,
Vouliagmeni GR-166 71, Greece

Received 8 July, 2005; accepted in revised form 16 August, 2005

Abstract: The resent studies of Bali et al. (2005) and Wei and Zhang (2004) showed that there is no relation between idiosyncratic risk and value-weighted portfolio returns and therefore asset specific risk does not matter in asset pricing. We propose a different approach of examining the forecasting ability of idiosyncratic risk by using switching regime models and therefore to be able to separate the stock returns to either high or low volatility periods in order to examine if idiosyncratic risk is related to either or both regimes. The empirical results suggest that for both markets (US and Japan) in the low variance regime there is a positive and statistically significant relation between future market returns and idiosyncratic risk. The proclaimed relation is not sample dependent as it holds before and after the 1999 during which it seems that a structural break has occurred.

Keywords: Idiosyncratic Risk, Stock Market Volatility and Switching Regime

1. Introduction

According to the Capital Asset Pricing Model (CAPM) investors are rewarded for bearing market risk. Even if the relation between risk and returns has been widely explored, the empirical evidence is controversial. The one strand of the literature supports the positive relation (see French et al. (1987), Bollerslev et al. (1988), Campbell and Hentschel (1992) and Santa-Clara (2005) among others), while Campbell (1987), Breen et al. (1989), Glosten et al. (1993), and Whitelaw (1994) documented a negative relation.

In equilibrium, idiosyncratic risk which in theory can be eliminated through diversification should get no reward. However, Goyal and Santa-Clara (2003) argued that idiosyncratic risk matters since the equally weighted total variance forecasts subsequent capitalisation weighted market returns. Given the fact that asset specific is the major component of total volatility, they substantiated that idiosyncratic risk drives this relation. Following their work, Bali et al. (2005) showed that the positive relation uncovered by Goyal and Santa-Clara (2003) disappears if the sample includes the more recent history of returns (from 2000 to 2001) and is partially driven by a liquidity premium. These findings were also confirmed by Wei and Zhang (2004) since they demonstrated that the positive relation between market returns and lagged idiosyncratic volatility is sample specific. When they extended the sample that Goyal and Santa-Clara (2003) had used (from 2000 to 2002), the positive relation between return and idiosyncratic risk was not statistically significant. Moreover, they showed that the investment strategy based on average risk that Goyal and Santa-Clara (2003) proposed does not hold, as it generates excess return only during the earliest years.

---

[*] Corresponding Author. Tel.: +30-210-8964-736.
E-mail address: taggel@unipi.gr
[**] Tel.: +30-210-8964-736.
E-mail address: ntessaro@alba.edu.gr. The usual disclaimer applies.

All previous studies were based on single state equations. However, since Hamilton (1989) modelled the U.S. business cycle as an outcome of a discrete-state Markov process, the switching regime models have become increasingly popular as it is widely stated that appear to be regime shifts in the structure of the markets. Hamilton and Susmel (1994) and Cai (1994) extended the simple regime switching model by introducing ARCH parameters in the conditional variances. Gray (1996) developed a regime switching GARCH model for the short term interest rate and demonstrated its superiority over single regime models in terms of statistical and economical tests. Brunetti et al. (2003) used Dueker's (1997) switching GARCH model and found out that real effective exchange rate, money supply relative to reserves, stock index returns and volatility are the major indicators of exchange rate instability. An e' and Ureche-Rangau (2004) developed an Asymmetric power GARCH switching regime model and estimated it for four Asian stock market indices. They stated that their model improved the simple switching GARCH model statistically and economically. Billio and Pelizzon (2000) introduced a multivariate switching regime model in order to calculate the VaR number for 10 Italian stocks. They tested the performance of their model by using two backtesting measures and concluded that the risk estimates based on the switching regime specification were more accurate than the other risk management techniques.

Timmermann (2000) derived the higher moments of a 2-state Markov regime switching model. He substantiated that the state means of the switching regime model create non-zero skewness, while excess kurtosis, relative to that of the standard normal distribution, is being produced mainly by the different regime volatilities. Due to the fact that it is common to observe these features (skewness and excess kurtosis) in financial markets, these models are expected to describe them better than the non-switching ones. Perez-Quiros and Timmermann (2000) analyzed the differences between small and large firms' stock returns over the economic cycle. They showed that during recessions, small firm's risk is mainly affected by the worsening credit conditions and therefore investors demand a higher risk premium in order to hold their shares. Guidolin and Timmermann (2003) analyzed the asset allocation problem under the framework of the switching regime models. They provided evidence of two regimes in the US stock market and derived the optimal portfolio weights that were depending on the state of the market. Ang and Bekaert (2001) argued that there are benefits from international diversification, as they found a high volatility – high correlation regime in three countries (US, UK and Germany).

In asset pricing studies, one potential limitation of the approaches that have been followed, is the assumption of a model with constant coefficients as there is evidence that betas, expected returns and volatilities are time-varying. Cooper et al. (2004) noted " models of asset pricing, both rational and behavioral, need to incorporate (or predict) such regime switches", and Ang and Chen (2002) argued that "asymmetric correlations may play a role in an asset-pricing model" and "regime switching models perform best in empirically explaining the amount of correlation asymmetry in the data". Kim et al. (2004a) used a log-linear present value framework under a Markov switching regime model and concluded that there is a positive relation between market risk and equity premium, while in Kim et al. (2004b) they showed that a structural break has occurred in the equity premium. Mayfield (2004) reached to the same conclusion, as he provided evidence of a structural change in the volatility process after the Great Depression which implies that the historical average risk premium is overstated.

Therefore, as the switching regime models have been applied in most of the areas of finance, it is interesting to investigate whether idiosyncratic risk matters under the framework of these models. Our study sheds a light on the issue as it partly explains the controversial results of the previous studies. Specifically, we examine the forecasting ability of idiosyncratic risk by assuming that the stock returns follow a two-state regime process. By using these models we will be able to separate the stock returns to high and to low volatility periods and to investigate if idiosyncratic risk is related to future stock returns in any of these two subperiods. For robustness purposes, we have used data from the two major stock exchanges (US and Japan) in order to avoid the results to be dependent on a specific financial market. The empirical results suggest that in the low variance regime and for both markets there is a positive and statistically significant relation between subsequent market returns and idiosyncratic risk. However, this relation does not hold during periods of high volatility which implies that asset specific risk matters only in periods of below average risk.

The structure of the paper is as follows. The second section describes the construction of idiosyncratic risk measure, while the third examines the forecasting ability of the idiosyncratic risk. Section 4 concludes the paper.

2. Idiosyncratic Risk Measure

Following Goyal and Santa-Clara (2003), we define the monthly variance of stock i based on daily returns as:

$$V_{i,t} = \sum_{d=1}^{D_t} r_{i,d}^2 + 2\sum_{d=2}^{D_t} r_{i,d} r_{i,d-1},$$ Eq. (1)

where D_t is the number of trading days in month t and $r_{i,d}$ is the return of stock i in day d. The second term of **Error! Reference source not found.** equation 1 adjusts the variance to the autocorrelation of stock returns, by employing the French et al. (1987) procedure. Similar to Goyal and Santa-Clara (2003) and Guo and Savickas (2003), we exclude stocks with less than 5 observations during month t, while we drop the $2\sum_{d=2}^{D_t} r_{i,d} r_{i,d-1}$ term from **Error! Reference source not found.** equation 1 if $V_{i,t} < 0$. TV_t^{Equal} is the average of the volatilities of all securities in month t, while TV_t^{Value} is the average value-weighted total variance.

Under the assumption that the betas of all securities against the market is one, (see Xu and Malkiel (2001)), the variance of stock i at time t, $V_{i,t}$, can be decomposed in two parts: a systematic part which equals to the variance of the market, MV_t and an idiosyncratic part which equals to the variance of the idiosyncratic return.

$$V_{i,t} = MV_t + IV_{i,t}$$ Eq. (2)

Therefore, the average aggregate idiosyncratic variance is calculated as [1] $IV_t^{Equal} = TV_t^{Equal} \quad MV_t^{Equal}$, while the corresponding value weighted asset specific risk is defined as $IV_t^{Value} = TV_t^{Value} \quad MV_t^{Value}$

3. Empirical Investigation of the Forecasting Ability of Idiosyncratic Risk

According to capital asset pricing theory, expected stock returns should be a function of systematic factors that affect stock prices. Idiosyncratic risk, which can be eliminated through diversification, should play no role in the pricing of stocks. We explore the relationship between idiosyncratic volatility and subsequent stock returns by regressing capitalization monthly market stock returns on various measures of lagged volatility (X_t).

$$r_{t+1} = \mu_j + b_j X_t + e_{jt}$$
$$e_{jt} = \varepsilon_{jt} \sigma_j$$ $, j = 1,2$ Eq. (3)

where r_{t+1} is the excess market return. Similar to Goyal and Santa-Clara (2003) and Bali et al. (2005) we also run regressions based on lagged standard deviations and log-variances, these transformations reduce both skewness and kurtosis and bring the distribution closer to the normal.

Table 2 presents the result of equation 3 for the US market[2]. In the single state case, value weighted idiosyncratic risk is negatively correlated with subsequent excess market returns[3]. However,

---

[1] For more information about the construction of the idiosyncratic risk measures, see Goyal and Santa-Clara (2003), Bali et al. (2005) among others.
[2] The corresponding results for the market of Japan are similar to those of US and are available from the authors upon request.
[3] Bali et al. (2005) reported a negative but statistically insignificant coefficient with a p-value close to 0.20.

when we apply the other two transformations the significant relation disappears, consistent with the findings of Bali et al. (2005). On the other hand, under the framework of the switching regime model, idiosyncratic risk, irrespectively of the transformation, is positively and statistically significant related to subsequent portfolio returns only when the market is in the low variance environment[4]. During volatile periods, the b_1 coefficient is negative but statistically insignificant, as the higher risk increases the error of the estimates.

For the equally weighted scheme and in the single regime case, idiosyncratic risk can not forecast market returns, a finding which is in line with the work of Bali et al. (2005) and Wei and Zhang (2004) who extended the sample that Goyal and Santa-Clara (2003) had used and contradicts the main result of Goyal and Santa-Clara (2003). As Figure 1 shows, both the market return and the idiosyncratic risk measures have switched the last four years to the high variance regime, a fact that may be the reason why Bali et al. (2005) and Wei and Zhang (2004) could not verify the positive and statistically significant relation between asset specific risk and subsequent stock returns. However, under the framework of the switching regime models, idiosyncratic risk is positively related to future portfolio return only when the market is in the low variance environment, as it was when we applied the value weighting scheme. In the high variance regime, idiosyncratic risk is negatively and statistically significant related to subsequent market returns only when we transform the risk measures. The insignificant relation in the single state case is due to the two opposite signs of b_j coefficient which are cancelled out when we combine the two regimes and hence the main findings of Goyal and Santa-Clara (2003), Bali et al. (2005) and Wei and Zhang (2004) can be attributed to the fact that they have used the "classical" regression approach. The magnitude of the estimated factors differs dramatically across regimes, which results to an either overestimation or underestimation of the parameters in the single state case.

4. Conclusions

Evidence that idiosyncratic volatility is a priced factor in asset pricing and that it can be used to forecast future stock returns are currently a topic of research and debate. The findings presented in this paper add to the conflicting and confusing results based on US data and shed new light to the issue. Even if the switching regime models have been proposed by the literature as alternatives to the single state models, they have not been applied under this framework yet. In this paper we have tried to investigate whether there are different states in the market which can be used to uncover the relation, if any, between subsequent market returns and asset specific risk.

Contrary to the findings of recent papers (Bali et al. (2005) and Wei and Zhang (2004)) we have shown that idiosyncratic risk is related to future market returns. Specifically, we have demonstrated that asset specific risk is positively and statistically significant related to subsequent market returns during periods of low volatility. Nevertheless, this is not the case for the high variance regime as in both countries asset specific risk can not forecast future stock returns. These results hold also during a shorter period (form 1973:01 to 1999:12) and therefore are not sample specific.

---

[4] The presented results are robust to different starting values.

Brill Academic Publishers
P.O. Box 9000, 2300 PA Leiden
The Netherlands

*Lecture Series on Computer
and Computational Sciences*
Volume 4, 2005, pp. 1320-1322

Household Portfolio Dynamics in the E.U. – 1994-2000

Angelos A. Antzoulatos[1]   Chris Tsoumas

Department of Banking and Financial Management,
University of Piraeus,
80 Karaoli & Dimitriou Street
Piraeus 18534
Greece

Received 18 July, 2005; accepted in revised form 16 August, 2005

Abstract: During the second half of the 1990s, for which comparable data exist in Eurostat's New Cronos database, household portfolios in thirteen E.U. countries underwent major changes, prominent among which were a significant increase and a convergence in the share of stocks, and a decrease and divergence in the share of safe assets (currency and deposits). The empirical finding that these changes can to a large extent be explained by the returns of the respective assets, together with the qualitative evidence on the composition of corporate and bank balance sheets, suggest that the trend towards a more market-oriented financial system in the E.U. has lost momentum and likely leveled off.

Keywords: Household Portfolios, E.U., Stocks, Bonds, Safe Assets.

1. Extented abstract

In the second half of the 1990s there was a dramatic rise of the share of stocks in household portfolios in E.U. countries, mostly at the expense of the share of the relatively safe assets, currency and deposits, and to a lesser extent at the expense of bonds – which were modest, anyway. These changes, the product of both institutional and market-driven developments, were not uniform across the sample countries – the share of stocks became more similar while those of the safe assets less so.

This paper attempts to quantify the contribution of risk-return considerations in the household portfolios in several E.U. countries (Austria, Belgium, Denmark, Finland, France, Germany, Italy, Netherlands, Norway, Portugal, Spain, Sweden, U.K.) and shed some light on the question whether the trend towards more market-oriented household portfolio will endure.

The data for household portfolios comes from Eurostat's *New Cronos* database, on a yearly basis. From the items *currency and deposits, securities other than shares* and *shares and other equity,* we calculate three ratios (s_{1t}, s_{2t}, s_{3t} respectively) that correspond to the shares of the three generic assets in household portfolios.

We assume that households use a mean-variance criterion defined over the one return period of their portfolios in order to decide the shares of the three generic assets. They choose at the end of period t-1 the shares s_{1t}, s_{2t} and s_{3t}. The households' optimization problem suggest the following regression equation for the portfolio shares s_{it} (i=1,2,3) in which the sign of the coefficients a_j (j=1,...,9) is not known à priori. In this equation, the additional subscript i denotes the country.

$$s_{i,t} = \alpha_0 + \alpha_1 E_{t-1} R_{i,1t} + \alpha_2 E_{t-1} R_{i,2t} + \alpha_3 E_{t-1} R_{i,3t}$$
$$+ \alpha_4 E_{t-1}\sigma^2_{i,1t} + \alpha_5 E_{t-1}\sigma^2_{i,2t} + \alpha_6 E_{t-1}\sigma^2_{i,3t}$$
$$+ \alpha_7 E_{t-1}\sigma_{i,12t} + \alpha_8 E_{t-1}\sigma_{i,13t} + \alpha_9 E_{t-1}\sigma_{i,23t} + \varepsilon_{i,t}$$

Two proxies are used for the expected returns and their variance/covariance terms: the realized during the previous period (lagged) values and the contemporaneous realized ones. A general-to-specific

---

[1] Corresponding author. E-mail: antzoul@unipi.gr

modelling approach is followed with pooled data in order to get enough degrees of freedom in estimation, paying attention to whether a fixed effects model (FEM) or a random effects model (REM) is more appropriate. The share of bonds (s_{2t}), being very low, is not examined further.
The results are summarized in tables 1 and 2:

Table 1. Lagged Determinants of the Share of Safe Assets (s_1) and of the Share of Stocks (s_3)

$$s_{i,1,3t} = \alpha_i + \alpha_1 R_{i,1t-1} + \alpha_2 R_{i,2t-1} + \alpha_3 R_{i,3t-1} + \alpha_4 \sigma^2_{i,1t-1} + \alpha_5 \sigma^2_{i,2t-1} + \alpha_6 \sigma^2_{i,3t-1} + \alpha_7 \sigma_{i,12t-1} + \alpha_8 \sigma_{i,13t-1} + \alpha_9 \sigma_{i,23t-1} + \varepsilon_{i,t}$$

	Return Coefficients		Variance Coefficients		Covariance Coefficients	Hausman's χ^2	FEM R^2	REM R^2
	α_1	α_3	α_4	α_6	α_7			
Basic Model								
S_1	1.111 (2.85)***	-0.077 (-4.49)***	-1.041 (-2.63)***			4.48	0.93	0.16
S_3	-1.903 (-3.43)***	0.103 (4.22)***	1.811 (3.22)***			19.92***	0.88	0.24
Variation #1: Sample Without the Southern European Countries (Italy, Portugal, Spain)								
S_1	1.068 (1.99)*	-0.069 (-3.91)***	-1.005 (-1.85)*			3.35	0.95	0.17
S_3	-2.130 (-2.88)***	0.087 (3.57)***	2.056 (2.74)***			15.19***	0.92	0.32
Variation #2: Sample Without the More Market-oriented U.K. and Netherlands								
S_1	0.577 (2.44)**	-0.079 (-4.29)***		-0.502 (-2.10)**		3.88	0.931	0.17
S_3	-1.606 (-2.70)***	0.103 (3.75)***			1.513 (2.51)**	19.55***	0.88	0.27

Table 2. Contemporaneous Determinants of the Share of Safe Assets (s_1) and of the Share of Stocks (s_3)

$$s_{i,1,3t} = \alpha_i + \alpha_1 R_{i,1t} + \alpha_2 R_{i,2t} + \alpha_3 R_{i,3t} + \alpha_4 \sigma^2_{i,1t} + \alpha_5 \sigma^2_{i,2t} + \alpha_6 \sigma^2_{i,3t} + \alpha_7 \sigma_{i,12t} + \alpha_8 \sigma_{i,13t} + \alpha_9 \sigma_{i,23t} + \varepsilon_{i,t}$$

	Return Coefficients		Variance Coefficients		Covariance Coefficients		Hausman's χ^2	FEM R^2	REM R^2
	α_2	α_3	α_5	α_6	α_7	α_8			
Basic Model									
S_1	0.192 (5.87)***	-0.025 (-1.99)**	1.414 (5.82)***	-1.209 (-3.98)***		5.044 (2.36)**	18.00***	0.96	0.21
S_3	-0.238 (-4.65)***	0.036 (1.95)*	-2.373 (-6.05)***	1.279 (2.86)***	4.329 (1.89)*	-5.466 (-1.76)*	17.09***	0.94	0.34
Variation #1: Sample Without the Southern European Countries (Italy, Portugal, Spain)									
S_1	0.154 (4.03)***	-0.030 (-2.34)**	1.492 (5.14)***	-1.135 (-3.82)***	-4.057 (-2.63)***	5.11 (2.47)***	16.90***	0.97	0.23
S_3	-0.223 (-4.08)***	0.033 (1.82)*	-2.160 (-5.19)***	1.465 (3.44)***	6.654 (3.01)***	-7.215 (-2.44)**	41.73***	0.96	0.37
Variation #2: Sample Without the More Market-oriented U.K. and Netherlands									
S_1	0.192 (5.02)***	-0.024 (-1.86)*	1.430 (5.55)***	-1.005 (-3.27)***	-2.976 (-1.87)*	4.531 (2.16)**	48.33***	0.97	0.23
S_3	-0.283 (-4.70)***		-2.618 (-6.89)***		5.100 (2.07)**		5.30	0.94	0.38

Notes:
1. One (*), (**) and three (***) asterisks denote significance at respectively the 10%, 5% and 1% level.
2. For the sake of comparison, the results shown are for the FEM model.
3. Sources: EUROSTAT, DATASTREAM and authors' calculations

The econometric results, together with the visual evidence of the composition of finance corporations' balance sheets and of the assets of non-finance corporations in the thirteen sample countries, suggests that the trend towards a more market-oriented financial system in the E.U. has lost momentum if not leveled off. In addition, despite the dramatic rise of the share of stocks in household portfolios, it is not a foregone conclusion that the role of banks in the E.U. has been diminished or is likely to be diminished in the foreseeable future.

Acknowledgments

The systematic risk is being correlated with the macroeconomic atmosphere (Wilson 1998). Sharpe and others (2002) proved that default and business risk are coincident risks. Macro-economists catch the idea of the systematic risk in view of the default risk and they call it business risk (William Goetzmann). Eilingsfeld and Schaetzler are driven into the same result. KMV (Peter Grosbie 1999) showed that the default risk, which is the risk that firms face up when they are not able to clear their liabilities that is the case that the asset value of a firm's capital are not enough to cover the book value of the liabilities, is called business risk or asset risk. So there is a strong connection between default risk and business risk. In other words, default risk and business risk are the same aspect of the same coin. Financial institutions define risk as the default risk and firms define risk as business risk, therefore we examine the same aspect from a different direction.

We focus on the examination of the default risk process into the European Union before and after the entrance of the new countries. The computation of the default risk will be a proxy of the business risk. In other words that examination minds to investigate the process and the effect of new countries' business risk on all EU business risk and on business risk of the countries that are already EU members. In the course, we will examine if there are conditions of convergence between the enlarged EU overall business risk on the business risk of old members as impact of the entrance of new countries.

The next step is about the estimation and the impact of the total business risk of the EU and on it's cost of capital. Here we have to point out that the systematic risk is calculated by the beta coefficient of the CAPM model. Beta offers a method of measuring the risk of an asset that cannot be diversified (Andre F. Perold 2004, *The Capital Asset Pricing Model*, Journal of Economic Perspectives, 18,3). The cost of capital is, in turn, the expected return of CAPM model (Fama & French 2004, *The Capital Asset Pricing Model: Theory and Evidence*, Journal of Economic Perspectives, 18,3). So we will focus on the CAPM model framework.

$$r = rf + \beta(rm - rf) \Rightarrow r - rf = \beta(rm - rf) \Rightarrow \beta = \frac{r - rf}{rm - rf} \Rightarrow \beta \cong \frac{r}{rm}$$

where we have r= cost of capital, β= business risk, rm= market portfolio, rf= risk free asset

Therefore, risk equals to the ratio between the cost of capital and portfolio market. One limitation of the CAPM model is that a low beta may be volatile. This happens because beta takes into account only the systematic risk and not the specific risk but we use the beta as measurement of the systematic risk. Therefore, there is no questioning that the result can be misleading.

Brill Academic Publishers
P.O. Box 9000, 2300 PA Leiden
The Netherlands

*Lecture Series on Computer
and Computational Sciences*
Volume 4, 2005, pp. 1325-1328

A Recursive Bootstrap Evaluation of Moving Average Trading Rules

R. Batchelor and N. Nitsas[1]

Cass Business School
City University
106 Bunhill Row
London EC1Y 8TZ

Received 18 July, 2005; accepted in revised form 16 August, 2005

Abstract: Technical trading rules aim to give investors timely forecasts of the direction of change of prices in financial markets. It is now well documented that the apparent ability of very simple technical trading rules to generate superior risk-adjusted profits in the US stock market has broken down in the last two decades. The question we seek to answer in this paper is whether an investor could have accounted in real-time for this breakdown and stopped utilizing these rules. We analyze the daily closing levels of the S&P500 for the period 29/7/1971 through to 31/12/2003. We show that a 'naïve' investor choosing among a set of 152 double Moving Average crossover rules can always finds superior in-sample performance throughout this period. We use a current best-practice bootstrap procedure to determine whether the in-sample out performance of certain rules is robust to data mining biases, given the full set of rules from which these particular rules were selected. Recursive application of this procedure captures in real-time the break-down of their performance after the mid 1980s.

Keywords: trading rules, technical analysis, data mining, stationary bootstrap

1. Introduction

Technical Analysis is a common tool of predicting asset prices over shot horizons. According to Taylor (1994), over 90% of foreign exchange dealers in London use some form of technical analysis to predict returns. Finance academics are very skeptical of technical analysis, employing the usual 'efficient market' arguments. However, after an influential study of Brock ct al. (1992) which showed that very simple technical trading rules vastly outperformed the risk-adjusted returns of the DJIA over the period 1897-1986, there has been renewed academic interest in the field and tens of studies for various asset classes. Unfortunately most of these studies suffer from two problems. First the 'best rules' are only identified ex-post and not in real time. Although it is possible to find a rule that could have produced superior returns ex-post we don't know whether the investor would have chosen this particular rule ex ante. Furthermore, the issue of data mining arises. If an arbitrarily large number of trading rules are applied to a financial time series one is bound to come up with a rule that will produce superior returns. However this could be the result of chance and not to any value inherent to the rule. In this study we check whether it is possible to produce superior risk adjusted returns in real-time on the S&P500 using crossover moving average rules. Furthermore, we use the Superior Predictive Ability (SPA) test proposed by Hansen (2005) to account for data mining affects in real time (recursively).

2. Moving Average rules and the SPA test

Our data (P_t) are the closing levels of the S&P500 for the period 29/7/1972 through to 31/12/2003. We utilize $k=152$ crossover moving average (MA) rules to produce trading rules recursively:

$$MA(P_t, m_1, m_2) \equiv 1, if \; \frac{\sum_{t=0}^{m_1} P_{t-1}}{m_1 + 1} > \frac{\sum_{t=0}^{m_2} P_{t-1}}{m_2 + 1}$$

---

[1] Corresponding author. E-mail: n.nitsas@city.ac.uk

$$o \quad if \quad \frac{\sum_{t=0}^{m_1} P_{t-1}}{m_1 + 1} \leq \frac{\sum_{i=0}^{m_2} P_{t-1}}{m_2 + 1}$$

$$m_{1,2} \equiv \{1,2,5,10,15,20,25,30,35,40,45,50,75,100,125,150,175,200,250\}, \quad m_1 < m_2$$

where l is go long and o be out of the market. An investing decision is made every 25 trading days. The investor uses the rule with the best in sample performance to decide over her investment decision for the next month:

$$\max_{m_{1,2} \in c} \sum_{i=t-N}^{t-1} MA \, (P_i, m_1 m_2) R_{t+1}$$

where $R_{t+1}=(P_t-P_{t-1})/P_{t-1}$ We set $N=500$ (approximately 2 years of data) and perform overall 308 recursive estimations.

In order to account for the search among the set of trading rules the investor uses the SPA test.

$$Let \quad L\,(R_t, \delta_{k,t-1} R_t) = -\delta_{k,t-1} R_t$$

be the loss of rule k, where $\delta_{k,t-1}$ a binary variable that instructs a trader to take either a short or a long position in an asset at time $t-1$. The relative performance of rule k with respect to the benchmark of always being in the market is:

$$d_{k,t} \equiv L\,(R_t, \delta_{0,t-1}) - L\,(R_t, \delta_{k,t-1})$$

where $\delta_{0,t-1}=1$ the benchmark of always being long in the market. The hypothesis of interest is that the benchmark is not inferior to any of the k alternatives, which is formulated as:

$$H_0 : \mu_k \leq 0, \quad with \quad \mu_k \equiv E[d_{k,t}]$$

Setting $d_{k,t}=X_{k,t}$ Hansen (2005) shows that the statistic of interest is:

$$T_n^{SPA} = \max_{k=1,...m} \frac{n^{1/2} \overline{X}_k}{\hat{\omega}_k}, \quad where \quad \overline{X}_k = \frac{1}{n}\sum_{t=1}^{n} X_{k,t} \quad and \quad \hat{\omega}_k^2 = var(n^{1/2} \overline{X}_k)$$

Unfortunately the distribution of the test statistic is not unique under the null. However Hansen shows that a consistent estimate of the p-value can be obtained by using the stationary bootstrap of Politis and Romano (1994).

If we define the vector of relative performances at time t as:

$$X_t \equiv (X_{1,t}, \dots X_{m,t})'$$

then using the stationary bootstrap (which resamples blocks of random length, geometrically distributed with parameter q) the bootstrap resamples are given by:

$$X_{b,t}^* = X_{\tau,b,t}, \quad t = 1 \dots n$$

which leads to sample averages:

$$\overline{X}_b^* = n^{-1}\sum_{t=1}^{n} X_{b,t}^*, \quad b = 1, \dots B$$

We define:

$$Z^*_{k,b,t} \equiv X^*_{k,b,t} - \overline{X}_k$$

and

$$\overline{Z}^*_{k,b} = n^{-1} \sum_{t=1}^{n} Z^*_{k,b,t}$$

Then the bootstrap statistics of interest is given by:

$$T^{SPA*}_{b,n} = \max_{k=1,\dots m} [n^{1/2} \overline{Z}^*_{k,b} / \widehat{var}(n^{1/2} \overline{X}_k)]$$

and we can obtain the p-value:

$$\hat{p}_{SPA} \equiv \sum_{b=1}^{B} \frac{1_{T^{SPA*}_{b,n} > T^{SPA}_n}}{B}$$

A small *p-value* is evidence against the null. We set *q=0.5, B=1000* and compute the *SPA p-value* recursively for a total of 308 estimations. The investor decides to invest according to the best performing trading rule only if it is below 0.1 (10% critical level).

3. Main Results

Figure 1 shows that the best moving average rule is time variant, especially after 1983. Figure 2 shows that the recursive p-value is below the 0.1 critical level only until February 1980. Therefore an investor who performed the SPA test would not have used the best performing rule after 1980. Table 1 shows the annualized return from the various investment strategies: ex post the best performing moving average rule is the MA (2,0) rule which produces an annualized return of 14.7% compared to 9.14% for the buy and hold strategy. However, more importantly in real time it is not possible to beat the buy and hold: the recursive MA strategy (which is the one an investor would have followed without the benefit of hindsight) produces an annualized return of only 6.49%. However if an investor had used the SPA test then he would have accounted in real time for the breakdown in the performance of the rules. In particular, the investor would have used these rules only up to 1980, with an annualized return of 11.06%

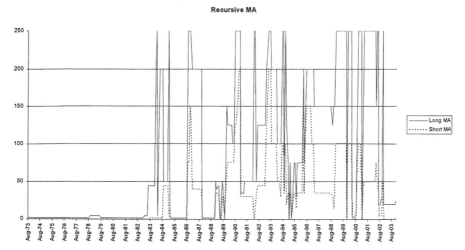

Figure 1: Best performing MA rule amongst the set of 152 rules, for the S&P500 calculated recursively every 25 trading days for the period 1/8/1973-31/12/2003

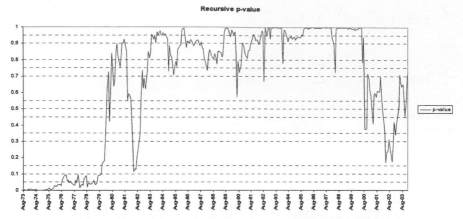

Figure 2: The SPA p-value calculated recursively every 25 trading days for the period 1/8/1973-31/12/2003

Buy and Hold	Best rule ex post MA (2,0)	Recursive MA	Data mining robust recursive MA
9.14%	14.7%	6.5%	11.06%

Table 1: Annualised returns of various investment strategies over the S&P500 for the period 1/8/1973-31/12/2003

4. References

[1] Brock W., Lakonishok J. and B. LeBaron(1992) 'Simple technical trading rules and the stochastic properties of stock returns', *Journal of Finance,* 47, 1731-1764

[2] Hansen, P. R. (2005), 'A test for superior predictive ability', forthcoming *Journal of Business and Economic Statistics*

[3] Politis, D.N., AND J. P. Romano (1994): "The Stationary Bootstrap," *Journal of the American Statistical Association*, 89, 1303–1313.

[4] Taylor, S.J. (1994) 'Trading futures using a channel rule: a study of the predictive power of technical analysis with currency examples', *Journal of Futures Markets*, 14:215-235

Brill Academic Publishers
P.O. Box 9000, 2300 PA Leiden
The Netherlands

Lecture Series on Computer
and Computational Sciences
Volume 4, 2005, pp. 1329-1331

Interactions Between Options and Stocks Within a VAR Framework: Evidence from Greece

George Filis[1]

University College Winchester,
Business School, Hampshire, UK

Costas Leon[2]

Democritus University of Thrace,
Department of International Economics and Development
Komotini 69100, Greece

Received 8 July, 2005; accepted in revised form 16 August, 2005

Abstract: VAR models are powerful devises to capture the interaction between variables. We use a VAR framework to examine the possible relationships between the Greek option and stock markets for the period September 2000 – September 2002, using daily data of returns and volumes from these two markets. Our tentative results show that "causality" runs from the option market to the stock market.

Keywords: VAR models, Options, Stocks, Cointegration Analysis, Athens Stock Exchange, Greece

1. Introduction

The interrelation between option and stock markets is an important issue, often examined from different angles. Some authors examine the relationship between the markets in general terms i.e. whether options provide efficiency to stock markets, or whether there is general interaction between both markets. Other authors study the option listing effects on individual stocks or on overall markets. Finally, others research which of the two markets is the primary market for investors, i.e. the one in which investors trade first. In these studies the methodologies usually employed are based on single-equation methods or, in fewer cases, on the traditional simultaneous equations. Despite the use of simultaneous equation systems, even in these few cases, the possible interaction between the two markets has not been thoroughly studied. The critic that one cannot be sure as to which variables are a-priori exogenous and what are endogenous is very well known, and in our opinion, applies equally here. A step forward to avoid the problem of the a-priori determination of the exogenous and endogenous variables is the development of VAR models in which all the involved variables are considered as endogenous and the exogeneity, if any, is not theory-imposed but, instead, data-driven. VAR models are very common in macroeconomics and monetary policy, but in finance are not frequently met. In particular, in the present issue of the possible interaction between option and stock markets, at least to our knowledge, only a few studies have examined this simultaneous interaction using a VAR approach, and regarding the Greek stock and option markets, such a study has not been conducted. The aim of this paper is to examine the possible relationships between the Greek option and stock markets using a VAR framework.

2. Relevant Literature

Many researchers have conducted studies regarding the interrelation between option and stock markets, drawing different conclusions. We present here three related lines of research but with different implications as to the direction of "causality". First, the stock market leads the option market. Second,

---

[1] Email: George.Filis@winchester.ac.uk.
[2] Corresponding Author. Email: kleon@ierd.duth.gr.

the reverse, i.e. the option market leads the stock market. Third, simultaneous interaction between the two markets. Summary of the relevant literature follows in the next Sections.

2.1. Stock Markets Lead Option Markets

Chan et al (1993) observed that the stock market leads the option market by 15 minutes. They used for their study a non-linear multivariate regression model, with the change of the call prices as the dependent variable. The independent variables were the change in stock price and the delta value. They examined stock and option prices for the 1st quarter of 1986 from NYSE, American Stock Exchange and CBOE. They explain this lead by the way the options and stock are traded. But, if the test performed had been examined with the bid-ask spread instead of the transaction costs, then this lead would disappear. Other studies, for example, Stephan and Whaley (1990), and Boyle et al (1999), support the view that the stock market leads the option market.

2.2. Option Markets Lead Stock Markets

On the other side, Manaster and Rendleman (1982) performed an ex-post and ex-ante test that contained closing stock and option prices to form portfolios in order to test whether there is more information in either the stock prices or the option prices. They concluded that the closing option prices contain information regarding the underlying stock that is not incorporated in the underlying price. Similar conclusions can be found in a study by Grenadier (1999). Other studies by Bhattacharya (1987), Anthony (1988) and Easley et al (1998) also concluded that option markets lead stock markets because they presented information, which predicts future stock price movements. Easley et al (1998) developed their own model of multimarket trading. It is a sequential trade model in which the traders make transactions in option and stock markets with risk neutral and competitive market makers. Their study contains 44 trading days from October to November 1990. The sample consisted of the first 50 firms ranked according to daily trading volume in the Market Statistics report of the CBOE. They suggested that only certain types of option trades have some predictive power for future stock price changes.

2.3. Simultaneous Interaction

Chan et al, in their later study (2002), investigate the interdependence of shares and their listed options in NYSE and CBOE, using a VAR framework. They observed that stock returns lead option returns. This indicates that order flows in stock markets are informative, that order flows in option markets are not, and that informed traders trade first in the stock market and then in the option market. A bivariate VAR model enhanced with GARCH errors to account for the autoregressive structure of the conditional variance is used by Berchtold and Norden (2004) who examine the information spillover effects between the Swedish OMX stock index and the index option market. Two types of information are examined. In the first type of information an informed investor knows the direction of the stock index, while in the second type of information, the direction is unknown, but an informed investor knows the trivial fact that the stock index either will increase or decrease. In their study, the authors detect significant conditional variance spillovers. It also turns out that today's options return shocks have an effect on tomorrow's conditional index returns variance but, on the other hand, stock index return shocks do not appear to influence the conditional option variance. This is in line with unidirectional information preceding directional information or information spillover from the option market to the stock market. Also, conditional stock index variance is affected by both types of information.

3. Methodology

We examine the possible relationships between the Greek option and stock markets within a VAR framework. The variables comprise of daily data for the period 11 September 2000 (the establishment of the Greek options market) up to 11 September 2002 and they are: index returns, index volumes, option returns and option volumes. The VAR model is defined as $\mathbf{X}_t = \mathbf{\mu} + \sum_{i=1}^{p} \mathbf{B}_i \mathbf{x}_{t-i} + \mathbf{C}\mathbf{Z}_t + \mathbf{\varepsilon}_t$ where $\mathbf{X}_t$ is the vector of the endogenous variables (i.e. the aforementioned variables), $\mathbf{\mu}$ is a vector of constants $\mathbf{X}_{t-i}$ are the vectors of endogenous random variables in lags $i = 1, 2, .. p$, $\mathbf{B}_i$ are the matrices corresponding to the various $\mathbf{X}_{t-i}$ in lags $i = 1, 2, .. p$, $\mathbf{Z}$ is a matrix of deterministic components such as dummies, exogenous variables or a linear time trend, and $\mathbf{C}$ is the corresponding parameters' matrix to Z. The vector ε_t is assumed to generate various shocks which are instantaneously orthogonal and also white noise processes. A series of testing procedures is the next step in a VAR analysis. Testing for unit roots, determination of the optimal lag length and testing for possible cointegration and determination of the cointegrating rank in the case of cointegration. According to the results of the last test, the VAR is estimated in levels, in differences or in a Vector Error Correction framework. After the estimation, Granger causality tests show the direction of the causality. Finally, impulse response functions show the response of each endogenous variable to simulated shocks. We have conducted a preliminary series for all of these tests and we have drawn the tentative conclusion that all series are stationary and the VAR can be estimated in levels. The tentative conclusion is that causality runs from the option market to the stock market.

4. Conclusion

We have attempted to test the interaction between the Greek option and stock markets using a VAR framework for the period 11 September 2000 up to 11 September 2002 within a VAR framework. The tentative conclusion is that causality runs from the option market to the stock market.

Acknowledgments

The authors wish to thank the anonymous referees for their careful reading of the manuscript and their fruitful comments and suggestions.

References

[1] Anthony, J., 1988, *The interrelation of stock and option market trading-volume data*, Journal of Finance, 43, pp.949-964.

[2] Berchtold F., Norden L., 2004, *Information spillover effects between stock and option markets*, Stockholm University, Working paper presented at the European Financial Management Association 2004 Meeting in Basel, Switzerland.

[3] Bhattacharya M., 1987, *Price changes of related securities: the case of call options and stocks*, Journal of Financial and Quantitative Analysis, 22, pp.1-15.

[4] Boyle P. P., Byoun S., Park H. Y., 1999, *Temporal price relation between stock and options markets and a bias of implied volatility in option prices,* Oxford Paper Number 99-07.

[5] Chan, K., Chung, P., H. Johnson, 1993, *Why option prices lag stock prices: A trading based explanation*, Journal of Finance, 48, no. 5, pp.1957-1968.

[6] Chan Y. K. Chung P., Fong W. M., 2002, *The Informational Role of Stock and Option Volume*, Review of Financial Studies, vol. 15, issue 4, pp.1049-1075.

[7] Easly D., O'Hara M., Srinivas P. S., 1998, *Option volume and Stock prices: Evidence on where informed traders trade,* Journal of Finance, vol. LIII, no.2, pp.431-465.

[8] Grenadier S., 1999, *Information revelation through option exercise*, Review of Financial Studies, vol. 12, no. 1, pp.95-129.

[9] Manaster S., Rendleman J., 1982, *Option prices as predictors of equilibrium stock prices*, Journal of Finance, vol. 37, no. 4, pp.1043-1057.

[10] Stephan J. A., Whaley R. E., 1990, *Intraday price change and trading volume relation in the stock and stock option markets*, Journal of Finance, vol. 45, pp.191-220.

Brill Academic Publishers
P.O. Box 9000, 2300 PA Leiden
The Netherlands

Lecture Series on Computer
and Computational Sciences
Volume 4, 2005, pp. 1332-1335

Bankruptcy Prediction for US Telecoms: Capital Expenditure and Profitability Effects

Dr. Fotios C. Harmantzis[1]   Angelo G. Christides

School of Technology Management
Stevens Institute of Technology
Hoboken 07030, NJ USA

Received 12 July, 2005; accepted in revised form 16 August, 2005

Abstract: The telecommunications market growth and downfall between 1996 and 2002, was characterized by excessive borrowing and spending, high inventories, over-capacity, and dismal returns. Numerous telecommunication companies have entered bankruptcy proceedings in the United States recently, with 2002 being a record year, e.g., WorldCom bankruptcy. This study demonstrates the effect of capital expenditure and profitability to corporate bankruptcy risk for the particular industries. Existing, widely used, bankruptcy prediction models are based on both accounting and market variables. In this paper, we draw a relationship between the bankruptcy models' signal, the capital expenditures and return on investment capital factors. The cross-sectional study covers a history of almost 20 years of publicly traded firms in the United States.

Keywords: Bankruptcy, Credit, Financial Risk, Profitability, Capital Expenditure, Multi-factor Regression

Mathematics Subject Classification: 91B30, 91B84, 62J05, 62H15, 62M20

1. Introduction

There has been extensive work in the area of corporate probability bankruptcy, in the last 35 years. This work has resulted in different types of bankruptcy prediction models, e.g., accounting-based, market-based, mathematical models, etc. Notably, Altman's (1968) Z-Score model [1] and Ohlson's (1980) Model-1 [6] relied primarily on accounting measures as proxies to bankruptcy risk. Altman's model is becoming an industry standard for financial analysts who want to access bankruptcy risk of a public company. A credit model based on Merton's (1974) option pricing theory [5], can be viewed as the starting point to associate probability of bankruptcy from market data. Merton's model is the premier of the so-called structural-type of credit models.

Recently, Chava and Jarrow demonstrated that simple classification, in financial and non financial firms, private and public firms, would not be sufficient to confirm the accuracy of these models [2]. Further classification of these firms is required, in industry groupings, in order to include industry effects in bankruptcy prediction. The Telecommunications (Telecom) industry that we deal with in this paper is one of those groupings. Since 1996 and for a period of at least seven years, the industry has seen an unprecedented rate of growth, both in revenue and spending. At the same time, with the turn of the century, the industry experienced an unprecedented rate of bankruptcy filings [4].

In our paper, we use four models, namely, Altman's, Ohlson's 1 and 2, and Merton's, as proxies for the bankruptcy risk of public telecom companies. The purpose of our research is not to access the validity of credit risk models. Since our purpose is to access the impact of certain financial statement items, the selection of accounting and market-based models is a natural one.

During the period between 1984 and 2003, we investigate the relationship between credit risk for these firms, and respective accounting and market variables for that period. We conduct a regression analysis in which we investigate the relationship between the bankruptcy scores and the Capital Expenditure (CapEx) (proxy for expenses) and Return On Invested Capital (ROIC) (proxy for profitability of the firm) as strong industry effect variables.

---

[1] Corresponding Author: Dr. Fotios Harmantzis, Assistant Professor, Head of Financial Analytics (FinA) Research Group. E-mail: fharmant@stevens.edu

2. Research Methodology

The data sample is consisted of an average of 200 public US telcos per year, classified into two groups: *Telecom Manufacturers* (e.g., Cisco, Lucent, Motorola, Nortel, etc.) and *Telecom Services* providers of local and long distance phone, service providers, wireless, etc., (e.g., AT&T, Verizon, SBC, Nextel, etc.). Accounting data and financial ratios have been collected from COMPUSTAT, starting in the year of 1984 and ending with the year 2003 (pre & post-1996 Telecom Act era). Our sample consists of selected firms with consistent reporting of their financials.

We fist start with Altman's Z-Score and Ohlson's Model-1 (for one and two years). The two models, utilize annual accounting variables as the primary prediction strength, with the Z-Score denoting financial strength and the Model-1 (O-Score) financial distress. The Z-Score results are inversely proportional to the probability of bankruptcy. Higher values of the Z-Score imply a lower probability of bankruptcy. The opposite relationship holds true for the Ohlson Models, in which higher scores imply higher probability of bankruptcy. In comparison, the two models are negatively correlated.

Similarly, Merton's option pricing model is used, taking into consideration market variables, e.g., stock price volatility (based on daily close), interest rates (10-year US Treasury) and market value of the firm (market capitalization plus debt minus cash).

Besides excessive capital expenditure and low profitability, a number of other factors, contributed to the downfall of well established Telecom firms during the dot-com era [4]. For example, for a typical telecommunications equipment manufacturing firm: high level of inventories, technology renewal/innovation, excessive capital and equipment lending by the firm to the customer, high merger and acquisition costs, etc. However for this paper, in order to demonstrate the strong effects of CapEX and ROIC in bankruptcy risk, we develop linear regression models between the output of the bankruptcy models (dependent variable), and the annual CapEX and ROIC figures of the firm (independent variables).

3. Cross-Sectional Regression Models

In this study we split the firms into two groups: the telecom manufacturing companies (voice, data, wireless) and telecom services companies (voice, data, wireless). Accounting data were collected for the period 1984-2003. The sample space in each year of observation ranged from a minimum of 38 firms in 1984 to a maximum of 233 in 1999 in the manufacturing sector, and from a minimum of 17 firms in 1984 to a maximum of 117 in 1999 in the services sector. Telecom sector has been a growth engine for the US economy for many years. After 1999, in both sectors, the sample space decreased by 24% and 22%, by 2003 respectively, as a result of the telecom fallout in US and worldwide. It can easily be seen the market growth in both sectors, especially after the Telecom Act was put in place in 1996, when the telecom market was open to pure competition.

At the same time interval of twenty years, CapEx followed the same trend. For both groups of companies, manufacturers and service providers, capital expenditure reached a positive maximum in the year 2000, and declined dramatically thereafter. This was inline with the market boom and bust in the late 90's. Low barriers to entry gave the opportunity to a large number of relatively small firms to enter the market. The market's perception of the Internet bandwidth, more than doubling every 100 days (a totally false assessment), forced CapEx spending to skyrocket – enough spending to just keep up with the competition.

In the telecom manufacturing sector, the return on invested capital numbers followed an inversely proportional to the CapEx pattern, reaching a negative peak in 2001. Most, if not all, of the sales during a period of four years, were primarily either on credit (bad credit) or on a loaner/free equipment basis. When the credit payments were due, the customer either had already filled bankruptcy, or simply could not re-pay the loan. In the services sector though, the situation was better. In particular, by the end of 1999, the telecom services market exhibited a positive ROIC, mostly attributed to optical and wireless networking services.

For the multi-regression analysis we define one dependent and two independent variables. We run the same regression equation to both groups of companies.

$$Bankruptcy\text{-}score_t = \beta_0 + \beta_1(CapEx_t) + \beta_2(ROIC_t) + e_i \qquad (1)$$

where: $Bankruptcy\text{-}score_t$ is the bankruptcy score for each firm per year t, $CapEx_t$ the capital expenditure of the firm at year t, and ROIC is the return on invested capital of the firm at year t.

This is a cross-sectional study. For each company in the sample, for each year, the bankruptcy score as well as the CapEx and ROIC figures were calculated, avoiding look-ahead bias. Some companies are present during the entire period, some they had a very short life. The CapEx and the ROIC for each company, for each year, were used as the two independent variables in each regression equation.

The results of the cross-sectional regressions are summarized in following tables. Those models constitute the contribution of our work. In the manufacturing sector, overall the coefficient of the positive CapEx was not significant in any bankruptcy model or regression equation. In the years between 1996 and 2003 CapEx was increasing at an alarming rate. Yet those expenditures were not justifiable by very poor profitability figures. This leads us to the second significance test and observation. On the other hand, the coefficient of the negative ROIC was significant and it provided a linear relationship with the bankruptcy scores in the first three models, Z-score, Ohlson (1), and Ohlson (2). For the Real-Options model, hardly any of the independent variables were significant. In the services sector, the situation is similar for the first three bankruptcy models. Out of the two independent variables only the ROIC is significant. However in the Real-Options model, CapEx now becomes significant and ROIC insignificant.

Z-Score: For both groups of companies the coefficient on ROIC is significantly positive while the coefficient on CapEx is insignificant.

Telecommunications Manufactures
Z-score $_t$ = -0.062$(CapEx_t)$ + 0.138$(ROIC_t)$

Telecommunications Service Providers
Z-score $_t$ = -0.075$(CapEx_t)$ + 0.179$(ROIC_t)$

One way of interpreting this could be that CapEx is typically a cash outlay, relatively small or large, which at a first glance does not indicate the firm's financial exposure or lack thereof. On the other hand, ROIC is a ratio of profits after taxes to the total assets minus the current liabilities. This ratio is more in line with the Z-score model's predicting ability through the use of selected financial ratios which focus on the firm's earning, sales, and liabilities. As it was mentioned previously, the Z-score indicates financial strength, and it is evident at this point, from the regression results, that the ROIC coefficient is tracking positively the Z-score results.

O-Score: Similar to the results from the regressions for the Z-score, the coefficients on ROIC, in both market sectors, and in both O-score models, are significant with a negative sign. Once again the coefficients on CapEx, in both cases respectively, are insignificant. The similarities between ROIC and the financial ratios utilized by the O-score models, are the same as in the previous section above, since the O-score model focuses on financial leverage, sales, and earnings.

Telecommunications Manufactures
O-score(1) $_t$ = 0.009$(CapEx_t)$ − 0.459$(ROIC_t)$

Telecommunications Service Providers
O-score(1) $_t$ = − 0.090$(CapEx_t)$ − 0.420$(ROIC_t)$

Telecommunications Manufactures
O-score(2) $_t$ = − 0.012$(CapEx_t)$ − 0.438$(ROIC_t)$

Telecommunications Service Providers
O-score(2) $_t$ = − 0.123$(CapEx_t)$ − 0.452$(ROIC_t)$

Attention should be paid to the negative sign of the coefficient on ROIC. As it was mentioned previously, the O-score indicates financial distress. Therefore, a higher bankruptcy score will be inversely proportional to a lower ROIC and visa versa. Therefore lower/higher returns on invested capital are tracking the O-score accordingly.

Options Model: For the telecommunications manufacturers industry both coefficients on CapEx and ROIC are insignificant. In the services sector however, the coefficient on CapEx is significant. The coefficient on ROIC is insignificant. Perhaps one possible explanation of the results is that under the Real-Options model we evaluate a call option on the market value of the firm's assets against its liabilities. CapEx in the services sector has more profound impact to the firm's financial strength than ROIC. CapEx is used primarily for the build-out of the telecom services and infrastructure which in turn may leave the firm with increased assets and liabilities.

Telecommunications Manufactures
Real-Options $_t$ $= -0.120(CapEx_t) - 0.094(ROIC_t)$
Telecommunications Service Providers
Real-Options $_t$ $= -0.339(CapEx_t) - 0.122(ROIC_t)$

Even so, someone would expect that the CapEx effect on the probability of bankruptcy under the Real-Options model would be significant during the latter years, between 1996 and 2003. Based on the regression results we find that this is not true. Instead, the coefficient on CapEx is significant during the years prior to 1996.

The coefficients on ROIC in both O-score models, in both market sectors, are highly significant compared to the ones in the Z-score. This can possibly be attributed to the slightly more accurate predicting ability of the O-score models. However, no noticeable differences between the 1- and 2-year models exist. The ROIC is equally significant to the probability of bankruptcy in both market sectors.

4. Conclusion

Our study is extensive as it covers a period of almost 20 years. Cross-sectional multi-factor regressions were conducted in the universe of almost all public telecom companies in the United States, in both services and equipment manufacturers. The frequency we used here to calculate financial items is annual (for stock price volatility daily prices are used).

There is a live debate in the credit risk community concerning the suitability and accuracy of credit risk models, i.e., accounting-based vs. market based, structural vs. reduced-form models. The purpose of our research is not to answer which model should be used for bankruptcy prediction. We use different models as proxies of bankruptcy risk. US telcos recorded bankruptcy filings in the beginning of the 21st century. We wanted to study the explanatory power of capital expenses and returns in credit risk indicators.

We found that capital expenditure alone, as a cash outlay, does not explain adequately the rise and fall of the telecom industry. Perhaps capital expenditure may need to be used in a ratio format combined with sales figures or used in conjunction with "excessive" inventories and/or assets, in order to be able to imply any type of bankruptcy risk. ROIC contains a debt term which is in line with the accounting based score models. Therefore, it performed as it was expected.

References

[1] E. Altman, Financial Ratios, Discriminant Analysis and the Prediction of Corporate Bankruptcy, *Journal of Finance*, 23 589-609 (1968).

[2] S. Chava and R. Jarrow, *Bankruptcy Prediction with Industry Effects, Market versus Accounting Variables, and Reduced Form Credit Risk Models*, Working Paper, Johnson Graduate School of Management, Cornell University (2001).

[3] D. Endicott, *Is Low Telecom ROIC here to Stay, and how Long will Investors Bear This?*, Working Paper, Columbia Institute for Tele-Information, Columbia University (2003).

[4] F. Harmantzis, *Inside the Telecom Crash: Bankruptcies, Fallacies and Scandals*, International Telecom. Society 15th Biennial Conference, Sept. 5-7 2004, Berlin, Germany (2004).

[5] R. Merton, On the Pricing of Corporate Debt: The Risk Structure of Interest Rates, *Journal of Finance*, 29 449-470 (1974).

[6] J. Ohlson, Financial Ratios and the Probability Prediction of Bankruptcy, *Journal of Accounting Research*, Vol.18, No.1, pp.109-131 (1980).

Brill Academic Publishers
P.O. Box 9000, 2300 PA Leiden
The Netherlands

*Lecture Series on Computer
and Computational Sciences*
Volume 4, 2005, pp. 1336-1336

Multivariate Long Memory Volatility Models: A Comparative Analysis with an Application to the Tokyo Stock Exchange

Kin-Yip Ho[1]

Department of Economics
Cornell University
Ithaca NY 14853 USA

Received 16 August, 2005; Accepted 16 August, 2005

Abstract: We analyze two competing approaches of modeling long-memory persistence and asymmetry in volatility in a multivariate framework: the fractionally integrated and the component approaches. Based on these two approaches, four main classes of multivariate models with time-varying correlations are developed, two of which are the varying-correlations fractionally integrated asymmetric power autoregressive conditional heteroskedasticity (VC-FIAPARCH) model and the varying-correlations component quadratic generalized ARCH (VC-COMPQGARCH) model. These models not only parsimoniously capture the features of long-memory volatility persistence, asymmetric volatility, and dynamic correlations typically exhibited by stock market returns, but also guarantee the positive-definiteness of the conditional variance-covariance matrix once parameter estimates are obtained. Our models are applied to the sectoral indices of the Tokyo stock market, which are constructed based on either the different sizes of market capitalization of the listed companies or the types of industry to which the component stocks belong.

In contrast to what is widely documented in the literature, asymmetric effects are not invariably present in the sectoral indices. Additionally, the conditional correlations are frequently highly positive and significantly time-varying. Our findings not only cast doubts on the "leverage effect" of equity returns, but also have bearing on the strategy of portfolio diversification among various sectors.

We also detect strong evidence of volatility persistence and long memory, and our two approaches of modeling volatility dynamics generally outperform the models without long-memory structures. The fractionally integrated approach also reveals that several sectors might share a common degree of fractional integration (long-memory persistence), whereas the component approach suggests that the weight of the integrated volatility component can vary remarkably across the sectors. However, the estimation results indicate that it is difficult to conclude which of the two approaches is deemed superior. We discuss some implications that these findings have on portfolio and risk management, asset allocation, and derivatives pricing.

Keywords: Component GARCH; Stock Market Volatility; Fractional Integration; Multivariate Asymmetric Long-Memory GARCH; Varying Correlations

Mathematics Subject Classification: 91B28, 91B84, 62H12, 62P20

---

[1] Corresponding author. PhD student of the Department of Economics, Cornell University. E-mail: kh267@cornell.edu

Brill Academic Publishers
P.O. Box 9000, 2300 PA Leiden
The Netherlands

*Lecture Series on Computer
and Computational Sciences*
Volume 4, 2005, pp. 1337-1339

The No Arbitrage Condition in Option Implied Trees:
Evidence from the Italian Index Options Market

V.Moriggia[a], S. Muzzioli[b1], C. Torricelli[b]

[a]Department of Mathematics, Statistics, Computer Science and Applications, Faculty of Economics,
University of Bergamo, Bergamo, Italy
[b]Department of Economics, Faculty of Economics , University of Modena and Reggio Emilia
41100 Modena, Italy

Received 12 August, 2005; accepted in revised form 16 August, 2005

Keywords: Binomial tree, implied volatility, calibration.
Mathematics Subject Classification: 91B28, 91B70

1. Extended Abstract

After the October 1987 crash, option markets exhibited implied volatilities that varied across different strikes (smile effect) and different times to expiration (term structure of the volatility), in contrast with the Black and Scholes assumption of constant volatility.

In order to capture the implied volatility dependence on strike and time to maturity, different smile-consistent no-arbitrage models have been proposed in the literature, which can be classified either as deterministic or stochastic volatility models[2]. Deterministic volatility models (see e.g. Derman and Kani (1994), Barle and Cakici (1998), Rubinstein (1994), Jackwerth (1997), Dupire (1994)) derive endogenously from European option prices the instantaneous volatility as a deterministic function of the asset price and time. Stochastic volatility models (see e.g. Derman and Kani (1997), Britten-Jones and Neuberger (2000), Ledoit and Santa Clara (1998)) allow for a no-arbitrage evolution of the implied volatility surface.

Deterministic volatility models have both theoretical and practical advantages: they preserve the no-arbitrage pricing property of the Black and Scholes model and are easily implementable. With the exception of Dupire (1994), which is developed in continuous time, most models are developed in discrete time. Among the latter, some (Derman and Kani (1994), Barle and Cakici (1998), Li (2001)) use forward induction in the derivation of the implied trees, others (Rubinstein (1994), Jackwerth (1997)), use backward induction[3]. The Rubinstein (1994) model is based on the assumption that different paths that lead to the same ending node have the same risk neutral probability, it captures only the smile effect and it is not useful for pricing path dependent options. The Jackwerth (1997) model, extend Rubinstein's by allowing the implied tree to fit intermediate maturity options, thus capturing both the smile effect and the term structure of the volatility. The main advantages of deriving implied trees by forward induction is that only observable data are used and, in contrast to backward induction, no estimation of ending risk neutral probabilities is needed.

A few papers empirically test the pricing performance of deterministic smile-consistent option pricing models (see among others, Dumas et al. (1998), Lim and Zhi (2002), Brandt and Wu (2002), Hull and Suo (2002), Linaras and Skiadopoulos (2005)), while, as underlined by Linaras and Skiadopoulos (2005), stochastic volatility smile-consistent models have not been tested yet, because of various computational limitations. The empirical tests compare different types of smile-consistent deterministic models w.r.t constant volatility models (such as Black and Scholes (1973), Cox-Ross-Rubinstein (1979)). The evidence on the pricing performance of deterministic smile-consistent models is mixed. Dumas et al. (1998) and Brandt and Wu (2002) find that they do not perform better than an ad

---

[1] Corresponding author. E-mail: muzzioli.silvia@unimore.it
[2] See Bates (2003) for a survey on the approaches taken in option pricing and Skiadopoulos (2001) for a taxonomy and an extensive survey on smile-consistent no arbitrage models.
[3] This paper departs here from the terminology used by Skiadopoulos (2001) in that forward induction models are meant as those that use also forward induction and backward induction ones are those that use only backward induction.

hoc procedure that smoothes Black and Scholes (1973) implied volatilities across strikes and time to expiration. By contrast, Hull and Suo (2002) find that they are superior to Black and Scholes in the pricing of exotic options. In Lim and Zhi (2002) and Linaras and Skiadopoulos (2005), the pricing performance of different types of deterministic-smile consistent models is shown to strongly depend on various factors (option class chosen, moneyness and time to expiration). No apparent superiority of one specific model w.r.t. the others emerges.

A major issue that negatively affects the pricing performance of implied trees based on forward induction is the occurrence of negative probabilities, which following Linaras and Skiadopoulos (2005) can be addressed to as "bad probabilities". Negative probabilities indicate the presence of arbitrage opportunities. Derman and Kani (1994) propose a methodology to override the nodes that violate the no arbitrage condition. Nonetheless negative probabilities are frequently found, questioning the correct replication of the observed smile. Barle and Cakici (1998) extend the Derman and Kani's algorithm, in order to increase its stability, in particular in the presence of high interest rates. Negative probabilities turn out less frequently, but in the presence of increasing interest rates and smile slopes, the fit to the smile is poor. In order to solve the problem, Li (2001) proposes to derive implied trees, by assuming constant nodal probabilities equal to 0.5. However, the Li's model strongly hinges on the assumption that the risk neutral measure exists. Moreover it is not appropriate for pricing path-dependent options, since all paths leading to the same node are equally likely (as in the Rubinstein's model).

In sum, focusing on deterministic volatility models based on forward induction, Derman and Kani (1994) remains comparatively the most suitable. In order to remove the problem of negative probabilities, the aim of this paper is to propose a modification of the no arbitrage test used to this purpose. The no arbitrage condition is examined by including dividends into the picture. In order to improve the fit to deep out of the money options, a no-arbitrage test for the nodes at the boundary of the tree is introduced. The proposed methodology is a modified Derman and Kani model and will be compared with the Barle and Cakici (1998) implied tree, both in the sample and out of sample. The empirical validation of the different implied trees is performed by using a data set, Italian index options over the period March 2000 - December 2003, which to our knowledge has not been yet used to the same purpose.

Overall findings support a better performance of the modified Derman and Kani's methodology. In particular, the better performance of the modified Derman and Kani can be attributed to the better pricing of out of the money put options, i.e. a better fit in the lower part of the tree. The better fit can be also explained by a lower number of no arbitrage violations for the modified Derman and Kani. This results in a lower number of stock price replacements and therefore in a better fit to traded option prices.

References

[1] Barle S., N. Cakici, 1998. How to grow a smiling tree. *Journal of Financial Engineering*, 7 (2) 127-146.

[2] Bates D. S., 2003. Empirical option pricing: a retrospection. *Journal of Econometrics*, 116, 387-404.

[3] Black F., Scholes M., 1973. The pricing of options and corporate liabilities. *Journal of Political Economy*, 81, 637,654.

[4] Brandt M. W., Wu T., 2002. Cross sectional tests of deterministic volatility functions. *Journal of Empirical Finance*, 9, 525-550.

[5] Britten-Jones M., Neuberger A., 2000. Option prices, implied price processes and stochastic volatility. *Journal of Finance*, 55 (2) 839-866 .

[6] Cox J., Ross S., Rubinstein M., 1979. Option pricing: a simplified approach. *Journal of Financial Economics*, 7, 229-263.

[7] Derman E., Kani I., 1994. Riding on a smile. *Risk*, 7 (2) 32-39.

[8] Derman E., Kani I., 1997. Stochastic implied trees: arbitrage pricing with stochastic term and strike structure of volatility. *International Journal of Theoretical and Applied Finance*, 1, 61-110.

[9] Dumas B., Fleming J., Whaley R.E., 1998. Implied volatility functions: empirical tests. *Journal of Finance*, 53, 2059-2106.

[10] Dupire B., 1994. Pricing with a smile. *Risk*, 7 (1) 18-20.

[11] Hull J.C., Suo W., 2002. A methodology for assessing model risk and its applications to the implied volatility function model. *Journal of Financial and Quantitative Analysis*, 37, 297-318.

[12] Jackwerth J., 1997. Generalized binomial trees. *Journal of Derivatives*, 5, 7-17.

[13] Ledoit O. Santa Clara P., 1998. Relative pricing of options with stochastic volatility. *Working Paper of the University of California, Los Angeles*.

[14] Li Y., 2001. A new algorithm for constructing implied binomial trees: does the implied model fit any volatility smile? *Journal of Computational Finance*, 4 (2) 69-95.

[15] Lim K., Zhi D., 2002. Pricing options using implied trees: evidence from FTSE-100 options. *Journal of Futures Markets*, 22, 601-626.

[16] Linaras H., Skiadopoulos G., 2005. Implied Volatility Trees and Pricing Performance: Evidence from the S&P100 Options, PP05-146 Working Paper, Financial Options Research Centre, University of Warwick.

[17] Rubinstein M., 1994. Implied binomial trees. *Journal of Finance* 49 (3) 771-818.

[18] Skiadopoulos G., 2001. Volatility Smile Consistent Option Models: A Survey. *International Journal of Theoretical and Applied Finance*, 4 (3) 403-437.

Brill Academic Publishers
P.O. Box 9000, 2300 PA Leiden
The Netherlands

*Lecture Series on Computer
and Computational Sciences*
Volume 4, 2005, pp. 1340-1344

The Predictive Content of Financial Variables:
Evidence from the Euro Area

Ekaterini Panopoulou[1]

National University of Ireland, Maynooth

Received 25 July, 2005; accepted in revised form 16 August, 2005

Abstract: In this paper we investigate the predictive ability of financial variables for real growth through bivariate and multivariate non-parametric Granger causality tests for the euro area. Apart from assessing the within-country forecasting ability of commonly-employed financial variables, such as the term spread, the stock market returns and the real growth of money supply, we also test for cross-country influences. In this way, we reveal the countries that are more useful in predicting growth in other member countries along with the ones that are more receptive to other countries' financial developments.

Keywords: Granger causality; Cross-correlations; Real growth; Financial variables

1. Introduction

A vast literature in finance and macroeconomics is devoted to the forecasting ability of financial variables for real economic activity. Empirical evidence is mixed and results are not robust with respect to model specification, sample choice and forecast horizon.[2]

To remedy potential caveats associated with the use of standard parametric techniques in the empirical investigation of the relationship between financial variables and output growth, we reinvestigate systematically this bivariate relationship by using the non-parametric methodology proposed by Cheung and Ng (1996).[3] To investigate the bivariate relationship between financial variables and industrial output growth in the context of these non-parametric methodologies we utilize monthly data from the euro area countries. As a second step, following Lemmens et al. (2005) we extend our bivariate testing procedure to a multivariate one by pooling together the information from the whole panel of the euro area countries. This multivariate testing procedure was introduced by El Himdi and Roy (1997) and enables us to investigate the general predictive content of a candidate financial variable for economic growth for the entire panel.

To the best of our knowledge, this multivariate testing methodology has been hardly employed in the literature. Specifically, El Himdi and Roy (1997), who proposed this methodology, applied this multivariate test to investigate the causal relations between money and income for Canada, as well as to study the causal directions between the Canadian and American economies. Lemmens et al. (2005) adapted the El Himdi and Roy test to jointly test the forecasting ability of multiple production expectation series for the members of the European Union. In this sense, they assessed whether part of the joint effect they found was due to cross-country influences and they determined the countries which have the most "clout" along with those with the most "receptivity".

---

[1]*Correspondence to:* Ekaterini Panopoulou, Department of Economics, National University of Ireland Maynooth, Co.Kildare, Republic of Ireland. E-mail: apano@nuim.ie. Tel: 00353 1 7083793. Fax: 00353 1 7083934.

[2]See Stock and Watson (2003) for a review of the empirical literature.

[3]The methodology put forward by Cheung and Ng (1996) extends the one proposed by Haugh (1976) to test for both causality in mean and variance. In the present text, we adopt the two-stage procedure of Cheung and Ng (1996), which filters out second-order effects prior to testing for causality in mean.

2. Econometric Methodology

In this section, we briefly describe the non-parametric techniques utilized in the present study which aim at detecting any Granger causality running from financial variables to output growth. In subsections 2.1 and 2.2 we describe the bivariate and multivariate methodology employed, respectively.

2.1. Bivariate within-country causality tests

Consider a bivariate stationary and ergodic stochastic process $Z_t = [y_t, x_t]^T$, $t = 1,2,\dots$. In our case, y_t, represents output growth and x_t, a financial variable. Cheung and Ng (1996) proposed a test based on the sample cross-correlations function of the standardized residuals and involves two stages. In the first stage, univariate time-series models are estimated for both the series under scrutiny, such as the typical ARMA(p,q)-GARCH(1,1). In our case the correct order of the ARMA(p,q) model for the mean of the series is determined by means of the Schwartz Information Criterion (SIC). In the second stage, we calculate the sample cross-correlations of the standardized residuals, typically defined as

follows: $\begin{aligned}\hat{u}_{yt} &= (y_t - \hat{\mu}_{y,t})/\hat{h}_{y,t}\\ \hat{u}_{xt} &= (x_t - \hat{\mu}_{x,t})/\hat{h}_{x,t}\end{aligned}$ where $\hat{\mu}_{y,t}, \hat{\mu}_{x,t}$ and $\hat{h}_{y,t}, \hat{h}_{x,t}$ are the estimated conditional means and

variances of output growth and real stock returns, respectively.

The sample cross-correlation function of u_{yt} and u_{xt} ($\hat{\tau}_{x,y}(k)$) is given by:

$$\hat{\tau}_{x,y}(k) = \frac{\hat{C}_{x,y}(k)}{\sqrt{\hat{C}_{x,x}(0)\hat{C}_{x,y}(0)}}$$ where $\hat{C}_{x,y}(k)$ is the sample cross-covariance, $\hat{C}_{x,x}(0)$, $\hat{C}_{y,y}(0)$ are the

sample variances of one of the financial variables and output growth, respectively, k is the lag length employed and T is the sample size. The test statistic, S, proposed by Cheung and Ng (1996) is given

by the following formula: $S_{x \to y} = T \sum_{k=1}^{M} \hat{\tau}_{x,y}^2(k)$ where M is a bandwidth parameter which under the

null hypothesis of no causality in mean from x_t to y_t follows asymptotically a X_M^2 distribution.

Similarly, when testing for causality in mean running from y_t to x_t, $S_{y \to x} = T \sum_{k=-M}^{-1} \hat{\tau}_{x,y}^2(k)$ is utilised.

We should also note that in order to obtain correct inference, M should be large enough to include all potential nonzero cross-correlations.

2.2. Multivariate causality tests

Let Y_t and X_t be two multivariate time series with $Y_t \in R^{d1}$ and $X_t \in R^{d2}$. In our case, the dimension of these multivariate time-series is $d = d_1 = d_2 = 12$, the number of the EMU countries under scrutiny. El Himdi and Roy (1997), extended the bivariate methodology of Haugh (1976) to the multivariate case. They proposed a test statistic for the hypothesis of no Granger causality between multivariate series. Similar to the bivariate case, the multivariate time series are prefiltered separately through Vector Autoregressive GARCH (VAR-GARCH) models.[4] In this respect, the residual series U_{yt} and U_{xt} are independent of the past of every single component of Y_t and X_t, respectively.[5] The estimated standardised residuals, U_{yt} and U_{xt} are cross-correlated with cross-correlation function,

$$\hat{R}_{xy}(k) = \begin{pmatrix} \hat{\tau}_{x1y1}(k) & \hat{\tau}_{x1y2}(k) & \cdots & \hat{\tau}_{x1yd1}(k) \\ \vdots & & & \vdots \\ \vdots & & & \vdots \\ \hat{\tau}_{xd2y1}(k) & \cdots & \cdots & \hat{\tau}_{xd2yd1}(k) \end{pmatrix} \in R^{d2 \times d1}, \text{ with } \hat{\tau}_{x_i y_i}(k) \text{ defined in Section 2.1.}$$

---

[4] The order of each VAR model is determined by the SIC criterion.
[5] These series differ from the ones employed in the bivariate tests, in the sense that the ones obtained by the within country VARs may still carry information with respect to the past of the series of the other countries.

Similarly, the corresponding auto-correlations are $\hat{R}_{x,x}(k) \in R^{d_2 \times d_2}$ and $\hat{R}_{y,y}(k) \in R^{d_1 \times d_1}$. The test statistic proposed by El Himdi and Roy (1997) is given by the following formula: $S_M = T \sum_{k=-M}^{M} \left[\left[vec(\hat{R}_{x,y}(k)) \right]^T A^{-1} \left[vec(\hat{R}_{x,y}(k)) \right] \right]$ where A is the asymptotic covariance matrix of $\sqrt{T} vec(\hat{R}_{x,y}(k))$ that is $A = \hat{R}_{x,x}(k) \otimes \hat{R}_{y,y}(k)$. When testing for causality in mean running from x_t to y_t, only positive values of the bandwidth parameter M should be employed. In this case the above quadratic form is shown to follow a $X^2_{Md_1 d_2}$ distribution. In the case that we are interested in revealing bidirectional causality between the series at hand, i.e. allowing for both positive and negative lags, the degrees of freedom are adjusted accordingly to $(2M+1)d_1 d_2$.

Naturally, this multivariate test is more powerful as opposed to the bivariate one introduced in Section 2.1 due to mainly two reasons. First, all countries are pooled together in order to find evidence of Granger causality and second, Granger causality across countries is also allowed. Moreover, this testing procedure can be modified so as to reveal more information with respect to the interdependencies within the euro area. Specifically, if we are interested in testing whether the developments in financial variables of one country affect real economic activity in the remaining countries, i.e. to discover a country's clout, we only include the financial variable x_i of country i and test whether it causes the variables y_j, with $j \neq i$. In such a case, $d_1 = 1$ and $d_2 = d-1$, with d the number of countries participating in the panel. Similarly, we can test for the receptivity of a country, i.e. discover the countries that are more likely to be led by developments in the financial variables of the remaining ones in the euro area. In this case, we test whether real economic activity in country j is Granger-caused by the financial variables x_i of the remaining countries, with $i \neq j$. The test again follows a X^2 distribution with $M(d-1)$ degrees of freedom.

3. Empirical Evidence

In this section we apply the techniques outlined in the previous section to examine the empirical relationship between growth and financial variables in the euro area.[6] To gauge this empirical relationship, we use existing measures of output, term spread, real stock price changes and real money supply growth for the 12 euro area countries. Our data set is monthly and covers the period from January 1988 to May 2005. As a measure of the growth rate of output we use the industrial production index (seasonally adjusted) from the OECD leading indicators (obtained by Datastream). Real stock price changes were obtained by use of Datastream-calculated composite indices, appropriately adjusted for the inflation rate of the countries under consideration. The term spread is calculated as the difference between a long-term bond and a short-term interest rate, mainly a three-month Treasury Bill obtained from the IMF, International Financial Statistics (Source: EcoWin). As regards the monetary aggregates, we employed real M3 money supply obtained by Datastream.

3.1. Bivariate predictive content

The within-country predictive content of financial variables was assessed through bivariate Granger causality tests reported in Table 1.

Table 1. Bivariate within-country analysis for testing
whether financial variables Granger cause growth

Country/variable	Term spread	Stock market	Money supply
Austria	0.095	0.240	0.032*
Belgium	0.008*	0.088	0.189

---

[6] All the reported results were obtained by programs written in E-views 4.1 and are available from the author upon request.

Country/variable	Term spread	Stock market	Money supply
Finland	0.456	0.001*	0.020*
France	0.072	0.002*	0.732
Germany	0.003*	0.005*	0.244
Greece	0.550	0.658	0.782
Ireland	0.521	0.420	0.041*
Italy	0.250	0.010*	0.977
Luxembourg	0.721	0.603	0.482
Netherlands	0.006*	0.221	0.210
Portugal	0.435	0.451	0.977
Spain	0.399	0.097	0.051

Note: p-values. An asterisk denotes significance at the 5% level. The bandwidth is 15 months

Our findings confirm the mixed evidence found in the literature. With respect to the term spread, our results suggest that this variable is a useful predictor of the real economic activity of Belgium, Germany and the Netherlands and only marginally (at the 10% level) for Austria and France. A similar picture emerges when the stock market returns are considered. Specifically, stock price developments Granger cause output growth in Finland, France, Germany and Italy. However, weaker evidence is found for Belgium and Spain. Monetary aggregates seem to have the lowest impact on output growth, with only Austria, Finland and Greece yielding significant dependencies.

3.2. Multivariate predictive content

Turning to the joint testing of Granger causality, our results paint a different picture. Table 2 reports the p-values from the El Himdi and Roy test for joint causality and for a variety of bandwidths ranging from 3 to 18 months.

Table 2. Multivariate cross-country analysis for testing
whether financial variables Granger cause growth

Variable/lag	Term spread		Stock market		Money supply	
	stat	p-value	stat	p-value	stat	p-value
3	404.17	0.828	426.18	0.570	492.99	0.022*
6	881.04	0.336	891.79	0.249	927.64	0.065
9	1309.37	0.392	1325.14	0.281	1384.50	0.043*
12	1731.49	0.472	1724.67	0.518	1878.21	0.006*
15	2175.33	0.404	2118.89	0.732	2308.62	0.013*
18	2611.56	0.390	2541.10	0.759	2720.88	0.038*

Note: p-values. An asterisk denotes significance at the 5% level. The bandwidth is 15 months.

Quite surprisingly, the only financial variable that could be a valuable indicator for economic growth for the euro area as a whole is the money supply growth. Jointly neither the term spread nor the stock market returns seem to Granger cause the output growth.

Table 3 provides a more in-depth analysis with repect to cross-country dependence. Specifically, the Table reports the relevant p-values for testing both the leading role (clout) of the euro area countries' financial variables and their receptivity.

Table 3. "Clout" and "Receptivity"

Country/variable	Term spread		Stock market		Money supply	
	Clout	Recepti vity	Clout	Recepti vity	Clout	Recepti vity
Austria	0.000*	0.061	0.059	0.050*	0.002*	0.055
Belgium	0.002*	0.000*	0.558	0.000*	0.001*	0.000*
Finland	0.560	0.000*	0.001*	0.000*	0.004*	0.013*
France	0.025*	0.051	0.538	0.000*	0.000*	0.000*
Germany	0.002*	0.000*	0.334	0.012*	0.099	0.000*
Greece	0.056	0.003*	0.207	0.012*	0.346	0.000*
Ireland	0.857	0.060	0.143	0.010*	0.169	0.000*
Italy	0.069	0.008*	0.023*	0.105	0.487	0.093
Luxembourg	0.056	0.237	0.369	0.001*	0.062	0.000*
Netherlands	0.260	0.001*	0.909	0.026*	0.211	0.000*
Portugal	0.202	0.966	0.622	0.615	0.352	0.683
Spain	0.076	0.497	0.898	0.816	0.595	0.495

Note: p-values. An asterisk denotes significance at the 5% level. The bandwidth is 15 months.

Developments in the term spread of Austria, Belgium, Germany and France appear to be significant for the euro area countries, which on the whole appear quite receptive. Only Luxembourg, Portugal and Spain are quite secluded and do not respond to changes in the term spread of the remaining member states. This finding is also confirmed for the receptivity of Portugal and Spain as far as stock market changes are concerned, which along with Italy appear to be indifferent to such developments. On the other hand, Italy and Finland appear to have a leading role in this respect. A similar picture emerges when monetary aggregates developments are considered. The receptivity of all countries, with the exception of Portugal and Spain appears high.

References

[1] Cheung Y.M. and L.K. Ng, 1996, "A causality-in-variance test and its application to financial market prices", *Journal of Econometrics*, 72, 33-48.

[2] El Himdi K. and R. Roy, 1997, "Tests for non-correlation of two multivariate ARMA time series", *Canadian Journal of Statistics*, 25, 233-256.

[3] Haugh L.D., 1976, "Checking the independence of two covariance-stationary time series: A univariate residual cross-correlation approach", *Journal of the American Statistical Association*, 71, 378-385.

[4] Lemmens A., C. Croux and M.G. Dekimpe, 2005, "On the predictive content of production surveys: A pan-European study", *International Journal of Forecasting*, 21, 363-375.

[5] Stock J.H. and M.W. Watson, 2003, "Forecasting output and inflation: the role of asset prices", *Journal of Economic Literature*, 41,788-829.

Brill Academic Publishers
P.O. Box 9000, 2300 PA Leiden
The Netherlands

*Lecture Series on Computer
and Computational Sciences*
Volume 4, 2005, pp. 1345-1348

Modelling Macroeconomic Effects in Central Eastern Economies Stock Returns

Aristeidis G. Samitas[1]

Department of Business Administration, Business School,
University of the Aegean,
6 Christou Lada Str.,
GR-105 61, Athens, Greece,

Dimitris F. Kenourgios

Faculty of Economics,
University of Athens,
5 Stadiou Street,
GR-10562 Athens, Greece

Received 5 July, 2005; accepted in revised form 16 August, 2005

Abstract: The present paper examines the long and short-run relationships between three Central European Economies stock returns (Poland, Hungary and Czech Republic) and their main western economic and trading partner, which is Germany. We obtain evidence of links between macroeconomic variables and stock returns in CE countries that are stronger than has previously been reported. The results from our empirical research are consistent in providing support to the integrated market hypothesis for the CE economies. Moreover, is proved that German economic activity has a significant influence on these stock markets portfolios returns in the long run but was less influential than domestic economic activity.

Keywords: Stock returns; macroeconomic policy; present value model; Central-Eastern stock markets.

Mathematical Subject Classification: 91B4, 62P20

1. Equity Markets Integration in CE European Economies

Recent finance literature reveals lot of research devoted to capital markets integration among Eastern European economies and the world's leading stock markets. Primary focus has been on whether the markets in CE countries are integrated with the international economy and, in particular the linkages between their capital markets and the major European Union equity markets. The last decade's economic reforms and the rapid transition to a market based economy attracted plenty of foreign direct investments in countries like Poland, Hungary and Czech Republic. The major share of these investments came from Germany, according to their trading exchanges (source: World Trading Organization) . As from May 1st 2004 the Czech Republic, Hungary and Poland became members of the EU, it means that they fulfilled the basic criteria for the accession as imposed by the Copenhagen Treaty. EU membership requires the reception of the *acquis communitaire* that includes some basic rules for the existence of a market economy. Following the Nineties' new economic orthodoxy, one of the bases of the *acquis* is the free circulation of capitals. According to the 2002 transition report, to some extent this has been already achieved. Once inside the EU, the very important issue is the introduction of the euro in these countries in due course. As it was for the countries of Western Europe, the new candidates will have to achieve the Maastricht criteria before joining the monetary union. This will include:

---

[1] Corresponding author. E-mail: asamitas@econ.uoa.gr

- Fiscal deficit not higher than 3% of GDP
- Government debt no more than 60% of GDP
- Inflation not higher than 1.5% the average of the inflation of the three countries with the lowest rate
- Long term interest rates not higher than 2% of the average of the three countries with the lowest rate

In addition to these requirements, the European Central Bank (ECB) imposed one more criterion on the accession countries that is to join the Exchange Rate Mechanism (ERM-II) for at least two years before the adoption of the single currency. It means that during that period exchange rate must be kept within a 30 percent-wide band (+/- 15%) around a fixed parity which may be revalued but not devalued. It is important to notice that the three countries are now adopting an inflation- targeting monetary policy. Hungary is already using a soft-peg with a 30% wide band, while Poland and Czech Republic allow a total free float of their currency. The introduction the ERM-II, even though it allows a wide fluctuation band, is likely to change the economic behaviour of the monetary institutions in these countries.

Long-run co movements between stock markets have important regional and global implications, as a domestic economy cannot be insulated from external shocks and the scope for independent monetary policy appears then limited. The present value model of stock prices suggests that stock markets should be a leading indicator of economic activity. The use of an aggregate proxy for interest rates (IR) and industrial production (IP) permits the relationship inherent in the present value model to be tested. The suggestion is that if current IR and IP are found to be significant explanators of price behavior, the present value model is violated. Many papers focus on European stock markets, [5], [11], [7], [15] and [12] among others. A body of research examines the relationships among international stock markets across regions, [1], [4], [13], [3], [2], [9] among others. Few only studies focus on the CE stock markets, such as [6], [8], [14], [10].

2. Data and Research Methodology

The sample periods for all countries are from 1994 till end 2004. The variables data for the present value model of share prices are cash flows (aggregate industrial production), interest rates (Government bond rate) and share prices (total return indexes). The total return share market indexes used are: the German DAX, the Polish WIG, the Czech PX50 and the Hungarian BUX. All indices were sourced from the Datastream International finance database. Interest rates (IR), and industrial production (IP) indexes for each country were sourced from the International Financial Statistics publication compiled by the International Monetary Fund (IMF).

Two models utilized to test the validity of the present value model and the relationship between economic variables and stock markets. The first model uses current industrial production to attempt to test for the relationship between a factor that represents current economic activity and stock prices:

$$SP_t = IP_t - IR_t,$$ (1)

where, SP denotes domestic stock prices, IP is industrial production, IR is a domestic interest rate series. The present value model is also tested using the relationship below in an identity which is more consistent with market efficiency:

$$SP_t = IP_{t+1} - IR_t,$$ (2)

where IP_{t+1} denotes domestic industrial production leading one quarter. According to the present value model, current share prices should be caused by future industrial production. As a proxy for future industrial production, share prices will be led by industrial production by one quarter. German industrial production and interest rates were used as external factors of influence. Germany is considered as the best representative of the European countries due to closer economic and trading ties and traditional neighboring with CE countries. In this case the equation is,

$$SP_t = GERIP_{t+1} - GERIR_t + IP_{t+1} - IR_t$$ (3)

where $GERIP_{t+1}$ is German industrial production leading one quarter ahead and $GERIR_t$ are the German interest rates. The theory of cointegration became the most sufficient method for testing the co-dependence between stock markets' indices and macroeconomic factors. The cointegration examines the existence of a long-run common stochastic trend among stock prices' returns, interest rates and industrial production. In sum, our paper includes a whole-set of time series analysis techniques including long run structural modelling (LRSM) of the cointegrating vectors, a vector error correction model (VECM) and a variance decomposition (VDC) analysis. After normalising share prices as the dependent variable, LRSM will used to determine the existence of a long run causal relationship by placing a restriction of zero on the variable in the cointegrating vector. The rejection of such a restriction implies the variable must enter the cointegrating vector significantly and a long run causal relationship is said to exist.

The vector error correction model (VECM) is a VAR where the non-stationary variables have been transformed into a stationary series by first differencing. Such tests can allow the researcher to examine the relative exogeneity and endogeneity of each variable in the system over the short run as well as examining the significance of the long run adjustment to the short run dynamics of the system.

Further more a Variance Decomposition (VDC) analysis is applied enhancing the above tests of causality by estimating the relative exogeneity and endogeneity of a system of variables in an out of sample test. This assists the comparison between domestic and international economic variables and their relative impact.

3. Empirical Results and Conclusions

CE capital markets have become increasingly integrated with the world markets. The main findings strongly suggest that the emerging CE European capital markets are macro economically cointegrated with the German economic influence. German macroeconomic variables have a significant influence on these stock markets portfolios returns in the long run but were less influential than domestic economic activity. Transition economies seem to become increasingly integrated with Europe. Such a situation was expected, especially with Germany, due to their very close trading, cultural and historical partnership.

Acknowledgements

The authors wish to thank the anonymous referees for their careful reading of the manuscript and their fruitful comments and suggestions.

References

[1] C.M Bilson, T.J Brailsford and V.J. Hooper, Selecting macroeconomic variables as explanatory factors of emerging stock market returns, *Pacific-Basin Finance Journal*, 9 401-426 (2001).

[2] D.A. Bessler and J. Yang, The structure of interdependence in international stock markets, *Journal of International Money and Finance*, 22 261-87 (2003).

[3] K. Chaudhuri and Y. Wu, Random walk versus breaking trend in stock prices: evidence from emerging markets, *Journal of Banking and Finance*, 27 575-92 (2003).

[4] G. Chen, M. Firth and O. Rui, 'Stock market linkages: evidence from Latin America', *Journal of Banking and Finance*, 26 1113-41 (2002).

[5] D.G. Dickinson, Stock market integration and macroeconomic fundamentals: an empirical analysis, *Applied Financial Economics*, 10 261-76 (2000).

[6] E. Dockery and F. Vergari, An investigation of the linkages between European Union equity markets and emerging capital markets: the East European connection, *Managerial Finance*, 27 (1/2) 24-39 (2001).

[7] R.J. Gerrits and A. Yuce, Short- and long-term links among European and US stock markets, *Applied Financial Economics*, 9 1-9 (1999).

[8] C.G. Gilmore and G.M. McManus, International portfolio diversification: US and Central European equity markets, *Emerging Markets Review*, 3 69-83 (2002).

[9] O. Ratanapakorn and S.C. Sharma, Interrelationships among regional stock indices, *Review of Financial Economics*, 11 91-108 (2002).

[10] A.G. Samitas and D.F. Kenourgios Macroeconomic factor's influence on "new" European stock returns: the case of four transition economies, *International Journal of Financial Services Management,* forthcoming (2005).

[11] P.L. Steely and J.M. Steely, Changes in the co-movement of European equity markets, *Economic Inquiry*, July 473-81 (1999).

[12] T. Syriopoulos, Modelling long run dynamics in transitional European equity markets, *European Review of Economics and Finance*, 3(4) 57-83 (2004).

[13] P.E. Swanson, The interrelatedness of global equity markets, money markets, and foreign exchange markets, *International Review of Financial Analysis*, 12 135-55 (2003).

[14] S. Voronkova, Instability in the long-run relationships: evidence from the Central European emerging stock markets, Discussion Paper, Symposium on International Equity Market Integration, Institute for International Integration Studies, Dublin (2003).

[15] J.J.Yang, I.Min and Q.Li, European stock market integration: does EMU matter?, *Journal of Business Finance and Accounting*, 30, forthcoming (2003).

Brill Academic Publishers
P.O. Box 9000, 2300 PA Leiden
The Netherlands

*Lecture Series on Computer
and Computational Sciences*
Volume 4, 2005, pp. 1349-1351

Evaluation of Correlation Forecasting Models for Risk Management

V. Skintzi[1]

Department of Economics,
School of Management and Economics,
University of Peloponnese,
GR 221 00 Tripolis, Greece

S. Xanthopoulos

Financial Engineering Research Centre
Department of Management Science & Technology
Athens University of Economics & Business
Evelpidon 47A & Lefkados 33
GR 113 62 Athens, Greece

Received 13 July, 2005; accepted in revised form 16 August, 2005

Abstract: Volatility and correlation forecasts are of paramount importance in modern risk management systems. The forecasting performance of volatility models in the context of risk management applications has extensively been investigated in the open literature. However, in spite of the plethora of correlation forecasting models, their impact on the VaR accuracy calculation has not yet been explicitly examined. In this paper, traditional and modern correlation forecasting techniques are compared using standard statistical loss functions, as well as, VaR based economic loss functions. Historical data on portfolios consisting of stocks, bonds and currencies are used for the purposes of this study.

Keywords: Correlation, Value-at-Risk, forecasting, GARCH

1. Extended Abstract

An important topic in financial econometrics that has received significant attention in the recent finance literature is the modelling of the second moments of asset returns. Accurate estimation and forecasting of asset return volatility and correlations are of paramount importance in most financial applications including asset pricing, capital allocation, risk management, derivatives pricing and hedging. The forecasting performance of volatility models in the context of various economic loss functions has been extensively investigated in the literature. Surprisingly, correlation estimation has not received significant attention in the finance literature until recently. Following the recognition of the time-variability of correlation and the success of univariate GARCH models, a number of multivariate volatility models have been proposed (e.g. Bollerslev, Engle and Wooldridge, 1988, Bollerslev, 1990, Engle and Kroner, 1995, Alexander, 2001). More recently, new multivariate GARCH models have been developed that focus on modeling the dynamic structure of conditional correlation (see Engle, 2002, Tse and Tsui, 2002). However, in spite of the increasing number of correlation forecasting models, only a limited number of studies have evaluated the forecasting performance of correlation models.

Over the past decade, the finance industry as well as the regulatory authorities have recognised the importance of measuring accurately financial risks and implementing sound risk management. In particular, the concept of Value-at-Risk (VaR) has received significant attention and is now established as a useful measure of risk. In the calculation of any parametric VaR model, accurate volatility and correlation forecasting is essential. The forecasting performance of volatility models in the context of risk management applications has been extensively investigated in the literature (see Brooks, 2003, Brooks and Persand, 2003, amongst others). However, only a limited number of studies compare VaR

---

[1] Corresponding author. E-mail: vikiski@uop.gr

Brill Academic Publishers
P.O. Box 9000, 2300 PA Leiden
The Netherlands

*Lecture Series on Computer
and Computational Sciences*
Volume 4, 2005, pp. 1352-1354

Optimal Prediction under Linlin Loss:
Empirical Evidence

Yasemin Ulu[1]

Department of Economics,
American University of Beirut,
Beirut, Lebanon

Received 12 July, 2005; accepted in revised form 16 August, 2005

Abstract: I compare the forecasts of returns from the mean predictor (optimal under MSE), with the pseudo-optimal and optimal predictor for an asymmetric loss function under the assumption that agents have asymmetric LINLIN loss function. I consider both univariate and multivariate cases. For the multivariate case, I generalize the LINLIN loss function to a multivariate LINLIN loss function and use a normal diagonal-BEKK GARCH(1,1) model to predict the time varying variances. The results strongly suggest not to use the conditional mean predictor under any kind of asymmetry. Forecasts can be improved considerably by the use of optimal predictor versus the pseudo-optimal predictor especially with the multivariate model.

Keywords: Symmetric Loss, Asymmetric loss, forecasting, optimal, pseudo-optimal predictors, volatility, GARCH, BEKK.

1. Multivariate Loss Function & Prediction

The theory of forecasting with asymmetric loss as originally presented by Granger (1969), and further developed by Christoffersen and Diebold (1996, 1997), only considered the prediction of a single variable based on its own passed values. In this section I extend the theory to a multivariate framework in which more than one series is to be forecasted. Let Y_{t+h} be an $n \times 1$ vector of variables to be forecasted at horizon h. $\hat{Y}_{t+h}$ be the $n \times 1$ vector of forecasts and $e_{t+h} = Y_{t+h} - \hat{Y}_{t+h}$ is the $n \times 1$ vector of forecast errors. I then have the following extension of Christoffersen and Diebold's (1997) Proposition 1.

Proposition 1: *If $Y_{t+h} | \Omega_t \sim N(\mu_{t+h|t}, \Sigma_{t+h|t})$ is conditionally multivariate normal and $L(e_{t+h})$ is any loss function defined on the vector of h-step-ahead prediction error, then the optimal predictor is of the form $\hat{Y}_{t+h} = \mu_{t+h|t} + \alpha_{t+h|t}$, where $\alpha_{t+h|t}$ depends only on the loss function and the conditional prediction error variance-covariance matrix $\Sigma_{t+h|t} = \mathrm{var}(Y_{t+h} | \Omega_t) = \mathrm{var}(e_{t+h} | \Omega_t)$.*

Proof: The optimal predictor $\hat{Y}_{t+h}$ minimizes the expected loss

$$E_t[L(Y_{t+h} - \hat{Y}_{t+h})] = \int_{-\infty}^{\infty} L(Y_{t+h} - \hat{Y}_{t+h})\phi(\Sigma_{t+h|t}^{-1/2}(Y_{t+h} - \mu_{t+h|t}) | \Omega_t) dY_{t+h}$$

where $\phi(\cdot)$ denotes the multivariate standard normal density and the integral sign denotes an n-fold integral over the elements of Y_{t+h}. Let $X_{t+h} = Y_{t+h} - \mu_{t+h|t}$ denote the observations deviation from its conditional mean. Changing variables, where determinant of the Jacobian of the transformation is one, the objective function can be expressed in deviation from mean as

$$E_t[L(X_{t+h} - \alpha_{t+h})] = \int_{-\infty}^{\infty} L(X_{t+h} - \alpha_{t+h})\phi(\Sigma_{t+h|t}^{-1/2} X_{t+h} | \Omega_t) dX_{t+h}$$

[1] Corresponding author: E-mail: yb06@aub.edu.lb

where $\alpha_{t+h} = \hat{Y}_{t+h} - \mu_{t+h|t}$ is chosen to be the optimal predictor of X_{t+h}. The objective function does depend on the conditional mean $\mu_{t+h|t}$, and therefore, the optimal predictor only depends on the loss function $L(\cdot)$ and the conditional variance-covariance matrix $\Sigma_{t+h|t}$. Given that α_{t+h} is the optimal predictor of X_{t+h}, the optimal predictor of $Y_{t+h|t}$ is $\mu_{t+h|t} + \alpha_{t+h}$.

Following Zellner (1986), it may be reasonable to assume the loss function is additively separable in the n prediction errors and can be written as

$$L(Y_{t+h} - \hat{Y}_{t+h}) = \sum_{i=1}^{n} L_i(y_{i,t+h} - \hat{y}_{i,t+h}).$$

Possible choices for $L_i(\cdot)$ are the linlin and linex loss functions. For the linlin loss function,

$$L(y_{i,t+h} - \hat{y}_{i,t+h}) = \begin{cases} a_i \, |y_{i,t+h} - \hat{y}_{i,t+h}| & \text{if } y_{i,t+h} - \hat{y}_{i,t+h} > 0 \\ b_i \, |y_{i,t+h} - \hat{y}_{i,t+h}| & \text{if } y_{i,t+h} - \hat{y}_{i,t+h} \le 0 \end{cases}.$$

The first order conditions are

$$\frac{\partial E[L(y_{t+h} - \hat{y}_{t+h})]}{\partial \hat{y}_{i,t+h}} = b_i \int_{-\infty}^{\hat{y}_{i,t+h}} f_i(y_{i,t+h}) dy_{i,t+h} - a_i \int_{\hat{y}_{i,t+h}}^{\infty} f_i(y_{i,t+h}) dy_{i,t+h} = 0 \quad i = 1, \dots, n$$

or

$$b_i F_i(\hat{y}_{i,t+h}) - a_i [1 - F_i(\hat{y}_{i,t+h})] = 0$$

which can be solved as

$$\hat{y}_{i,t+h} = F^{-1}(a_i / (a_i + b_i))$$

If $y_{i,t+h}$ is conditionally normal, then $F_i(y_{i,t+h}) = \Phi((y_{i,t+h} - \mu_{i,t+h}) / \sigma_{ii,t+h})$ and the optimal predictor vector can be written as

$$\hat{y}_{i,t+h} = \mu_{i,t+h} + \sigma_{ii,t+h} \cdot \Phi^{-1}(a_i / (a_i + b_i)).$$

The coefficients a_i and b_i might differ for each coordinate. However, for convenience I set all $b_i = 1$ as in the univariate case.

Acknowledgments

The author wishes to thank the anonymous referees for their careful reading of the manuscript and their fruitful comments and suggestions.

References

[1] Bollerslev, T. (1986), "Generalized Autoregressive Conditional Heteroscedasticity", *Journal of Econometrics*, 31, 307-327.

[2] Bollerslev, T., Engle, R.F., Wooldridge, J.M. (1988), "A capaital Asset Pricing Model with timevarying covariances", *Journal of Political Economy*, 96, 116-131.

[3] Chiristoffersen, P. F. and Diebold F. X. (1996), "Further results on Forecasting and Model Selection Under Asymmetric Loss", *Journal of Applied Econometrics*, Vol. 11, No. 5, Special Issue: Econometric Forecasting (Sep.-Oct., 1996), 561-571.

[4] Chiristoffersen, P. F. and Diebold F. X. (1997), "Optimal Prediction Under Asymmetric Loss", Econometric Theory, 13, 808-817.

[5] Diebold, F. X. and Nason J. A (1990), "Nonparametric Exchange Rate Prediction" *Journal of International Economics*, 28, 315-332.

- a time and space efficient algorithm for locating multirepeats in a set of strings. This problem is a string manipulation problem which is biologically motivated,
- in silico evaluation of bioactive compounds: Docking Simulations based Enzyme-Inhibitor Interaction compared with X-ray models,
- an interesting simulation of nuclei images using spatial point pattern models.
- a study concerning the way the usage of synonymous codons in gene expression can be explored in order to act as a potential coding mechanism for tissue differentiation.

We would like to thank all the authors for their contribution and the members of the Program Committee for their help.

Program Committee

1. Athanasios Tsakalidis (University of Patras and RACTI)
2. Christos Makris (University of Patras and RACTI)
3. Yannis Panagis (University of Patras and RACTI)
4. Katerina Perdikuri (University of Patras)
5. Evangelos Theodoridis (University of Patras and RACTI)

Professor Athanasios Tsakalidis

Department of Computer Engineering and Informatics
School of Engineering, University of Patras, 26500 Patras – GREECE

Athanasios Tsakalidis was born in Katerini, Greece. Professor Tsakalidis completed his Diploma of Mathematics, University of Thessaloniki in 1973. In 1980 he completed his studies of Informatics in University of Saarland, Germany. He completed his Ph. D in Informatics in 1983 in the same University. During the period 1983-1989, he was a researcher in the University of Saarland, student and cooperator (12 years) of Prof. Kurt Mehlhorn (Director of Max-Planck Institute of Informatics in Germany). Prof. Tsakalidis is the Research & Development Coordinator of Research Academic Computer Technology Institute and the Scientific Coordinator of Research Unit 5 of the same Institute. Since 1993 he is Professor in the Dept. of Computer Engineering and Informatics of the University of Patras and its Chairman since 2001. He is one of the contributors of the "Handbook of Theoretical Computer Science" (Elsevier and MIT-Press 1990). Professor's main interests are Data Structures, Graph Algorithms, Computational Geometry, Multimedia, Information Retrieval and Bioinformatics. Professor has authored three books, two chapters and numerous publications in international journals having an especial contribution to the solution of elementary problems in the area of data structures. He has also contributed to national and international conference proceedings.

Assistant Professor Christos Makris

Christos Makris was born in Greece, 1971. He graduated from the Department of Computer Engineering and Informatics, School of Engineering, University of Patras, in December 1993. He received his Ph.D. degree from the Department of Computer Engineering and Informatics, in 1997. Today he works as an Assistant Professor in the Department of Computer Engineering and Informatics, University of Patras. His research interests include Data Structures, Computational Geometry, Data Bases and Information Retrieval. He has published over 50 papers in various scientific journals and conferences

Brill Academic Publishers
P.O. Box 9000, 2300 PA Leiden
The Netherlands

*Lecture Series on Computer
and Computational Sciences*
Volume 4, 2005, pp. 1357-1361

Chaotic Dynamics in a Tryptophan Repressible Operon Model

L.B. Drossos[1] H. Isliker[2] T.C. Bountis[3] S.Parthasarathy[4]

[1]Dept. of Applied Informatics, Technical Educational Institute of Messologhi, Greece
[2]Department of Physics, University of Thessaloniki, 54006 Thessaloniki, Greece
[3]Department of Mathematics, University of Patras Patras 26110, Greece
[4]Centre for Cellular and Molecular Biology Uppal Road, Hyderabad 500 007, India

Received 31 July, 2005; Accepted in revised form 10 August, 2005

Abstract: A repressor-mediated repression process, often observed in bacteria, is modelled using a gene enzyme-end-product control unit. In this paper, we study a periodically driven repressible tryptophan operon model describing such a process. We find that, when unperturbed, our model typically possesses a 3-dimensional **limit cycle**, which is a global attractor, for all initial conditions of the system. However, under the effect of external periodic forcing, this simple periodic behavior becomes considerably more complex, as the limit cycle turns to a **strange attractor**, with a broadband spectrum, positive largest Lyapunov exponent, and a non integer fractal dimension, displaying chaotic dynamics in its vicinity.

Keywords: Prokaryotic Systems, Gene Regulation, Operon Models, Proteins, Dynamical Systems, Strange Attractors, Power Spectrum, Lyapunov Exponents, and Fractal Dimension.
Mathematics Subject Classification: 9240
PACS : 31.15-p

1. Introduction

The operon hypothesis was introduced to describe the process of regulation of gene expression in prokaryotic systems, [16]. It was applied, first, to the so called lac operon and later to all other classical inducible and repressible operons, (such as trp and arg) in which the endproduct of a particular metabolic pathway is controlled by the operon. The above hypothesis was formulated mathematically, [7, 8] by using a genetic control circuit, which describes a gene-enzyme-endproduct unit. These models have been thoroughly studied from the stability point of view. Some perturbations have also been introduced to the model to induce oscillations and instability, [2, 6, 17, 18, 19, 27]. Finally, the operon model has been used to describe a number of periodic phenomena observed in cellular systems, [4, 6, 28].

According to the operon hypothesis, a small molecule (e.g. protein) called "inducer" or "repressor" mediates, in many cases, processes of induction and repression. This molecule represents a gene which does not reside in the operon, but is used to control the amount of the produced endproduct. In each of these molecules, there are two binding sites which correspond to the endproduct of the pathway and the operator **DNA**.

The first binding affects the configuration of this molecule and thus influences its attachment to the operator **DNA**. These two interactions have binding constants which differ considerably. The evolution of these interactions in time determines the behaviour of the system i.e. the synthetic profile of the endproduct. Note that mutations of operon, repressor or inducer genes strongly affect the dynamics of the system through structural changes in the molecules of interest, which, in turn lead to changes in the evolution of the interactions.

In this paper we study the dynamical properties of a periodically driven tryptophan repressible operon, described in [21], of the form

Table 1: Correlation dimension $D^{(2)}$ estimates of the chaotic attractor, for different ε values using the Maximum-Likelihood-method, with error Δ

ε	$D^{(2)}$	Δ
0.0	1.10	± 0.02
0.017	1.28	± 0.03
0.095	1.62	± 0.03
0.25	1.46	± 0.03
0.495	1.35	± 0.03
0.75	1.02	± 0.02
0.99	1.31	± 0.03
1.75	1.59	± 0.03
2.75	1.79	± 0.04
3.75	1.69	± 0.03
5.75	1.66	± 0.03

Table 2: Lyapunov exponents for different values of ε: Λ_{max} estimated from time series, whereas λ_i, $i = 1,..., 4$, are the four Lyapunov exponents estimated directly from the equations of motion

ε	Λ_{max}	λ_1	λ_2	λ_3	λ_4
0.0	-9.9e-6	5.1e-4	-1.9e-2	-1.5	0.0
0.017	8.0e-3	7.0e-3	-2.6e-2	-1.5	0.0
0.095	2.3e-2	2.2e-2	-4.1e-2	-1.5	0.0
0.25	3.1e-2	2.6e-2	-4.3e-2	-1.5	0.0
0.495	3.4e-2	2.8e-2	-6.1e-2	-1.4	0.0
0.75	2.1e-2	1.5e-2	-5.6e-2	-1.4	0.0
0.99	3.1e-2	3.1e-2	-5.9e-2	-1.4	0.0
1.75	3.7e-2	4.2e-2	-6.8e-2	-1.5	0.0
2.75	4.0e-2	5.2e-2	-7.5e-2	-1.5	0.0
3.75	4.2e-2	4.7e-2	-7.6e-2	-1.4	0.0
5.75	4.1e-2	5.4e-2	-8.2e-2	-1.4	0.0

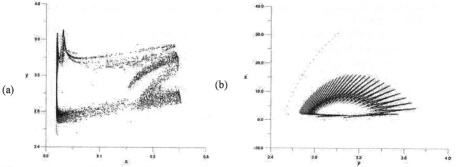

Fig. 1: Projection of the "strange" attractor, at $\varepsilon = 2.750$, (a) on the x, y plane, (b) on the y, z plane.

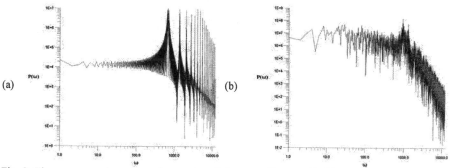

Fig. 2: The power spectrum for a time series of the z-coordinate, with (a) $\varepsilon = 0$, (b) $\varepsilon = 1.75$. The frequencies are in units of $2\pi / (N_a \tau)$, where the length of the analyzed time series is $N_a = 20\ 000$.

Acknowledgements

This work was supported by grants of research European program "Archimedes I".

References

[1]   Abarbanel H.D.I., Brown R., Sidorowich J.J., Tsimrind L.S., (1993), Rev. Mod. Phys . **65** , 1331.

[2]   Allwright D.J., (1977), J.Math. Biol. 4, 363.

[3]   Badii R., Politi A., (1984) : 'Intrinsic Oscillations in Measuring the Fractal Dimension', *Phys. Lett. A* 104, 303.

[4]   Bliss, R.D., Painter, P.R., Marr, A.G., (1982), J. Theor. Biol. 97, 177.

[5]   [8] Ellner S., (1988) : 'Estimating Attractor Dimensions from Limited Data: a New Method, with Error Estimates', *Phys. Lett. A* 133, 128.

[6]   Goldbeter A., Nicolis G., (1976), Progr. Theor. Biol. 4, 65.

[7]   Goodwin B.C., (1965), Adv. Enz. Regul. 3, 425.

[8]   Goodwin B.C., (1966), Nature 209, 479.

[9]   Goodwin B.C., (1976), Analytical Physiology of Cells and Developing Organisms. London: Academic Press.

[10]   Grassberger P., Procaccia I., (1983a): 'Characterization of Strange Attractors', *Phys. Rev. Lett.* 50, 346.

[11]   Grassberger P., Procaccia I., (1983b): 'Measuring the Strangeness of Strange Attractors', *Physica D* 9, 189.

[12]   Guckenheimer J., Holmes P., (1983): 'Nonlinear Oscillations, Dynamical Systems and Bifurcations of Vector Fields' (Springer, Berlin) .

[13]   Isliker H., (1992): 'A Scaling Test for Correlation Dimensions', *Physics Letters A* 169, 313.

[14]   Isliker H., Kurths J., (1994) : 'A Test for Stationarity: Finding Parts in Time Series Apt for Correlation Dimension Estimate', *Internat. J. of Bifurcation and Chaos*, in press.

[15]   Isliker H., Benz A.O., (1994) : 'Nonlinear Properties of the Dynamics of Bursts and Flares in the Solar and Stellar Coronae', *Astron. Astrophys.*, in press.

[16]   Jacob F., Monod J., (1961), J. Mol. Biol. 3, 318.

[17]   MacDonald N., (1977), J. Theor. Biol. 67, 549.

[18]   Rapp P. E., (1975), Math. Biosci. 25, 165.

[19]   Sanglier M., Nicolis, G., (1976), Biophys. Chem. 4, 113.

[20]   Schevitz R.W., Otwinwski Z., Joachimiak A., Lawson C.L., Sigler P.B., (1985), Nature 317, 782.

[21]   Sinha S., Ramaswamy R., (1988), J. Theor. Biol. 132, 307.

[22]   Smith L.A., (1988) : 'Intrinsic Limits on Dimension Calculations', *Phys. Lett. A* 133, 283.

[23]   Takens F., (1981) : 'Detecting Strange Attractors in Turbulence', in: Dynamical Systems and Turbulence, Lecture Notes in Mathematics, 898, p. 366 (Springer, Berlin.

[24]   Takens F., (1984) : 'On the Numerical Determination of the Dimension of an Attractor', in: Dynamical Systems and Turbulence, Lecture Notes in Mathematics, 1125, p. 99 (Springer, Berlin.

[25]   Theiler J., (1986) : 'Spurious Dimension from Correlation Algorithms Applied to Limited Time-series Data', *Phys. Rev. A* 34, 2427.

[26]   Theiler J., (1991) : 'Some Comments on the Correlation Dimension of $1 = f$® Noise', *Phys. Lett. A* 155, 480.

[27]   Tyson J.J., Othmer, H. G. (1978), Prog. Theor. Biol. 5, 1.

[28]   Tyson J.J. (1983), J. Theor. Biol. 103, 313.

[29]   Wolf A., Swift J.B., Swinney H.L., Vastano J.A., (1985) : `Determining Lyapunov Exponents from a Time Series', *Physica D* 16, 285-317.

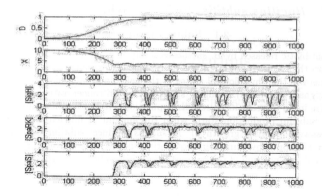

Figure 1: Simulation results (stochastic: red, deterministic: blue)

X_0 and D_{max} are constants of the model; the latter represents constraints on the population because of space limitations and competition within the population.

The continuous dynamics for x_2 are governed by:

$$\dot{x}_2 = -k_1 x_1 + k_2 x_3$$

where k_1 denotes the rate of nutrient consumption per unit of population and k_2 the rate of nutrient production due to the action of subtilin. In reality, the second term is proportional to the average concentration of SpaS, but for simplicity we follow [1] and assume that the average concentration is proportional to the concentration of SpaS for a single cell.

The continuous dynamics for the remaining three states depend on the discrete state, i.e. the state of the three switches. In all three cases the equations take the form:

$$\dot{x}_i = \begin{cases} -l_i x_i & \text{if } S_i \text{ is OFF} \\ k_i - l_i x_i & \text{if } S_i \text{ is ON.} \end{cases}$$

It is easy to see that the concentration x_i decreases exponentially toward zero whenever the switch S_i is OFF and tends exponentially toward k_i/l_i whenever S_i is ON.

2.3 Simulation

A simulation program was developed in Matlab to simulate the stochastic hybrid dynamics outlined in the previous section. Figure 1 shows the results of a simulation based on this program. To estimate the validity of the results, we also provide simulation results generated by a deterministic approximation for the stochastic model developed in [1].

3 Genetic algorithms

3.1 How we use Genetic algorithms

The evolution of the *B. Subtilis* model depends on the values of five synthesis rates k_1 to k_5, three degradation rates l_3 to l_5, the constants r, c, X_0, D_{max} and the threshold η. Our objective is to find

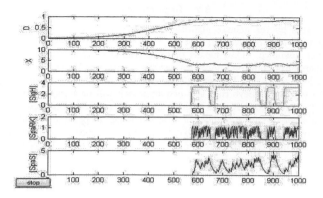

Figure 2: Simulation results for the B.subtilis system with GA identificated parameters

parameter values such that the predictions of the model match the experimental data as closely as possible. The search for these optimal parameter values can naturally be formulated as an optimization problem, which, due to the stochastic, hybrid nature of the dynamics is very difficult to solve by conventional optimization methods [8], [9]. In this section, we propose to use Genetic Algorithms to approximate the solution of this optimization problem.

As an initial test of the potential of randomized optimization methods for parameter identification in stochastic hybrid systems we applied genetic algorithms to the model *B. Subtilis* outlined above. Our intention is to estimate a set of parameters such that the trajectories generated by the model match the real cell function. To be more precise, what we are trying to do is to infer all the parameters of the model keeping its structure unchanged.

Since experimental data was not available at this stage, we tested the genetic algorithms on synthetic data. In other words, we run simulations of the model using known values for the parameters and then pretended the values were unknown and tested whether the genetic algorithm could estimate them. The genetic algorithm uses a population of candidate parameter values. For each of the candidates simulations were executed. The fitness of each candidate was then evaluated as the distance (mean square) between the trajectories for the states x_1 and x_2 generated by the candidate and the trajectories for x_1 and x_2 generated by the "real" parameter values. These two states were chosen since they should be the most easily observable in biological data. We underline the fact that only the trajectories representing food evolution and population evolution are used in the fitness function. Our goal is therefore to utilize only obtainable biological data in order to infer unknown (to date) parameters.

The simulation was held on the MATLAB environment for ease of use [11]. The Genetic Algorithm toolbox provides useful conveniences, making it possible to obtain interesting results. These results include different sets of parameters that reproduce satisfactorily the trajectories observed. It must be clarified,however, that the fidelity of the set of parameters generated was only tested by qualitative criteria. In example if all the substances present inside the cell took part in the control system and if the curves generated appeared to be close to their original counterparts.

3.2 Importance

The results generated by the genetic algorithm demonstrate that a reasonable mechanism describing the operation of a cell can support the identification of reasonable parameters. These parameters are yet unknown. That means that the results obtained by genetic algorithms can drive our experiments from now on in order to test the accuracy of these numbers. What we hope, is to get a closer look of what is happening inside the cell only by observing what is happening outside of it. We strongly believe that this new approach will probably produce a clear insight of the quantitative way that substances interact inside the cell. Furthermore, the technique just described, is capable of serving a number of alternative sets of parameters all leading to a reasonable operation of the cell. This is desirable, because it can reveal a completely different set of rates. Easily, this method can be applied to numerous cases, where only partial knowledge of a mechanism is gained, so as to reveal the rest of it.

Acknowledgment

Research supported by the European Commission under the project HYGEIA, NEST-4995. The authors wish to thank the anonymous referees for their careful reading of the manuscript and their fruitful comments and suggestions.

References

[1] J. Hu and W.C. Wu and S. Sastry,*Modeling Subtilin Production in Bacillus subtilis Using Stochastic Hybrid Systems*, 2004.

[2] M.B. Elowitz and A.J. Levine and E.D. Siggia and P.S. Swain, *Stochastic Gene Expression in a Single Cell*, Science Vol. 297, 2001, 1183–1186.

[3] J. Hu and J. Lygeros and S. Sastry, *Towards a theory of stochastic hybrid systems*, Hybrid Systems: Computation and Control pp. 160-173, 2000.

[4] J. Lygeros and G. Pappas and S. Sastry, *An Introduction to Hybrid System Modeling, Analysis and Control*, First Nonlinear Control Network (NCN) Pedagogical School Vol.8, 1999, 307-329.

[5] T. Stein and S. Borchert and P. Kiesau and S. Heinzmann and S. Kloss and M. Helfrich and K.D. Entian, *Dual Control of Subtilin Biosynthesis and Immunity in Bacillus Subtilis*, Molecular Microbiology Vol. 44, 2002, 403-416.

[6] J. Holland, *Adaptation In Natural and Artificial Systems*, The University of Michigan Press, 1975.

[7] M. Mitchell, *An Introduction to Genetic Algorithms*, MIT Press, 1996.

[8] D. Goldberg, *Genetic Algorithms in Search, Optimization, and Machine Learning*, Addison-Wesley, 1989.

[9] http://lancet.mit.edu/ mbwall/presentations/IntroToGAs/

[10] http://www.rennard.org/alife/english/gavintrgb.html

[11] http://www.mathworks.com/access/helpdesk/help/techdoc/matlab.shtml

Brill Academic Publishers
P.O. Box 9000, 2300 PA Leiden
The Netherlands

*Lecture Series on Computer
and Computational Sciences*
Volume 4, 2005, pp. 1367-1370

Metabolism of arene substrates on iron site in cytochrome P450: Quantum chemical DFT modeling

E. Broclawik*[1,2], S. Abdul Rajjak[1], M. Ismael[1], H. Tsuboi[1], M. Koyama[1], M. Kubo[1], C. A. Del Carpio[1] and A. Miyamoto[1]

[1]Department of Applied Chemistry and New Industry Creation Hatchery Centre, Tohoku University, Aoba, Aramaki, Aoba-ku, Sendai 980-8579, Japan
[2]Institute of Catalysis, Polish Academy of Sciences, ul. Niezapominajek, 30-239 Kraków Poland.

Received 31 July, 2005; accepted in revised form 10 August, 2005

Abstract: CYP3A4 mediated metabolism of (S)-N-[1-(3-morpholin-4-ylphenyl)ethyl]-3-phenylacrylamide is studied by DFT modeling; the important issue is substantial lowering of the activation energy for the rate-determining step by hyperconjugation of lone electron pair on nitrogen of o-substituent in phenyl ring.

Keywords: DFT modeling, CYP3A4 drug metabolism, reaction mechanism, C-H activation

Mathematics Subject Classification: 92C40

PACS: 31.10.-z; 31.15.Ew

1. Introduction

Cytochrome P450 (CYP) enzymes are membrane bound proteins that catalyze primary oxidations of endobiotics and xenobiotics. CYP3A4 is a major CYP450 isoform and contributes extensively to human drug metabolism due to its high level of expression in liver and broad capacity to oxidize structurally diverse substrates [1]. CYP450 enzymes metabolize the majority of drugs thus metabolism pathways and prospective drug-drug interactions are of primary importance. Hydroxylation of the C-H bond is one of major metabolism steps and it can influence drug bioavailability by transforming the substrate either to active form or to toxic compounds. A detailed understanding of this metabolism step and prediction of metabolites is thus a major challenge in drug design.

Since the advent of a robust computational techniques based on Density Functional Theory, QM modeling has provided extensive information on elementary steps in catalytic reactions of CYP450: transformation of the initial, inactive form of the enzyme into the active oxyferyl form, model ligand binding and subsequent metabolism. Shaik et al have recently published comprehensive review on the subject [2]. The target substrate of this study (actual bulky drug molecule), however, has not yet been examined at molecular level by QM calculations. (S)-N-[1-(3-morpholin-4-ylphenyl)ethyl]-3-phenylacrylamide [3] is the novel KCNQ2 potassium channel opener with significant activity in a cortical spreading depression model of migraine. From comparative mechanistic considerations it is believed to metabolize mainly by hydroxylation of the phenyl ring on *N*-phenylmorpholine moiety. Our study utilizes the existing model of the oxidized active enzyme form and focuses on the ligand transformation. DFT calculations are done for a gas phase model composed of the active oxyferyl Cpd I form of the iron – porphyrine site [4] and the ligand molecule.

Our gas-phase QM model has been validated by comparison with docking studies of the substrate in CYP3A4 structure. All quantum chemical calculations were done by means of DMol[3] (Accelrys) [5], with DNP basis set and PW91 GGA exchange-correlation. Transformation mechanism between stable structures was followed by minimum energy pathway calculations based on geometrically selected reaction

---

* Corresponding author. Permanent address: Inst. of Catalysis, Polish Academy of Sciences; E-mail: broclawi@chemia.uj.edu.pl

coordinate (DRC). Aryl carbon hydroxylation proceeds primarily via the formation of a strong σ complex and proton shuttle mediated by a porphyrin ring; however, other possible reaction routes have also been considered and tested by calculating prospective intermediates.

2. Results and discussion

Scheme 1 shows general route for a metabolism reaction with branching pathways for the formation of alcohol, ketone or epoxide product. Table 1 summarizes properties of stable structures, intermediates: weakly and strongly bonded complex or proton-shuttle, and final metabolites. All steps in the metabolism proceed on the doublet potential energy surface, with antiferromagnetic coupling of Cpd I preserved in the encounter complex.

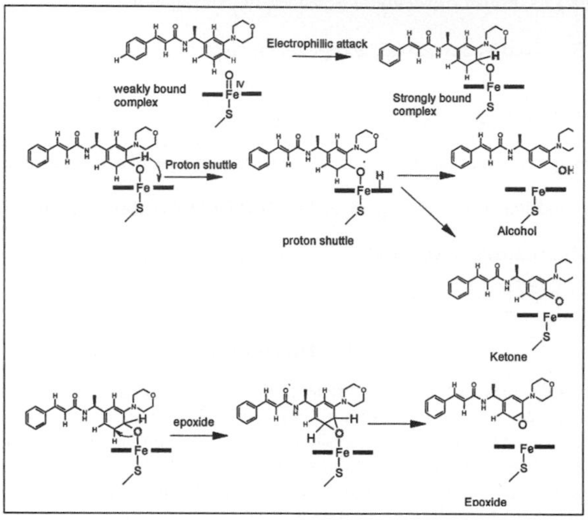

Scheme 1: Proposed mechanism for the substrate metabolism by Cytochrome P450 3A4.

Figures 1, 2 and 3, show optimized model structures for the σ-bonded complex and weakly bonded products. Hydrogen-bonded complex of Cpd I and the substrate (not shown) has GGA stabilization energy of about -7.6 kcal/mol that comes predominantly from electrostatic interaction and from two hydrogen bonds with the oxyferryl. After crossing the energy barrier of only 6 kcal/mol the substrate forms strong covalent complex with the site, stabilized by additional -8.3 kcal/mol with respect to weak association. Neither the stabilization of the active site – substrate complexes nor significant lowering of the energy barrier was expected after inspecting energetical parameters of this elementary step for benzene hydroxylation [2, 6]. Therefore critical analysis of structural and electronic factors influencing the bonding of the actual drug molecule to the ferric active site in CYP450 3A4 was of prime interest.

Table 1: Electronic properties of stable structures along transformation pathway: stability with respect to isolated fragments, charge and spin on substrate and spins on crucial fragments of Cpd I.

Complex	Interaction energy kcal/mol	Substrate		Spin		
		Charge	Spin	Fe	O	S
Weakly bound	-21.424	0.199	-0.186	1.151	0.816	-0.535
Strongly bound	-29.738	0.795	0.069	0.81	0.047	0.121
Proton-shuttle	-62.997	0.216	0.001	0.758	-0.002	0.272
Alcohol	-63.946	0.551	0.008	-0.001	0.765	0.255
Ketone	-51.818	0.610	-0.004	0.739	0.00	0.305

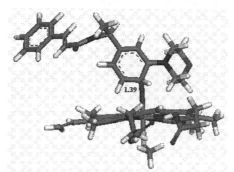

Figure 1: Strongly bonded σ-complex between the substrate and Cpd I (with C-O bond distance in Å). Color captions: gray – C; white – H; red – O; violet – Fe.

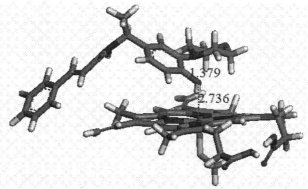

Figure 2: Hydroxylated substrate weakly bonded to the active site (C-O and Fe-O distances in Å)

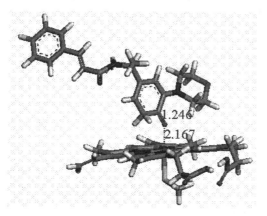

Figure 3: Ketone product weakly bound to Cpd I ((C-O and Fe-O distances in Å, colors as in Fig. 1).

The transition state structure for σ-bond formation reveals the importance of the morpholine nitrogen bound in ortho position to the phenyl ring. In a free substrate this nitrogen atom has clear sp³ hybridization and thus pyramidal structure; in transition state this fragment becomes nearly planar with sp^2 hybridization. This indicates that the nitrogen lone pair forms double bond with the phenyl carbon thus regaining three conjugated double bonds lost on formation of the σ-complex with the oxyferryl site.

In the σ-complex the C-H bond on tetrahedral carbon is already substantially activated, r_{CH}=1.15 Å as compared to regular bond length of 1.09 Å, and points towards the nearest porphyrine nitrogen. Larger activation of the ipso hydrogen and the shorter N-H distance than in the case of benzene oxidation indicates that the H abstraction by proton shuttle to porphyrine nitrogen should have very low barrier and be distinctly kinetically favored over epoxide or direct ketone formation. Indeed, the calculated proton-shuttle reaction pathway gave very low energy barrier, below 1 kcal/mol at GGA. Only approximate transition state could be located here due to the flatness of potential energy surface and large size of the system, with r_{CH}=1.23 Å and r_{NH}=1.64 Å (early transition state). From the stable structure with the proton positioned on porphyrine nitrogen low energy pathways lead to the two products: alcohol and ketone precursors. Both low activation energies and high product stabilities (Table 1) indicate fast reaction with some preference for alcohol formation in the metabolism of a prospective drug molecule on CYP3A4 active site.

3. Conclusions

Quantum chemical modeling for elementary steps in the transformation of the (S)-N-[1-(3-morpholin-4-ylphenyl)ethyl]-3-phenylacrylamide catalyzed by the active site in CYP3A4 revealed new aspects of the metabolism. In conclusion we propose that the presence of an electron pair donating substituent in the ortho position to the oxidized phenyl carbon will substantially lower the energy barrier for the rate-determining step in the drug metabolism. The o-morpholine substituent studied here was found to allow the π-electron deficient fragment to increase the number of conjugated π bonds and restore the aromaticity. From the σ-bonded strong complex the substrate transformes along low-energy pathways to the final metabolites: alcohol and ketone, with the former one indicated as a favorable one.

References

[1] G. N. Kumar, S. Surapaneni, Role of drug metabolism in drug discovery and development , _Medicinal Research Reviews_ **21** 397-411(2001) .

[2] S. Shaik, D. Kumar, S. P. de Visser, A. Altun and W. Thiel, Theoretical Perspective on the Structure and Mechanism of Cytochrome P450 Enzymes, _Chemical Reviews_ **105** _2279-2328_ (2005).

[3] Y. J. Wu, C. D. Davis, S. Dworetzky, W. C. Fitzpatrick, D. Harden, H. He, R. J. Knox, A. E. Newton, T. Philip, C. Polson, D. V. Sivarao, L. Sun , S. Tertyshnikova, D. Weaver, S. Yeola, M. Zoeckler and M. W. Sinz, Fluorine substitution can block CYP3A4 metabolism-dependent inhibition: identification of (S)-N-[1-(4-fluoro-3-morpholin-4-ylphenyl)ethyl]-3-(4-fluorophenyl) acrylamide as an orally bioavailable KCNQ2 opener devoid of CYP3A4 metabolism-dependent inhibition, _Journal of Medicinal Chemistry_ **46** 3778-3781(2003).

[4] S. P. de Visser and S. Shaik, A proton-shuttle mechanism mediated by the porphyrin in benzene hydroxylation by cytochrome p450 enzymes, _Journal of American Chemical Society_ **125**, 7413-7424(2003).

[5] a) B. Delley, An All-Electron Numerical Method for Solving the Local Density Functional for Polyatomic Molecules, _Journal of Chemical Physics_ **92** 508-517(1990), b) B. Delley, From molecules to solids with the DMol3 approach, _Journal of Chemical Physics_ **113** 7756-7764(2000).

[6] C. M. Bathelt, L. Ridder, A. J. Mulholland and J. N. Harvey, Mechanism and structure-reactivity relationships for aromatic hydroxylation by cytochrome P450, _Organic and Biomolecular Chemistry_ **2** 2998-3005(2004).

Brill Academic Publishers
P.O. Box 9000, 2300 PA Leiden
The Netherlands

*Lecture Series on Computer
and Computational Sciences*
Volume 4, 2005, pp. 1371-1374

A protein classification engine based on stochastic finite state automata

F.E. Psomopoulos[1] and P.A. Mitkas[2]

Dept. Electrical and Computer Engineering, Aristotle University of Thessaloniki, GR-541 24 Thessaloniki, Greece	and	Informatics and Telematics Institute, Centre for Research and Technology Hellas, GR-570 01 Thessaloniki, Greece

Received 31 July, 2005; accepted in revised form 10 August, 2005

Abstract: Accurate protein classification is one of the major challenges in modern bioinformatics. Motifs that exist in the protein chain can make such a classification possible. A plethora of algorithms to address this problem have been proposed by both the artificial intelligence and the pattern recognition communities. In this paper, a data mining methodology for classification rules induction in proposed. Initially, expert – based protein families are processed to create a new hybrid set of families. Then, a prefix tree acceptor is created from the motifs in the protein chains, and subsequently transformed into a stochastic finite state automaton using the ALERGIA algorithm. Finally, an algorithm is presented for the extraction of classification rules from the automaton.

Keywords: protein classification, data mining, bioinformatics, motifs, finite state automata

Mathematics Subject Classification: 68T30, 68Q45, 68R10

PACS: 87.15.Cc, 87.14.Ee

1. Protein Classification

Protein classification is currently one of the most interesting problems in computational biology. The biological action of proteins is traditionally identified by time consuming and expensive in-vitro experiments. However, recent developments in bioinformatics have enabled the use of computational tools and techniques towards this end [1]. Clustering algorithms [2], artificial neural networks [3], decision trees [4, 5] and statistical models [6] are few of the methods currently employed. *Motifs* that can be found in a protein chain have provided a higher level of abstraction to the problem, since protein properties are mainly defined by them. Motifs can be either *patterns*, which are short aminoacid chains with a specific order, or *profiles*, which are computational representations of multiple sequence alignments derived by the use of hidden Markov models.

On the other hand, proteins can be assorted into families, each family containing proteins with similar functions. Those families can either be *expert–based*, meaning that they have been experimentally specified and their significance is biologically meaningful, or *computer–generated*, meaning that they have been created with the use of unsupervised protein classification algorithms. In the latter case the major disadvantage is that the protein classes will not necessarily have any biological meaning or significance. In the former, expert – based classes are often overlapping, thus adding to the complexity of the classification algorithms.

2. Core Engine

In [7] a technique for extraction of classification rules using finite state automata was introduced, the classes being expert – based protein families. The technique is outlined in Figure 1. First, the training set is constructed from a set of known proteins. The next step is the creation of the Prefix Tree

---

[1] E-mail: fpsom@auth.gr

[2] E-mail: mitkas@eng.auth.gr

Acceptor (*PTA*) using the protein chains of the train set. Using the ALERGIA algorithm [8], the PTA is converted to an equivalent Stochastic Finite State Automaton (*SFSA*), in which every transition is associated with a probability. Since the SFSA is a generalized representation of the protein structure, information can be obtained directly from it. As a result, certain probabilities can be calculated, such as the probability of a protein chain to contain a certain motif or a specific subset of motifs. Using these probabilities, rules can be extracted to better describe the form of the protein chains. In this paper we extend this technique by introducing both a hybrid form of classification and a new algorithm for the extraction of classification rules from the automata.

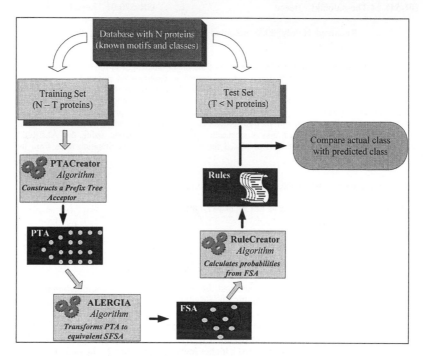

Figure 1: FSA based technique for protein classification rules induction

3. Methodology Outline

The proposed methodology for rule induction consists of three sequential parts: a) Preprocessing, b) Model Creation and c) Rule Extraction.

During the Preprocessing phase, all the protein classes in the training set are recombined to create *Virtual Classes*. These constitute a hybrid form of the original partitioning, where the overlapping areas between classes are assigned to new classes, as shown in Figure 2. In this case, the classes that arise still maintain their biological meaning, but they also represent more closely the proteins they contain. On an algorithmic level, this step creates disjoint sets of proteins, which can be independently processed for classification rules.

For each ensuing Virtual Class, during the Model Creation phase, a SFSA is created using the algorithm described in [7], modified as follows: before the construction of the PTA, the motif list is transformed in order to move the most frequently appearing motifs at the beginning of each list. This transformation ensures that the order of appearance of the motifs in the lists will have no impact on the classification rules.

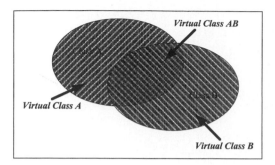

Figure 2: Virtual protein classes

The final step in the methodology is the extraction of classification rules from the SFSA. This is accomplished using the *Max_Probability_Path* algorithm, presented in Box 1. This algorithm is a modification of the shortest path: it finds the maximum probability path from the starting vertex to any other vertex in the SFSA. However, only the paths to the terminating vertices are important due to the fact that they define the conditions necessary for a protein to belong to a class. For each vertex the algorithm stores its predecessor, thus allowing tracing the path from the starting vertex to any other vertex in the SFSA.

```
algorithm Max_Probability_Path
input:
   A: Stochastic Finite State Automaton
output:
   p[]           : probability to reach each vertex from the starting vertex
   predecessor[]: the previous vertex following the max probability path
begin
   s: starting vertex in A
   for each vertex u ∈ A
      p[u] = 0;
   end for
   p[s] = 1;
   predecessor[s] = null;
   Q = Queue with all vertices in A;

   while(Q NOT empty)
      u = first vertex in Q;
      for (each vertex v that connects with u)
         if (p[u]*probability(u, v) > p[v])
            p[v] = p[u]*probability(u, v);
            remove vertex v from Q;
            predecessor[v] = u;
         end if
      end for
   end while

end algorithm
```

Box 1: Max_Probability_Path algorithm

The concept behind the algorithm is that the probability of any path from the starting node is equal to the product of the probabilities of the edges that comprise the path. A classification rule is derived from a path simply by reading the edge "types", i.e. the motifs that exist in the path. The probability of the path is a quantitative measure of the interest of the rule: the higher the probability of the path the stronger the rule will be.

An overview of the complete methodology is presented in Figure 3.

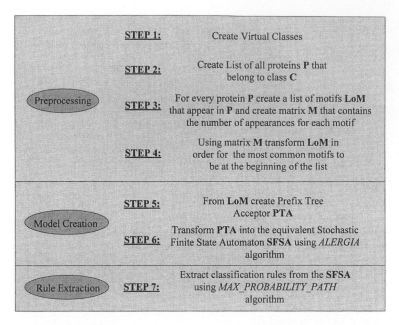

Figure 3: Methodology Outline

References

[1] P.F. Baldi, S. Brunak, *Bioinformatics: A Machine Learning Approach*, The MIT Press, Cambridge, MA, 2001.

[2] C. Makris, Y. Panagis, K. Perdikuri, E. Theodoridis, A. Tsakalidis, Algorithms for Protein Clustering, In Proceeding of the 2nd International Greek Biotechnology Forum, pp 38-44, July 1-3, 2005.

[3] C. Bishop, *Neural Networks for Pattern Recognition*, Oxford University Press, New York, 1995.

[4] D. Wang, X. Wang, V. Honavar, D. Dobbs, Data-driven generation of decision trees for motif-based assignment of protein sequences to functional families, *In: Proceedings of the Atlantic Symposium on Computational Biology*, Genome Information Systems & Technology, 2001.

[5] J.R. Quinlan, *Programs for Machine Learning*, Morgan Kaufmann, San Mateo, CA, 1992.

[6] R. Duad, P. Hart, *Pattern Classification and Scene Analysis*, Wiley, New York, 1973.

[7] F. Psomopoulos, S. Diplaris, P.A. Mitkas, A Finite State Automata Based Technique for Protein Classification Rules Induction, *In Proceedings of the Second European Workshop on Data Mining and Text Mining for Bioinformatics*, pp. 54-60, ECML/PKDD 2004, Pisa, Italy, September 20-24, 2004.

[8] R.C. Carrasco, J. Oncina, Learning Stochastic Regular Grammar by means of a State Merging Method, *In Proceedings of the Second International Colloquium on Grammatical Inference (ICGI '94)*, Alicante, Spain, Lecture Notes in Artificial Intelligence, pp 139-152, Springer – Verlag, 1994.

Brill Academic Publishers
P.O. Box 9000, 2300 PA Leiden,
The Netherlands

*Lecture Series on Computer
and Computational Sciences*
Volume 4, 2005, pp. 1375-1378

Finding Multirepeats in a Set of Strings

C. Makris, S. Sioutas, E. Theodoridis, K. Tsichlas
Department of Computer Engineering and Informatics, University of Patras, 26500 Patras, Greece
makri,sioutas,theodori,tsihlas@ceid.upatras.gr
A. Bakalis
Department of Computer Science, King's College, Strand, London WC2R 2LS, England
bakalis@dcs.kcl.ac.uk

Received 31 July, 2005; accepted in revised form 10 August, 2005

Abstract: A multirepeat in a string is a substring that appears many times. A multirepeat is maximal if it cannot be extended either to the right or to the left and produce a multirepeat. In this paper, we present an algorithm for the problem of finding maximal multirepeats in a set of strings. We propose an algorithm with $O(N^2 n + \alpha)$ time complexity, where N is the number of strings, n is the mean length of each string, m is multiplicity of the multirepeat and α is the number of reported occurrences.

Keywords: Computational Biology, Tandem Arrays, Maximal Repeats, Suffix Trees

Mathematics Subject Classification: 68W40, 68Q25

PACS: 02.10.Ox

1 Introduction

In this paper we investigate the problem of locating maximal multirepeats in a set of strings. A multirepeat in a string is a substring that appears many times. The distance (number of intermediary characters) between the occurrences of the same substring is called a gap. When the gap is equal to zero, then the multirepeat is in fact a tandem array.

A tandem repeat (square) is a string of the form $s's'$, where s' is a non-empty string. Tandem repeats have been widely investigated. In [2, 5] two algorithms with $O(n \log n)$ time for finding perfect tandem repeats (no errors are allowed) are given. Gusfield [3] proposed an $O(n + \alpha)$ algorithm for finding all maximal pairs in a string with no restrictions on the gap between the two substrings, where α is the number of reported pairs. A maximal pair is a substring that occurs twice in a string. The multirepeats are a generalization of maximal pairs. We extend his results by investigating the problem in the case of a set of strings.

The problem of finding common regularities among a set of strings is very important (see [3] Chapt. 14 and 15). In biological strings (DNA, RNA, or protein) the problem of finding repeats in a set of strings arises in many contexts, like database searching and sequence allignment. In particular, as stated in [3] (pp. 127):

> *Less directly, the problem of finding (exactly matching) common substrings in a set of distinct strings arises as a subproblem of many heuristics developed in the biological literature to align a set of strings.*

Figure 1: A maximal multirepeat of multiplicity 3 for two strings.

An example of such use is given in [4], where the algorithm of Gusfield for finding maximal pairs is employed in order to find approximate repeats in a biological string. In fact, the basic routine of their repeat analysis software is the algorithm to find exact maximal pairs. Moreover, as stated in [1], the search for maximal pairs with small gaps could somehow compensate for errors in tandem repeats or general repeats.

Assume a set S of N strings, where each string in S has mean length n. The problem is to find all repeats that occur at least m times in each of at least q strings of set S with no restrictions imposed on the length of the gaps. For this problem, we propose an algorithm with $O(N^2 n + \alpha)$ time complexity, where α is the number of reported occurrences.

2 Preliminaries

Consider an alphabet $\Sigma = \{1, 2, \ldots, \sigma\}$. A string s of length n is represented by $s[1..n] = s[1]s[2]\cdots s[n]$, where $s[i] \in \Sigma$ for $1 \le i \le n$. The empty word is the empty sequence (of zero length) and is denoted by ε; we write $\Sigma^* = \Sigma^+ \cup \{\varepsilon\}$. Moreover, a word is said to be *primitive* if it cannot be written as v^e with $v \in \Sigma^+$ and $e \ge 2$.

We define as *factor* f of length p at position i of a string s, a sequence of length p of consecutive characters occurring at position i in string s; that is $f = s[i, \cdots, i+p-1]$. A string has a repetition when it has at least two equal factors. In the case where these factors are consecutive then this occurrence is called a *square* or *tandem repeat*.

A multirepeat of multiplicity m is an occurrence of p that happens m times. A multirepeat is said to be *left-maximal* (*right-maximal*) if the characters to the immediate left (right) of the m occurrences of p are different and so we cannot extend the multirepeat. A multirepeat is *maximal* if it is both left-maximal and right-maximal. In Figure 1 an example of a maximal multirepeat is given for two strings.

Consider a set S of N strings $s_1, s_2, \ldots, s_N$ whose mean length is n. We say that a maximal multirepeat p occurs in i strings, if each of these i strings contains the same maximal multirepeat. We define the *quorum* q to be the minimum number of strings such that a maximal multirepeat must occur, in order to be considered valid. In this paper we consider the following problem:

Problem 1 *Given a set of strings $S = \{s_1, s_2 \ldots, s_N\}$, where each string s_i has length n, and integers $q \le N$ and $m \ge 2$ we must find all maximal multirepeats of multiplicity m that occur in at least q strings of set S.*

We propose algorithms that make heavy use of suffix trees and generalized suffix trees (see [3] for a good introduction to suffix trees). The Generalized Suffix Tree (GST) for a set of strings $S = \{s_1, s_2, \ldots, s_N\}$, $N \ge 2$, contains all suffixes of all strings in this set. The leaves of the GST contain the position in the string where the suffix occurs and in addition an identifier of the string in which the suffix belongs. Thus, the pair (i, j) stored in a leaf corresponds to the occurrence at position j of a suffix of a string s_i. Note that a leaf may store multiple pairs that correspond to

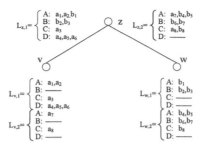

Figure 2: For an alphabet $\Sigma = \{A, C, G, T\}$, an example is given of the sublists in a suffix tree.

the same suffixes in different strings. The substring defined between the root of the suffix tree and a node v is called the *path label* of v.

The internal nodes of the GST have degree at least 2 and at most σ. However, the algorithms in this paper make the silent assumption that the suffix tree is a binary tree. In order to binarize the GST, we replace each node v with out-degree $|v| > 2$ by a binary tree with $|v|$ leaves and $|v| - 1$ internal nodes with $|v| - 2$ edges. Each new edge is labelled with the empty string ε so that all new nodes have the same path-label as node v that they replace. The size of the initial GST was $O(Nn)$. The size of the binary GST is also $O(Nn)$ since for each edge in the initial GST we add two nodes in the binary GST. Finally, we make the assumption that the size of the alphabet Σ is $O(1)$ (in molecular biology usually $\Sigma = \{A, C, G, T\}$).

3 Identifying Maximal Multirepeats

First, a generalized suffix tree T on the set of strings S is constructed. This suffix tree is binarized by the simple procedure we described in 2, resulting in tree T'. Then, we traverse the tree to report all maximal multirepeats. For each pair of nodes v and w with father z in T' we first report the maximal multirepeats (if any) between nodes v and w (the reporting stage). Next, we prepare their father z for the reporting stage between z and its brother in T' (the construction stage). As a result, the algorithm is divided into two stages, the construction stage and the reporting stage for each pair of nodes. First, we focus on the construction stage.

The Construction Stage

Our algorithm traverses the suffix tree in a bottom-up fashion. Each internal node v is attached a set of leaf-lists. There is a leaf-list for every string in set S and each leaf-list stores all occurrences of suffixes of the respective strings that occur in the leaves of the subtree T_v. The leaf-list $L_{v,i}$ corresponds to occurrences of suffixes of s_i in the subtree T_v. Each leaf-list consists of sublists $L_{v,i}^{\sigma_j}$ that stores occurrences of suffixes of string s_i in T_v such that they are preceded by symbol σ_j in string s_i. The leaf-lists of the leaves consist of the set of suffixes they represent. An example of sublists is depicted in Figure 2.

Assume the brother nodes v and w and their respective father z as in Figure 2. The problem is to construct the leaf-lists of z by the leaf-lists of v and w. This can be accomplished by the merge of the leaf-lists of v and w. Each sublist $L_{v,i}^{\sigma_j}$ is concatenated with the sublist $L_{w,i}^{\sigma_j}$ and this concatenation results in $L_{z,i}^{\sigma_j}$ (see Figure 2). The concatenation is a simple list operation that can be realized in $O(1)$ time since the occurrences of the suffixes need not be ordered. The binary GST contains $O(Nn)$ nodes and each node carries at most N leaf-lists. Each leaf-list has at most σ sublists. As a result, the construction procedure of the sublists is realized in total $O(N^2n)$ time.

The Reporting Stage

At this point we will focus on the reporting stage. Assume two nodes v and w. We want to generate all maximal multirepeats from the leaf-lists $L_{v,i}$ and $L_{w,i}$. This is accomplished by combining the positions of two or more sublists and then combine the results for at least q leaf-lists. If there are at least $N - q$ leaf-lists for which the combination is empty, then no output can be generated from these two nodes (the quorum constraint).

The design of a good output-sensitive algorithm for this problem relies heavily on whether we are able to find out efficiently if there are multirepeats to report for the nodes v and w. We accomplish this by a simple check as follows. The combination of sublists of leaf-lists $L_{v,i}$ and $L_{w,i}$ is empty if one of the following holds:

1. Each of the leaf-lists have at most one non-empty $L_{v,i}^{\sigma_j}$ and $L_{w,i}^{\sigma_j}$ for the same symbol

2. The number of positions stored in leaf-lists $L_{v,i}$ and $L_{w,i}$ is $< m$ and (1) does not hold.

If the first criterion holds, then we have nothing to combine, so there is no output from these leaf-lists. The second criterion is related to multiplicity. So, if the first criterion does not hold then we must have at least m positions in each string in order to satisfy the multiplicity constraint.

We can make this check by a simple traversal of all leaf-lists checking whether their sublists are empty or not while keeping the total number of positions stored. This is accomplished in $O(N)$ time. Since the GST has $O(Nn)$ nodes, this check costs $O(N^2 n)$ in total.

If at least $N - q$ of them are empty, then no valid multirepeat can be reported. Otherwise, the reporting of multirepeats is performed. Either we report the path label to z as a multirepeat or if we want the exact positions we make all possible combinations of m elements between $L_{v,i}$ and $L_{w,i}$ such that in each reported multirepeat there is at least one position from each leaf-list. Obviously, this can be accomplished in time linear to the number of reported multirepeats.

An example of this procedure can be given with the help of Figure 2. Assume that $q = 2$ and $m = 2$ and $S = \{s_1, s_2\}$. Then, the path label of node z will be reported as a multirepeat since: (a) the leaf-lists $L_{v,i}^{\sigma_j}$ and $L_{w,i}^{\sigma_j}$ for $i = 1, 2$ are non-empty for more than one σ_j and (b) the number of positions stored in these lists is $\geq m$. For example, a maximal pair (since $m = 2$) occurs at positions (a_1, b_2) and (a_8, b_7) in strings s_1 and s_2 respectively. However, if we set $m = 8$ then there is no occurrence to report because the size of lists $L_{v,2}$ and $L_{w,2}$ is 7. As a result, only one string satisfies the multiplicity constraint which further means that the quorum constraint is violated.

Theorem 1 *The computation of maximal multirepeats in a set of strings S, uses linear space and its time complexity is $O(N^2 n + \alpha)$, where α is the size of the output.*

References

[1] G.S. Brodal, R.B. Lyngsø, C.N.S. Pedersen and J. Stoye. Finding Maximal Pairs with Bounded Gaps. *Journal of Discrete Algorithms*, 0(0):1-27, 2000.

[2] M. Crochemore. An Optimal Algorithm for Computing the Repetitions in a Word. *Information Processing Letters*, 12(5):244-249, 1981.

[3] D. Gusfield. Algorithms on Strings, Trees, and Sequences. Cambridge University Press, New York, USA, 1997.

[4] S. Kurtz, J.V. Choudhuri, E. Ohlebusch, C. Schleiermacher, J. Stoye and R. Giegerich. REPuter: the Manifold Applications of Repeat Analysis on a Genomic Scale. *Nucleic Acids Research*, 29(22):4633-4642, 2001.

[5] M.G. Main and R.J. Lorentz. An $O(n \log n)$ Algorithm for Finding all Repetitions in a String. *Journal of Algorithms*, 5:422-432, 1984.

Brill Academic Publishers
P.O. Box 9000, 2300 PA Leiden
The Netherlands

*Lecture Series on Computer
and Computational Sciences*
Volume 4, 2005, pp. 1379-1382

In silico evaluation of bioactive compounds: Docking Simulations based Enzyme-Inhibitor Interaction compared with X-ray models.

Georgios Vlachopoulos[*], Athanasios Papakyriakou[+], George Dalkas[*], Georgios A. Spyroulias[1*], Paul Cordopatis[*]

[*1]Department of Pharmacy, University of Patras, GR-26504, Patras, Greece
[+]Institute of Physical Chemistry, NCSR "Demokritos," 153 10 Ag. Paraskevi Attikis, Greece.

Received 31 July, 2005; accepted in revised form 10 August, 2005

Abstract: Modern drug design approaches are based on accurate prediction of the protein-ligand binding interface and binding properties/modes of the ligand, even if experimentally determined (through X-ray or NMR) protein-ligand complex model is not available. The knowledge of the structure and physicochemical determinants of protein-substrate recognition and binding is of fundamental importance in structure-based drug design. What is highlighted herein is the procedure which makes use of tools of bioinformatics in order to test the binding properties of a ligand to a biomacromolecule. Since our research activities oriented in the quest of bioactive compounds as inhibitors towards zinc metallopeptidases, such us Angiotensin-I Converting Enzyme (ACE) [1] and Anthrax Lethal Factor (ALF) [2], we are studying the enzyme's catalytic sites and/or inhibitors conformational characteristics through Nuclear Magnetic Spectroscopy (NMR). Furthermore, we exploit the acquired structural data in an attempt to screen various compounds according their binding affinity *in silico*, through docking simulations methodology. Possible lead compounds will be optimized, synthesized and their binding properties would be then determined experimentally (using Xray or NMR).
In this procedure application of docking simulations approaches is a prerequisite and the evaluation of the binding modes of a ligand to a protein target should be performed. To this effect, we implement docking simulations to the study of enzyme-inhibitor complexes and the results are compared to already known enzyme-inhibitor crystal structures. The potential for a docking algorithm to be used as a virtual screening [3][4]tool is based on both speed and accuracy .
Keywords: Structure-based Drug Design, Docking Simulations, Structural Bioinformatics, Angiotensin-I Converting Enzymes, Anthrax Lethal Factor.

Mathematics SubjectClassification: 9240
PACS: 31.15-p

1. Methods and Materials

1.1 Simulation Method

AutoDock version 3.0.5 and AutoDock Tools [5] software was used for the docking simulation. In this procedure the Lamarckian Genetic Algorithm (LGA) [5] was applied for ligand conformational search. The target-molecule (protein, enzyme, etc) is considered as a rigid body and the ligand/substrate poses conformational freedom in terms of torsion angle rotation. The first step was the target preparation. All crystallographic waters were removed. Polar hydrogens were added and all histidine residues were made neutral. Kollman [5] charges were assigned to all atoms. As a docking site the Zn-catalytic site was chosen. We used two kinds of 3D grid boxes. In the case of ACE complexes the selected dimensions for the first grid box (no localized box) was 84x52x82 and for the second grid box (localized box) was 40x30x40. Similarly in the case of ALF complex the corresponding grid boxes had dimensions 88x60x84 and 44x30x842. In both cases the grid boxes centered on the active site, with 0,375 Å spacing where the energies of each grid point calculated for each of the following atom types C, A (aromatic C), N, O, S, H, Zn and e (electrostatic).

For the ligands/substrates (Lisinopril, Enapatrilat, NSC 12155), all Hydrogens were added and Gasteiger [5] charges were assigned. The rotatable bonds assigned via AutoTors routine. For each of

---

[1] Corresponding author. E-mail: G.A.Spyroulias@upatras.gr

the above algorithms the docking parameters were the following: Trials of 100 dockings, population size of 150, random starting position and conformation, translation step ranges of 1.5 Å, rotation step ranges of 5°, elitism of 1, mutation rate of 0.02, crossover rate of 0.8, local search rate of 0.06, and 1,5 million energy evaluations.

1.2 Evaluation Method

The success of a program [6] in predicting ligand's binding pose is usually measured by the root-mean-square deviation (RMSD) between the experimentally-determined heavy atom positions of the ligand and those of the models predicted by the program (more specifically, the top-ranked solution from the program). Given docking results for a set of test complexes, the RMSD values can then be summarized in different ways. The most frequently used statistic is the percentage of test complexes for which the docking solution has RMSD 2 Å (termed here the success rate). The 2Å threshold is arbitrary but rather commonly used[7]-[9]. Some authors count docking-derived ligand conformers whose RMSDs lie between 2and 3 Å as partial successes.[10][11]

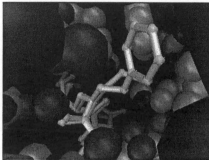 

Figure 1a: *Ligand conformers in ACE–Lisinopril complex, with low RMSD (0,89 Å). The light grey structure represents the experimental structure, and the yellow represents the simulated one with the lowest energy. The similarity between the two structures is high, and the aromatic rings fit perfectly one each other.*

Figure 1b: *Ligand conformers in ACE–Lisinopril Complex, with high RMSD (8,81 Å). The light grey structure represents the experimental structure, and the yellow one represents the simulated structure. The difference between the two conformations is apparent.*

Friesner et al [12] used the average RMSD over all their test complexes as an overall success measure. However, if two programs are run on a test complex and produce solutions with RMSDs of 4 Å and 6 Å, it could be concluded that both lack precision and accuracy in yielding or predicting the binding pose of the ligand than to consider that one is twice as accurate as the other. In contrast, if these programs give predictions with RMSDs of 0.3 Å and 0.6 Å among the conformers (ligand), it might be argued that both yield rather precise and accurate binding modes of the ligand, especially when the resolution of the experimental results is taking into account (see Goto et al.[13] for an interesting discussion of this point). Additionally, since the distribution of RMSDs is positively skewed, the average is a poor estimate of central location.

According to literature data RMSD should be treated and used with caution in result evaluation, since it is flawed as a success measure while it is possible to get solutions that have good RMSDs but form different interactions with the protein than these which are observed experimentally.

Results and Discussion

As it was stated above, we used two kinds of grid boxes. When the not localized grid box with the large dimensions is selected, the ligand is allowed to move around a large area near the catalytic site. This freedom of the ligand to occupy any position into the frames of grid box, gives the permeation to the ligand to find other binding spots except for the active site, and to adopt different conformations than that in which is crystallized in solid state (as determined by x-ray).

Even though, the structures with the lowest docking energies, which are the structures with the higher binding ability, are the structures with the most successful values of RMSD. The percentage success in all three cases hurdles 95%.

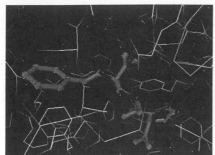 

Figure 2a: *Ligand conformers in ACE–Enapatrilat Complex, with low RMSD (1,3). The blue structure represents the experimental structure, while the green one represents the simulated structure with the lowest energy.*

Figure 2b: *Ligand conformers in ACE–Enapatrilat Complex, with high RMSD (8,81). The blue structure represents the experimental structure, while the green one represents the simulated structure.*

As it is illustrated by the data presented above when the localized box is used, the results are significantly better in terms of RMSD, due to space constrains applied to the ligand, and force the molecule to move in a short area around the active site. Therefore, the vast majority of the conformers look alike the X–ray structures, and the success percentages are higher.

ACE-Lisinopril

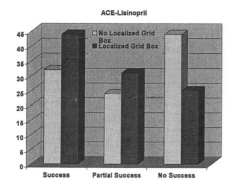

	Success	*Partial Success*	*No Success*
No Localized Grid Box	*32*	*24*	*44*
Localized Grid Box	*44*	*31*	*25*

ACE-Enapatrilat

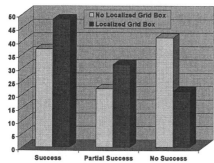

	Success	*Partial Success*	*No Success*
No Localized Grid Box	*37*	*22*	*41*
Localized Grid Box	*48*	*31*	*21*

ALF-NSC 12155

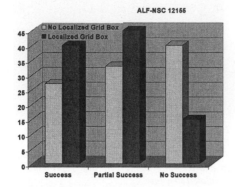

ALF-NSC 12155

	Success	Partial Success	No Success
No Localized Grid Box	27	33	40
Localized Grid Box	40	45	15

The above analysis leads to the conclusion that the docking simulation is a powerful and reliable computational tool. It can be safely used for predicting the interaction of ligands with biomacromolecular targets, with apparent interest in the field of computer-aided drug design. Finally, molecular docking is often used in virtual screening methods [3] [4], whereby large virtual libraries of compounds are evaluated according to their binding ability. This process leads to a manageable subset of compounds, constituted by molecules with high binding affinities to a target receptor, thus reducing the number of compounds that would be further optimized and evaluated *in vitro* and *in vivo*.

References

[1]   G.A. Spyroulias, P. Nikolakopoulou, A. Tzakos, I.P. Gerothanassis, V. Magafa, E. Manessi-Zoupa, P. Cordopatis. Comparison of the solution structures of angiotensin I & II. Implication for structure-function relationship. *Eur. J. Biochem.* 270, 2163-2173(2003)

[2]   Hicks RP, Hartell MG, Nichols DA, Bhattacharjee AK, van Hamont JE, Skillman DR. The medicinal chemistry of botulinum, ricin and anthrax toxins. Curr Med Chem. 2005;12(6):667-90. Review.

[3]   Paul D. Lyne, Structure-based virtual screening: an overview, *DDT Vol. 7, No. 20 October 2002*

[4]   Ingo Schellhammer and Matthias Rarey Fast, Structure-Based Virtual Screening, *PROTEINS: Structure, Function, and Bioinformatics 57:504–517 (2004)*

[5]   G.M. Morris, D.S. Goodsell, R.S. Halliday, R. Huey, W.E. Hart, R.K. Belew and A.J. Olson Automated Docking Using a Lamarckian Genetic Algorithm and and Empirical Binding Free Energy Function J. Computational Chemistry 19, 1639-1662(1998).

[6]   Jason C. Cole, Christopher W. Murray, J. Willem M. Nissink, Richard D. Taylor, and Robin Taylor, Comparing Protein–Ligand Docking Programs Is Difficult PROTEINS: Structure, Function, and Bioinformatics 60:325–332 (2005)

[7]   . Kontoyianni M, McClellan LM, Sokol GS. Evaluation of docking performance: comparative data on docking algorithms. *J Med Chem 2003;47:558–565.*

[8]   Kramer B, Rarey M, Lengauer T. Evaluation of the FlexX incremental construction algorithm for protein-ligand docking. *Proteins 1999;37:228–241.*

[9]   Gohlke G, Hendlich M, Klebe G. Knowledge-based scoring functions to predict protein-ligand interactions. *J Mol Biol 2000;295: 337–356.*

[10] Vieth M, Hirst JD, Kolinski A, Brooks III CL. Assessing energy functions for flexible docking. *J Comp Chem 1998;19:1612–1622.*

[11] Bursulaya BD, Totrov M, Abagyan R, Brooks III CL. Comparative study of several algorithms for flexible ligand docking. J Comput Aided Mol Des 2003;17:755–763.

[12] Friesner RA, Banks JL, Murphy RB, Halgren TA, Klicic JJ, Mainz DT, Repasky MP, Knoll EH, Shelley M, Perry JK, and others. Glide: a new approach for rapid, accurate docking and scoring. 1. Method and assessment of docking accuracy. *J Med Chem 2004;47:1739–1749*

[13] Goto J, Kataoka R, Hirayama N. Ph4Dock: pharmacophore-based protein-ligand docking. *J Med Chem 2004;47:6804–6811.*

Brill Academic Publishers
P.O. Box 9000, 2300 PA Leiden
The Netherlands

Lecture Series on Computer
and Computational Sciences
Volume 4, 2005, pp. 1383-1386

Simulation of nuclei cells images using spatial point pattern models

S. Zimeras[1]

Department of Statistics and Actuarial-Financial Mathematics,
University of the Aegean,
83200 Karlovassi, Samos, Greece

Received 31 July 2005; accepted in revised form 10 August 2005

Abstract: Markov random fields (MRF) modelling are a popular pattern analysis method with many applications in image restoration, texture analysis, classification and image segmentation. They are used to model spatial interaction on lattice system. The important characteristic of MRF is that they form the global pattern as conditional probabilities of local interaction. Models based on Markov random fields are widely used to model spatial processes, especially in biological spatial patterns. A particular subclass of MRF is the auto-models, introduced in Besag (1974) and further studied in Cross and Jain (1983), Besag (1986), Aykroyd et. al. (1996) and Zimeras (1997). In order to estimate MRF parameter efficiently an MRF parameter estimation method based on MCMC could be applied. The fundamental idea is to use an algorithm, which generates a discrete time Markov chain converging to the desired distribution. The most commonly algorithms include the Gibbs sampler (Geman and Geman, 1984) and the Metropolis-Hasting algorithm (Metropolis et. al., 1953; Hastings, 1970). In this work spatial properties of the auto-Poisson model was studied and a deterministic univariate iterative scheme, which can be used to predict these properties, has been proposed. Realizations from auto-Poisson model have been generated using the Gibbs sampler. As a result, the parameter space can be divided into regions, each with distinct spatial behavior. The iterative procedure classifies each region as either stable or unstable. The auto-Poisson model is fitted to a real data example from he area of medical biology. A variety of image structures of nuclei cells created by the proposed model is presented.

Keywords: Markov random fields, spatial point patterns, medical biology, spatial point models.

Mathematics Subject Classification: 62M40

PACS: 02.50.-r

1. Introduction

Data in the form of sets of points, irregular distributed in a region of space could be identified in varies biological applications for examples the cell nuclei in a microscope section of tissue. These kinds of data sets are defined as spatial point patterns and the presentation of the positions in the space are defined as points. The spatial pattern generated by a biological process, can be affected by the physical scale on which the process is observed. With these spatial maps, the biologists will usually want a detailed description of the observed patterns. One way to achieve this is by forming a parametric stochastic model and fitting it to the data. If a model can be found which fits the data well, the estimated values of its parameters provide statistical measures, which can be used to compare similar data sets. Also a fitted model can provide an explanation of the biological processes. Model fitting especially for large data sets is difficult. For that reason, statistical methods can apply with main purpose to formulate a hypothesis for the implementation of biological process.

Classification of the patterns could be regular, random or aggregated based on the scale of the natural environment. The approach taken to reconstruct these spatial patterns will be to develop

---

[1] Corresponding author. Lecturer- Grad. Stat. E-mail: zimste@aegean.gr

methods for the analysis based on **stochastic models**, assuming randomness inside the process. The main characteristic of these models is that the spatial patterns can occur at any spatial location in some 2D Euclidean space. Special classes of spatial models known as **Markov random fields** (MRFs) (Besag, 1974; 1986) have been developed for situations where the set of all possible spatial locations is discrete. A MRF is a discrete stochastic process whose global properties are controlled by conditions of local properties. This collection of such sites is defined as lattice. There are regular and irregular lattices: regular lattices have neighbors that are often defined by adjusting sites and irregular often have neighbors defined by Euclidean distances. The important characteristic of MRF is that they form the global pattern as conditional probabilities of local interaction. Models based on Markov random fields are widely used to model spatial processes, especially in biological spatial patterns. A particular subclass of MRF is the **auto-models**, introduced in Besag (1974) and further studied in Cross and Jain (1983), Besag (1986), Aykroyd et. al. (1996), Zimeras (1997), Zimeras and Georgiakodis (2005). Especially, the family of auto-Poisson point pattern models (Figure 1) has been proposed by Diggle (1983) as an appropriate to model biological structures or treatment diseases.

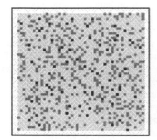
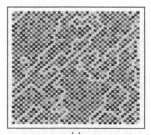

Figure 1. Realizations of the auto-Poisson model for different parameter combinations

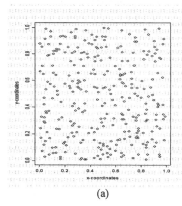
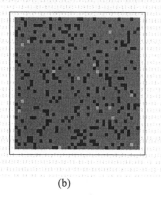

(a)   (b)

Figure 2. (a) Location of the nuclei cells. (b) Image showing intensities of data points.

Key components of any statistical analysis using such models are the choice of an appropriate model and the estimation of prior model parameters. In many applications, appropriate values of these parameters will be found be *trial- and- error*. In order to estimate MRF parameter efficiently an iterative estimation method based on MCMC could be applied. The fundamental idea is to use an algorithm, which generates an MRF model converging to the desired distribution. The most commonly algorithms include the **Gibbs sampler** (Geman and Geman, 1984) and the **Metropolis-Hasting algorithm** (Metropolis et. al., 1953; Hastings, 1970). In this work, investigation of the properties of auto-Poisson models is considered implementing a phase transition iterative process (Chandler 1978; Zimeras, 1997). Estimation of the model parameters has been proposed using the nearest-neighborhood process and investigation for the modelling of the spatial point patterns has been done using K-functions.

Conclusions

In this work spatial properties of the auto-Poisson model was studied and realizations from auto-Poisson model have been generated using the Gibbs sampler; for appropriate combinations of models parameters different Markov random fields can be introduced and for certain combinations the behavior is not straightforward. As a result, the parameter space can be divided into regions, each with distinct spatial behavior. The iterative procedure classifies each region as either stable or unstable leading to phase transition process.

Estimation of the model parameters have been introduced using the nearest-neighborhood process combined with modified maximum likelihood method. Results of the proposed methodology have been illustrated in Figure 3. The data represents the positions of the centers of nuclei of certain cells in an approximatelly 0.25mm square histolgical section of tissue from a laboratory metastastic lymphoma in the kidney of a hamster (Diggle 1983, page 109). The data region is split into a grid with width 0.025m and the frequencies of the points in the pixels are calculated. From the real data set, very few high values towards image are apparent as well as large smooth areas. Both smooth areas and high values pixels occur using the fitted model justifying the effectiveness of the process. Comparison between the two images shows that the results are reasonable and the method works satisfactory.

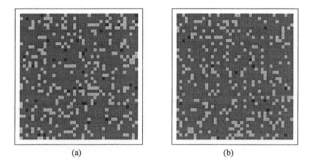

(a) (b)

Figure 3. (a) Fitted model; (b) Data set

References

[1] Besag J. (1974): Spatial interaction and the statistical analysis of lattice systems, J. Royal Statistical Society, Series B, 36, 192-236.

[2] Cross G. R. and Jain A. K. (1983): Markov random field texture models, IEEE Trans. Pattn. Anal. Mach. Intell., 5(1), 25-39.

[3] Besag J. (1986): On the statistical analysis of dirty pictures, J. Royal Statistical Society, Series B, 48, 259-302.

[4] Aykroyd R. G., Haigh J. G. B. and Zimeras S. (1996): Unexpected spatial patterns in exponential familly auto-models, Graphics, Models and Image processing, Vol 58, No 5, 452-463.

[5] Zimeras S. (1997): Statistical models in medical image analysis, Ph.D. Thesis, Leeds University, Department of Statistics.

[6] Geman S. and Geman D. (1984): Stochastic relaxation, Gibbs distribution and the Bayesian restoration of images, IEEE Trans. Pattern Anal. Mach. Intell., 6, 721-741.

[7] Hastings W. K. (1970): Monte Carlo sampling methods using Markov chains and their applications, Biometrika 57, 97-109.

[8] Metropolis N., Rosenbluth A. W., Rosenbluth M. N., Teller A. H., Teller E. (1953): Equations of state calculations by fast computing machines, J. Chem. Phys., 21, 1087-1091.

[9] Diggle P.J. (1983): Statistical analysis of spatial pattern point, Academic Press, London.

[10] Chalmond B. (1989): An Iterative Gibbsian Technique for Simultaneous Structure Estimation and Reconstruction of M-ary Images, Pattern Recognition, 22 (6), 747-761.

[11] S. Zimeras and F. Georgiakodis: Bayesian models for medical image biology using Monte Carlo Markov Chain techniques, Mathematical and Computer Modeling (in press).

Brill Academic Publishers
P.O. Box 9000, 2300 PA Leiden
The Netherlands

Lecture Series on Computer
and Computational Sciences
Volume 4, 2005, pp. 1387-1395

Exploring the Way Synonymous Codons Affect the Distribution of Naturally Occurring Amino Acids

G. Anogianakis, A. Anogeianaki[1]

Department of Physiology, Faculty of Medicine,
University of Thessaloniki,
GR-54124 Thessaloniki, Greece

K. Perdikuri, A. Tsakalidis

Computer Engineering & Informatics Department,
University of Patras,
GR-26500 Patras, Greece
AND
Research Academic Computer Technology Institute,
N. Kazantzaki Str., University of Patras,
GR-26500 Patras, Greece

N. Kozeis

Eye Department, Ippokrateio Hospital,
GR-54642 Thessaloniki, Greece

Received 31 July 2005; accepted in revised form 10 August 2005

Abstract: In this work we expand on our previous exploration of automata which search for protein patterns in nucleic acid sequences. The purpose of our work is to explore the usage of synonymous codons in gene expression and to investigate if it is a potential coding mechanism for tissue differentiation.
Keywords: Synonymous Codons, Biological Sequences, Tissue Differentiation

Mathematics SubjectClassification: 93B40
PACS: 02.10.0x

1. Introduction

DNA and protein sequences can be seen as long texts over specific alphabets encoding the genetic information of living beings. In the case of DNA sequences this alphabet consists of the four nucleotides, $\Sigma_{DNA} = \{a, c, g, t\}$, while in the case of protein sequences, the corresponding alphabet consists of the twenty amino acids. Codons, constitute triplets of nucleotides that code the amino acids, the "words" in this 4-letter language. One of the corollaries of the DNA coding mechanism is that DNA sequence alterations in a gene can change the structure of the protein that it codes for.

One of the earliest explored features of the genetic code is its redundancy: Given that the genetic alphabet uses four letters and each word is three letters long one expects to come up with a sixty four different words "language". But the language of genetics contains twenty one "semantically different words'" (i.e.: the twenty amino acids and a "terminating codon" which serves as the punctuation mark). Thus it is evident that there must be many different ways of "spelling" for at least some, of the words of the language of genetics.

This is best illustrated by the fact that despite drastic variations between the base compositions of the DNA for different species, the amino acid composition of the proteins DNA codes for does not reflect a wide range of variation. Extreme examples of this are the organism Tetrahymena pyriformis,

---

[1] Corresponding author. A. Anogeianaki. E-mail: anogian@auth.gr

on one hand, where only 25% of its double stranded DNA consists of GC pairs and Micrococcus lysodeikticus where 72% of its DNA consists of GC pairs [1].

The actual correspondences between codons and amino acids (at the DNA replication level) are tabulated in Table 1. For example the codon TCC codes for the amino acid Serine, CCT for Proline etc, but also CCC, CCA and CCG all code for Proline. Since the codons: {CCT, CCC, CCA, CCG} code the same amino acid they are called **Synonymous Codons**. According to the recent literature synonymous codons have significant implications in tissue differentiation and evolution.

The transformation of information content from the form of a DNA coded message to a protein structure is accomplished in two steps: DNA is first transcribed into messenger RNA (mRNA) and then mRNA is translated into protein by the ribosome. During this step, transfer RNA (tRNA) acts as a ferry of amino acids to the translation site and at the same time as label for the ribosome to identify the correct amino acid before it places it in the amino acid sequence that will eventually give rise to the DNA coded protein.

At the DNA to mRNA translation level, T is substituted by U and, the corresponding pairing between bases thus becoming AU instead of AT. The RNA alphabet, thus, consists of the four nucleotides $\Sigma_{RNA} = \{a, c, g, u\}$. The same alphabet holds during the mRNA transcription into protein (mRNA – tRNA – ribosome interaction).

From what is known about the amino acid composition of proteins in general, without taking into account the different organisms from which they derive, the observed relative amino acid frequency in protein is as presented in Table 2.

The genomes of species from bacteria [2] to Drosophila [3] show unique biases for particular synonymous codons and, recently, it was shown that such codon preferences exist in mammals [4]. Systematic differences in synonymous codon usage between genes selectively expressed in six adult human tissues were reported, while the codon usage of brain-specific genes is, apparently, selectively preserved throughout the evolution of human and mouse from their common ancestor [4]. In particular, when genes that are preferentially expressed in human brain, liver, uterus, testis, ovary, and vulva were analyzed, synonymous codon biases between gene sets were found. The pairs that were compared were brain-specific genes to liver-specific genes; uterus-specific genes to testis-specific genes; and ovary-specific genes to vulva-specific genes. All three pairs differed significantly from each other in their synonymous codon usage raising the possibility that codon biases may be partly responsible for determining which genes are expressed in which tissues. Such a determination may, of course, take place at the level of transcriptional control. However, given the relatively low number of genes identified in the human genome, "vis-à-vis" our preconceptions about human structure and function, one is tempted to explore whether other control mechanisms operate (alone, in tandem or in parallel with transcriptional level mechanisms) in tissue differentiation [5].

Table 1: The 64 possible combinations of the four bases and the codons they represent. X stands for the terminating codon.

1ST BASE	2ND BASE				3RD BASE
	T	C	A	G	
T	PHE	SER	TYR	CYS	T
	PHE	SER	TYR	CYS	C
	LET	SER	X	X	A
	LET	SER	X	TRP	G
C	LET	PRO	HIS	ARG	T
	LET	PRO	HIS	ARG	C
	LET	PRO	GLN	ARG	A
	LET	PRO	GLN	ARG	G
A	ILE	THR	ASN	SER	T
	ILE	THR	ASN	SER	C
	ILE	THR	LYS	ARG	A
	MET	THR	LYS	ARG	G
G	VAL	ALA	ASP	GLY	T
	VAL	ALA	ASP	GLY	C
	VAL	ALA	GLT	GLY	A
	VAL	ALA	GLT	GLY	G

Assuming that some kind of parsimony principle governs the operation of control mechanisms at the DNA transcription level (a very strong but necessary assumption that is required in order to limit and focus the subsequent discussion), there are two obvious mechanisms that can be used to link synonymous codon choice and tissue-specific gene expression:

- The first mechanism depends on local tRNA abundance. The tRNA pools in the brain, e.g., may differ from the pools in liver, and so if the codon usage of a gene is calibrated to the tRNA pools that exist in the brain, that gene will be translated more efficiently in brain.
- The second mechanism would make use of the different chemical affinities, on one hand between different tRNAs coding for the same amino acid and the mRNA they help transcribe and between mRNA and the underlying DNA structure, on the other. In other words, if certain codons have greater affinities for their corresponding tRNAs than other synonymous codons, it stands to reason that they will be expressed more readily. A corollary of this argument is that differentiation will occur when the appropriate genes find themselves in an energetically appropriate environment to be expressed.

Table 2: Observed relative amino acid frequency in naturally occurring proteins (from B. Lewin, based on data of King and Jukes [1]).

a.a.	% of total a.a.'s
TRP	1,29
MET	1,81
HIS	2,92
TYR	3,35
CYS	3,36
GLN	3,70
ILE	3,78
PHE	3,96
ARG	4,21
ASN	4,39
PRO	4,99
GLU	5,76
ASP	5,93
THR	6,19
VAL	6,79
LYS	7,22
LEU	7,39
ALA	7,40
GLY	7,48
SER	8,08
	100,00

In order to resolve the question of whether the second mechanism that was proposed as the link between synonymous codon choice and tissue-specific gene expression has any theoretical (or practical) merit, it is necessary to associate each codon with a value reflecting its potential for expression. To illustrate this point let us assume, e.g., that a gene is coding a peptide with the sequence:

ALA-ALA-ALA-ALA-ALA-ALA-ALA-ALA-ALA-ALA.

Let us, further, assume that codon GCU has twice the affinity for its corresponding tRNA than codon GCC has for its own corresponding tRNA. Similarly, codon GCC has twice the affinity for its corresponding tRNA than codon GCA has for its own corresponding tRNA and, finally that codon GCA has twice the affinity for its corresponding tRNA than codon GCG has for its own corresponding tRNA. It is evident that, if chemical affinities alone were the determining factor and if gene expression was a linear function of the product of its codons' affinities for their corresponding tRNAs, then a gene represented by the sequence:

GCUGCUGCUGCUGCUGCUGCUGCUGCUGCU,

would be expressed 2^{30} times more readily than the gene represented by the sequence:

GCGGCGGCGGCGGCGGCGGCGGCGGCGGCG,

despite that these two hypothetical genes are synonymous.

The example above serves as an illustration of the approach that is required to resolve the problem of amino acid coding redundancies, codon synonyms and their possible involvement in tissue differentiation.

In the present work we concentrate on further exploring the way that the differential transcriptional (i.e., DNA to mRNA) or translational (mRNA – tRNA – ribosome interaction) affinities for synonymous codons may influence the conversion of DNA sequences into proteins.

The structure of the paper is as follows: In Section 2 we develop some basic principles needed throughout the paper while in Section 3 we present an application of these concepts to determining the different affinities for translation of synonymous codons and their influence on differential gene expression, along with some arguments about the kinetics of DNA translation. Finally in Section 4 we conclude and discuss our future research in the area.

2. The process of DNA Translation

In order to gain an insight as to how we may calculate the differential affinities for translation of synonymous codons we must first explore the stereochemistry of the DNA and RNA molecules respectively. Indeed, as early as 1953 [6, 7] the DNA structure was known (Figure 1), consisting of two covalently linked chains of deoxyribose - phospate which are oriented in opposite (antiparallel) directions (i.e., one running 5' to 3' and the other 3' to 5') and are held together solely by hydrogen bonds between their always complementary base pairs. Other physical characteristics of the double stranded DNA inside the cell are as follows:

- G forms three hydrogen bonds with C while A forms two hydrogen bonds with T.
- The angle of interaction between base pairs results in major and minor grooves
- Helix diameter is 2.0 nanometers.
- Helix rise per base pair is 0.34 nanometers.
- The helix pitch (distance along the axis per 360 degree turn) is 3.4 nanometers.
- Phosphates are on the outside of the molecule.
- Base pairs are always in the inside of the molecule stacked close to each other.
- Inside cells, DNA completes an average of 10.5 base pairs per helical turn.

Only four base pairs were possible either in the double stranded DNA molecule or during its replication: TA, CG, GC and AT. It is interesting to note that the aggregate molecular weight of each of these pairs is

$$TA = AT = 259.80 \text{ Dalton and}$$
$$CG = GC = 260.35 \text{ Dalton.}$$

Given that this represents a variation in the density of matter (analogous to the aggregate molecular weight (agMW) of the base pairs) within the volume enclosed by the backbones of the DNA of at most 0.6% per base pair, it is reasonable to conclude that the distance (Figure 2) across each base pair is constant throughout the entire DNA strand, which is confirmed by crystallography. Furthermore, assuming that the process of DNA replication is irreversible, at the nanometer level the probability of one base matching its opposite one within the limited space available (i.e., the two deoxyribose – phospate, *sugar*, backbones constricted by the cavity of the enzyme catalyzing the reaction) will be approximately constant.

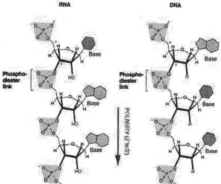

Figure 1: Structure of DNA and RNA strands[1]

---

[1] http://oregonstate.edu/instruction/bb492/lectures/RegulationIV.html

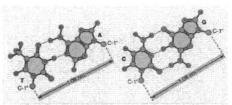

Figure 2: Schematic of the TA (AT) and CG (GC) interactions (hydrogen bonding) in the DNA double helix[1].

During translation of a DNA single strand to mRNA, an arrangement similar to that of DNA replication takes place, i.e. mRNA forms by complementing the single DNA strand which will eventually be translated into protein, by matching the DNA bases and forming DNA – RNA base pairs, with the exception that U is now substituting for T. Therefore, the base pairs that are formed during DNA translation into an mRNA strand will represent A to U, C to G, G to C and T to A pairings. It is interesting to note that deoxyribose differs from ribose in one position only (Figure 1). It does so in a way that does not affect the helix rise per base pair of the sugar backbone of the DNA strand. On the other hand, it is important to note that the agMW of each AU (or UA) pair is 245.65 suggesting a variation (of approximately 6% across their entire length) in the density of matter within the volume enclosed by the backbones of the DNA template and the mRNA strand that is forming and, therefore, in the probability of one base matching its opposite one. This variation depends on whether an AU (or UA) or either a GC (or CG) or a TA (but not AT) pair is being formed. Also, keeping in mind that the translation process takes place one triplet at a time, the probability of one base matching its opposite one will not be constant but it will depend on which triplet is being translated (Table 3).

An equivalent situation holds during the translation of the mRNA coded information to a peptide chain at the ribosome, given the pairing that has to take place between the A's of the mRNA and the U's of the tRNA and vise versa (Table 3).

Table 3: Variation of density of matter in the volume enclosed by the nucleic acid sugar backbones											
A.A.	DNA replication (Triplet)			AgMW	DNA to mRNA (Triplet)			AgMW	mRNA – tRNA interaction (Triplet)			AgMW
PHE	T	T	T	777	A	A	A	777	U	U	U	735
PHE	T	T	C	778	A	A	G	778	U	U	C	750
LEU	T	T	A	777	A	A	U	763	U	U	A	735
LEU	T	T	G	778	A	A	C	778	U	U	G	750
SER	T	C	T	778	A	G	A	778	U	C	U	750
SER	T	C	C	779	A	G	G	779	U	C	C	765
SER	T	C	A	778	A	G	U	764	U	C	A	750
SER	T	C	G	779	A	G	C	779	U	C	G	765
TYR	T	A	T	777	A	U	A	763	U	A	U	735
TYR	T	A	C	778	A	U	G	764	U	A	C	750
X	T	A	A	777	A	U	U	749	U	A	A	735
X	T	A	G	778	A	U	C	764	U	A	G	750
CYS	T	G	T	778	A	C	A	778	U	G	U	750
CYS	T	G	C	779	A	C	G	779	U	G	C	765
X	T	G	A	778	A	C	U	764	U	G	A	750
TRP	T	G	G	779	A	C	C	779	U	G	G	765
LEU	C	T	T	778	G	A	A	778	C	U	U	750
LEU	C	T	C	779	G	A	G	779	C	U	C	765
LEU	C	T	A	778	G	A	U	764	C	U	A	750
LEU	C	T	G	779	G	A	C	779	C	U	G	765
PRO	C	C	T	779	G	G	A	779	C	C	U	765
PRO	C	C	C	780	G	G	G	780	C	C	C	780
PRO	C	C	A	779	G	G	U	765	C	C	A	765
PRO	C	C	G	780	G	G	C	780	C	C	G	780
HIS	C	A	T	778	G	U	A	764	C	A	U	750
HIS	C	A	C	779	G	U	G	765	C	A	C	765
GLN	C	A	A	778	G	U	U	750	C	A	A	750
GLN	C	A	G	779	G	U	C	765	C	A	G	765
ARG	C	G	T	779	G	C	A	779	C	G	U	765
ARG	C	G	C	780	G	C	G	780	C	G	C	780

---

[1] http://oregonstate.edu/instruction/bb492/lectures/RegulationIV.html

ARG	C	G	A	779	G	C	U	765	C	G	A	765
ARG	C	G	G	780	G	C	C	780	C	G	G	780
ILE	A	T	T	777	U	A	A	763	A	U	U	735
ILE	A	T	C	778	U	A	G	764	A	U	C	750
ILE	A	T	A	777	U	A	U	749	A	U	A	735
MET	A	T	G	778	U	A	C	764	A	U	G	750
THR	A	C	T	778	U	G	A	764	A	C	U	750
THR	A	C	C	779	U	G	G	765	A	C	C	765
THR	A	C	A	778	U	G	U	750	A	C	A	750
THR	A	C	G	779	U	G	C	765	A	C	G	765
ASN	A	A	T	777	U	U	A	749	A	A	U	735
ASN	A	A	C	778	U	U	G	750	A	A	C	750
LYS	A	A	A	777	U	U	U	735	A	A	A	735
LYS	A	A	G	778	U	U	C	750	A	A	G	750
SER	A	G	T	778	U	C	A	764	A	G	U	750
SER	A	G	C	779	U	C	G	765	A	G	C	765
ARG	A	G	A	778	U	C	U	750	A	G	A	750
ARG	A	G	G	779	U	C	C	765	A	G	G	765
VAL	G	T	T	778	C	A	A	778	G	U	U	750
VAL	G	T	C	779	C	A	G	779	G	U	C	765
VAL	G	T	A	778	C	A	U	764	G	U	A	750
VAL	G	T	G	779	C	A	C	779	G	U	G	765
ALA	G	C	T	779	C	G	A	779	G	C	U	765
ALA	G	C	C	780	C	G	G	780	G	C	C	780
ALA	G	C	A	779	C	G	U	765	G	C	A	765
ALA	G	C	G	780	C	G	C	780	G	C	G	780
ASP	G	A	T	778	C	U	A	764	G	A	U	750
ASP	G	A	C	779	C	U	G	765	G	A	C	765
GLU	G	A	A	778	C	U	U	750	G	A	A	750
GLU	G	A	G	779	C	U	C	765	G	A	G	765
GLY	G	G	T	779	C	C	A	779	G	G	U	765
GLY	G	G	C	780	C	C	G	780	G	G	C	780
GLY	G	G	A	779	C	C	U	765	G	G	A	765
GLY	G	G	G	780	C	C	C	780	G	G	G	780
	Av. AgMW			**778,5**	**Av. AgMW**			**768**	**Av. AgMW**			**757,5**

3. Determining relative affinities for synonymous codons

To estimate the velocity of reaction in the case of DNA translation, depending on which codon is being translated, we will follow the rather informal but elegant thinking of Miranda [8]. We will assume that, the DNA strand that is copied interacts with one triplet to elongate either its complementary DNA or its complementary mRNA strand and that the critical step in this process is the stabilization of the bases added to the elongating chains in place between the two sugar backbones. In this way, the formation of hydrogen bonds between the template and the free bases that will be used for the elongation of the complementary strand becomes a question of probability: How many ways are there for the volume of the reactants to fit in the space allotted to them in the presence of the triplet that has to be copied. A large number of ways implies a small probability of the reactants to be at the right place to form the prerequisite hydrogen bonds and the opposite.

In our treatment, we consider that the number of ways in which the reactant molecules fit in the space allotted to them is inversely proportional to the volume that their atoms occupy, i.e., proportional to their MW (measured in Daltons) multiplied by the volume of the hydrogen atom. Consequently, the probability of the reactants to be at the right place to form the prerequisite hydrogen bond is proportional to the agMW of the triplet being added as well as proportional to the agMW of the triplet being copied. We proceed therefore to tabulate:

- The volumes (in nm^3) of the equivalent spheres that correspond to the agMW of all possible codons for the mRNA strand,
- The volumes (in nm^3) of the equivalent spheres that correspond to the agMW of all possible codons for the DNA strand from which the mRNA was copied,
- The volumes (in nm^3) of the equivalent spheres that correspond to the agMW of all possible anticodons for the tRNA – amino acid complexes (that give rise to the polypeptide chain) to the mRNA strand agMW of the translated codon (translation),
- The volume (in nm^3) enclosed between the two sugar backbones of the double helix for a distance of 1 nm (or rise of three steps for the helix, corresponding to the space occupied by a triplet) under

the assumption that this volume corresponds to the space available for the reaction leading to base pair formation.

The implicit assumption, in this case, is that the replicated, translated or transcribed chain of codons represents the optimum reactant–enzyme–product spatial arrangement (reaction volume) since it has to be preserved intact throughout the corresponding processes.

Table 4: Velocity per stage of transformation (arbitrary units)			
Amino Acid	DNA to mRNA	mRNA to tRNA	Synonym Affinity (normalized)
PHE	0,682019961	0,999998526	0,682020966
PHE	0,690078629	0,870280467	0,600563722
LEU	0,784667603	0,981008315	0,769767712
LEU	0,682019961	0,843793234	0,575485525
SER	0,690078629	0,870280467	0,600563722
SER	0,704880074	0,762227431	0,537280512
SER	0,786571818	0,847964026	0,666986572
SER	0,680422074	0,786571818	0,535202406
TYR	0,784667603	0,981008315	0,769767712
TYR	0,786571818	0,847964026	0,666986572
X	0,847964026	0,981008315	0,831862213
X	0,737949016	0,847964026	0,625756063
CYS	0,682019961	0,843793234	0,575485525
CYS	0,680422074	0,786571818	0,535202406
X	0,737949016	0,847964026	0,625756063
TRP	0,646311025	0,737949016	0,476945991
LEU	0,690078629	0,870280467	0,600563722
LEU	0,704880074	0,762227431	0,537280512
LEU	0,786571818	0,847964026	0,666986572
LEU	0,680422074	0,786571818	0,535202406
PRO	0,704880074	0,762227431	0,537280512
PRO	0,71496395	0,71496395	0,511174957
PRO	0,737949016	0,737949016	0,544570356
PRO	0,685229534	0,685229534	0,469540898
HIS	0,786571818	0,847964026	0,666986572
HIS	0,737949016	0,737949016	0,544570356
GLN	0,843793234	0,843793234	0,711989121
GLN	0,786571818	0,786571818	0,618697049
ARG	0,680422074	0,786571818	0,535202406
ARG	0,685229534	0,685229534	0,469540898
ARG	0,786571818	0,786571818	0,618697049
ARG	0,685229534	0,685229534	0,469540898
ILE	0,784667603	0,981008315	0,769767712
ILE	0,786571818	0,847964026	0,666986572
ILE	0,847964026	0,981008315	0,831862213
MET	0,737949016	0,847964026	0,625756063
THR	0,786571818	0,847964026	0,666986572
THR	0,737949016	0,737949016	0,544570356
THR	0,843793234	0,843793234	0,711989121
THR	0,786571818	0,786571818	0,618697049
ASN	0,847964026	0,981008315	0,831862213
ASN	0,843793234	0,843793234	0,711989121
LYS	0,999998526	0,999998526	1
LYS	0,870280467	0,870280467	0,757390324
SER	0,737949016	0,847964026	0,625756063
SER	0,786571818	0,786571818	0,618697049
ARG	0,870280467	0,870280467	0,757390324
ARG	0,762227431	0,762227431	0,580992369
VAL	0,682019961	0,843793234	0,575485525
VAL	0,680422074	0,786571818	0,535202406
VAL	0,737949016	0,847964026	0,625756063
VAL	0,646311025	0,737949016	0,476945991
ALA	0,680422074	0,786571818	0,535202406
ALA	0,685229534	0,685229534	0,469540898
ALA	0,786571818	0,786571818	0,618697049
ALA	0,685229534	0,685229534	0,469540898
ASP	0,737949016	0,847964026	0,625756063
ASP	0,786571818	0,786571818	0,618697049
GLU	0,870280467	0,870280467	0,757390324

Brill Academic Publishers
P.O. Box 9000, 2300 PA Leiden
The Netherlands

*Lecture Series on Computer
and Computational Sciences*
Volume 4, 2005, pp. 1396-1396

Computational Methods In Chemical Engineering

Organizer:

Dr. George D. Verros

Description of the topic of the session

In the past decades many significant insights have been made in several areas of Computational Chemical Engineering. The aim of this session is to bring together chemical engineers from several disciplines in order to share methods, methodologies and ideas.

Dr. George D. Verros

Dr. George D. Verros was born in 1965 in Thessaloniki, Greece. He graduated from the Chem. Eng. Dep., Polytechnic School, Aristotle University of Thessaloniki (AUTH) in 1989, following his family tradition in Chemistry and Engineering He gained a Doctorate   in polymer reaction engineering in 1995 from the same department under the supervision of Prof. C. Kiparissides. During his studies he was awarded by several fellowships and he participated as a researcher in programs sponsored by EXXON. His primary interests focused in the area of polymer science & technology with emphasis on process simulation and   include a) polymer reaction engineering (mathematical modeling, on line optimization, calculation of Molecular Weight Distribution in polymerization reactors, diffusion controlled reactions, polymerization in supercritical solvents, etc) b) membrane & coating formation (estimation of diffusion coefficients, multi-component diffusion, prediction of membrane & coating morphology as a function of the process conditions, etc). He is author or co-author of fifteen publications in scientific journals such as Polymer and he has written more than twenty publications in Proceedings of Conferences such as International Conference on Computational Methods in Science and Engineering (ICCMSE), AIChE Annual Meetings, ISCRE, etc. He is a member of several Scientific Organizations including the European Society of Computational Methods in Science and Engineering (ESCMSE). Dr. Verros also serves on the Editorial Board of Res. J. Chem. Env.   and he acts as a reviewer in several journals including Polymer and Macromol. Theory & Simul. He has received many invitations to publish in research journals; George and his wife Olga-Maria Plakaki live in the wonderful city of Lamia in the central Greece. He has been working in environmental applications for the Greek Government since 1999. He has also been teaching as Adjunct Professor in the Department of Electrical Engineering, Technological & Educational Institute of Lamia since 2000.

Brill Academic Publishers
P.O. Box 9000, 2300 PA Leiden
The Netherlands

*Lecture Series on Computer
and Computational Sciences*
Volume 4, 2005, pp. 1397-1404

Determination of Chaos in the Direct Simulation of Two Dimensional Turbulent Flow over a Surface Mounted Obstacle

V. P. Fragos[1*], S. P. Psychoudaki[1], and N. A. Malamataris[2†]

[1] Department of Hydraulics, Solid Sciences and Agricultural Engineering
Aristotle University of Thessaloniki
GR-54124 Thessaloniki

[2] Department of Mechanical Engineering
TEI W. Macedonia
GR-50100 Kila, Kozani

Received 23 July, 2005; accepted in revised form 2 August, 2005

Abstract. *The quasi two dimensional turbulent flow over a surface mounted obstacle is studied as a numerical experiment by directly solving the transient Navier Stokes equations with Galerkin finite elements. The Reynolds number defined with respect to the obstacle height is 1304. Energy and enstrophy spectra yield the dual cascade of two dimensional turbulence and the -1 power law decay of enstrophy. Other statistical characteristics of turbulence such as Eulerian autocorrelation coefficients, longitudinal and lateral coefficients are also computed. Finally, oscillation diagrams of computed velocity fluctuations yield the chaotic behavior of turbulence*

Keywords: Quasi two dimensional turbulence, energy and enstrophy spectra, calculation of auto-correlation and cross-correlation functions, Galerkin Finite Elements, Chaos.

1 Introduction

The turbulent flow over a surface mounted obstacle is a fundamental problem in fluid mechanics having a wide range of applications in all domains of engineering science, as recently reviewed by Fragos et al[3],[4] Although the flow has received a lot of attention in the engineering community, it is still an open ended problem partly due to its complicated geometry and partly due to the unresolved issues of the nature of turbulence. The term flow over a surface mounted obstacle is used ambiguously in the literature. In this work, it is examined the flow over a cubic or a prismatically shaped obstacle having a width that extends up to the walls of a wind tunnel, where the obstacle is placed. This case is a quasi two dimensional flow, that takes place in the two dimensional space, where any three dimensional effects are generated from the existence of walls or from turbulence. For this particular flow, there is some recent experimental work in the turbulent regime conducted by Acharya et al[1] and Larichkin and Yakovenko[9] for obstacles with rectangular cross section of aspect ratio 1:1. There are also attempts to study this flow computationally by Acharya et al[1], Hwang et al[7] who used k-ε models with a finite difference method. In this work, the quasi two dimensional flow over a surface mounted obstacle is studied computationally solving the unsteady Navier Stokes equations in primitive variable formulation with standard Galerkin finite elements. This approach is used for the first time for this flow. The experimental set up and the process parameters of the work of Acharya et al[1] are taken for comparison with the numerical results of this work.

In this work, the issues of both quasi two dimensional turbulence and direct numerical simulation of turbulent flows are addressed in the study of two dimensional turbulent flow over a surface mounted obstacle with square cross section. In the following, the governing equations are presented along with the computational domain and the parameters of the flow. The issues of initial condition and inflow as well as outflow boundary condition are examined next, followed by the finite

---

* Corresponding author: e-mail: fragos@agro.auth.gr , web page: http://www.auth.gr/agro/eb/hydraul/intex.htm
† e-mail: nikolaos@eng.auth.gr

Table III. Correlation Matrix between Physicochemical Descriptors, Indicator Parameter and Biological Activity of phenol derivatives used in present study

	log 1/C	MR	MV	Pc	η	ST	D	α	I
log1/C	1.00000								
MR	−0.31563	1.00000							
MV	−0.46260	0.95216	1.00000						
Pc	−0.38700	0.99568	0.96683	1.00000					
η	0.51715	0.19128	−0.11544	0.12843	1.00000				
ST	0.38721	0.10890	−0.18880	0.06487	0.96341	1.00000			
D	0.09015	0.30311	0.03197	0.27727	0.85050	0.90864	1.00000		
α	−0.31524	1.00000	0.95210	0.99564	0.19147	0.10899	0.30307	1.00000	
I	−0.54595	0.85037	0.82651	0.87377	0.06935	0.08005	0.39152	0.85037	1.00000

References

[1] Burper A., Medicinal Chemistry, third edition, part I (1969)

[2] Colbom T., dumanoski D., Mayer J.P., Our Stolen Future, Dutton Publishing (1996)

[3] Henderson B.E., Ross R., Bemstein L., *Cancer Res.*, 48, 246 (1988)

[4] Asadawa E.M., Hagiwata A., Takahoshi S., Ito N., *Int. J. Cancer*, 56, 146 (1994)

[5] Hansch C. Kurup A., Garg R., Hua Gao, *Chem. Rev.*, 101, 619 (2001)

[6] Thakur A., Thakur M. and Khadlikar P.V., *Bioorg. Med. Chem.*, 11, 23, 5203 (2003)

[7] Thakur M., Thakur A. and Khadikar P.V., *Bioorg. Med. Chem.* 12, 4, 825 (2004)

[8] Thakur A., Thakur M. and Vishwakarma S., *Bioorg. Med. Chem.*, 12, 5, 1209 (2004)

[9] Thakur A., Thakur M., *Bioinformatics India,* 1, 63 (2003)

[10] Thakur M. Thakur A. and Sudele P., *Indian Journal of Chemistry*, 43 B, 976 (2004)

[11] Thakur A. and Thakur M., *Res. J. Chem. Environ.,* 7, 1, 51 (2003)

[12] Thakur A., Agrawal A., Thakur M. and Sudele P., *Bioinformatics India,* 1, 3, 18 (2003)

[13] Thakur M. Agrawal A., Thakur A. and Khadikar P.V., *Biorganic and Medicinal Chemistry,* 12, 9, 2287 (2003)

[14] Computer software Chemsketch 5.0 (from ACD labs), http/www/ACDlabs.com

[15] Chaterjee S., Hadi A.S. Price B., Regression Analysis by Examples, 3rd ed. Wiley VCH: New York (2000)

Brill Academic Publishers
P.O. Box 9000, 2300 PA Leiden
The Netherlands

*Lecture Series on Computer
and Computational Sciences*
Volume 4, 2005, pp. 1409-1412

A Computational Approach to Mass Transfer in Hollow Fiber Membrane Contactors with Linear and Nonlinear Boundary Conditions

G. Pantoleontos[1], S. P. Kaldis[2], D. Koutsonikolas[1], G. Skodras[*,1,2,3] and G. P. Sakellaropoulos[1,2,3]

[1] Chemical Process Engineering Laboratory, Department of Chemical Engineering, Aristotle University of Thessaloniki, P.O. Box 1520, 54006, Thessaloniki, Greece
[2] Chemical Process Engineering Research Institute, Centre for Research and Technology Hellas, 6th Km Harilaou-Thermis , P.O. Box 361, 57001, Thermi, Thessaloniki, Greece
[3] Institute for Solid Fuel Technology and Applications, Centre for Research and Technology Hellas , 4th Km Ptolemaidas-Kozanis , P.O. Box 95, 50200, Ptolemaida, Greece

Received 29 July, 2005; accepted in revised form 5 August, 2005

Abstract: Mass transfer in fully developed, laminar flow in a hollow fiber membrane contactor occurs in a variety of many important applications, such as supported gas and liquid membrane, reverse osmosis, pervaporation, membrane reactors and biological systems. The complexity of the partial differential equation which describes the concentration profile in the lumen with the associated linear or nonlinear boundary conditions at the fiber wall is simplified by means of analytical and numerical methods using current computational tools.

Keywords: hollow fiber membrane contactor, laminar flow, continuity equation

1. Problem formulation

Hollow fiber membrane contactors are a quite novel technology found in many membrane applications because of their high mass transfer area (known *a priori*), modular design, easy scale-up and other advantages [1]. Mathematical models for hollow fiber membrane processes have made use of two different approaches: (a) Plug flow and lumped mass transfer effects into a film-type mass transfer coefficient [2] and (b) Concentration profiles along the fiber by means of the continuity equation [3]. In the present study only the latter case is considered.

The following assumptions are made in order to describe the fluid flow within the fiber and the transport of the gas species through the membrane pores [4]: (a) steady state and isothermal operation; (b) Newtonian fluids with constant physical properties; (c) fully developed, laminar flow in the lumen and (d) applicability of Henry's law. When the velocity profile is fully developed, velocity term in radial direction becomes zero [5]. Furthermore, for short tubes diffusion occurs mainly in the tube walls, so axial molecular diffusion is neglected [6]. Considering the assumptions above and making some rearrangements, continuity equation inside a single fiber becomes in dimensionless form [7]:

$$\left(1-r^{*2}\right)\frac{\partial C^*}{\partial z^*} = \frac{2}{r^*}\cdot\frac{\partial C^*}{\partial r^*} + 2\cdot\frac{\partial^2 C^*}{\partial r^{*2}} \tag{1}$$

$$C^*(0,r^*) = 1 \tag{2}$$

$$\frac{\partial C^*}{\partial r^*}_{z^*,0} = 0 \tag{3}$$

where C^*, z^* and r^* are concentration, axial and radial distance in dimensionless form, respectively.

For certain cases the flux of the diffusing component through the membrane pores is given by the linear boundary condition at the lumen interface [7]:

$$\frac{\partial C^*}{\partial r^*}_{z^*,1} = -\frac{Sh_W}{2}\cdot C^*(z^*,1) \tag{4}$$

where Sh_W is wall Sherwood number based on shell and membrane mass transfer.

---

* Corresponding author (skodras@vergina.eng.auth.gr).

Dimensionless mixed-cup concentration is equal to [8]:

$$C_z^* = 4 \cdot \int_0^1 r^* \cdot \left(1 - r^{*2}\right) \cdot C^*\left(z^*, r^*\right) dr^* \tag{5}$$

2. Analytical solution

The set of equations (1-4) can be solved analytically by the separation of variables method, which yields an infinite series solution [9]:

$$C^*(z^*, r^*) = \sum_{n=1}^{\infty} c_n \cdot \exp\left(-2 \cdot \Lambda_n^2 \cdot z^*\right) \cdot R(r^*) \tag{6}$$

where

$$R(r^*) = \exp\left(-\frac{\Lambda_n \cdot r^{*2}}{2}\right) \cdot M\left(\frac{1}{2} - \frac{\Lambda_n}{4}, 1, \Lambda_n \cdot r^{*2}\right) \tag{7}$$

and

$$c_n = \frac{\int_0^1 r^* \cdot \left(1 - r^{*2}\right) \cdot R(r^*) dr^*}{\int_0^1 r^* \cdot \left(1 - r^{*2}\right) \cdot R^2(r^*) dr^*} \tag{8}$$

$M(a,b,x)$ is the Kummer function, which is a library routine in the mathematical package Maple V 5.00. For every Sh_W the eigenvalues, Λ_n, are the zeros of the equation:

$$\left(\frac{Sh_W}{2} - \Lambda_n\right) \cdot M\left(\frac{1}{2} - \frac{\Lambda_n}{4}, 1, \Lambda_n\right) + 2 \cdot \Lambda_n\left(\frac{1}{2} - \frac{\Lambda_n}{4}\right) \cdot M\left(\frac{3}{2} - \frac{\Lambda_n}{4}, 2, \Lambda_n\right) = 0 \tag{9}$$

Despite the seemingly complexity of expanding the solution in infinite series, for $z^* > 0.01$ only 6 eigenvalues must be determined for the evaluation of the dimensionless concentration, C^*. The coefficients, c_n, and the eigenvalues, Λ_n, can be easily evaluated in a few lines code in Maple commercial package. In Table 1 below an example is presented in Maple notation, with $Sh_W=1$. We suspect that one root lies in the domain {21, 22}, and then we define the precision of the output (here is 14). Then we define $R(r^*)$ and evaluate Fourier coefficient with the desired precision (here is 13) for the eigenvalue found. The greatest convenience of Maple is the incorporation of Kummer function as a library routine (KummerM).

Table 1: Evaluation of eigenvalues and Fourier coefficients in Maple V 5.00

```
> L:=evalf(RootOf((1/2-L)*KummerM(1/2-L/4,1,L)+2*L*(1/2-L/4)*KummerM((6-L)/4,2,L),21 ..
22),14); R(r) := exp(-L*r^2/2)*KummerM(1/2-L/4,1,L*r^2):
> c:=int(r*(1-r^2)*R(r),r=0..1)/int(r*(1-r^2)*(R(r))^2,r=0..1): evalf(c,13);
```

3. Numerical solution of the general case of nonlinear boundary condition

A very common situation in membrane processes is a nonlinear boundary condition at the fiber wall, e.g. a reaction takes place or the solubility of the diffusing component depends on its concentration. In that case superposition principle does not apply and numerical integration must be implemented. Equation (4) is substituted with the general case:

$$\frac{\partial C^*}{\partial r^*}\bigg|_{z^*, 1} = -Sh_W \cdot f\left(C^*(z^*, 1)\right) \tag{10}$$

where $f(C^*(z^*, 1))$ is a function of C^* in the fiber wall. Method of lines is used by discretizing in terms of r^* with finite difference formulas of 2nd order, thus resulting in a system of ordinary differential equations. If we choose to have m equally spaced subdomains in r^* direction, equations (1, 3) become:

$$\left(1 - r_i^{*2}\right) \cdot \frac{dC_i^*}{dz^*} = \frac{1}{r_i^*} \cdot \frac{C_{i+1}^* - C_{i-1}^*}{h_r} + 2 \cdot \frac{C_{i+1}^* - 2 \cdot C_i^* + C_{i-1}^*}{h_r^2} \text{, for i=2...m} \quad (11)$$

$$\frac{dC_1^*}{dz^*} = \frac{1}{3} \cdot \left[4 \cdot \frac{dC_2^*}{dz^*} - \frac{dC_3^*}{dz^*} \right] \text{, for i=1} \quad (12)$$

with $h_r = 1/m$ and $r_i^* = (i-1) \cdot h_r$.

Discretization of (10) yields $\qquad \dfrac{3 \cdot C_{m+1}^* - 4 \cdot C_m^* + C_{m-1}^*}{2 \cdot h_r} = -Sh_W \cdot f\left(C_{m+1}^*\right)$ \qquad (13)

After rearrangement and differentiation in respect to z*we get:

$$\left(2 \cdot h_r \cdot Sh_W \cdot f'\left(C_{m+1}^*\right) + 3\right) \cdot \frac{dC_{m+1}^*}{dz^*} = \left[4 \cdot \frac{dC_m^*}{dz^*} - \frac{dC_{m-1}^*}{dz^*} \right] \text{, for i=m+1} \quad (14)$$

Initial conditions are $C_i^* = 1$ for i=1...m. For point m+1 one has to solve equation (13) (with $C_m^* = C_{m+1}^* = 1$) to define initial condition at that point. For linear (and certain nonlinear) cases this root is explicitly evaluated, but for the general case a numerical method must be used, such as the Newton-Raphson, in order to find the zero of the nonlinear equation. In order to evaluate the dimensionless mixed-cup concentration Simpson's rule may be used.

Equations (11-12, 14) represent an initial value problem with m+1 ordinary differential equations, which was solved by Gear's BDF method taken from IMSL Math Library and compiled in Fortran programming language. Figures below show concentration profiles in the hollow fiber for various z* and r* derived from the linear case, when a numerical solution was attempted.

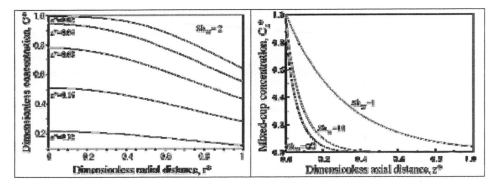

Figures 1, 2: Numerical solution of dimensionless concentration profiles for various z*, r* with only 17 discretization points.

In the case of a reaction taking place inside the membrane, the boundary condition in dimensionless form in the fiber wall might be of the type [10]:

$$\frac{\partial C^*}{\partial r^*}\bigg|_{z^*,1} = -\frac{Sh_W}{2} \cdot \frac{C^*(z^*,1)}{1 + 100 \cdot C^*(z^*,1)} \quad (15)$$

Values of dimensionless mixed-cup concentration are shown in Table 2, which were obtained for various discretization points. We see that for m=256 values are identical to Qin and Cabral [10], who used finite elements for discretization in r* direction.

Table 2: Dimensionless mixed-cup concentration for equations (1-3, 15) ($Sh_w = 1000$) and comparison with literature data [10]

z*	m=64	m=128	m=256	m=512	Values from [10]
0.001	0.961668320	0.961270575	0.961162744	0.961134779	0.9611
0.004	0.871340307	0.871065217	0.870992209	0.870973420	0.8710

0.050	0.400382367	0.400297576	0.400275113	0.400269243	0.4002
0.090	0.222842521	0.222781434	0.222765377	0.222761247	0.2275
0.160	0.080298426	0.080266486	0.080258117	0.080256084	0.0825
0.250	0.021617481	0.021605423	0.021602234	0.021601511	0.0216
0.300	0.010427673	0.010420901	0.010419014	0.010418741	0.0104

4. Conclusion

In the previous lines a very simple numerical scheme was presented by using finite difference formulas of 2nd order for the evaluation of concentration profiles in a hollow fiber with fully developed, laminar flow. Spatial discretization in r^* direction resulted in an initial value problem with a system of ordinary differential equations, which was solved by a stiff integrator, namely the Gear's BDF method. The above approach requires the incorporation of a root-finding routine in the case of a nonlinear boundary condition for the diffusing component through the membrane pores. A case study is presented and comparison with literature data shows a very good agreement. For a linear boundary condition at the fiber wall an analytical solution in terms of infinite series expansion exists, for which, apart from the entrance region ($z^* < 0.01$), only 6 eigenvalues are needed for an accurate evaluation of C^*. Current computational tools, such as Maple, are easily implemented in a few lines code for the evaluation of the eigenvalues and Fourier coefficients, an example of which is presented in the analytical solution section.

References

[1] A. Gabelman and S.-T. Hwang, Hollow fiber membrane contactors, *J. Membr. Sc.* **159** 61-106 (1999).

[2] R. D. Noble, Two-dimensional permeate transport with facilitated transport membranes, *Sep. Sc. Tech.* **19** 469-478 (1984).

[3] J. M. Kooijman, Laminar heat or mass transfer in rectangular channels and in cylindrical tubes for fully developed flow: comparison of solutions obtained for various boundary conditions, *Chem. Eng. Sc.* **28** 1149-1160 (1973).

[4] M. Mavroudi, S. P. Kaldis and G. P. Sakellaropoulos, Reduction of CO_2 emissions by a membrane contacting process, *Fuel* **82** 2153-2159 (2004).

[5] A. H. P. Skelland, *Diffusional Mass Transfer*, Wiley: New York, 1974.

[6] E. L. Cussler, *Diffusion*, Cambridge University Press, 1984.

[7] Y. Qin and J. M. S. Cabral, Lumen mass transfer in hollow-fiber membrane processes with constant external resistances, *A.I.Ch.E. J.* **43** 1975-1988 (1997).

[8] D. O. Cooney, S. - S. Kim and E. J. Davis, Analyses of mass transfer in hemodialysers for laminar blood flow and homogeneous dialysate, *Chem. Eng. Sc.* **29** 1731-1738 (1974).

[9] G. Pantoleontos, Aristotle University of Thessaloniki, Chemical Engineering Department, Diploma Thesis (in greek), 2000.

[10] Y. Qin and J. M. S. Cabral, Lumen mass transfer in hollow-fiber membrane processes with nonlinear boundary conditions, *A.I.Ch.E. J.* **44** 836-846 (1998).

Brill Academic Publishers
P.O. Box 9000, 2300 PA Leiden
The Netherlands

*Lecture Series on Computer
and Computational Sciences*
Volume 4, 2005, pp. 1413-1415

Computational Approaches to Artificial Intelligence: Theory, Methods, Applications

M.N. Vrahatis[1]

Computational Intelligence Laboratory, Department of Mathematics,
University of Patras, GR-26110 Patras, Hellas

G.D. Magoulas[2]

School of Computer Science and Information Systems,
Birkbeck College, University of London,
Malet Street, London WC1E 7HX, UK

Abstract: The Session on "Computational Approaches to Artificial Intelligence: theory, methods, applications" is part of the *International Conference of Computational Methods in Sciences and Engineering 2005* (ICCSME 2005) that was held at Loutraki, Greece, 21-26 October 2005. The Session aimed at providing an up-to-date view of computational and mathematical approaches to artificial; intelligence fields, covering both theoretical and applied works.

Keywords: cellular automata, cryptosystems, genetic algorithms, financial forecasting, fuzzy cognitive maps, memetic algorithms, neural networks, nonlinear mappings, particle swarm optimization, unsupervised clustering.

Mathematics Subject Classification: 03B52, 14G50, 37C25, 37M10, 62M10, 62M20, 62M45, 65P40, 90C30, 91B84,93C10, 94A60.

Artificial Intelligence (AI) was founded in the 1950s in an attempt to understand and model human intelligence and reasoning in order to build intelligent systems. Currently AI comprises a wide field of activities which focus on reasoning methods as well as learning and intelligent data and information processing. Other subfields are related to particular tasks such as medical diagnosis, time series prediction, "intelligent" optimization and adaptive modeling.

In recent years computational approaches have made their way from mathematical theory and practice into areas of Artificial Intelligence. Nonlinear optimization, nonlinear systems and global search methods are perhaps the most widely known approaches, complemented by applications in various settings such as cryptography, financial prediction, and control systems.

The session on "Computational approaches to artificial intelligence: theory, methods, applications"' presents the current capabilities of computational approaches as well as ideas on how these methods can be applied to real-world problem solving. Below we provide an overview of the papers of this session.

The first paper entitled "Genetic algorithm evolution of cellular automata rules for complex binary sequence prediction" proposes hybrid evolutionary algorithms for the prediction of complex binary sequences. The authors' algorithm incorporates cellular automata with sets of rules that are suitably coded in order to be evolved by a genetic algorithm.

In the paper "Optimal rural water distribution design using Labye's optimization method and linear programming optimization method" the authors apply "intelligent" optimization to water distribution design. In this context it is particularly important to calculate the optimal head of the pump station as well as the corresponded optimal pipe diameters so that the minimal total cost of the irrigation network is obtained. The paper comparatively evaluates an approach based on the Labye's optimization method and the conventional linear programming method.

---

[1] Email: vrahatis@math.upatras.gr
[2] Email: gmagoulas@dcs.bbk.ac.uk

The next work entitled "Computational intelligence methods for financial forecasting" concerns the prediction of the daily spot exchange rate of the Euro against the Japanese Yen. This challenging problem is treated using the recently proposed k-windows unsupervised clustering algorithm to automatically approximate the number of clusters in the data, and feedforward neural networks to approximate the unknown nonlinear relationships. The reported results illustrate the advantages of this methodology in capturing short-term dynamical behaviour when compared against a dedicated feedforward neural network and nearest neighbour regression.

The paper entitled "Fuzzy cognitive maps learning using memetic algorithms" proposes a new learning algorithm, which is based on Memetic Algorithm (MAs) for determining the weight matrix of a fuzzy cognitive map. MAs are hybrid search schemes which combine a global optimization algorithm and local search. The proposed approach is applied to an industrial process control problem and the results are particularly promising.

The next work entitled "Unsupervised clustering under parallel and distributed computing environments" presents an algorithmic framework for the k-windows clustering algorithm that tries to minimize communication in a distributed computing environment, without sacrificing its efficiency.

Lastly, the paper entitled "New orbit based symmetric cryptosystem" focuses on the application of chaotic systems to cryptography. The authors contribute a new symmetric key cryptosystem based on the dynamical systems theory. This cryptosystem exploits the idea of nonlinear mappings and their fixed points to encrypt information.

Michael N. Vrahatis

Professor Michael N. VRAHATIS is with the Department of Mathematics at University of Patras, Greece. He received the Diploma and Ph.D. degrees in Mathematics from the University of Patras, in 1978 and 1982, respectively. He was a visiting research fellow at the Department of Mathematics, Cornell University (1987-1988) and a visiting professor to the INFN (Istituto Nazionale di Fisica Nucleare), Bologna, Italy, (1992, 1994 and 1998); the Department of Computer Science, Katholieke Universiteit Leuven, Belgium, (1999); the Department of Ocean Engineering, Design Laboratory, MIT, Cambridge MA, USA, (2000), and the Collaborative Research Center "Computational Intelligence" (SFB 531) at the Department of Computer Science, University of Dortmund, Germany (2001). He was a visiting researcher at CERN (European Organization of Nuclear Research), Geneva, Switzerland, (1992), and at INRIA (Institut National de Recherche en Informatique et en Automatique), France (1998, 2003 and 2004). He is the author of more than 250 publications (more than 110 of which are published in international journals) in his research areas, including computational Mathematics, optimization, neural networks, evolutionary algorithms and artificial intelligence. His research publications have received more than 600 citations. He has been principal investigator of several research grants from the European Union, the Hellenic Ministry of Education and Religious Affairs and the Hellenic Ministry of Industry, Energy and Technology. He is among the founders of the "University of Patras Artificial Intelligence Research Center" (UPAIRC), established in 1997, where currently he serves as director. He is the founder of the "Computational Intelligence Laboratory" (CI Lab), established in 2004 at the Department of Mathematics of University of Patras, where currently he serves as director.

George D. Magoulas

Dr George D. MAGOULAS is Reader in Computer Science in the School of Computer Science and Information Systems, Birkbeck College, University of London, UK. He was educated at the University of Patras, Greece, in Electrical and Computer Engineering (BEng/MEng, PhD). He held R&D positions in industry, AMBER S.A (1990-1993) and SYNDESIS Ltd (1997-1998), where he worked in several national and international projects on the development of embedded systems employing soft computing and computational intelligence methodologies. Dr Magoulas held Lecturer and Senior Lecturer posts at the Department of Information Systems and Computing, Brunel University, UK. He has secured research grants from the UK Engineering and Physical Sciences Research Council, the Arts and Humanities Research Board and the Joint Information Systems Committee, and has published more than 130 articles in leading international journals and conferences, and edited two books in the area of adaptive web-based systems. Dr. Magoulas is on the Board of Special Reviewers of the archival journal User Modeling and User-Adapted Interaction, a member of the IEEE, the User Modeling Inc., the Technical Chamber of Greece, and the Hellenic Artificial Intelligence Society. More details on his research activities can be found at http://www.dcs.bbk.ac.uk/~gmagoulas/

Brill Academic Publishers
P.O. Box 9000, 2300 PA Leiden,
The Netherlands

Lecture Series on Computer
and Computational Sciences
Volume 4, 2005, pp. 1416-1419

Computational Intelligence Methods for Financial Forecasting

N.G. Pavlidis[1], D.K. Tasoulis[1], V.P. Plagianakos[1],
C. Siriopoulos[2] and M.N. Vrahatis[1]

[1]**Department of Mathematics and Computational Intelligence Laboratory,**
University of Patras, GR-26110 Patras, Greece.

[2]**Department of Business Administration,**
University of Patras, GR-26110 Patras, Greece.

Received 10 July, 2005; accepted in revised form 20 July, 2005

Abstract: Forecasting the short run behavior of foreign exchange rates is a challenging problem that has attracted considerable attention. High frequency financial data are typically characterized by noise and non–stationarity. In this work we investigate the profitability of a forecasting methodology based on unsupervised clustering and feedforward neural networks and compare its performance with that of a single feedforward neural network and nearest neighbor regression. The experimental results indicate that the proposed combination of the two methodologies achieves a higher profit.

Keywords: Financial Forecasting, Unsupervised Clustering, Neural Networks, Nearest Neighbors

Mathematics Subject Classification: 62M10, 62M45, 91B84

1 Introduction

After the inception of floating exchange rates in the early 1970s, economists have attempted to explain and forecast the movements of exchange rates using macroeconomic fundamentals. Empirical evidence suggests that although these models appear to be capable of explaining the movements of major exchange rates in the long run and in economies experiencing hyperinflation, their performance is poor when it comes to the short run and out-of-sample forecasting [2]. Conventional time series models forecasting on global approximation models, employing techniques such as linear and non-linear regression, polynomial fitting and artificial neural networks. Such models are better suited to problems with stationary dynamics [13]. In the analysis of real–world systems two of the key problems are non–stationarity (often in the form of switching between regimes) and overfitting (which is particularly important for noisy processes) [13]. To improve forecasting performance numerous researchers have proposed methodologies that attempt to identify regions of the input space exhibiting similar dynamics and subsequently employ a local model for each region [5–8, 10, 13].

In [6, 7] the application of unsupervised clustering for the segmentation of the input space, and feedforward neural networks (FNNs) acting as local predictors for each identified cluster, was proposed. In this work we investigate the profitability of this methodology combined with a simple trading rule, and compare its performance with two other widely known forecasting methodologies, namely FNNs and the nearest neighbor regression. The profitability of these methodologies is evaluated on the time series of the daily spot exchange rate of the Euro against the Japanese Yen.

2 Algorithms

In this section we outline the unsupervised k-windows (UKW) algorithm, the nearest neighbors algorithm, and FNNs.

2.1 Unsupervised k-windows Algorithm

UKW generalizes the original k-windows algorithm [12] by approximating the number of clusters. The UKW algorithm employs a windowing technique to identify the clusters present in a dataset. Specifically, assuming that the dataset lies in d dimensions, UKW initializes a number of d–dimensional windows over the dataset. Next it moves and enlarges these windows so as to enclose all the patterns that belong to one cluster within a window. The movement and enlargement procedures are guided by the points that lie within each window at each iteration. As soon as the movement and enlargement procedures do not increase significantly the number of points within each window they terminate. The final set of windows defines the clustering result of the algorithm. UKW approximates the number of clusters by employing a sufficiently large number of initial windows, and at the final stage of its execution assigning windows that enclose a high proportion of common points to the same cluster [11].

2.2 Nearest Neighbors

Assume that the time series $\{z_t\}_{t=1}^T$ has been embedded in an n-dimensional space, yielding pattern vectors of the form $z_t^n = (z_t, z_{t-1}, \ldots, z_{t-n+1})$. To generate a prediction for z_t from the information available up to time $(t-1)$, first the k nearest neighbors of the pattern z_t^n are identified, and subsequently, an estimator of $E(x_t | x_{t-1}, \ldots, x_{t-n})$ by $\sum_{i=1}^k \omega_{ti} x_i$ is computed where ω_{ti} represents the weight assigned to the ith nearest neighbor. Alternative configurations of the weights are possible but we employed uniform weights as they are the most frequently encountered configuration in the literature.

2.3 Feedforward Neural Networks

In the context of time series modeling the inputs to the FNN typically consist of a number of delayed observations, while the target is the next value of the series. The *universal myopic mapping theorem* [9] states that any shift–invariant map can be approximated arbitrarily well by a structure consisting of a bank of linear filters feeding an FNN. An immediate implication of this theorem is that, FNNs alone can be insufficient to capture the dynamics of a non–stationary system. The supervised training process is a data driven adaptation of the weights that propagate information between the neurons. Learning in FNNs is achieved by finding a minimizer $w^* = (w_1^*, w_2^*, \ldots, w_n^*) \in \mathbb{R}^n$, such that $w^* = \min_{w \in \mathbb{R}^n} E(w)$, where E is the batch error measure of the FNN. The error function is based on the squared difference between the actual output value and the target value.

3 Experimental Results

The time series that was studied is the daily spot exchange rate of the Euro against the Japanese Yen. The 1682 available observations cover the period from 12/6/1999 to 29/6/2005. The first 1482 observations were used as a training set, while the last 200 observations were used to evaluate the profit generating capability of the different forecasting methodologies. Using the *false nearest neighbors* [4] method on the observations forming the trainig set we selected an embedding dimension of 5. Firstly we employed a global FNN with architecture 5-5-4-1 to model the time series. The FNN was trained for 200 epochs using the Improved Resilient Propagation algorithm [3] and subsequently its profit generating capability on the test set was evaluated. In particular, we assume that on the first day we have 1000 Euros available. The trading rule that we considered is the following: if the system at date t, holds Euros and $\widehat{x_{t+1}} > x_t$ (where $\widehat{x_{t+1}}$ is the predicted price for date $t + 1$ and x_t is the actual price at date t) then the entire amount available is converted to Japanese Yen. On the other hand, if the system holds Japanese Yen and $\widehat{x_{t+1}} < x_t$, then the entire amount is converted to Euros. In all other cases, the holdings do not change currency at date t. The last observation of the series is employed to convert the final holdings to Euros. In all transactions we assume a cost of 0.25% [1]. The profitability from trading based on the predictions of the global FNN model is depicted with the red line (Global FNN) in Fig 1. A perfect predictor, i.e. a predictor that correctly predicts the direction of change of the spot exchange rate at all dates, achieves a total profit of approximately 9716 Euros excluding transactions costs, while including transactions costs reduces total profit to approximately 7301 Euros.

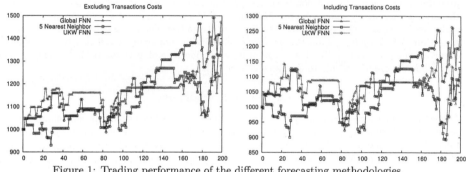

Figure 1: Trading performance of the different forecasting methodologies.

As can be seen from Fig. 1, excluding transactions costs the FNN is capable of achieving a profit of 282.94 Euros over the 200 days of the test set, while including transactions costs reduces total profit to 109.57 Euros. Next the results obtained by the k nearest neighbor regression method are presented. We experimented with all the integer values of k in the range $[1, 20]$. The best results were obtained for $k = 5$, and are illustrated with the black line (5 Nearest Neighbor) in Fig 1. The 5 nearest neighbor method achieved a profit of 379.51 excluding transctions costs and 129.16 including transactions costs. Finally, the performance of the forecasting system based on the segmentation of the input using the UKW algorithm and utilizing an FNN to act as a local predictor for each cluster, is illustrated with the blue line (UKW FNN) in Fig. 1. This approach achieved the highest profit: in the absence of transactions costs, 472.16 Euros, and 235.54 including transactions costs.

4   Concluding Remarks

In this paper, we report results concerning the profitability of a forecasting methodology that employs the UKW clustering algorithm and FNNs, acting as local predictors for each cluster, to predict the direction of movement of the daily spot exchange rate of the Euro against the Japanese Yen. The advantages of this methodology are that UKW automatically approximates the number of clusters present in a dataset, and that FNNs are capable of approximating an unknown nonlinear relationship. The profit generating performance of this methodology compares favorably with that of a single FNN and nearest neighbor regression, which implies that it is capable of capturing more accurately the short run dynamics of the time series.

Acknowledgment

The authors thank the European Social Fund (ESF), Operational Program for Educational and Vocational Training II (EPEAEK II) and particularly the Program PYTHAGORAS, for funding this work. This work was supported in part by the "Karatheodoris" research grant awarded by the Research Committee of the University of Patras.

References

[1] F. Allen and R. Karjalainen, *Using genetic algorithms to find technical trading rules*, Journal of Financial Economics **51** (1999), 245–271.

[2] J. A. Frankel and A. K. Rose, *Empirical research on nominal exchange rates*, Handbook of International Economics (G. Grossman and K. Rogoff, eds.), vol. 3, Amsterdam, North-Holland, 1995, pp. 1689–1729.

[3] C. Igel and M. Hüsken, *Improving the Rprop learning algorithm*, Proceedings of the Second International ICSC Symposium on Neural Computation (NC 2000) (H. Bothe and R. Rojas, eds.), ICSC Academic Press, 2000, pp. 115–121.

[4] M. B. Kennel, R. Brown, and H. D. Abarbanel, *Determining embedding dimension for phase–space reconstruction using a geometrical construction*, Physical Review A **45** (1992), no. 6, 3403–3411.

[5] R. L. Milidiu, R. J. Machado, and R. P. Renteria, *Time-series forecasting through wavelets transformation and a mixture of expert models*, Neurocomputing **28** (1999), 145–156.

[6] N. G. Pavlidis, D. K. Tasoulis, and M. N. Vrahatis, *Financial forecasting through unsupervised clustering and evolutionary trained neural networks*, Proceedings of the Congress on Evolutionary Computation (CEC 2003), 2003, pp. 2314–2321.

[7] N.G. Pavlidis, D.K. Tasoulis, V.P. Plagianakos, and M.N. Vrahatis, *Computational intelligence methods for financial time series modeling*, International Journal of Biffurcation and Chaos (accepted for publication) (2005).

[8] J. C. Principe, L. Wang, and M. A. Motter, *Local dynamic modeling with self–organizing maps and applications to nonlinear system identification and control*, Proceedings of the IEEE **86** (1998), no. 11, 2240–2258.

[9] I. W. Sandberg and L. Xu, *Uniform approximation of multidimensional myopic maps*, IEEE Transactions on Circuits and Systems I: Fundamental Theory and Applications **44** (1997), no. 6, 477–485.

[10] A. Sfetsos and C. Siriopoulos, *Time series forecasting with a hybrid clustering scheme and pattern recognition*, IEEE Transactions on Systems, Man, and Cybernetics Part A: Systems and Humans **34** (2004), no. 3, 399–405.

[11] D. K. Tasoulis and M. N. Vrahatis, *Unsupervised distributed clustering*, IASTED International Conference on Parallel and Distributed Computing and Networks, 2004, pp. 347–351.

[12] M.N. Vrahatis, B. Boutsinas, P. Alevizos, and G. Pavlides, *The new k-windows algorithm for improving the k-means clustering algorithm*, Journal of Complexity **18** (2002), 375–391.

[13] A. S. Weigend, M. Mangeas, and A. N. Srivastava, *Nonlinear gated experts for time series: Discovering regimes and avoiding overfitting*, International Journal of Neural Systems **6** (1995), 373–399.

Brill Academic Publishers
P.O. Box 9000, 2300 PA Leiden,
The Netherlands

*Lecture Series on Computer
and Computational Sciences*
Volume 4, 2005, pp. 1420-1423

Fuzzy Cognitive Maps Learning using Memetic Algorithms

**Y.G. Petalas[1], E.I. Papageorgiou[2], K.E. Parsopoulos[1], P.P. Groumpos[2],
and M.N. Vrahatis[1]**

[1]**Computational Intelligence Laboratory, Department of Mathematics,
University of Patras Artificial Intelligence Center(UPAIRC)
University of Patras, GR-26110 Patras, Greece.**

[2]**Department of Electrical and Computer Engineering,
Laboratory of Automation and Robotics,
University of Patras, GR-26500 Patras, Greece.**

Received 10 July, 2005; accepted in revised form 20 July, 2005

Abstract: Memetic Algorithms (MAs) are proposed for learning in Fuzzy Cognitive Maps
(FCMs). MAs are hybrid search schemes, which combine a global optimization algorithm
and a local search one. FCM's learning is accomplished through the optimization of an
objective function with respect to the weights of the FCM. MAs are used to solve this
optimization task. The proposed approach is applied to a well-established process control
problem in industry and the results are promising.

Keywords: Fuzzy Cognitive Maps, Memetic Algorithms, Particle Swarm Optimization,
Local Search

Mathematics Subject Classification: 03B52 90C30

1 Introduction

Fuzzy Cognitive Maps (FCMs) are a soft computing methodology developed by Kosko as an expansion of cognitive maps which are widely used to represent social scientific knowledge [1]. They belong to the class of neuro–fuzzy systems, which are able to incorporate human knowledge and adapt it through learning procedures. FCMs are designed by experts through an interactive procedure of knowledge acquisition, and they have a wide field of application, including modeling of complex and intelligent systems, decision analysis, and extend graph behavior analysis. They have also been used for planning and decision–making in the fields of international relations and social systems modeling, as well as in management science, operations research and organizational behavior [1, 2].

An FCM consists of nodes–concepts, C_i, $i = 1, \ldots, N$, where N is the total number of concepts. Each node–concept represents a key–factor of the system, and it is characterized by a value $A_i \in [0, 1]$, $i = 1, \ldots, N$. The concepts are interconnected with weighted arcs, which imply the relations among them. Each interconnection between two concepts C_i and C_j has a weight W_{ij}, which is proportional to the strength of the causal link between C_i and C_j. At each step, the value, A_i, of the concept C_i is influenced by the values of the concepts–nodes connected to it, and it is updated according to the scheme [2]: $A_i(k+1) = f\left(A_i(k) + \sum_{\substack{j=1 \\ j \neq i}}^{n} W_{ji} A_j(k)\right)$, where k stands for the iteration counter; and W_{ji} is the weight of the arc connecting the concept C_j to the concept C_i. The function f is the sigmoid function.

A few algorithms have been proposed for FCM learning [3]. The main task of the learning procedure is to find a setting of the FCM's weights, that leads the FCM to a desired steady state. This is achieved through the minimization of a properly defined objective function.

We propose a new approach for FCM learning that is based on Memetic Algorithms (MAs) [4]. MAs are hybrid search schemes, integrating an evolutionary algorithm and a local search method, and they have been used with success in many difficult optimization problems. Their efficiency can be attributed to the exploitation of the advantages of both the global and the local search schemes. Global search algorithms can explore the whole search space but they are not efficient in locating the optimum of the objective function with high accuracy. On the other hand, local search methods can compute the optimum with high accuracy if they are initialized in its basis of attraction, but they are prone to getting stuck to local minima. MAs are used for the determination of optimum weight matrices for the system through the minimization of a properly defined objective function [3].

The rest of the paper is organized as follows: In Section 2, the proposed learning algorithm is presented, while results from the application of the proposed method in a industrial control problem are reported in Section 3. Section 4 concludes the paper.

2 The Proposed Approach

The main goal in FCM learning is to determine the values of the weights of the FCM that produce a desired behavior of the system. The determination of the weights is of major significance and it contributes towards the establishment of FCMs as a robust methodology. The desired behavior of the system is characterized by output concept values that lie within desired bounds prespecified by the experts.

The computation of the FCM's weights is accomplished through the minimization of a problem–dependent objective function. For this purpose, the following objective function was employed [3]:

$$F(W) = \sum_{i=1}^{m} H\left(A_{\text{out}_i}^{\min} - A_{\text{out}_i}\right) \left|A_{\text{out}_i}^{\min} - A_{\text{out}_i}\right| + \sum_{i=1}^{m} H\left(A_{\text{out}_i} - A_{\text{out}_i}^{\max}\right) \left|A_{\text{out}_i}^{\max} - A_{\text{out}_i}\right|, \qquad (1)$$

where H is the well–known Heaviside function and $A_{\text{out}_i}^{\min}$, $A_{\text{out}_i}^{\max}$, are bounds of the output concepts' values. This function has been used with success in the past, combined with a swarm intelligence algorithm, for FCM learning [3]. In the proposed approach, we use MAs to solve this optimization problem.

The proposed MA, called MemeticPSO (MPSO), consists of the Particle Swarm Optimization (PSO) algorithm as the global search component, and the Hooke and Jeeves (HJ) algorithm as local search component. PSO is a stochastic global optimization algorithm and it has been applied successfully for FCM's learning [3]. More specifically, it belongs to the class of *swarm intelligence* algorithms, which are inspired from the social dynamics and emergent behavior that arise in socially organized colonies. A brief description of PSO is provided below.

Assume a D–dimensional search space, $S \subset \mathbb{R}^D$, and a swarm consisting of N particles. Let $X_i = (x_{i1}, x_{i2}, \ldots, x_{iD})^\top \in S$, be the i–th particle and $V_i = (v_{i1}, v_{i2}, \ldots, v_{iD})^\top \in S$, be its velocity. Let also the best previous position (i.e., the position that has the lowest function value) encountered by the i–th particle in S be denoted by $P_i = (p_{i1}, p_{i2}, \ldots, p_{iD})^\top$. Assume g_i to be the index of the particle that attained the best previous position among all the particles in the neighborhood of the i–th particle, and G to be the iteration counter. Then, the swarm is manipulated by the equations [5]:

$$V_i(G+1) = \chi\left[V_i(G) + c_1\, r_1\left(P_i(G) - X_i(G)\right) + c_2\, r_2\left(P_{g_i}(G) - X_i(G)\right)\right], \qquad (2)$$

$$X_i(G+1) = X_i(G) + V_i(G+1), \qquad (3)$$

Table 1: Pseudo code for the Memetic algorithm.

Input: N , χ, c_1, c_2, w_{min}, w_{max} (lower & upper bounds)	
Step 1	**Set** $t = 0$.
Step 2	**Initialize** $w_i(t), v_i(t) \in [w_{min}, w_{max}]$, $p_i(t) \leftarrow w_i(t)$, $i = 1, \ldots, N$.
Step 3	**Evaluate** $F(w_i(t))$. **Determine** the indices g_i, $i = 1, \ldots, N$.
Step 4	**While** (stopping criterion is not satisfied) **Do**
Step 5	**Update** the velocities $v_i(t+1)$, $i = 1, \ldots, N$, according to Eq. (2).
Step 6	**Set** $w_i(t+1) = w_i(t) + v_i(t+1)$, $i = 1, \ldots, N$.
Step 7	**Constrain** each particle w_i in $[w_{min}, w_{max}]$.
Step 8	**Evaluate** $f(w_i(t+1))$, $i = 1, \ldots, N$.
Step 9	**If** $f(w_i(t+1)) < f(p_i(t))$ **Then** $p_i(t+1) \leftarrow w_i(t+1)$
	Else $p_i(t+1) \leftarrow p_i(t)$.
Step 10	**Update** the indices g_i.
Step 11	**While** (local search is applied) **Do**
Step 12	**Choose** a best position, $p_q(t+1)$, $q \in \{1, \ldots, N\}$.
Step 13	**Apply** local search on $p_q(t+1)$ and obtain a new solution y.
Step 14	**If** $F(y) < F(p_q(t+1))$ **Then** $p_q(t+1) \leftarrow y$.
Step 15	**End While**
Step 16	**Set** $t = t + 1$.
Step 17	**End While**

where $i = 1, \ldots, N$; χ is a parameter called *constriction coefficient*; c_1 and c_2 are two parameters called *cognitive* and *social* parameters, respectively; and r_1, r_2, are random vectors with components uniformly distributed within $[0, 1]$ (all vector operations in Eqs. (2) and (3) are assumed to be performed componentwise).

HJ is a direct search algorithm that uses function evaluations solely, without computing any derivative information [6]. Therefore, it can be applied in problems with non–differentiable or discontinuous objective functions. A pseudocode of the proposed methodology is provided in Table 1, where F denotes the objective function, and w denotes the matrix W of the FCM's weights, represented as a vector that contains its rows in turn.

3 Experimental Results

The proposed method has been applied to the industrial control problem investigated in [3]. The ranges of the weights implied by the fuzzy regions, as they were suggested by experts, were: $-0.50 \leqslant W_{12} \leqslant -0.30$, $-0.40 \leqslant W_{13} \leqslant -0.20$, $0.20 \leqslant W_{15} \leqslant 0.40$, $0.30 \leqslant W_{21} \leqslant 0.40$, $0.40 \leqslant W_{31} \leqslant 0.50$, $-1.0 \leqslant W_{41} \leqslant -0.80$, $0.50 \leqslant W_{52} \leqslant 0.70$, $0.20 \leqslant W_{54} \leqslant 0.40$. Since the consideration of all eight constraints on the weights prohibits the detection of a suboptimal matrix, some of the constraints were omitted. More specifically, the constraints for the weights W_{15}, W_{52}, and W_{54}, for which, the experts' suggestions regarding their values varied widely, were omitted. The corresponding weights were allowed to assume values in the range $[-1, 0]$ or $[0, 1]$, in order to avoid physically meaningless weight matrices [3]. The results obtained through MPSO were compared with that of PSO that are reported in [3]. We performed 100 independent experiments. The error goal for the optimization problem was set to 10^{-8}, and the swarm size of MPSO was set to 20. The HJ algorithm was applied on the best particle of the swarm with probability 0.05 at each iteration. In all cases, the local version of PSO was used, with neighborhood size 3. The performance of PSO and MPSO is analyzed statistically in Table 2, with respect to the required number of function evaluations. In Table 3, statistics regarding the weights are reported. The results suggest that MPSO is a very promising approach for FCM learning.

Table 2: Statistical analysis for the function evaluations required by PSO and MPSO.

	Max	Min	Mean	Stdev
PSO	760	240	491.20	104.16
MPSO	1071	40	315.51	155.07

Table 3: Statistical analysis for the weights obtained with PSO and MPSO.

	PSO				MPSO			
	Max	**Min**	**Mean**	**Stdev**	**Max**	**Min**	**Mean**	**Stdev**
W_{12}	−0.3000	−0.5000	−0.3369	0.0655	−0.3000	−0.5000	−0.3444	0.0705
W_{13}	−0.2000	−0.3068	−0.2100	0.0231	−0.2000	−0.3000	−0.2165	0.0306
W_{15}	1.0000	0.7163	0.902348	0.0900	1.0000	0.7166	0.9323	0.0789
W_{21}	0.4000	0.3903	0.399828	0.0011	0.4000	0.3991	0.3999	0.0000
W_{31}	0.5000	0.4843	0.4998	0.0015	0.5000	0.5000	0.5000	0.0000
W_{41}	−0.8000	−0.8000	−0.8000	0.0000	−0.8000	−0.8000	−0.8000	0.0000
W_{52}	1.0000	0.8272	0.9309	0.0619	1.0000	0.8224	0.9525	0.0594
W_{54}	0.1591	0.1000	0.1064	0.0136	0.1614	0.1000	0.1068	0.0138

4 Conclusions

A new learning algorithm, which is based on MAs, was proposed for determining the weight matrix of an FCM. MPSO proved to be very efficient, providing promising results. Future work will include the application of MAs on more complex problems, as well as the investigation of different memetic schemes for FCM learning.

Acknowledgment

This work was supported by the "Pythagoras II" research grant co funded by the European Social Fund and National Resources.

References

[1] B. Kosko. Fuzzy cognitive maps. *Int. J. Man–Machine Studies*, **24** 65–75(1986).

[2] B. Kosko. *Fuzzy Engineering*. Prentice Hall, New York, 1997.

[3] E. I. Papageorgiou, K.E Parsopoulos, C.D. Stylios, P. P. Groumpos, and M. N. Vrahatis. Fuzzy cognitive maps learning using particle swarm optimization. *Journal of Intelligent Information Systems*, 2004. accepted for publication.

[4] D. Corne, M. Dorigo, and F. Glover. *New Ideas in Optimization*. McGraw-Hill, 1999.

[5] M. Clerc and J. Kennedy. The particle swarm–explosion, stability, and convergence in a multidimensional complex space. *IEEE Trans. Evol. Comput.*, **6(1)** 58–73(2002).

[6] R. Hooke and T. A. Jeeves. Direct search solution of numerical and statistical problems. *Journal of the Association for Computing Machinery*, **8** 212–229(1961).

Brill Academic Publishers
P.O. Box 9000, 2300 PA Leiden,
The Netherlands

Lecture Series on Computer
and Computational Sciences
Volume 4, 2005, pp. 1424-1427

Genetic Algorithm Evolution of Cellular Automata Rules for Complex Binary Sequence Prediction

A.V. Adamopoulos[a],[1], **N.G. Pavlidis**[b] and **M.N. Vrahatis**[b]

[a] **Medical Physics Laboratory, Department of Medicine,
Democritus University of Thrace, GR-681 00 Alexandroupolis, Hellas**

[b]**Computational Intelligence Laboratory, Department of Mathematics,
University of Patras, GR-26110 Patras, Hellas.**

Received 10 July, 2005; accepted in revised form 20 July, 2005

Abstract: Complex binary sequences were generated by applying a simple threshold, linear transformation to the *logistic iterative function,* $x_{n+1} = r x_n (1 - x_n)$. Depending primarily on the value of the *non-linearity parameter* r, the logistic function exhibits a great variety of behavior, including stable states, cycling and periodical activity and the period doubling phenomenon that leads to high-order chaos. Binary sequences of length $2L$ were used in our computer experiments. The first L bits (first half) were given as input to Cellular Automata (CA) with the task to regenerate the remaining L bits (second half) of the binary sequence in less than L evolution steps of the CA. To perform this task a suitably designed Genetic Algorithm (GA) was developed for the evolution of CA rules. Various complex binary sequences were examined, for a variety of initial values of x_0 and a wide range of the non-linearity parameter, r. The proposed hybrid prediction algorithm, based on a combination of GAs and CA proved quite efficient.

Keywords: Cellular Automata, Genetic Algorithms, Complex Binary Sequence Prediction

Mathematics Subject Classification: 37M10, 62M20

1 Introduction

Cellular Automata (CA) are decentralized structures of simple and locally interacting elements (cells). A CA starts with a given initial configuration that refers to the initial state of its cells. A CA evolves following a set of rules that incorporate [1, 2]. This set of rules yields the derived cell states (values) at the next evolution step, for all the possible combinations of cell states. CA have been proposed as a novel approach for a large number of problems. Among others, CA have been proposed as models for physical, biological and social systems, games and pattern recognition [3, 4]. As systems, CA have proved capable of parallel and emergent computation [5, 6]. In this work, simple, bistable, one-dimensional CA were used for the purpose of prediction of binary sequences of high complexity. The considered binary sequences were derived by applying a linear threshold transformation, proposed in [7], on data obtained from the logistic function. The task under consideration for the CA was to use the first half of a given binary sequence as input (initial configuration), in order to reproduce the second half of that sequence in a given number of evolution steps of the CA. For that purpose, a population of CA was evolved using a

---

[1]Corresponding author. Email adam@med.duth.gr

suitably designed Genetic Algorithm (GA) [8, 9]. Namely, the GA evolved the set of rules of the CA [10–12]. Results show that the proposed method can provide CA rules that give in a small number of evolution steps a 100% prediction of long binary sequences.

2 Materials and Methods

The main task of this work is to utilize a GA to evolve CA rules, so that the CA is capable of predicting a complex binary sequence in a certain number of evolution steps. Complex binary sequences were generated using a two step procedure. At the first step, the logistic function:

$$x_{n+1} = r\, x_n\, (1 - x_n), \tag{1}$$

was used, for the generation of sequences of real numbers of length $2L$. The choice of the logistic function to obtain data was not arbitrary; on the contrary, it was based on the highly complex behavior it exhibits. Specifically, for values of the non-linearity parameter r in the range $[0, 3)$ the system reaches a single-state stable value. For r in the range $[3, 3.57)$ the period doubling phenomenon occurs, and the system exhibits cycling (periodical) behavior with increasing cycling period as the value of r increases. This results to a fully chaotic behavior for even larger values of r in the range $[3.57, 4]$.

After the generation of the chaotic data, x, using Eq. (1), at a second step, the binary sequence $b_n(x_0, r)$ is generated by applying the transformation [7]:

$$b(x_0, r) = \begin{cases} 0, & \text{if } x_n \leqslant 0.5 \\ 1, & \text{if } x_n > 0.5 \end{cases} \tag{2}$$

Taking a binary sequence $b_n(x_0, r)$ of length $2L$ bits, the first half (first L bits) were used as input to the CA. In other words, these L bits were used for the construction of the initial configuration of the CA, which corresponds to the evolution step zero. The task was to let the CA evolve using the set of rules for a number of evolution steps less than or equal to L and to investigate if the CA was able to regenerate the second half of $b_n(x_0, r)$, namely the last L bits of $b_n(x_0, r)$. The number of rules of a specific CA depends on the order R (radius) of cell neighborhood. This parameter determines the number $2R+1$ of cells that a specific cell interacts with in a local manner. Thus, for $R = 1$, the neighborhood of each cell consists of three cells. This is shown in Table 1. A three-cell neighborhood, with each cell considered as a bistable element, may present $C = 2^{2R+1} = 2^3 = 8$ distinct combinations. Thus, the adaptation of a set of 8 rules is necessary. In the general case, these C combinations are ordered and numbered from 0 to $C - 1$ following the representation of integer numbers in the binary arithmetical system.

Rule Nr. C:	0	1	2	3	4	5	6	7
	000	001	010	011	100	101	110	111
	1	0	0	1	1	0	1	0

Table 1: An example of a set of rules for $R = 1$, which results to $C = 8$ distinct rules.

To accomplish our task, the sets of rules of a population of N CA were evolved using a GA. Each set of rules was represented as the chromosome of the individuals of the GA, with each gene being a binary digit. The length of the chromosome was C, as this is the number of rules that comprise the corresponding set of rules. The binary representation of the GA individuals allowed the use of well known and widely used genetic operators for selection, crossover and mutation [8]. In particular, as selection operator we used the *roulette-wheel selection operator*; as crossover operator we used the *one-point crossover operator*, and finally, as mutation operator we used the *bit-flip operator*. Recalling our goal, as fitness function for the evaluation of the individuals of the GA we used the number of bits that were successfully predicted after L evolution steps of the CA.

The necessity of utilizing a GA must be emphasized. As it is shown in Table 2, the size V of the search space is enormously expanding, even for small values of R.

Radius R	Number of neighbors H	Number of rules C	Size of search space V
1	3	8	$2^8 = 256$
2	5	32	$2^{32} = 4294967296$
3	7	128	$2^{128} = 3.40282 \cdot 10^{38}$
4	9	512	$2^{512} = 1.34078 \cdot 10^{154}$

Table 2: Number of possible combinations of CA rules for various values of R.

3 Results

The proposed method was tested for various binary sequences which were generated for a large number of combinations of the parameters x_0 and r of the logistic equation in Eq. (1). The values of the non-linearity parameter, r, were selected in the range $[3.57, 4]$ for which chaotic behavior is exhibited by Eq. (1).

The values of half-length L of the binary sequence reached up to 50. That is, a sequence of 50 bits was given as input (initial configuration) to a CA, in order to regenerate the next 50 bits of the given binary sequence. Typically, the GA employed 50 to 200 individual, the crossover probability was in the range $[0.9, 1.0]$, the mutation probability was in the range $[0.05, 0.2]$, and the GA evolved for 20 up to 500 consecutive generations.

The evolution for 40 GA generations of the fitness function of the individual with the highest fitness function value at each generation for an experiment with $L = 20$ is shown in Fig. 1. As it is shown, the GA was able to track very quickly (in less than 40 GA generations) a suitable set of rules that regenerate the desired output in less than 20 evolutionary steps of the CA.

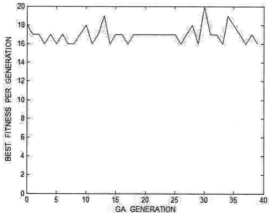

Figure 1: Evolution of the best fitness value of each GA generation, for a binary sequence with $L = 20$.

4 Discussion

In the present work a hybrid evolutionary algorithm was proposed for the prediction of compex binary sequences. The algorithm incorporates Cellular Automata with sets of rules that are suitably codified in order to be evolved by a Genetic Algorithm. Our purpose was to test the method in binary sequences of high complexity, like the ones that are obtained by applying a simple, linear threshold transformation on the iterative logistic function of Eq. (1), which is a well-known system that through period doubling reaches to chaotic behavior. The proposed algorithm was given an L number of bits as input and its task was to evolve the sets of rules of the CA in order to regenerate the next L bits of the binary sequence in less than L evolution steps of the CA. The obtained results for values of L up to 50 and for a variety of values of the non-linearity parameter r of Eq. (1) indicated that the proposed algorithm was able to find the proper sets of CA rules in a small number of GA generations.

Acknowledgment

This work was partially supported by the Hellenic Ministry of Education and the European Union under research Program PYTHAGORAS-89203.

References

[1] S. Wolfram: *Theory and Applications of Cellular Automata*, World Scientific, 1986.

[2] S. Wolfram: *Nature*, **311**, 419, (1984).

[3] S. Wolfram: *Cellular Automata and Complexity*, World Scientific, Singapore, 1994.

[4] N. Ganguly, B.K. Sikdar, A. Deutsch, G. Canright, P.P. Chaudhuri: A Survey on Cellular Automata, Technical report, Centre for High Performance Computing, Dresden University of Technology, December 2003.

[5] M. Mitchell. Computation in cellular automata: A selected review. In T. Gramss et al., (ed.), Nonstandard Computation, pp. 95-140, Wiley-VCH, Weinhcim,1998. Available at http://www.cs.pdx.edu/~mm/publications.tml

[6] J.P. Crutchfield and M. Mitchell, Proceedings of the National Academy of Sciences, USA, **92**(23), 10742, (1995).

[7] N.H. Packard, Complex Systems **4**, 543 (1990).

[8] M. Mitchell: *Introduction to Genetic Algorithms*. MIT Press (1996).

[9] J.R. Koza, *Genetic Programming: On the Programming of Computers by Means of Natural Selection*, MIT Press, 1992.

[10] M. Mitchell, J. Crutchfield, and R. Das, Evolving Cellular Automata with Genetic Algorithms: A Review of Recent Work, In: First International Conference on Evolutionary Computation and its Applications, 1996.

[11] M. Mitchell, J.P. Crutchfield, and P.T. Hraber, Physica D 75, 361–391, 1994.

[12] R. Das, J.P. Crutchfield, M. Mitchell and J.E. Hanson, Evolving globally synchronized cellular automata, In: L.J. Eshelman (ed.), Proceedings of the Sixth International Conference on Genetic Algorithms 336–343. San Francisco, CA Morgan Kaufmann, 1995.

Brill Academic Publishers
P.O. Box 9000, 2300 PA Leiden,
The Netherlands

Lecture Series on Computer
and Computational Sciences
Volume 4, 2005, pp. 1428-1431

Unsupervised Clustering under Parallel and Distributed Computing Environments

D.K. Tasoulis[1], L. Drossos[2] and M.N. Vrahatis[1]

[1]Department of Mathematics and Computational Intelligence Laboratory,
University of Patras, GR-26110 Patras, Greece.

[2]Department of Applications of Informatics
to Economics and Business Administration, TEI Messologi, Greece.

Received 10 July, 2005; accepted in revised form 20 July, 2005

Abstract: The urge to discover, potentially useful, implicit information in gross amounts of data has rendered the development of clustering algorithms a necessity. Clustering algorithms aim to discover groups of objects in a way that maximizes the intra-group similarity. However, the more recent technological status has formed a new database environment. In detail, while traditional datasets where located in a single database, the lately come settings consider data to be spread in different sites. These facts should be endorsed by the clustering research, towards the development of parallel and distributed algorithms. In this work, we present an algorithmic framework for the k-windows clustering algorithm that tries to minimize communication in a distributed computing environment, without sacrificing its efficiency.

Keywords: Unsupervised Clustering, Parallel Algorithms, Distributed Computing

Mathematics Subject Classification: 91C20, 68W15, 68W10

1   Introduction

Clustering is one of the fundamental processes of Knowledge Discovery. It refers to the partitioning of a set of elements into disjoint groups (clusters) in a way that elements in the same cluster are more similar to each other than elements in different clusters. Clustering techniques have a very broad application domain ranging from data mining [5], and statistical data analysis [1], to global optimization [12] and web personalization [8].

Through the exponential growth of the Internet new kinds of computing environments, have emerged. A typical such environment includes a collection of computing units, with different characteristics, connected through a network infrastructure. Data on these new settings are distributed among the different computing sites. Thus performing searches and similarity queries involves the sending of messages among the nodes. This fact, introduces apart from the computation cost of the algorithm, a communication cost. Efficient parallel and distributed clustering algorithms aim to minimize the communication cost by taking under consideration the limitations of the underlying environment.

Although several approaches have been introduced for parallel and distributed Data Mining [6], parallel and distributed clustering algorithms have not been extensively studied. In [13] a parallel version of DBSCAN [9] and in [4] a parallel version of k-means were introduced.

In this paper, we adapt the k-windows unsupervised clustering algorithm [11] to this kind of environments. In detail, we present an algorithmic framework that aims to minimize the communication cost by discovering when it is necessary for the algorithm to communicate information, and avoid it otherwise.

2   The Unsupervised k-windows Clustering Algorithm (UKW)

Here we briefly describe the basic concepts of the unsupervised k–windows algorithm (UKW). Suppose that we have a set of points in the $\mathbb{R}^d$ space. Intuitively, the k–windows algorithm for every cluster present in the dataset tries to place a d–dimensional window containing patterns that belong to a single cluster.

At first, k-windows of a certain size are randomly selected initialized over the data. Next the algorithm, employs three fundamental procedures: *movement*, *enlargement* and *merging*. The movement procedure aims at positioning each window as close as possible to the center of a cluster. During this procedure each window is centered at the mean of the patterns that it includes. On the other hand, the enlargement procedure tries to augment each window so as to include as many patterns from the cluster over which it is positioned, as possible. Thus, the range of each window, for each coordinate separately is enlarged, as long as, a significant number of additional points is enclosed by the window. Finally, the merging operation, by computing the proportion of data objects in the intersection of any two overlapping windows with respect to the total number of data object they include, either: (a) ignores one of the windows if this number is very high, (b) considers the windows to contain parts of the same cluster for a relatively high number, or (c) considers the windows to capture different clusters.

By iteratively executing the movement and enlargement procedures, until they cease to alter any of the windows, and subsequently executing the merging procedure, the algorithm identifies the clusters and provides an approximation to their number. In Fig. 1 the three processes are illustrated.

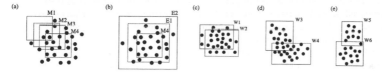

Figure 1: (a) Sequential movements M2, M3, M4 of initial window M1. (b) Sequential enlargements E1, E2 of window M4. (c) W_1 and W_2 have a very high overlapment proportion thus W_1 is discarded. (d) W_3 and W_4 have a significant overlapment and are considered to belong to the same cluster. (e) W_5 and W_6 have a small overlapment and capture two different clusters.

The computationally demanding step of the k–windows clustering algorithm is the determination of the points that lie in a specific window. This is the well studied *orthogonal range search* problem [7]. Numerous Computational Geometry techniques have been proposed [2, 7] to address this problem.

3   Parallel and Distributed Unsupervised Clustering

Under the assumption that a pool of densely interconnected computing nodes is the underlying environment, parallel computation can be exploited for the most computationally demanding steps. For k-windows algorithm, this is the answering of the range queries. In this case, each computer node is considered to have a part of the database, so the range queries are answered by simultaneously querying all the nodes. In [3], the authors have thoroughly analyzed this approach and provided bounds for the speedup that can be obtained.

On the other hand, a different assumption would enforce minimal communication among the sites. This could be due to privacy issues, or very slow and expensive network connections. In this case it is possible to modify the k-windows algorithm to distribute locally the whole clustering procedure, as described in [10]. In more detail, the k-windows algorithm is executed over each locally stored dataset. This step results in a set of windows for each site. To obtain the final clustering result over the whole dataset, all the final windows are collected and merged through the merging and similarity procedure of the k-windows algorithm.

However, today's settings include mixed environments. To this end we propose a modification of the k-windows algorithm that is able to operate under these circumstances. The underlying environment is assumed to be composed of a number of sites that although connected with a communication network, they are considered distant. Moreover, to comply with privacy concerns no specific record exchange is allowed among the sites. The proposed procedure at first involves the execution of a local k-windows algorithm at each site individually. If each site contains a collection of computing units then the parallel version of the algorithm could also be employed. The next step, involves the gathering of the set of final windows that constitute the clustering result, from each site at a central node, that plays the role of the coordinator. This role could be performed by any of the participating sites. At this point, the characteristics of the k-windows are exploited, to guide the clustering procedure. In detail, the proposed algorithmic framework aims to instruct a parallel clustering procedure only to specific regions of the data domain for which more than one sites contain information. This can be performed by examining the overlapment of windows originating in different sites. If windows from different sites are found to overlap each other, then a new window is initialized, in their overlapment area. This window is processed by the movement and enlargement procedures of the k-windows algorithm, but the answering of the range queries is performed by simultaneously querying only the involved sites. After these procedures terminate this window replaces the initial overlapping windows, in the final clustering result.

The proposed procedure has the advantage not performing a total parallel clustering procedure over all the sites, but using parallel computation only when and where it is necessary. In Fig. 2, an example of the operation of this framework is demonstrated. In Fig. 2(a), it is assumed that two sites A and B participate. The filled circles and the dashed-line windows (WA1, WA2), represent the data and the final windows after the local clustering procedure respectively, of site A. The empty circles and the solid-line windows represent the data and final windows (WB1, WB2), of site B. Examining all four windows reveals, that only windows WA1 and WB2 overlap. Thus a new window WAB1 is initialized in their overlapment. The final position of WAB1 after the movement and enlargement procedures using information from both sites A and B is demonstrated in Fig. 2(b). This is also the final clustering result.

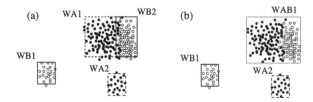

Figure 2: The operation of the proposed framework.

4   Discussion

The development of efficient distributed clustering algorithms has attracted considerable attention in the past few years. In this work, we present an algorithmic framework for the k-windows clustering algorithm that aims to minimize the communication cost among the involved sites, by analyzing local clustering results and determining when co-operation among the sites is necessary. Further experimental analysis could provide accurate speedup measurements over traditional approaches.

References

[1] M.S. Aldenderfer and R.K. Blashfield, *Cluster analysis*, Quantitative Applications in the Social Sciences, vol. 44, SAGE Publications, London, 1984.

[2] P. Alevizos, *An algorithm for orthogonal range search in $d \geqslant 3$ dimensions*, Proceedings of the 14th European Workshop on Computational Geometry, Barcelona, 1998.

[3] P. Alevizos, D.K. Tasoulis, and M.N. Vrahatis, *Parallelizing the unsupervised k-windows clustering algorithm*, Lecture Notes in Computer Science (R. Wyrzykowski, ed.), vol. 3019, Springer-Verlag, 2004, pp. 225–232.

[4] I.S. Dhillon and D.S. Modha, *A data-clustering algorithm on distributed memory multiprocessors*, Large-Scale Parallel Data Mining, Lecture Notes in Artificial Intelligence, 2000, pp. 245–260.

[5] U.M. Fayyad, G. Piatetsky-Shapiro, and P. Smyth, *Advances in knowledge discovery and data mining*, MIT Press, 1996.

[6] H. Kargupta, W. Huang, K. Sivakumar, and E.L. Johnson, *Distributed clustering using collective principal component analysis*, Knowledge and Information Systems **3** (2001), no. 4, 422–448.

[7] F. Preparata and M. Shamos, *Computational geometry*, Springer Verlag, New York, Berlin, 1985.

[8] M. Rigou, S. Sirmakessis, and A. Tsakalidis, *A computational geometry approach to web personalization*, IEEE International Conference on E-Commerce Technology (CEC'04) (San Diego, California), July 2004, pp. 377–380.

[9] J. Sander, M. Ester, H.-P. Kriegel, and X. Xu, *Density-based clustering in spatial databases: The algorithm GDBSCAN and its applications*, Data Mining and Knowledge Discovery **2** (1998), no. 2, 169–194.

[10] D.K. Tasoulis and M.N. Vrahatis, *Unsupervised distributed clustering*, IASTED International Conference on Parallel and Distributed Computing and Networks, Innsbruck, Austria, 2004, pp. 347–351.

[11] D.K. Tasoulis and M.N. Vrahatis, *Novel approaches to unsupervised clustering through the k-windows algorithm*, Knowledge Mining, Studies in Fuzziness and Soft Computing, 2005.

[12] A. Törn and A. Žilinskas, *Global optimization*, Springer-Verlag, Berlin, 1989.

[13] X. Xu, J. Jgerand, and H.P. Kriegel, *A fast parallel clustering algorithm for large spatial databases*, Data Mining and Knowledge Discovery **3** (1999), 263–290.

Brill Academic Publishers
P.O. Box 9000, 2300 PA Leiden,
The Netherlands

*Lecture Series on Computer
and Computational Sciences*
Volume 4, 2005, pp. 1432-1435

New Orbit Based Symmetric Cryptosystem

M.N. Vrahatis[†,♭1], **G.A. Tsirogiannis**[‡,♭2], **E.C. Laskari**[†,♭3]

†Computational Intelligence Laboratory, Department of Mathematics,
University of Patras, GR-26110 Patras, Greece

‡Department of Engineering Sciences, University of Patras, GR-26110 Patras, Greece

♭University of Patras Artificial Intelligence Research Center (UPAIRC),
University of Patras, GR-26110 Patras, Greece

Received 10 July, 2005; accepted 20 July, 2005

Abstract: In this contribution a new symmetric key cryptosystem is presented. This cryptosystem exploits the idea of nonlinear mappings and their fixed points to encrypt information.

Keywords: Cryptosystems, symmetric key, nonlinear mappings, fixed points, stability

Mathematics Subject Classification: 94A60, 14G50, 93C10, 37C25, 65P40

1 Introduction

Recently, chaotic systems and its application to cryptography have received considerable attention [1, 2, 3]. In this contribution a new symmetric key cryptosystem based on dynamical systems is proposed. This cryptosystem exploits the idea of nonlinear mappings and their fixed points to encrypt information.

A *symmetric key cryptosystem* can be defined as follows [4]. Consider an encryption scheme consisting of the sets of encryption and decryption transformations $\{E_e : e \in \mathcal{K}\}$ and $\{D_d : d \in \mathcal{K}\}$, respectively, where $\mathcal{K}$ denotes the key space. The encryption scheme is said to be symmetric key if for each associated encryption–decryption key pair (e, d) it is computationally "easy" to determine d knowing only e, and to determine e from d. A large variety of such cryptosystems exists [5, 6, 7, 8].

The original message is usually called *plaintext* and the transformed (encrypted message) *ciphertext*. In the proposed cryptosystem the keys e, d are identical. Moreover, the plaintext consists of strings of symbols (characters) from a known alphabet (e.g. ASCII) and the cipher text consists of fixed points of nonlinear mappings. The cryptosystem is based on two-dimensional non-linear mappings. In general, these mappings are of the following form:

$$\Phi : \begin{cases} \widehat{x}_1 & = \phi_1(x_1, x_2), \\ \widehat{x}_2 & = \phi_2(x_1, x_2). \end{cases} \tag{1}$$

In a nonlinear mapping of the form (1) we can find points which are invariant or fixed under the mapping; these points are commonly referred to as *periodic orbits* of the mapping. We say that

---

[1]E-mail: vrahatis@math.upatras.gr
[2]E-mail: gtsirog@ceid.upatras.gr
[3]E-mail: elena@math.upatras.gr

$X = (x_1, x_2)^\top$ is a *fixed point* of Φ if $\Phi(X) = X$ and a *fixed point of order p* or a *periodic orbit of period p* if

$$X = \Phi^p(X) \equiv \underbrace{\Phi(\Phi \cdots (\Phi(X)) \cdots))}_{p \text{ times}} \qquad (2)$$

A typical example of a non linear mapping is the following Hénon's quadratic area-preserving two-dimensional mapping [9, 10]:

$$\Phi : \begin{pmatrix} \widehat{x}_1 \\ \widehat{x}_2 \end{pmatrix} = R(a) \begin{pmatrix} x_1 \\ x_2 + g(x_1) \end{pmatrix}, \qquad (3)$$

where $(x_1, x_2)^\top \in \mathbb{R}$ and

$$R(a) = \begin{pmatrix} \cos a & -\sin a \\ \sin a & \cos a \end{pmatrix},$$

where $a \in [0, \pi]$ is the rotation angle and $g(x_1) = -x_1^2$. Using the CHABIS package [11, 12] for the Hénon's mapping with $a = \cos^{-1}(0.24)$, we can compute the following periodic fixed point:

$$x_1^5 = (0.5672405470221847, -0.1223202134278941)^\top,$$

with period $p = 5$. If we iterate the mapping we obtain the remaining fixed points of the same orbit:

$$
\begin{aligned}
X_2^5 &= \Phi(X_1^5) = (0.5672405470221847, 0.4440820516139216)^\top, \\
X_3^5 &= \Phi(X_2^5) = (0.0173925844399303, 0.5800185952239573)^\top, \\
X_4^5 &= \Phi(X_3^5) = (-0.5585984457571741, 0.1560161118011652)^\top, \\
X_5^5 &= \Phi(X_4^5) = (0.0173925844399305, -0.5797160932304572)^\top.
\end{aligned}
$$

Thus, if we know the mapping and a fixed point with period p, we can easily produce all the remaining $(p-1)$ fixed points of the same periodic orbit. Furthermore, to each periodic orbit corresponds a *rotation number* $\sigma = \nu/(2\pi) = i_1/i_2$, where ν is the *frequency of the orbit* and i_1, i_2, are two positive integers. From the sequence with which the above points are created on the (x_1, x_2) plane, we can infer the rotation number of this orbit $\sigma = 1/5$, indicating that it has produced $i_2 = 5$ points, by rotating around the origin $i_1 = 1$ times.

There is a large variety of such mappings that we be can used including the Standard Map [13], the Gingerbreadman Map [14], the Predator-Prey Map [15] as well as higher dimensional maps including the Lorenz Map [16], the Rössler Map [17], the Hénon's 4-Dimensional Symplectic Map [18] among others. Using the CHABIS package one is able to compute stable and unstable periodic orbits with periods up to hundreds of thousands [10, 18].

2 The Basic Idea of Orbit Based Cryptosystems

The central idea of the proposed cryptosystem is hiding information in fixed points of an orbit (denoted by O). Let $X_0, \Phi_1^p(X_0), \Phi_2^p(X_0), \ldots, \Phi_{p-1}^p(X_0)$ the fixed points of orbit O. The order of the elements is very important. We store them following the order of their production by the iteration of the mapping Φ for $(p-1)$ times using the initial fixed point X_0. Equivalently, we can consider the group $G_o = \langle \mathbb{Z}_p, \circ \rangle$, with $k \longrightarrow \Phi_k^p(X_0)$, where $\langle \mathbb{Z}_p, \circ \rangle \longrightarrow X_0, \Phi_1^p(X_0), \Phi_2^p(X_0), \cdots, \Phi_{p-1}^p(X_0)$, and the operation 'o' can be defined as $(\Phi_k^p \circ \Phi_l^p)(X_0) = \Phi_{k+l}^p(X_0)$.

We can compute such an orbit O using the Characteristic Bisection method [10, 11, 12], computational intelligence methods [19], or other methods.

The central idea is the following. If we know the nonlinear mapping Φ, the orbit O and a fixed point X_0 of O (which is assumed to be the initial fixed point of O), we are able to count how many times we must iterate the nonlinear mapping Φ starting from the initial fixed point X_0 so as to obtain any other fixed point of the same orbit. The important observation is that the elements of O which are placed on a linked–list, have a very specific order which is known only if the nonlinear mapping Φ is known. As will be shown, the encryption–decryption procedures are based on this order of the elements.

In the basic form the cryptosystem accepts a message m as input (e.g. a sequence of ASCII characters) and produces a matrix $C \in \mathbb{R}^{lenght(m) \times 2}$ as output. This means that every character of the plaintext is encrypted to a fixed point of a specific orbit. The **key** of the cryptosystem consists of the following.

- A nonlinear mapping Φ with the following characteristics: (i) the value of $\alpha \in [0, \pi]$ and (ii) nonlinear term $g(x_1)$,

- A fixed point X_0,

- A positive integer p (large enough i.e. $p \geqslant 300$, we prefer p to be a prime) which indicates the order of the orbit O in which X_0 belongs to.

In order to construct such a key we follow the procedure given below.

1. Choose at random a rotation angle $\alpha \in [0, \pi]$

2. Choose a nonlinear term $g(x_1)$

3. Choose a quite large positive integer p (i.e. $p \geqslant 300$) (we prefer p to be a prime)

4. Choose a region of the plain in which the fixed point X_0 must lie.

5. Use a method (e.g. CHABIS [12]) to locate a fixed point X_0.

6. Check if p is the smaller integer that satisfies $X_0 = \Phi_p^p(X_0)$, terminate.

The **encryption algorithm** takes as an input a message m with n characters (i.e. ASCII form) and a key k, gives as output a ciphertext $C \in \mathbb{R}^{n \times 2}$, and can be described by the following steps.

1. Initialize $C_0 = X_0$, $i = 1$

2. Iterate the nonlinear mapping Φ (which is described in key k) $ASCII(m_i)$ times starting from the fixed point C_{i-1}. We write $C_i = \Phi_{ASCII(m_i)}^p (C_{i-1})$

3. Proceed to the next character by setting $i = i + 1$

4. **If** $i < n$ **then** goto step 2, **else** terminate.

In reverse, the **decryption algorithm** takes as an input a ciphertext $C \in \mathbb{R}^{n \times 2}$ and a key k and provides as output the original plaintext m, by the following procedure.

1. Initialize $C_0 = X_0$, $i = 1$

2. Iterate the nonlinear mapping Φ (which is described in key k) so $ASCII(m_i)$ times on the fixed point C_{i-1} so as to ensure that the equality, $C_i = \Phi_{ASCII(m_i)}^p (C_{i-1})$, holds for the first time.

3. Proceed to the next character by setting $i = i + 1$.

4. If $i < n$ **then** goto step 2, **else** terminate.

The security of this system is mainly based to the difficulty to sort the elements of a given orbit when the nonlinear mapping is not known. Any kind of brute force attack is completely inefficient because small changes of the parameters of the nonlinear mapping (rotation angle α and non linear term $g(x_1)$) lead to very different results. Furthermore, an evaluation of the cryptographical properties of the designed system will be considered along with improvements to its basic form.

References

[1] T. Yang, C.W. Wu, L.O. Chua, Cryptography based on chaotic systems, *IEEE Trans. Circ. Sys.*, **44**, 469–472, 1997.

[2] G. Grassi, S. Mascolo, A System Theory Approach for Designing Cryptosystems Based on Hyperchaos, *IEEE Trans. Circ. Sys.*, **46**(9), 1135–1138, 1999.

[3] F. Dachselt, W. Schwarz, Chaos and Cryptography, *IEEE Trans. Circ. Sys.*, **48**(12), 1498–1509, 2001.

[4] A. Menezes, P. Van Oorschot, S. Vanstone, *Handbook of applied cryptography*, CRC Press, 1996.

[5] Data Encryption Standard (DES), U.S. Dept. of Commerce, Dec. 30, 1993, FIPS PUB 46-2 (C13.52).

[6] Advanced encryption Standard (AES) Fact Sheet, October 3, 2000.

[7] X. Lai, On the Design and Security of Block Ciphers, *ETH Series on Information Processing*, Verlag, 1992.

[8] A. Shimizu, S. Miyaguchi, Fast Data Encipherment Algorithm FEAL, *IEICE*, July 1987.

[9] M. Hénon, Numerical study of quadratic area-preserving mappings, *Quart. Appl. Math.*, 27, 291–311, 1969.

[10] M.N. Vrahatis, An efficient method for locating and computing periodic orbits of nonlinear mappings, *J. Comput. Phys.*, **119**, 105–119, 1995.

[11] M.N. Vrahatis, Solving systems of nonlinear equations using the nonzero value of the topological degree, *ACM Trans. Math. Soft.*, **14**(4), 312–329, 1988.

[12] M.N. Vrahatis, CHABIS: A mathematical software package for locating and evaluating roots of systems of nonlinear equations, *ACM Trans. Math. Soft.*, **14**(4), 330–336, 1988.

[13] S.N. Rasband, *Chaotic Dynamics of Nonlinear Systems*, Wiley, New York, 1990.

[14] R.L. Devaney, A piecewise linear model for thezones of instability of an area preserving map, *Physica D*, 10, 387–393, 1984.

[15] M.J. Smith, *Mathematical Ideas in Biology*, Cambridge University Press, London, 1968.

[16] E.N. Lorenz, Deterministic nonperiodic flow, *J. Atmos. Sci.*, 20, 130–141, 1963.

[17] O.E. Rössler, An equation for continuous chaos, *Phys. Lett. A.*, 57, 397–398, 1976.

[18] M.N. Vrahatis, H. Isliker, T.C. Bountis, Structure and breakdown of invariant tori in a 4-D mapping model of accelerator dynamics, *Inter. J. Bifurc. Chaos*, 7, 2707–2722, 1997.

[19] K.E. Parsopoulos, M.N. Vrahatis, Computing periodic orbits of nondifferentiable /discontinuous mappings through particle swarm optimization, in *Proc. IEEE 2003 Swarm Intelligence Symposium*, 34–41, 2003.

Brill Academic Publishers
P.O. Box 9000, 2300 PA Leiden
The Netherlands

*Lecture Series on Computer
and Computational Sciences*
Volume 4, 2005, pp. 1436-1441

Optimal Rural Water Distribution Design using Labye's Optimization Method and Linear Programming Optimization Method

M.E. Theocharis[1*] , C.D. Tzimopoulos [2] , M. A. Sakellariou - Makrantonaki [3] , S. I. Yannopoulos [2], and I. K. Meletiou[1]

[1] Department of Crop Production
Technical Educational Institution of Epirus
GR- 47100 Arta, Greece

[2] Department of Rural and Surveying Engineers
Aristotle University of Thessaloniki
GR- 54006 Thessaloniki, Greece

[3] Department of Agricultural Crop Production and Rural Environment
University of Thessaly
GR- 38334 Volos, Greece

Received 10 July, 2005; accepted 20 July, 2005

Abstract: The designating factors in the design of branched irrigation networks are the cost of pipes and the cost of pumping. They both depend directly on the hydraulic head of the pump station. It is mandatory for this reason to calculate the optimal head of the pump station as well as the corresponded optimal pipe diameters, in order to derive the minimal total cost of the irrigation network. The certain calculating methods in identified the above total cost of a network, that have been derived are: the linear programming optimization method, the non linear programming optimization method, the dynamic programming optimization method and the Labye's method. All above methods have grown independently and a comparative study between them has not yet been derived. In this paper a comparative calculation of the pump station optimal head as well as the corresponded economic pipe diameters, using the Labye's optimization method and the linear programming method, is presented. Application and comparative evaluation in a particular irrigation network is also developed. From the study it is being held that the two optimization methods in fact conclude to the same result and therefore can be applied with no distinction in the studying of the branched hydraulic networks.

Key words: Irrigation, head, pump station, network, cost, optimization, linear programming, Labye

1. Introduction

The problem of selecting the best arrangement for the pipe diameters and the optimal pumping head as like as the minimal total cost to be produced, has received considerable attention many years ago by the engineers who study hydraulic works. The knowledge of the calculating procedure in order that the least cost is obtained, is a significant factor in the design of the irrigation networks and, in general, in the management of the water resources of a region. The classical optimization techniques, which have been proposed so long, are the following: a) The linear programming optimization method [1,7,8,9,10,11,13], b) the nonlinear programming optimization method [9,10,13], c) the dynamic programming method [9,13], and d) the Labye's optimization method [2,3,4,5,6,9,12,13]. The common characteristic of all the above techniques is an objective function, which includes the total cost of the network pipes, and which is optimized according to specific constraints. The decision variables that are generally used are: the pipes diameters, the head losses, and the pipes' lengths. As constraints are used: the pipe lengths, and the available piezometric heads in order to cover the friction losses. In this study, a systematic calculation procedure of the optimal pipe diameters using the Labye's method and the linear programming method is presented. Application and comparative evaluation in an irrigation network is also developed.

---

* Corresponding author. e-mail: theoxar@teiep.gr

2. Methods

2.1 The Labye's optimization method

According to this method [2,3,4,5,6,9,10,11], the optimal solution of hydraulic networks is obtained considering that the pipe diameters can only be chosen in a discrete set of values corresponding to the standard ones considered. It consists of the tracing of a zigzag line in a coordinates diagram, from which the minimal cost of the network can be obtained as a function of the total piezometric losses of the network.

2.1.1 A network with pipes in sequence

For every pipe of the network [2,3,4,5,6,9,12] the available commercial size of diameters are selected and then calculated: **i.** the frictional head losses per meter ,J_{ij} ; **ii.** the pipe cost per meter, c_{ij}; and **iii.** the various gradients $\varphi_{ij} = \left| \dfrac{\Delta c_{ij}}{\Delta J_{ij}} \right|$ witch are classified in decreased order. After that, the graph

P – H, (figure 1) is constructed witch is a convex zigzag line witch is called *" the characteristic"* of the network. The gradients of the zigzag line various parts, φ_{ij}, are progressively decreased from left to right. The terminal right point of the characteristic, A, corresponds to the minimal diameters for all the pipes of the network with the maximal total frictional losses, H_A, and the minimal total cost of the network P_A. Similarly the terminal left point of the characteristic, F, corresponds to the maximal diameters for all the pipes of the network with the minimal total frictional losses, H_F, and the maximal total cost of the network P_F.

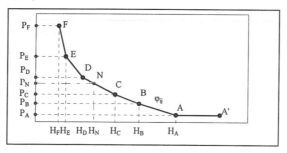

Figure 1: A network with pipes in sequence

After that, the total cost of the network, P_N, corresponded to the available total frictional head loss, H_N, is calculated. If the point N lies in the part of the characteristic with gradient φ_{ij}, it means that only the ith pipe must be constructed with two different diameters. On the passing from the point N to the point F, linear parts with progressively increasing gradients corresponding to the various pipes of the network are detected. Each pipe is constructed with the lower diameter corresponding to the gradient φ_{ij}. Similarly any pipe, the gradients of which are detected on the right of the point N, is constructed with the higher diameter corresponding to the gradient φ_{ij}. It is concluded that only one pipe of the network is possible to be constructed with two different diameters.

2.1.2. A network with two branched pipes

At first the characteristic lines of the two branches are constructed. Then the cumulative characteristic line is constructed, which is produced by adding the ordinates of the contributing branches. This characteristic line is also a convex zigzag line similar to the characteristic lines of the contributing branches [2,3,4,5,6,9,12].

2.1.3 Branched networks

The following steps (figure 2) are followed [2,6,9,12]: a. The characteristic lines of the branches i.g. BC , BD , AB, AE and OA are constructed according to the paragraph 2.1.1 (case of pipes in sequence). b. The characteristic line of the branch BCD is constructed according to the paragraph 2.1.2. c. The characteristic line of the branch BCD – AB is constructed, adding the characteristic lines of the branches BCD and AB according to the paragraph 2.1.1. d. The characteristic line of the branch $(BCD - AB) - AE$ is constructed, according to the paragraph 2.1.2. The procedure is continued until the head , O , of the network. Finally the total cost of the network, P_N, corresponded to the available total frictional head loss, H_N, is calculated and then the economic pipe diameters are selected.

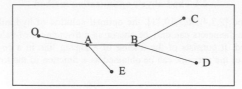

Figure 2: A branched irrigation network

2.2 The linear programming optimization method

According to this method [1,7,8,9,10,11,13], the search for optimal solutions of hydraulic networks is carried out considering that the pipe diameters can only be chosen in a discrete set of values corresponding to the standard ones considered. Due to that, each pipe is divided into as many sections as there are standard diameters, the length of these sections thus being adapted as decision variables. The least cost of the pipe network, P_N, is obtained from the minimal value of the objective function, meeting the specific functional and non negativity constraints.

2.2.1 The objective function

The objective function is expressed by [9,10,11]:

$$f(X) = CX \tag{1}$$

where C is the vector giving the cost of the pipes sections in Euro per meter and X is the vector giving the lengths of the pipes sections in meter. The vectors C and X are determined as:

$$C = (C_1 .... C_i ... C_n) \quad ; \quad C_i = (\delta_{i1} ...... \delta_{ij} ... \delta_{ik}) \quad ; \quad X = (X_1 .... X_i ... X_n)^T ;$$

$$X_i = (x_{i1} ...... x_{ij} ... x_{ik})^T \quad \text{for } i = 1, 2, ..., n \quad \text{and} \quad j = 1, 2, ..., k$$

where $x_{11}, x_{12}, ..., x_{nk}$, are the decision variables in meter ; δ_{ij} is the cost of jth section of ith pipe in Euro per meter ; n is the total number of the pipes in the network ; and k the total number of each pipe accepted diameters (= the number of the sections in which each pipe is divided).

2.2.2 The constraints of the problem

The constraints of the problem are specific functional and non negativity constraints [9,10,11]. The functional constraints are length constraints and friction loss constraints. The length constraints are expressed by $L_i = \sum_{j=1}^{k} x_{ij}$ .The friction losses constraints are expressed by: $\sum_{i=1}^{i} \Delta h_i \leq H_A - h_i$ for all the nodes ,i, where H_A is the piezometric head of the water intake, h_i is the minimum required piezometric head at each node i. The sum $\sum_{i=1}^{i} \Delta h_i$ is taken along the length of every route i, of the network. The non negativity constraints are expressed by: $x_{ij} \geq 0$.

2.2.3 The variance of the head of the pump

The calculation of the optimal pump head is achieved through the following process: a) the variance of the pump station head are determined, b) using one of the known optimisation methods, the optimal annual total cost of the project for every head of the pump station is calculated, c) The graph P_{ET}.- H_A is constructed and its minimum point is defined, which corresponds in the value of H_{man}, This value constitutes the optimal pump station head. [9,10,11]

2.2.4 The least cost and the economic pipe diameters of the network

The least cost of the pipe network, PN, is obtained from the minimal value of the objective function. The minimal value of the objective function, min $f(X) = C X$, is obtained using the simplex method.

3. Application

The optimal cost of the irrigation network, which is shown in Figure 3, is calculated. The material of the pipes is PVC 10 atm and the available total head is $H_N = 60.00$ m.

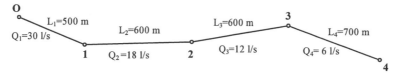

Figure 3: The under solution network

3.1 Selecting the acceptable commercial pipe diameters

Using the continuity equation the acceptable commercial diameters, as well as the cost per meter for every pipe of the network are selected. From the Darcy- Weisbach equation the head losses are calculated. From the values of head losses and pipe cost the various gradients φ_{ij} are calculated. The results are presented in Table 1.

Table 1. The geometric and hydraulic characteristics off the pipes

Pipe	Internal Diameter [mm]	Head Losses [%]	Cost [€/m]	Gradient φ	Pipe	Internal Diameter [mm]	Head Losses [%]	Cost [€/m]	Gradient φ
1	144.6	1.935	21.31	0.7923	2	113.0	2.524	13.82	0.2809
	180.8	0.640	31.57	2.5587		126.6	1.436	16.88	0.6418
	203.4	0.359	38.76	8.3432		144.6	0.745	21.31	2.0681
	253.2	0.123	58.45			180.8	0.249	31.57	6.6010
3	99.4	2.231	11.33	0.2378		203.4	0.140	38.76	
	113.0	1.184	13.82	0.6036	4	99.4	0.619	11.33	0.7428
	126.6	0.677	16.88	1.3673		113.0	0.331	13.82	
	144.6	0.353	21.31						

3.2. Solving the network according to the linear programming method

The objective function is expressed by :
$Z = 21.31X_1 + 31.57X_2 + 38.76X_3 + 58.45X_4 + 13.82X_5 + 16.88X_6 + 21.31X_7 + 31.57X_8 + 38.76X_9 + 11.33X_{10} + 13.82X_{11} + 16.88X_{12} + 21.31X_{13} + 11.33X_{14} + 13.82X_{15}$
Subject to:

$X_1 + X_2 + X_3 + X_4 = 500$; $X_5 + X_6 + X_7 + X_8 + X_9 = 600$

$X_{10} + X_{11} + X_{12} + X_{13} = 600$; $X_{14} + X_{15} = 700$

$1.935X_1 + 0.640X_2 + 0.359X_3 + 0.123X_4 \le H_A - h_1 = 1800$

$1.935X_1 + 0.640X_2 + 0.359X_3 + 0.123X_4 + 2.524X_5 + 1.436X_6 + 0.745 X_7 + 0.249X_8 + 0.140X_9 \le 2300$

$1.935X_1 + 0.640X_2 + 0.359X_3 + 0.123X_4 + 2.524X_5 + 1.436X_6 + 0.745 X_7 + 0.249X_8 + 0.140X_9 + 2.231X_{10} + 1.184X_{11} + 0.677X_{12} + 0.353X_{13} \le 2700$

$1.935X_1 + 0.640X_2 + 0.359X_3 + 0.123X_4 + 2.524X_5 + 1.436X_6 + 0.745X_7 + 0.249X_8 + 0.140X_9 + 2.231X_{10} + 1.184X_{11} + 0.677X_{12} + 0.353X_{13} + 0.619X_{14} + 0.331X_{15} = 3000$

The minimization of the objective function is obtained using the simplex method. The results are presented in Table 2. From the Table is produced that min Z=36929 €

Brill Academic Publishers
P.O. Box 9000, 2300 PA Leiden
The Netherlands

*Lecture Series on Computer
and Computational Sciences*
Volume 4, 2005, pp. 1442-1444

Preface to the Symposium: Explicit Density Functional of the Kinetic Energy in Computer Simulations at Atomistic Level

T.A. Wesolowski[1]

Department of Physical Chemistry, University of Geneva,
30, quai Ernest-Ansermet, 1211 Geneva, Switzerland

Received 10 August, 2005; accepted in revised form 12 August, 2005

Keywords: kinetic energy functional, density functional theory, computer

The Kohn-Sham formulation [1] of Density Functional Theory (DFT) is the basis of the most successfull computational strategy which emerged from decades of intensive research dating from the early works of Thomas and Fermi [2]. Its success can be attributed to the fact that even very simple approximations to the exchange-correlation functional, the object of unknown analytic form of its dependence on electron density, appear sufficiently accurate to be used in a great variety of computer modelling studies dealing with: large chemical molecules, materials, biomolecules, liquids, nanostructures, etc. Owing to the introduction of the fictitious system of non-interacting electrons, which is the key element in the Kohn-Sham formalism, the kinetic energy of the reference system of non-interacting electrons of the electron density ρ ($T_s[\rho]$), which is one of the largest components of the total energy, can be calculated exactly. $T_s[\rho]$ is defined in Levy constrained search [3] performed among single determinantal wavefunctions (Ψ_s) as (in atomic units):

$$T_s[\rho] = \min_{\{\Psi_s\} \to \rho} = \left\langle \Psi_s \left| -\frac{1}{2}\nabla^2 \right| \Psi_s \right\rangle \tag{1}$$

In the Kohn-Sham formalism, the orbitals which minimize the right-hand side of Eq. (1) are obtained from one-electron equations. Orbitals are, therefore, indispensable to evaluate $T_s[\rho]$ exactly. As a consequence, the analytic expression (exact or approximate) for $T_s[\rho]$ as a functional of ρ is not needed at all.

The symposium *Explicit density functional of the kinetic energy in computer simulations at atomistic level* deals, however, not with Kohn-Sham formulation of DFT but with other theoretical frameworks, in which explicit analytic expression to evaluate $T_s[\rho]$ is indispensable. A universally applicable approximate explicit functional $T_s[\rho]$ of a reasonable accuracy, would revolutionize the computer simulations of polyatomic systems because the use of orbitals could be completely eliminated. This requires, however, approximating $T_s[\rho]$ by means of an explicit analytic expression depending on ρ. Unfortunately, such an approximation which would be *universally applicable* has not been developed yet. However, approximations of applicability limited to specific types of problems have been developed. The recent review by Wang and Carter [4] provides complete overview of these developments until 2000. In 2002, we organized a three-day workshop at CECAM in Lyon (France) aimed mainly at summarizing the efforts concerning *i*) mathematical properties of $T_s[\rho]$, *ii*) development of approximations to $T_s[\rho]$, *iii*) practical applications of approximate functional of $T_s[\rho]$, and *iv*) outlining the perspectives for the future [5]. The symposium in taking place in Lutraki in October 2005, is planned to be a similar event but smaller in scale. It was conceived as a forum for presenting the newest developments relevant for the four aforementioned topics. The presented papers provide a representative overview of such computer simulations which use explicit functional $T_s[\rho]$ in different formal frameworks and in studies dealing with various systems. They can be divided in two categories.

---

[1] Corresponding author. E-mail: Tomasz.Wesolowski@chiphy.unige.ch

One, which can be seen as the modern realization of the ideas of Thomas and Fermi does not involve any orbitals (it is orbital-free therefore). Instead of performing the minimization of Eq. (1) an analytic approximate expression $T_s[\rho]$ is used. The paper by B. Zhou and Y.A. Wang provides a concise formal description of the orbital-free methodology and an example of its application in studies of the structure of cubic diamond Si crystal. The application of the orbital-free methodology to metal clusters is reported in the paper by J.A. Alonso which also provides a representative example of recent efforts to construct approximations to $T_s[\rho]$. The paper by D.Garcia-Aldea and J.E. Alvarellos collects benchmark results to asses the accuracy of known approximations to $T_s[\rho]$ and also provides a new approximation which is thoroughly tested in model systems.

The second group of papers concerns methods, which do not use $T_s[\rho]$, but only a closely related object – its non-additive component. In the case of two subsystems, it is a bi-functional:

$$T_s^{nad}[\rho_A, \rho_B] \equiv T_s[\rho_A + \rho_B] - T_s[\rho_A] - T_s[\rho_B] \qquad (2)$$

Approximating only this component of the whole kinetic energy is a key element in subsystem formulation density functional theory introduced by Cortona in 1993 [6]. The paper by Cortona, reviews the formalism and provides several examples of its applications to study cohesive properties and elastic properties of alkali-earth chalcogenide solids. Two remaining papers deal with yet another type of modelling in which many levels of description are used to describe complex polyatomic systems, in which only one subsystem is described by means of orbitals (*embedded orbitals*) whereas the other one uses only electron density derived from simplified methods (orbital-free embedding). The formal basis of this type of simulations, introduced in one of our earlier works [7], can be seen as a special case of Cortona's formulation of DFT. The complete overview of recent applications of this embedding formalism can be found elsewhere [8]. Recent application of the orbital-free embedding in studies of electronic structure of the manganese impurities in perovskites will be presented by J.-M. Garcia-Lastra. The efficient computer implementation of the orbital-free embedding formalism is also of key importance in studies of solvated systems such as our recent studies of solvatochromism [9]. The paper presented by M. Dulak addresses the issue of efficient grid integration in such cases.

This work was supported by the Swiss National Scientific Foundation.

References

[1] W. Kohn, L.J. Sham, *Phys. Rev.* **1965**, *140*, A1133.

[2] E. Fermi, *Z. Phys.*, **1928**, *48*, 73 ; L.H. Thomas, *Proc. Camb. Phil. Soc.*, **1927**, *23*, 542.

[3] M. Levy, Proc. Natl. acad. Sci. USA, **1979**, *76*, 6062.

[4] Y.A. Wang and E.A. Carter, *Orbital-Free Kinetic Energy Density Functional Theory* In: *Theoretical Methods in Condensed Phase Chemistry*, Ed. Schwartz, p. 117-184, Kluwer, Dordrecht, 2000.

[5] CECAM workshop: *Computer of atoms, molecules, and materials using approximate functionals of the kinetic energy* at CECAM, July 31-August 2, 2002, Lyon (France) organized by T.A. Wesolowski and H. Chermette. The final report of the workshop is available at: http://lcta.unige.ch/~tomek/cecam2002/cecam2002.html

[6] P. Cortona, *Phys. Rev. B* **1991**, *44*, 8454.

[7] T.A. Wesolowski, A. Warshel, *J. Phys. Chem.* **1993**, *97*, 8050.

[8] T.A. Wesolowski, *Chimia* **2002**, *56*, 707; T.A. Wesolowski, *One-electron equations for embedded electron density: challenge for theory and practical payoffs in multi-level of complex polyatomic systems*, In: *Current Trends in Computational Chemistry*, vol. XI, J. Leszczynski Ed., World Scientific **2005**, *in press*.

[9] J. Neugebauer, M.J. Louwerse, E.J. Baerends, T.A. Wesolowski, *J. Chem. Phys.* **2005**, *122*, 09411; J. Neugebauer *et al*, *J. Chem. Phys.* **2005**, *in press*.

Tomasz A. Wesolowski

Tomasz A. Wesolowski, studied Physics and Biophysics at University of Warsaw (PhD in 1991). He was appointed Assistant Professor at the Department of Physics (in Biophysics) in 1991. Between 1992 and 2000, he conducted research concerning mainly development and applications of the orbital-free embedding formalism for multi-level computer simulations at University of Southern California (Los Angeles) in the lab of Prof. Arieh Warshel and at the University of Geneva in the Applied Theoretical Chemistry Lab. He joined the faculty of the Department of Chemistry at University of Geneva in 2000. His current research involves applications of this formalism in computer simulations studies of the electronic structure of condensed matter. His other interests concern description of weak intermolecular forces in density functional theory and modelling chemical systems at the quantum mechanical level.

Brill Academic Publishers
P.O. Box 9000, 2300 PA Leiden
The Netherlands

*Lecture Series on Computer
and Computational Sciences*
Volume 4, 2005, pp. 1445-1449

Study of Mn^{2+}-doped fluoroperovskites by means of the Kohn-Sham Constrained Electron Density embedding formalism

J. M. García-Lastra[1], T. Wesolowski[2], M. T. Barriuso[1], J. A. Aramburu[3], and M. Moreno[3]

[1]Departamento de Física Moderna, Universidad de Cantabria, Avda. de los Castros s/n. 39005 Santander, Spain
[2]Depártement de Chimie, Université de Genève, 30, quai Ernest-Ansermet, CH-1211 Genève 4, Switzerland
[3]Departamento de Ciencias de la Tierra y Física de la Materia Condensada, Universidad de Cantabria, Avda. de los Castros s/n. 39005 Santander, Spain

Received 20 July, 2005; accepted 5 August, 2005

Abstract: The local structure and optical and vibrational properties associated with Mn^{2+}-doped cubic fluoroperovskites are studied by means of DFT calculations using the orbital-free Kohn-Sham with Constrained Electron Density embedding formalism. The influence of chemical and hydrostatic pressures upon the luminescence is discussed. The nonequivalence of both pressures is also shown.

Keywords: MnF_6^{4-} complexes; luminescence; Stokes shift; DFT calculations; embedding effects
PACS: 71.55.-i; 78.55.Hx; 31.15.Ew; 31.70.Dk

1. Introduction

Photoluminescence is one of the most attractive phenomena associated with transition metal impurities in insulators. In this process an optical absorption band peaked at energy E_{abs} gives rise to an emission band whose maximum is placed at energy $E_{em} < E_{abs}$. In the case of doped insulating lattices, the Stokes shift $E_S = E_{abs} - E_{em}$ arises basically from the linear coupling of the electronic excited state responsible for the emission with local vibrational modes, and thus reflects an equilibrium geometry in the excited state different to that of the ground state [1]. Due to the localized character of active electrons in both ground and excited states such modes involve basically the distortion of ligands.

Luminescence in Mn^{2+} impurities in cubic AMF_3 fluoroperovskites comes from the $^4T_{1g} \rightarrow {}^6A_{1g}$ transition of octahedral MnF_6^{4-} complexes, involving the first $^4T_{1g}$ crystal field state [2]. For an octahedral complex a T state can be coupled linearly only to e_g and t_{2g} modes, in addition to the symmetric a_{1g} mode which is always allowed. Working in the $\{xz, yz, xy\}$ basis of the triplet T state, the coupling with the *stretching* a_{1g} and e_g modes is pictured by the following effective Hamiltonian [3]:

$$H_{eff} = V_a I Q_a + V_e (U_\theta Q_\theta + U_\varepsilon Q_\varepsilon) \qquad (1)$$

Here I means the identity 3×3 matrix, U_θ, U_ε are Pauli matrices, while V_i ($i = a_{1g}, e_g$) are linear coupling constants. The coupling with the i mode induces a decrease of the energy minimum corresponding to the $^4T_{1g}$ state, given by [1]

$$E_i = V_i^2 / 2M_L \omega_i^2 = S_i \hbar \omega_i \qquad (2)$$

where M_L stands for the ligand mass, and S_a and S_e denote the Huang-Rhys factors related to a_{1g} and e_g modes, respectively.

It is worth noting that if only the a_{1g} and e_g local modes are important the coupling with the non-symmetric Jahn-Teller mode does not modify the shape of an optical transition from a singlet to a triplet state. Nevertheless, it produces additional vibrational progressions to those coming from the a_{1g} mode. In this situation the Stokes shift can simply be written as [1]:

$$E_S = E_S^0 + 2\hbar\omega_u \tanh(\hbar\omega_u / 2k_B T) \qquad (3)$$

$$E_S^0 = 2(S_a \hbar\omega_a + S_e \hbar\omega_e) \qquad (4)$$

The second term in (3) simply reflects that d-d transitions for an octahedral complex are parity forbidden. Thus, progressions at $T = 0$ K start not at zero-phonon lines but at the so called false origins,

3   Results for the CD Si

The OO-DFT and KS-DFT results are obtained from a modified *ABINIT* code [10]. Both the NLPS [11] and the BLPS [5] are employed. In our OF-DFT calculations, the Wang-Govind-Carter (WGC) KEDF [6] and the BLPS [5] are used to compute the kinetic energy and nuclear-electron interaction energy. Local density approximation (LDA) [12] is used in all the DFT calculations. The computational details are reported in Refs. [5] and [7].

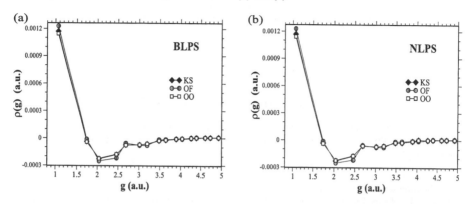

Figure 1: The electron densities in reciprocal space from a unit cell of the CD Si with lattice constant 5.40 Å, (a) from the BLPS and (b) from the NLPS, using KS (solid diamond), OF (opaque circles), and OO (open squares) DFT methods.

Fig. 1 depicts the electron density in reciprocal space, $\rho(\mathbf{g})$, of the CD Si with the BLPS and the NLPS. Discrepancies between $\rho_{KS}(\mathbf{g})$ and $\rho_{OF}(\mathbf{g})$ are obvious. In both cases, the value of $\rho_{OF}(\mathbf{g})$ at the first Bragg vector is too high, while those at some intermediate $\mathbf{g}$ vectors are too low. The density with the BLPS is notably improved by the one-iteration OO-DFT and $\rho_{OO}(\mathbf{g})$ is getting very close to $\rho_{KS}(\mathbf{g})$. Even when the NLPS is used, the error in $\rho_{OO}(\mathbf{g})$ is significantly reduced. This illustrates that the error in OF-DFT density is mainly due to the less optimal KEDF.

The equation of state (EOS) for the CD Si is computed in KS-DFT, OO-DFT , and OF-DFT. In order to obtain the static structural properties, we fit the EOS data to Murnaghan's EOS [13] for an arbitrary volume V,

$$E_{\text{tot}}(V) = \frac{B_0 V}{B_0'} \left[\frac{(V_0/V)^{B_0'}}{B_0' - 1} + 1 \right] + \text{constant} , \qquad (7)$$

where B_0 and B_0' are the bulk modulus and its pressure derivative at the equilibrium volume V_0, respectively.

Table 1 displays the resulting equilibrium volumes V_0, equilibrium bulk moduli B_0, and equilibrium energies E_0. The combination of the WGC KEDF and the BLPS in OF-DFT generates results containing large errors, with V_0 too large by *ca.* 20%, B_0 too small by *ca.* 70%, and E_0 too low by *ca.* 0.1%. In contrast, OO-DFT produces results very close to those from KS-DFT with the errors of the three properties reduced to 0.4%, 1.5%, and 0.01%, respectively. We further use the NLPS to test the accuracy of OO-DFT. The data in Table 1 shows that the KS-DFT results are very well reproduced by OO-DFT.

Table 1: OF (BLPS), OO (NLPS and BLPS), and KS (NLPS and BLPS) LDA predictions of the CD Si equilibrium bulk properties: V_0, B_0, and E_0.

	Property	OF	OO	KS
	V_0 (Å^3)	23.966	19.356	19.431
BLPS	B_0 (GPa)	25.8	97.0	95.6
	E_0 (eV/atom)	-110.345	-110.228	-110.236
	V_0 (Å^3)		19.393	19.489
NLPS	B_0 (GPa)		94.8	92.6
	E_0 (eV/atom)		-108.059	-108.067

4   Summary and Conclusions

We have developed an effective linear-scaling OO-DFT method by combining OF-DFT and KS-DFT via a density connection. OO-DFT offers remarkable improvements over the OF-DFT results and achieves comparable accuracy as KS-DFT at the cost of performing only a single non-self-consistent iteration of KS calculation under a fixed KS effective potential. We anticipate OO-DFT to venture into domains beyond the limits of other first-principles quantum mechanical methods.

Acknowledgment

The financial support from the Natural Sciences and Engineering Research Council (NSERC) of Canada is gratefully acknowledged.

References

[1] P. Hohenberg and W. Kohn, Phys. Rev. **136**, B864 (1964).

[2] Y. A. Wang and E. A. Carter, in *Theoretical Methods in Condensed Phase Chemistry*, Chapter 5 (Kluwer, Dordrecht, 2000), pp. 117-184.

[3] S. C. Watson and P. A.Madden, Phys. Chem. Commun. **1**, 1 (1998).

[4] M. Fago, R. H. Hayes, E. A. Carter, and M. Ortiz, Phys. Rev. B **70**, 10010 (2004).

[5] B. Zhou, Y. A. Wang, and E. A. Carter, Phys. Rev. B **69**, 125109 (2004).

[6] Y. A. Wang, N. Govind, and E. A. Carter, Phys. Rev. B **60**, 16350 (1999).

[7] B. Zhou, V. L. Lignères, and E. A. Carter, J. Chem. Phys. **122**, 044103 (2005).

[8] J. F. Annett, Comput. Mater. Sci. **4**, 23 (1995).

[9] S. Goedecker, Rev. Mod. Phys. **71**, 1085 (1999).

[10] X. Gonze *et al.*, Comput. Mater. Sci. **25**, 478 (2002).

[11] N. Troullier and J. L. Martins, Phys. Rev. B **43**, 1993 (1991).

[12] J. P. Perdew and A. Zunger, Phys. Rev. B **23**, 5048 (1981).

[13] F. D. Murnaghan, Proc. Nat. Acad. Sci. USA **30**, 244 (1944).

Brill Academic Publishers
P.O. Box 9000, 2300 PA Leiden,
The Netherlands

Lecture Series on Computer
and Computational Sciences
Volume 4, 2005, pp. 1462-1466

A Study of Kinetic Energy Density Functionals: a New Proposal

D. García-Aldea and J. E. Alvarellos

Departamento de Física Fundamental. Universidad Nacional de Educación a Distancia.
Apartado 60141, E-28080 Madrid, Spain.

Received ...

Abstract: The aim of this paper is to carry out a brief study of orbital-free kinetic energy density functionals, which has not attracted as much attentions as the exchange-correlation density functionals. Defining a kinetic energy density we compared the exact one with those generated by the semilocal functionals, with the help of the local relative error. We have found that commonly used corrections to the Thomas-Fermi functional do not improve the performance of the local density approximation. Finally we propose a new kinetic energy density functional based on the von Weizsäcker functional instead of the more traditional nonlocal functionals derived from a Thomas-Fermi approach.

Keywords: Density functional Theory, Orbital-free, Kinetic Energy Density Functional, von Weizsäcker, Thomas-Fermi.

1 Introduction

Density Functional Theory [1] is nowadays a most preferred method for electronic structure calculations. The Kohn-Sham (KS) method [2] provides an efficient way to perform *ab-initio* calculations in both extended and localized systems of electrons and nuclei. But the KS method requires the use of orbitals whereas the original Hohenberg-Kohn formulation gives the entire protagonism to the electron density. In the KS method the total energy of the system is divided into four terms: the non-interacting kinetic energy density functional (KEDF, $T_s[n]$), the energy due to the external potential ($V[n]$), the classical electrostatic energy or Hartree energy ($J[n]$) and the exchange-correlation energy ($E_{xc}[n]$):

$$E[n] = T_S[n] + V[n] + J[n] + E_{xc}[n].$$

Much of the effort has been focused on the development of exchange correlation density functionals. However the development of approximations to the kinetic energy based only on the density is enough interesting by itself: an efficient approximation to the so-called *orbital-free* kinetic energy density functional (OF-KEDF) will provide a better scaling with the system size than the Kohn-Sham method in electronic structure calculations.

The oldest kinetic energy density functional is even previous to the Hohenberg-Kohn formulation. The local density approximation, or Thomas-Fermi (TF) [3], functional gives the exact energy of the free electron gas. Another "classical" functional was developed by von Weizsäcker (vW) [4] to provide the exact energy for one or two particle systems in the fundamental state. When applied to other systems, that are far away from this two limits, both functionals provide poor values of the kinetic energy. Later on, a large amount of combinations of this two functionals were formulated. The GGA functionals (see studies in references cited in [5]) are modifications of the TF functional that include the gradient of the electron density through the *enhancement factor F*:

$$T_S^{GGA}[n] = \int d\mathbf{r} \; t_s^{TF}(n(\mathbf{r})) \; F\left(\frac{|\nabla n|}{n^{4/3}}\right).$$

The semilocal KEDF's, when applied over electron densities obtained from Kohn-Sham or Hartree-Fock calculations, give values of the kinetic energy that differ only about 1% from the exact results. But the same functionals, when applied in a variational scheme which minimizes the total energy, give density profiles that are physically wrong. In atoms no shell structure is obtained and in low dimensional systems, like electrons in an infinite well potential, no quantum oscillations are found.

In the semilocal KEDF's the functional only takes into account the electron density and its gradient, but more sophisticated functionals can be formulated. Non-local KEDF's take into account the electron density in the whole space to evaluate the contribution to the kinetic energy coming from a certain region of the space. There has been two main approaches for the development of nonlocal KEDF's. The first one is the so-called Weighted Density Approximation (WDA) [6]; in this approximation the nonlocal form of the functional allows to give the correct normalization of the exchange-correlation hole. The second one is the Average Density Approximation (ADA) [7,8,9] scheme, where the functional is constructed to reproduce the exact linear response function of the free electron gas (i.e. the Lindhard response function). Several different approaches has been developed within the ADA approximation. Nowadays the functionals included in the ADA scheme are considered the most sophisticated and accurate *orbital-free* KEDF's. This family of functionals, when working in a variational scheme, are able to produce some atom shell structure as well as quantum oscillations [10].

2   A Kinetic Energy Density Study of Semilocal Functionals

The kinetic energy density (KED) [11] is a function that gives the exact kinetic energy of a system when integrated over all the space. In the KS theory the exact KED distribution can be obtained using the KS orbitals ϕ_i. Usually two different definition of KED are used:

$$t_S^I(\mathbf{r}) = \frac{1}{2}\sum_{i=1}^{N} |\nabla \phi_i(\mathbf{r})|^2$$

$$t_S^{II}(\mathbf{r}) = -\frac{1}{2}\sum_{i=1}^{N} \phi_i^*(\mathbf{r})\nabla^2 \phi_i(\mathbf{r}).$$

These two definitions are useful because they have different advantages in different contexts. The first one is always positive in the whole space and the second one has the form of the kinetic energy operator in the KS equations. Both expressions only differ locally and are related to each other by a function that is proportional to the laplacian of the electron density:

$$\frac{1}{4}\nabla^2 n(\mathbf{r}) = t_S^I(\mathbf{r}) - t_S^{II}(\mathbf{r}).$$

On the other hand, the KED distribution is not unique; one can choose a specific definition but can add any function (e.g. the laplacian of the electron density) that integrates to zero in the whole space to obtain another valid distribution for the KED.

As every functional approximation is also an approximation to the KED, a study of the KED's generated by the functionals compared to the exact definitions will be useful to characterize the failure of the semilocal functionals. As the definitions of the KED are not unique a comparison of the approximate KED's with only one the infinite choices would not be adequate.

In this work we present an exhaustive study of the KED of the semilocal KEDF's, discussing for each case the approximated and exact KED's. In order to allow the comparison with many definitions we have defined the exact KED as

$$t_S^{exact}(\mathbf{r}) = t_S^l(\mathbf{r}) + a\nabla^2 n(\mathbf{r}),$$

where a is a parameter that can have any value. In this way, with a simple expression we have infinite choices of the KED that include the two typical for values of $a = 0$ and $a = -1/4$. Comparison has been made possible by defining a parameter σ able to describe the quality of the KED related with the semilocal functionals:

$$\sigma = \frac{\int d\mathbf{r} \left| t_S^{exact}(n; \mathbf{r}) - t_S^{func}(n; \mathbf{r}) \right|}{T_S[n]}.$$

This parameter is essentially the local difference between the exact and approximate KED's, integrated over the whole space, divided by the exact kinetic energy. This definition gives a measure of the error regardless of the size of the system. We choose the value of a that minimizes the parameter σ for a given functional. In this way we are testing every functionals in the best conditions for itself.

In Table I we summarize the performance of different functionals, showing the values of σ for several atoms and their averages. We note that corrections to the TF functional do not improve the values of sigma. One can interpret this result saying that the corrections of TF functional modify the local KED profile in non suitable way. In fact, they correct the TF functional only in terms of the total energy but they do not locally improve the approximate KED generated by TF functional.

Table I.- Values of σ for several atoms.

	He	Be	N	Ne	Avg.
TF	0.166	0.168	0.162	0.160	**0.164**
TF+$\frac{1}{9}$vW	0.187	0.190	0.187	0.186	**0.188**
TF+$\frac{1}{5}$vW	0.225	0.228	0.224	0.221	**0.225**
DK	0.218	0.208	0.200	0.194	**0.205**
vW	0.000	0.064	0.179	0.275	**0.130**

As the GGA corrections do not improve the KED of the TF functional, it should be a good idea to explore a new strategy in the development of KEDF's. We will propose to use the vW approach as the foundation of new nonlocal functionals.

3 Proposal of New Nonlocal Functionals based on the von Weizsäcker One

Nonlocal functionals usually have three terms. The first one is a local term that is the TF functional multiplied by a prefactor. The second one is always the full von Weizsäcker functional, which is indispensable to reproduce the correct limit of the linear response for large values of the momentum (in both WDA and ADA schemes). The third term is properly the nonlocal term, that allows the functionals to exactly reproduce certain physical properties. For the WDA this property is the normalization of the exchange-correlation hole and for the ADA is the Lindhard response function.

All the nonlocal terms of the functionals previously formulated are based on the modification of the TF functional, and can be usually written as:

$$T_{TF,nl}\left[n(\mathbf{r})\right] = \int d\mathbf{r} \int d\mathbf{r}'\, n^{\alpha}(\mathbf{r})\, n^{5/3-\alpha}(\mathbf{r}')\, \Omega\zeta\left((\mathbf{r},\mathbf{r}'),|\mathbf{r}-\mathbf{r}'|\right).$$

Instead of this TF based modification, we propose a new approach based on a modification of the von Weizsäcker functional, which includes $n^{1/2}$ and the gradient operator. We develope two different nonlocal vW functionals:

$$T^1_{vW,nl}[n] = \frac{1}{2}\int d\mathbf{r} \int d\mathbf{r}'\, \vec{\nabla}\varphi(\mathbf{r})\cdot\vec{\nabla}\varphi(\mathbf{r}')\,\Omega\left(\zeta(\mathbf{r},\mathbf{r}'),|\mathbf{r}-\mathbf{r}'|\right)$$

$$T^2_{vW,nl}[n] = -\frac{1}{2}\int d\mathbf{r} \int d\mathbf{r}'\, \varphi(\mathbf{r})\,\nabla^2\varphi(\mathbf{r}')\,\Omega\left(\zeta(\mathbf{r},\mathbf{r}'),|\mathbf{r}-\mathbf{r}'|\right),$$

where $\varphi(\mathbf{r}) = n^{1/2}(\mathbf{r})$, $\Omega\left(\zeta(\mathbf{r},\mathbf{r}'),|\mathbf{r}-\mathbf{r}'|\right)$ is a universal weight function constructed in the same way than the ADA scheme (i.e. to reproduce the linear response function of the free electron gas) and finally $\zeta(\mathbf{r},\mathbf{r}')$ is the two-body Fermi wavevector that is an average of the Fermi wave vectors in two different points of space. A parameter γ is introduced, in order to determine how the average is made.

The form of the universal weight function Ω is obtained from a differential equation. We want to point out that this differential equation is much simpler than in the case of the ADA nonlocal functionals.

We have applied these new functionals to some simple systems, electrons in an infinite potential well for $T^1_{vW,nl}[n]$ and noble gas atoms for $T^2_{vW,nl}[n]$. In the next tables we summarize the results we have obtained.

In Table II we present the relative errors in the total energy obtained for the $T^1_{vW,nl}[n]$ functional when applied to N electrons in a infinite potential well. As shown by the averages, the functionals perform better for systems far way from the one particle limit. Note that values for $\gamma = 1$ and $\gamma = 3/4$ yield the best results for the total kinetic energy calculations.

Table II.- Relative errors in the total kinetic energy of the potential well for $T^1_{vW,nl}[n]$.

N / γ	Geom	1	3/4	1/2	1/4	−1/4
2	0.180	0.046	0.066	0.101	0.145	0.205
4	0.074	0.009	0.018	0.032	0.054	0.088
6	0.044	0.002	0.007	0.016	0.030	0.053
8	0.030	0.002	0.003	0.010	0.020	0.037
10	0.023	0.003	0.000	0.006	0.014	0.028
12	0.018	0.004	0.001	0.004	0.011	0.023
14	0.015	0.004	0.001	0.003	0.009	0.019
16	0.012	0.004	0.002	0.002	0.007	0.016
18	0.011	0.004	0.002	0.002	0.006	0.014
20	0.009	0.004	0.002	0.001	0.005	0.012
Avg.	0.041	0.008	0.010	0.018	0.030	0.050

Table III shows the relative errors in total kinetic energy obtained with the $T^2_{vW,nl}[n]$ functional, and their averages, when applied to noble gas atoms. We have found small errors, and in the one particle limit the smallest ones are those obtained with the geometric average for the two-body Fermi wavevector.

verify the indirect to direct transition of Ge nanoparticles as their size decreases, showing that the HOMO-LUMO transition is both spin and symmetry allowed.

Acknowledgments

We thank the European Social Fund (ESF), Operational Program for Educational and Vocational Training II (EPEAEK II), and particularly the Program PYTHAGORAS, for funding the above work

References

[1] Canham L T 1990 *Appl. Phys. Lett.* **57** 1046

[2] Vasiliev I, Öğüt S and Chelikowsky J R 2001 *Phys. Rev Lett*. **86**, 1813

[3] Garoufalis C S, Zdetsis A D and Grimme S 2001 *Phys. Rev. Lett.* **87 276402**

[4] Wilcoxon J P, Provencio P P and Samara G A 2001 *Phys. Rev. B* **64** 035417

[5] Heath J R, Shiang J J and Alivisatos A P 1994 *J. Chem. Phys.* **101** 1607

[6] Palummo, Onida G and Del Sole R 1999 *Phys. Stat. sol.* **175** 23

[7] Weissker H, Furthmüller J and Bechstedt F 2002 *Phys. Rev. B* **65** 155328

[8] Melnikov D V and Chelikowsky J R 2003 *Solid State Commun.* **127** 361

[9] Casida M E, in *Recent Advances in Density Functional Methods,* edited by D. P. Chong (World Scientific, Singapore,1995).

[10] Stephens P J, Devlin F J, Chabalowski C F and Frisch M J 1994 *J. Phys. Chem.* **98**, 11 623

[11] Muscat J, Wander A and Harrison N M 2001 *Chem. Phys. Lett.* **342** 397

[12] TURBOMOLE (Version 5.3), Universitat Karlsruhe, 2000.

[13] Schäfer A, Horn H and Ahlrichs R 1992 J. Chem. Phys. 97 2571

[14] Zdetsis A D, Garoufalis C S and Grimme S 2003 *NATO Advanced Research Workshop on "Quantum Dots: Fundamentals, Applications, and Frontiers"*

[15] Zdetsis A D, Optical properties of small size semiconductor nanocrystals and nanoclusters, *Reviews on Advanced Materials Science (RAMS)* (2005)

Brill Academic Publishers
P.O. Box 9000, 2300 PA Leiden
The Netherlands

Lecture Series on Computer
and Computational Sciences
Volume 4, 2005, pp. 1473-1476

Electron Behavior in Si nanocrystals

X.Zianni[†]
Department of Applied Sciences,
Technological Institution of Chalkida,
34 400 Psachna,, Greece

and

A.G. Nassiopoulou
IMEL/NCSR 'Demokritos'
153 10 Aghia Paraskevi, Greece

Received 11 July, 2005; accepted in revised form 14 August, 2005

Abstract: We present a continuum model for the electron states in Si dots that allows for the effects of size, shape and crystallographic orientation of the dots. This formalism has been used to study the behavior of the photoluminescence (PL) lifetime in Si nanodots. Due to the anisotropy of Si band structure, confinement causes in dots a bunch of energy levels coming from the different valleys that although are close in energy they have very different recombination rates. It is concluded that dispersion in the magnitude of the PL lifetimes should be the case in Si dots samples.

Keywords: electron states, silicon nancrystals, effective mass approximation, photoluminescence lifetime

PACS: 78.67 Hct, 78.55.Mb, 73.21.Hb

1. Electron states

Experiment has shown that the optical properties of Si nanostructures are much different than those of bulk Si [1,2]. The optical properties are closely related with the electronic structure of these systems [3-6]. We consider free standing homogeneous dots and we use the Effective Mass approximation (EMA) to describe the electron states. The wavefunction is written as a product of the Bloch periodic function and an envelope function. The envelope function and the energy eigenvalues are determined by the solution of an effective mass equation, to which are applied the appropriate boundary conditions. This method has been described for electrons and for holes in Si quantum wires [7,8] and for Si quantum dots [9,10].

In bulk Si, electrons occupy at the minimum of the conduction band three pairs of equivalent valleys along the three main crystallographic directions. These anisotropic valleys are ellipsoids with two transverse masses, $m_t = 0.19\ m_e$, and a longitudinal mass, $m_l = 0.98\ m_e$. As it becomes evident below, the anisotropic character of the conduction band of bulk Si reduces to a rich electronic structure, in Si nanostructures, that is responsible for their distinct optical behavior compared to for example the III-IV semiconductors.

The distance of the center of the ellipsoids along the [001], [010] and [100] directions from the Γ-point is denoted by α. We denote these six valleys in the x, y and z axes as $[\pm\alpha,0,0]$, $[0,\pm\alpha,0]$ and $[0,0,\pm\alpha]$ respectively. We define the dot crystallographic direction in a system of coordinates (X,Y,Z) as follows: the Z axis is along the z direction and the X and Y axes are rotated anticlockwise by an angle θ relative to the x and y directions respectively. Electron eigenstates are obtained by solving Schröndinger's equation. A three dimensional infinitely deep confining potential, $V(X,Y,Z)$ is assumed and the electron envelope function, $\varphi_c(\mathbf{r})$, and the energy states are determined by solving the following effective mass equations:

[†] Corresponding author. E-mail: xzianni@teihal.gr

Brill Academic Publishers
P.O. Box 9000, 2300 PA Leiden
The Netherlands

*Lecture Series on Computer
and Computational Sciences*
Volume 4, 2005, pp. 1484-1487

Optical Properties of Oxygen Contaminated Si Nanocrystals

C. S. Garoufalis and A. D. Zdetsis[1]

Department of Physics,
University of Patras,
GR-26500 Patras, Greece

Received 15 July, 2005; accepted in revised form 10 August, 2005

Abstract: We report accurate high level calculations of the optical gap and absorption spectrum of small oxygen contaminated Si nanocrystals. Our calculations have been performed in the framework of time dependent density functional theory (TDDFT) using the hybrid nonlocal exchange and correlation functional of Becke and Lee, Yang and Parr (B3LYP). The accuracy of these calculations has been verified by high level multi-reference second order perturbation theory. For diameters smaller than 20 Å, the optical gap which is significantly lower than the corresponding gap of oxygen-free nanocrystals, is determined by oxygen induced states. We find that, for small amounts of oxygen, the size of the optical gap is practically insensitive to the exact number of oxygen atoms, but depends strongly on their relative distribution and bonding type.

Keywords: Si nanocrystals, optical properties, quantum dots,

PACS: 71.15.Mb, 71.35.Cc, 73.22.-f,78.67.Bf, 78.67.-n

1. Introduction

The optical properties of silicon nanocrystals, have been a very promising field of research over the last decade. A large number of experimental and theoretical approaches have been carried out in order to explore the properties and resolve the origins of the observed visible photoluminescence (PL). However, although most of the main issues have been resolved, for reasons which we have summarized elsewhere [6], there are still conflicting results in the literature. The central issue of dispute is the variation of the optical gap as a function of diameter, and in particular the critical range of diameters for the observation of visible PL. The majority of the earlier experimental work gives diverse results as for the size of the Si nanocrystals capable of emitting in the visible. The results of Wolkin et al [3] revealed optical gaps as small as 2.2 eV, for nanoclusters with a diameter of 18Å. For nanoclusters of about the same size Wilcoxon et al [2] obtained a similar result (2.5 eV) together with a much larger gap of about 3.2 eV for highly purified samples of the same dot diameter. Furthermore, Schupler et al [5] have estimated the critical diameter for visible PL to be less than 15 Å.

The main source of discrepancies in the experimental results is oxygen contamination, coupled, in many times, with difficulties in size determination. Recent studies about the role of surface oxygen on the optical properties of silicon nanoclusters report conflicting levels of importance, ranging from minimal to crucial. In many experiments, the presence of oxygen is considered as a means to effectively passivate the surface dangling bonds or as a means to reduce (through oxidation) the size of the silicon nanoparticles. In most of the cases oxygen is just a contaminant, which is very difficult to remove.

The sources of discrepancies in the theoretical results have been reviewed and summarized elsewhere [6]. The present calculations are based on time dependent density functional theory

---

[1] Corresponding author. Professor at the Physics Department of the University of Patras. E-mail: zdetsis@upatras.gr, zdetsis@physics.upatras.gr

(TDDFT)[7] employing the hybrid nonlocal exchange-correlation functional of Becke, Lee, Yang and Parr (B3LYP)[13], as well as the non-hybrid functional of Becke and Perdew BP86 [14], for comparison. We have verified the accuracy of the TDDFT/B3LYP calculations for the optical gap, which is better than 0.3 eV, by sophisticated multi-reference second order perturbation theory (MR-MP2 [8])

2. Outline of calculations

The oxygen "contaminated" nanocrystals have been prepared by a special substitution procedure of surface hydrogen atoms in order to maintain the Td symmetry of the initial oxygen-free nanocrystals. The number of the surface oxygen atoms is dictated by the need to maintain the symmetry. Thus, we have considered here only the $Si_{17}H_{12}O_{12}$, $Si_{29}H_{12}O_{12}$, $Si_{35}H_{24}O_6$, $Si_{47}H_{48}O_6$, $Si_{71}H_{72}O_6$, $Si_{99}H_{76}O_{12}$ and $Si_{147}H_{52}O_{24}$ "doubly bonded" nanocrystals. However, we also constructed two additional nanoparticles ($Si_{47}H_{36}O_{12}$, $Si_{71}H_{60}O_{12}$), for which the surface oxygen atoms instead of being doubly bonded, they form a bridge between two Si atoms. For the first one ($Si_{47}H_{36}O_{12}$) we also constructed a variant ($Si_{47}H_{24}O_{18}$) which contains both bridging bonds (Si-O-Si) and double bonds (Si=O).

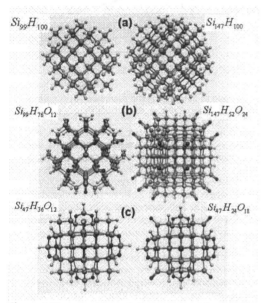

The DFT and the TDDFT calculations were performed with the TURBOMOLE [9] suite of programs using Gaussian atomic orbital basis sets of split valence [SV(P)]: [4s3p1d]/2s [10] quality. The TDDFT calculations have been performed as described in detail in Ref. [5,12] using the B3LYP functional consistently for both, the self-consistent solution of the Kohn-Sham equation for the ground state, and the solution of the linear response problem. As we have shown shown by comparison to multi-reference second-order perturbation theory MR-MP [8] calculations, the partially exact Hartree-Fock (HF) exchange that is included in the B3LYP method is crucial for the correct description of the optical properties. This has been verified by performing additional TDDFT calculations using the well known functional of Becke and Perdew (BP86) [14], which does not include exact (or partially exact) exchange. The inclusion of exact HF exchange remedies the well-known deficiency of local-density approximation (LDA) to underestimate the band gap. As a result, the TDDFT/B3LYP method has an estimated accuracy of about 0.3 eV for the excitation energies as is demonstrated in tables I and II in Ref. [5] and in agreement with earlier experience [11, 12].

Figure 1.The structure of the nanocrystals: (a) The oxygen-free $Si_{99}H_{100}$ and $Si_{147}H_{100}$ (b) the corresponding (doubly bonded) Si=O oxygen-rich nanocrystals and (c) the $Si_{47}H_{36}O_{12}$ and $Si_{47}H_{24}O_{18}$ nanocrystals. $Si_{47}H_{36}O_{12}$ includes only bridging oxygen atoms whereas $Si_{47}H_{24}O_{18}$ includes both bridging and doubly bonded oxygen atoms.

3. Results and discussion

In figure 2 we present our TDDFT/B3LYP calculations for the optical gap of both oxygen-free and doubly-bonded oxygen "contaminated" Si nanocrystals. We have plotted the optical gap as a function of the nanocrystal diameter and compare the present oxygen-free [5] and oxygen-rich results with analogous experimental results of Wilcoxon et al [2] (inset in fig. 2b). The shaded area has been designated by Wilcoxon et al as the region of oxygen contaminated samples. For diameters 10Å < d < 20Å the average calculated optical gap for oxygenated nanoparticles ranges from ≈2.5 eV to ≈1.85 eV, while the experimental data range from ≈ 2.2 eV to ≈ 1.7 eV. As we can see, there is an excellent agreement between experiment and theory for both oxygen-free and oxygen-rich samples (the latter fall

Brill Academic Publishers
P.O. Box 9000, 2300 PA Leiden
The Netherlands

*Lecture Series on Computer
and Computational Sciences*
Volume 4, 2005, pp. 1488-1491

Optical and Electronic Properties of Mixed SiGe:H Nanocrystals

A.D. Zdetsis[1], C.S. Garoufalis and E.N. Koukaras
Department of Physics
University of Patras
26500 Patras, Greece

Received 14 July, 2005; accepted in revised form 11 August, 2005

Abstract: The optical and electronic properties of hydrogenated mixed semiconductor (Si, Ge) nanocrystals are calculated. All calculations are performed in the framework of the Desity Functional Theory (DFT) and Time-Dependent Density Functional Theory (TDDFT), using the hybrid non-local exchange-correlation functional of Becke, Lee, Parr and Yang (B3LYP). Our results show that by proper adjustment of the relative concentrations we can obtain nanocrystals with the desired electronic and optical properties.

Keywords: Nanocrystals, Nanoparticles, Optical Gap, Density Functional Theory

PACS: 73.22.-f, 61.46.+w, 71.15.Pd, 78.67.-n

1. Introduction

The possibility of tunable photoluminescence from Si and Si-like nanocrystals as well as from porous silicon (composed of Si nanocrystals) has attracted the interest and research on this type of materials over the last few years[1-7]. A large portion of this work has been devoted to understanding the visible photoluminescence of these materials and its dependence on the diameter of the nanoparticles (or equivalently, the porosity of p-Si). It is widely accepted and well established by now [4-6] that the luminescence of oxygen-free Si nanocrystals (of well defined diameter), is mainly due to quantum confinement of the corresponding nanoparticles. This is also true for Ge nanoparticles [7].

By varying the diameters of the nanocrystals (or equivalently the porosity of porous silicon) intense PL can be obtained across the visible spectrum, which could never have taken place for bulk crystalline silicon with a band gap of 1.2 eV. Quantum confinement is responsible for the opening of the gap from the bulk value of 1.2 eV to values of 2-3 eV (for larger nanocrystals) up to 6-7 eV for smaller nanocrystals [4-5]. These large values of the gap for small size nanocrystals obviously is equally undesirable as the shrinking of the gap (in bulk Si).

It is anticipated that at small sizes of nanocrystals would not be so strong for Ge nanocrystals [7] because of the smaller band gap of bulk Ge, compared to Si. Even in this case, the possibilities of adjusting the optical gap (and the band gap) are limited only to proper size selection. The possibility of combining the advantages of Si (not only in the optical, but also in the electronic and technologically useful properties) with those of Ge is an intriguing and very promising project. Needless to say, that, not only the minimum critical diameter of the nanocrystals for visible PL is important, but also the maximum possible diameter.

With this in mind, we have examined the optical and electronic properties of mixed nanocrystals of the form $Si_xGe_y:H_z$, where all variables x, y and z have been varied within the symmetry restrictions and the current limitations of our computational system. It should be noted that up to date no such high level calculations exist for such systems. It should be mentioned that even for the optical properties of pure Ge nanocrystals, besides our recent work [7], only semiempirical methods, or local density (LDA) ground state calculations are known in the literature (without gradient corrections) [7].

2. Some technical details of the calculations

[1] Corresponding author. Professor at the Physics Department of the University of Patras. E-mail: zdetsis@upatras.gr, zdetsis@physics.upatras.gr

All calculations in this work are based on Time Dependent Density Functional Theory (TDDFT) employing the nonlocal exchange-correlation functional of Becke, Lee, Yang and Parr (B3LYP) [4-7]. The accuracy of these calculations (TDDFT/B3LYP) for the optical gap has been tested before by high level multireference second-order perturbation theory (MR-MP2) for the case of Si nanocrystals, with excellent results [4].

The size of the nanocrystals considered here ranges from 5 to about 20 Å. This corresponds to values of x and y from 5 to 99 Si or Ge atoms and to values of z between 12 and 100 H atoms (a total of about 199 atoms). The symmetry of the nanocrystals is Td and their geometries have been fully optimized within this symmetry constrain using the hybrid B3LYP functional. Representative geometries of Si_xGe_y:H_z nanocrystals for x+y=71 are shown in figure 1. The Ge atoms are placed in sites corresponding to successive layers such that the Td symmetry is maintained. The bulk of our calculations were performed with the TURBOMOLE [6] suite of programs using Gaussian atomic orbital basis sets of split valence [SV(P)]: [4s3p1d]/[2s] quality [9].

3. Results and discussion

In figure 2 we display the fundamental optical gap for Si and Ge nanoparticles of diameters up to around 20 Å. As we can see in these figures there is an inverse correlation of the optical gap with the size of the nanoparticle, as is well known [1] (the same trend is followed by the HOMO-LUMO gap, not shown here). Also shown in figure 2 are the optical gaps of Si_xGe_y:H_z nanocrystals for a given diameter (and thus, for a given total number of atoms x+y). We see that the values for the optical gap of these nanocrystals are intermediate of the pure Si and Ge ones. For this figure we have chosen $x \cong y$. For nanoclusters of fixed size (x+y=constant), the values of the optical gap vary up to 0.4 eV, depending on the consentration (relative x,y values) of Ge atoms in the nanocrystal. This enables adjustment of the optical gap for a given size nanocrystal.

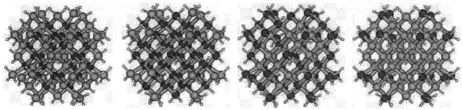

Figure 1. A succession of Ge containing layer depositions corresponding to the Si_xGe_y:H_z nanocrystal with x+y=71. The Td symmetry is maintained.

Calculations of the HOMO-LUMO gap of mixed Si_xGe_y:H_z nanocrystals have also been recently performed by Yu et al. [10], specifically for nanocrystals with a total number of Si and Ge atoms of 71 (x+y=71). These calculations are based on density-functional theory (DFT) in the local-density approximation (LDA). The resulting HL gaps range from 3.3 – 4.1 eV corresponding to the pure Ge and pure Si nanocrystals, while our results give 3.8 eV for the pure Ge nanocrystal and 4.2 eV for the pure Si nanocrystal. The differences are attributed mainly to the use of LDA approximation which is known [4,11] to underestimate both the HOMO-LUMO and the optical gap. The B3LYP functional used in our calculations includes a partially exact Hartree-Fock (HF) exchange which is very important for the correct description of the optical properties[4].

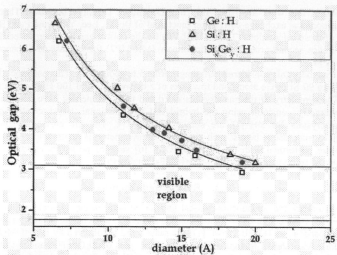

Figure 2. Comparison of the optical gap of Si_xGe_y:H (solid circles) for approximately equal concentrations of Si and Ge atoms ($x \cong y$).

Finally, to facilitate the comparison of the electronic structure ("band structure") of the nanocrystals, for different concentrations of Ge atoms, we have plotted in figure 3, the electronic density of states (DOS) for three typical nanocrystals (one Si-rich, one Ge-rich and one of about equal concentrations of Si and Ge). The DOS curves were generated from the eigenstates of the ground state calculations with a suitable gaussian broadening. As we can see, the largest variation with the Ge concentration occurs in the valence band edges, while the conduction band edge is relatively insensitive.

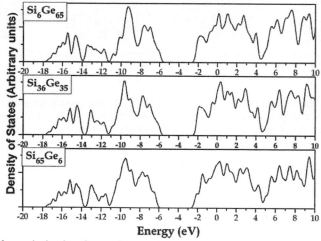

Figure 3. The electronic density of states for the $Si_{65}Ge_6$:H, $Si_{36}Ge_{35}$:H, and Si_6Ge_{65}:H nanocrystals.

4. Conclusions

We have shown that, indeed, the mixed SiGe nanocrystals have optical and electronic properties intermediate between those of pure Si and Ge nanocrystals. The large variety of optical and band gaps depends, not only on the relative concentrations, but also on the relative spatial distribution of the Ge atoms with respect to the surface of the nanocrystals. This work is in progress.

Acknowledgments

We thank the University of Patras Research Committee and particularly the basic research program "K. KARATHEODORI 2003" for funding the above work.

References

[1]   Canham L T, *Appl. Phys. Lett.* **57** 1046 (1990)

[2]   Wilcoxon J P, Provencio P P and Samara G A, *Phys. Rev. B* **64** 035417 (2001)

[3]   Weissker H Ch, Furthmuller J and Bechstedt F, *Phys. Rev. B* **65** 155328 (2002)

[4]   Garoufalis C S, Zdetsis A D and Grimme S, *Phys. Rev. Lett.* **87** 276402 (2001)

[5]   Zdetsis A D, Garoufalis C S and Grimme S, proceedings of *NATO Advanced Research Workshop on "Quantum Dots: Fundamentals, Applications, and Frontiers"* *Crete, Greece*, 317-332, Kluwer-Springer, 2005, and references therein

[6]   Garoufalis C S, Zdetsis A D, *J. Phys. Conf. Ser.* **10** 69 (2005)

[7]   Garoufalis C S, Skaperda M S and Zdetsis A D, *J. Phys. Conf. Ser.* **10** 97 (2005)

[8]   TURBOMOLE (Version 5.6), Universitat Karlsruhe, 2002

[9]   Schäfer A, Horn H and Ahlrichs R, *J. Chem. Phys.* **97** 2571 (1992)

[10]   Yu M, Jayanthi C S, Drabold D A and Wu S Y, *Phys. Rev. B* **68** 035404 (2003)

[11]   A. D. Zdetsis, Optical properties of small size semiconductor nanocrystals and nanoclusters, *Reviews on Advanced Materials Science (RAMS)* (2005)

Brill Academic Publishers
P.O. Box 9000, 2300 PA Leiden
The Netherlands

Lecture Series on Computer
and Computational Sciences
Volume 4, 2005, pp. 1492-1495

Structural and Electronic Properties of the Ni@Si$_{12}$ Nanocluster.

E.N. Koukaras, C.S. Garoufalis and A.D. Zdetsis[1]

Department of Physics
University of Patras
26500 Patras, Greece

Received 15 July, 2005; accepted in revised form 8 August, 2005

Abstract: The structural and electronic properties of the Ni@Si$_{12}$ clusters are studied in the framework of the density functional theory using the hybrid functional B3LYP. A distorted hexagonal structure of Cs symmetry and a structure of S4 symmetry are identified as the best candidates for the ground state of Ni@Si$_{12}$. The structures resulting by substitution of a Si atom in low lying Si$_{13}$ isomers or by insertion of a Ni atom in an energetically low lying Si$_{12}$ isomers, after a geometry optimization, are not energetically the lowest ones. We report energetic properties such as binding energies, ionization potentials, electron affinities and embedding energies.

Keywords: Metal encapsulated clusters; Nanocluster; Silicon; Transition metal; Ab initio calculations

PACS: 61.46.+w, 73.22.-f, 71.15.Nc, 36.40.Cg, 31.15.Ew, 82.30.Nr

1. Introduction

In the past 20 years there has been a considerable amount of research, both theoretical and experimental, on silicon based materials. The intense interest on TM encapsulated Si clusters is initiated by both the promising applications in nanotechnology and nanoscale electronics as well as its pure scientific significance. For example, advancements in the electronics industry have led to silicon based integrated circuits with elements (MOSFETS) as small as a few nanometers. Furthermore, stable silicon structures may serve as building blocks for the construction of novel nanoscale silicon based materials with a tunable band gap depending on the encapsulated TMA.

Within this perspective, a subject of great concern is the chemistry that occurs at the interface between a metal and a silicon surface. An example of such a process is the introduction of energy states, that lie between the valence and conduction band energy in semiconductors, from the metal-semiconductor reaction products. This type of energy states impede our ability to control Schottky barrier heights. The study of small and medium size silicon clusters reveals information related to the reaction at the interface. It has been speculated [1] that metal-containing stable silicon clusters may represent the earliest products formed at the interface.

Experimentally, the production and study of metal atom-silicon clusters was first reported by Beck[1]. In this work Beck et. al. observed the formation of metal-containing silicon clusters for three types of transition metals atoms (TMA), specifically Tungsten (W), Molybdenum (Mo) and Chromium (Cr). The produced metal-containing silicon clusters exhibited increased stability compared to similar sized pure silicon clusters when subjected to photofragmentation. Hiura et al. report[2] that TMAs react with silane (SiH$_4$) producing Si clusters with an encapsulated TM atom. The TMA, considered to be endohedral, stabilizes the Si cage. The TMAs that react with SiH$_4$ to form stable cages have a partially filled d shell with *d* electrons ≥ 2 in the ground state. The TM@Si$_n$ clusters that were formed lost their reactivity to SiH$_4$ when n reached 12 suggesting that TM@Si$_{12}$ constitute stable clusters.

From a theoretical point of view, the structural stability and electronic properties of TM@Si$_n$ clusters have been heavily investigated the last few years with a large variety of theoretical methods. The difficulties entailed in such attempts can be readily appreciated when one considers that even for

[1] Corresponding author. Professor at the Physics Department of the University of Patras. E-mail: zdetsis@upatras.gr, zdetsis@physics.upatras.gr

small pure Si clusters (which have been under investigation for many years) there are still numerous fundamental issues to be resolved [8]. The incorporation of TMAs, such as Ni, introduce partially filled d-electrons (and possible different spin states) that makes the situation even more complex.

Up to date, there is little research done concerning specifically the Ni@Si$_{12}$ cluster. Menon et al.[3] have performed a series of TBMD as well as DFT calculations in an effort to identify the energetically more stable isomer of Ni@Si$_{12}$. The results of both theoretical approximations suggest that the lowest energy Ni-encapsulated Si structures is a cage of C5v symmetry as the one shown in figure 1(h). Thus far, this conclusion has only been questioned by Kumar et al. [4] who suggest that the structure of C5v symmetry of Menon et al. is the third lowest in energy, while a hexagonal prism and a chair structure are energetically more favorable than the former. Moreover, based on calculations on a double C5v structure they claim that icosahedral packing is not favorable for Ni doping.

In this work we examine the geometric structure and electronic properties of TM encapsulated Si clusters, with a particular emphasis on the Ni@Si$_{12}$ cluster, which we present here.

2. Calculations

Several different approaches were adopted for the construction of the initial candidate geometries. We started by considering ideal fcc and hcp cells, with up to second order neighbors from a center Ni atom, as well as two additional structures of hexagonal and icosahedral symmetry. In all cases we performed symmetry unconstrained (C1) geometry optimizations, using the gradient corrected BP86[] functional with the SVP basis set [4s3p1d] for Si and [5s3p2d] for Ni. In this stage of the calculations the resolution of the identity (RI)[12] approximation for the two-electron integrals was consistently employed. In order to characterize the resulting structures, we defined their symmetry using loose symmetry criteria. The final symmetric species were then reoptimized in the framework of the hybrid three parameter, non-local correlation functional Becke-Lee, Parr and Yang (B3LYP) with the large triple-zeta quality split valence basis set, TZVP[5s4p1d] for Si and [6s4p3d] for Ni. Based upon prior experience[5], the B3LYP functional gives exceptionally accurate results for both electronic[5] and structural[8] properties for Si. Moreover, for the case of silicon clusters[8], it has been shown that the quality of B3LYP results is comparable to more sophisticated and computationally demanding methods, such as CCSD(T). For the case of TMA containing structures, the B3LYP functional (but even for other hybrid functionals) yields accurate binding energies, but rather short bond lengths[13]. Specifically for Ni, the bond lengths are rather elongated.

The geometric and electronic configuration of the clusters was determined by additional stability calculations, by performing calculations on different spin states (singlet, triplet, etc.), by employing occupation number optimization procedures using (pseudo-Fermi) thermal smearing and by comparing to other high level methods when it was considered necessary.

The bulk of our calculations were performed using the Turbomole[6] program.

Table 1: Energetic properties of the Ni@Si$_{12}$ isomers, Binding Energies (per atom) and Embedding Energies (see text for difference in definitions). The isomers are characterized by their symmetry group.

Sym	Spin State	BE / atom (eV)	EE (eV)	EE2 (eV)
Cs	s	3.170	5.435	3.3244
S4	s	3.166	6.017	3.2828
basket	s	3.131	5.432	2.8183
C2h	s	3.129	5.443	2.8015
C2v	s	3.109	5.687	2.5405
C2~	s	3.034	5.106	1.5637
C5v (g)	s	3.034	5.132	1.5596
C5v (h)	s	3.011	4.486	1.2639
C2v	t	3.117	5.443[†]	2.6371

[†]This value corresponds to the pure Si cluster being in a triplet occupation.

3. Results and discussion

Our calculations reveal two distinct, nearly isoenergetic, low-energy structures for the Ni@Si$_{12}$ cluster of Cs and S4 symmetry. The Cs structure, shown in figure 1(a), is proposed by Kumar et al.[4] as the lowest energy structure for the Ni@Si$_{12}$ cluster. The S4 structure is not previously mentioned, to our knowledge, in any bibliography as a stable isomer of the Ni@Si$_{12}$ cluster. The Cs isomer can be regarded as a hexagonal prism where a single Si atom has elongated bond lengths. The S4 isomer, shown in figures 1(b) and 1(b'), can be regarded as four distorted neighboring pentagons. Figures 1(b) and 1(b') correspond to a relative 90 degrees rotation.

The S4 isomer was obtained using as an initial geometry an hcp structure with a central Ni atom and up to second order neighboring Si atoms. A subsequent frequency calculation revealed no imaginary values. In the case of the Cs isomer the structure used as the initial geometry was a hexagonal prism of tight C2h symmetry, as shown in figure 1(d). This (hexagonal) structure in itself is significant as it exhibits remarkable stability regardless of the TMA embedded[7]. As can be seen in Table 1, this hexagonal prism has a HL gap of 1.02 eV (using the TZVP basis set) which is in complete agreement with the value given by Sen et al.[7] (1.03 eV). However, a frequency calculation revealed three imaginary values. By performing a continuous cyclic procedure of distorting the structure in accordance to the imaginary frequencies, relaxing the new structure and reevaluating frequencies, in every case the final structure obtained was the distorted hexagonal prism (of Cs symmetry).

The procedure mentioned above that leads to the Cs structure, via a number of intermediate structures, suggests a flat potential energy surface (PES), and indicates that the stability of the structure may be a dynamic[8,14] process rather than a static one. The differences in energy between the intermediate structures are less than 0.38 eV.

Another isomer obtained from the initial hexagonal geometry is shown in figure 1(i) of C2v symmetry. The difference in this case is the assignment of a triplet configuration. As can be seen in table 1, this is an energetically low structure. This structure is comprised of two distorted pentagons (not shown on figure 1(i) as they are on the left and right of the structure), neighboring with six distorted rhombuses, and caped by a single silicon atom. By omitting this single silicon atom, the remaining structure is non-convex.

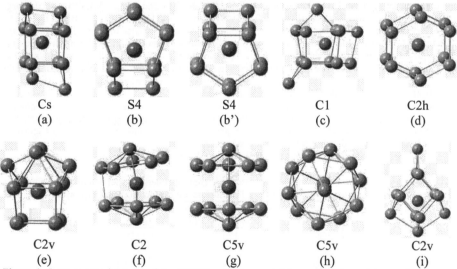

Figure 1: Low energy isomers of the Ni@Si$_{12}$ cluster, sorted by spin state and increasing energy. Structures (a),(b) are isoenergetic and correspond to the lowest energy isomers. Figures (b) and (b') correspond to the same isomer rotated by 90 degrees. Ni-Si bonding has been omitted wherever necessary for clarity.

The ability of a transition metal to stabilize a Si cluster of a specified size can be partially quantified by the Embending Energy (EE). The EE is defined by two different ways in modern bibliography, given by the following relations:

$$EE = \left[E(\text{Si}_n)_u + E(\text{TM}) \right] - E(\text{TM@Si}_n) \tag{1}$$

and

$$EE2 = \left[E(\text{Si}_n)_{lowest} + E(\text{TM}) \right] - E(\text{TM@Si}_n) \tag{2}$$

In both relations the Embedding Energy is given by the difference in energy of the transition-metal embedded silicon cluster from the energy of the single TMA and of the pure Si$_n$ cluster. The difference in the two definitions is in regard to the Si$_n$ structure. In one case the coordinates of the Si atoms in Si$_n$ are that of the TM@Si$_n$ cluster (i.e. simply remove the TM), whereas in the other case the Si$_n$ is the lowest-energy isomer of the Si$_n$ cluster[11]. Our calculations correspond to both definition, the former being EE and the later EE2. The former definition gives an insight on the stabilizing effects of the TMA for the specific structure and thus is true to its designation. An advantage of the later definition is that the lowest-energy (Si$_n$)$_{lowest}$ cluster acts as a common reference, thus providing a means of comparing the stabilization effect of the TM atom in each TM@Si$_n$ isomer. We present calculations based on both definitions as both of these quantities are found in current literature.

In addition to the properties listed in table 1, we have also calculated other electronic, energetic and structural properties. Our results show that the structures are not suited for a charge-transfer-type acceptor to other substances with the exception perhaps of structure C5v (h)[10].
This work is currently in progress.

Acknowledgments

We thank the University of Patras / Research Committee and particularly the basic research program "K. KARATHEODORI 2003" and the European Social Fund (ESF), Operational Program for Educational and Vocational Training II (EPEAEK II), and particularly the program PYTHAGORAS, for funding the above work.

References

[1] S. M. Beck, *J. Chem. Phys.*, **90**, 6306 (1989); S. M. Beck, *J. Chem. Phys.*, **87**, 4233 (1987).

[2] H. Hiura, T. Miyazaki and T. Kanayama, *Phys. Rev. Lett.*, **86**, 1733 (2001).

[3] M. Menon, A. Andriotis, G. Froudakis, *Nano Lett.*, **2**, 301 (2002); A. Andriotis, G. Mpourmpakis, G. Froudakis and M. Menon, *New J. Phys.*, **4**, 78 (2002).

[4] A. K. Singu, T. M. Briere, V. Kumar and Y. Kawazoe, *Phys. Rev. Lett.*, **91**, 146802 (2003).

[5] C. S. Garoufalis, A. D. Zdetsis and S. Grimme, *Phys. Rev. Lett.*, **87**, 276402 (2001).

[6] TURBOMOLE (Version 5.6), *Universitat Karlsruhe*, 2000.

[7] P. Sen, L. Mitas, *Phys. Rev. B*, **68**, 155404 (2003).

[8] A. D. Zdetsis, *Phys. Rev. A*, **64**, 023202 (2001).

[9] R.G. Parr and W. Yang, *Density-Functional Theory of Atoms and Molecules*, Oxford University Press, New York, 1989.

[10] T. Miyazaki, H. Hiura and T. Kanayama, *Eur. Phys. J. D*, **24**, 241 (2003).

[11] C. Xiao, F. Hagelberg, *Phys. Rev. B*, **66**, 075425 (2002).

[12] K. Eichkorn, O. Treutler, H. Öhm, M. Häser, R. Ahlrichs, *Chem. Phys. Lett.*, **240**, 283 (1995).

[13] K. Yanagisawa, T. Tsuneda and K. Hirao, *J. Chem. Phys.*, **112**, 545 (2000).

[14] A. Zdetsis, to appear in *Reviews on Advanced Materials Science (RAMS)* (2005)

Brill Academic Publishers
P.O. Box 9000, 2300 PA Leiden
The Netherlands

*Lecture Series on Computer
and Computational Sciences*
Volume 4, 2005, pp. 1496-1497

Ab initio MRD-CI Investigation of the Electronic Spectrum of Glyoxal (CHO)₂

Margret Gruber-Stadler[1,2], Max Mühlhäuser[1], Claus J. Nielsen[2]

[1]Studiengang Verfahrens- und Umwelttechnik, MCI – Management Center Innsbruck Internationale
Fachhochschulgesellschaft mbH, Egger-Lienz-Straße 120, A-6020 Innsbruck, Austria
[2]Department of Chemistry, University of Oslo, Blindern, N-0315 Oslo, Norway

Received 5 August, 2005; accepted in revised form 11 August, 2005

Abstract: High level *ab initio* calculations of the electronic spectrum of Glyoxal have been performed. The results are in excellent agreement with experimental.

Keywords: Ab Initio calculations, oscillator strengths

PACS: 31.15.Ar, 32.70.Cs

Global change depends critically on the atmospheric cycling of Methane. Prominent species in these processes are Formaldehyde CH_2O and Glyoxal $(CHO)_2$. While Formaldehyde has been the subject of numerous studies, much less is known about the electronic spectrum of Glyoxal. Consequently we employed multi-reference configuration interaction (MRD-CI) calculations to compute the electronic spectrum of Glyoxal. The calculations are very sensitive to the inclusion of s-, p- and d-Rydberg functions into the AO basis.

The first dipole-allowed transition $(1^1A_u \leftarrow X^1A_g)$ is calculated at 2.80 eV [1] in excellent agreement to experimentally obtained values of 2.72 eV [2,3] and 2.97 eV [4]. Its oscillator strength is computed with f = 0.0002. A much stronger transition (f = 0.38) is obtained at 8.51 eV $(3^1B_u \leftarrow X^1A_g)$. This 3^1B_u state corresponds to the transition $1b_g \rightarrow 2a_u$, which can be characterized as $\pi(CO) \rightarrow \pi^*(CO)$ and $\pi^*(CC) \rightarrow \pi(CC)$ type.

Due to the limited energetic range (below 5 eV) of the experimental investigations [2-4] carried out up to now, this excitation has not been reported in literature yet. On the other hand it is very likely that this transition is of importance for the photochemistry of atmospheric Glyoxal due to its large transition probability. Thus the present *ab initio* multi-reference MRD-CI study provides values that can be an appropriate guideline for future experimental measurements in the energetic region around 8.5 eV to provide new insight into the photolysis of Glyoxal under atmospheric conditions.

Table 1: Calculated electronic transition energies ΔE (eV) and oscillator strengths f of Glyoxal.

State	Excitation	ΔE (singlet) [eV]	f	ΔE (triplet) [eV]
X^1A_g	$(5a_g)^2(1a_u)^2(4b_u)^2(1b_g)^2$	0.0	–	–
$1\,A_u$	$5a_g \rightarrow 2a_u$	2.80	0.0002	2.37
$3\,B_u$	$1b_g \rightarrow 2a_u$	8.51	0.38	5.26

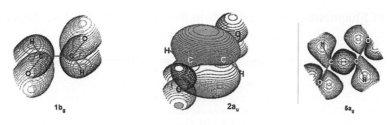

Figure 1: Characteristic orbitals (HOMO $5a_g$, LUMO $2a_u$ and virtual $1b_g$) of Glyoxal $(CHO)_2$.

References

[1] GRUBER-STADLER, M., MÜHLHÄUSER, M. and NIELSEN, C. J., *J. Mol. Phys. submitted* (2005).

[2] HOROWITZ, A., MELLER, R., and MOORTGAT, G. K., *J. Photochem. Photobiol. A*, **146** 19 (2001).

[3] VOLKAMER, R., SPIETZ, P., BURROWS, J., and PLATT, U., *J. Photochem. Photobiol. A*, in press (2005).

[4] ZHU, L., KELLIS, D., and DING, C.-F., *Chem. Phys. Lett.*, **257** 487 (1996).

Brill Academic Publishers
P.O. Box 9000, 2300 PA Leiden
The Netherlands

*Lecture Series on Computer
and Computational Sciences*
Volume 4, 2005, pp. 1498-1501

Effects of Fluctuating Electric Fields in the Dynamics of a Two-Electron Quantum Dot Molecule

Andreas F. Terzis[a,1], Antonios Fountoulakis[a] and Emmanuel Paspalakis [b]

[a]Physics Department,
School of Natural Sciences,
University of Patras,
Patras 265 04, Greece

[b]Materials Science Department,
School of Natural Sciences,
University of Patras,
Patras 265 04, Greece

Received 3 August, 2005; accepted in revised form 12 August, 2005

Abstract: We study the dynamics of two interacting electrons confined in a symmetric double quantum dot structure under the influence of time-dependent external electric fields. In particular, we investigate the case of an AC-DC electric field accompanied by a fluctuating electric field. By numerical solutions of the time-dependent Schrödinger equation for a realistic GaAs quantum dot structure it is shown that the fluctuating field, depending on its strength, can have a significant influence on the electrons localization.

Keywords: two-electron quantum dot structure, localization, AC-DC field, fluctuating field

PACS: 73.40.Gk, 73.63.Nm, 73.23.Ad, 73.20.Fz

1. Introduction

In the recent years the quantum mechanical control of the dynamics of two-electron quantum dot structures such as the two-electron quantum dot molecule [1-10] has been the subject of several investigations. Most of these studies concentrated on the localization effect and on the generation of entangled states of two electrons. In particular, in the two-electron quantum dot molecule it has been shown that localization can be achieved by applying an AC electric field [1,2,4-6] or by applying an AC-DC driving field [3,7,8] for specific values of the strength and the frequency of the applied fields. The localization conditions involve ratios of the Rabi frequency of the field-matter interaction to the angular frequency of the external driving field corresponding to zeroes of ordinary Bessel function of order determined by the DC field.

In the present work we study the dynamic behavior of two electrons in a symmetric double quantum dot structure under the influence of an external AC-DC electric field and a fluctuating electric field and give particular emphasis to effect of the fluctuating electric field in the dynamics of the system and on the localization.

2. Theoretical model for the two-electron quantum dot molecule

First, we derive the parameters for the two-electron quantum dot molecule. The Hamiltonian of the two interacting electrons in a coupled quantum dot driven by an external electric field can be written as

$$H(t) = \sum_{i=1,2} h(\vec{r}_i, \vec{p}_i, t) + \frac{e^2}{\varepsilon \mid \vec{r}_1 - \vec{r}_2 \mid}, \tag{1}$$

---

[1] E-mail: terzis@physics.upatras.gr, Corresponding Author

where $h(\vec{r}, \vec{p}, t)$ represent the single electron Hamiltonian and the last term expresses the Coulomb repulsion, with ε being the static dielectric constant of the quantum dot structure. The Hamiltonian $h(\vec{r}, \vec{p}, t)$ describing one electron confined in a double dot structure is

$$h(\vec{r}, \vec{p}, t) = \frac{p^2}{2m} - ezE(t) + V_l(\vec{r}) + V_v(\vec{r}).$$ (2)

The first term describes the kinetic energy of the electron with m being the effective mass of the electron. The second term describes the dipolar interaction when an external time-dependent electric field is applied to the structure. Finally, the confinement potentials (last two terms) are described in harmonic oscillator terms [4]

$$V_l(\vec{r}) = V_l(x, y) = \frac{m\omega^2}{2} C^2(x^2 + y^2)$$
$$V_v(\vec{r}) = V_v(z) = \frac{m\omega^2}{8a^2}(z^2 - a^2)^2$$ (3)

where C determines the strength of the vertical confinement ($V_v(\vec{r})$) relative to the lateral confinement($V_l(\vec{r})$). In this representation the two quantum dots are centered at $\pm a$, i.e. the distance between the quantum dots is $2a$.

In order to describe the ground state of our system we apply the Hund-Mulliken method known from molecular physics. We define the states with double occupation, $|\Psi_\pm^d\rangle$, where both electron belonging to the same quantum dot and the states with single occupation, $|\Psi_\pm^s\rangle$, where each electron belongs to different quantum dots. Using both spatial and spin states we generate six basis vectors with respect to which we diagonalize the two-particle Hamiltonian H (Equation (1)). The basis vectors are constructed such that to ensure the anti-symmetry of the total wave function. We ignore the spin-triplet part of the Hamiltonian, as the number of the electrons on each quantum dots is always one, independently on the applied electric fields.

Then, in the spin-singlet part, the time evolution of any initial state $|\Psi(0)\rangle$ can be expressed as

$$|\Psi(t)\rangle = c_1|\Psi_-^d\rangle + c_2|\Psi_+^s\rangle + c_3|\Psi_+^d\rangle,$$ (4)

where $c_1(t)$, $c_2(t)$, $c_3(t)$ are the corresponding probability amplitudes. The probability amplitudes are determined by the following time-dependent Schrödinger equation

$$i\frac{d}{dt}\begin{bmatrix} c_1 \\ c_2 \\ c_3 \end{bmatrix} = \begin{bmatrix} v(t) & \sqrt{2}k & 0 \\ \sqrt{2}k & -W & \sqrt{2}k \\ 0 & \sqrt{2}k & -v(t) \end{bmatrix} \begin{bmatrix} c_1 \\ c_2 \\ c_3 \end{bmatrix}.$$ (5)

Here, $W \equiv U_1 - U_2$, is the effective Coulomb interaction with U_1 and U_2 being the intradot and the interdot Coulomb interaction of the electrons and k denotes the single-electron tunneling amplitude. Also, $v(t)$ describes the coupling between the electrons and the external electric field(s). In the above equations the co-tunneling matrix element was omitted. An approximation that is satisfied in the regime of the parameters of interest.

In the following we will investigate the dynamics for a GaAs system consisted of two identical quantum dots. For this system $\varepsilon = 13.1$ and $m = 0.067m_0$, where m_0 is the mass of the electron. We assume vertical confinement of energy $\hbar\omega = 16meV$ and $C = 0.5$ which determines the lateral confinement. For inter-dot distance at 40nm the parameters read W=5.6meV and k= -0.15meV.

3. Effects of fluctuating electric field

In this section we study the dynamics of a two-electron quantum dot molecule in the presence of an AC-DC external field and a stochastic electric field. The latter field is described by a time-dependent randomly fluctuating external field. We assume that the applied fields are such that the field-quantum dot coupling term $v(t)$ can be described by $2\hbar v(t) = \Omega_0 + \Omega_1 \cos \omega t + \tilde{\Omega} f_n(t)[\cos \omega t]$ where the square brackets [] means that we study cases with the cosine term present (AC stochastic field) and cases without the cosine term present (DC stochastic field). The function $f_n(t)$ is a random variable which describes the stochastic field. We study the case of Gaussian noise, so for the random variable we require that $\langle f_n(t) \rangle = 0$ and $\langle f_n(t) f_n(t') \rangle = \delta(t - t')$, such that it has a zero mean value and there are no correlations at different times.

The dynamical evolution of our system is described by the time-dependent Schrödinger equation, written in the form of equation (5). But now the presence of the stochastic term in the potential $v(t)$ makes the equations stochastic differential equations. We reside to numerical solution of these equations for the effect of the stochastic field. As it is customary, the solution is an average of several stochastic evolutions (trajectories) [9]. By using several iterations, each with a different set of random numbers, we ensure that the numerical results are not affected by small number statistics. Moreover, the accuracy of the numerical results is tested by using several time steps.

First we investigate cases where the two electrons undergo synchronous oscillations between the two quantum dots with period modified by a monochromatic AC electric field. We also assume that the Rabi frequency of the AC field is $\Omega_1 = 3.831\omega$, with $\omega = 0.5meV$ [8]. In this case the period of oscillations is estimated by $T = \dfrac{\pi W}{2\left| J_0\left(\dfrac{\Omega_1}{\omega}\right) \right| k^2} = 638.93$ psec.

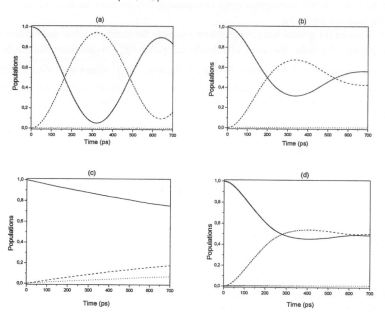

Figure 1. The time evolution of the populations in left localized state (solid curve), right localized state (dashed curve) and delocalized state (dotted line), for parameters $\hbar W$ =5.6meV, $\hbar k$ =-0.15meV (interdot distance at 40nm). In all cases $\Omega_0 = 0$, $\Omega_1 = 3.831\omega$. (a) $\tilde{\Omega} = \omega/15$, (b) $\tilde{\Omega} = \omega/5$, (c) $\tilde{\Omega} = \omega$ and (d) $\tilde{\Omega} = \omega/5$, but now without the cosine term in the stochastic field. In all plots $\hbar\omega$ =0.5meV.

We now study the effect of the stochastic field to the system. In Figure 1, we plot the populations for localized and delocalized states. Initially, when the strength of the fluctuating field is weak, the influence on the oscillations of the two electrons is small (see figure 1(a)). When the stochastic field gets larger values we observe in figure 1(b) an increased influence on the dynamic evolution of the system. Now the synchronous oscillations of the two electrons seems to disappear very soon. The system rapidly evolves to a state where the two localized stated are equally populated and the delocalized state does not appear. In figure 1(c) we observe slow evolution of the system and weak effective localization in the initially occupied quantum dot. In addition, there is a small population transfer in the delocalized state. The cases studied in figures 1(a) –(c) correspond to stochastic terms in which the cosine oscillatory term is present. By systematically investigated cases with the same strength parameters but without an oscillatory term, we have found the two-electron system is driven faster in the mixed state where both localized states are equally populated (see figure 1(d)).

Next, we investigate the effect of the stochastic fields on the localization of the electrons achieved by the application of AC-DC external electric field. In figure 2, we plot results for investigations performed for DC $\Omega_0 = \omega$, Rabi frequency of the AC field $\Omega_1 = 3.831\omega$ and $\omega = 0.5 meV$. This is a case that leads to localization in the system [8]. We observe that the influence of the fluctuating field is small for relatively low values of $\tilde{\Omega}$ ($\tilde{\Omega} < \omega/4$) (not shown here). Effect on the localization is found for $\tilde{\Omega} = \omega/2$, as can be seen from figure 2(a). In this case the localization is gradually destroyed and there is population transfer to the delocalized state and to the other two-electron localized state. The effect is more pronounced as we increase the strength of the stochastic field such that $\tilde{\Omega} = 2\omega$ (see figure 2(b)).

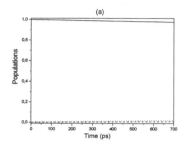 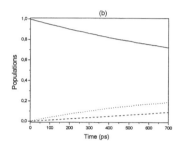

Figure 2: The same as in figure 2 but with $\Omega_0 = \omega$. (a) $\tilde{\Omega} = \omega/2$ and (b) $\tilde{\Omega} = 2\omega$.

4. Summary

In summary, in this work we studied the effects of combined AC-DC and fluctuating electric fields on the dynamics of a two-electron quantum molecule and presented numerical solutions of the time-dependent Schrödinger equation for a GaAs quantum dot structure.

Acknowledgements

We thank the European Social Fund (ESF), Operational Program for Educational and Vocational Training II (EPEAEK II), and particularly the Program PYTHAGORAS II, for funding the above work.

References

[1] G. Platero and R. Aguado, 'Photon-assisted transport in semiconductor nanostructures', *Physics Reports* **395**, 1-157 (2004).

[2] P. Zhang and X.-G. Zhao, 'Quantum dynamics of a driven double quantum dot', *Physics. Letters* A **271**, 419-428 (2000).

[3] P. I. Tamborenea and H. Metiu, 'Localization and entanglement of two interacting electrons in a quantum-dot molecule', *Europhysics Letters* **53**, 776-782 (2001).

[4] P. Zhang, Q.-K. Xue, X.-G. Zhao, and X. C. Xie, 'Coulomb-enhanced dynamic localization and Bell-state generation in coupled quantum dots', *Physical Review* A **66**, 022117 (2002).

[5] C. E. Creffield and G. Platero, 'Ac-driven localization in a two-electron quantum dot molecule' *Physical Review B* **65**, 113304 (2002).

[6] E. Paspalakis, 'Localizing two interacting electrons in a driven quantum dot molecule' *Physical Review B* **67**, 233306 (2003).

[7] Z. Jiang, D. Suqing and X.-G. Zhao, 'Dynamical localization of quantum system: long-time averaged occupation probability method' *Physics Letters A* 330, 260-266 (2004).

[8] E. Paspalakis and A.F. Terzis, 'Localization effects in a two-electron quantum dot molecule: The case of AC-DC driving fields' *Journal of Applied Physics* **95**, 1603-1605 (2004).

[9] G. Vemuri and D. M. Wood, 'Lasing without inversion with a fluctuating pump: Gain dependence on pump noise and frequency' *Physical Review A* 50, 747-753 (1994).

From our calculations, it turns out that in order to interpret tunneling experiments in structures with active elements silicon nanocrystals, one has to take into account the structure characterization and appropriate electronic structure calculations. To this should further contribute the inclusion of the holes transport as it is pointed out in Ref. [10].

Acknowledgments

The present work has been funded by the European Commission and by the Greek Ministry of Education, O.P. 'Education' (E.P.E.A.E.K.), under the 'Archimides' programme.

References

[1] L.W. Yu, K.J. Chen, M. Dai, W. Li, X.F. Huang, Collective behavior of single electron effects in a single layer Si quantum dot array at room temperature, *Physical Review B* l **71**, No 24305 (2005).

[2] N.T. Bagraev, A.D. Bouravlev, L.E. Klyachkin, A.M. Malyarenko, W. Gehlhoff, Y.I. Romanov, S.A. Rykov, Local tunneling spectroscopy of silicon nanostructures, *Semiconductors* **39** 685-696 (2005).

[3] Z.H. Zhong, Y. Fang, W. Lu, C.M. Lieber, Coherent single charge transport in molecular-scale silicon nanowires, *Nano Letters* 5 1143-1146 (2005).

[4] Z.W. Pei, A.Y.K. Su, H.L. Hwang, H.L. Hsiao, Room temperature tunneling transport through Si nanodots in silicon rich silicon, *Applied Physics Letters* **86** art. no. 063503 (2005).

[5] V. Ioannou-Sougleridis, A.G. Nassiopoulou and A. Travlos, *Nanotechnology* **14** 1 (2003).

[6] D.V. Averin, A.N. Korotkov and K.K. Likharev, *Physical Review B* **44** 6199 (1991).

[7] C.W.J. Beenakker,Theory of Coulomb-blockade oscillations in the conductance of a quantum dot, *Physical Review B* **44** 1646 (1991).

[8] X. Zianni and A.G. Nassiopoulou, Optical Properties of Si quantum wires and dots, *Handbook of Theoretical and Computational Nanotechnology*, American Scientific Publishers, in press (2005).

[9] X. Zianni and A.G. Nassiopoulou, Photoluminescence lifetimes in Si quantum dots, submitted for publication.

[10] Y.M. Niquet, C. Delerue, M. Lannoo and G. Allan, Single-particle tunneling in semiconductor quantum dots, *Physical Review B* **64** 113305 (2001).

Brill Academic Publishers
P.O. Box 9000, 2300 PA Leiden
The Netherlands

Lecture Series on Computer
and Computational Sciences
Volume 4, 2005, pp. 1505-1506

Computer Graphics and Computer Geometry Symposium 2005

Jiawan Zhang,
Associate Professor,
Department of Computer Science,
Tianjin University, Tianjin, China

Computer graphics and computer geometry has played an important role in various computational tasks because they help people to gain the essence of the problem by graphics and images, which are more comprehensible and informative than other forms.

As a symposium of International Conference of Computational Methods in Sciences and Engineering 2005, the aim of Computer Graphics and Computer Geometry Symposium (CGCG 2005) is to provide a forum for computer graphics and computer geometry related scientists and engineers to share ideas with computational experts.

In order to promoting the discovery of new computational methods both in traditional computations, CGCG fields and cross-disciplinary field, this year the main interests of the symposium include: Computer Graphics Algorithms, Geometric Modeling, Computer Geometry, Virtual and Augmented Reality, Real-time Modeling, Real-time Rendering, GPU-based Computing, GPU-based Rendering, Shading Language Techniques, Photorealistic Rendering, Non-Photorealistic Rendering, Computer-Aided Design, Scientific Visualization, Information Visualization, Mobile Graphics, and related Applications. We received 15 papers and finally 8 papers are accepted and included in the proceedings. These papers deal with topics varying from nonphotorealistic rendering (Yuan Luo and Marina L.Gavrilova, "Morphing Facial Expressions from Artistic Drawings"), Modeling and geometric rendering (Quanyu Wang, "Research of Modeling in the COMPUROBOT Platform", Qin Zheng, "Vectors based Fault-Tolerant Routing in Hypercube Multi-computers", Xiaolin Lu, "Constructing Smooth Connecting Surfaces with High Order Geometry Continuity for Polyhedron Corner Rendering"), to Information Visualization(Tae-Dong Lee, "Real-time Visualization for Distributed Military Application on RTI-G - RunTime Infrastructure on Grid") and 3D volume visualization(Jiawan Zhang, "An Accelerated Ray Casting Algorithm Using Segment Composition" and S. Zimeras, "An efficient 3D shape reconstruction technique for CT images using volume definition tools", Dai Wenjun, "3D Model Retrieval Based on Volume Distribution").

We hope all participants to contribute to a constructive and fruitful meeting, where exchanging views and opinions on the issues surrounding computer graphics and computer geometry and related computation methods will lead to more clarity of the issues involved and the current status of scientific research. We also hope that this symposium will be the first step for related domain experts to foster new computational methods in computer graphics and computer geometric.

Finally, we thank Professor Dr. T.E. Simos, chairman of ICCMSE 2005, and other staff members of the scientific and organizing committee for the excellent logistical support they provided for the symposium.

Jiawan Zhang,
Associate Professor,
Department of Computer Science,
Tianjin University, Tianjin, China
Organizer of Computer Graphics and Computer Geometry (CGCG) Symposium

}

In order to compare the fault-tolerant performances of the routing algorithms based on *SV*, *MMSPV*,we simulate their ability to send messages along optimal paths in the 8 dimensional and 10 dimensional hypercube interconnection networks separately. For a given fault links number, we select 100 kinds of faulty distributed models, and send messages from source node to the destination node for one time. By considering all of the messages sent between fault-free nodes as 100%, we obtain the percentages of messages sent along optimal paths according to *SV* and *MMSPV* as following table2.

Table 2 Comparative results of *SV* and *MMSPV* (8-D hypercube with fault links)

Total numbers of fault links	OP in *SV*	OP in *MMSPV*	*OP Exist*
15	99.9774	99.9962	99.9969
30	99.5550	99.9931	99.9938
40	97.2506	99.9907	99.9914
50	86.4614	99.9879	99.9884
60	62.3946	99.9853	99.9861

From the simulation results of table 2, it is easy to know that the fault-tolerant ability of algorithm based on *MMSPV* is better than *SV*.

5. Conclusions

Aiming at handling the link-faults that exist in hypercube multi-computers networks, this paper proposes the concept of *MMSPV*. And in addition, algorithm on how to construct *MMSPV*, and fault-tolerant algorithm based on *MMSPV* are also presented. Due to that excellent characteristics of *MMSPV*, we solve the problem arisen during the research of n-dimensional hypercube interconnection networks: for any node A in an n-dimensional hypercube network, with the storage cost of $n^2/2$ bits, how to record the optimal paths information as much as possible by the n-1 rounds of information exchanges among neighboring nodes only.

References

[1] Wang Lei, Lin Ya-ping, Chen Zhi-ping, *Fault-tolerant Routing for Hypercube Multicomputers Based on Maximum Safety-Path Matrix*. Journal of Software 994-1004(2004).

[2] Wu J. *Adaptive fault-tolerant routing in cube-based multicomputers using safety vectors*. IEEE Trans Parallel and Distributed Systems 321-334(1998).

[3] J. Al-Sadi, K. Day, M. Ould-Khaoua. *Unsafety vectors: A new fault-tolerant routing for k-ary n-cubes*. Microprocessors and Microsystems 239-246(2001).

[4] J. Al-Sadi, K. Day, and M. Ould-Khaoua. *Unsafety vectors: A new fault-tolerant routing for the binary n-cube*. Journal of Systems Architecture 783-793(2002).

[5] JJ.Narraway. *Alternative shortest paths in n-cubes*. IEEE Transactions on Electronics Letters, 1916-1918 (2000).

[6] D. Xiang. *Fault-Tolerant Routing in Hypercube Multicomputers Using Local Safety Information*. IEEE Transactions on parallel and Distributed Systems, 942-951(2001).

Brill Academic Publishers
P.O. Box 9000, 2300 PA Leiden
The Netherlands

Lecture Series on Computer
and Computational Sciences
Volume 4, 2005, pp. 1525-1528

Real-time Visualization for Distributed Military Application on RTI-G (RunTime Infrastructure on Grid)

Tae-Dong Lee, and Chang-Sung Jeong[1]

School of Electrical Engneering in Korea University 1-5ka, Anam-Dong,
Sungbuk-Ku, Seoul 136-701, Republic of Korea

Received 18 July, 2005; accepted in revised form 27 July, 2005

Abstract: In this paper, we firstly describe the RTI-G (RunTime Infrastructure on Grid) which is made up of three components: libRTI, Fedexec and RTIexec. We designed and implemented RTI-G which supports dynamic, many-to-many communication in a distributed environment, and a large amount of computing resources at different geographical locations, solving the problems of dynamic coordination concerns and fault tolerance among the communication nodes using Grid computing. Many military applications on RTI-G need the graphic visualization for information representation. However, the use of graphic visualization makes the performance of system worse because military applications require the synchronization among them. For the solution of degradation cause for synchronization, after explaining the algorithm for relationship between visualization and time management, and designing and implementing each part(Inner/Inter FTM and visualization) with active object pattern, we describe the military system composing of six federates using the algorithm, and then evaluate the performance.

Keywords: Grid, RTI, HLA, Visualization, Time Management, Military System

PACS: 89.75.-k, 89.20.Dd, 89.20.Ff

1. Introduction

Running a large-scale distributed simulation may need a large amount of computing resource at different geographical locations. In particular, the Grid[1] provides a platform for large scale resource sharing, and thus an obvious candidate running large-scale simulations. On the other hand, the High Level Architecture (HLA)[2] was developed by the Defense Modeling and Simulation Office (DMSO) to provide a common architecture that facilitates simulation interoperability and reusability across all classes of simulations. HLA has been adopted as IEEE standard 1516 in September 2000. The Runtime Infrastructure (RTI)[3] is a collection of software that provides commonly required services to simulation systems using HLA. When many distributed federates(applications) on RTI need real-time graphic visualization, the relation between the logical time in time management in RTI and the real time of visualization is a important factor for the performance of system because the object information is continuously represented in real-time. Time management services in RTI encompass two aspects of transportation services and time advance services[4]. The categories of transportation service are distinguished by reliability of event delivery and event ordering, and the time advance services provide a protocol for the federate and RTI to jointly control the advancement of logical time. Also, when the fault in the federate among distributed federates happens, the time management is broken, and then the system is stopped. For solving these problems, we designed and implemented RTI-G(RunTime Infrastructure on Grid) using Grid computing which makes both the communication efficiency and the visualization performance better, and provides the solution of fault by resource sharing. In this paper, after we first describe the architecture of RTI-G, and then explain and evaluate the scheme of graphic visualization of federates using time management in RTI-G.

---

[1] Corresponding author. E-mail: lyadlove@snoopy.korea.ac.kr, csjeong@charlie.korea.ac.kr

2. Architecture of RTI-G

None of the existing RTIs[5] consider the problem of coordinating and managing the resources for a distributed simulation to be completed efficiently and effectively. Moreover, neither the HLA nor its RTI define the support needed for the access controls providing necessary protection levels. The HLA does not currently support the authentication mechanism required for the federates which join the architecture. Grid technology can make up for the weak points of the HLA in terms of resource integration, resource sharing and security. Based on this idea, we designed and implemented the RTI-G which is a grid-enabled implementation of the RTI. That is, using the services obtained from the Grid, RTI-G makes the necessary provision for a dynamic configuration, dynamic execution and security communication, and is described in [5] in detail. To apply Grid technologies to the RTI, we use the multi-layered architecture which can provide a well-defined model of a system that reflects the scale and depth of the application-level services and separates the application models into discrete tiers.

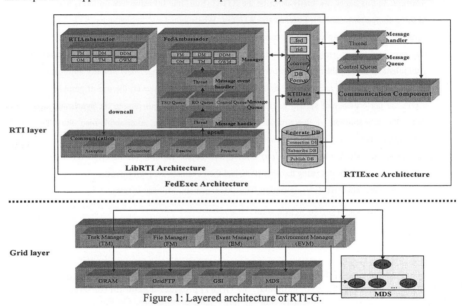

Figure 1: Layered architecture of RTI-G.

3. Model and Algorithm of Time Synchronization

Figure 2(a) explains the time relationship within a federate with a graphic visualization, which consists of Inner Federate Time Management (Inner-FTM), Inter Federate Time Management (Inter-FTM), and real time of visualization. For synchronization schemes, we designed and implemented the visualization part and the time management(Inner/Inter-FTM) in RTI respectively using active design pattern that each object retains its own thread (or even multiple threads) and uses this thread for execution. Inner-FTM considers a sequential simulation and achieves synchronization by coordinating the local time of objects and the real time of visualization in each federate. There may exist many entities in one federate. Each entity has its local time. Inner-FTM advances time by checking each entity's local time, synchronized with real time. Inter-FTM achieves synchronization with real time by coordinating federates using RTI time management API which is to establish or associate events with logical time, interactions, attribute updates, object reflections or object deletion by federate time scheme, support the causal behavior within a federation and the interaction among federates using different timing schemes. Figure 2(b) shows the time-stepped algorithm in a graphic mode, where timer( ) is called at regular interval in real time to perform visualization while processing local and remote events. The algorithm is composed of Inner-FTM, Inter-FTM and visualization. Inner-FTM processes the events of local objects and advances local object's time. Inter-FTM updates information of local objects and advances time synchronized with other federates and reflects information of remote objects. find_earliest_object( ) is to find an object having earliest local time and trigger_event( ) is to process an event and advance the object's local time.

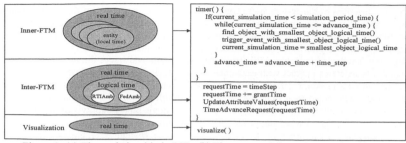

Figure 2: (a) Time relationship in RTI   (b) Time-stepped time advance algorithm.

4. Experiments

This experiment (see Figure 3(a)) was conducted in order to evaluate the efficiency of graphic visualization through the communication and fault tolerance of distributed military application on RTI-G with 50 updates/sec sending 128 bytes per update, comparing RTI-NG(a representative in RTI field). The test environment is composed of 10 servers (S1-S10) and 10 clients, which can be regarded as one virtual organization as shown in Figure 3(b). The scenario involves six clients which are connected to S1, and RTIexec is created and causes Fedexec to be generated on the same host in the experiments for RTI-NG. In the case of RTI-G, the six clients are also connected to S1. Once these connections are established, RTIexec is created and causes Fedexec to be generated through a fork operation. However, in this case, Fedexec is generated on the best server (S10 in this experiment) by components in the RTI-G. The communication hub is Fedexec, which six clients connect to and communicate through.

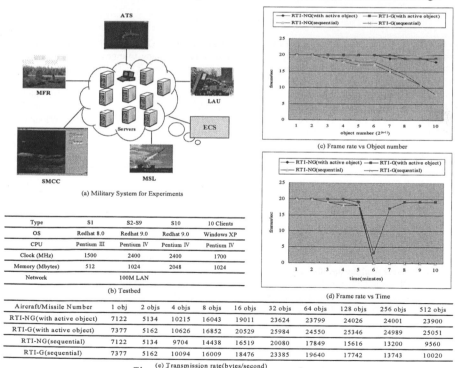

(a) Military System for Experiments

(c) Frame rate vs Object number

(d) Frame rate vs Time

(b) Testbed

Type	S1	S2-S9	S10	10 Clients
OS	Redhat 8.0	Redhat 9.0	Redhat 9.0	Windows XP
CPU	Pentium III	Pentium IV	Pentium IV	Pentium IV
Clock (MHz)	1500	2400	2400	1700
Memory (Mbytes)	512	1024	2048	1024
Network		100M LAN		

(e) Transmission rate(bytes/second)

Aircraft/Missile Number	1 obj	2 objs	4 objs	8 objs	16 objs	32 objs	64 objs	128 objs	256 objs	512 objs
RTI-NG(with active object)	7122	5134	10215	16043	19011	23624	23799	24026	24001	23900
RTI-G(with active object)	7377	5162	10626	16852	20529	25984	24550	25346	24989	25051
RTI-NG(sequential)	7122	5134	9704	14438	16519	20080	17849	15616	13200	9560
RTI-G(sequential)	7377	5162	10094	16009	18476	23385	19640	17742	13743	10020

Figure 3: Distributed simulation experiment.

For the evaluation, we have designed and implemented military system for air defense, which is composed 6 components: SMCC(Surface Missile Control Center), ATS(Air Target Simulator), MFR(Multi Function Radar), ECS(Engagement Control Simulator),LAU(LAUncher), MSL(MiSsiLe).

The design objectives of the system are to construct the synthesis environment for organization of anti-air arms. SMCC monitors and controls the system. ATS creates the air targets and sends their information. MFR searches the position of air targets and missiles. ECS evaluates the threat degree of air targets and informs priority of threat of missiles. LAU transfers information of ECS into MSL and simulates the function of launcher. MSL represents the missile and traces the air targets by the threat evaluation. Six federates interact with each other through RTI. Among these federates, SMCC renders the 3D terrain data (10km x 10km) through OpenGL using DEM(digital elevation model) data on geometric data represented by a finite set of data points with associated elevation values storing each 100m data by unit about Korea peninsula[6]. We use the triangle binary subdivision which the index of children is each (2i), (2i+1) when the parent index is (i) with error limit 30m (the number of triangles: 6642). The data have the information: triangle index, maximum triangle error, and normal vector.

For the performance evaluation, we evaluate the frame rate with respect to the object number and the frame rate with respect to time respectively. The latter is related to fault tolerance. For the first test, after the generation of Fedexec, a object are tested initially, and the number of object is increased twice per 60 seconds. We have measured the frame rate with respect to the object number on SMCC. For RTI-NG test, Pentium III was chosen, and Pentium IV for the case with RTI-G, since RTI-G selects the host allocating the resource with higher performance according to the information collected by MDS in Grid. Therefore, as shown in Fig. 3(c), the latter case shows higher frame rate than the former case. Also, we performed the data transmission rate on Fedexec about the first test, which the fluctuation ratio of transmission is almost proportional with that of frame as shown in Fig. 3(e). In the second experiment, we measured the frame rate with respect to time on SMCC when each ATS and MSL sends 128bytes/update. Suppose a fault occurs in SMCC after 5 minutes, and then only RTI-G detects the fault by checkpoint mechanism using Grid services. After the detection of fault, RTI-G selects the new host and configures to restore SMCC. The result is shown in Fig. 3(d).

5. Conclusions

In this paper, we described the RTI-G (RunTime Infrastructure on Grid) which is made up of three components: libRTI, Fedexec and RTIexec, supporting dynamic, many-to-many communication in a distributed environment, and a large amount of computing resources. RTI-G solves the problems of dynamic coordination concerns and fault tolerance among the communication nodes using Grid computing. Many military applications need the graphic visualization for information representation. However, the use of graphic visualization makes the performance of system worse because military applications require the synchronization among them. After explaining the algorithm for relationship between visualization and time management, designing and implementing each part(Inner/Inter FTM and visualization) with active object pattern, we used the algorithm on military system composing of six federates, and then evaluated the performance. The performance evaluation showed that the performance using our algorithm is better than that using sequential method because dynamic coordination makes the transportation of data faster, and the fault tolerance provides the availability, reliability, safety, and maintainability to the system.

References

[1] I. Foster, C. Kesselman, J. Nick, S. Tuecke, "The Physiology of the Grid: An Open Grid Services Architecture for Distributed Systems Integration," June 22, 2002.

[2] IEEE Standard for Modeling and Simulation (M&S), "High Level Architecture(HLA) Federate Interface Specification," IEEE Std 1516.1-2000.

[3] T.B. Stephen, J.R. Noseworthy, J.H. Frank, "Implementation of the Next Generation RTI," 1999 Spring Simulation Interoperability Workshop.

[4] T. McLean, R. Fujimoto, "Predictable time management for real-time distributed simulation," Parallel and Distributed Simulation, 2003. (PADS 2003). Proceedings. Seventeenth Workshop on 10-13 June 2003 Page(s):89 - 96

[5] T.D. Lee, C.S. Jeong, "Service-oriented RunTime Infrastructure on Grid," GCC 2004 Workshops, LNCS 3252, pp736-743

[6] K. Xu, X. Zhou, X. Lin, "Direct mesh: a multiresolution approach to terrain visualization," Data Engineering, 2004. Proceedings. 20th International Conference on 30 March-2 April 2004 Page(s):766 - 776

Brill Academic Publishers
P.O. Box 9000, 2300 PA Leiden
The Netherlands

*Lecture Series on Computer
and Computational Sciences*
Volume 4, 2005, pp. 1529-1532

An Efficient 3D Shape Reconstruction Technique for CT Images Using Volume Definition Tools

S. Zimeras[1] and G. Karangelis[2]

[1]Department of Statistics and Actuarial-Financial Mathematics,
University of the Aegean,
83200 Karlovassi, Samos, Greece

[2] Medintec GmbH, Feldstrasse 26a, 44867 Bochum, Germany.

Received 15 July, 2005; accepted in revised form 27 July, 2005

Abstract: The definition of structures and the extraction of organ's shape are essential parts of medical imaging applications. These might be applications like diagnostic imaging, image guided surgery or radiation therapy. The aim of the volume definition process is to delineate a specific shape of an organ on a digital image as accurate as possible especially for 3D rendering, radiation therapy, and surgery planning. This can be done, either by manual user interaction or applying imaging processing techniques for the automatic detection of specific structures in the image. In this work we present a set of tools that are implemented on several computer based medical application. Central focus of this work, are techniques used to improve time and interaction needed for a user when defining one or more structures. These techniques involve interpolation methods for the manual volume definition and methods for the semi-automatic organ shape extraction. Finally different segmentation techniques would be proposed for the particular organs of interests (lungs, skin and spine canal) and a 3D shape reconstruction of these regions would be illustrate the efficiency of the segmentation techniques. Finally, the proposed technique would be compared with the manual segmentation obtained from the doctor experts using quantitative (shape matching measures) and qualitative (visual comparison) measures.

Keywords: Shape reconstruction, CT medical data, volume definition, rendering techniques, segmentation techniques.

1. Introduction

Volume definition aims to localise areas of interest as well the surrounding to them tissues, focusing on several medical-imaging applications. Methods for performing segmentations vary widely depending on the specific application, imaging modality, and other factors. This might involve treatment (e.g. cancer treatment, image-guided surgery and invasive techniques) or diagnostic imaging [1-3]. In general there is currently no single segmentation method that yields acceptable results for every medical imaging. Nowadays medical imaging devices give us the possibility to acquire large amount of digital data in short time providing high-resolution information. Although this improves the visualisation results, in addition it increases the working effort during volume definition since the user has to go through a large number of images step-by-step. Traditionally this is done on the original acquired image plane. This means that on each original volume plane a number of contours are defined, relative to the number of structures we want to have. The user usually has the possibility to modify and edit the contour or each contour point. Eventually, the sequence of the contoured objects is triangulated creating surface objects that can be reconstructed in 3 dimensions. The target definition can be used for calculations and measurements or as orientation indicator. For example in case of diagnostic imaging, the physician needs to know the size of a structure on a specific or arbitrary plane, the volume it occupies and its 3D shape.

---

[1] Corresponding author. Lecturer- Grad. Stat. E-mail: zimste@aegean.gr
[2] Medintec GmbH, Feldstrasse 26a, 44867 Bochum, Germany.

In this work we will present a number of methods that improve the time and interaction one needs to define one or more structures. These techniques involve interpolation methods for the manual volume definition and methods for the semi-automatic organ shape extraction (segmentation techniques). Once an accurate segmentation is obtained, this information may be used by the doctors to compare the volume and morphology characteristics of each region against known anatomical norms, other regions in the same image set, and corresponding regions in related image sets.

Many methods have been proposed to detect and segment 2D shapes, the most of which is template matching. However their low speed has prevented its wide spread use. Other techniques called snakes or active contours have been used, but the main drawbacks associated with their initialization and poor convergence to boundary concavities limit their utility. Substantial computational and storage requirements become especially acute when object orientation and scale have to be considered. Therefore automated or semi-automated segmentation techniques are essential if these software applications are ever to gain widespread clinical use. An effective semi-automatic method, based on the *boundary tracking technique* [4-7], which improves the time when one or more structures are in use. The implemented algorithms can segment within a few seconds the complete volume of specific organs e.g. lungs, skin, spine. The only interaction of the user is to select the starting point in the region of interest and the algorithm will track the object boundaries in 3 dimensions. For the visual comparison of the semi-automatic techniques, the EXOMIO[3] [8, 10] virtual simulation package was used (Figure 1). The above tools are already in use and involve applications of the InViVo family. InViVo is a 3D visualization system of medical volume data that has been developed over several years in Fraunhofer-IGD, applied in ultrasound diagnostic imaging, interstitial brachytherapy and virtual simulation of external beam radiotherapy.

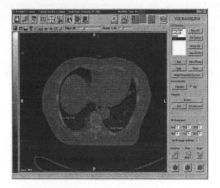

Figure 1: EXOMIO contouring of the structures user interface

Conclusions

In this work different volume definition techniques has been presented. In the first part we present interpolation techniques aim to accelerate the manual contouring process. That includes the well-known linear interpolation technique (Figure 2) implemented for simple bisection organ shapes. In addition we introduce a new interpolation method, the orthogonal interpolation (Figure 3), which enables the user to define a volume, drawing only a small number of contours. The main drawback of this method is the weakness to handle complex organ shapes. For optimum results, the produced interpolated contours can be used as reference for the ACM algorithm.

---

[3] http://www.medcom-online.de/

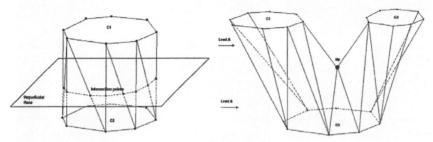

Figure 2. On the left side a simple case where an interpolated contour is created from the plane intersection with the triangles. On the right side an example of contour bisection.

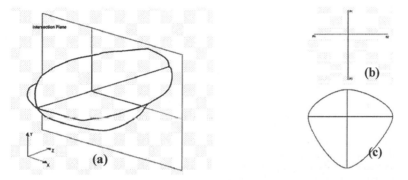

Figure 3. The orthogonal contour interpolation. In (a), the two drawn contours and the intersection plane in Z direction. The result of the intersection is shown in (b). After applying the cubic Spline interpolation a new contour is created (c)

At the second part, an effective semi-automatic method was presented, based on the *boundary tracking technique* that improves the time when one or more structures are in use. Also an implementation of an active contour method is taking place [11, 12]. Resulting images are compared for the benefit of the first segmentation algorithm. The implemented algorithm can segment within a few seconds the complete volume of specific organs e.g. lungs, skin, spinal canal. The only interaction of the user is to select the starting point in the region of interest and the algorithm will track the object boundaries in 3 dimensions. For each particular organ of interests (lungs, skin and spine canal), a different segmentation technique is proposed, but all of them are based on the *boundary tracking technique.* The 3D-shape is reconstructed out of the segmented regions, in order to illustrate the efficiency of the segmentation techniques. The benefit of the 3D illustration is the visual presentation of the organs (Figure 4). In many medical cases, illustration of the organ involves an anomaly, clinical problem or generally artifacts. Visual representation of the particular organ, in addition with the clinical examinations, could be a powerful tool to the doctors for diagnosis, medical treatment or surgery.

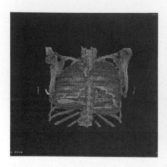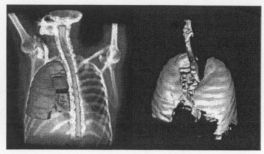

Figure 4. 3D segmentation images for different organs: Lungs and Spinal canal, Spinal canal with left lung and tumor, Spinal canal and Lungs (different view).

Acknowledgments

The authors would like to thank Prof George Sakas and Prof. Nikos Zamboglou for the useful scientific help and comments about the progress of this work. Also many thanks to MedCom Company and Städtisches Klinikum Offenbach whom provided equipments and medical data sets for the implementation of the above work. This work was supported by a Marie Curie Industry Host Fellowship Grant no: HPMI-CT-1999-00005 and the first author is a MCFA member.

References

[1] Strassman G., Kolotas C., Heyd R.: Navigation system for interstitial brachytherapy, Radiotherapy & Oncology, Vol 56, pp:49-57, 2000.

[2] Hoegemann D., Stamm G.: Volumetric evaluation and 3D-visualization of the liver before living-related donation, CARS'99, France, pp: 259-252, 1999.

[3] Meyer JL, Purdy JA (eds): 3D Conformal Radiotherapy, Front Radiat Ther Oncol. Basel, Karger, vol 29, pp 31-42, 1996.

[4] S. Zimeras, G. Karangelis: Semi-automatic Segmentation Techniques for CT Medical Data, 3rd caesarium Computer Aided Medicine November 12-13, Bonn, Germany.

[5] G. Karangelis, S. Zimeras: An Accurate 3D Segmentation Method of the Spinal Canal Applied on CT Images, BVM 2002, Conference Proceedings Meiler M, Saupe D, Kruggel F, Handels H, Lehmann TM, (Hrsg) Bildverarbeitung für die Medizin 2002, Springer-Verlag, Berlin, 2002, 366-369, Germany.

[6] S. Zimeras, G. Karangelis, and E. Firle: Object segmentation and shape reconstruction using computer- assisted segmentation tools, Medical Imaging 2002, San-Diego, USA.

[7] G. Karangelis and S. Zimeras: 3D segmentation method of the spinal cord applied on CT data, Computer Graphics Topics, 1/2002, Vol 14, 28-29

[8] Cai, W., Sakas, G., Karangelis, G. Volume Interaction Techniques in the Virtual Simulation of Radiotherapy Treatment Planning. International Conference on Computer Graphics and Vision (Graphicon) Moscow, 1999.

[9] G. Sakas. Interactive volume rendering of large fields. The Visual Computer, 1993, 9(8): 425-438.

[10] Karangelis G, Zamboglou N: "EXOMIO: A 3D Simulator for External Beam Radiotherapy.", Volume Graphics 2001- Proceedings of the Joint IEEE TCVG and Eurographics Workshop in Stony Brook, New York, USA, Springer-Verlag Wien New York, 2001 351-362.

[11] Terzopoulos D.: Dynamic 3D models with local and global deformations: deformable superquadrics, IEEE Trans. Pattern Anal. Mach. Int., vol 13, 703-714, 1987.

[12] Eviatar H. and Somorjai R. L.: A fast, simple active contour algorithm for biomedical images, Pattern Recognition Letters, vol 17, 969-974, 1996.

Brill Academic Publishers
P.O. Box 9000, 2300 PA Leiden
The Netherlands

Lecture Series on Computer
and Computational Sciences
Volume 4, 2005, pp. 1533-1536

An Accelerated Ray Casting Algorithm Using Segment Composition

Jiawan Zhang[1], Yi Zhang, Jizhou Sun

Visualization and Image Processing Group, SRDC,
School of Electronic and Information Engineering,
Tianjin University, 300072, China

Received 12 July, 2005; accepted in revised form 28 July, 2005

Abstract: Volume rendering has been a key technology in the visualization of datasets from various disciplines. In volume rendering, ray casting is popular because of its ability to generate high-quality images. But it is computational expensive. In this paper, a segment-based ray casting algorithm is proposed to speed up the composition stage and improve the parallel property of traditional ray casting algorithm. The new algorithm divides each ray into several segments each of which includes some sequential samples. The samples in the same segment have similar optical attributes. These segments substitute for the samples to be the basic composition units. Therefore, the number of composition operations is efficiently reduced. The experiments results show that the segment-based ray casting algorithm improves the rendering performance by at least 30% compared to traditional ray casting while still produces appropriate quality images.

Keywords: ray casting, segment, composition

1. Introduction

Volume rendering is a technique for directly displaying a sampled 3D scalar field without first fitting geometric primitives to the samples. It has been widely used in various disciplines. Ray casting, which was first proposed by Kajiya [1] and Levoy [3], is the conventional and most widely used method for volume rendering. The necessary stages of ray casting are resampling, classification and composition. It casts the rays of sight through the volume until the rays leave the volume. In the reampling stage, the discrete volume data is resampled at equispaced intervals along the ray. After resampling, the scalar data value of the sample is mapped to optical properties via transfer function table in the classification stage. These optical properties are integrated using the emission-absorption volume rendering equation in either back-to-front or front-to-back order in the composition stage.

The popularity of ray casting is due to its high-quality results and good parallel performance. However, ray casting method also suffers from a serious weakness: it is computational expensive. One reason of this weakness is that the sequential composition operations cost much. In the meanwhile, the ordinal character of the composition stage destroys the parallel capacity which behaves excellently in the other rendering stages.

Much work has been done to improve the original ray casting algorithm. The challenge of the accelerated methods is to take advantage of data's coherence. A famous accelerated algorithm is Lacroute's Shear-Warp Factorization algorithm [2]. It exploits coherence in both the volume data and the image. Many data structures and compression methods such as k-trees、octrees、BSP trees、wavelet and run-length coding are used efficiently in practice. However, these techniques are always used to skip the samples which have no contributions to final result before composition stages. The coherence of the samples which are allowed to enter the composition stage is not used. Thus no performance enhancement takes place in the composition stage.

In this paper, an accelerated ray casting algorithm is presented to speed up the composition stage and to improve the parallel performance. Our solution extends the basic composition unit from a sample to a segment which consists of several sequential samples with similar scalar value.

We implement original ray casting algorithm with only early ray termination optimization and improve it by our accelerated algorithm. The results show that segment-base ray casting algorithm can improve the rendering performance by at least 30% compared to original ray casting algorithm. Although many

---

[1] Corresponding author. E-mail: jwzhang@tju.edu.cn, zhangjiawan@hotmail.com

comparing of two and more graphs always is a time-consuming task. Hence, the efficient implementation of approximate matching methods is a current research issue.

In this paper, we propose a new approach of 3D model feature extraction, which represents 3D model by volume distribution characteristic curves along three principal axes. And we implement this technique in our 3D model retrieval system. Experiments show the efficiency and discriminative power of this approach. And this approach can process various 3D models with arbitrary geometry and topology, even models with holes inside.

The outline of the rest of the paper is as follows: Feature extraction is proposed in section 2 and matching algorithm in section 3. In section 4 we present experiment results and conclusion.

2   Feature Extraction

In a preprocessing step, feature of each model in the database is extracted and stored. 3D models have arbitrary scale, orientation and position in the 3D space. So models must be placed into a canonical coordinate frame first. There are three problems: normalizing for translation, scale and rotation. We solve these problems in the following section, and then come to feature extraction in section 2.2.

2.1   Normalization of Models

First, models are normalized for rotation by aligning the model with PCA[6]. And then we normalize the models for translation by transforming the model's center of gravity to the origin. Our algorithm of calculating center of gravity is as below:

If an object is divided into several segments, the weight of segment i is P_i, the center of gravity of the object is (x_i, y_i, z_i) $x_c = \dfrac{\sum P_i x_i}{\sum P_i}$ $y_c = \dfrac{\sum P_i y_i}{\sum P_i}$ $z_c = \dfrac{\sum P_i z_i}{\sum P_i}$ [7]. So we slice the model along three axes to

calculate x_c, y_c and z_c. The following is the demonstration of getting x_c by slicing model along x-axis.

Slice the model along x-axis into n segments. When n is large enough, the two cross sections of one segment are almost the same. X-coordinate value of the center of gravity of each segment x_i is the median of X-coordinate value of two slice planes. And the volume of each segment can be calculated.

The weight of each segment P_i can also be calculated, so we can get x_c by $x_c = \dfrac{\sum P_i x_i}{\sum P_i}$. Similarly, y_c

and z_c can be calculated. So far we get the approximating position of center of gravity (x_i, y_i, z_i), and transform it to the origin for normalization. Scale normalization is in the process of feature extraction.

2.2   Extraction of Volume Distribution

Each model is represented by three volume distribution curves along three principal axes. First we slice the model along three principal axes with a set of parallel planes. From cross section in each plane, a polygon set composed of one or more polygons can be obtained and the area of each cross section can be consequently computed. The volume distribution curves can be calculated according to these areas. The following is the definition of volume distribution characteristic curve:

Definition: Volume distribution curve of a certain direction is represented in $y = f(x)$, and x-axis indicates the sequence of sliced segments, The value of y-axis is the ratio of the sliced segment volume to the whole volume.

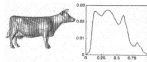

Figure 1: A cow model and its volume distribution curve

Figure 1 shows the demonstration of volume distribution curve of a cow model along the primary direction of principal axis.

Without losing of generality, we only discuss cutting models along z-axis. We name the z-axis value of the highest and lowest slice planes of the set of slice planes z_{max} and z_{min}, and the number of

planes is n. So we get $n-1$ segments and the interval of planes is $\dfrac{z_{max} - z_{min}}{n-1}$. The areas of cross

sections are calculated and named $S_1, S_2, \cdots, S_n$. The approximate value of the model's volume is:

$$V \approx \frac{z_{max} - z_{min}}{3 \times (n-1)} \sum_{i=1}^{n-1} (S_i + \sqrt{S_i S_{i+1}} + S_{i+1})$$

When n is large enough, every two adjacent cross sections are almost the same. The formula can be simplified to $V \approx \frac{z_{max} - z_{min}}{n-1} \sum_{i=1}^{n} S_i$ (Evolution operation is avoided). The volume of segment i is:

$$V_i \approx \frac{z_{max} - z_{min}}{n-1} \times \frac{(S_i + S_{i+1})}{2}$$

(When n is large enough, volume can be calculated by multiplying area of cross section and the thickness of the segment) Then the volume distribution curve at segment i $f(\frac{i}{n-1} - \frac{1}{2(n-1)})$ is:

$$f(\frac{i}{n-1} - \frac{1}{2(n-1)}) = \frac{V_i}{V} = \frac{\frac{z_{max} - z_{min}}{n-1} \times \frac{(S_i + S_{i+1})}{2}}{\frac{z_{max} - z_{min}}{n-1} \sum_{t=1}^{n} S_t} = \frac{(S_i + S_{i+1})}{2 \sum_{t=1}^{n} S_t} (i = 1, 2, \cdots n-1)$$

Other place can be calculated by linear interpolation. Because the curve is continuous, formula is simplified to:

$$f(\frac{i}{n-1}) = \frac{S_{i+1}}{\sum_{t=1}^{n} S_t} (i = 0, 1, 2 \cdots n-1)$$

To deal with the normalization for scale, we make $z_{max} = R$ and $z_{min} = -R$. R is distance between the center of gravity and the farthest vertex to the center of gravity. Then we can slice the model along three principal axes and get three characteristic curves.

Here we can reuse the result when center of gravity is calculated. We slice the model along three principal axes when calculating the center of gravity, the same as we slice along here, the only difference is the value of z_{max} and z_{min}. We record the area of cross sections when we calculate the center of gravity and reuse them by linear interpolation. Because the range of slice axis here include the range of slice process in calculating the center of gravity, so this will not lose information and the time of slicing is reduced to 3.

3   Matching Algorithm

As in Figure 2(a) each model has three characteristic curves. The distance between two models is the sum of the distance of three pairs of characteristic curves. We have investigated one method usually used to compare function f and g : Minkowski-form distance L_p : $D(f, g) = (\int |f - g|^p)^{\frac{1}{p}}$ [1].

And finally we decided the distance between two curves f and g is measured by L_1 :

$$D(f, g) = \int |f - g|$$

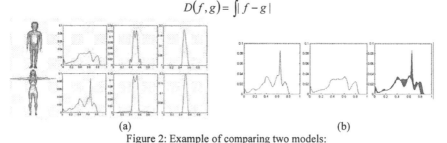

(a)   (b)

Figure 2: Example of comparing two models:
(a) two models and their three characteristic curves   (b) distance between two curves

Figure 2(b) shows that the distance between the first characteristic curves of models of woman and child is area of the shaded region.

When we align 3D models with PCA, the direction of principal axes may be reverse, so the two characteristic curves may be reverse. So when we calculate the distance of two models, we must consider all the situations and get the shortest distance.

Where Φ is a set of m sigmoidal functions, W in $\mathfrak{R}^{m \times n}$ is the weight (free-parameters) matrix associated with the label space, and $\hat{W}^j$ in $\mathfrak{R}^{m \times m}$ is the weight (free-parameters) matrix associated with the j-th subgraph space. This process can be graphically described unfolding the encoding process through the input structures, as shown in Figure 1 for two molecular structures. For each input structure the encoding begins at the leaves (shown by arrows in Figure 1) and follow a bottom-up topological order. At the root, this process computes a code of the whole molecular structure. The code is then mapped to the output property value. A single neural unit performing a linear regression realizes the mapping function. The *encoding* and *mapping* free parameters of the neural network are adapted to the task through the learning algorithm on the basis of the training examples. In particular, we used a constructive approach, which automatically determines the number of hidden units, to realize the model, by a Recursive Cascade Correlation method [4]. Summarizing, RecNN models a direct and adaptive relationship between molecular structures and target properties by the encoding and mapping functions. In particular, the recursive model can learn an encoding of the input structured representations according to the given QSPR/QSAR task. Hence, RecNN can automatically discover by learning the specific structural descriptors (x) for the particular task to be solved. As a result, no *a priori* definition and/or selection of properties by an expert are needed.

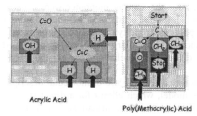

Figure 1: Unfolding the encoding process through structures. Each box includes the subtree progressively encoded by the recursive neural network.

3. Overview of the results

Finding a thorough representation of polymer structures as RecNN input constituted the first goal of this research. The basic idea was to extend the representation developed for mono- and poly-functional compounds to macromolecules [5]. In agreement, the representation of each polymer was based on the 2D graph of its repeating unit treated as *k-ary* tree. To build unique chemical trees, the repeating unit is divided into a number of defined atomic groups. Each group corresponds to a vertex of the tree and each bond between groups corresponds to an edge. With respect to the small molecule representation, the most relevant innovation was the positioning of the tree root. Indeed, the root was not placed on the highest-priority chemical group, but on an additional super-source vertex (the group "Start"), not related to the molecular graph. The super-source conveys information on the average macromolecule characteristics through its label. In the present study, the label of the super-source encoded the information of polymers tacticity. Since a repeating unit is an "open" structure, the Start group is also used to close one of its sides. In particular, the Start is linked to the atomic group with the highest priority at the end of the repeating unit. The structural unit is capped at the other end by a "Stop" group with the only purpose of closing the molecule. Neither the Start nor the Stop affect the chemical features of the groups they are linked to. This representation can be easily extended to deal equally with homo- and co-polymers. Indeed, two or more repeating units can be rooted on the super-source that may contain information on their distribution. The representation of poly(methacrylic acid) as a chemical three is reported in Figure 2 as an example.

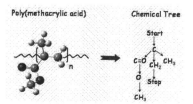

Figure 2: Representation of poly(methacrylic acid) as a chemical three.

The mean absolute Tg error computed by our method was satisfactory and below the errors obtained by the majority of standard methods over data sets of only amorphous polymers. Additionally, the RecNN model learned the Tg dependence from tacticity and then correctly predicted new data. Noteworthy, the model found the Tg-tacticity relationships by treating simultaneously polymers with either only one or different tacticity forms. It must be stressed that the ability of accounting for the extent and type of stereoregularity of polymer chains is of paramount importance because of the impact of this feature on several material properties. For instance, very often stereoregular polymers are highly crystalline, whereas atactic ones are amorphous. On the other hand, methods able to correlate the Tg of polymers with their tacticity are lacking in literature. For instance, the most widely referenced model in this field, is that of Karasz and MacKnight [6]. The authors rationalize the influence of polymer stereoregularity on the Tg without any prediction purpose.

The RecNN model resulted also to be a powerful tool for data cleaning. In fact, analysis of the training set outliers was useful to pick the most uncertain Tg values out of the data set. In particular, the RecNN found the highest terms of homologous series as outliers. In agreement, recent literature on polyacrylics and polymethacrylics confirmed that these values are indeed melting points and not glass transition temperatures.

4. Conclusions

The reported results highlight the greater generality and flexibility of the proposed method and molecular structure representation with respect to standard literature techniques. The RecNN method is able to deal in the same way with small molecules and polymers. In the latter case, the method allows for taking into account also the average macromolecule characteristics and for the simultaneous handling of polymers having one or more values of the considered average property. Moreover the molecular representation can be extended to the treatement of all kinds of copolymers (random, alternating, block).

Acknowledgments

The financial support by MIUR Cofin 2003 Project "*Integrated multidisciplinary design and realization of biologically active systems for biomedical applications*" is gratefully acknowledged.

References

[1] a) A. Micheli, A. Sperduti, A. Starita, A. M. Bianucci, *Soft Computing Approaches in Chemistry* (Editors: L. M. Sztandera and H. M. Cartwright), Springer-Verlag: Heidelberg (2003) 265-296; b) A.M. Bianucci, A. Michel, A. Sperduti, A. Starita, *Appl. Int. J.*, **12** (2000) 117-146; c) A. Micheli, A. Sperduti, A. Starita, A. M. Bianucci, *J. Chem. Inf. Comput. Sci.* **41** (2001) 202-218.

[2] a) A. R. Katritzky, P. Rachwal, K. W. Law, M. Karelson, V. S. Lobanov, *J. Chem. Inf. Comput. Sci.* **36** (1996) 879-884.

[3] a) J. Bicerano, *Prediction of polymer properties*, 3rd ed., Marcel Dekker, New York, Basel (2002); b) D.W. Van Krevelen, *Properties of Polymers-Their Estimation and Correlation with Chemical Structure*, 2nd ed., Elsevier: New York, 1976; c) M. G. Koehler and A. J. Hopfinger, *Polymer* **30** (1989) 116-126; d) A.R. Katritzky, S. Sild, V.S. Lobanov, M. J. Karelson, *Chem. Inf. Comput. Sci.* **38** (1998) 300-304; e) P. Camelio, C. C. Cypcar, V. Lazzeri, B. J. Waegel, *J. Polym. Sci.: Part A: Polym. Chem.* **35** (1997) 2579-2590; f) S. J. Joyce, D. J. Osguthorpe, J. A. Padgett, G. J. Price, *J. Chem. Soc., Faraday Trans.* **91** (1995) 2491-2496; g) B. G. Sumpter and D. W. Noid, *J. Thermal Anal.* **46** (1996) 833-851; i) B.E. Mattioni and P.C. Jurs, *J. Chem. Inf. Comput. Sci.* **42** (2002) 232-240.

[4] A. Micheli, Recursive Processing of Structured Domains in Machine Learning. *PhD Thesis TD-13/03*, Department of Computer Science, University of Pisa (2003).

[5] C. Duce, Physical chemical methods in the rational design of new materials: QSAR and calorimetric approaches, *PhD Thesis*, University of Pisa (2005).

[6] F. E. Karasz and W. J. MacKnight, *Macromolecules* **1** (1968) 537-540.

2. Examples and Applications

We demonstrate in Fig. 1 and Table 1 that the introduced descriptors of topological complexity describe adequately the patterns of increasing complexity in acyclic graphs having up to 7 vertices.

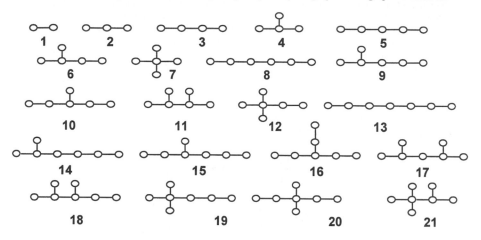

Figure 1: Acyclic graphs depicting hydrocarbon molecules with two to seven carbon atoms

Table 1: Overall complexity values of graphs **1 – 21**. Included are the subgraph count, *SC*, the overall connectivity, *OC*, the overall Wiener index, *OW*, the overall Zagreb indices *OM1* and *OM2*, and the overall Hosoya index, *OZ*

Graphs	SC	OC	OW	OM1	OM2	OZ
1	3	4	1	1	1	4
2	6	14	6	22	10	10
3	10	32	21	56	26	21
4	11	39	24	87	27	23
5	15	60	56	110	60	40
6	17	76	67	168	67	46
7	20	100	80	292	68	52
8	21	100	126	188	130	72
9	24	127	154	277	149	84
10	25	136	161	300	161	89
11	28	164	188	404	172	100
12	30	181	197	505	168	103
13	28	154	252	294	272	125
14	32	194	311	418	315	147
15	34	214	333	468	351	159
16	36	234	354	516	390	172
17	37	246	384	584	366	173
18	40	276	411	668	410	191
19	41	284	414	762	370	185
20	44	314	440	850	412	202
21	49	369	510	1075	433	225

The applicability of the sets of six overall topological indices to QSPR (Quantitative Structure-Property Relationships) was tested against ten physicochemical properties of alkanes: boiling points, critical temperatures, critical pressures, critical volume, molar volume, molecular refraction, surface tension, the heat of formation in gaseous state, the heat of vaporization, and the heat of atomization. All five-variable models with overall connectivity indices showed lower standard deviations than those produced by molecular connectivity indices, the difference being considerable for heats of atomization, boiling point, and critical temperature. The standard deviations of the models are very low: boiling point - 1.60 °C, heat of formation − 1.02 kJ/mol, heat of vaporization − 0.67 kJ/mol, heat of atomization − 0.17 kcal/mol, critical temperature − 3.23 °C, critical pressure − 0.37 atm, critical volume − 0.0079 L/mol, molar volume − 0.23 cm³/mol, molar refraction − 0.041 cm³/mol. In eight of the ten models the best results are obtained with overall connectivity indices, and in two of them (molar volume and critical temperature) − with the overall Wiener indices. This indicates the considerable potential for applications in QSPR and QSAR.

References

[1]   S. Bertz and W. C. Herndon, Similarity of Graphs and Molecules. *Artificial Intelligence Applications in Chemistry,* American Chemical Society, Washington, D. C., (1986) 169-175.

[2]   S. H. Bertz, and T. J. Sommer, Rigorous Mathematical Approaches to Strategic Bonds and Synthetic Analysis Based on Conceptually Simple New Complexity Indices. *Chem. Commun.* (1997) 2409-2410.

[3]   S. H. Bertz and W. F. Wright, The Graph Theory Approach to Synthetic Analysis: Definition and Application of Molecular Complexity and Synthetic Complexity. *Graph Theory Notes New York Acad. Sci.* (1998) 32-48.

[4]   D. Bonchev, Kolmogorov's Information, Shannon's Entropy, and Topological Complexity of Mmolecules. *Bulg. Chem. Commun.* **28** (1995) 567-582.

[5]   D. Bonchev, Novel Indices for the Topological Complexity of Molecules, *SAR QSAR Environ. Res.* **7** (1997) 23-43.

[6]   D. Bonchev, Overall Connectivities /Topological Complexities: A New Powerful Tool for QSPR/QSAR, *J. Chem. Inf. Comput. Sci.* **40** (2000) 934-941.

[7]   D. Bonchev and G. A. Buck, Quantitative Measures of Network Complexity. *Complexity in Chemistry, Biology and Ecology* (Editors: D. Bonchev and D. H. Rouvray), Springer, New York, (2005) 191-235.

[8]   D. Bonchev, The Overall Wiener Index - A New Tool for Characterization of Molecular Topology, *J. Chem. Inf. Comput. Sci.* **41** (2001) 582-592.

[9]   D. Bonchev and N. Trinajstić, Overall Molecular Descriptors. 3. Overall Zagreb Indices. *SAR QSAR Environ. Res.*, **12** (2001) 213-235.

[10] D. Bonchev, Overall Connectivity - A Next Generation Molecular Connectivity, *J. Mol.Graphics Model.* **5271** (2001) 1-11.

[11] H. Hosoya, Topological Index. A Newly Proposed Quantity Characterizing the Topological Nature of Structural Isomers of Saturated Hydrocarbons. *Bull. Chem. Soc. Jpn* **44** (1971) 2332-2339.

[12] L. B. Kier and L. H. Hall, *Molecular Connectivity in Chemistry and Drug Research.* Academic Press, New York, 1976.

Brill Academic Publishers
P.O. Box 9000, 2300 PA Leiden
The Netherlands

*Lecture Series on Computer
and Computational Sciences*
Volume 4, 2005, pp. 1558-1561

Graph Theory and Structure-Aromaticity Relationships

R. Bruce King[*]

Department of Chemistry
University of Georgia
Athens, Georgia 30602, USA

Received 6 April, 2005; accepted in revised form 10 July, 2005

Abstract: Three-dimensional aromaticity is now well-established in the deltahedral boranes as an extension of the familiar two-dimensional aromaticity in benzene and related aromatic hydrocarbons. Möbius aromaticity, in which the π-electron network is twisted as in a Möbius strip, can be postulated to occur in the five-center four-electron bond in the equatorial pentagon of pentagonal bipyramidal $RhBi_7(\mu-Br)_8$. This delocalized bonding in $RhBi_7(\mu-Br)_8$ can be contrasted with the localized bonding in the ions $[RhBi_6(\mu-X)_{12}]^{3-}$ (X = Br, I). Another type of aromaticity is the σ-aromaticity found in cyclopropanes. σ-Aromaticity can also account for the stability of triangular metal carbonyls such as $M_3(CO)_{12}$ (M = Fe, Ru, Os) relative to square metal carbonyls such as $Os_4(CO)_{16}$.

Keywords: Aromaticity, graph theory, polyhedra, Möbius strip

Mathematics Subject Classification: graphs and matrices, 3D polytopes, chemistry: molecular structure

PACS: 05C50, 52B10, 92E10

1. Introduction

A graph can be used to model a chemical structure with the vertices corresponding to atoms or atomic orbitals and the edges corresponding to chemical bonds or other interactions. Thus consider a graph G with v vertices and e edges. Its adjacency matrix $\mathbf{A}$ is a $v \times v$ matrix with $A_{ii} = 0$, $A_{ij} = 1$ if vertices i and j are connected by an edge, and $A_{ij} = 0$ if vertices i and j are not connected by an edge. The eigenvalues λ_i ($1 \leq i \leq v$) of G are solutions of the determinantal equation $|\mathbf{A} - \lambda\mathbf{I}| = 0$ where $\mathbf{I}$ is the "identity matrix" ($I_{ii} = 1$ and $I_{ij} = 0$ for $i \neq j$). The spectrum of a graph is the set of all of its eigenvalues. These eigenvalues relate to molecular orbital energy parameters for the chemical system modelled by the graph G using simple Hückel theory. Thus the "topological resolution" of the secular equation $|\mathbf{H} - E\mathbf{S}| = 0$ in terms of $\mathbf{A}$ and $\mathbf{I}$ gives $\mathbf{H} = \alpha\mathbf{I} + \beta\mathbf{A}$ and $\mathbf{S} = \mathbf{I} + S\mathbf{A}$. This relates to the following equation for the relationship of molecular orbital energy levels to the eigenvalues λ of the adjacency matrix $\mathbf{A}$:

$$E = \Sigma\varphi\acute{\alpha}\lambda\mu\alpha! \tag{1}$$

Since the energy unit $\beta < 0$, positive or negative values of λ correspond to bonding and antibonding orbitals, respectively. In Hückel theory S is taken to be zero and α is the energy zero so $E = \lambda\beta$.

2. Three-dimensional Aromaticity in Deltahedral Boranes

This graph theoretical approach [1,2] can be used to demonstrate the analogy between the aromaticity in two-dimensional planar polygons such as benzene, cyclopentadienide, and tropylium and three-dimensional deltahedral boranes, where a deltahedron is a polyhedron where all faces are triangles. In both cases the vertex atoms are assumed to contribute three orbitals to the skeletal bonding. Two of these orbitals, namely the twin internal orbitals, are equivalent whereas the third internal orbital is unique. In a two-dimensional system with n vertices such as benzene ($n = 6$) the twin internal orbitals are trigonal sp^2 hybrids and overlap pairwise around the circumference of the polygon leading to the σ-bonding network with n bonding and n antibonding orbitals; graph-theoretically this corresponds to n disconnected K_2 graphs. The unique internal orbitals in a polygonal system overlap cyclically

---

[*] Corresponding author. E-mail: rbking@sunchem.chem.uga.edu.

corresponding to a C_n cyclic graph which has $2k + 1$ positive eigenvalues corresponding to the π-bonding. This leads to the famous $4k + 2$ π-electron rule for aromatic hydrocarbons.

Consider now the deltahedral boranes, $B_nH_n^{2-}$, as three-dimensional analogues of the planar hydrocarbons. The deltahedra involved are the so-called "most spherical" deltahedra where all of the vertices are as similar as possible (Figure 1). In the case of an n-vertex deltahedron the twin internal orbitals are p orbitals, also called tangential orbitals. They overlap pairwise in the surface of the deltahedron to give n bonding and n antibonding orbitals similar to the σ-bonding in the planar polygons discussed above. The unique internal orbitals are sp hybrids, also called radial orbitals, which form an n-center bond in the core of the deltahedron. The corresponding overlap topology can be described by the K_n complete graph, which has one positive eigenvalue and $n-1$ negative eigenvalues leading to one core bonding orbital regardless of the value of n. Combining the n surface bonding orbitals and the single core bonding orbital leads to $n + 1$ bonding orbitals for an n-vertex deltahedron in accord with the Wade-Mingos rules [3] of $2n + 2$ skeletal electrons for deltahedral boranes with n vertices.

| 6 vertices: Octahedron | 7 vertices: Pentagonal Bipyramid | 8 vertices: Bisdisphenoid ("D_{2d} Dodecahedron") | 9 vertices: 4,4,4-Tricapped Trigonal Prism | 10 vertices: 4,4-Bicapped Square Antiprism | 11 vertices: Edge-coalesced Icosahedron | 12 vertices: Icosahedron |

Figure 1: The borane deltahedra.

3. High-Spin Versus Low-Spin Aromaticity: A Vanadoborate as "High Spin Benzene"

The aromaticity in both planar aromatic hydrocarbons and deltahedral boranes involves interaction between orbitals on adjacent atoms, i. e., atoms connected by an edge of the polygon or deltahedron. In such systems the energy parameter β (e.g., equation 1) is large compared with the electron pairing energy so low-spin systems are obtained. However, aromaticity can also be observed in metal oxide structures where the metal atoms contributing the orbitals for delocalization are separated by oxygen atoms in M–O–M linkages. In such cases β can be much smaller so that high-spin aromatic systems can be found in certain metal oxide structures. The distinction is similar to low-spin and high-spin coordination complexes, e.g., $Fe(CN)_6^{3-}$ versus FeF_6^{3-}.

An interesting case is the macrohexagonal vanadoborate $V_6B_{20}O_{50}H_6^{8-}$ in which a vanadium macrohexagon is imbedded into an electronically inert borate matrix [4]. The paramagnetism of this species corresponds approximately to the four unpaired electrons expected if electron pairing in the hexagon molecular orbitals is incomplete (Figure 2). Thus $V_6B_{20}O_{50}H_6^{8-}$ may be regarded as a "high spin" analogue of benzene [5].

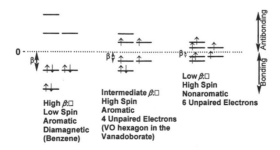

Figure 2: High-spin versus low spin aromaticity.

4. Hückel versus Möbius Aromaticity: A Rhodium Bismuth Bromide as a Möbius Aromatic System

The so-called Hückel aromaticity in the cyclopentadienide anion, $C_5H_5^-$, like that in benzene, arises from the π-bonding involving cyclic overlap of p orbitals without any phase change (Figure 3a). However, there is a pentagonal bipyramidal rhodium bismuth bromide $RhBi_7(\mu\text{-}Br)_8$ where there is a phase change in the overlapping orbitals of the equatorial pentagon atoms analogous to the twist in a Möbius strip [6]. Delocalization of this type leads to Möbius aromaticity with a molecular orbital pattern inverted from the standard Hückel molecular orbital pattern. Using the graph-theory derived model in equation (1), the phase-change or "twist" in the Möbius aromatic system corresponds to a "–1" entry in the adjacency matrix **A**. The interaction between the five orbitals in a Möbius pentagon (e.g., Figure 3) can equivalently be considered as a five-center four-electron (5c-4e) bond.

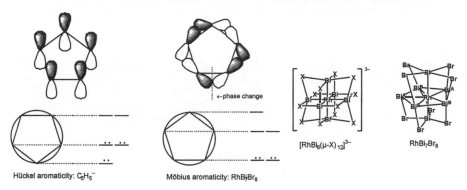

Hückel aromaticity: $C_5H_5^-$

Möbius aromaticity: $RhBi_7Br_8$

$[RhBi_6(\mu\text{-}X)_{12}]^{3-}$

$RhBi_7Br_8$

Figure 3: (a) Hückel versus Möbius aromaticity in a pentagon; (b) A comparison of octahedral $[RhBi_6(\mu\text{-}X)_{12}]^{3-}$ and pentagonal bipyramidal $RhBi_7(\mu\text{-}Br)_8$.

Compare the two rhodium bismuth bromides in Figure 3b. The structure of the octahedral anion $[RhBi_6(\mu\text{-}X)_{12}]^{3-}$ with non-bonding Bi–Bi distances of 3.83 Å can be constructed exclusively from 6 Rh–Bi and 24 Bi–X 2c-2e bonds and corresponds to an octahedral d^6 Rh(III) complex with six BiI_2 ligands. However, the structure of the pentagonal bipyramidal $RhBi_7(\mu\text{-}Br)_8$ has potentially bonding Bi–Bi distances of 3.2 to 3.3 Å and requires a 5c-4e Bi_5 bond from the Bi_5 Möbius pentagon in addition to 7 Rh–Bi 2c-2e bonds and 16 Bi–Br 2c-2e bonds.

5. σ-Aromaticity in Metal Carbonyl Triangles: Relationship to Möbius Aromaticity

The metal carbonyl triangles $M_3(CO)_{12}$ (M = Fe, Ru, Os) are very stable compounds and the Ru and Os derivatives, without bridging CO groups, may be considered as metal-carbonyl analogues of cyclopropane. Thus the metal atoms in $M_3(CO)_{12}$ have the favored 18-electron configurations if the M_3 metal triangles consist of 2c-2e bonds. However, no similar metal carbonyl analogues of higher cycloalkanes are known except for $Os_4(CO)_{16}$, an analogue of cyclobutane. In contrast to the very stable $Os_3(CO)_{12}$ the cyclobutane analogue $Os_4(CO)_{16}$ is unstable at room temperature.

The difference in stability between triangular $Os_3(CO)_{12}$ and square $Os_4(CO)_{16}$ can arise from σ-aromaticity [7] similar to that used to explain some special properties of cyclopropanes relative to higher cycloalkanes. In this connection the metal-metal bonding in the Os_3 triangle of $Os_3(CO)_{12}$ is not formulated simply as consisting of three 2c-2e bonds in an edge-localized bonding model but instead as one 3c-2e bond and one 3c-4e bond in a σ-aromatic model (Figure 4). Both bonding schemes use six orbitals and six electrons. In the σ-aromatic model for $Os_3(CO)_{12}$ the 3c-2e bond in the center of the Os_3 triangle is analogous to the 3c-2e B–B–B bonds found in some triangles of borane deltahedra and fragments thereof. The 3c-4e bond around the periphery of $Os_3(CO)_{12}$ in the σ-aromatic model is a Möbius triangle with a single phase change similar to the Möbius pentagon depicted in Figure 3a.

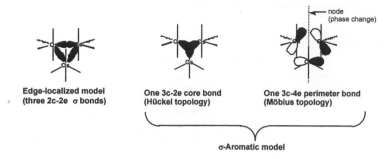

Figure 4: A comparison of the edge-localized bonding model and the σ-aromatic model for $Os_3(CO)_{12}$.

Acknowledgments

The author is indebted to the National Science Foundation for partial support of this work under Grant CHE-0209857.

References

[1] R. B. King and D. H. Rouvray, Chemical Applications of Group Theory and Topology. VII. A Graph-Theoretical Interpretation of the Bonding Topology in Polyhedral Boranes, Carboranes, and Metal Clusters, *J. Am. Chem. Soc.*, **99** (1977) 7834–7840.

[2] R. B. King., Three-dimensional Aromaticity in Polyhedral Boranes and Related Molecules, *Chem. Revs.* **101** (2001) 1119–1152.

[3] K. Wade, Structural Significance of the Number of Skeletal Bonding Electron-pairs in Carboranes, Higher Boranes, and Borane Anions, and Various Transition Metal Carbonyl Cluster Compounds, *Chem. Commun.* (1972) 792–793.

[4] D. J. Rose, J. Zubieta, and R. Haushalter, Hydrothermal Synthesis and Characterization of an Unusual Polyoxovanadium Borate Cluster: Structure of $Rb_4[(VO)_6\{B_{10}O_{16}(OH)_6\}_2]\cdot0.5\ H_2O$, *Polyhedron* **17** (1998) 2599

[5] R. B. King, Aromaticity in Transition Metal Oxide Structures, *J. Chem. Inf. Comput. Sci.* **41** (2001) 517–526.

[6] R. B. King, Möbius Aromaticity in Bipyramidal Rhodium-Centered Bismuth Clusters, *J. Chem. Soc., Dalton Trans.* (2003) 395–397.

[7] R. B. King., Metal Cluster Topology. 21. Sigma Aromaticity in Triangular Metal Carbonyl Clusters, *Inorg. Chim. Acta* **350** (2003) 126–130.

Brill Academic Publishers
P.O. Box 9000, 2300 PA Leiden
The Netherlands

Lecture Series on Computer
and Computational Sciences
Volume 4, 2005, pp. 1562-1565

Back-of-Envelope Prediction of Aromaticity of Non-Benzenoid Conjugated Hydrocarbons

Haruo Hosoya[1]

Emeritus Professor
Ochanomizu University,
Bunkyo-ku, Tokyo 112-8610, Japan

Received 4 July, 2005; accepted 10 July, 2005

Abstract: For deciphering the secret of the Hückel's $(4n+2)$-rule and for extending it to polycyclic systems the aromaticity index, ΔZ, is introduced based on the graph-theoretical molecular orbital method. All the information either stabilizing or destabilizing the π-electronic system of a given graph **G** is contained in the characteristic polynomial, $P_G(x)$, obtained by expanding the secular determinant of HMO theory. The present author derived the general expression of $P_G(x)$ in terms of the non-adjacent number, $p(G,k)$, for **G**, the sum of which gives the topological index, Z_G of **G**. By using ΔZ mathematical origin of the Hückel's rule was clarified and expanded to the "extended Hückel's rule" for polycyclic conjugated systems.

Keywords: conjugated hydrocarbon, π-electronic structure, topological index, aromaticity index, graph-theoretical molecular orbital method

Mathematics Subject Classification: 92E99
PACS: 05C75, 68R10, 81V55

1. Aromaticity

Benzene and cyclobutadiene are respectively called typical aromatic and anti-aromatic hydrocarbon molecules [1]. Both from experimental and theoretical standpoints so many different characteristic quantities representing their aromatic character or aromaticity have been assigned to these two molecules, between which all the conjugated hydrocarbon molecules find their respective positions with respect to aromatic character. However, as the number of rings and the variety of the size of the component rings increase, say from three to eight or nine, discussion on their aromaticity becomes more complicated. The motivation of this study is to obtain the mathematical foundation for the secret of this Hückel's $(4n+2)$-rule and to derive an extended version of this rule applicable to polycylic network systems.

Aromaticity is not an observable quantity but is a quality which can freely be defined for various π-electronic properties of those molecules originated from or related to benzene and cyclobutadiene. Thus for clarifying the nature and origin of aromaticity theoretical studies have big advantage of discussing a variety of hypothetical molecules as in Fig. 1 which may or may not exist. Actually **665** is a stable molecule known as acenaphthylene, while **755** is an olefinic substance known as one of Hafner's hydrocarbons [2]. All others in Fig. 1 have not been synthesized yet or have no chance of their existence. With the Hückel molecular orbital (HMO) method their relative stabilities of these π-electronic systems can be predicted semi-quantitatively under the assumption that they take planar structure. Namely, the order of their stability is calculated as in the following order,

$$665.> 953 > 755 > 863 >> 764 > 854 > 944 > 773. \tag{1}$$

---

[1].E-mail: hosoya@is.ocha.ac.jp

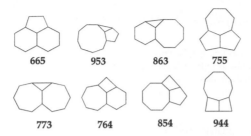

Figure 1: Carbon atom skeletons of the isomers of 12π-3-ring conjugated hydrocarbons.

Note that the molecules containing a three-membered ring and **944** with two tetragons are anticipated to be highly strained. Then these four molecules can be eliminated from our target to be scrutinized. Actually *ab initio* molecular orbital calculation of HF/6-31G(d,p) level [3] has shown that **953, 863, 773**, and **944** are non-planar, and the relative stability of our target molecules is predicted as in the following order,

$$665 \gg 755 \gg 854 > 764. \tag{2}$$

Contrary to these sophisticated molecular orbital calculations the present author defined the aromaticity index, ΔZ [4], using the topological index, Z_G, which has been proposed by him in 1971 [5] for characterizing the topological nature of a graph representing the carbon atom skeleton of a hydrocarbon molecule. With the aid of the recursion formulas the ΔZ values for the graphs in Fig. 1 can quite easily be obtained by back-of-envelope calculation. The results are given in the following order,

$$665 \gg 755 \gg 764 > 854. \tag{3}$$

The purpose of the present paper is to introduce this aromaticity index and to clarify why this simple index can sort out rather correctly the relative π-electronic stability of conjugated hydrocarbon molecules.

2. Topological Index and Characteristic Polynomial

First define the non-adjacent number, $p(G,k)$, as the number of ways for choosing k disjoint edges from **G**, with $p(G,0)$ being unity for any graph. Then the Z-index (originally coined as "topological index" by the present author [5]) is defined as the sum of all the $p(G,k)$'s for **G**, or

$$Z_G = \sum_{k=0}^{m} p(G,k). \tag{4}$$

where m is the maximum number of $k=[N/2]$ for **G** with N vertices.

The Z-indices for several series of graphs are found to be related to important series of numbers. Namely, the Z values of alkanes (or alkenes), or path graphs, $\{S_N\}$, form the family of the Fibonacci numbers ($F_N = 1, 1, 2, 3, 5, ...$), while those of cyclic alkanes (or alkenes) or monocyclic graphs, $\{C_N\}$, are Lucas numbers ($L_N = 2, 1, 3, 4, 7, ...$).

By using $p(G,k)$'s the matching polynomial [6], acyclic polynomial [7], or reference polynomial [8], $M_G(x)$, was proposed by several groups to be defined using the $p(G,k)$'s as

$$M_G(x) = \sum_{k=0}^{m} (-1)^k p(G,k) x^{N-2k}. \tag{5}$$

In HMO theory the (molecular) orbital energies of a given π-electronic system **G** composed of N carbon atoms are the zeroes of the characteristic polynomial, $P_G(x)=0$, which is defined in terms of the adjacency matrix A and unit matrix E of **G** as

$$P_G(x) = (-1)^N \det(A - x E). \tag{6}$$

The present author has shown [5] that for a tree graph $P_G(x)$ is identical to $M_G(x)$. This means that $P_G(x)$ can be obtained just from the counting of $p(G,k)$ numbers without decomposing the secular determinant.

$$P_G(x) = \sum_{k=0}^{m} (-1)^k p(G,k) x^{N-2k} = M_G(x) \quad (G \in \text{tree}). \tag{7}$$

On the other hand for non-tree graphs the $P_G(x)$ is shown to be expressed in terms of the $p(G,k)$ numbers of G and its subgraphs as in Eq. (8) [9], where $G \ominus R_r$ is the subgraph of G obtained by deleting ring R_r together with all the edges incident to R_r.

$$P_G(x) = \sum_{k=0}^{m} (-1)^k p(G,k) x^{N-2k}$$

$$+ (-2) \sum_{r}^{\text{Ring}} \sum_{k=0}^{m_r} (-1)^{k+n_r} p(G \ominus R_r, k) x^{N-n_r-2k}$$

$$+ (-2)^2 \sum_{r>s}^{\text{Ring}} \sum_{k=0}^{m_{rs}} (-1)^{k+n_r+n_s} p(G \ominus R_r \ominus R_s, k) x^{N-n_r-n_s-2k}$$

$$+ \cdots$$

$$= M_G(x) - 2 \sum_{r}^{\text{Ring}} M_{G \ominus R_r}(x) + 4 \sum_{r>s}^{\text{Ring}} M_{G \ominus R_r \ominus R_s}(x) - \cdots \tag{8}$$

The first term is the matching polynomial $M_G(x)$ of G with no ring contribution, and the second summation term counts the contribution caused by the isolated rings, R_r's. The third term is the contribution caused by the pairs of disjoint rings, R_r and R_s, composed, respectively, of n_r and n_s vertices, and so on. Note that $P_G(x)$ of a polycyclic graph is thus expressed by the set of the matching polynomials of G and the subgraphs obtained by deleting a ring or set of disjoint rings from G.

The total π-electron energy, E_π, of a given hydrocarbon with N (usually even) carbon atoms is the double of the half sum of the solutions of $P_G(x)=0$ as

$$E_\pi = 2 \sum_{k=1}^{[N/2]} x_k. \tag{9}$$

From Eq. (8) one can analyze how the component rings and their combinations stabilize and destabilize the π-electronic state of polycyclic hydrocarbon molecules. The secret of aromaticity is hidden in Eq. (8).

3. Aromaticity Index

By taking into account the meaning of Eq. (8) the aromaticity index, ΔZ, is defined as the weighted sum of the Z_G's of the set of subgraphs obtained by deleting even-membered rings as follows:

$$\Delta Z_G = 2 \sum_{\substack{r=1 \\ n=4k+2}}^{R} Z_{G \ominus R} \quad \bigcirc, \; \bigcirc\!\!\bigcirc, \; \bigcirc\!\!\bigcirc\!\!\bigcirc, \; \cdots$$

$$-2 \sum_{\substack{r=1 \\ n=4k}}^{R} Z_{G \ominus R} \quad \square, \; \bigcirc, \; \bigcirc\!\!\bigcirc, \; \cdots$$

$$+4 \sum_{\substack{r=2 \\ n=4k}}^{R} Z_{G \ominus R} \quad \triangle + \bigcirc, \; \square + \square, \; \triangle + \bigcirc\!\!\bigcirc, \\ \square + \bigcirc, \; \bigcirc + \bigcirc, \; \bigcirc + \bigcirc, \; \cdots$$

$$-4 \sum_{\substack{r=2 \\ n=4k+2}}^{R} Z_{G \ominus R} \quad \triangle + \triangle, \; \triangle + \bigcirc, \; \square + \bigcirc, \\ \bigcirc + \bigcirc, \; \square + \bigcirc\!\!\bigcirc, \; \cdots$$

$$+8 \sum_{\substack{r=3 \\ n=4k+2}}^{R} Z_{G \ominus R} \quad \triangle + \triangle + \square, \; \square + \square + \bigcirc, \\ \bigcirc + \bigcirc + \bigcirc, \; \bigcirc + \bigcirc + \bigcirc, \; \cdots$$

$$+ \cdots \tag{10}$$

This equation shows two important factors governing the π-electronic stability of a ring network, namely, the value of the Z-index of subgraph $G \ominus R$ and the sign given to each term. The first two lines give the well-known Hückel's rule. Namely, a $(4n+2)$-membered ring stabilizes, while $(4n)$-membered ring destabilizes the π-electronic conjugated system. The third line tells us that a pair of disjoint rings with a total of $4n$ vertices stabilizes, while the fourth line shows a destabilization contribution by a pair of disjoint rings with a total of $4n+2$ vertices.

The selection rule for the combined contribution from odd number of rings goes back to the case of a single ring, while for the case with even number of disjoint rings the selection rule is the same as the case with a pair of disjoint rings.

We have seen in the above discussion that the value of $\Delta\square$ for tree graphs is always zero, while for $(4n+2)$-membered monocyclic graph +2, and for $(4n)$-membered monocyclic graph −2 (Hückel's rule).

4. Calculation

Finally the values of $\Delta\square$ are compared with the E_π of HMO and ΔE by *ab initio* MO calculations in Table 1. By checking the contribution from each term to $\Delta\square$ in Eq. (10) one can easily understand the reason why each hydrocarbon becomes aromatic or not. This is the biggest advantage of the aromaticity index, $\Delta\square$.

Table 1: Comparison of aromaticity index with molecular orbital calculations for molecules in Figure 1.

Compound	$\Delta\square$	$E_\pi\ \square\square\square$	ΔE (kcal/mole)
665	+ 56	16.6189	0
755	− 2	16.3660	17.81
764	−52	15.9662	46.62
854	−74	15.8617	35.43

Acknowledgments

The author thanks Professor Kichisuke Nishimoto for executing *ab initio* molecular orbital calculations.

References

[1] E. Hückel: Quanten theoretische Beiträge zum Benzolproblem. I, *Zeitschrift für Physik* **70** (1931) 204–272

[2] K. Hafner and J. Schneider: Vorbereitung und Eigenschaften der Pentalen und Heptalen Derivate, *Annalen der Chemie* **624** (1959) 37–47.

[3] K. Nishimoto: private cmmunication.

[4] H. Hosoya, K. Hosoi, and I. Gutman: A Topological Index for the Total π-Electron Energy, *Theoretica Chimica Acta* **38** (1975) 37–47.

[5] H. Hosoya: Topological Index. A Newly Proposed Quantity Characterizing the Topological Nature of Structural Isomers of Saturated Hydrocarbons, *Bulletin of the Chemical Society of Japan* **44** (1971) 2332–2339.

[6] E. J. Farrell: An Introduction to Matching Polynomial, Journal of Combinatorial Theory **B27** (1979) 75–86.

[7] I. Gutman, M. Milun, and N. Trinajstic: Non-parametric Resonance Energies of Arbitrary Conjugated Systems, Journal of American Chemical Society **99** (1977) 1692–1704.

[8] J. Aihara: A New Definition of Dewar-type Resonance Energy, Journal of American Chemical Society **98** (1976) 2750–2758.

[9] H. Hosoya: Graphical Enumeration of the Coefficients of the Secular Polynomials of the Hückel Molecular Orbitals, *Theoretica Chimica Acta* **25** (1972) 215–222.

Brill Academic Publishers
P.O. Box 9000, 2300 PA Leiden
The Netherlands

*Lecture Series on Computer
and Computational Sciences*
Volume 4, 2005, pp. 1566-1569

Molecular Modeling of Polyphenols from Croatian Wines

M. Medić-Sarić[*,1], A. Mornar[1], V. Rastija[2], I. Jasprica[1] and S. Nikolić[3]

[1]Department of Pharmaceutical Chemistry,
Faculty of Pharmacy and Biochemistry,
University of Zagreb,
HR-10 000 Zagreb, A.Kovacica 1, Croatia
[2]Faculty of Agriculture, University J.J. Strossmayer
HR-31000 Osijek, Trg Sv. Trojstva 3, Croatia
[3]The Rugjer Bošković Institute, P.O.B. 180, HR-10002 Zagreb, Croatia

Received 7 July, 2005; accepted 12 July, 2005

Abstract: Because of health benefits of phenolic compounds in wine, it is important to investigate their structure-property relationship. We have investigated linear and nonlinear (polynomial) relationships between given topological indices and lipophilicity of polyphenols, main pharmacological active components of wine.

Keywords: Polyphenols, Topological Indices, Lipophilicity, QSPR

Mathematics Subject Classification: 92E10

1. Introduction

Phenolic compounds are secondary plant metabolites and they have been implicated in number of varied roles including UV protection, pigmetation and disease resistance. Wines, especially red wines contain a wide range of phenolic compounds that include phenolic acids, the trihydroxystilbene resveratrol, the flavonols (e.g. quercetin and myricetin), flavan-3-ols (e.g. catehin and epicatehin), as well as polymers of the latter, defined as procyanidins and anthocyanins that are the pigments responsible for the colour of red wines. The processes of viticulture and vinification determine the content and profile of phenolic compounds in wine. Vineyard factors are variety, quality, climate, ageing, geographical origin and disease and vinification factors are lenght of skin contact and temperature [1]. White wines are usually made from the free running juice, without the grape mash, having no contact with the grape skins, and this is main reason for relatively low phenolic content and lower antioxidant acitivity of white wine in comparison to red wine [2]. Phenolic compounds play a major role in wine quality. They are responsible for sensory characteristic such as color and bitter flavour of wine. Most of them show beneficial physiological properties including cardioprotective, anticarcinogenic, anti-inflammatory and antioxidant activities [3,4].

Molecular structure of polyphenols is the basis of their molecular properties – from chemical and physical properties to the certain biological activity. Molecular structure is described with number of parameters that can be calculated from molecular topology, *e.g.* topological indices. The topological indices, as nonempirical structural parameters, are convenient tools to formulate direct relationships between chemical structure and physical, chemical and biological properties of molecules [5].

Lipophilicity is a physicochemical property of primary interest for the medicinal chemist determining pharmacokinetic and pharmacodinamic behavior of drugs. Its quantitative descriptor, log P (the logarithm of octanol-water partition coefficient), is one of the most important pharmacokinetic parameter, which describes oral absorption, cell uptake, protein binding, blood-brain penetration, metabolism and toxicity (ADME/Tox processes) of the bioactive substances. Traditionally partition coefficient is determined by the tedious and time-consuming "shake flask" method. The first attempts to

•   Corresponding author. Head of Department. E-mail: bebamms@pharma.hr predict log *P* values were developed by Hansch and Rekker. Since than, several calculative procedures, based on different theoretical approaches, have been developed [6].

2. Materials and Methods

Nine red and three white wines commercially available in Croatia were analyzed by TLC and HPLC methods [7]. List and structures of investigated polyphenols are given in Figure 1.

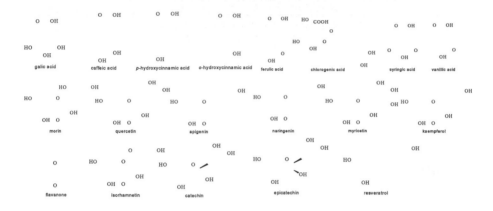

Figure 1: Structures and full names of studied polyphenols.

In this paper we have investigated whether five topological indices (Wiener index, valence connectivity index, Balaban index, information-theoretic index and Schultz index) are applicable to QSPR studies of polyphenols from wines. All indices used in our work were calculated using TAM program [8]. Molecular lipophilicity was calculated using three computer programs based on different theoretical approach (ALOGPS v2.1 – based on e-state indices and neural networks, CLOGP – based on fragmental method and MLOGP(DRAGON) – based on 13 parameters) [9-11]. Statistical analysis was performed using STATISTICA v6.0 (StatSoft).

3. Results and Discussion

Selected physico-chemical properties (molecular weight - *M.w.*, polar surface area - *PSA* and partition coefficient - log *P*) and calculated topological indices (Wiener index - *W*, valence connectivity index - χ^v, Balaban index – *J* and information theoretic index - *I*) for studied polyphenols are shown in Table 1.

Table 1: Physico-chemical properties (*M.w.*, *PSA*, and log *P*) and calculated topological indices (*W(G)*, $\chi^v(G)$, *J(G)*, *I(G)*, and *MTI(G)*) for investigated compounds.

Compound	N(G)	W(G)	J(G)	I(G)	$^0\chi(G)$	$^1\chi(G)$	M.w.	ALOGPs	CLOGP	MLOGP	PSA
gallic acid	12	246	3.371	2.774	5.696	2.925	170.12	1.17	0.43	0.03	17.07
caffeic acid	13	417	3.079	3.453	6.336	3.201	180.16	1.67	0.98	1.12	17.07
p-hydroxycinnamic acid	12	307	2.771	3.255	6.189	3.344	164.16	1.74	1.57	1.66	17.07
o-hydroxycinnamic acid	12	292	2.861	3.135	6.189	3.35	164.16	1.90	1.57	1.66	17.07
ferulic acid	14	444	3.018	3.309	7.52	3.873	194.16	1.58	1.42	1.42	26.30
chlorogenic acid	25	1960	2.215	3.786	12.707	7.225	354.31	0.16	-1.88	-0.04	43.37
syringic acid	14	368	3.498	2.872	7.696	3.742	198.18	1.55	1.07	0.68	35.53
vanillic acid	12	251	3.232	2.83	6.365	3.213	168.15	1.71	1.36	0.88	26.30
morin	22	1226	2.29	3.381	10.744	6.026	302.24	2.23	1.13	-0.75	26.30
quercetin	22	1252	2.269	3.417	10.744	6.026	304.26	1.07	0.77	-0.75	26.30

Table 1:continued

apigenin	20	1006	2.162	3.407	10.082	5.778	270.25	3.07	2.91	0.79	26.30
naringenin	20	952	2.105	3.341	10.29	6.033	272.27	2.47	2.45	0.90	26.30
myricetin	23	1403	2.303	3.434	11.075	6.147	318.24	1.66	0.84	-1.49	26.30
kaempferol	21	1111	2.26	3.392	10.413	5.905	286.24	1.99	1.37	0.01	26.30
flavanone	17	612	2.052	3.162	9.297	5.682	224.26	3.10	3.48	2.80	26.30
isorhamnetin	23	1510	2.353	3.587	11.586	6.201	316.27	1.96	1.95	-0.50	35.53
resveratrol	17	795	2.179	3.573	8.921	5.076	228.25	2.57	2.83	2.63	0.00
catechin	21	1057	2.109	3.353	10.698	6.269	290.27	1.02	0.53	0.25	9.23
epicatechin	21	1057	2.109	3.353	10.698	6.269	290.27	1.02	0.53	0.25	9.23

We have investigated linear and nonlinear (polynomial) relationships between given topological indices and selected properties of polyphenols, main pharmacological active components of wine. To test the quality and accuracy of derived models, following statistical parameters were used: number of data points (n), correlation coefficient (r), mean squares residual (MS) and F-ratio between the variances of observed and calculated values (F).

Preliminary QSPR modeling has shown that each group of polyphenols (flavonoids and phenolic acids) sholud be investigated separately. Linear regression analysis was found suitable to describe the relationship between MLOGP and Wiener index, $W(G)$:

$$MLOGP = 5.6518 \, (\pm 1.7156) - 0.0048 \, (\pm 0.0015)W \qquad (1)$$
$$n = 11, r = -0.9218, MS = 0.3060, F = 50.8556.$$

A plot of MLOGP *vs.* Wiener index, $W(G)$, is given in Figure 2.

The phenolic acids have shown excellent correlation between MLOGP values and information-theoretic indices, $I(G)$, as shown in Figure 3. and equation (2):

$$MLOGP = -58.7478 \, (\pm 20.8329) + 37.1698 \, (\pm 12.9312)I - 5.7275 \, (\pm 1.9906)I^2 \qquad (2)$$
$$n = 8, r = 0.9572, MS = 0.0527, F = 27.3524.$$

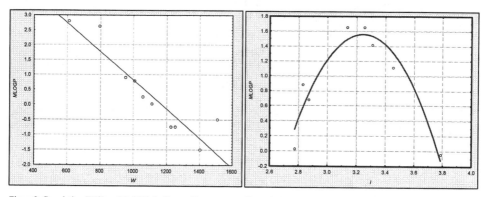

Figure 2: Correlation $W(G)$ *vs.* MLOGP for flavonoids.   Figure 3: Correlation $I(G)$ *vs.* MLOGP for phenolic acids.

4. Conclusion

The best QSPR models and statistical correlations of examined physicochemical properties and certain indices are based on linear or polynomial regression analysis. Lipophilicity is useful parameter for structure-property analysis that can be easily calculated for a polyphenols from wines and used for prediction of their pharmacological properties.

Acknowledgments

This work was supported in part by Grants No. 0006541 (M. M.-S.) and No. 0098034 (S. N.) awarded by the Ministry of Science, Education and Sport of the Republic of Croatia.

References

[1] J. Burns, P.T. Gardner, D. Matthews, G.G. Duthie, M.E.J. Lean and A. Crozier, Extraction of Phenolics and Changes in Antioxidant Activity of Red Wines during Vinification, *Journal of Agriculture and Food Chemistry* **49** (2001) 5797-5808.

[2] B. Fuhrman, N. Volkova, A. Suraski and M. Aviram, White Wine with Red Wine-like Properties: Increased Extraction of Grape Skin Polyphenols Improves the Antioxidant Capacity of the Derived White Wine, *Journal of Agriculture and Food Chemistry* **49** (2001) 3164-3168.

[3] G. Mazza, L. Fukumoto, P. Delaquis, B. Girard and B. Ewert, Anthocyanins, Phenolics, and Color of Cabernet Franc, Merlot, and Pinot Noir Wines from British Columbia, *Journal of Agriculture and Food Chemistry* **47** (1999) 4009-4017.

[4] N. Al-Awwadi, J. Azay, P. Poucheret, G. Cassanas, M. Krosniak, C. Auger, F. Gasc, J.-M. Rouanet, G. Cros and P.-L. Teissedre, Antidiabetic Activity of Red Wine Polyphenolic Extract, Ethanol, or Both in Streptozotocin-Treated Rats, *Journal of Agriculture and Food Chemistry* **52** (2004) 1008-1016.

[5] N. Trinajstić, *Chemical Graph Theory.* CRC Press: Boca Raton, FL, 1992.

[6] V. Pliska, R. Mannhold, H. van de Waterbeemd and H. Kubinyi, *Lipophilicity in Drug Action and Toxicology.* VCH, Weinheim, 1996.

[7] V. Rastija, A. Mornar, I. Jasprica, G. Srečnik and M. Medić-Šarić, Analysis of Phenolic Components in Croatian Red Wines by Thin-Layer Chromatography, *Journal of Planar Chromatography* **17** (2004) 26-31.

[8] M. Vedrina, S. Marković, M. Medić-Šarić and N. Trinajstić, TAM: a program for the calculation of topological indices in QSPR and QSAR studies, *Computers Chemistry* **21** (1997) 355-361.

[9] I. V. Tetko, V. Y. Tanchuk, T. N. Kasheva and A. E. P. Villa, Internet software for the calculation of the lipophilicity and aqueous solubility of chemical compounds, *Journal of Chemical Information & Computer Sciences* **41** (2001) 246–252.

[10] J. T. Chou and P. C. Jurs, Computer-assisted computation of partition coefficients from molecular structures using fragment constants,. *Journal of ChemicalInformation & Computer Sciences* **19** (1979) 172–178.

[11] I. Moriguchi, S. Hirono, I. Nakagome and H. Hirano, Comparison of reliability of log P values for drugs calculated by several methods, *Chemical and Pharmaceutical Bulletin* **42** (1994) 976–978.

Brill Academic Publishers
P.O. Box 9000, 2300 PA Leiden
The Netherlands

*Lecture Series on Computer
and Computational Sciences*
Volume 4, 2005, pp. 1570-1573

Nonlinear Multivariate Polynomial Ensembles in QSAR/QSPR

Damir Nadramija[1,a], Bono Lučić[b]

[a]SciBase International LLC,
21 Diablo Creek Pl, Danville 94506, USA

[b]The Rudjer Bošković Institute,
P. O. Box 180, 10002 Zagreb, Croatia,

Received 10 July, 2005; accepted in revised form 10 July, 2005

Abstract: In this study, which is part of Damir Nadramija's PhD thesis developed in collaboration with The Rudjer Bošković Institute, we demonstrate use of ensembles of linear and nonlinear multivariate regression models, based on multivariate polynomials of initial descriptors, in QSAR/QSPR modeling. Data sets, which varied significantly in size regarding number of variables and number of points, were all previously referenced in literature and molecular structures were either obtained from authors of these publications or generated in our laboratories. All data sets were encoded as SMILES and converted to 3D structures (SD files) by the CORINA program (www2.chemie.uni-erlangen.de/software/corina/). All descriptors were computed by the program DRAGON 2.1 (http://www.disat.unimib.it/chm/). Linear ensembles were built with multiple linear regression models (MLR) and nonlinear ensembles consisted of multivariate polynomials, which were constructed as controlled subsets selected among linear descriptors, their two-fold cross-products and squares, as well as cubic potencies of (only) single descriptors. Ensemble responses were computed as mean or median or weighted values of all intrinsic models. Models and ensembles discussed in this paper were constructed with the application NQSAR, a Windows console application, which is available upon request. Results obtained show clear advantage of nonlinear ensembles over linear counterparts when data sets contain 4 to 5 times more points than model coefficients. On the other side linear ensembles, which in general exhibit higher robustness and stability, are better suited for small data sets with many variables outperforming nonlinear ensembles in predicting values of data points from external data set. This can be explained by the fact that the linear models are less affected by small variations than nonlinear models while they equally benefit from the key ensemble features. Primarily, we note the impact of the inclusion of more variables spread across optimized variable subsets, which are used in ensembles' intrinsic models that individually satisfy before mentioned rule on over-fitting. The overall ensemble responses are more stable and robust with higher predictive powers than single models.

Keywords: QSAR/QSPR modeling, selection of the most relevant molecular descriptors, ensembles of multivariate regression models, linear and nonlinear models

Mathematics SubjectClassification: AMS-MOS: 62J02, 62J10, 62J12, 62H25, 90C29, 62–07
PACS: 02.50.Sk, 02.60.Ed, 89.75.Kd

1. Introduction

As the imperative to produce safe and effective orally available drugs continuously grows in line with stronger market demands, so the importance of the ADMETox modeling rises to the ever high levels. The high failure rate of drug candidates in the late development stage is often (about 50% failures) caused by the ADMETox (absorption, distribution, metabolism, excretion and toxicity)

---

[1] Corresponding author. *Damir Nadramija*, Head of [a]SciBase International LLC; E-mail: damir.nadramija@scibase.com

deficiencies, as well as adequate solubility of a molecule (drug candidate) is extremely important [1]. For this reason, development of effective computational approaches/models for early screening of ADMETox properties is a very important and promising field of research [2-4]. Here we discuss our experiences in developing new highly robust and prediction focused modeling methods based on ensembles of linear and nonlinear (polynomial) models.

2. Methods

Four data sets, containing experimental values of properties or activities of molecules, used in this study are: (1) solubility data set, with 1297 compounds [1], (2) toxicity data set, with 295 compounds [5], (3) log BBB data set, with 106 compounds [6], and (4) Selwood data set with 31 compounds and 53 descriptors [4]. Selwood data set was used to verify findings related to nonlinear ensembles behavior in sets with relatively small number of data points. In case of the solubility data set, we performed initial partitioning into training (1039 molecules) and test (258 molecules) sets, as it was done by Liu and So in their neural network study [1]. Partitions for other sets were set as 4:1 in favor of training sets vs. external (prediction) sets.

Initial structures of molecules were encoded as SMILES and converted into 3D structures (SD file) by the CORINA program (www2.chemie.uni-erlangen.de/software/corina/) and more than 1000 initial descriptors were computed by the Dragon 2.1 program (http://www.disat.unimib.it/chm/). The primary set of descriptors was filtered in order to remove non-significant and highly inter-correlated descriptors (123 descriptors were left after filtering in solubility data set, 186 in toxicity data set, 337 in log BBB data set, and for Selwood data set filtering was not performed and 53 initial descriptors were used).

Ensemble construction used a quasi-genetic algorithm [7] to select best possible combinations of initial descriptors for respective linear or nonlinear models. In each search step we used either a generalized Levenberg-Marquardt technique for fitting nonlinear polynomials or a standard MLR routine for fitting linear models [2-4]. The best models (corresponding to the number of ensemble elements) were retained after the search stage as ensemble models. Ensemble responses were computed based on the ensemble type (mean, median or weighted) and residual statistics and the goodness of fit statistics were performed. The cross-validation (CV) of all ensemble models as well as of single models was performed using a standard leave-one-out (LOO) procedure on the respective training set. As the aim of this study was to create more robust models with higher predictive power, at the end of the processing of each data set, an external set of molecules was used to calculate predicted values. Subsequent residual statistics (fit, LOO cross-validated, predicted) yielding the correlation coefficient (R) between the calculated (or predicted) and observed values, as well as the standard error (S). The same process was repeated with all other sets used in this study.

Nonlinear polynomials were defined as

$$P_n^m(x_n) = \prod_i^n P^m(x_i) \tag{1}$$

where m = (1,2,3) and n=(number of variables in the model). However, as only free term, linear, quadratic, cubic and two-fold cross-products of initial (linear) terms were allowed the number of coefficient was kept at the reasonable level. For example, for a multivariate polynomial of quadratic order a number of produced coefficients was equal to

$$1 + 2 \bullet n + (n \bullet (n-1)/2) \tag{2}$$

3. Results

Correlation coefficient (R), standard error of fit, standard error of LOO CV and standard error of prediction of linear and nonlinear ensembles are displayed in Table 1. In general, ensemble responses were better and more robust than single models' behavior; although in nonlinear ensembles this is somewhat dependent on the over fitting problems of intrinsic nonlinear models for data set with fewer data points. Automatic removal of outliers produced much more favorable statistics not presented in this extended abstract.

Table 1: Correlation coefficients (R) and standard errors (S) for data sets and ensembles displaying main features of linear and nonlinear ensembles dependent on the set size and ensemble parameters

Data Set/Ensemble	Ensemble R / S	Best model R / S	LOO CV R / S	Predicted R / S
Solubility, 100 linear models with 4-7 vars.	0.8902/0.9315	0.8947/0.9122	0.8880/0.9401	0.8920/0.9062
Solubility, 100 linear models with 9 vars.	0.9131/0.8442	0.9122/0.8652	0.9112/0.853	0.8876/0.8881
Solubility, 100 nonlinear models with 4-7 vars.	0.9548/0.5931	0.9502/0.6209	0.9508/0.6179	0.9576/0.6309
Solubility, 100 nonlinear models with 9 vars.	0.9628/0.4255	0.9610/0.4351	0.9568/0.4573	0.9454/0.6481
Toxicity, 100 linear models with 4-7 vars.	0.9239/0.546	0.9155/0.5736	0.9156/05731	0.9444/0.5289
Toxicity, 100 nonlinear models with 4-7 vars.	0.9739/0.3272	0.9685/0.3581	0.9358/0.5079	0.9159/0.4278
Log BBB, 100 linear models with 4-7 vars.	0.8637/0.4228	0.8447/0.4467	08354/04591	0.6517/0.5958
Log BBB, 100 nonlinear models with 4-7 vars.	0.984/0.1501	0.9675/0.2066	0.7109/0.6582	0.5879/1.0143
Selwood, 50 linear models with 4-7 vars.	0.9618/0.2308	0.9561/0.2474	0.8978/0.3713	0.8922/0.9391
Selwood, 50 nonlinear models with 3-5 vars.	1.00/0.0004	0.9997/0.0026	0.7158/10175	0.6405/0.838

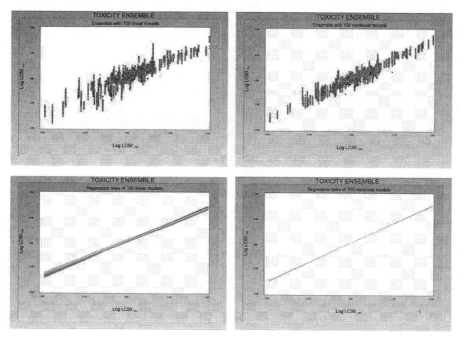

Figure 1: Illustration of main differences between linear and nonlinear ensembles developed on the toxicity data set. Spread of models in linear ensemble vs. practical colinearity of nonlinear models in a nonlinear ensemble.

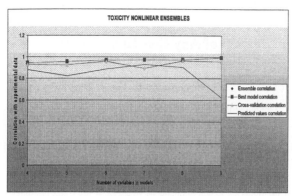

Figure 2: Effect on the stability of nonlinear ensemble correlations based on number of variables

4. Conclusions

As this paper is focused on a method of developing such models and ensembles and not on the models themselves, a number of data sets that were used as well as respective models developed were not further discussed nor compared to previously published models. Results obtained show that ensembles, in general, exhibit higher robustness and more stable predictive power than single models, but that they are still prone to the same maladies related to over fitting that originate from the problems inherent to their intrinsic models. Nevertheless, nonlinear ensembles seem as an excellent choice for modeling activities/properties of large data sets.

Acknowledgments

The authors wish to thank Prof. Dr. Nenad Trinajstić (Zagreb), Dr. Dragan Amić (Osijek) and Dr. Ivan Bašic (Zagreb), all from Croatia, for their help and guidance in preparing data sets, designing experiments, as well as for their comments and suggestions.

References

[1] R. Liu, S.-S. So, Development of quantitative structure-property relationship models for early ADME evaluation in drug discovery. 1. Aqueous solubility. *J. Chem. Inf. Comput. Sci.* **41** (2001) 1633-1639.

[2] Lučić B, Trinajstić N, Multivariate regression outperforms several robust architectures of neural networks in QSAR modeling. *J. Chem. Inf. Comput. Sci.* **39** (1999) 121-132.

[3] Lučić B, Amić D, Trinajstić N, Nonlinear multivariate regression outperforms several concisely designed neural networks on three QSPR data sets. *J. Chem. Inf. Comput. Sci.* **40** (2000) 403-413.

[4] Lučić B, Nadramija D, Bašic I, Trinajstić N, Toward generating simpler QSAR models: nonlinear multivariate regression versus several neural network ensembles and some related methods. *J. Chem. Inf. Comput. Sci.* **43** (2003) 1094-1102.

[5] Katritzky, A. R.; Tatham, D. B.; Maran. U. Theoretical Descriptors for the Correlation of Aquatic Toxicity of Environmental Pollutants by Quantitative Structure-Toxicity Relationships, *J. Chem. Inf. Comput. Sci.* **41** (2001) 1162-1176.

[6] Rose K, Hall LH, Kier LB. Modeling blood-brain barrier partitioning using the electrotopological state. *J Chem Inf Comput Sci.* **42** (2002) 651-666.

[7] Dunn, W.; Rogers, D. Genetic Partial Least Squares in QSAR. In *Genetic Algorithms in Molecular Modeling*; Devillers, J. Ed.; Academic Press: London 1996; pp 109-130.

Brill Academic Publishers
P.O. Box 9000, 2300 PA Leiden
The Netherlands

Lecture Series on Computer
and Computational Sciences
Volume 4, 2005, pp. 1574-1577

Simple Approximation of Any Constants of Homologues Using Single Recurrent Function

I.G. Zenkevich[1]

Chemical Research Institute, St. Petersburg State University,
Universitetsky pr., 26, St. Petersburg 198504, Russia

Received 28 May, 2005; accepted in revised form 10 July, 2005

Abstract: The dependencies of various physicochemical constants of organic compounds (A) *vs.* number of carbon atoms in the molecule within homologous series [A = f(n_C)] are non-linear. However the simplest recurrent equation $A(n+1) = a\,A(n) + b$, connecting any A-values for homologues with the values of the same constants for previous members of series, indicates practically "ideal" linear character for most of all known properties of organic compounds. This fact permits us to approximate (or to extrapolate) any physicochemical data within any homologous series using the standard approach without special search of complex algebraic functions.
Principal mathematical properties of the function $A(n+1) = a\,A(n) + b$ are considered.

Keywords: Organic compounds; Homologous series; Physicochemical constants; Approximation; Recurrent function; Mathematical properties; Applications

Mathematical Subject Classification: 92E99
PACS: 03.65.Fd, 82.80-d

1. Introduction

Various physicochemical constants of organic compounds are used in the characterization and identification of organic compounds up to present [1]. However, not a total multitude of these objects is exhaustively characterized by all known constants. Many compounds have been obtained only once many years ago [2] and since that time never been re-synthesized again. Various series are characterized by constants of only simplest homologues owing to the increase the number of possible isomers for higher members of them [3]. These objectives explain us the importance of theoretical precalculation of various physicochemical constants.

Numerous methods for the prediction of the values of different properties have been elaborated in physical chemistry [4]. Most of them are based on the using of other available constants of the same compounds. Another principal approach implies the comparison of the set of constants for compounds of one series with the set of the same constants for objects of another series (structural analogues) [5-7].

The approach of the third type implies the estimation of relationships connected the values of any constants of organic compounds with their position within corresponding homologous series, i.e., functional dependencies A = f(n_C), where n_C is the number of carbon atoms in the molecule.

An objective difficulty in the solution of this problem is non-linear character of these dependencies. As an example two plots A = f(n_C) are presented on Figs 1 and 2, namely for: boiling points (T_b, °C) of perfluoro-*n*-alkanes (Fig. 1); relative densities (d_4^{20}, non-dimensional constant) of *n*-alkyl bromides (Fig. 2).

These non-linear dependencies can be both prominent and concave, having asymptotes or not. Very often the choice of these functions has been fulfilled empirically. The example of that is the most precise equation (named after Kreglewsky) for the approximation of boiling points of alkanes $C_1 - C_{100}$ [8]. Similarly, the function $A = a\,n_C^k + b$ was proposed for the approximation of refractive indices and

---

[1] Corresponding author. Head of Gas Chromatographic laboratory of Chemical Research Institute of St. Petersburg State University. E-mail: igor@IZ6246.spb.edu

relative densities [9]. Other examples are non-linear three- and four-parameters dependencies for precalculation of GC RIs using reference data on boiling points [10-12].

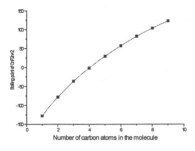

Figure 1: Typical non-linear dependence of boiling points (T_b, °C) of perfluoro-*n*-alkanes *vs.* number of carbon atoms in the molecule (has no limit at $n_C \rightarrow \infty$).

Figure 2: Typical non-linear dependence of refractive indices (n_D^{20}) of *n*-1-alkanols *vs.* number of carbon atoms in the molecule (tends to a limit approx. 1.457 ± 0.003 at $n_C \rightarrow \infty$).

To simplify the choice of non-linear approximation functions for various properties of organic compounds the special approach has been recommended [13]. It implies the selection of approximation function not for initial set of data [i.e. $A = f(n_C)$], but for second numerical differences of A-values [i.e. $A'' = g''(n_C)$], followed by its integrating twice. The last (fourth) approach in the precalculation of physicochemical constants is additive schemes. As an example of them the methods of boiling points precalculation using so-called "boiling numbers" can be mentioned [14,15].

Comparing all known methods of the approximation of any constants of organic compounds, it is interesting to note that there were no attempts to evaluate any constants using the values of the same properties for the previous members of the same series up to present. Generally, it means the application of recurrent relationships, namely the simplest of them:

$$A(n+1) = a\,A(n) + b \qquad (1)$$

The coefficients a and b are calculated by LSM.

Surprisingly, this simplest relationship provides an excellent linear approximation practically for all known physicochemical constants within homologous series.

2. Some mathematical properties of recurrent relationship $A(n+1) = a\,A(n) + b$

The application of function (1) for the properties of homologues means that the values of argument belong to the equidistant set of data (the number of carbon atoms in the molecules). Obviously, the recurrent equation (1) at $a = 1$, $b \neq 0$ generalizes the arithmetical progressions, whilst at $a > 1$ and $b = 0$ – geometrical progressions. It can be used for exponential dependence $y = e^x$, when $a = 2.71828...$ and $b = 0$, like for other algebraic functions as well. The application of equation (1) for the approximation of Fibonacci numbers (F) [16]: 1, 1, 2, 3, 5, 8, 13, 21, 34, 55, 89, ... [$F(i+2) = F(i+1) + F(i)$] means the existence of linear dependence $F(n+1) = a\,F(n) + b$. The coefficient a tends to the value $1.61803...$ (golden section), whilst $b \rightarrow 0$.

The function (1) has the following algebraic solution:

$$A(n) = k\,a^n + b\,(a^n - 1)\,/\,(a - 1), \qquad \text{where } a, b, k - \text{constants} \qquad (2)$$

Formula (2) permits us to conclude noticeable feature on the application of this approach for the approximation of the properties of homologues. Namely, depending on the values of coefficient a, at the hypothetical increasing the number of carbon atoms in the molecules of homologues ($n_C \rightarrow \infty$), the values of function $A(n_C)$ can tend to infinity (at $a > 1$), or to a limit $A(n_C \rightarrow \infty) = b\,/\,(1 - a)$ (at $a < 1$). If $a = 1$ the function $A(n_C)$ describes the linear dependence $A(n_C) = ka + bn_C$. This mathematical property of function (2) seems very important, because all constants of organic compounds can be sub-divided onto two types: having no limits at $n_C \rightarrow \infty$ and tending to a non-infinity limits within homologous series:

Instead of application of general recurrent relationship (1), another recurrent procedure can be proposed for calculation of constants of higher homologues using the data for previous members of series. We can write equation (1) for any two consecutive pairs of homologues:

$$A(n) = a\,A(n\text{-}1) + b; \qquad A(n\text{-}1) = a\,A(n\text{-}2) + b,$$

That gives the expressions for coefficients a, and b:

$$a = [A(n) - A(n\text{-}1)] / [A(n\text{-}1) - A(n\text{-}2)]; \quad b = [A(n\text{-}1)^2 - A(n) \times A(n\text{-}2)] / [A(n\text{-}1) - A(n\text{-}2)]$$

Finally we receive the alternative recurrent formula for calculation of any properties of homologues using the data for three previous members of series:

$$A(n+1) = \{A(n\text{-}1)^2 + A(n) \times [A(n) - A(n\text{-}1) - A(n\text{-}2)]\} / [A(n\text{-}1) - A(n\text{-}2)] \qquad (3)$$

3. Examples of application

So far as the recurrent function $A(n+1) = a\,A(n) + b$ possesses the extremely great "approximation power", it is not surprising that it provides the linearization of the dependencies of any physicochemical constants *vs.* number of carbon atoms in the molecules of organic compounds within homologous series. Some examples are presented in the graphical form on Figs 3 and 4, namely for: boiling points of perfluoro-*n*-alkanes (Fig. 3; compare with Fig. 1); refractive indices of *n*-alkanes (Fig. 4; compare with Fig. 2).

The similar practically "ideal" linear plots are observed for other properties of organic compounds (densities, dynamics viscosities, surface tensions, vapor pressures, dielectric constants, ionization potentials, partition coefficients in heterophaseous systems, solubilities, etc.), as well as for various functions of these properties. Obviously, the constants indicating significant alternating effects for homologues with even and odd carbon number in the molecules (e.g., melting points) cannot be approximated by this manner.

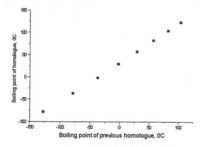

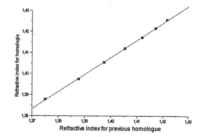

Figure 3: Linear dependence $T_b(n+1) = a\,T_b(n) + b$ for boiling points (^{0}C) of perfluoro-*n*-alkanes C_nF_{2n+2}; $a = 0.872 \pm 0.005$; $b = 31.8 \pm 0.4$; $\rho = 0.9999$, $S_0 = 1.1$.

Figure 4: Linear dependence $n_D^{20}(n+1) = a\,n_D^{20}(n) + b$ for refractive indices (non-dimensional constants) of *n*-alkanes C_nH_{2n+2}; $a = 0.804 \pm 0.008$; $b = 0.154 \pm 0.006$; $\rho = 0.9997$, $S_0 = 0.0007$.

So far as correlation coefficients of the dependencies of general type (1) usually exceed 0.999, it means the easy way to precalculate the values of any constants for higher homologues using available data for previous members of series *without special selection of approximation functions*.

The number of practical applications of this approach can be extremely large. Moreover, if the equation $A(n+1) = a\,A(n) + b$ seems the "universal" approximation function in nature, it can be applied *not only in chemistry or other natural sciences*. One of the most curious examples is the linearization of the typically non-linear dependence of the time of wine (as well as that for other beverages or technical liquids) dropping from empty bottles. Fig. 5 illustrates the plot of this dependence on the example of red Crimean dessert wine "Massandra" (35 drops during the time of about five minutes were registered). Really it is impossible to propose theoretically any reasonable approximation function for so complex process. However, in the accordance with the statement mentioned above, it is not necessary to choose any special approximation function for processing of these data, because the equation $t_{drop}(n+1) = a\,t_{drop}(n) + b$ should be correct even in this case. The linear plot presented on Fig. 6 confirm this conclusion unambiguously ($\rho = 0.9995$). It means that we can predict the time of next drop formation for any liquids using the information about previous dropping times with high reliability.

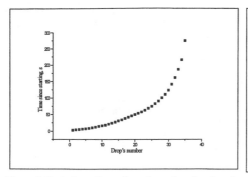
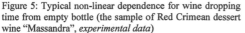
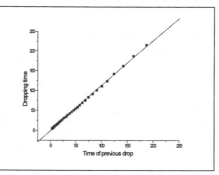

Figure 5: Typical non-linear dependence for wine dropping time from empty bottle (the sample of Red Crimean dessert wine "Massandra", *experimental data*)

Figure 6: Linear dependence $t_{drop}(n+1) = a \, t_{drop}(n) + b$ for wine dropping time from empty bottle (compare with Fig. 5); $a = 1.139 \pm 0.007$; $b = -0.85 \pm 0.48$; $\rho = 0.9995$, $S_0 = 1.8$.

The existence of simple recurrent function describing the variations of different physicochemical constants of organic compounds within homologous series can be interpreted by different manners, namely as universal mathematical regularity (chemometrics aspect of the problem), as a new "hyper-property" of organic compounds (should be important in organic and physical chemistry), or even as the general law of nature: *the values of most properties of organic compounds are in the linear dependence upon the values of the same properties for previous members of homologous series.*

Acknowledgements

The author wishes to thank Dr. Anatoly N. Marinichev (Chemical Research Institute of St. Petersburg State University) who pointed out the existence the analytical solution of general recurrent equation and Eng. Andrew A. Malov (C-MAP Corp., St. Petersburg) for his assistance in experiments with wine dropping from empty bottles.

References

[1] *Organikum. Organisch-chemisches Grundpraktikum.* Berlin, VEB Geutscher Verlag, 1976.

[2] *Beilstein Handbuch der Organishen Chemie.* 4-te Auflage, Berlin, 1918 – present time, Bd. I - XXXI.

[3] Z. Slanina, *Teoreticke aspekty fenomeny chemicke isomerie.* Praha, Akademia, 1981.

[4] R.C. Reid, T.K. Sherwood, *The Properties of Gases and Liquids.* 2nd Edn., New York, McGraw-Hill Book Co, 1966.

[5] M.Kh. Karapet'yants, *Methods of Comparative Calculation of Physico-Chemical Properties,* Moscow, Nauka Publ. House, 1965, 404 p.

[6] I.G. Zenkevich, *Fresenius' J. Anal. Chem.* **365** (1999) 305-309.

[7] I.G. Zenkevich, *Zh. Org. Khim. (Rus.)* **30** (1994) 1441-1447.

[8] A. Kreglewsky, B.J. Zwolinsky, *J. Phys. Chem.* **65** (1961) 1050-1053.

[9] B.V. Ioffe, I.G. Zenkevich, *Zh. Phys. Khim. (Rus.)* **74** (2000) 2101-2106.

[10] K. Heberger, *Anal. Chim. Acta.* **223** (1989) 161-174.

[11] K. Heberger, *Chromatographia.* **29** (1990) 375-384.

[12] I.G. Zenkevich, L.M. Kuznetsova, *Collect. Czech. Chem. Commun.* **56** (1991) 2042-2054.

[13] I.G. Zenkevich, B. Kranics, *Chemometr. & Intel. Lab. Systems.* **67** (2003) 51-57.

[14] C.R. Kinney, *J. Amer. Chem. Soc.* **60** (1938) 3032-3035.

[15] I.B. Sladkov, *Zh. Prikl. Khim. (Rus.)* **72** (1999) 1266-1271.

[16] M. Gardner, *Mathematical Games. From Scientific American.* (in Russian transl.). Moscow, MIR Publ. House, 1974.

Brill Academic Publishers
P.O. Box 9000, 2300 PA Leiden
The Netherlands

Lecture Series on Computer
and Computational Sciences
Volume 4, 2005, pp. 1578-1581

QSTR with Extended Topochemical Atom Indices. 7.
QSAR of Substituted Benzenes to *Saccharomyces cerevisiae*

Kunal Roy[1] and Indrani Sanyal

Drug Theoretics and Cheminformatics Lab, Division of Medicinal and Pharmaceutical Chemistry,
Department of Pharmaceutical Technology
Faculty of Engineering and Technology,
Jadavpur University,
Kolkata 700 032, INDIA

Received 7 July, 2005; accepted in revised form 10 July, 2005

Abstract: In continuation of our recent efforts to explore quantitative structure - toxicity relationships (QSTRs) of compounds of diverse chemical classes using extended topochemical atom (ETA) indices, the present paper deals with the modeling of the nonspecific toxicity of 51 substituted benzenes to the yeast *Saccharomyces cerevisiae*. The results of the study suggest that ETA parameters are sufficiently rich in chemical information to encode the structural features contributing significantly to the nonspecific toxicity of the substituted benzenes to *Saccharomyces cerevisiae*.

Keywords: QSAR, QSTR, ETA, Topological indices, *Saccharomyces cerevisiae*.

Mathematics Subject Classification: 62J05

PACS: 02.50.–r

1. Introduction

The effect of hazardous chemicals and pollutants on the ecosystem is a matter of great concern considering that though large number of chemical compounds (in the tune of tens of thousands) are in commercial use, relatively few of these have been subjected to adequate assessment for their hazardous environmental properties. [1] In order to evaluate environmentally safe levels of dangerous chemicals, there is the need for a set of toxicological data on organisms representative of the ecosystems, which is often unavailable or inadequate. [2] With the advancement of computational efficiency and toxicological understanding, quantitative structure-activity relationships (QSARs) are evolving as popular and indispensable tool in assessing potential toxic effects of organic chemicals. Considering the ever-increasing production of new chemicals, and the need to optimize resources to assess thousands of existing chemicals in commerce, regulatory agencies have turned to QSARs as essential tools to help prioritize tiered risk assessments when empirical data are not available to evaluate toxicological effects. [3] Progress in designing scientifically credible QSARs is intimately associated with the development of empirically derived databases of well-defined and quantified toxicity endpoints, which are based on a strategic evaluation of diverse sets of chemical structures, modes of toxic action, and species.

2. Materials and Methods

In continuation of our recent efforts to explore quantitative structure - toxicity relationships (QSTRs) of compounds of diverse chemical classes using extended topochemical atom (ETA) indices, [4-9] the present paper deals with the modeling of the nonspecific toxicity of 51 substituted benzenes to the yeast *Saccharomyces cerevisiae*. [10] Principal component factor analysis (FA) was used as the data-

---

[1] Corresponding author. Email: kunalroy_in@yahoo.com, kroy@pharma.jdvu.ac.in ;
URL: http://www.geocities.com/kunalroy_in

preprocessing step for the selection of independent variables for the subsequent multiple linear regression (MLR) analysis. Besides FA-MLR, stepwise regression analysis and partial least square (PLS) analysis were used as additional statistical tools. Attempts were made to model the data set with three different sets of descriptors : ETA, non-ETA (topological indices including Wiener, Hosoya Z, molecular connectivity, kappa shape, Balaban J and E - state parameters apart from physicochemical parameters like AlogP98, *MolRef, H_bond_acceptor* and *H_bond_donor*) and a combined set including both ETA and non-ETA parameters. The values for the non-ETA topological descriptors for the compounds have been generated by QSAR+ and Descriptor+ modules of the Cerius 2 version 4.8 software. [11] The statistical quality of the MLR equations was judged by the parameters like *explained variance* (R_a^2, i.e., adjusted R^2), *correlation coefficient* (*r* or *R*), *standard error of estimate* (*s*) and *variance ratio* (*F*) at specified *degrees of freedom* (*df*). All the accepted equations have regression constants and *F* ratios significant at 95% and 99% levels respectively, if not stated otherwise. All the developed models were cross-validated [12] using "leave-one-out" technique, and *leave-one-out cross-validation* R^2 (Q^2) values along with *predicted residual sum of squares* (*PRESS*) values were reported.

2. Results and Discussion

FA-MLR
The statistical qualities of the best models obtained from FA-MLR are given below:
1. ETA model: *n*= 51, Q^2 = 0.851, R_a^2 = 0.874, *R*=0.940, *F*= 87.9 (*df* 4, 46), *s* = 0.235, *PRESS*=3.3.
2. Non-ETA model: *n* = 51, Q^2= 0.837, R_a^2= 0.855, *R* = 0.929 , *F*=98.9 (*df* 3,47), *s* = 0.253, *PRESS* =3.6.
3. Combined model: *n* = 51, Q^2=0.824, R_a^2 =0.852, *R* = 0.940, *F*=73.0 (*df* 4,46), *s* =0.255, *PRESS* = 3.9.

Stepwise regression
The statistical qualities of the best models obtained from stepwise regression analysis are given below:
1. ETA model: *n* = 51, Q^2=0.851, R_a^2 =0.874, R^2 =0.884, *F*= 87.9 (*df* 4, 46), *s* = 0.236, *PRESS* = 3.3
2. Non-ETA model: *n* = 51, Q^2 =0.847, R_a^2=0.860, R^2 =0.869, *F*=103.4 (*df* 3,47), *s* = 0.248, *PRESS* =3.4.
3. Combined descriptors: *n* = 51, Q^2 = 0.875, R_a^2= 0.888, R^2 = 0.894, *F*=132.7 (*df* 3,47), *s* = 0.222, *PRESS* =2.7.

The equation obtained using ETA descriptors from stepwise regression is same as that obtained from FA-MLR.

PLS
The statistical qualities of the best models obtained from PLS analysis are given below:
1. ETA model: *n* = 51, Q^2 = 0.863, R^2 =0.893, *F*=96.0 (*df* 4,46), *PRESS* = 3.0.
2. Non-ETA model: *n* = 51, Q^2 =0.867, R^2 =0.884, *F*=371.9(*df* 1,49), *PRESS* = 2.9.
3. Combined model: *n* = 51, Q^2=0.875, R^2 = 0.909, *F* = 90.9(*df* 5,45), *PRESS* = 2.7.

PCRA
Using factor scores as independent variables, principal component regression analysis (PCRA) was performed and the derived relations were of the following statistical qualities: Q^2 values being 0.926, 0.878 and 0.869 while R^2 values being 0.942, 0.903 and 0.899 for factor scores derived from ETA, non-ETA and combined matrices respectively.

Interpretation of the ETA models
The use of the ETA indices suggested negative contributions of functionalities of amino and carboxylic acid substituents on the benzene ring and the presence of the electronegative atoms, and positive contributions of branching and functionality of chloro substituent.

Overview
Table 1 shows observed toxicity values of 51 diverse functional benzene derivatives and calculated values from the ETA models. Table 2 shows comparison of statistical quality of different models. The ETA models are comparable to the non-ETA models in statistical quality, and addition of non-ETA descriptors does not significantly increase the quality of the ETA models. This study suggests that ETA

parameters are sufficiently rich in chemical information to encode the structural features contributing significantly to the nonspecific toxicity of the substituted benzenes to *Saccharomyces cerevisiae*. This indicates that ETA indices merit further assessment to explore their potential in QSAR / QSPR / QSTR modeling.

Table 1: Observed and calculated toxicity values of 51 benzene derivatives against *Saccharomyces cerevisiae*.

Sl. No.	Compounds	Y_{OBS} (ref. 10)	Y_{CALC} ETA model (FA-MLR or stepwise)	Y_{CALC} ETA model (PLS)
1	Chlorobenzene	1.18	1.26	1.25
2	Bromobenzene	1.40	1.69	1.63
3	1,2-Dichlorobenzene	1.96	2.04	1.96
4	1,4-Dichlorobenzene	1.96	2.03	1.95
5	1,3-Dichlorobenzene	2.32	2.51	2.45
6	4-Bromo-chlorobenzene	2.08	2.31	2.22
7	1,2,3-Trichlorobenzene	2.41	2.63	2.58
8	1,2,4-Trichlorobenzene	2.54	2.63	2.57
9	2,5-Dichlorotoluene	2.33	2.35	2.17
10	2,4,5-Trichlorotoluene	2.91	2.96	2.82
11	3-Chlorobenzoic acid	1.72	1.47	1.43
12	4-Chlorobenzoic acid	1.85	1.46	1.42
13	3-Bromobenzoic acid	1.94	1.74	1.69
14	4-Fluorobenzoic acid	1.37	0.96	1.05
15	4-bromobenzoic acid	1.95	2.12	1.91
16	2-Aminobenzoic acid	0.79	0.96	0.97
17	3-Aminobenzoic acid	0.32	0.56	0.75
18	4-Aminobenzoic acid	0.23	0.92	0.95
19	Pentachlorophenol	3.83	3.59	3.82
20	2,4-Dichlorophenol	2.43	2.27	2.43
21	2-Methylphenol	1.38	1.23	1.15
22	2-Chlorophenol	1.43	1.59	1.53
23	4-Chlorophenol	1.63	1.58	1.52
24	2,6-Dimethylphenol	1.35	1.69	1.49
25	Phenol	0.86	0.64	0.74
26	Nitrobenzene	1.01	0.96	1.03
27	2-Chloronitrobenzene	1.65	1.73	1.75
28	3-Chloronitrobenzene	1.64	1.73	1.74
29	4-Chloronitrobenzene	1.65	1.72	1.73
30	4-Bromonitrobenzene	2.13	2.00	1.99
31	2-Methylnitrobenzene	1.52	1.42	1.34
32	3-Methylnitrobenzene	1.52	1.42	1.34
33	4-Methylnitrobenzene	1.50	1.42	1.34
34	3,4-Dichloronitrobenzene	2.20	2.28	2.33
35	2,4-Dichloronitrobenzene	2.24	2.29	2.34
36	*o*-Dinitrobenzene	1.41	1.44	1.54
37	*m*-Dintrobenzene	1.45	1.44	1.54
38	2,4-Dinitrobromobenzene	2.47	2.21	2.36
39	2,4-Dinitrochlorobenzene	1.90	2.03	2.16
40	2,4-Dintrotoluene	2.02	1.77	1.76
41	2,3-Dintrotoluene	1.97	1.77	1.76
42	2,6-Dintrotoluene	1.61	1.77	1.76
43	3-Chloroaniline	1.80	1.43	1.48
44	4-Chloroaniline	1.80	1.44	1.48
45	4-Bromoaniline	1.91	1.79	1.81
46	2,4-Dichloroaniline	2.40	2.11	2.13
47	2,4,6-Tribromoaniline	3.12	2.95	3.19
48	2,4,6-Trichloroaniline	2.45	2.63	2.69
49	4-Methylaniline	0.77	1.10	1.14
50	*p*-Phenylenediamine	0.89	0.78	1.01
51	2-Chloro-4-nitroaniline	1.42	1.81	1.38

Y_{OBS} and Y_{CALC} represent observed and calculated toxicity values.

Table 2: Comparison of statistical quality of different models

Model	Statistical Tools	R^2	R_a^2	Q^2	PRESS
ETA	FA – MLR	0.884	0.874	0.851	3.3
	Stepwise	0.884	0.874	0.851	3.3
	PLS	0.893	0.878	0.863	3.0
Non – ETA	FA - MLR	0.863	0.855	0.837	3.6
	Stepwise	0.869	0.860	0.847	3.4
	PLS	0.884	0.868	0.867	2.9
Combined	FA - MLR	0.884	0.852	0.824	3.9
	Stepwise	0.894	0.888	0.875	2.7
	PLS	0.909	0.886	0.875	2.7

Acknowledgments

One of the authors (KR) thanks the All India Council for Technical Education (AICTE), New Delhi for financial assistance under the Career Award for Young Teachers scheme.

References

[1]   D. Mackay, J. Hubbarde, E. Webster, The role of QSARs and fate models in chemical hazard and risk assessment. *QSAR Comb. Sci.,* **22** (2003) 106-112.

[2]   M. Vighi, P. Gramatica, F. Consolaro, R. Todeschini, QSAR and chemometric approaches for setting water quality objectives for dangerous chemicals. *Ecotoxicol. Environ. Saf.* **49** (2001) 206-220.

[3]   S. P. Bradbury, C. L. Russom, G. T. Ankley, T. W. Schultz, J. D. Walker, Overview of data and conceptual approaches for derivation of quantitative structure-activity relationships for ecotoxicological effects of organic chemicals. *Environ. Toxicol. Chem.* **22** (2003) 1789-1798.

[4]   K. Roy, G. Ghosh, Introduction of extended topochemical atoms (ETA) indices in the valence electron mobile (VEM) environment as tool for QSAR/QSPR studies. *Internet Electron. J. Mol. Des.* **2** (2003) 599-620; http://www.biochempress.com

[5]   K. Roy, G. Ghosh, QSTR with extended topochemical atom indices. 2. Fish toxicity of substituted benzenes. *J. Chem. Inf. Comput. Sci.* **44** (2004) 559-567.

[6]   K. Roy, G. Ghosh, QSTR with extended topochemical atom indices. 3. Toxicity of nitrobenzenes to *Tetrahymena pyriformis*. *QSAR Comb. Sci,* **23** (2004) 99-108.

[7]   K. Roy, G. Ghosh, QSTR with Extended Topochemical Atom Indices. 4. Modeling of the Acute Toxicity of Phenylsulfonyl Carboxylates to Vibrio fischeri Using Principal Component Factor Analysis and Principal Component Regression Analysis. *QSAR Comb. Sci.* **23** (2004) 526-535.

[8]   K. Roy, G. Ghosh, QSTR with Extended Topochemical Atom Indices. Part 5. Modeling of the Acute Toxicity of Phenylsulfonyl Carboxylates to *Vibrio fischeri* Using Genetic Function Approximation. *Bioorg. Med. Chem.* **13** (2004) 1185-1194 http://dx.doi.org/10.1016/j.bmc.2004.11.014

[9]   K. Roy, G. Ghosh, QSTR with extended topochemical atom (ETA) indices. VI. Acute toxicity of benzene derivatives to the tadpoles (*Rana japonica*). *J. Mol. Model.* (under revision).

[10]   Y.-Y. Liao, L.-S. Wang, Y.-B. He, H. Yang, *Bull. Environ. Contam. Toxicol.* **56** (1996) 460-466.

[11]   Cerius 2 version 4.8 is a product of Accelrys Inc., San Diego, CA.

[12]   S. Wold, L. Eriksson . Statistical validation of QSAR results. In: H. van de Waterbeemd (Ed.) *Chemometric Methods in Molecular Design*, VCH, Weinheim, 1995, pp. 312-317.

Brill Academic Publishers
P.O. Box 9000, 2300 PA Leiden
The Netherlands

*Lecture Series on Computer
and Computational Sciences*
Volume 4, 2005, pp. 1582-1585

A Novel Method for Selecting Clusters in Cluster Analysis

G. Restrepo[1,a], E. J. Llanos[b] and A. Bernal[b]

[a]Laboratorio de Química Teórica, Universidad de Pamplona, Pamplona, Colombia
[b]Observatorio Colombiano de Ciencia y Tecnología, Bogotá, Colombia

Received 4 June, 2005; accepted in revised form 10 July, 2005

Abstract: We developed a mathematical method for selecting clusters in a dendrogram (tree). The procedure considers a cluster as a subset of a set Q of objects. The elements that belong to a cluster are considered as an equivalence class where the equivalence relation is a similarity relationship. We showed that the searching for clusters in a dendrogram produces a partition on the set Q and we developed a mathematical method for build up partitions according to an integer number, where $|Q|$ means the cardinality of the set Q. This method is called "chemotopology" and n means the maximum number of elements that can be contained in a subset of a partition of Q. This methodology produces a collection of $|Q|$ possible ways to select the clusters in the dendrogram. The procedure for selecting the partition that shows the similarity relationships among the elements in Q is based on the number of subsets of the partition and their population. We developed a mathematical definition of these criteria and build up the *selection number* (S) that combines the population and number of subsets in order to determine the partition that produces the clusters. We showed that the value of n producing the maximum value of S is the one that offers many very populated clusters (the criterion to select clusters in the dendrogram). Finally, we applied this methodology to the set of 72 chemical elements (Z=1-86, omitting Z=58-71) and we found a value of n=4 to select the clusters. We found a close relationship among the number of elements in these clusters and the number of elements shown in the periodic table.

Keywords: Chemotopology, Cluster analysis, Partitions, Chemical elements, Periodic Table

Mathematics SubjectClassification: 05A18, 00A69

1. Extended Abstract

A general problem of the studies of Cluster Analysis (CA) is the one related to the selection of the clusters in a dendrogram (tree). There are several methods for selecting clusters in a tree, some of them are the "Phenon line" [1], the amalgamation coefficient [2], the Taylor-Butina [3,4] method and some others. Those selection methods are based on the dendrogram as it and do not consider the procedure used to build up such a dendrogram. It means that those methods are independent of the similarity function [5,6] and grouping methodology [5,6] used to obtain the dendrogram. All these methods consider the dendrogram as a mathematical binary tree. In general, these methods try to find a way to choose the appropriate number of clusters that can be defined on a set Q of elements.

Taking into consideration the uses of CA as a tool for determining similarity relationships among the elements of Q, we consider that a criterion for selecting the clusters of a dendrogram should include the similarities among the elements [7]. It means that the elements that belong to a particular cluster should be similar according to the properties used to define each one of the elements in Q. But the question is, how can we determine the enough degree of similarity among the elements to build up the clusters? On the other hand, all the methods for selecting clusters in a tree produce partitions of the set Q [8]. Thus, our aim is to develop a method for selecting the appropriate partition of the set Q that shows the similarity relationships among the elements in Q.

We showed in recent works a mathematical method for partitioning a set Q according to the similarities shown by the dendrogram [5-7,9-12]. This method was called "chemotopology" [12] due to it

---

[1] Corresponding author. Professor at Universidad de Pamplona, Colombia. E-mail: grestrepo@unipamplona.edu.co

combines the chemometric results (CA) with topology. The chemotopological approach defines the dendrogram as a graph [5,6,10,12] and partitions the graph in subgraphs called subtrees [5,6,10,12]. These subtrees correspond to equivalence classes where the equivalence relation is a similarity relationship [13]. The method is based on the size of subtrees or branches of the dendrogram, in such a way that the subtrees can have from one element to $|Q|$ elements. The size of these subtrees or clusters is defined according to an integer number $1 \leq n \leq |Q|$ [5-7,9-12]. It means that for each n we can obtain a particular partition of Q [12]. Thus, if we have a set of 100 elements then we have 100 different ways of partitioning the set. We show an example of application of the chemotopological partition for the set $Q=\{a,b,c,d,e,f\}$, according to the dendrogram shown in Figure 1. The set partitions produced by each value of $1 \leq n \leq 4$ appear in Table 1.

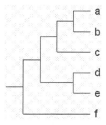

Figure 1: A dendrogram of the set $Q=\{a,b,c,d,e,f\}$.

Table 1: Chemotopological set partitions of Q according to the dendrogram of Figure 1, where n means the maximum number of elements in each cluster, $|TSPn|$ the number of set partitions, GPn the geometrical population and S the selection number

n	Set partition	$|TSPn|$	GPn	S
1	$\{\{a\},\{b\},\{c\},\{d\},\{e\},\{f\}\}$	6	1	6
2	$\{\{a,b\},\{c\},\{d,e\},\{f\}\}$	4	4	16
3	$\{\{a,b,c\},\{d,e\},\{f\}\}$	3	6	18
4	$\{\{a,b,c\},\{d,e\},\{f\}\}$	3	6	18
5	$\{\{a,b,c,d,e\},\{f\}\}$	2	5	10
6	$\{\{a,b,c,d,e,f\}\}$	1	6	6

In the particular example of $Q=\{a,b,c,d,e,f\}$ we have 6 ways for partitioning the tree. If we observe the clusters generated with $n=1$ we have that all the elements are different or dissimilar. This result is a trivial case and it is not important in the searching for similarities [7]. The other extreme case, $n=|Q|$, offers a partition of the set with only one element, the whole set. This is another trivial case since it establishes that all the elements are similar, which is not important in terms of similarity [7]. Then, we need a value of n that does no consider the extreme cases. Although, we have $|Q|-2$ possibilities for selecting the value of n. This number is still a big number of possibilities. In order to select the value of n for doing a partition on the set Q, we can study the effect of $n=1$ in the partition. In this case we have $|Q|$ subsets of only one element, which means the maximum number of subsets and also with the minimum population each subset. On the other hand, when $n=|Q|$, it produces only one subset (minimum number of subsets) with maximum population. Thus, what we need for selecting the value of n is to consider the population of the subsets (clusters) and the number of them. In order to take this into account we called $|TSPn|$ the number of clusters generated by means of the use of a particular n. On the other hand, we called GPn the geometrical population of the clusters under the selection of a particular n. GPn is the multiplication of the cardinality of each one of the clusters generated by n. Finally, the multiplication of $|TSPn|$ with GPn produces a measure of the number and population of the clusters generated under a particular value of n. This multiplication is called S, *selection number*. The values of $|TSPn|$, GPn and S for the dendrogram of Figure 1 appear in Table 1. According to the meaning of S, "many cluster very populated", we can observe that the maximum value of S produces at least one partition with several subsets and also very populated. In the case of the dendrogram of Figure 1 we have that S is maximum when $n=3=4$, and both values of n produce the same partition in Q according to the chemotopological method.

We applied this methodology to the set of 72 chemical elements ($Z=1-86$, omitting $Z=58-71$), each one defined by 128 properties [7,11]. We used 4 similarity functions and 5 grouping methodologies. Thus, we obtained 20 dendrograms and we calculated the values of S for every one of the 72 different values of n. The most frequent maximum value of S occurred for $n=4$, which is in agreement with the population of the groups of chemical elements showed by the Periodic Table. This result reinforces the idea that the chemical elements that belong to a group are the most similar elements according to their properties [6].

Acknowledgments

We thank Dr. Mesa from the Universidad del Valle for his mathematical support. G. Restrepo thanks the Universidad de Pamplona in Colombia for its financial support and especially, Dr. A. González, Head Dean of this University, for his interest in the development of the mathematical chemistry.

References

[1] P. H. A. Sneath and R. R. Sokal: *Numerical Taxonomy: The Principles and Practice of Numerical Classification*. W. H. Freeman Company, San Francisco, 1973.

[2] M. S. Aldenderfer and R. K. Blashfield: *Cluster Analysis (Sage University Paper series on Quantitative Applications in the Social Sciences, N°. 44)*. Sage University, Newbury Park, 1984.

[3] R. Taylor, Simulation Analysis of Experimental Design Strategies for Screening Random Compounds as Potential New Drugs and Agrochemicals, *Journal of Chemical Information and Computer Sciences* **35** (1994) 59-67.

[4] D. Butina, Unsupervised Data Base Clustering Based on Daylight's Fingerprint and Tanimoto Similarity: A Fast and Automated Way To Cluster Small and Large Data sets, *Journal of Chemical Information and Computer Sciences* **39** (1999) 747-750.

[5] G. Restrepo, H. Mesa, E. J. Llanos and J. L. Villaveces, Topological Study of the Periodic System, *Journal of Chemical Information and Computer Sciences* **44** (2004) 68-75.

[6] G. Restrepo, H. Mesa, E. J. Llanos and J. L. Villaveces, Topological Study of the Periodic System, In: B. King and D. Rouvray, *The Mathematics of the Periodic Table*. Nova, New York, 2005 (in press).

[7] G. Restrepo, E. J. Llanos and J. L. Villaveces, Trees (Dendrograms and Consensus Trees) and their topological information, In: S. Basak,; D. K. Sinha, (Ed.), Proceedings of the Fourth Indo-US Workshop on Mathematical Chemistry, University of Pune, Pune, India, 2005.

[8] S. Lipschutz: *General Topology*. McGraw-Hill, New York, 1965.

[9] G. Restrepo, H. Mesa, E. J. Llanos and J. L. Villaveces, Topological Study of the Chemical Elements, In: P. Willett, (Ed.), Proceedings of Third Joint Sheffield Conference on Chemoinformatics, The University of Sheffield, Sheffield, United Kingdom, 2004.

[10] G. Restrepo and J. L. Villaveces, From Trees (Dendrograms and Consensus Trees) to Topology, *Croatica Chemica Acta*, In press.

[11] G. Restrepo, E. J. Llanos and H. Mesa, Chemical Elements: A Topological Approach, In: T. Simos and G. Maroulis, (Ed.), Proceedings of The International Conference of Computational Methods in Sciences and Engineering 2004, VSP, Athens, Greece, 2004.

[12] G. Restrepo, H. Mesa and J. L. Villaveces, On the Topological Sense of Chemical Sets, *Journal of Mathematical Chemistry*, In press.

[13] G. Restrepo and R. Brüggemann, Ranking regions through cluster analysis and posets, Accepted for being presented at the 9th WSEAS International Conference on COMPUTERS, Athens, Greece, 2005.

Brill Academic Publishers
P.O. Box 9000, 2300 PA Leiden
The Netherlands

*Lecture Series on Computer
and Computational Sciences*
Volume 4, 2005, pp. 1586-1588

A Short Survey of Lunn-Senior's Model of Isomerism
in Organic Chemistry

Valentin Vankov Iliev[1]

Section of Algebra,
Institute of Mathematics and Informatics,
Bulgarian Academy of Sciences,
1113 Sofia, Bulgaria

Received 2nd June, 2005; accepted in revised form 10 July, 2005

Abstract: The paper contains a brief survey of the present state of Lunn-Senior's mathematical theory of isomerism in organic chemistry.

Keywords: substitution isomerism, stereoisomerism, structural isomerism, groups, genetic (substitution) reactions, chiral pairs

Mathematics Subject Classification: 05E25, 20B35, 92E10

The papers [1] and [2] continue, generalize and trace out the framework of the famous theory of isomerism developed by A. C. Lunn and J. K. Senior, see [5]. In Lunn-Senior's mathematical model of isomerism in organic chemistry the molecule under consideration is divided into skeleton and d univalent substituents whose valences are numbered $1,2,...d$, and the structural formula of the substituents part can be identified with a tabloid with d nodes $A = (A_1, A_2,...)$, where A_i is the set of numbers of valences of the substituents of type x_i, $i=1,...,d$. Here, if $\lambda_i = |A_i|$, then $\lambda_1 \geq \lambda_2 \geq ...$, $\lambda = (\lambda_1, \lambda_2,...)$ is a partition of d, and $(x_1)_{\lambda 1}(x_2)_{\lambda 2}...$ is the empirical formula of the substituent part. The univalent substitution isomerism, stereoisomerism, and the structural isomerism, respectively, determine three equivalence relations on the set T_λ of all structural formulae $A = (A_1, A_2,...)$ with the above empirical formula. On the other hand, the symmetric group S_d and any of its subgroups acts naturally on the set T_λ. Lunn-Senior's thesis asserts that the equivalent classes of each one of these equivalence relations are the orbits of three groups $G \leq G' \leq G''$, respectively, that are subgroups of S_d, and $|G':G| \leq 2$. Thus, the G-orbits in T_λ represent the univalent substitution isomers, the G'-orbits represent the stereoisomers, and the G''-orbits represent the structural isomers. Moreover, among the stereoisomers there are chiral pairs if and only if G has index 2 in G', and in this case any G'-orbit that contains two G-orbits represents a chiral pair, whose members are represented by these G–orbits; the G'-orbits that contain one G-orbit represent the dimers. The set T_d of all structural formulae is disjoint union of T_λ for all partitions λ of d, the orbit spaces $T_{d;W}$, where $W \leq S_d$, are disjoint unions of the orbit spaces $T_{\lambda;W}$, so the elements of the orbit spaces $T_{d;G}$, $T_{d;G'}$, $T_{d;G''}$ represent the corresponding type of isomers of a molecule divided into skeleton and d univalent substituents.

The substitution (or, genetic) reactions among the derivatives of the given molecule are reflected by the model in the following way:

(a) on the level of empirical formulae a simple substitution reaction has the form
$(x_1)_{\mu 1}...(x_i)_{\mu i}...(x_j)_{\mu j}... \rightarrow (x_1)_{\lambda 1}...(x_i)_{\lambda i}...(x_j)_{\lambda j}...$,
where $\mu_1 = \lambda_1,..., \mu_i = \lambda_i +1,..., \mu_j = \lambda_j -1,..., \mu_d = \lambda_d$, that is, a replacement of a ligand of type x_i by a ligand of type x_j, and in this case we write $\mu = \rho_{i,j}\lambda$ and $\lambda < \mu$;

(b) on the level of structural formulae the simple substitution reaction from (a) has the form

---

[1] Partially supported by Grant MM-1106/2001 of the Bulgarian Foundation of Scientific Research;
Corresponding author. E-mail: viliev@math.bas.bg, viliev@aubg.bg

B = (B$_1$, B$_2$,..., B$_i$,..., B$_j$,...) → A = (A$_1$, A$_2$,..., A$_i$,..., A$_j$,...),
where B can be obtained from A by moving an element s ∈ A$_j$ to the set A$_i$, and in this case we write B = R$_{i,s}$A and A < B;

(c) in general, we write λ ≤ μ if λ can be obtained from μ via a finite sequence of simple replacement of type (a) – this is the famous dominance order in the set P$_d$ of all partitions of *d*.; we write A ≤ B if B can be obtained from A via a finite sequence of movements of type (b); an equivalent definition of the latter partial order is: A ≤ B if A$_1$ ∪...∪ A$_k$ ⊂ B$_1$ ∪..∪ B$_k$ for all *k* = 1,..., *d*;

(d) we factor out the partial order from (c) and obtain a partial order on the set T$_{d;G}$ of G-orbits via the rule: *a* ≤ *b* if there exist A ∈ *a* and B ∈ *b* such that A ≤ B.
The relation *a* < *b* means that the product that corresponds to *a* can be obtained from the product that corresponds to *b* via several consecutive simple substitution reactions. Thus, the partially ordered set (T$_{d;G}$, ≤) represents the univalent substitution isomers and all possible genetic reactions among them.
Given *a*, *b* ∈ T$_{d;G}$ with *a* < *b*, in [1, Theorem 4.2.3] necessary and sufficient conditions are presented for *b* to cover *a*, that is, for the interval [*a*, *b*] to contain two elements.
Given a permutation group W ≤ S$_d$ and a one-dimensional character χ of W with kernel W$_χ$ ≤ W, we can define unambiguously a subset T$_{λ;χ}$ ⊂ T$_{λ;W}$ consisting of all W-orbits that contain |W: W$_χ$| in number W$_χ$-orbits. In particular, when W = G', and χ is the canonical homomorphism G' → G'/G = {1,-1}, so W$_χ$ = G, the elements of set T$_{λ;χ}$ represent all chiral pairs and the elements of the difference T$_{λ;W}$ − T$_{λ;χ}$ represent the dimers. Our hypothesis is that in general the set T$_{λ;χ}$ for various one-dimensional characters χ of G, G', and G″, also represents chemically interesting type property of certain molecules – it can be called generalized chirality. If χ is the unit character of W, then we have T$_{λ;χ}$= T$_{λ;W}$ and get all isomers. In [1, Theorem 5.2.7] we find a counting formula for the numbers n$_{λ;χ}$ = |T$_{λ;χ}$| and the same considerations yield Ruch's formula from [6] (see also [1, Corollary 5.2.10]). The former counting formula implies the well known Kauffmann's formulae for the number of the derivatives of naphthalene C$_{10}$H$_8$.- see [1, Section 6]. In [1, Theorem 5.3.1] we prove that the map λ → n$_{λ;χ}$ is decreasing, and as a corollary we obtain a result of E. Ruch of special beauty: if a distribution of ligants according to the partition μ amounts to a chiral molecule and λ < μ, then also a distribution according to λ yields a chiral molecule.

The complete identification of the di-substitution homogeneous derivatives of benzene C$_6$H$_6$ (para, ortho, and meta) and its tri-substitution homogeneous derivatives (asymmetrical, vicinal, and symmetrical) via the famous Körner's relations is the archetype of theory that we develop in [2]. The intrinsic reason for this identification is that the corresponding automorphism group of the mathematical structure in this case is trivial. For any set D ⊂ P$_d$ of partitions of *d* (that is, empirical formulae) we set T$_{D;G}$ = ∪$_{λ∈D}$ T$_{λ;G}$. A bijection *u*: T$_{D;G}$ → T$_{D;G}$ is said to be automorphism of T$_{D;G}$ if the following four conditions are satisfied: (1) *a* ≤ *b* if and only if *u*(*a*) ≤ *u*(*b*); (2) *u*(T$_{λ;G}$) = T$_{λ;G}$ for any λ ∈ D; (3) *u* maps any G'-orbit onto a G'-orbit; (4) *u* maps any G″-orbit onto itself. All automorphisms of T$_{D;G}$ form a group Aut(T$_{D;G}$) and the products that correspond to *a*, *b* ∈ T$_{D;G}$ are said to be indistinguishable via substitution reactions among the elements of T$_{D;G}$ if *u*(*a*) = *b* for some *u* ∈ Aut(T$_{D;G}$). Otherwise, *a* and *b* are called distinguishable via the same reactions. Since the bijection of T$_{D;G}$ that permutes the members of any chiral pair and leaves the dimers invariant is an involution of the group Aut(T$_{D;G}$), we obtain the indistinguishability of the members of a chiral pair via substitution reactions. – see [2, Theorem 8.2.6]. Using the sets T$_{λ;χ}$ for various χ in order to separate pairs *a*, *b* ∈ T$_{D;G}$, we define indistinguishability via characters and prove that the members of a chiral pair can not be distinguished via characters, too – see [2, Corollary 9.2.2]. As a result we obtain that the members of the two structurally identical di-substitution products of ethene C$_2$H$_4$ can not be distinguished either by substitution reactions or by characters (see [2, Corollaries 10.1.3, 10.1.5]). In accord to the results in [3], we can establish Körner's relations in a more general situation – see [2, Remark 10.2.3]. Moreover, the comprehensive results about all di-substitution and the tri-substitution derivatives of cyclopropane C$_3$H$_6$, which we prove in [2, Corollary 10.3.6], can be generalized at the same level – see [2, Remark 10.3.7]. In the unpublished paper [4] we describe completely all univalent substitution isomers, stereoisomers, and structural isomers of ethane C$_2$H$_6$ as well as all substitution reactions among them. We also list the structurally identical products of ethane with the same empirical formula that can and that can not be distinguished via substitution reaction among the elements of certain sets T$_{D;G}$.

References

[1] V.V. Iliev, On Lunn-Senior's Mathematical Model of Isomerism in Organic Chemistry. Part I, *MATCH Commun. Math. Comput. Chem.* **40** (1999) 153-186.

[2] V.V. Iliev, On Lunn-Senior's Mathematical Model of Isomerism in Organic Chemistry. Part II, *MATCH Commun. Math. Comput. Chem.* **50** (2004) 39-56.

[3] V.V. Iliev, On certain organic compounds with one mono-substitution and at least three di-substitution homogeneous derivatives, *J. Math. Chem.* (2003) 137-150.

[4] V.V. Iliev, The genetic reactions of ethane, *J. Math. Chem.* (submitted for publication).

[5] A.C. Lunn and J.K. Senior, Isomerism and Configuration, *J. Phys. Chem.* **33** (1929) 1027-179.

[6] E. Ruch, W. Hässelbarth, B. Richter, Doppelnebenclassen als Klassenbegriff und Nomenklaturprinzip für Isomere und ihre Abzählung, *Theoret. Chim. Acta (Berl.)* **19** (1970) 288-300.

Brill Academic Publishers
P.O. Box 9000, 2300 PA Leiden,
The Netherlands

Lecture Series on Computer
and Computational Sciences
Volume 4, 2005, pp. 1589-1592

Computer Simulation Studies of Molecular Systems and their Reactions

D. Janežič[1], M. Penca, K. Poljanec, M. Hodošček

National Institute of Chemistry,
Hajdrihova 19,
SI-1000 Ljubljana, Slovenia

Received 10 July 2005; accepted in revised form 20 July, 2005

Abstract:
The survey of our past and present work on computer simulation studies of molecular systems and their reactions will be presented. In particular, new integration algorithms for molecular dynamics simulations and quantum chemical calculations for determination of the energy levels of the main intermediate products of the isocianide multicomponent reactions will be described.

Keywords: molecular dynamics simulation, normal mode analysis, symplectic integration methods, Hamiltonian systems, Lie algebra, vibrational modes, large systems, multi component reactions, isocyanides, quantum chemistry

Mathematics Subject Classification: 65C20, 65L20, 70H05, 70H15

PACS: 31.15.Qg

1   Introduction

Many physical problems, particularly in chemical and biological systems, involve processes that occur on widely varying time scales. Such problems have motivated the development of new methods for computer simulation studies of molecular systems and their reactions [1].

Among the main theoretical methods of investigation of the dynamic properties of molecular systems are computer simulations and quantum chemical computatations. The technique of molecular dynamics clearly is a powerful tool for simulating molecular motion [2, 3, 4]. A complementary computational method to examine molecular behavior is a normal mode analysis which examines motion in the harmonic limit [5, 6, 7].

Computer-based calculations are now used generally to supplement experimental technics. For several decades they have been developed and refined so that it is now possible to analyse the structure and properties of matter in detail. Quantum chemistry is today used within all branches of chemistry and molecular physics. As well as producing quantitative information on molecules and their interactions, the theory also affords deeper understanding of molecular processes that cannot be obtained from experiments alone. Today, for example, quantum chemistry calculations can be used to explain how multicomponent reactions of isocyanides occur.

Multicomponent reactions (MCRs) are reactions in which more than two starting compounds react to form the product or intermediate in one step. Three fundamental types of MCRs exist [8]:

---

[1] Corresponding author. E-mail: dusa@cmm.ki.si

the type I of MCRs presents an equilibrium of starting compounds, intermediates and product. In type II of MCRs, the starting materials and intermediates equilibrate, but their last reaction step to product is irreversible. The type III of MCRs correspond to a sequence of irreversible elementary reactions.

Isocyanides are unusual organic compounds with extraordinary functional group contains divalent carbon atom which is in exothermic reactions oxidized to tetravalent carbon atom [8]. For that reason it is quite natural that they react in a different way than the rest of organic compounds.

2   Methods

The Passerini reaction (P-3CR) [9, 10] was introduced as the first three component reaction of carbonyl compounds, carboxylic acids, and isocyanides. Their products are formed via the hydrogen bridged intermediate with its carbonyl compounds and carboxylic acid in suitable solvents [9, 11]. The alpha-addition of the isocyanide can be considered as a three component reaction since hydrogen bridged intermediate can be also considered as the two components.

In 1959, Ugi et al. described the most important variations of four-component condensation, the U-4CRs [12, 13, 14]. The classic version of the U-4CRs is the reaction of a primary amine, an oxo component (aldehyde or ketone), an isocynide, and a carboxylic acid, to form an alpha-aminoalkyl-cations, and the anions of the acid components. In suitable polar solvents these ions are solvated and they are mutually attracted so they can be close to each other.

The cations, anions and isocyanides can directly form the alpha-adducts and subsequently rearrange into the final product. However, the intermediate can be formed by two steps: the intermediate could come from cation with isocyanide and in the next step the anion could be added into alpha-addition. And similarly from anion with isocyanide the intermediate could be formed and subsequently react with cation into the alpha-adduct.

To determine which reaction mechanism have energetic advantages over the other possibilities, the energetic levels of the reactants, intermediates and the isocyanide were calculated. A simple model of this reaction forming alpha-adduct was investigated. By HF/6-31G* and B3LYP/6-31G* quantum chemistry calculations [15, 16] the preferred reaction mechanism of the U-4CR was indicated.

3   Results

The energy minimization of eight possible conformations of the alpha-adduct structure were performed. These conformations were obtained by rotation of three torsional angles respectively, so all of the conformational space of the intermediate structure was explored. The conformation with the lowest energy was chosen for further analysis of the system.

The total energy and the difference between the total energy and zero energy were calculated to asses the stability of the intermediate structure. The results for the HF/6-31G(d) and B3LYP/6-31G(d) of these calculations using GAMESS program, interfaced with CHARMM utilizing the ABNR minimization routine are presented in Figure 1.

The starting models for calculations were obtained by breaking the covalent bonds between the subsystems in the alpha-adduct structure by setting the distance to 3 Å. When the starting models are fully minimized, they either transform into the alpha-adduct structure (the upper right structure on Figure 1) or some other structure on the potential energy surface, but never into the structures shown on the bottom left and right on Figure 1. To establish if structures shown on the bottom left and right on Figures 1 are even possible we also calculated the isolated intermediates of these two structures, but found they disintegrate. Since these two structures do not posses a minimum on the potential energy surface, we can conclude that the alpha-adduct (the upper right structure on Figure 1) is formed by a three component reaction [17].

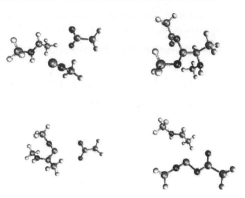

Figure 1: The figure shows a three-component reaction of isocyanides. The reactants are in the upper left. The alpha-adduct structure is in the upper right. Two intermediates are shown on the bottom left and right. We have shown with a quantum-mechanical calculation that the alpha-adduct forms directly from the three reactants as well as that the two intermediates do no exist.

4   Conclusions

The present work provides an overview of a variety of methods for computer simulation studies of molecular systems in the case when all degrees of freedom were taken into account, and their reactions.

Acknowledgment

This work was supported by the Ministry of Higher Education, Science and Technology of Slovenia under grants No. P1-0002 and J1-6331.

References

[1] D. Janežič, M. Penca, Molecular Simulation Studies of Fast Vibrational Modes. *Lect. Ser. Comput. Computat. Sci.* 2004, 1, 756-759.

[2] D. Janežič, M. Praprotnik, F. Merzel, Molecular dynamics integration and molecular vibrational theory. I. New symplectic intgrators. *J. Chem. Phys.* 2005, 122, 17, art.no. 174101.

[3] M. Praprotnik and D. Janežič, Molecular dynamics integration and molecular vibrational theory. II. Simulation of non-linear molecules. *J. Chem. Phys.* 2005, 122, 17, art.no. 174102.

[4] M. Praprotnik and D. Janežič, Molecular dynamics integration and molecular vibrational theory. III. The IR spectrum of water. *J. Chem. Phys.* 2005, 122, 17, art.no. 174103.

[5] B.R. Brooks, D. Janežič, M. Karplus, Harmonic analysis of large systems. I. Methodology. *J. Comput. Chem.* 1995, 16, 1522-1542.

[6] D. Janežič and B.R. Brooks, Harmonic analysis of large systems. II. Comparison of different protein models. *J. Comput. Chem.* 1995, 16, 1543-1553.

[7] D. Janežič, B.R. Brooks, and R.M. Venable, Harmonic analysis of large systyems. III. Comparison with molecular dynamics. *J. Comput. Chem.* 1995, 16, 1554-1566.

[8] A. Dolming, I. Ugi, Multicomponent Reactions with Isocyanides, *Angew. Chem. Int. Ed. Engl.* 2000, 39, 3169-3210

[9] I. Ugi, Isonitrile Chemistry, Academic Press, New York, 1971

[10] M. Passerini, G. Ragni, Sopra gli isonitrili - XIX Reazoni con acidi aldehidi e chetonoci, *ibid.* 1931, 61, 964.

[11] R.H. Baker, L.E. Linn, The Passerini Reaction III. Stereochemistry and Mechanism *J.Am. Chem. Soc.* 1951, 73, 699-702.

[12] R. Meyr and U. Fetzer, C.Steinbruckner, Versuche mit Isonitrilen, *Angew. Chem.* 1959, 71, 386.

[13] I. Ugi, Neuere Methoden der praparativen organishen Chemie IV. Mit sekundar-reationen gekoppelte alfa-additionen von Immonium-ionen und anionen an isonitrile, *Angew. Chem. Int. Ed. Engl.* 1962, 1, 8.

[14] I. Ugi, S. Lohberger and R. Karl, The Passerini and Ugi reactions, Comprehensive Organic Synthesis: Selectivity for Synthetic Efficiency, vol. 2, chap. 4.6, B.M. Trost, C.H. Heathcock (eds), Pergamon, Oxford 1991, p. 1083.

[15] K.P. Eurenius, D.C. Catfield, B.R. Brooks and M. Hodošček, Enzyme Mechanisms with Hybrid Quantum and Molecular Mechanical Potentials. I. Theoretical Considerations, *Int.J.Quant. Chem.* 1996, 60, 1189-1200.

[16] B.R. Brooks, R.E. Bruccoleri, B.D. Ofafson, D.J. States, S.Swaminathan and M Karplus, CHARMM: A Program for Macromolecular Energy, Minimization and Dynamics Calculations, *J. Comput. Chem.* 1983, 4, 187-217.

[17] D. Janežič, M. Hodošček, and I. Ugi, The simultaneous [alpha]-addition of a cation and an anion onto an isocyanide. *Internet Electron. J. Mol. Des.* 2002, 1, 6, 293-299.

VSP International
Science Publishers
P.O. Box 346, 3700 AH Zeist
The Netherlands

Lecture Series on Computer
and Computational Sciences
Volume 4, 2005, pp. 1593-1595

Approaches and Methods of Security Engineering

Tai-hoon Kim[1]

Sansung Gongsa,
San 7, Geoyeo-Dong, Songpa-Gu, Seoul, Korea

Abstract: The general systems of today are composed of a number of components such as servers and clients, protocols, services, and so on. Systems connected to network have become more complex and wide, but the researches for the systems are focused on the 'performance' or 'efficiency'. While most of the attention in system security has been focused on encryption technology and protocols for securing the data transaction, it is critical to note that a weakness (or security hole) in any one of the components may comprise whole system. Security engineering is needed for reducing security holes may be included in the software. There are very many approaches or methods in software development or software engineering. Therefore, more security-related researches are needed to reduce security weakness may be included in the software and complement security-related considerations of general software engineering.

Keywords: Security engineering, Security hole, Security weakness.

1. Security Engineering for Information Assurance

There are many standards, methods and approaches which are used for assuring the quality of software. ISO/IEC TR 15504, the Software Process Improvement Capability Determination (SPICE), provides a framework for the assessment of software processes [1-2]. But, in the ISO/IEC TR 15504, considerations for security are relatively poor to others. For example, the considerations for security related to software development and developer are lacked.

When we are making some kinds of software products, ISO/IEC TR 15504 may provide a framework for the assessment of software processes, and this framework can be used by organizations involved in planning, monitoring, controlling, and improving the acquisition, supply, development, operation, evolution and support of software. But, in the ISO/IEC TR 15504, considerations for security are relatively poor to other security-related criteria such as ISO/IEC 21827, the Systems Security Engineering Capability Maturity Model (SSE-CMM), or ISO/IEC 15408, Common Criteria (CC) [3-6]. Security-related software development is concerned with many kinds of measures that may be applied to the development environment or developer to protect the confidentiality and integrity of the IT product or system developed.

It is essential that not only the customer's requirements for software functionality should be satisfied but also the security requirements imposed on the software development should be effectively analyzed and implemented in contributing to the security objectives of customer's requirements. Unless suitable requirements are established at the start of the software development process, the resulting end product, however well engineered, may not meet the objectives of its anticipated consumers. The IT products like as firewall, IDS (Intrusion Detection System) and VPN (Virtual Private Network) which made to perform special functions related to security are used to supply security characteristics. But the method using these products may be not the perfect solution. Therefore, when making some kinds of software products, security-related requirements must be considered.

[1] Corresponding author. Sansung Gonsa, San 7, Geoyeo-Dong, Songpa-Gu, Seoul, Korea, E-mail: taihoonn@empal.com

2. Approaches and Methods of Security Engineering

Assurance methods are classified in Fig.1 according to the three assurance approach categories. Depending on the type of assurance method, the assurance gained is based on the aspect assessed and the lifecycle phase. Assurance approaches yield different assurance due to the deliverable (IT component or service) aspect examined. Some approaches examine different phases of the deliverable lifecycle while others examine the processes that produce the deliverable (indirect examination of the deliverable). Assurance approaches include facility, development, analysis, testing, flaw remediation, operational, warranties, personnel, etc. These assurance approaches can be further broken down; for example, testing assurance approach includes general testing and strict conformance testing assurance methods [3].

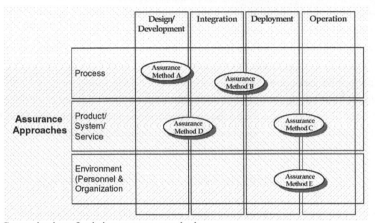

Figure 1: Categorization of existing assurance methods.

References

[1] ISO. ISO/IEC TR 15504-2:1998 Information technology – Software process assessment – Part 2: A reference model for processes and process capability
[2] ISO. ISO/IEC TR 15504-5:1998 Information technology – Software process assessment – Part 5: An assessment model and indicator guidance
[3] ISO. ISO/IEC 21827 Information technology – Systems Security Engineering Capability Maturity Model (SSE-CMM)
[4] ISO. ISO/IEC 15408-1:1999 Information technology - Security techniques - Evaluation criteria for IT security - Part 1: Introduction and general model
[5] ISO. ISO/IEC 15408-2:1999 Information technology - Security techniques - Evaluation criteria for IT security - Part 2: Security functional requirements
[6] ISO. ISO/IEC 15408-3:1999 Information technology - Security techniques - Evaluation criteria for IT security - Part 3: Security assurance requirements
[7] Tai-Hoon Kim, Byung-Gyu No, Dong-chun Lee, Threat Description for the PP by Using the Concept of the Assets Protected by TOE, ICCS 2003, LNCS 2660, Part 4, pp. 605-613
[8] Tai-hoon Kim and Haeng-kon Kim: The Reduction Method of Threat Phrases by Classifying Assets, ICCSA 2004, LNCS 3043, Part 1, 2004.
[9] Tai-hoon Kim and Haeng-kon Kim: A Relationship between Security Engineering and Security Evaluation, ICCSA 2004, LNCS 3046, Part 4, 2004.
[10] Eun-ser Lee, Kyung-whan Lee, Tai-hoon Kim and Il-hong Jung: Introduction and Evaluation of Development System Security Process of ISO/IEC TR 15504, ICCSA 2004, LNCS 3043, Part 1, 2004

Dr. Tai-hoon Kim

Dr. Tai Hoon Kim received his M.S. degrees and Ph.D. in Electrics, Electronics & Computer Engineering from the Sungkyunkwan University, Korea. After working with Technical Institute of Shindoricoh 2 years as a researcher from December 1st 1996 and working at the Korea Information Security Agency as a senior researcher 2 years and 6 months from January 1st 2002, he is currently working at the DSC (Defense Security Command) SERG (Security Engineering Research Group).

He wrote sixteen books about the software development, OS such as Linux and Windows 2000, and computer hacking & security. And he lectured at the Sungkyunkwan University. He was a program committee of SNPD (International Conference on Software Engineering, Artificial Intelligence, Networking, and Parallel/Distributed Computing) 2004 and SERA (International Conference on Software Engineering Research, Management and Applications) 2004, and he is a program committee of SERA 2005.

He is a Guest Editor of AJIT (Asian Journal of Information Technology) Special Issue on: Multimedia Services and The Security in the Next Generation Mobile Information Systems.

He was a technical session chair in ICCSA 2004, PARA 2004, SCI 2004, ICCMSE 2004, PCM 2004, ICCSA 2005, KES 2005, RSFDGrC 2005, KES 2005, and he is a mini-symposium chair in ICCMSE 2005 and ICIC 2006.

He researches security engineering, the evaluation of information security products or systems and the process improvement for security enhancement. The results of his work were presented in many conferences and about 20 papers were listed in SCI.

Brill Academic Publishers
P.O. Box 9000, 2300 PA Leiden,
The Netherlands

*Lecture Series on Computer
and Computational Sciences*
Volume 4, 2005, pp. 1596-1599

A Role-based Process Security Model in Business Process Management

Kwanghoon Kim[1]

Collaboration Technology Research Lab,
Department of Computer Science,
KYONGGI UNIVERSITY,
Suwon-si, Kyonggi-do, 442-760, South Korea

Changmin Kim[2]

Division of Computer Science,
SUNGKYUL UNIVERSITY,
Anyang-si, Kyonggi-do, 430-742, South Korea

Received 20 June, 2005; accepted in revised form 30 July, 2005

Abstract: This paper formally defines a role-based security model of a business process in order eventually to provide a theoretical basis for realizing the role-based access control in enacting business processes. That is, we propose a graphical representation and formal description of the mechanism that generates a set of role-based security models from a business process modeled by the information control net (ICN) modeling methodology that is a typical business process modeling approach for defining and specifying Business processes.

Keywords: security and role-based access control, secured business process management

1 Introduction

In the business process management literature, the current set of security policies, access control guidelines, and mechanisms has little grown out of research and development efforts. But, according to the growths of the business process management market, the security issue is taking the immediate attention in the literature. Today the best known computer-related security standard is the Role-Based Access Control (RBAC) [1,2,3] that has been accepted as a secured access control model being appropriate and central to the secure processing needs within industry and civilian government. However, without any modifications and extensions, it is unreasonable for the role-based access control model to be directly applied in order to realize a secured business process management system. Therefore, for realizing the secured business process management, in this paper we propose a new business process security model, which is called a role-based business process security model, by analyzing and reflecting the properties of business processes, and the security model will be possibly extended to the role-based access control standard in the future.

---

[1]Corresponding author. E-mail: kwang@kyonggi.ac.kr
[2]E-mail: kimcm@sungkyul.ac.kr

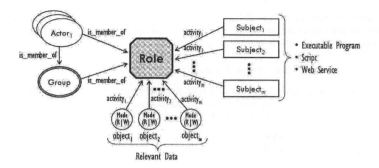

Figure 1: Graphical Representation of the Role-based Security Model

2 Motivation and Related Work

So far, in terms of the security criteria, standards, and guidelines in commercial and civilian government organizations, the Trusted Computer System Evaluation Criteria (TCSEC) must be the best known standard [1]. The TCSEC specifies three types of access controls: Discretionary Access Controls (DAC), Mandatory Access Controls (MAC), and Role-Based Access Control (RBAC) [1,2,3]. DAC requirements have been perceived as being technically correct for commercial and civilian government security needs, as well as for single-level military systems. MAC is used for multi-level secure military systems, but its use in other applications is rare. And, as stated in [1,2,3], the Role-Based Access Control (RBAC) is more appropriate and central to the secure processing needs within industry and civilian government than that of DAC. On the web site [10], we can find a lot of documents and materials about the RBAC model and its technologies. Especially, the site [6] reads that there is a patent of workflow management employing RBAC, which seems to be surely related with this paper's approach. Even though we couldn't read the contents of the patent through the web site, it is manifest that the approach is completely difference from our approach because our approach is based upon not the RBAC model but our own model that is automatically generated from a business process model. In the next section, we define our own model - the role-based business process security model.

3 Role-based Business Process Security Model

In this section, we formally define the role-based business process security model (RB^2PSM) and propose an algorithm that automatically generates a set of the underlining role-based security models from a business process model.

3.1 Graphical Representation of RB^2PSM

Based upon the ICN-based business process model in the previous section, we define and clarify the notion of RB^2PSM by giving a graphical representation as well as a simple formal description. The graphical representation of RB^2PSM is shown in Fig. 1. As seen in the figure, each role defined in the business process model is associated with a set of subjects that is assigned into the corresponding role through its assigned activities, and it also has a set of member actors reflecting the organizational structure. Finally, a set of objects with permission modes (read or/and write) is assigned into the role. Note that the notion of the subject is same to the application entity in the ICN-Based business process model, and the object is same to the notion of the relevant data entity. Also, as you can intuitively guess, the number of role-based security models to be generated from a business process model is exactly same to the number of roles in it.

3.2 Formal Description of RB^2PSM

In order to clarify the graphical representation of the role-based business process security model, we give a formal description of the model and also simply use a way of describing, in terms of sets and relations, the formal description of RBAC [1], as followings:

- For each activity, the active role is the one that the activity is currently using:
  $AR(a : activity) = \{the\ active\ role\ for\ activity\ a\}$.

- Each activity may be authorized to perform one or more roles:
  $RA(a : activity) = \{the\ authorized\ roles\ for\ activity\ a\}$.

- Each role may be assigned to one or more activities:
  $AA(r : role) = \{the\ authorized\ activities\ for\ role\ r\}$.

- Each activity may be authorized to perform one or more subjects:
  $AS(a : activity) = \{the\ authorized\ subjects\ for\ activity\ a\}$.

- Each role may be assigned to one or more actors (or groups):
  $RC(r : role) = \{the\ authorized\ actors\ for\ role\ r\}$.

- Each activity may be authorized to read-access or write-access to one or more objects:
  $AO(a : activity) = \{the\ authorized\ objects\ for\ activity\ a\}$.

- For each activity, the active actor (or group) is the one that is currently performing the activity:
  $AC(a : activity) = \{the\ active\ actor\ or\ group\ for\ activity\ a\}$.

- Activities may execute subjects. The predicate exec(a,s) is **true** if *activity* a can execute *subject* s at the current time, otherwise it is **false**:
  exec($a : activity, s : subject$) = {**true** *iff* activity a can execute subject s}.

Also, four basic rules are required in the rule-based business process security model as followings:

1. Role assignment: An activity can execute a subject only if the activity has selected or been assigned a role:
   $\forall a : activity, s : subject(\text{exec}(a, s) \Rightarrow AR(a) \neq \emptyset)$.

   Like in the RBAC model, the identification and authentication process (e.g. login) is not considered an activity. All other user activities on the business process management system are conducted through subjects. Thus all active actors are required to have some active role.

2. Role authorization: An activity's active role must be authorized for the activity:
   $\forall a : activity(AR(a) \subseteq RA(a))$.

   With (1) above, this rule ensures that actors can take on only roles for which they are authorized.

3. Subject authorization: An activity can execute a subject only if the subject is authorized for the activity's active role:
   $\forall a : activity, s : subject(\text{exec}(a, s) \Rightarrow s \in AS(AA(AR(a))))$.

   With (1) and (2), this rule ensures that actors can execute only subjects for which they are authorized. Note that, because the conditional is "only if", this rule allows the possibility that additional restrictions may be placed on subject execution. That is, the rule does not guarantee a subject to be executable just because it is in AS(AA(AR(a))), the set of subjects potentially executable by the activity's active role.

4. Object authorization: A subject can access objects only if the subject is authorized for the objects with access modes:

$\forall a : activity, s : subject, o : object(\text{exec}(a, s) \Rightarrow access(\text{AR}(a), s, \text{AO}(a), x)$.

This rule is defined using a subject to object access function access(r, s, o, x) which indicates if it is permissible for a subject in *role* r to access *object* o in *mode* x using *subject* s, where x is taken from some set of modes such as *read* and *write*.

Like the RBAC model [1], access control decisions of RB^2PSM, such as duties, responsibilities, and qualification, are determined by the roles. But the difference is on the way of determination. That is, in the business process management it is determined at the build-time of business processes. A RBAC policy bases access control decisions on the functions of organization-specific protection guidelines within an organization, while a RB^2PSM policy is based upon access control decisions on the functions of business process-specific authorization within an organization. This is a fundamental difference between RBAC and RB^2PSM.

4   Conclusions

So far, the role-based access control model has been accepted as a promise solution and standard that is able to accomplish the central administration of an organizational specific security policy and to meet the secure processing needs of many commercial and civilian government organizations. In spite of these facts, the RBAC model should be inapplicable to the business process management systems without further modifications and extensions. So, we proposed the role-based process security model that is directly applicable to the business process management system. As future research issues, there might be several extensions that we should investigate how the concept of role-based security model is incorporated into the other families of coordination models and systems.

References

[1] David F. Ferraiolo and D. Richard Kuhn, "Role-Based Access Controls," Proceedings of the 15th NIST-NSA National Computer Security Conference, Baltimore, Maryland, October 13-16, (1992)

[2] David F. Ferraiolo and et al., "AN INTRODUCTION TO ROLE-BASED ACCESS CONTROL", NIST/ITL Bulletin December, (1995)

[3] David F. Ferraiolo, Janet A. Cugini, and D. Richard Kuhn, "Role-Based Access Control (RBAC): Features and Motivations," Proceedings of the 11th Annual Computer Security Applications, (1995)

[4] Clarence A. Ellis, "Formal and Informal Models of Office Activity", Proceedings of the 1983 Would Computer Congress, Paris, France, (1983)

[5] Kwanghoon Kim and et al., "Role-based Model and Architecture for Workflow Systems", International Journal of Computer and Information Science, Vol. 4, No. 4, (2003)

[6] http://csrc.nist.gov/rbac

Brill Academic Publishers
P.O. Box 9000, 2300 PA Leiden,
The Netherlands

*Lecture Series on Computer
and Computational Sciences*
Volume 4, 2005, pp. 1600-1603

Correlation Analysis System using
VA data, IDS alerts

Jong-Hyouk Lee and Tai-Myung Chung[1]

Internet Management Technology Laboratory,
Dept. of Computer Engineering, Sungkyunkwan University,
300 Cheoncheon-dong, Jangan-gu,
Suwon-si, Gyeonggi-do, 440-746, Korea

Received 27 June, 2005; accepted in revised form 2 August, 2005

Abstract: We propose a novel framework named Correlation Analysis System (CAS) using vulnerability assessment data, IDS alerts. Traditional IDS focus on low-level attacks or anomalies, and raise alerts independently, though there may be logical connections between them. In situations where there are intensive intrusions, not only will actual alerts be mixed with false alerts, but the amount of alerts will also become unmanageable. We propose CAS using vulnerability assessment data, IDS alerts to reduce false positives and negatives. This provides an effective mean to lower the number of false positives and negatives that an administrator has to deal with. It also improves the results of alert correlation systems by cleaning their input data from spurious attacks. We have developed an active verification system based on Snort and Nessus. As the current implementation stands, it is a useful tool for reducing the false alarm rate of Snort.

Keywords: Correlation, vulnerability, IDS

1 Introduction

Intrusion detection systems (IDS) have become widely available in recent years, and are beginning to gain acceptance in enterprises as a worthwhile improvement on security. However, traditional IDS focus on low-level attacks or anomalies, and raise alerts independently, though there may be logical connections between them. In situations where there are intensive intrusions, not only will actual alerts be mixed with false alerts, but the amount of alerts will also become unmanageable. As a result, it is difficult for network security administrators or intrusion response systems to understand the alerts and take appropriate actions. Several alert correlation techniques have been proposed to facilitate the analysis of intrusion alerts, including those based on the similarity between alert attributes [1, 2], previously known attack scenarios [3, 4]. However, most of these correlation methods focus on IDS alerts.

In order to solve these problem, we propose CAS (Correlation Analysis System) using VA (Vulnerability Assessment) data, IDS alerts to reduce false positives and negatives.

The rest of this paper is organized as follows. The next section discusses related work. In section 3, we define the CAS and describe how it works and how it is organized. In Section 4, we evaluate the performance of the CAS. At last, section 5 summarizes our work.

---

[1]Corresponding author: Jong-Hyouk Lee. E-mail: jhlee@imtl.skku.ac.kr

2 Background and Motivation

2.1 Vulnerability Assessment

Every year, the Computer Emergency Response Team (CERT) publishes more and more vulnerability disclosures. These disclosures come from product vendors, security research companies and many other sources. In 2002, CERT reported 4,129 vulnerabilities. In 2003, this number rose to 3,784 vulnerabilities and In 2004, over 3780 have been reported [5]. Each one of these vulnerabilities is not present on every network, and each vulnerability is not as serious as the next one, but there is no way to know the impact of the vulnerabilities on a given network without looking at how they effect the given network directly. For example, a simple vulnerability in a web service application which allows the remote browsing of the contents of a directory may seem innocuous, but combined with a server that processes credit card information, this vulnerability could disclose end-user private information.

To VA, automated tools (e.g., Nessus, SATAN) are used to catalog each of the hosts on a network, and then to interrogate the discovered hosts for any known vulnerabilities. Typically these automated tools are run from a single host and have a database of vulnerabilities. When run, they produce reports about the types of hosts discovered, and lists of the potential vulnerabilities found. Once the VA is completed, the information is communicated to those who can fix the problems discovered.

2.2 IDS alerts

As more IDSs are developed, network security administrators are faced with the task of analyzing an increasing amount of alerts resulting from the analysis of different event streams. In addition, IDSs are far from perfect and may produce both false positives and non-relevant positives, which are alerts that are not representative of a successful attack. Therefore, there is a need for tools and techniques that allow the administrator to aggregate and combine the outputs of multiple IDSs, filter out spurious or incorrect alerts, and provide a succinct, high-level view of the security state of the protected network.

Many of the false positives associated with an IDS can be mitigated by considering the vulnerabilities of the protected network. At a high level, if an IDS knows that a system is vulnerable to a particular vulnerability, then it should only concern itself for attacks against that particular vulnerability. If such a system existed, then we can expect high quality alerts to be generated. The system knows what the vulnerabilities are, and it knows that a particular vulnerability is being exploited. This level of information results in a higher level of confidence that a system is under immediate threat. Because of this correlation, better decisions can be made in an automated fashion. These include firewall rule changes to drop the attacker and mailing the security staff to notify them in real time of the attack.

3 Implementation of CAS

The CAS system known as an extension of Snort uses Nessus Attack Scripting Language (NASL) scripts to scan the Nessus vulnerabilities. Deployment of the combined system does not cost as much as a stand-alone Snort sensor does. The modifications mainly consist of an addition to Snorts alert processing pipeline which intercepts alerts to be passed to enabled alert plug-ins. These alerts are queued for verification by a pool of verification threads. This allows Snort to continue processing events while alert verification takes place in the background. Because our verification system is implemented as a part of the Snort sensor, they are located together. This is not a requirement of active verification, however, and it would also be possible to have a single verification system that

Brill Academic Publishers
P.O. Box 9000, 2300 PA Leiden
The Netherlands

Lecture Series on Computer
and Computational Sciences
Volume 4, 2005, pp. 1604-1607

A Multilayered Neural Network Based Computer Access Security System: Effects of Training Algorithms

A.S. Anagun[1]

Department of Industrial Engineering,
Osmangazi University,
Eskisehir 26030, Turkey

Received 10 June, 2005; accepted in revised form 20 July, 2005

Abstract: In this paper, effects of training algorithms applicable to multilayered neural networks (NNs) are examined for a multilayered NN based computer access security system designed in order to differentiate an appropriate person from an intruder. Five training algorithms are taken into consideration for such system in terms of recognition accuracy. The algorithms studied are Backpropagation (BP), Quickprop (QP), Delta-Bar-Delta (DBD), Extended-Delta-Bar-Delta (EDBD), and Resilient Prop (RP). The designed system is trained using the data obtained from time intervals between successive characters while entering a password via keyboard. The performances of the algorithms are compared with each other in terms of classification accuracy.

Keywords: Security system, multilayered neural networks, training algorithms, keystroke dynamics.

1. Introduction

A computer security system should not only be able to identify a person and let him/her access to the system if he/she has a correct security code or deny the access otherwise - *preventive security*, but also be capable of identifying the person whether he/she is indeed the right person - *detection of violations* [1]. In the area of computer access security, a number of researchers have carried out studies related to user identification based on individual's typing pattern using NNs trained using BP algorithm as powerful tools for pattern recognition and classification applications [1-5]. In these studies, since the same password has been entered by a group of people, this situation may be classified as *multiple users-single password* and *preventive security*, basically classifying people into two groups without evaluating whether they are indeed authorized.

However, due to the developments in computer technology and the complexity of information systems, which the organizations might have, there may be a different situation such as *multiple users-multiple passwords*. As discussed in [6], *multiple users-multiple passwords* may be applied to the computer access security systems considering passwords with different lengths depending on his/her preferences or system's requirements, if applicable. In order to identify users and differentiate valid users from invalid ones (intruder), the NN was trained with BP algorithm using a large set of data consisted of keystroke patterns of the participants. In the study of [7], a NN based security system has been designed and applied to the cases of *multiple users-one password* and *multiple users-multiple passwords* with different lengths for *preventive security* and *detection of violation* purposes, mentioning the importance of questioning both password and user identification simultaneously. Two critical issues, password-dependent identification and password-independent identification were evaluated in terms of recognition accuracy.

Here, a neural network based intelligent computer access security system is developed for a situation of *multiple users-multiple passwords* considering both password and user identification codes. In order to differentiate an authorized person from an intruder, the data composed of time intervals between keystrokes typed via a keyboard are obtained by means of a data collection structure designed. In addition to the BP algorithm, four more training algorithms are taken into consideration and the

---

[1] Corresponding author. E-mail: sanagun@ogu.edu.tr

performances of the algorithms are compared with each other in terms of reliability of the security system.

2. Training Algorithms for Multilayered NNs

The BP algorithm gives the change $\Delta w_{ji}(k)$ in the weight of the connection between neurons i and j at iteration k as [8]:

$$\Delta w_{ji}(k) = -\alpha \frac{\partial E}{\partial w_{ji}(k)} + \beta \Delta w_{ji}(k-1) \qquad (1)$$

where α is called the learning coefficient, β the momentum coefficient and $\Delta w_{ji}(k-1)$ the weight change in the immediately preceding iteration. Training a multilayered NN by BP algorithm involves presenting it sequentially with all training patterns. Difference between the target output $y_d(k)$ and the actual output $y(k)$ of the multilayered NNs are propagated back through to adapt its weights. A training iteration is completed after a pattern in the training data has been presented to the network and the weights updated.

The QP algorithm was developed as a method of improving the rate of convergence to a minimum value of $E(w_{ji})$ by using information about the curvature of the error surface $E(w_{ji})$ [9]. The QP algorithm employs the gradient $\partial E / \partial w_{ji}$ at two points $w_{ji}(k)$ and $w_{ji}(k-1)$ to find the minimum of E. The weight change $\Delta w_{ji}(k)$ is computed as follows:

$$\Delta w_{ji}(k) = -\alpha \frac{\partial E}{\partial w_{ji}(k)} + \beta \Delta w_{ji}(k-1) \qquad (2)$$

where

$$\beta = \frac{\partial E / \partial w_{ji}(k)}{[\partial E / \partial w_{ji}(k-1)] - [\partial E / \partial w_{ji}(k)]} \qquad (3)$$

Since the momentum coefficient β is variable and thus QP algorithm can be viewed as a form of BP algorithm that employs a dynamic momentum coefficient.

The DBD algorithm is developed to improve the speed of convergence of the connection weights in the multilayered NNs [10]. By assigning a learning coefficient to each weight and permitting it to change over time, more degrees of freedom are introduced to reduce the to convergence towards a minimum value of $E(w_{ji})$. The weight change is given by:

$$\Delta w_{ji}(k) = -\alpha_{ji}(k) \frac{\partial E}{\partial w_{ji}(k)} \qquad (4)$$

where $\alpha_{ji}(k)$ is the learning coefficient assigned to connection from neuron i to j.

The EDBD algorithm is an extension of DBD algorithm and also aims to decrease the training time for multilayered NNs [11]. The changes in weights are calculated as:

$$\Delta w_{ji}(k) = -\alpha_{ji}(k) \frac{\partial E(k)}{\partial w_{ji}(k)} + \beta_{ji}(k) \Delta w_{ji}(k-1) \qquad (5)$$

where $\alpha_{ji}(k)$ and $\beta_{ji}(k)$ are learning and momentum coefficients, respectively. On the other hand, the use of momentum in EDBD algorithm is the major difference between it and DBD algorithm.

The RP is an adaptive learning rate method where weight updates are based only on the sign of the local gradients, not their magnitudes. This algorithm generally provides faster convergence than most other algorithms [12]. When the update-value for each weight is adapted, the delta weights are changed as follows:

compared to the single SMPS case, even with no error correction. The reason for this is some of the periodic impulse noise generated by one SMPS was canceled by the other SMPS.

Also, as shown in Figure 6, we simulated the cases of channels with single and triple SMPSs using the proposed interleaving method. In each case, we see improved performance of about 2dB compared to the respective cases without error correction.

6. Conclusions

In this paper, we introduced an interleaving method to improve impulse noise techniques that is important in Home Power Line Communication for overcoming poor communication channel environments.

Impulse noise(especially home appliance and load generated impulse noise) has already been proved as the most influential noise that degrades bit error rate properties because impulse components of voltage and current waveforms occur in wide frequency bands widely due to switching of semiconductor devices in home appliances. We can confirm the results that communication performances are improved by implementing a block interleaver before modulating and after demodulating the burst noise caused by impulse noise.

Future work includes 1) investigation of transmission channels in various environments to implement more efficient Home Power Line Communication Network Systems and 2) improvement of noise reduction with the interleaving method using reliable error correcting.

Acknowledgement

This work was supported by KESRI (R-2004-B-228), which is funded by MOCIE (Ministry of Commerce, Industry and Energy).

References

[1] Y.K. Choi, G.H. Park, J.W. Seo, J.S. Cha, H.S. Seo, M.C. Shin, "A Study of Broadband Indoor Power Line Communication Channel Modeling", KIEE Summer Annual Conference, Vol. A, pp.271-273, 2003

[2] Michel ROUSSEAU, Patrick MOREAU, "Characterization and optimization of multicarrier-technologies over PLC channel", ISPLC2001, pp299-304, 2001

[3] J.H. Lee, J.S. Cha, J.J. Lee, M.C. Shin, R.H. Sung, "A Study of Channel Property Analysis for Power Line Communication", KIEE Summer Annual conference, Vol. A, 558-560, 2000

[4] Beranrd Sklar, "Digital Communication - Fundamentals and Applications", Second Edition, pp.463-466, Prentice hall, 2000

[5] G.H. Park, Y.K. Choi, B.G. Lee, H.M. Kim, J.S. Cha, M.C. Shin, "A Study of High-Speed Power Line Communication", KIEE Summer Annual Conference, Vol. A, pp.274-276, 2003

Brill Academic Publishers
P.O. Box 9000, 2300 PA Leiden
The Netherlands

Lecture Series on Computer
and Computational Sciences
Volume 4, 2005, pp. 1613-1616

Home Security System in a Web-based Networking Environment

Kyung-Bae Chang[1], Jae-woo Kim, Il-Joo Shim, and Gwi-Tae Park

Department of Electrical Engineering,
Korea University,
Anam-dong 5ga, Seongbuk-gu 136-713 Seoul, Korea

Received 30 July, 2005; accepted in revised form 10 August, 2005

Abstract: In a Ubiquitous environment, people want control over their homes every moment and from any location.[1] If the environment that directly controls and surveys a home uses the web, this becomes possible. Instead of using expensive imaging equipment, cheap and reliable monitoring and control can be accomplished by the use of a USB camera. We used a LabVIEW program to verify this paper, and the presented method can be easily applied to other applications.[2]

Keywords: Home Security, LabVIEW, Web-based Control

1. Introduction

A representative application field for security and monitoring is the home network service, which is a emerging topic in the IT field. As the development to a knowledge information society becomes more rapid, based on the outstanding information communication network infra leaded by the government, demand for fast and reliable information has increased and 'quality of life' has become an important factor. The development of information technology has even influenced industrial fields and also everyday livelihoods. Based on this, home networks and home automation systems which increase the safety, convenience and pleasure of the living environment, and security solutions which has considered the secure and economic values of the industrial fields are being highlighted.[3] As the quality of life increases and lifestyles change based on information, the need for more efficient control and facilities that improve the safety and convenience of the residents increase. In order to fulfill the peoples aspirations of maintaining safety, convenience and pleasantness of the residential area, services that provide surveillance monitoring at any location through computers, PDA and cell phones became possible by the use of an open architecture using the internet rather than the closed CCTV(Closed Circuit Television) method. Also, based on the cameras type, movement control or enlarging and reducing the image become possible for the user in any location. A security solution can be built by using only a camera and software with the use of a personal PC, without any separate servers or equipment. Therefore, cheap and convenient surveillance easily becomes possible by various industrial fields. Also, with the development of home automation technology for the common household, the ON/OFF of household electrical items such as HVAC(heating, ventilating, and air conditioning) systems, lighting, front door can be controlled from a remote location through a PC, cell phone or PDA, utilizing the home automation system. For such purpose, usually CCTV is employed that uses a screen capture grab board and an expensive camera. However, it is very passive, showing the image on the screen while recording it to the video tape. Usually, CCTV is used by the apartment building administration or some other controlling entity, since it is expensive to buy and install the system for the personal use by an individual. However, when the digital home service becomes popular and hence more affordable, we can use such inexpensive equipment to monitor our home for intruders or nursery, operating room in the hospital, building entrance and so on. It can also make our life more convenient.[4] For example, to inform the home owner about a person entering or trying to enter the home and even allow entrance via a certification process through a mobile device such as a cell-phone or PDA. Further, it can allow to remotely monitor young children or a patient in a house or in the hospital while being away enhancing personal safety and comfort.[5]

---

[1] Corresponding author. Kyungbae Chang(E-mail: lslove@korea.ac.kr)

of voice, image and remote control data and data communications, thus becoming a more complete security solution.[10][11]

5. Conclusion

In this paper, we produced a home security system on a web based network environment, with the use of LabVIEW. A system more cheaper than using common visual devices, but equally effective in a household can be produced by this method. Also, we proved that a more reliable surveillance system can be produced this way by of the use of the web, giving the ability to actually see and control whenever and wherever the user is. As the market for home security and automation systems grows, the proposed method will become more applicable.

References

[1] Schulzrinne, H. Xiaotao Wu Sidiroglou,S. Berger, "*Ubiquitous computing in home networks*" Volome: 41, Issue: 11 128 - 135 Communications Magazine, IEEE 20

[2] www.ni.com */LabVIEW*

[3] Renato Jorge Caleira Nunes, *A web-Based Approach to the Specification and Programming of Home Automation Systems*, IEEE MELECON,pp693-696 , Croatia 2004

[4] Yuan-Hsiang Lin   I-Chien Jan "*A wireless PDA-based physiological monitoring system for patient transport*" Volume: 8, Issue: 4 439 - 447 IEEE Transactions 2004

[5] M. Joler and C. G. Cristodoulou, "*Virtual laboratory instruments and simulations remotely controlled via the Internet,*" in Proc. IEEE Antennas Propagat. Soc. Int. Symp., vol. 1, pp. 388–391 Boston, MA, July 2001

[6] Kwnag Yeol Lee and Jae Weon Choi, *Remote-controlled Home Automation System via Bluetooth Home Network* , SCIE Annual Conference pp 2824-2829, 2003.

[7] S.kuo,Z.Salcic and U.Madawala *A Real-time Hybrid Web-client Access Architecture for Home Automation* ,ICICS-PCM pp1752-1756,Singapore 2003

[8] R..J.C.Nunes, and J.C.M. Delgado *An Internet Application for Home Automation* IEEE Transactions consumer Electronics, vol 43.no.4,pp1063-1069,1997

[9] Jeffrey Travis, *Internet applications in LabVIEW* ,1997

[10] St Moscibrodzki,W., Kwiatkowski,and S.,Malinowski, A. *A reference implementation of the Internet-supported Home Security System* ,IEEE Conference IECON'01 vol 3.pp1838-1843

[11] Sin-Min Tsai , Po-ching Yang and Shya-Shiow Sun *A service of home security system on intelligent network* ,IEEE Consumer Electronics, vol 44, issue 4,pp 1360-1366, 1998

Brill Academic Publishers
P.O. Box 9000, 2300 PA Leiden
The Netherlands

Lecture Series on Computer
and Computational Sciences
Volume 4, 2005, pp. 1617-1620

Integration and Control of Building System using Embedded Web Server and Wireless LAN

Kyung-Bae Chang[1], Tae-Kook Kim, Il-Joo Shim, and Gwi-Tae Park

Department of Electrical Engineering,
Korea University,
Anam-dong 5ga, Seongbuk-gu 136-713 Seoul, Korea

Received 30 July, 2005; accepted in revised form 10 August, 2005

Abstract: The servers that are used in the existing building systems use several PC grade servers for the functional control of DDC(Direct Digital Controller). Because of this, there are problems like space limitation or cost increases. Also, because of the companies using individual protocols and wiring, flexible use and compatibilities become problems. In this paper we propose a method by replacing a PC which controls each facility, with a Windows CE installed Embedded Web Server. Many Embedded Web Servers that control each facility, with their security strengthened, has good compatibilities and can be managed conveniently by the integration with the SI server. Also it has concise wiring and it can be controlled by wireless equipments like PDAs.

Keywords: EWS(Embedded Web Server), Embedded system, System integration, Intelligent building

1. Introduction

Building systems are formed by many facilities like HVAC (heating, ventilating, and air conditioning) control system, Light control system, Power control system and Parking control system. And as the system became more intelligent, it started having even more control systems.
Control networks that integrate the building systems now have many protocols [1]. Figure 1 shows the current building systems integration.

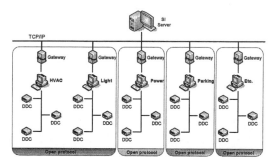

Figure 1: Building system at a present
In the current building system integration, between each facility server and the SI server, a data protocol TCP/IP is used. Between each facility and the DDC, control protocols like BACnet, LonWorks, ModBUS are used. And each facility uses a PC grade server.
There cannot be a perfect integration with the use of TCP/IP and BACnet, LonWorks, ModBUS together. Also in order to connect the different protocols, expensive network devices like Gateway

---

[1] Corresponding author. Kyungbae Chang(E-mail: lslove@korea.ac.kr)

must be used [2]. Therefore the cost goes higher, the wiring becomes more complex and the additional installing space becomes more limited.

The reason of the use of control protocols like BACnet or LonWorks instead of data protocols like DDC in each of the facilities is because there were problems like data loss and transmission delay. But after a lot of research, these problems have been solved [3,4,5].

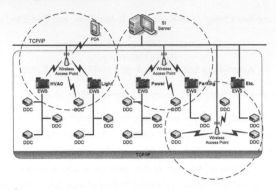

Figure 2: System integration using EWS and wireless LAN

This paper proposes the advantages and proves the possibilities of system integration through Embedded Web Server and wireless LAN.

2. Embedded Web Server Design

Embedded Web servers are micro controllers that contain internet software and operation codes that monitor and control the system. In order to obtain the Embedded Web Servers functions, the following are needed. First, because the embedded system has a small memory, it needs a divided memory block structure that can manage the memory with more efficiency. Second, a dynamic page must be created in order to send the information, with the use of Ethernet TCP/IP protocol, from the subsystem (DDC, Embedded device, sensor) to the web browser. Therefore the information on the web browser matches the information on the actual subsystem.

2.1. Hardware Concept

Hardware is composed of three parts, ARM (Advanced RISC Machines) processor, memory and interface expanded I/O controller. Figure 3 shows the hardware structure.

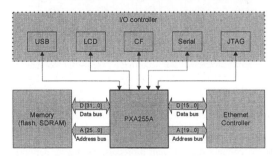

Figure 3: Hardware structure of Embedded Web Server
2.2. Software Concept

The application is composed of independent control tasks and communication tasks [6].
Figure 4 shows the programs structure.

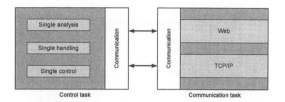

Figure 4: Software structure of Embedded Web Server

3. EWS running and wireless communication Tests

In order to assemble and test this system, there must be a DDC controller that is capable of Ethernet TCP/IP communications. To make a desktop computer obtain DDC functions, control functions have been applied by a LabVIEW^{TM} program.
Figure 5 shows the GUI (Graphical User Interface) of DDC.

Figure 5: GUI of DDC

When the Embedded Web Server sends a control signal, this program controls the sensors and controllers. The sensors and controllers that should be controlled, uses a simple device as shown in Figure 6.

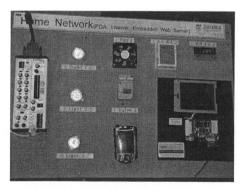

Figure 6: DDC, Sensor, PDA, EWS, and etc.

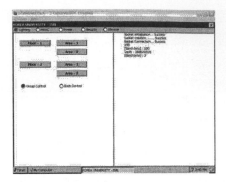

Figure 7: Integration software of Embedded Web Server

Figure 7 shows the executed EWS's integration software. If the connections between EWSs, between EWS and DDC, and between SI server and EWS are normal, the control of HVAC systems, Lights systems and power systems become possible on EWS. The figure below shows the controlling of the Light system and HAVC. The left side of the screen indicates the control devices and the right side indicates the communication status of the control signal, which is functioning normally.

4. Conclusion

This paper surveyed the problems in the existing building system integrations, and to solve these problems we proposed the Embedded Web Server and conformed its possibilities.
The proposed system has the following advantages over the existing system.
- lower costs
- simple wiring
- the possibility of various linked services
- no limitation to installation space
- flexible network structure

Research on the performance distinction between the Embedded Web Server and a PC grade server, research on the security of wireless communication will be required in the future.

References

[1] Sun Jun-ping, Sheng Wan-xing, Wang Sun-an and WU Ke-gong, Substation Automation High speed Network Communication Platform Based on MMS + TCP/IP + Ethernet, *T Power System Technology*. 2(2002).

[2] John Petze, Powering Smart Energy Networks Now, *VP product Development*. Article, 2000, [online] Available: http://automatedBuilding.com

[3] Mike Donlon, Using Standard Internet protocols in Building Automation, *Networked controls white Papers,* Director of Reesearch and Development Computrols inc, [online] Available: http://automatedBuilding.com.

[4] Dave VanGompel, Peter Cleaveland, Moving popular industrial protocols to Ethernet TCP/IP, 2001, [online] Available: http://controlsolutionsmag.com

[5] Understanding and Evaluating Ethernet and TCP/IP Technonlogies for Industrial automation, 2000, [online] Available: http://www.ManagetherealWorld.com

[6] Christian Eckel, Georg Gaderer and Thilo Sauter, Implementation Requirements for Web-enabled Appliances a Case Study: Emerging Technologies and Factory Automation Vol. 2, 2000.

Brill Academic Publishers
P.O. Box 9000, 2300 PA Leiden
The Netherlands

*Lecture Series on Computer
and Computational Sciences*
Volume 4, 2005, pp. 1621-1624

Smart Home Control and Security System with PDA, USB Web Camera using LabVIEW™

Kyung-Bae Chang[1], Jae-Woo Kim[1], Il-Joo Shim[1] and Gwi-Tae Park[1]

1 ISRL, Korea Univ, 1, 5-ka Anam-dong Sungbuk-ku 136-701 Seoul
South Korea

Received 7 July, 2005; accepted in revised form 15 August, 2005

Abstract: People want to control various facilities regardless location in the ubiquitous era. In this paper a method to establish a security monitoring system using an USB web camera is proposed and we implemented a control program in a PDA that uses Wireless LAN among other wireless protocols. Proposed method can be applied to other applications (smart home control, monitoring system, approaches of security system etc.) [1] For implementation LabVIEW™ was chosen as software along with USB web camera, a wireless client (PDA)

Keywords: LabVIEW™, LabVIEW™ PDA Module, Smart Home, Ubiquitous, Monitoring System

1. Introduction

Ubiquitous computing aims to "enhance computer use by making many computers available throughout the physical environment, but making them effectively invisible to the user". Thus, it is a desirable system for home networks where every home owner acts as an administrator. An example usage of such a system in a home environment is to automatically change communication devices when the talker roams from one room to another. The talker would always use the wired devices of the room he/she is in. Compared with using wireless devices for roaming in home, using wired devices can get better conversation quality since they usually have higher bandwidth and bigger displays. [2]

In a home environment, some devices might be controlled by a PDA and Web Server for Smart home. In efficient aspect, a user can use his/her PDA as a universal communication. A user can use his/her PDA as a universal communication and control agent, not only for communication, but also as a controller of network-connected appliances. Furthermore a representative application field for security and monitoring is the home network service, which is an emerging topic in the IT field.

We use inexpensive web camera instead of expensive devices, and are going to propose and verify a scenario that can embody home monitoring system through control signal of wireless devices. The digital home service becomes popular and hence more affordable. We can use such in-expensive equipment to monitor our home for intruders or nursery, operating room in the hospital, building entrance and so on. [3] It can also make our life more convenient. For example, to inform the home owner about a person entering or trying to enter the home and even allow entrance via a certification process through a mobile device such as a cell-phone or PDA. After analyzed information, we can do the second control. Further, it can allow to remotely monitoring young children or a patient in a house or in the hospital while being away enhancing personal safety and comfort. Furthermore, it may be possible to take a photograph through sounds (crash spot) and to monitor financial institution or jail with web server, and to analyze degree of congestion of complicated place. Therefore, in this paper, we embodied program that can send control signal as real time after accept the information of someone in far away place using by USB web camera and also can watch with direct control to Internet.

We implemented home automation using LabVIEW™ in home network technology (using centered home server that control, utilize, and monitoring some electronic devices) and are going to propose digital home system that uses web server as well.

2.1 Architecture discusses advantages and drawbacks of using technologies for communication with LabVIEW™ and LabVIEW™ PDA module and discusses to analysis front panels and block diagrams

---

[1] Corresponding author. Kyung-Bae Chang
E-mail: lslove@korea.ac.kr

in a server and a client program. [4] We confirmed proposed home model through 2 experiments in 2.2 home controls. Lastly, 2.3 applications discuss about useful functions in a building and about home controls and conclude the paper.

2. Implementation and proposal of home control

2.1 Architecture

Figure 1 shows overview of technologies for communication with LabVIEW™ over a network

Because LabVIEW™ is a graphical program language, the user can make programs easy and fast. It is easy to learn and understand. LabVIEW™ displays strong power when is used to control. In the communication field, LabVIEW™ is developing to be worth comparing with functions which are realizable in the C++. LabVIEW™ PDA Module is one of options among other application modules and it was chosen for the development. Figure 1 shows overview of technologies for communication with LabVIEW™ over a network. [5]

2.2 Home Control

There are three different experiments of the home control. First control is made by using a PDA, second control by using the PDA or a cell phone after monitoring program operated by USB web camera, lastly a controlling through the Internet by using program that is consisted of below three parts (monitoring part, sound part, and control part). We need to experiment equipment such as Table 1.

Table 1. Experiment appliance specification

Computer - P4 2.4G, RAM 512M
PDA - HP 5450
Samsung USB web camera(*OS Windows Pocket PC 2002*)
AP(Access Point) - LINKSYS WRT54G
DAQ – NIDAQ (*PCI 6040E*)(*Data Acquisition System*) Valve-valcon AT12-3T , Lamp-300w , Fan

2.2.1 Home control using PDA

We composed a model of the home network consisting of the lighting, a fan, a temperature indicator and a valve. So we created a PDA client program that has five on-off is switches available in order to enable the control. And PDA program enables the control of the each device pictured in the Figure 2

Figure 2. PDA client program for experiment (*left*) and Operating Home Control System (*right*)

2.2.2 Home control using monitoring system

Second experiment was done by using an USB web camera. Web camera is connected to server and when motion is detected, program developed in LabVIEW™. Since, when the E-mail arrives, using a service that sends such message to portable phone in case when we are not at the house, we can be notified when someone enters our house. Since we can confirm someone's identity by looking at the picture, a user can control device in user's house (door, window, temperature) using client program through wireless LAN. Therefore, we are proposing a scenario as in Figure 3. Program that is stored on a server sends the email, and when the email arrives, we can confirm someone visit in real time using service that sends the messages. Figure 3 shows downloading JPEG files from Internet in a PDA, and log in pages. After log-in process, we can control other device like in the first experiment (*send control signal by PDA*).

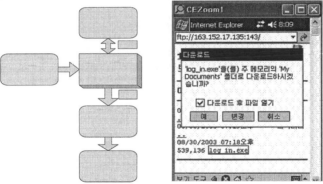

Figure 3. Scenario and Connecting pages after downloading the picture

2.2.3 Home control through the web-based interfaces

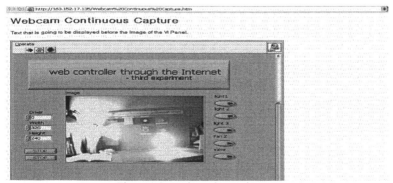

Figure 4 Control panel on web pages

It is important for us to enable the control by a web interface, to allow the control to be possible anytime and anywhere. So we developed the application for the control in run-time using the LabVIEW™ Run-time Module. Figure 4 shows the control using the Web.[6] It can be controlled from Web environment in the world.

2.3 Applications

For example, to inform the home owner about a person entering or trying to enter the home and even allow entrance via a certification process through a mobile device such as a cell-phone or PDA. After analyzed information, we can do the second control. Further, it can allow to remotely monitoring young children or a patient in a house or in the hospital while being away enhancing personal safety and comfort. Furthermore, it may be possible to take a photograph through sounds (*crash spot*) and to monitor financial institution or jail with web server, and to analyze degree of congestion of complicated place. And with log files, it is possible to predict the lighting, fan, motor's total time of use, etc. So we can know time of exchange and equipment and this program is also capable of informing administrator about the time of managing.

3. Future Works

For future work, we will enforce program's security and check its stability in a long time period. Further, when multiple users connect to this program, there is a priory problem that has to be settled. And need research about link and implementation of embedded web server in order to apply in additional application.

4. Conclusion

For home networks, many network appliances that can be controlled from mobile environment. We made a control program in a server and a wireless client (PDA) using WLAN and LabVIEW™. And we also made a smart home model to propose useful functions in building and home controls. Since control program is always processing signals of sensors, to make a program using LabVIEW™ with the support of measurement for safety and accuracy is profitable as communication program, since when we make programs in a PDA, to use LabVIEW™ helps us to shorten a developing time and gains in accuracy measurement values. Lastly, we implemented useful functions in building and home controls using PDA and web. Proposed monitoring system can be used conveniently without expensive image acquisition equipment and can be applied to other applications as well.

References

[1] Alheraish, A. "Design and implementation of home automation system" Volume: 50, Issue: 4 1087 - 1092 IEEE Transactions Nov. 2004
[2] Schulzrinne, H. Xiaotao Wu Sidiroglou,S. Berger, "Ubiquitous computing in home networks" Volome: 41, Issue: 11 128 - 135 Communications Magazine, IEEE 2003
[3] Nikunja K. Swain, "Remote data acquisition, control and analysis using LabVIEW™ front panel and real time engine", Proceedings IEEE southeast Con, 2003
[4] Kuo, S. Salcic, Z. Madawala, U. "A real-time hybrid Web-client access architecture for home automation" 1752 - 1756 vol.3 Processing 2003
[5] Jeffrey travis "INTERNET APPLICATION IN LabVIEW™", PHPTR
[6] M. Joler and C. G. Cristodoulou, "Virtual laboratory instruments and simulations remotely controlled via the Internet," in Proc. IEEE Antennas Propagat. Soc. Int. Symp., vol. 1, pp. 388–391 Boston, MA, July 2001

Brill Academic Publishers
P.O. Box 9000, 2300 PA Leiden
The Netherlands

Lecture Series on Computer
and Computational Sciences
Volume 4, 2005, pp. 1625-1629

Software Audit System for Ubiquitous Computing

Hyun-kyu Cho[1], Moo-hun Lee[2], Young-jo Cho[3], Hyeon-sung Cho[4], Jung-hoo Cho[5], Eui-in Choi[6]

Received 9 July, 2005; accepted in revised form 11 August, 2005

Abstract: We can use information and software of various forms without being restricted for place and time if ubiquitous computing age comes. However, its dysfunction is causing security problems such as outflow of personal information, hacking, diffusion of virus. Specially, dissemination of software that has malicious purpose in ubiquitous computing environment causes serious damage. We have studied about malicious code detection and software vulnerability detection tool to prevent this. But, existent detection tools are not suited to general software, because they are limitative in specification area. In addition, they can not detect a newly appeared malicious code. Because they use a simplicity pattern matching technique that must update pattern of new malicious code. In this paper, we propose rule-based software audit system that analyzes structure of target code to solve these problems, define this as rule, and detect. Proposed system can construct secure ubiquitous computing environment, because it will be used by a common software audit tool that detects malicious codes and software vulnerabilities at the same time.

Keywords: Ubiquitous Computing, Malicious Code, Software Vulnerability

1. Introduction

Accessing information are not restricted by time and place in ubiquitous computing environment. However, if we distribute malicious software to device that is used in ubiquitous computing, damage may be fairly serious. Therefore, we need a tool that can verify security for all software that is used in ubiquitous computing. Also, we need automatic audit tool that examine software vulnerability that is created by developer's carelessness. Existent detection tools can not evaluate general security of software because malicious code, virus, and vulnerability detection were separate into each other. Also, existent techniques are depending mostly on pattern matching of signature base. Therefore, such techniques can not detect new malicious code. Because they must update signature of malicious code. In this paper, we propose audit system that can verify safety for software used in ubiquitous computing environment. Proposed audit system detects malicious codes(virus, trojan) in source code inside of software and software vulnerability(buffer overflow, format string bug, race condition). Users will trust and use safely software in ubiquitous computing environment.

The rest of the paper is organized as follows. Section 2 reviews several detection tools with related works. The software audit system we propose are described in Section 3. Section 4 compares there results with existent tools. Section 5 describes our research conclusion.

2. Related Works

ITS4 stand for It is The Software Stupid (Security Scanner). It statically scans C and C++ code for vulnerabilities, but it does not do so by parsing the actual source code used in a single build configuration. Instead, ITS4 looks at several files to check for vulnerabilities in multiple builds of the

This work was supported by Electronics and Telecommunication Research Institute under grant No. 2004-S-007

[1] Intelligent robot research division, Electronics and Telecommunications Research Institute, 161 Gajeong-Dong, Yuseong-gu, Daejeon, Korea, E-mail : hkcho@etri.re.kr

[2] Corresponding author. Department of Computer Engineering, Hannam University, 133 Ojeong-Dong, Daedeok-Gu, Daejeon, Korea, E-mail : mhlee@dblab.hannam.ac.kr

[3] Intelligent robot research division, Electronics and Telecommunications Research Institute, 161 Gajeong-Dong, Yuseong-gu, Daejeon, Korea, E-mail : youngjo@etri.re.kr

[4] Intelligent robot research division, Electronics and Telecommunications Research Institute, 161 Gajeong-Dong, Yuseong-gu, Daejeon, Korea, E-mail : hsc@etri.re.kr

[5] Department of Computer Science, University of California, Los Angeles, CA 90095, E-mail : cho@cs.ucla.edu

[6] Department of Computer Engineering, Hannam University, 133 Ojeong-Dong, Daedeok-Gu, Daejeon, Korea, E-mail : eichoi@dblab.hannam.ac.kr

software. This is done for many reasons. First, it reduces the false negatives to almost zero. Second, it avoids the complexities of real parsing that add no value to the security scanning requirement. Third, it allows ITS4 to be used in an integrated development environment to highlight potential errors from within an editor[1].

MOPS stands for Model Checking Program for Security Properties. It checks for security vulnerabilities from a sequence of operations viewpoint. It uses model checking together with specific rules to detect the violation of temporal safety properties. A user describes the rules in the form of a finite state machine. If the software finds any problems related to a property, it prints out the offending path found in the source code. Such techniques can find potential issues with buffer overflow, user privileges, and array index [2,3,4].

Blast (Berkeley Lazy Abstraction Software Verification Tool) is a verification system for checking safety properties of C programs. Blast implements an abstract refine loop to check for reachability of a specified label in the program. The abstract model is built on the fly using predicate abstraction. This is model checked. If there is no path to the specified error label, Blast reports that the system is safe. Otherwise, it checks if the path is feasible using symbolic execution of the program. If the path is feasible, Blast outputs the path as an error trace, otherwise, it uses the infeasibility of the path to refine the abstract model[5,6,7].

3. Software Audit System

3.1 Architecture of software audit system

Proposed software audit system consists of analysis module, rule process, and check module. Analysis module analyzes target code. Rule process analyzes malicious code and software vulnerability and creates rule. Also, check module examines using analyzed Intermediate code and rule. Check module consists of malicious code detector module, vulnerability detector module, and reporter. It separates and examines malicious code and software vulnerability that is included on source code inside. It evaluates security of software through two examinations. Architecture of software audit system describes in figure 1.

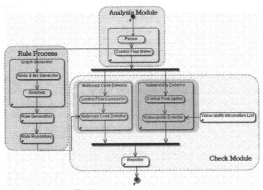

Figure 1: Architecture of Software Audit System

3.2 Analysis module

Analysis module is consisted of parser and control flow maker. It creates Intermediate code that is used in check module and rule process. Parser identifies program internal expression, statements, declarations, blocks, and procedures, and source code is compiled by GCC compiler. Each statements generated in parsing process is delivered to control flow maker and it describes processing flow.

Intermediate code that Parser is used as it is GCC compiler's compile technique and compiles does tokening for control flow maker's argument creation. In case of grammatical mistake happens in source code, it displays error information, and analyzes Intermediate code(compiled assembly language) and describes processing flow of code. It flows tokening process and examines token type and divides each token by expression, statement, declaration, block, and procedure for processing flow analysis of code. Control flow maker uses identifier who correspond to token and token type, and describe flow of program using them. If token identifier is decided, it analyzes inclusion relation with previous identifier. If relation analysis is completed, it inserts token and identifier that is responded and describes processing flow about one token. If all tokens are described by control flow maker, intermediate code that describes source code's processing flow is created. Parser and control flow maker are applied equally to analysis of malicious code for rule definition as well as target code.

3.3 Rule process

We need predefined rule to detect malicious code and vulnerability in target code. Analysis module analyzes malicious code and generates intermediate code. Rule process analyzes generated intermediate code and defines rule. Proposed system divided from process to 2 processes according to malicious code analysis' processing flow, and is same figure 2. Malicious code analysis process is achieved before create rule, and rule constitute process defines and stores and manages rule.

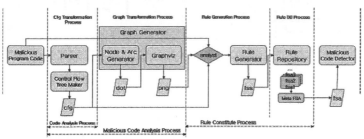

Figure 2: Malicious Code Analysis and Rule constitution Process

Analysis module compiles malicious code and creates intermediate code. Rule process changes created intermediate code to dot file and changes in graph structure for analysis. Auditor or analyst who have special knowledge about malicious program analyzes its code, analysis graph, and intermediate code and defines rule through FSA(Finite State Automata) grammar. First code of figure 3 is su code that is kind of malicious code. The second code is intermediate code that is created by analysis module. 12 lines and 18 lines are statements that cause actuality malicious behavior. The third figure is analysis graph that analyze structure of su code. analyst defines rules using three products.

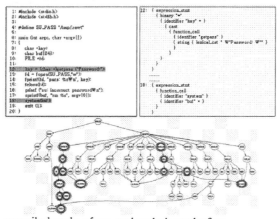

Figure 3: su code, compiled code of su, and analysis graph of su

3.4 Check module

Check module consists of malicious code detector, vulnerability detector, and reporter as describe in figure 1. Malicious code detector detects malicious code that is included to source code, and vulnerability detector detects software vulnerability that is created by developer's carelessness. And reporter estimates results and verifies software's security. Malicious code detector and vulnerability detector use intermediate code that is created by analysis module. Malicious code detector and vulnerability detector use intermediate code that is created by analysis module. Intermediate code is refined by control flow compacter and control flow splitter. Refined codes examine with rule that is stored to rule repository, and examines with function that is stored to Vulnerability information list. Check module executes two examinations, and conveys the result to reporter and evaluates software. Figure 5 is describing process of malicious code detector and vulnerability detector. Refined intermediate code uses PDA(PushDown Automata) to grasp target code's structure in malicious code detector. Malicious code detector converts intermediate code and rule file to internal data model. And it finds program architecture such as malicious code by comparing two internal data models.

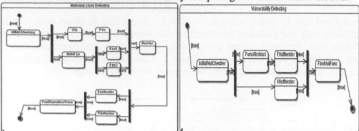

Figure 4: Processing procedure of Malicious Code Detector and Vulnerability Detector

4. Comparative Analysis

Proposed software audit system can verify security of software that is used in ubiquitous computing environment combining existent vulnerability audit tool and malicious code detection tool. ITS4, MOPS, and BLAST are a tool that is used in specification area. However, proposal system can verify security of software effectively than other tool. Because it can evaluate security of software generally over malicious code and software vulnerability. Moreover, it differs from simplicity pattern matching tool in many respects, and detection of new malicious code is easy because based on rule. Table 1 is result that compare existent detection tools with proposed software audit system.

Table 1: Compare proposed system with existent system

	ITS4	MOPS	BLAST	Proposed system
Subject of detection	Software vulnerability	Security property	Security of C program	Malicious code and software vulnerability
Detection technique	Pattern matching based detection	Model checking based detection	Abstract model based detection	Rule based detection
Detection time	Before compile	After compile	Before compile	After compile

5. Conclusion

In this paper, we proposed software audit system that can detect malicious code and software vulnerability that is included by malicious purpose on source code inside and can verify security. Proposed system analyzes structure of target code, and verifies rules. Therefore, it solved existing detection tool's problem that can not detect new malicious code by simplicity pattern matching. Also, it can verify security of software more effectively by integrating vulnerability audit tool and malicious code detection tool. We can construct confidential ubiquitous computing environment by taking advantage of proposed system. Future research in this paper needs the study of agent which can collect new malicious code, and analyze structure of malicious code, and create rule automatically.

References

[1] John Viega, J.T. Bloch, Tadayoshi Kohno, Gary McGraw Reliable Software Technologies, "ITS4: A Static Vulnerability Scanner for C and C++ Code", http://www.rstcorp.com.
[2] Hao Chen, David Wagner and David Schultz at Computer Science Division, UC Berkeley, "MOPS User's Manual"
[3] Hao Chen and David Wagner University of California at Berkeley "MOPS: an Infrastructure for Examining Security Properties of Software"
[4] Hao Chen, David Wagner, and Robert Johnson, "Model Checking Software for Security Violations", Short talk at 2001, IEEE Symposium on Security and Privacy.
[5] Thomas A. Henzinger, Ranjit Jhala, Rupak Majumdar and Gregoire Sutre, "Lazy Abstraction", In ACM SIGPLAN-SIGACT Conference on Principles of Programming Languages, pages 58-70, 2002.
[6] Thomas A. Henzinger, Ranjit Jhala, Rupak Majumdar, and Gregoire Sutre, "Software Verification with Blast", Proceedings of the 10th SPIN Workshop on Model Checking

Software (SPIN), Lecture Notes in Computer Science 2648, Springer-Verlag, pages 235-239, 2003.

[7] See the Tutorial Introduction section in the "BLAST User's Manual", May 16, 2004. http://www-cad.eecs.berkeley.edu/~tah/blast/

Brill Academic Publishers
P.O. Box 9000, 2300 PA Leiden
The Netherlands

*Lecture Series on Computer
and Computational Sciences*
Volume 4, 2005, pp. 1630-1633

Rule-based System for Vulnerability Detection of Software

Sung-Hoon Cho, Chang-Bok Jang, Moo-Hun Lee, Eui-In Choi [*]
Department of Computer Engineering, Hannam University, 133 Ojeong-Dong, Daedeok-Gu, Daejeon, Korea

Received 3 July, 2005; accepted in revised form 19 July, 2005

Abstract: IT infrastructure is expanding since computing and communication technologies are improved. However, this expansion takes place the security vulnerability because the reason of this problem is the software vulnerability. This problem is researched to resolve the security problem at many projects. But the perfect solution do not announce yet. The fundamental solution of software vulnerability is the development of secure software without the vulnerability. Thus this paper proposes the system for the detection of the software vulnerability. This system doesn't perform the simple pattern matching. Instead this system defines the vulnerability into a rule, analyses the structure of software and seeks the violation.

Keywords: Rule, Vulnerability Detection, Security, Buffer-Overflow, Software Vulnerability

1. Introduction

The improvement of computer and internet techniques is various services and information to user. However it may become dangerous elements at the security between networked computers, and an individual's privacy. These security problems take place at network, software and other fields now. Software vulnerability causes the software's problems. The best solution to resolve software vulnerability is the development of the secure software without any defeats. Although to make the perfectly secure software is mostly impossible, many research work to solve these problems.

The major software vulnerability is buffer-overflow[1, 2]. To resolve buffer-overflow, many challenges go on. Conventionally the manner for the detection and prevention of buffer-overflow scan the source code, and then find the vulnerable function. Associated projects are ITS4 and CQUAL[3, 4]. The other manner is that the modified compiler that added the canary checks buffer-overflow. Associated projects are Stack Guard, Point Guard[5, 6]. ITS4 performs the detection about C source code. But ITS4 takes many times to detect vulnerability about a large number of source code unfortunately. And Stack Guard and Point Guard checks only the exploit that the buffer-overflow happened. It doesn't handle other vulnerability. Thus in this paper, we propose the system for the detection of the software vulnerability. This system defines the vulnerability into a rule, analyses the structure of software and finds the violation. In the paper is organized as follows: Section 2 explores related works in the techniques of software vulnerability detection Section 3 discusses our proposal to detect the vulnerability. And finally Section 4 describes our future works and conclusions.

2. Related Works

2.1 Signature Recognition

Signature recognition is able to rapidly diagnose the exploit and explicitly identify the kind of malicious code. However it doesn't handle about unknown malicious code. Also users are exposed to malicious behavior until antivirus company distributes the signature of the malicious code and the solution. Thus the recently research that detects the unknown malicious code apply heuristic analysis and behavior blocking. The projects as ITS4, Cqual, Splint and others employ signature recognition.

2.2 Heuristic Analysis

This technique is divided two manners. One is used the static heuristic of the code in malicious code. Others are used dynamic heuristic about result through the emulation at runtime. Static heuristic analysis stores code fragments that are frequently used at malicious behavior into database. And then it

This work was supported by a grant No.R12-2003-004 -03002-0 from Ministry of Commerce, Industry and Energy
[*] E-mails: {shcho, chbjang, mhlee, eichoi}@dblab.hannam.ac.kr

scans target codes to decide the existence of malicious codes and to analyze the appearance frequency. These steps are static heuristic's procedure. This procedure is relatively fast. It has also the high detection ratio. However it has a defeat that false-positive is frequently occurred. Dynamic heuristic analysis performs target code at the emulator on virtual machine. And it monitors system call and the change of system resource at the runtime. These steps are dynamic heuristic's procedure. However it has defeats that must implement the virtual machine. Also emulation cannot trace all program's control-flow. Thus hybrid heuristic analysis is proposed. It detects dangerous code through the string search and analyses the program's control-flow for avoiding malicious code's logic trick.

2.3 Behavior Blocking

Behavior blocking technique is similar with dynamic heuristic technique except the target code is executed at the target system. Emulation can distinguish the maliciousness of target code through behavior monitoring without side effects. But the execution of malicious code at the system and the monitoring may cause the disk formatting or the change of system files. If these inappropriate behavior are detected, the process will have to stop the execution immediately. Thus behavior blocking is difficult to monitor the behavior pattern for a long time like emulation. Once dangerous behaviors happen, behavior blocking must notify the caution. This operation may occur very frequent false-positive.

2.4 Integrity Checking

Integrity checking stores file information such as checksum or hash value about all or some files on the local file system. Then it checks the changes of file information after a certain period. This manner is indirect. Since it only checks the modification of the specified files, a valid change may decide an inappropriate change. Thus it has a false-positive problem.

3. The rule based system for the detection of buffer-overflow

The mostly security vulnerability is caused by software vulnerability which is related to buffer-overflow problem. Many researches performed in order to prevent buffer-overflow problem, but the perfect prevention is yet none. Thus in this paper, we propose the rule based system for the detection of buffer-overflow.

3.1 The structure and function of system

The proposed system is consist of RC compiler, Rule Generation Process, VRR(Vulnerability Rule Repository), Checker and Alerter. RC compiler is used at MOPS project. It is able to parse the source code and to describe the structure of program and control-flow into ".cfg" file. Software vulnerability has the unique feature. VRR is a rule which converted this feature. Checker module refers to this VRR in checking the vulnerability. Checker implements a role of the detection of buffer-overflow. Alerter prints the violation list which is transmitted by Checker. Finally Rule Generation Process is not implemented by C code. It means a process to generate a rule. Figure 1 is the architecture of this system.

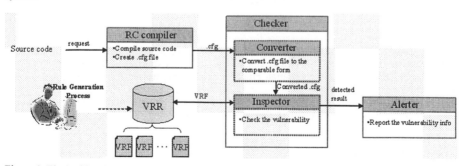

Figure 1: The architecture of the proposed system

3.1.1 A role of RC Compiler

RC compiler is based on gcc compiler. And it is able to describe the program's structure and control flow. It is described by indicator set. This indicator set has 94 indicator. And it is also able to represents the notations such as datatype, operator, function, pointer, array and values of variables in C language. For example, "function_call" means a function call and "function_decl" means function declaration. Others are indicated by RC. And "{" and "}" in .cfg" file can represent the relation of invocation and implication. Thus it is able to describe the program's structure and control flow. A right side of figure 2 is compiled left side by RC compiler. A right side contains various indicators and several braces. The statements of right side are able to convert to graph using the brace.

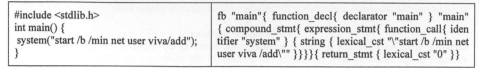

| #include <stdlib.h>
 int main() {
   system("start /b /min net user viva/add");
 } | fb "main"{ function_decl{ declarator "main" } "main"
 { compound_stmt{ expression_stmt{ function_call{ iden <br> tifier "system" } { string { lexical_cst "\"start /b /min net <br> user viva /add\"" }}}}{ return_stmt { lexical_cst "0" }} |

Figure 2: The target code and the .cfg file's contents.

3.1.2 VRR(Vulnerability Rule Repository)
VRR reposit a rule that is created by Rule Generation Process, That is, it performs a role such as database. VRR is consisted of a VRL(Vulnerability List File) and VRFs. A VRL records the list of VRFs. In checking buffer-overflow through Checker, it first refers to a VRL and loads the related VRFs. And then VRFs are used to check the possibility of buffer-overflow.

3.1.3 Rule Generation Process
Rule Generation Process creates a rule as a name. A rule is another form of malicious code. It express malicious behavior by states and transitions. Buffer-overflow can be occurred by various techniques and method. Therefore we must create rules as many as possible. This work is manual. And it needs the knowledge of C language for the rule generation and the analysis. In this paper, we divided rule creation process into following details step. it is difficult to apply to all the cases, because malicious code don't have always the same behaviors.

Step 1: Confirm runtime environment.
Step 2: Find the vulnerable functions of buffer-overflow in source code
Step 3: Analyze the statements related to vulnerable function.
Step 4: Analyze the statements that invade the return address of the buffer.
Step 5: The initial state is a state that will execute the vulnerable function. Establish the initial state.
Step 6: The terminal state is a state of the completed execution. Establish the terminal state.
Step 7: Establish the next state of the current state. And each state is added to state set.
Step 8: The input is transit the current state into the next state. Establish the transition and add the input in transition set.
Step 9: To generate a rule, the current state, the next state and the input notate as "[t] [current state] [next state] [input]".
Step 10: Store the notation of step 9 in VRF and add the VRF's file name to VRL.
Step 11: Repeat step 1 - 10 to process the several vulnerable function.

3.1.4 Checker for the detection of buffer-overflow
Checker analyzes the program's structure and checks the vulnerability of buffer-overflow within C source code. It receives ".cfg" file that is created the program's structure and control flow by RC compiler and VRF that is Rule Generation Process's result. The received data are the input data of Checker.
Checker needs the converter and the inspector. The converter converts the ".cfg" file to the comparable form with rule. The inspector compares the converter's results with rules. The convert steps is as following:

Step 1: Converter loads a ".cfg" file.
Step 2: Find the notation of "functon_call" in ".cfg" file. "function_call" is the separator of the relation of invocation. A function contained the "function_call" establishes in "caller". And a invoked function establishes in "callee".
Step 3: The transmitted parameter's value from "caller" to "callee" is established by "input".
Step 4: Record "caller", "callee" and "input" at a file.Step 5: Repeat the step 1 - 4 to process others "function_call".

If Checker implements above steps, ".cfg" data will be similar with rules. That is, mutual comparison is available. The converted data and rules are loaded to Inspector. Inspector examines that the states and transitions that are described to Rule were also appeared to ". cfg". If the same states and transition is occurred at ".cfg", we will be able to detect the vulnerability. The examined results are delivered to Alerter.

3.1.5 The operation of Alerter

Alerter provides the processed information that is generated through Inspector to user. The processed information contains the replacement information of vulnerable functions. The contained information is that the function such as strncpy, strncat and snprintf have to replace the function such as strcpy, strcat and sprintf.

4. Conclusion

In this paper, we introduced the rule based detection system to prevent buffer-overflow. This system generates the rule of vulnerability and detects the buffer-overflow vulnerability. The legacy detection systems examined the vulnerability through the simple pattern matching. However, Our system examines the vulnerability through the rule and program's control-flow. Thus it is able to provide more precise detection than the legacy. Also It can distinguish whether a function causes the malicious behavior or does not. In the case of new malicious behavior, a new rule is able to create through Rule Generation Process.

Future works are to define rule about the more various vulnerability functions and to automate Rule Generation Process.

References

[1] Crispin Cowan, Perry Wagle, Calton Pu, Steve Beattie, and Jonathan Walpole, "Buffer Overflows: Attacks and Defenses for the Vulnerability of the Decade", SANS 2000.

[2] Crispin Cowan, Calton Pu, David Maier, Heather Hinton, Peat Bakke, Steve Beattie, Aaron Grier, Perry Wagle, and Qian Zhan, "Automatic Detection and Prevention of Buffer-Overflow Attacks", the 7th USENIX Security Symposium.

[3] John Viega, J.T. Bloch, Tadayoshi Kohno, Gary McGraw, "ITS4: A Static Vulnerability Scanner for C and C++ Code", ACM Transactions on Information and System Security, 2000.

[4] Jeffrey S. Foster, Tachio Terauchi, and Alex Aiken. "Flow-Sensitive Type Qualifiers ", In ACM SIGPLAN Conference on Programming Language Design and Implementation (PLDI'02), pages 1-12. Berlin, Germany. June 2002.

[5] C.Cowan et.al, Stackguard "Automatic adaptive detection and prevention of buffer-overflow attacks.", In proc. Of the 7th USENIX Security Conference, January, 1988.

[6] Crispin Cowan, Steve Beattie, John Johansen and Perry Wagle, "PointGuard: Protecting Pointers From Buffer Overflow Vulnerabilities", the 12th USENIX Security Symposium, Washington DC, August, 2003.

Click Nodes *Click*: this node models a student's action of clicking the spot on the game board to move there. Each node has two states: Correct and Wrong. Correct denotes that the student has clicked on a correct spot, that is a number from the arithmetic expression which is composed with the given three random numbers.

Support Node *Hint* and *Help*: these nodes denote an agent's action of giving the hint or help window on the arithmetic expressions and game strategies. *Hint* are initiated by an agent, whereas *Help* is initiated by student. These have two states, Yes and No. Yes denotes that the agent gives the hint or help window to help the student. These nodes are also evidence nodes. The left picture of Figure 2 shows the basic dependencies among the nodes which encode the assumptions to be mentioned below.

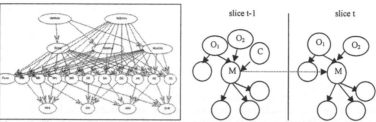

Figure 2. The dependency between nodes and two slices of an example DBN

The Conditional Probability Table(CPT) for each node is set up by taking the assumptions mentioned above into consideration. If there are multiple parent nodes for a particular node X, its conditional probability table is defined in the following way: assume that this node has n parent nodes. For each assignment of the parent node values, if there are m parent nodes ($0 \leq m \leq n$) which have the state Unmastered, then the corresponding probability in the conditional probability table for this node to be Mastered is calculated using the equation which is a little bit modified suggested in [7]: $P(X=\text{Mastered})= p-\{(p-(1-p)/n\}*m$. In this equation, p is designed by hand to denote a high probability as "Mastered". The value of p is determined by considering the student's prior knowledge about the relevant knowledge domain. Traditionally, Dynamic Bayesian Networks (DBN) [5] [6] are extensions of BNs specifically designed to model worlds that change over time. The right picture of Figure 2 shows two slices of an example DBN. After action C (i.e. Click)occurs, a new evidence node C is added to the network. After the network is updated, and before a new action node is taken in, the evidence node C is removed in slice t+1. The CPT for the node M is changed according to the value of C in slice t by $P(M=T|C=T, O_1, O_2) = P(O_1, O_2| M=T) +d$, where d is the weight the action C brings to the assessment of the corresponding knowledge nodes, and it can be either positive or negative, depending on the type of action. If the M represents the arithmetic expression nodes E_x, the increment weight d_m is determined as follows.

$$d_m = \frac{1}{n}\sum_{i=1}^{N} p_i - p_m ,$$

where p_m is the probability of the node M and p_i is the probabilities of the occurrences of number i which are computed by applying the expression M. The probability of knowledge nodes M not directly affected by Click node don't remains unchanged, because we could be assumed to slowly forget the unfired knowledge. The S-shaped membership function is utilized to guess the weight to be applied to decay the unfired knowledge.

$$f(x) = \frac{1}{1+e^{-\alpha x}}$$

α is the coefficient for controlling the slope. The range of x can be scaled as 0 to s. The scaling weight is calculated by $\beta = f(P(\cdot | M = F)/s)$, where s is scale factor for $f(x)$ and d is the given weight mentioned above. Hence, a node M may be decayed as the following simple rule.

If $P(M=F|OM=T) > \theta$, then $P(\cdot |M=F) = P(\cdot |M=F) + \beta d,$

θ is the given threshold value for making decision of decay and is used to determine the timing for Help event. OM stands for the root node, OptimalMove, which has dual meanings, that is, a player chooses the optimal move currently and a student has mastered the ways to solve the arithmetic expressions as well.

4. Simulation

The simulation is set up for the test of plausibility of overall performance effectiveness. The main goal of simulation is to show that a pedagogical agent that provides hints to students enhance student learning in the game environment. The 200 simulation sessions are set up and the related parameters are determined such as d=0.01, θ=0.05, α=1.0, s=20. The figure 3 show that the resultant transitions of several CPTs after 200 simulation sessions reflect relatively correct assessment of the suggested student model. Figure (a) shows that three CPTs, P(OptimalMove=T|Paren=T), P(OptimalMove=T|Paren=T, Bump=T), and P(OptimalMove=T|Paren=T, Bump=T,MD=T) give the meaningful results that if a student master the Bump strategy and the usage of multiplication and division operators(MD), he may master the optimal moves in the game board. We set up the parameters for OptimalMove node to increase the true value monotonically. From this figure, we can not tell that the strategy node, Bump gives more evidence for the optimal move, but we can see the overall conditional values correctly generated, that can promise highly plausible structure for future usage.

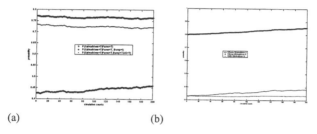

(a)   (b)

Figure 3. (a) +:P(OptimalMove=T|Paren=T), *: P(OptimalMove=T|Paren=T, Bump=T), -: P(OptimalMove=T|Paren=T, Bump=T,MD=T), (b) -:P(Paren=T|OptimalMove=T), *:P(Bump=T|OptimalMove=T), -:P(MD=T|OptimalMove=T)

Figure 3(b) shows three CPTs, P(Paren=T|OptimalMove=T), P(Bump=T|OptimalMove=T), P(MD=T|OptimalMove=T). From this figure, we can assume that if a student knows how to apply the game strategy such as Bump, he might have mastered the optimal move on the game board. This fact implies that he may increase the knowledge on the application of MD (the possible combination of multiplication and division in arithmetic expressions).

5. Conclusion and Future Works

This paper presented research on using an intelligent pedagogical agent in the Yut-nori educational game to enhance student learning by integrating gaming strategies into problem solving framework to enhance the problem solving performance. The agent's actions are based on both some simple strategy and the assessments from a probabilistic student model based on Bayesian Networks. To reduce the computational cost of updating a large BN at run-time, we can design different student models depending on the game levels using adding the Mals and numbers additionally. In this case, special framework should be designed to reflect the lower level strategies and domain knowledge. Currently, the numbers and weights in the CPTs are assigned using our subjective estimates. If possible, the empirical model could be used to refine these estimates to get more accurate assessments.

References

[1]. A.S. Gertner, C. Conati, K. VanLehn," Procedural help in Andes: Generating hints using a Bayesian network student model," http://www.cs.ubc.ca/~conati/my-papers/Gertner-AAAI98.pdf

References

[1]   N. Al-Dhahir, "Single-carrier frequency-domain equalization for space-time block-coded transmissions over frequency-selective fading channels," IEEE Commun. Lett., vol. 16, no. 8, pp. 1451–1458, Oct. 1998.

[2]   S. L. Ariyavisitakul, A. B. B. Eidison, B. Falconer, "Frequency domain equalization for single-carrier broadband wireless systems," [Online]. Available: www.sce.carleton.ca/bbw/papers/Ariyavisitakul.pdf

[3]   S. M. Alamouti, "A simple transmit diversity technique for wireless communications," IEEE J. Select. Area Commun., vol. 16. pp. 1451–1458, Oct. 1998.

[4]   D. Falconer et. al., "Frequency domain equalization for single-carrier broadband wireless systems," IEEE Commun. Mag., vol. 40, no. 4, pp. 58–66, Apr. 2002.

Brill Academic Publishers
P.O. Box 9000, 2300 PA Leiden
The Netherlands

Lecture Series on Computer
and Computational Sciences
Volume 4, 2005, pp. 1653-1658

ANCS Technique for Replacement and Security of Execution Code on Active Network Environment[1]

Chang-Bok Jang[2], Moo-Hun Lee, Sung-Hoon Cho, Eui-In Choi

Department of Computer Engineering,
Hannam University,
Daejeon, Republic of Korea

Received 3 July 2005; accepted in revised form 2 August, 2005

Abstract: As developed internet and computer capability, many users take a lot of information through the network. So requirement of user that use to network was rapidly increased and become various. But it spent much time to accept user requirement on current network, so studied such as active network for solved it. This active node on active network has the capability that stored and processed execution code aside from capability of forwarding packet on current network. So required execution code for executed packet arrived in active node, if execution code should not be in active node, have to take by request previous action node and code server to it. But if this execution code takes from previous active node and code server, bring to time delay by transport execution code and increased traffic of network and execution time. So, as used execution code stored in cache on active node, it need to increase execution time and decreased number of request. Also unexpected increasing of execution code should be cause of network attack such as DOS(Denial Of Service). Our paper suggest ANCS technique that able to decrease number of execution code request and time of execution code prevent from DOS attack by efficiently store execution code to active node. Therefore, ANCS technique that suggested by us is able to decrease cost of network management due to decrease network traffic. Also because execution time of program was decreased, we can more than fast execute program.

Keywords: Active Network, Network Security, Cache Policy, DOS

1. Introduction

As developed computer performance and network, many users take the information through network. So demand of user that using Internet become varies and complicate. But it spent much time to accept new technology and standard in current network, cause of institute standard of technology and replacement of device. And as it was relatively slow change of network system then user's demand that rapidly changed, it was impossible to apply user's demand at the proper time. So, to solve the problem, active network was studied with activity. Active network was environment that could dynamically process packet to execute program in active node. It could execute the program in router or switch[1, 16, 17, 18]. Study of active network is ANTS, Switchware, PAN, CANE, FAIN[4, 10, 19, 20]. Also it provided flexibility and merit that is process various task in node of active network, different from current node that set up path of packet and pass it [2, 8, 11]. The middle node in active network is classified packet and judge whether it is execute or not, when arrived packet. And then if execute the packet, it was passed to the next node by current node. Therefore, execution code must be in active node and need the execution code, data for process packet that passed by previous node[22]. If not exist the execution code to process in node, should be request the code from previous and code server. And as network user increases, network security became important. Therefore network attack such as DOS and increase of request of execution code should be cause of serious obstacle such as network fault [24, 25].

So research to decrease execution time to store execution code in cache and to prevent from DOS needed. Therefore our paper suggested the ANCS(Active Network Cache and Security) technique that used the reference time and time limit, Access Ratio, PEV for decreasing execution time and number of

[1] This work was supported by a grant No.R12-2003-004-03002-0 from Korea Ministry of Commerce, Industry and Energy.
[2] Corresponding author. Hannam University. E-mail: chbjang@dblab.hannam.ac.kr

code request and prevent from DOS. ANCS technique that suggested by us is able to decrease cost of network management due to decrease network traffic. Also because execution time of program was decreased, we can more than fast execute program.

2. Related Works

2.1 Active Network

Switch and router(active network node) have not only simple function that pass the packet, but also function that store/process/pass the execution code. After users include program code to packet, send to the destination node. Users use the function of network by execute the program code that received, or call the program that already installed in active node[1, 2, 7, 8, 11, 13, 15]. So Execution code must store in node, if not exist the code to need in node, should be receive from other active node. In this case, required technique is code request technique, studying two method: that received from previous node, or code server[3].

2.2 DOS Attack

In an environment where a considerable fraction of the traffic will be continuous media traffic, security must include resource management and protection with an eye to preserving timing properties. In particular, a pernicious from of attack is the so-called DOS(Denial of Service) attack.

3. ANCS Techniques

There are many execution code in active network, it execute as needed in active node. If not exist execution code in active node, because of receiving the it to request from previous active node, bring to delay of execution time and increase of network traffic. So, if execution code efficiently maintains in cache, it could decrease the frequency of request and delay time of packet process and increase rate of packet process. And unexpected increasing of execution code should be cause of network attack such as DOS. In our paper, for execution code more efficiently stored and prevent from network attack, we suggest ANCS technique that stored and managed frequency of reference that execution code was how referenced and time limit that could know to reference recently, Access ratio that how many the code used, PEV(Previous Execution no Value) in code information table, code information table that showed in table 1 was consisted of code id about execution code, execution no that has frequency of code execution, and time limit no that could know to reference recently, access ratio, PEV.

Table 1. Code Information Table

Code ID	Execution No	Time Limit No	PEV	Access Ratio
A	5	1	4	0.1
B	1	1	1	0.0
C	7	0	0	0.0
D	8	8	5	0.3
E	7	7	5	0.2

Code id is for identify the execution code, stored in unique identifier. Execution no has value how code many executed. When codes are executed, its have 1 value, if they are executed again, the value was increased each 1. Time limit no store to value how long not used after first used, if time goes on fixed period, the value was updated. Namely, code has fixed value(10) about time limit after it was executed, and then decrease value of the time limit no each 1 per fixed time(320 sec) after code is executes. Time limit no has initial value(10), when it executed again. PEV is value of previous execution no. So it stored previous execution no value for fixed time(320 sec), and then After 320 second, it is update. Access ratio is value for how many codes are used. If we assume that execution no is E, PEV is P, access ratio is A, value of A compute as $A = (E-P)/100$.

For storage space of cache limited and prevent from DOS, we suggested ANCS technique for check and delete, store execution code in cache. It processes as follow. First, as time goes for a fixed period, execution code that has o value in time limit no not immediately delete, it maintain in cache. If need the replacement, it first delete thing which execution no was the least low in one of execution code that has 0 value in time limit no. This method solves a problem that execution code was deleted directly to get 0 values in time limit no, in spite of high frequency of reference.

Second, in case of one thing that execution code had 0 values in time limit no when requests the replacement of execution code, not used execution no, delete the code that had 0 values in time limit no.

This technique increases efficiency of cache space to delete the execution code that not referenced for a long time, only referenced for a shot time. If value of time limit has not 1, first action was applied.

And third, we check the value of access ratio for whether execution code is used for DOS attack or not. Access ratio value was computed by method as mentioned above. If new value of access ratio more increase than previous value of it unexpectedly, we could consider that this code are used for DOS attack. So we can properly act against DOS attack such as stop of execution code or deny of execution code. For example, such as code information table in table 1, execution code A, B, C, D, E was stored in cache, if execution code F is going to store in cache, because cache space was lacked, execution code must delete. So, execution code C was deleted and stored F by first method that we suggested, because the case that time limit no has 0 value in A, B, C, D, E is only one. If time goes for 320 second, value of time limit no was decreased each 1. In case of new execution code G is going to store, because code that had 0 value in time limit no is A, B(assumption: any code not executes in active node for 320 second), deleted code is B that had lowest value of execution no by second method that we suggested. In result, code information table will be such as table 2.

Table 2. Code information table after delete execution code

Code ID	Execution No	Time Limit No	PEV	Access Ratio
A	5	0	5	0.0
G	1	10	0	0.1
F	1	9	0	0.1
D	8	7	8	0.0
E	7	6	7	0.0

If code A was used unexpectedly(assumption : used 100 times for 320 second), and then access ratio should be 0.95(A = (100 − 5)/100). So it will probably be DOS attack. Therefore security manager should monitor this code.

In our paper, for prevent directly deletion of code that executes frequently after time goes on fixed period, applied different technique of cache replacement by value in time limit no. It was decreased the frequency of code request and use efficiently cache space by maintains execution code which used frequently in cache and delete the execution code that not used for a long time. Also as monitors code that used unexpectedly, could prevent from DOS attack.

4. Comparative Analysis

In this chapter, we explain the result and simulation for estimate performance of ANCS technique that we were suggested. Our simulation focused on whether network traffic could decrease or not through decreasing frequency of code, stop or deny execution of code. We set the argument that used in simulation to number of request, frequency of code execution, and code execution time, deny of code execution.

4.1 Experimentation strategy and assumption

In ANCS technique, if codes not exist after check the execution code in cache, requested code from previous node. At this time, in case of not received code from previous node, except from experimentation. Also, we should stored code in cache as maximum 5, assume that size of buffer to store code that arrived in node is 20. Code execution time in active node has random value within constant range. And we assume that execution code made of same size, so not consider the method of code deployment by different size of execution code. Finally, we experiments execution time of code that increased unexpectedly. And then we compare ANCS with other technique whether code safely executes or not.

4.2 Result of analysis

4.2.1 Analysis about unexpected use of code

For we compute the ratio of code execution, we assumed that each codes(A, B, C, D, E) stored in cache and Only A code used unexpectedly. And we experiments that A code was executed for short time period. Figure 1 is ratio of code execution when A code was used unexpectedly. In part of FIFO, LFU, LRU, because other codes not stored in buffer due to increase of A code unexpectedly, A code was executed about 90% and other code was not executed. But As ANCS technique that suggested denies execution of A code by monitoring increase of execution of A code, more other codes are executed.

Therefore when occur the network traffic or attack such as DOS, because we efficiently cope with attacks by ANCS, it was able to more safety management of network.

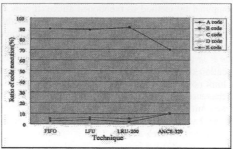

Fig. 1. Ratio of code execution when A code was used unexpectedly

4.2.2 Analysis in redundancy ratio

In our experimentation, we define redundancy rate 1, 2, 3, 4, 10, 15 for estimate frequency of code request, frequency of code execution, code execution time. As number increasing, number of redundancy was decreased and various codes are existed. Value of time limit no in redundancy ratio 2, 3, 5, 10, 15 except 1 in ANCS that we suggested applied case of ANCS-320(320 second) and ANCS-80(80 second) to experimentation, LRU only applied LRU-200(200 second) to experimentation. Result of estimation about each technique is follow. Figure 6 shows how code many executed directly until redundancy ratio 2 to 15 every cache technique. Figure 2 shows that compared frequency of code execution in ANCS and frequency of code execution in LFU by redundancy ratio. In case of high redundancy ratio, they have similar performance, but ANCS-320 that our suggested have more than better frequency of code execution than LFU in other redundancy ratio. But ANCS-80 is rather worst frequency of code execution than LFU, due to unsuitable value of Time Limit No.

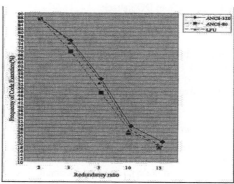

Fig. 2. Comparison of LFU, ANCS-320, ANCS-80

In table 3, it shows result that we compare ANCS with other cache technique. And (++) represents very good performance, (--) represents very worst performance. (0) represents average performance. As showed in table 3, ANCS-320 that suggested by us is very good performance in all field. But ANCS-80 has average performance, due to choice wrong value of time limit no. And LFU has good performance in our experimentation.

Table 3. Comparison of each cache technique

	FIFO	LRU	LFU	ANCS-80	ANCS-320(Proposed Technique)
Frequency of code request	--	-	+	-	++
Frequency of code execution	--	-	+	0	++
Time of code execution	--	-	++	0	++
Ratio of code execution	--	--	--	-	++

5. Conclusions

In our paper, we suggest ANCS technique that prevents from DOS and stored execution code in cache for decreasing frequency of code request. General nodes are executed code regardless of whether code that used for DOS attack or not. But ANCS technique that suggested denies execution of code by monitoring increase of execution of it, more other codes are executed. Therefore when occur the network traffic or attack such as DOS, because we efficiently cope with attacks by ANCS, it was able to more safety management of network. Execution code that stored in cache was process by request of packet, if it not exists in cache, it should be request from previous node. Also, ANCS technique can decrease frequency of code request by stored efficiently execution code in limited cache. As used ANCS technique that suggested, could decrease delay time of packet process, frequency of code request, waste of cache space, and increase the velocity of packet process. Hence, ANCS technique is able to decrease cost of network management due to decrease network traffic. Also because execution time of program was decreased, we can more than fast execute program.

Future works in our paper needs the study of time limit no for efficient replacement regardless of redundancy ratio of execution code.

References

[1] D. L. Tennenhouse, J. M. Smith, W. D Sncoskie, D. J. Wetherall, and G. J. Minden. : A Survey of Active Network Research, IEEE Communications Magazine, Vol. 35, No. 1, pp. 80-86, 1977.

[2] D. L. Tennenhouse, D. J. Wetherall. : Towards an Active Network Architecture, Multimedia Computing and Networking, January 1996.

[3] David J. Wetherall, John V. Guttang and David L. Tennenhouse. : ANTS : A Toolkit for Building and Dynamically Deploying Network Protocols, IEEE OPENARCH, 1998.

[4] D.S Alexander, et. al. : The SwitchWare Active Network Architecture : IEEE Network Specail Issue on Active and Controllable Networks, vol.12 no.3, 1998.

[5] Jack Jensen. : A Guide to Business Decision-Making Using Visual SLAM II and AweSim, 1999

[6] K. L. Calvert. : Architectural Framework for Active Networks Version 1.0, Active Network Working Group, DRAFT July 27, 1999

[7] Konstantinos Psounis. : Active Networks: Application, Security, Safety, And Architectures, IEEE Communications Surveys 1999

[8] Beverly Schwartz, Alden W. Jackson, W. Timothy Strayer, et. al. : Smart Packets: Applying Active Networks to Network Management, ACM Transactions on Computer Systems, Vol. 18, No. 1, February 2000, Pages 67-88.

[9] Thomas Becker, et al. : Initial Active Network and Active Node Architecture ver 1.0, May 2001

[10] M. Hicks, et. al. : PLANet: An Active Internetwork, IEEE INFOCOM, 1999

[11] T. Wolf, et. al. : A Scalable High-Performance Active Network Node, IEEE Network, 1999

[12] S. Merugu, et. al. : Bowman: A node OS for Active Networks, Proceedings of IEEE Infocom 2000, March 2000

[13] G. Alex, et. al. : A Flexible IP Active Networks Architecture, IWAN 2000 Conference, 2000

[14] L. Peterson.(Editor) : NodeOS Interface Specification, DARPA AN NodeOS Working Group, 1999

[15] D. Wetherall, et. al. : The Active IP Option, 7th ACM SIGOPS European Workshop, 1996

[16] Tal Lavian, Phil Yonghui Wang. : Active Networking On A Programmable Networking Platform, IEEE OPENARCH 2001, 2001

[17] Danny Raz, Yuval Shavitt. : Active Networks for Efficient Distributed Network management, IEEE Communications Magazine, March 2000

[18] Danny Raz, Yuval Shavitt. : An Active Network Approach to Efficient Network management, Proceedings of the First International Working Conference on Active Networks (IWAN '99), 1999

[19] Erik L. Nygren, Stephen J. Garland, and M. Frans Kaashoek. : PAN: A High-Performance Active Network Node Supporting Multiple Mobile Code Systems, IN PROCEEDINGS IEEE OPENARCH'99, MARCH 1999

[20] Jonathan T. Moore Scott M. Nettles. : Towards Practical Programmable Packets, Technical Report MS-CIS-00-12, Department of Computer and Information Science, University of Pennsylvania, May 2000 (switchware)

[21] S.Merugu S.Bhattacharjee et al. : Bowman and CANEs : Implementation of an Active Network. In Proceedings of 37th Annual Allerton Conference, Monticello, IL,September 1999

[22] Samrat Bhattacharjeey Martin W. McKinnon. : Performance of Application-Specific Buffering Schemes for Active Networks, Technical Report GIT-CC-98-17, College of Computing, Georgia Tech

[23] Ying Shi, Edward Watson, Ye-sho Chen. : Model-driven simulation of world-wide-web cache policy, Proceedings of the 1997 Winter Simulation Conference, 1997

[24] Konstantinos Psounis, : ACTIVE NETWORKS: APPLICATIONS, SECURITY, SAFETY, AND ARCHITECTURES, IEEE Communications Surveys, 1999

[25] Scott Alexander, : Security in Active Networks, Secure Internet Programming: Issues in Distributed and Mobile Object Systems, 1999L.Gr. Ixaru and M. Micu, *Topics in Theoretical Physics*. Central Institute of Physics, Bucharest, 1978.

Brill Academic Publishers
P.O. Box 9000, 2300 PA Leiden
The Netherlands

*Lecture Series on Computer
and Computational Sciences*
Volume 4, 2005, pp. 1659-1661

A Study on CVMP (Competing Values Model about Privacy) in Ubiquitous Computing Environment

Jang Mook, Kang[1]

Dept. Of Computer Engineering,
Seokyeong University,
16-1 Jung-Ryung dong,
Seoul, South Korea

Received 7 July, 2005; accepted in revised form 20 July, 2005

Abstract:
It is said that the ubiquitous computing environment will be come by the technological developments in the future. This paper points out Privacy in ubiquitous computing environment and presents CVMP (Competing Values Model in Privacy). This research will focus on the following: first, to check out the mechanism by which the ubiquitous computing environment is distinguishable from the IT environment and draw out from it. Second, it will focus on the following to provide the CVMP (competing values model in Privacy) in ubiquitous computing environment.
Keywords: UC, Ubiquitous Computing, Privacy, CVMP, Competing Values Model

1. Introduction

It is said that the ubiquitous computing environment will be come by the new technological revolution in the near future. This paper points out the problem that engineer thinks that UC(ubiquitous computing) connected to network have become more complexes and distributed, but most system-related research is still focused on the 'performance' or 'efficiency'. And that most studied papers in privacy have been focused on encryption technology and protocols for securing the personal information data transaction. This paper secondly points out the problem that law-related researcher thinks that law can handle technology beyond human control. These days, due to the complexity of problem, new privacy-related research based on total solution is integrative needed to consider laws, technology, economy, norms and so on.

This paper will focus on analysis about Privacy in ubiquitous computing environment and total protection considerations on privacy. So, this research will focus on the following: first, to check out the mechanism by which the ubiquitous computing environment is distinguishable from the IT environment and draw out from it. Secondly, this research will focus on the following to provide the CVMP(competing values model in Privacy) in ubiquitous computing environment.

2. The characteristics and significance for ubiquitous computing era

The ubiquitous computing technology creates the ubiquitous computing space, by connecting the 'Cyber Space & Physical Space, the Limited time & Real time' organically. Supposing the IT-based space is just an extension into a mere cyber one, the ubiquitous space equals to a new space created from the real networking between the cyber space and the physical space(as shown in Figure 1). Cyber space, Physical space, Ubiquitous computing space is created beyond the time and space. In consequence ubiquitous computing environment means an arena, technologically based by real time mobility and pervasiveness of computing functions. So much of the technology needed to protect ubiquitous computing systems exits. Cryptographic techniques can be applied in support of authentication, authorization, integrity, confidentiality and so on[1]. Technology can also be applied in conjunction with legal, norm, commerce to provide protection for personal information or individual privacy. To be useful, however, the infrastructure supporting these technologies must be put in

---

[1] Corresponding author. Active member of the Korea Security Association. E-mail: mooknc@paran.com, mooknc@naver.com

space(cyber-space, real-space) and the technology must be integrated with social law, norm, and commerce to give protection for individual privacy.

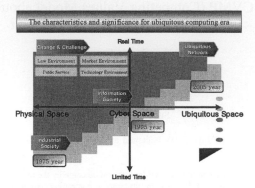

Figure 1: The characteristics and significance for ubiquitous computing era

3. In Definition of Privacy

3.1 Definition of privacy

The debate on privacy has initiated from the publication of the thesis by Samuel D. Warren and Louis D. Brandeis in 'Harvard Law Review'[2]. They claimed that, the state of mind, emotion, intelligence can be protected by suing against the libel disgrace of honor and ownership of a copyright law, however, the privacy law should be advocated to protect the 'right to be let alone' as an individual.

3.2 Changing scope of privacy

We can extension for scope of privacy as shown fig. 2. Ubiquitous Computing Technology will make new space which is augmented real space. So scope of privacy necessary change and adapt new technology environment. New protection scopes of privacy are physical space, cyber space and augmented space.

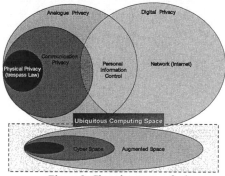

Figure 2: Changing scope of privacy

4. Introduction

4.1 CVP(Competing Values in Privacy)

The values of privacy in ubiquitous computing systems are to be classified as Public values, Private values, Commercial values and Law and Normative values and each (value) has both collusive and cooperative relationship with one another. First, the values of privacy reserves public and private values as well as protection and check relations. Secondly, the values of privacy reserves competition and charge relation as well as Commercial values and Law & Normative values. Third, Public and law & normative values in privacy reserves Control and Charge Relation values. Fourth, Law & Normative values and commercial values in privacy reserves Protection and Check relation values. Fifth, the technological values representative of mobility and pervasiveness in Ubiquitous computing system [3] threatens the cost and Benefit towards the diverse values.

4.2 CVMP(Competing Values Model in Privacy)

CVMP (Competing values Model in Privacy) are to be found by classifying the correlation between CVP (Competing Values in Privacy). The higher the personal information level goes up, the more protective action it should be taken the lesser the danger it entails, the higher charge (economic and administrative convenience) is required. The higher requirements in regard of stability in personal information, it should be in control and competition when flexibility is required emphasized, instead of stability. The balance of all values can be achieved through technological values.

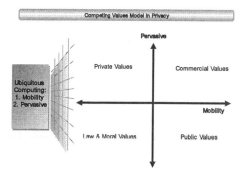

Figure 3: CVMP(Competing Values Model in Privacy)

5. Conclusion

As important as privacy has been for digital society (1980-2005), it will be even more important for the acceptance of ubiquitous computing environment (2005-2020)[4]. Ubiquitous computing will exacerbate a problem already present and critical in CEO, bureaucrat, citizens, lawyer and so on. Stubborn idea that Ubiquitous computing system might bring about atrophy of other values, by expansion, should be avoided. There might be negative relationships between each other (where the various values collide each other) in CVMP(Competing Values Model in Privacy), it also opens up the possibility of positive relations where synergy effect is created through it. Therefore it is a model based on the technological feature, which enables the limit and the degree of invasion of privacy definable. CVMP is adaptive to lots situation for new privacy problem. So CVMP is available to understanding about complexity of privacy problem as integrated privacy model(Law, Norm, Technology, Commerce, Public and Private)

Acknowledgments

The author wishes to thank the anonymous referees for their careful reading of the manuscript and their fruitful comments and suggestions.

References

[1] Clifford Neuman, *Security, Payment, and Privacy for Network Commerce,* IEEE Journal on selected areas in communications. Vol. 13. no.8 1530(1995)
[2] Samuel Warren, Louis Brandeis.: The Right to Privacy. Harvard Law Review, Vol. 4. 193(1890)
[3] Lyytinen., Kelly., Youngjin Yoo., *Issues and Challenges in Ubiquitous Computing,* Communications of ACM. Vol 45. no12. 12-13(2002).
[4] Mark Weiser, John seely brown, *The Coming age of calm technology,* Xerox PARC, (1996) available at http://www.ubiq.com/hypertext/weiser/acmfuture2endnote.htm

Brill Academic Publishers
P.O. Box 9000, 2300 PA Leiden
The Netherlands

Lecture Series on Computer
and Computational Sciences
Volume 4, 2005, pp. 1662-1665

An Authentication Mechanism using HMAC-based Security Association Token for Smart Space

Myung-Hee Kang, Hwang-Bin Ryou

1 Dept. of Computer Science, Kwangwoon University,
447-1, Wolgye-dong, Nowon-gu, Seoul, 139-701, KOREA
{mhkang, ryou}@kw.ac.kr

Received 12 July, 2005; accepted in revised form 18 August, 2005

Abstract: The spread of smart gadgets, appliance, mobile devices, PDAs and sensors has enabled the construction of ubiquitous computing environments, transforming regular physical spaces into "Smart space" augmented with intelligence and enhanced with services. However, the deployment of this computing paradigm in real-life is disturbed by poor security, particularly, the lack of proper authentication and access control techniques. In this paper, we propose efficient authentication and access control mechanism for smart space by using HMAC-based pair-wise security association token.

1. Introduction

Ubiquitous computing environments or Smart spaces promote the spread of embedded devices, smart gadgets, appliance and sensors. We envision a Smart space to contain hundreds, or even thousands, of devices and sensors that will be everywhere, performing regular tasks. Providing new functionality, bridging the virtual and physical worlds, and allowing people to communicate more effectively and interact seamlessly with available computing resources and the surrounding physical environment.

However, the real-life deployment of Smart spaces is disturbed by poor and inadequate security measures, particularly, authentication and access control techniques.

These mechanisms are inadequate for the increased flexibility required by distributed environments. Moreover, traditional authentication methods using symmetric key techniques either do not well in massively distributed environments, with hundreds or thousands of embedded devices like Smart space. Authentication methods using public key techniques would clearly be a better solution compared to authentication methods using symmetric key techniques. But, it may inappropriate to use public key techniques in smart space with the computational constrained devices.

In this paper, we propose efficient authentication and access control mechanism for distributed environments with the computational constrained devices by using HMAC-based pair-wise security association token.

We organize the paper as follows. In section 2, we describe authenticated key agreement protocol, Kerberos system and secure device association mechanism. In section 3, we propose authentication and access control mechanism using HMAC-based pair-wise security association token. Experimental results and analysis will be stated in section 4, and finally the conclusions are given in section 5.

2. Related Work

Leighton-Micali Scheme[4] focuses on the efficient, authenticated key agreement protocol based symmetric techniques. Key management using public key techniques would clearly be a better solution compared to key management using symmetric techniques from the standpoint of both ease of management and scalability. However, implementation of public key management techniques requires wide-spread deployment of public key infrastructure. Also, it may inappropriate to use public key techniques in smart space with the computational constrained devices. Leighton-Micali Scheme may be a good solution in distribution environments with the computational constrained devices.

While Kerberos[1] was a success in meeting authentication challenges in early distributed system, it has serious limitations that hinder its effectiveness in ubiquitous computing environments. First, it is mainly based on passwords, and as such is prone to password-guessing attacks. Second, Kerberos assumes that every user in the system accesses services through a designated workstation. In other words, a user has to log into some workstation on the network and only from that workstation the user can access the distributed services. On the contrary, in a ubiquitous computing environment there is no notion of a single machine that the user can access the services through any of the hundreds of machines that populate the smart space. Third, because it is required a lot of encryption/decryption and network transactions (i.e.: the number of encryption/decryption operation is 8 and 6-way transaction per one service access), it is not scalable in massively distributed environments with hundreds or thousands of devices like smart space.

Kindberg and Zhang[2] have described protocols for validating secure associations setup spontaneously between devices. The secure device association method employs lasers, which provide relatively precise physically constrained channels. The initiator device is equipped with a laser whose output can be rapidly switched to provide a data stream. The target device is equipped with a securely attached light sensor that can read the data emitted by the initiator device. The initiator device creates new session key K_S and sends it to target device using laser. Kindberg and Zhang scheme has merit to enable nomadic users to securely associate their devices without communication with trusted third parties. However, the scheme has potential security vulnerabilities, particularly, denial of service attacks, because K_S is shared between initiator and target device without authentication.

3. Proposed Authentication and Access Control Mechanism

3.1 3-Party Authentication and Access Control Model

This paper proposes efficient authentication and access control mechanism for distributed environments with the computational constrained devices by using HMAC-based pair-wise security association token. The security association token is similar to Kerberos's ticket and employs concept of Leighton-Micali Scheme[4]'s authenticated key agreement mechanism.

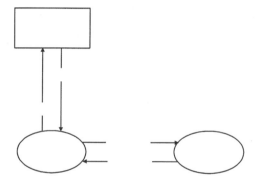

Fig. 1. 3-Party Authentication and Access Control Model

(1) $U \Rightarrow AAS$: REQ_SAT = {U, T}
The user U requests security association token to access to the target T by sending the user and target T identifiers to Authentication and Access Control Server(AAS).

(2) $AAS \Rightarrow U$: RES_SAT = {SATInfo, R_{AAS}, PK_{UT}, MAC_U, MAC_T}
The AAS searches it's key database for the shared keys K_U (with U) and K_T (with T), generates a random number R_{AAS} and computes PK_{UT} , MAC_U and MAC_T as

$$KM_U = hmac(K_U, R_{AAS}),$$
$$KM_T = hmac(K_T, R_{AAS}),$$
$$PK_{UT} = (KM_U \oplus KM_T),$$
$$MAC_U \quad = hmac(K_U, SATInfo \parallel KM_T),$$
$$MAC_T \quad = hmac(K_T, SATInfo \parallel KM_U)$$

Then it sends the response token RES_SAT back to U. Here note that PK_{UT} plays the role of a pair-wise public key between U and T in the sense that from this value each party can obtain a secret (KM_U for U , KM_T for T) only computed by the other party(and AAS). Also note that MAC_U / MAC_T has a similarity to a digital signature. The SATInfo is described in subsection 3.2

(3) $U \Rightarrow T$: REQ_SERV = {SATInfo, R_{AAS}, PK_{UT}, MAC_T}

U first computes KM_U = hmac(K_U, R_{AAS}) using it's secret K_U shared with *AAS* and checks that the received MAC_U equals the computed value hmac(K_U, SATInfo $\parallel$ $KM_U \oplus PK_{UT}$). If they are not equal, *U* sends an appropriate error message to *AAS* restarts the protocol. If the check succeeds, *U* sends service request REQ_SERV to the intended target *T*.

(4) $T \Rightarrow U$: RES_SERV = {Accept or Reject}

On receiving the token, *T* first computes KM_T = hmac(K_T, R_{AAS}) using it's secret K_T and the received random number R_{AAS}. Next it computes hmac(K_T, SATInfo $\parallel$ $KM_T \oplus PK_{UT}$) and checks that it is equal to the received value MAC_T. If the check fails, *T* sends a service reject message to *U* and terminates the protocol. If it succeeds, *T* is assured that the token is a valid for *U* and *T* generated by *AAS* and sends a service accept message to *U*.

3.2 Security Association Token Profile

The HMAC-based pair-wise security association token is similar to tickets of Kerberos system. But, it is not used encryption algorithm but used MAC(e.g.; HMAC) algorithm. Table 1 shows security association token profile proposed in this paper.

Table 1. Security Association Token Profile

Filed		Descriptions
SATInfo	SAT_Id	Identifier of a security association token
	Issuer_Name	Issuer Name of a security association token(i.e.; *AAS*)
	Holder_Name	Owner Name of a security association token(i.e.; *U*)
	TargetDevice_Name	Target Device Name of a security association token(i.e.; *T*)
	MAC_AlgId	MAC algorithm Identifier
	Validity	Validity period of a security association token
Random Number(R_{AAS})		Secure random number which is generated by *AAS*
Pair-wise public key(PK_{UT})		Pair-wise public key value between user *U* and target device *T*
MAC_holder(MAC_U)		MAC value which is generated by *AAS* with shared secret key between user and *AAS* (The holder can be assured that the token is a valid signature generated by *AAS*.)
MAC_target(MAC_T)		MAC value which is generated by *AAS* with shared secret key between target device *T* and *AAS* (The target device *T* can be assured that the token is a valid signature generated by *AAS*.)

4. Experimental Results

For performance evaluation, we use AES algorithm for Kerberos system and HMAC-SHA1 algorithm for our scheme. Table 2 and 3 respectively shows the cryptographic algorithm performance and the experimental result between Kerberos system and our scheme.

Table 2. Performance of cryptographic algorithms

AES	Encryption (Decryption)	Round key generation for encryption	Round key generation for decryption
	224(Mbps)	0.286(micro-sec.)	1.441(micro-sec.)
HMAC-SHA1	HMAC Generation		
	337.5(Mbps)		

* Pentium-III 1Ghz, Windows 2000, Microsoft Visual C++ 6.0

Table 3. Experimental Results between Kerberos system and our scheme

	Kerberos system		Our scheme	
Communication transaction	6-way		4-way	
The number of cryptographic operation	Encryption	6	HMAC generation	8
	Decryption	2		
	Enc. round key generation	6		
	Dec. round key generation	2		
Ticket(Security Association Token) generation	39.48(micro-sec.)		23.12(micro-sec.)	

We suppose that the length of the security association token(or Kerberos ticket) is 128 bytes.

Our scheme is simpler and better than Kerberos system in sense of the number of communication transaction and cryptographic computation processing. Moreover, if the AAS(Authentication and Access Control Server) share a key among homogenous target devices, our scheme has big advantage to reduce the load of it. For instance, if 3 printers have a shared key between AAS and provide printing service of same quality for user, he(she) can access some printers by using a association token supporting multiple device association.

5. Conclusion

In this paper, we propose efficient authentication and access control mechanism for distributed environments with the computational constrained devices by using HMAC- based Security Association Token. We are now investigating a key update protocol and association token supporting multiple device association as extension of our scheme.

References

1. B. Neumann and T. Ts'o, "Kerberos: An Authentication Service for Computer Networks", IEEE Communications Magazine, 32(9): 33-38, September 1994.
2. Kindberg, T. and Zhang, K., "Secure Spontaneous Device Association", Proceedings of UbiComp 2003: Ubiquitous Computing,, vol. 2864 of Lecture Notes in Computer Science, Seattle, WA, USA, October 12-15, 2003. Springer Verlag
3. Jalal Al-Muhtadi, Anand Ranganathan, Roy Campbell and M. Dennis Mickunas, "A Flexible, Privacy-Preserving Authentication Framework for Ubiquitous Computing Environments", Proceedings of the 22nd International Conference on Distributed Computing Systems Workshops(ICDCSW'02), 2002.
4. Tom Leighton and Silvio Micali, "Secret-Key Agreement without Public-Key Cryptography", Advances in Cryptology CRYPTO 1993, 456-479, 1994.
5. Kindberg, T. and Zhang, K., "Validating and Securing Spontaneous Associations between Wireless Devices", Proceedings of 6th Information Security Conference(ISC'03), October 2003.
6. C.H. Lim, "Authenticated Key Distributed for Security Services in Open Networks", Technical Report, May 1997.
7. Yuh-Min Tseng, "A scalable key-management scheme with minimizing key storage for secure group communications", International Journal of Network Management, 13:419-425, 2003.
8. T.Kindberg and A. Fox, "System Software for Ubiquitous Computing", IEEE Pervasive Computing, January-March, 2002, pp. 70-81.
9. Weiser, M., "The Computer for the Twenty-First Century", Scientific American, Vol.256, No. 3, pp. 94-104, September 1991.
10. Yuh-Min Tseng, "A scalable key-management scheme with minimizing key storage for secure group communications", International Journal of Network Management, 13: 419-425, 2003.
11. Harney H, Muchenhirn C., "Group key management protocol(GKMP) specification",RFC 2093, July, 1997.
12. Li M, Poovendran R, Berenstein C., "Design of secure multicast key management schemes with communication budget constraint", IEEE Communication Letters 2002; 6(3): 108-110.
13. Ha, W., et at. "Confusion of Physical Space and Electronic Space: Ubiquitous IT Revolution and the Third Space", Korean Elctronic Times, 2002.
14. G. Banavar and A. Bernstein, "Software infrastructure and design challenges for ubiquitous computing applications", Communications of the ACM, vol. 45(12), pp. 92-6, 2002.
15. J. Undercoffer, F. Perich, A. Cedilnik, L.Kagal and A. Joshi, "A secure infrastructure for service discovery and access in pervasive computing", Mobile Networks and Applications, vol. 8(2), pp. 113-25, 2003.
16. Sakamura, K., "Ubiquitous Computing", Dongbang Media Books, 2003.

Brill Academic Publishers
P.O. Box 9000, 2300 PA Leiden
The Netherlands

Lecture Series on Computer
and Computational Sciences
Volume 4, 2005, pp. 1666-1669

A Covert Channel-Free Validation Algorithm Using Timestamp Order for Multilevel-Secure Optimistic Concurrency Control

Sukhoon Kang [1]

Department of Computer Engineering, Daejeon University
96-3 Yongun-Dong, Dong-Gu, Daejeon, Korea 300-716

Received 29 June 2005; accepted in revised form 26 July, 2005

Abstract: Although multilevel secure optimistic concurrency control (MLS/OCC) is capable of providing much higher parallelism compared to the locking-based scheme, the use of transaction aborts as the main mechanism for maintaining consistency is a serious disadvantage in the sense that it can induce *unnecessary aborts problem* due to clumsy definition of conflict and *starvation problem* due to indefinite delay of high-level transactions in order to avoid covert timing channel. From this line, the refinements of MLS/OCC will take a closer look at the following questions: (1) Are timestamps useful to refine validation of transactions in order to reduce the risk of restart among equal-level transactions. (2) Can starvation due to prevention of covert channel be solved in a more elegant way than has been suggested so far?

In order to reap the potential advantage of high parallelism and delay-secureness, this sort of unnecessary aborts problem should be alleviated. The basic refinement philosophy for the solution on unnecessary aborts problem is to incorporate the advantage of timestamp ordering into MLS/OCC. To this end, timestamp ordering is employed mainly for transaction validation. In this sense, we call this approach MLS/OTSO. This variant is shown to preserve the security semantics of the MLS/OCC while significantly reducing the cost of its validation phase. We should note at this point that the MLS/OTSO algorithm presents an opportunity for reducing the amount of work wasted by restarted transactions. Write timestamps can be tested when items are read as well as during validation, thus allowing transactions to detect certain inevitable restarts earlier in their execution. We will pursue this possibility further in MLS/OTSO, although the resulting algorithm would not really be an optimistic one, but we note that this modification may indeed be worth considering for performance reasons.

Keywords: Optimistic Concurrency Control, Multilevel Security, Databases, Covert Channel

1. Introduction

A *multiple-level-secure* database management system (MLS/DBMS) is a secure database manager which is shared by users of more than one clearance level and contains data of more than one classification level. An MLS/DBMS is different from a conventional DBMS in two respects: (1) every data item controlled by an MLS/DBMS has associated with it, perhaps indirectly, a unique classification level, and (2) a user's access to data must be controlled on the basis of clearance and classification.

The concurrency control requirements for transaction processing in an MLS/DBMS are different from those in conventional transaction processing systems with respect to inclusion of covert-channel freeness. In particular, there is the need to coordinate transactions at different security levels avoiding both potential *covert timing channels* and the *starvation* of transactions at high security levels. For instance, suppose that a low-level transaction attempts to write a data item that is being read by a higher-level transaction. A covert timing channel arises if the low-level transaction is either *delayed or aborted* by the transaction scheduler. In addition, the high-level transaction may be subject to an indefinite delay if it is forced to abort repeatedly. The user responsible for initiating the aborted transaction must be notified of its unsuccessful termination. But this notification constitutes a flow of information from the DBMS to a low-level user based on the activity of a high-level transaction, and such an information flow may be readily exploited to divulge sensitive information between conspired transactions.

---

[1] Corresponding author. Fax: +82-42-284-0109. E-mail: shkang@dju.ac.kr (S. Kang).

2. Motivation for Multilevel-Secure Optimistic Concurrency Control

2.1. Multilevel-Secure Optimistic Concurrency Control: MLS/OCC

If lower-level transactions were somehow allowed to continue with its execution in spite of the conflict of high-level transactions, *covert timing-channel freeness* would be satisfied. This sort of optimistic approach for conflict insensitivity is the basic principle behind the set of multilevel-secure optimistic concurrency control (MLS/OCC) schemes [1]. An advantage of the optimistic concurrency control (OCC) schemes [2] is their potential to allow a higher level of concurrency. Optimistic concurrency control for multilevel-secure databases can be made to work by ensuring that whenever a conflict is detected between a higher-level transaction (T_j) in its validation phase and a lower-level transaction (T_i), T_j is aborted, while T_i is not affected. Ideally, OCC has the properties of non-blocking and deadlock freedom. These properties make the OCC scheme especially attractive to multilevel-secure transaction processing. However, the original OCC innately possess the clumsy definition of conflict, and thus some transactions can be aborted unnecessarily. This sort of unnecessary abort problem that is caused due to clumsy definition of conflict should be eliminated. Consider a set of transactions that are concurrently executed as shown in Example 1.

Example 1 (Unnecessary Aborts Problem Due to Clumsy Definition of Conflict):

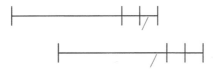

Figure 1: Execution Schedule for Concurrent Transactions

In this execution schedule, transaction $T_2(S)$ conflicts with $T_1(U)$, since its read-set overlaps with the write-set of $T_1(U)$. Using the validation scheme of original OCC, $T_2(S)$ should be aborted during its validation. However, since $T_2(S)$ reads x after $T_1(U)$ has written it, the serializability between $T_1(U)$ and $T_2(S)$ is guaranteed in the order of $T_1(U)$ (r) $T_2(S)$. Restart of $T_2(S)$ is unnecessary.

2.2. Refinements to MLS/OCC

The basic refinement philosophy for the solution on unnecessary aborts problem is to incorporate the advantage of timestamp ordering [3] into MLS/OCC mainly for transaction validation. In this sense, we call this approach MLS/OTSO. This variant is shown to preserve the security semantics of the MLS/OCC while significantly reducing the cost of its validation phase. Write timestamps can be tested when items are read as well as during validation, thus allowing transactions to detect certain inevitable restarts earlier in their execution.

Although MLS/OCC is capable of providing much higher parallelism compared to the locking-based scheme [4], the use of transaction aborts as the main mechanism for maintaining consistency is a serious disadvantage in the sense that it can induce unnecessary aborts problem due to clumsy definition of conflict and starvation problem due to indefinite delay of high-level transactions in order to avoid covert timing channel.

Following this line, the refinements of MLS/OCC will take a closer look at the following questions: (1) Are timestamps useful to refine validation of transactions in order to reduce the risk of restart among equal-level transactions. (2) Can starvation due to prevention of covert channel be solved in a more elegant way than has been suggested so far?

In order to reap the potential advantage of high parallelism and delay-secureness, this sort of unnecessary aborts problem should be alleviated. We should note at this point that the MLS/OTSO algorithm presents an opportunity for reducing the amount of work wasted by restarted transactions. Write timestamps can be tested when items are read as well as during validation, thus allowing transactions to detect certain inevitable restarts earlier in their execution. We will pursue this possibility further in MLS/OTSO, although the resulting algorithm would not really be an optimistic one, but we note that this modification may indeed be worth considering for performance reasons.

3. A Covert Channel-Free Validation Algorithm Using Timestamp Order for MLS/OCC

In the MLS/OCC, a transaction that wishes to commit is validated by comparing its *readset* with *writesets* of all transactions in recently committed transactions, $RC(T)$. If the number of $RC(T)$ is large, this validation test will be expensive. Thus we need to explore an equivalent validation test with a complexity that is independent of the $RC(T)$, making it more suitable for use in large transaction processing environments. The basic algorithm used in our design of MLS/OTSO for secure database transactions is as follows: (Note that transaction at a given security level cannot be aborted due to activity at a higher-level, or else a signaling covert channel results. This means that a transaction cannot modify the read timestamp on any lower-level data item.)

Algorithm 1: Covert Channel-Free Validation Phase Algorithm in MLS/OTSO

[Global Sets for System]
active_set is the set of transactions that have finished the read phase but have neither been committed nor aborted.
submitted_set   is the set of transactions which have been submitted but are still in the read phase.
[Sets and Variables for Each Transaction]
T is the current transaction.
$RS(T)$ is the readset of transaction T.
$WS(T)$ is the writeset of transaction T.
$cp\_active\_set(T)$ is a copy of active_set maintained for transaction T.
$conflict\_cnt(T)$ is the number of the transactions that can be affected by a conflict with T.
$TS(T)$ is a timestamp of T.

1. $< cnt := cnt + 1;\ T_j := cnt;$
2. **for each** $T_i \in active\_set$ **do**
3. $conflict\_cnt(T_i) := conflict\_cnt(T_i) + 1;$
4. $cp\_active\_set(T_j) := active\_set;\ conflict\_cnt(T_j) := 0;$
5. $RS(T_j) := \emptyset;\ WS(T_j) := \emptyset;$
6. $submitted\_set := submitted\_set \cup \{T_j\} >$
7.
8. **READ-PHASE:**
9.
10. $< submitted\_set := submitted\_set - \{T_j\};$
11. **for each** $T_i \in submitted\_set$ **do**
12. $cp\_active\_set(T_i) := cp\_active\_set(T_i) \cup \{T_j\};$
13. $conflict\_cnt(T_j) := conflict\_cnt(T_j) + |submitted\_set|;$
14. $active\_set := active\_set \cup \{T_j\};$
15. $cnt' := cnt' + 1;\ TS(T_j) := cnt' >$
16.
17. **procedure** validate(T_j);
18. **begin**
19. valid:=true;
20. **for all** $T_i \in cp\_active(T_j)$ **do**
21.   for all $x \in RS(T_j) \cap WS(T_i)$ **do**
22.   if $TS(x) < TS(T_i)$ **then** valid:=false; /* $TS(x)$ is copy of timestamp(x) stored locally by T_j */
23.
24. **if** valid **then** /* commit */
25.   **begin**
26.   **for each** $x \in WS(T_j)$ **do** /* commit $WS(T)$ to database */
27. $< $ **if** $TS(T_j) > TS(x)$ **then** { write $x;\ TS(x) := TS(T_j);$ } $>$
28. $< active\_set := active\_set - \{T_j\}; >$
29.   **if** $conflict\_cnt(T_j) = 0$ **then** discard $WS(T_j)$ due to Thomas's Write Rule;
30.   **end;**
31. **else** /* abort */
32.   **begin**
33. $< active\_set := active\_set - \{T_j\}; >$
34.   **if** $conflict\_cnt(T_j) = 0$ **then** discard $WS(T_j)$
35.   **else** $WS(T_j) := \emptyset;$
36.   **end;**
37.
38.   **for each** $T_i \in cp\_active\_set(T_j)$ **do**
39. $< conflict\_cnt(T_i) := conflict\_cnt(T_i) - 1;$
40.   **if** $conflict\_cnt(T_i) = 0$ **then** discard $WS(T_i) >$
41. **if** valid **then** cleanup
42.   **else** backup (high-level) $T;$

43. **end**.

In MLS/OTSO scheme, each transaction is assigned a startup timestamp $S\text{-}TS(T)$ at startup time, and each transaction receives a commit timestamp $C\text{-}TS(T)$ when it enters its commit processing phase. A write timestamp, $TS(x)$, is maintained for each data item x as the commit timestamp of most recent committed writer of x. A transaction T is allowed to commit iff $S\text{-}TS(T) > TS(x_r)$ for each data item x_r in its readset. Each T which successfully commits will update $TS(x_w)$ to be equal to $C\text{-}TS(T)$ for all data items x_w in its writeset. Write timestamps for recently written data items can be maintained in main memory, just as the writesets of recent transactions can be in the MLS/OCC.

Compared to previous approaches and MLS/OCC, the MLS/OTSO has the following advantages:

(1) Write-write conflict in the same security level do not lead to restarts and are resolved by using timestamp ordering with Thomas write rule. This allows more concurrency and makes the optimistic approach more attractive to both different security level interaction and the same level interaction environments with a higher fraction of update transactions.

(2) Only writing single data item at a time is needed to be in a critical section rather than the entire write-phase. Thus transactions could be concurrently writing data items in their writesets even if they have write-write conflicts. Since write transactions in low-level or equal-level are the main cause of rollbacks, reducing restarts of update transactions will have a positive effect on read-only transactions and overall system performance. We believe that this improvement represents an important step towards making the implementation of MLS/OTSO practically acceptable.

(3) MLS/OTSO does not require affixing a permanent timestamp to each data item in MLS/DB. Timestamps are created in main memory only for those data items accessed by current active transactions. Thus, although MLS/OTSO uses secure timestamps, it does not require changing the structure of the MLS/DB. Using main memory to store the timestamps of the subset of data items needed by active transactions can be affordable overhead.

(4) One of the problems of the OCC scheme is its discrimination against transactions having long read-phase because of higher probability of restart. MLS/OTSO alleviates the severity of this problem in the following two ways. First, reducing restarts of updating transactions created an environment in which transactions with long read-phase would have a better chance for not being interfered with. Second, timestamps are used to resolve read-write conflicts in which the read-only transaction has read a consistent state of the MLS/DB. Thus, it is possible for a long read-only transaction to finish execution correctly in spite of the presence of conflicting update transactions.

4. Conclusions

If lower-level transactions were somehow allowed to continue with its execution in spite of the conflict of high-level transactions, covert timing-channel freeness would be satisfied. This sort of optimistic approach for conflict insensitiveness and the properties of non-blocking and deadlock freedom make the optimistic concurrency control scheme especially attractive to multilevel-secure transaction processing. MLS/OCC with timestamp ordering (MLS/OTSO), in order to eliminate unnecessary aborts among equal-level transactions, is the performance refinement to MLS/OCC. Write timestamps can be tested when items are read as well as during validation, thus allowing transactions to detect certain inevitable restarts earlier in their execution. To solve the starvation problem, it is necessary to maintain multiple versions of low-level data.

References

[1] S. Kang and S. Moon, "Read-Down Conflict-Preserving Serializability as A Correctness Criterion for Multilevel-Secure Optimistic Concurrency Control: *CSR/RD*," Journal of System Architecture, Vol. 46, pp. 889-902, 2000.

[2] H. T. Kung and J. T. Robinson, "On Optimistic Methods for Concurrency Control," ACM Trans. Database System, Vol. 6, No. 2, pp. 213-226, June 1981.

[3] M. Carey and M. Stonebraker, "The Performance of Concurrency Control Algorithm for Database Management Systems," Proc. of the 10th VLDB Conf., pp. 107-118, 1984.

[4] J. McDermott and S. Jajodia, "Orange-Locking: Channel-Free Database Concurrency Control via Locking," C. E. Landwehr, Database Security VI: Status and Prospects, North-Holland, pp.262-274, 1993.

Brill Academic Publishers
P.O. Box 9000, 2300 PA Leiden
The Netherlands

*Lecture Series on Computer
and Computational Sciences*
Volume 4, 2005, pp. 1670-1678

A Genetic Approach to LQR Design for Thyristor Controlled Series Compensation Application in Power Grid Network

Hak-Man Kim*, Jea-Sang Cha**, Myong-Chul Shin***[1]

*Korea Electrotechnology Research Institute,
Euiwang, South Korea
**Seokyeong University,
Seoul, South Korea
***Sungkyunkwan University,
Suwon, South Korea

Received 5 July 2005; accepted in revised form 28 July, 2005

Abstract: When the Linear Quadratic Regulator (LQR) is applied to power grid network, one of the difficult problems is the selection of the weighting matrices Q and R. The performance of LQR depends on the selection of Q and R. The selection of Q and R is usually carried out by a trial and error manner, which is not systematic. A genetic approach provides a simple structure to implement optimization problems with high probability to find global optimal values. This paper presents a genetic approach to the selection of Q and R of LQR for Thyristor Controlled Series Compensation (TCSC) application. The proposed approach was applied to LQR design of TCSC in the WSCC 3-machine, 9-bus grid network.

Keywords: Linear Quadratic Regulator (LQR) Design, Genetic Algorithms (GAs), Selection of weighting matrices, Thyristor Controlled Series Compensation (TCSC), Power Grid Network

1. Introduction

Thyristor controlled series compensation (TCSC) which has a powerful ability to change its reactances by controlling the firing angle of thyristors has been developed in recent years. However, research on how to control TCSC to improve the stability of power grid network is still at a challenging stage. Linear quadratic regulator (LQR) is a very effective and widely used technique of linear control system design [1]. In power grid network, LQR has been studied for control of excitation, governor and power system stabilizer (pss) [2,3]. When LQR is applied to power systems, one of main difficulties is the selection of the weighting matrices Q and R [4]. The selection of Q and R is usually carried out by a trial and error manner, which is not systematic [1].

  Genetic Algorithms (GAs) are search algorithms based on the mechanism of natural selection and natural genetics [5,6]. GAs provide simple structures to implement optimization problems with high probability to find global optimal values. In recent years, GAs have been used to the design of various controllers [7,8].

 In this paper, a genetic approach to LQR design of TCSC for stability enhancement of a multi-machine grid network is proposed. The main design parameters Q and R are selected by a genetic approach.

 The proposed approach was applied to LQR design of TCSC in the WSCC 3-machine, 9-bus grid network. The performance of LQR of TCSC was evaluated by eigenvalue analysis and time domain simulation.

2. Introduction

Each synchronous machine and its excitation system in the test grid network are modelled using a set of nonlinear dynamic equations (1)-(7) [10,11].

$$T_{do}'\dot{E_q}' = -E_q' - (x_d - x_d')i_d + E_{FD} \tag{1}$$

---

[1]   Corresponding author. Faculty of Information Technology School, Sungkynkwan University. E-mail: mcshin@yurim.skku.ac.kr

$$T_{qo}' \dot{E}_d' = -E_d' + (x_q - x_q') i_q \tag{2}$$

$$\dot{\delta} = \omega_0 (\omega - 1) \tag{3}$$

$$2H \dot{\omega} = -(E_q i_q + E_d i_d) - (\omega - 1)D + T_M \tag{4}$$

$$T_a \dot{V}_R = -V_R - (K_a K_f / T_f) E_{FD} + K_a R_f + K_a (V_{ref} - V_t) \tag{5}$$

$$T_e \dot{E}_{FD} = -(K_e + A_{sat}(1 + B_{sat} E_{FD})) E_{FD} + V_R \tag{6}$$

$$T_f \dot{R}_f = -R_f + (K_f / T_f) E_{FD} \tag{7}$$

For TCSC in the test grid network, the relationship between the controlled reactance and the input to the TCSC controller can be expressed by (8) [11]. The supplementary input is controlled to stabilize the disturbed power grid network.

$$T_C \dot{X}_C = -X_C + X_{ref} + X_{sup} \qquad X_{Cmin} \le X_C \le X_{Cmax} \tag{8}$$

Where

$$X_C = \text{reactance controlled by TCSC controller}$$
$$X_C = \text{reactance controlled by TCSC controller}$$
$$X_{ref} = \text{reference input}$$
$$X_{sup} = \text{supplementary input}$$
$$X_{Cmin}, X_{Cmax} = \text{limits of TCSC output}$$
$$T_C = \text{time constant of TCSC}$$

In this paper, a genetic approach to the LQR design for TCSC application is studied in the discrete state-space. For the purpose of study, the system of equations described in (1)-(7) can be linearized as shown in (9). The subscript *d* of (9) stands for discrete state-space.

$$\Delta x(k+1) = A_d \Delta x(k) + B_d \Delta u(k) \tag{9}$$

3. Linear Quadratic Regulator

LQR is known as a very effective and widely used technique of linear control systems. The LQR problem is to minimize *J* with respect to the control input *u(k)*.

$$J = \frac{1}{2} \sum_{k=0}^{\infty} [\Delta x^T(k) Q \Delta x(k) + u^T(k) R u(k)] \tag{10}$$

$$u(k) = -K \Delta x(k) \tag{11}$$

The control design consists of picking up a proper control gain vector *K*. Since *K* is obtained by solving (12),(13), the selection of *Q* and *R* in the algebraic Riccati equation is very important in LQR problems. However, usually the selection of *Q* and *R* is carried out by a trial and error manner, which is not systematic.

$$S - A_d^T S A_d + A_d^T S B_d (R + B_d^T S B_d)^{-1} B_d^T S A_d - Q = 0 \tag{12}$$

$$K = (R + B_d^T S B_d)^{-1} B_d^T S A_d \tag{13}$$

Fig.1 shows the block diagram of the TCSC controlled by LQR.

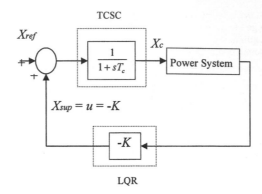

Figure 1: The block diagram of the TCSC controlled by LQR.

4. Genetic Algorithm

GAs do not use real parameters but use chromosomes composed of string coded genotype such as Fig.2.

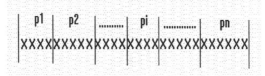

Figure 2: A chromosome composed of parameters.

The basic operators of GAs are reproduction, crossover and mutation. Reproduction is a process in which individual strings are copied according to their objective function (fitness function) values. After reproduction, crossover and mutation operators alter chromosomes in the new population. Under the crossover operator, two structures in the new population exchange portions of their genes. After a crossover operator, a mutation operator may be applied to explore other patterns. Fig.3 and Fig.4 show operation process of crossover and mutation, respectively in GAs [5,6].

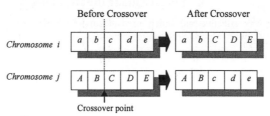

Figure 3: Crossover operation.

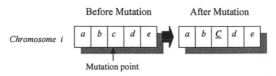

Figure 4: Mutation operation.

GAs explore the optimal solution with the following procedure.

Step 1: Set the initial population of N chromosomes randomly. (N = the number of chromosomes)

Step 2: Evaluate the fitness of each chromosome using the fitness function. If termination condition is satisfied then go Step 5.
Step 3: Reproduce offspring.
Step 4: Alter chromosomes with crossover and mutation operators. Go to Step 2.
Step 5: End.

5.   Proposed Genetic Approach to LQR Design

5.1. Chromosome Composition

For the genetic approach to the selection of Q and R, Q and R were assumed to be diagonal matrices. Chromosomes were composed by diagonal elements of Q and R as shown in Fig.5.

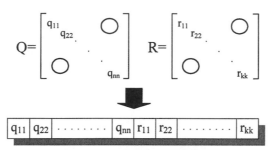

Figure 5:A chromosome composed by diagonal elements of Q and R.

It is required that Q be symmetric positive semi-definite and R symmetric positive definite for a meaningful optimization problem. Using (14) and (15), those constraints are satisfied during the genetic searching procedure. In the (14) and (15), $q_{ii\ lower}$ is the lower limit of the range of diagonal elements of Q and $r_{jj\ lower}$ is the lower limit of the range of diagonal elements of R.

$$q_{ii\ lower} \geq 0 \qquad\qquad i = 1,.....,n \qquad\qquad (14)$$
$$r_{jj\ lower} > 0 \qquad\qquad j = 1,.....,k \qquad\qquad (15)$$

5.2. Fitness Function

It is important in this genetic approach to construct an efficient fitness function. The distinction should be made between the objective function of the LQR problem and the fitness function. Conventional LQR problem is to find the optimal solution K which minimizes the value of the cost function (10), where Q and R are design parameters. If the performance is not satisfactory, the process is repeated with different Q and R until it is satisfactory.
In applying GAs to Q and R selection problem, however, Q and R are not design parameters, but control variables to be determined by some cost function. Using the cost function of LQR (10) as a fitness function always gives only the trivial solution. That is, (10) will be minimized to zero with zero Q and R matrices.
This paper proposes the fitness function considering the following points.

(1)   The performance of a controller mainly depends on the control of dominant oscillation modes. Using participation factors, some states that have the most influence on dominant modes, can be identified [12]. Those states can be considered in the fitness function as a performance measure. Also, another state which is related to the control input may be considered as a design constraint. Consequently, the general fitness function can be formulated as (16) based on the time domain deviations of the dominant modes in the finite time interval.
(2)   The only parameters given by the designer are α and β as shown in (16). The ratio of them is related to the transient response of the TCSC, which should be determined based on the capacity of TCSC.

$$Fitness(v_i) = \sum_{}^{M}\left\{\sum_{}^{L}\alpha(\Delta X_{dom})^2 + \beta\Delta X_C^{\ 2}\right\} \qquad\qquad (16)$$

where

$$v_i = \text{i th chromosome}$$

$$\Delta X_{dom} = \text{deviation of the states which have the most influence on dominant modes}$$

$$\Delta X_c = \text{deviation of the states related to the TCSC}$$

$$\alpha, \beta = \text{weights}$$

$$M = \text{evaluation time} \div \text{step size}$$

$$L = \text{the number of the states which have the most influence on dominant modes}$$

5.3. The proposed procedure of LQR design

Fig.6 shows the proposed procedure of LQR design, which consists of the procedure of LQR design and the procedure of the genetic search. In this paper, elitism [5] was adopted in the genetic search.

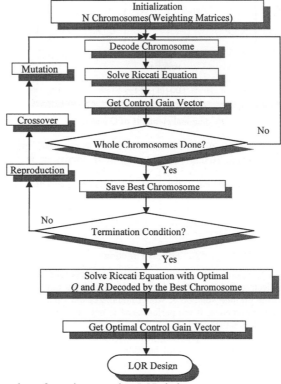

Figure 6: The procedure of genetic approach to LQR design.

6. Case Study

6.1. Test grid network

The Western System Coordinating Council (WSCC) 3-machine, 9-bus grid network is considered as the test grid network. TCSC is installed on line 6-9 to improve the damping of all poorly damped oscillation modes in a multi-machine power grid network [13]. The test grid network and its parameters are shown in the appendix.

(17) shows state variables of the test grid network and (18) is the control input.

$$x = [E'_{q1} \; E'_{q2} \; E'_{q3} \; R_{f1} \; R_{f2} \; R_{f3} \; E'_{d1} \; E'_{d2} \; E'_{d3} \; \delta_{12} \; \delta_{13}$$

$$\omega_1 \; \omega_2 \; \omega_3 \; V_{R1} \; V_{R2} \; V_{R3} \; E_{FD1} \; E_{FD2} \; E_{FD3} \; X_C]^T \qquad (17)$$

$$u = X_{sup} \qquad (18)$$

Table 1 shows the mode characteristics of the test grid network. In table 1, modes 1 and 2 exhibit poor damping. These two modes represent dominant modes that govern system dynamics after disturbance. State variables δ_{12} and δ_{13} have the most influence on dominant modes. Especially δ_{12} has the most influence on mode 2 and δ_{13} has the most influence on mode 1. To increase the damping of poorly damped oscillation modes with TCSC, the damping associated with these two modes must be improved by TCSC control. For the control of two dominant modes, the TCSC controller aims at minimizing the rotor angle deviations $\Delta\delta_{12}$ and $\Delta\delta_{13}$. In this case, the fitness function for the genetic approach was expressed such as (19).

$$Fitness(v_i) = \sum^{M} \left\{ \alpha(\Delta\delta_{12}^2 + \Delta\delta_{13}^2) + \beta\Delta X_C^2 \right\} \qquad (19)$$

Table 1: Mode characteristics of the test grid network.

Mode	Eigenvalues	Damping Ratio	Freq. (rad/sec)	Participation Factor	
				$\Delta\delta_{12}$	$\Delta\delta_{13}$
1	-0.9268 ±i13.2265	0.0699	13.2589	0.1698	0.8651
2	-0.2096 ±i7.6745	0.0273	7.6774	0.8378	0.1562
3	-8.6094 ±i8.1908	0.7245	11.8832	0.0008	0.0006
4	-8.6813 ±i8.3372	0.7213	12.0363	0.0002	0.0000
5	-8.7011 ±i8.4275	0.7183	12.1133	0.0018	0.0040
6	-6.5576	1.0000	6.5576	0.0093	0.0186
7	-4.2284	1.0000	4.2884	0.0095	0.0030
8	-2.0183	1.0000	2.0183	0.0026	0.0007
9	-0.3654 ±i0.7957	0.4317	0.8756	0.0010	0.0002
10	-0.2406 ±i0.5409	0.4064	0.5920	0.0012	0.0000
11	-0.1752 ±i0.3738	0.4244	0.4129	0.0002	0.0006
12	-0.0003	1.0000	0.0003	0.0000	0.0000
13	-10.0000	1.0000	10.0000	0.0000	0.0000

6.2. LQR Design

Genetic parameters and controller design parameters are as follows.

No. of generations = 100
No. of population = 100
Mutation probability = 0.1
Crossover probability = 0.5
Precision = 10^{-4}
Ranges of diagonal elements of Q and R = $0.0000 \leqslant q_{ii} \leqslant 1.5000, 0.0001 \leqslant r_{jj} \leqslant 1.5001$
Evaluation time = 1[sec]
Step size = 1.3899[msec]
$\alpha = 1$
$\beta = 1$

Although the number of possible combination of Q and R exists $7.4818 \times 10^{91} (=15,000^{22})$ cases in the feasible solution space, a genetic algorithm searches less than $10,000 (100 \times 100)$ cases. Fig.7 shows the evolution of fitness function index through each generation.

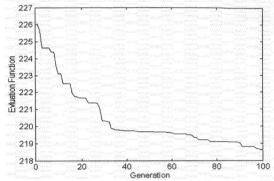

Figure 7: The evolution of fitness function index.

Q and R selected by the proposed genetic approach were as follows.

$$Q=\text{diag}[0.1298\ 0.0106\ 0.1399\ 0.6781\ 0.7563\ 0.7980\ 0.2552$$
$$0.1535\ 0.6080\ 0.0160\ 1.4683\ 0.3147\ 0.6798\ 0.6602$$
$$0.0015\ 0.0059\ 0.8274\ 0.5988\ 0.4136\ 0.6564\ 1.3520] \qquad (20)$$
$$R=0.9063 \qquad (21)$$

The optimal control gain vector obtained by selected Q and R was as follows.

$$K=[-2.2786\ -1.0520\ -3.3851\ 0.0071\quad 0.3640\quad 7.0272\ -0.5700$$
$$-1.2332\ -1.9460\quad 0.3766\ -0.4533\ -32.1501\ 13.3455\quad 8.8772$$
$$0.0002\quad 0.0023\quad 0.1099\ -0.0090\quad 0.0045\ -0.7064\quad 1.0299] \qquad (22)$$

6.3. Eigenvalue Analysis

Fig.8 shows the pole positions of open-loop and closed-loop transfer functions. By TCSC with designed LQR, two dominant modes, mode 1 and mode 2, shift to left more and less respectively. Table 2 shows pole positions and damping ratios of two dominant modes. Damping ratios of dominant modes increase about $2 \sim 4$ times by the designed controller.

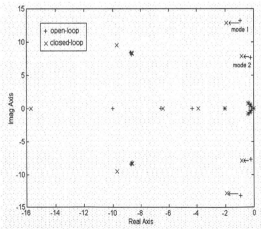

Figure 8: Pole positions with respect to open-loop and closed-loop transfer function.
Table 2: Pole positions and damping ratios of dominant modes.

Mode	Open-loop		Closed-loop	
	Eigenvalue	Damping Ratio	Eigenvalue	Damping Ratio
1	-0.9268± i13.2265	0.0699	-1.9069± i12.8792	0.1465
2	-0.2096± i7.6745	0.0273	-0.8421± i7.8537	0.1066

6.4. Time simulation

To evaluate the performance of LQR designed by the proposed approach, a three-phase short-circuit fault was applied at bus 7. This fault was cleared after 3 cycles without grid network configuration change. Fig.9 shows the simulation results, which demonstrate that the damping of poorly damped oscillation in the test grid network was improved satisfactorily.

6.5. Results

As shown in the results of eigenvalue analysis and time simulation, damping ratios of dominant modes increased about 2-4 times and LQR damped out dominant modes.

Although about 10,000 chromosomes of 7.4818×10^{91} chromosomes were searched by the proposed approach, the performance of designed controller was satisfactory.

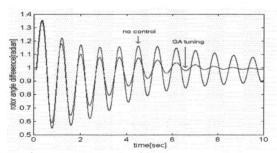

(a) Rotor angle difference between generators 1 and 2.

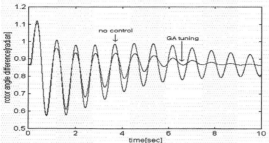

(b) Rotor angle difference between generators 1 and 3.
Figure 9: Rotor angle difference.

7. Conclusion

The genetic approach to LQR design for TCSC application has been studied in this paper. The designed LQR by the proposed genetic approach showed good performance to the enhancement of the damping of poorly damped oscillations in the test grid network. Key contributions of this paper are

(1) a genetic approach to LQR design of TCSC in multi-machine grid network
(2) presentation of the systematic procedure for the genetic approach to LQR design.

The case study showed that LQR was systematically designed by the proposed approach. The proposed genetic approach can be applied to other state feedback control techniques such as LQG (Linear Quadratic Gaussian), LQG/LTR (Linear Quadratic Gaussian with Loop Transfer Recover), etc..

Acknowledgments

Authors wish to thank Samsung Electro-Mechanics for their support .

References

[1] G.F. Franklin, et. al., Feedback Control of Dynamic Systems, 3rd Ed., Addison-Wesley Publishing Company, 1994.

[2] H.A. Mousa and Y. Yu, "Optimal Power System Stabilization through excitation and/or governor control", IEEE Trans. on PAS, pp. 1166-1174, May/June, 1972.

[3] Y. Yu and H.A. Mousa,"Optimal Stabilization of Multi-Machine System", IEEE Trans. on PAS, pp. 1174-1182, May/June, 1972.

[4] Y. Yu, Electrical Power System Dynamics, Academic Press, 1983.

[5] D.E. Goldberg, Genetic Algorithms in Search, Optimization & Machine Learning", Addision Wesley, 1989.

[6] Zbigniew Michalewicz, Genetic Algorithms + Data Structures = Evolution Programs, Spring-Verlag, 1992.

[7] B. Chen, Y. Cheng, C. Lee, "A Genetic Approach to Mixed Optimal H_2/H_∞ PID Control", IEEE Control System, Vol. 15, No. 5, pp. 51-60, Oct., 1995.

[8] F. Klawonn, J. Kinzel, R. Kruse, "Modification of Genetic Algorithms for Designing and Optimizing Fuzzy Controllers", IEEE   ECEC'94, pp. 28-33, 1994.

[9] P.M. Anderson, A.A. Foud, Power System Control and Stability, Iowa State Univ. Press, 1977.

[10] IEEE Committee Report, "Computer Representation of Excitation Systems", IEEE Trans. on PAS, Vol. 87, pp. 1460-1464, 1968.

[11] J.J. Paserba, N.W. Miller, E.V. Larsen and R.J. Piwko, "A Thyristor Controlled Series Compensation Model for Power System Stability Analysis", IEEE Transactions on Power Delivery, Vol. 10, No.3, pp.1471-1478, July, 1995.

[12] Power System Engineering Committee, Eigenanalysis and Frequency Domain Methods for System Dynamic Performance, 90TH0292-3-PWR, The Institute of Electrical and Electronics Engineers, Inc. 1989.

[13] X.R. Chen, N.C. Pahalawaththa, et. al., "Controlled Series Compensation for Improving the Stability of Multi-Machine Power Systems", IEE Proc. Gener. Transm. Distrib., Vol. 142, No. 4, July, 1995.

Brill Academic Publishers
P.O. Box 9000, 2300 PA Leiden
The Netherlands

Lecture Series on Computer
and Computational Sciences
Volume 4, 2005, pp. 1679-1685

Enhanced Blind MVDR Beamforming Implementation Using Real Toeplitz-Plus-Hankel Approximation Scheme

KyungSeok Kim[1] and Jae-Sang Cha[2]

[1] School Electrical and Computer Eng., Chungbuk National Univ., Korea
[2] Dept. of Information and Communication Eng. Seokyeong Univ., Korea

Received 25 June 2005; accepted in revised form 21 July, 2005

Abstract: The efficient scheme for enhancing the performance of the MVDR beamforming is proposed. Generally, the computational complexity of the MVDR beamforming is shorter than that of blind cyclostationary beamforming on applying neural network based signal processing. The main focus of the proposed scheme is to enhance the practical estimation of an array response vector for the MVD beamforming on the arbitrary geometry structure of antenna arrays. Thus, the proposed scheme used the real Toeplitz-plus-Hankel Approximation (RTHA) scheme to obtain the covariance matrix having the theoretical property of an overall noise-free signal. Computer simulation results are presented to show the effectiveness of the proposed scheme and the possibility for network-based beamforming.

Keywords: Adaptive array, network-based beamforming, DS/CDMA.

1. Introduction

Most of conventional algorithms have been used at antenna array structure such as uniform linear arrays (ULAs) to reduce the multiple access interference (MAI) efficiently at the base station in the DS/CDMA uplink scenario. However, the solution using ULAs can not be applied at nonuniform linear arrays (NLAs). NLAs are frequently encountered in practice because it is consequently important to have methods that can detect and estimate the DOA independently of the array configuration. Therefore, we introduce a new approach that enhances the performance of antenna arrays under an arbitrary linear arrays environment, which is NLAs as well as ULAs, on DS/CDMA systems. The overall idea behind the proposed scheme is to modify the calculation of a weight vector for the purpose of the performance improvement of the minimum variance distortionless response (MVDR) beamforming, which is the optimum beamforming. Specifically, the proposed scheme for calculating of a weight vector is to use the real Toeplitz-plus-Hankel Approximation (RTHA) method which has the theoretical property of a overall noise-free signal.

2. System model

For simplicity, we consider a DS/CDMA system employing BPSK and a single cell. After the received signal is converted to the baseband signal, we can write the complex baseband signal vector $\mathbf{x}(t)$ of an array with K antenna elements at the base station as

$$\mathbf{x}(t) = \sum_{l=1}^{L} \sum_{i=1}^{K_u} \varphi_i \alpha_{l,i}(t) S_{l,i} e^{-\phi_{l,i}(t)} \mathbf{a}(\theta_{l,i}) + \mathbf{n}(t) \tag{1}$$

$$S_{l,i} = \left\{ b_i \left(\frac{t - \tau_{l,i}}{T_b} \right) c_{i,t}(t - \tau_{l,i}) + c_{i,p}(t - \tau_{l,i}) \right\} \tag{2}$$

---
[1] E-mail: kseokkim@chungbuk.ac.kr
[2] E-mail: chajs@skuniv.ac.kr

where $\alpha_{l,i}(t)$, $\tau_{l,i}$, and $\phi_{l,i}$ are the channel gain, the path delay, and the time-varying phase shift for the lth multipath signal of the ith user. L is the number of multipath signal, K_u is the number of users, φ_i is the amplitude of the ith user, and $b_i(\cdot)$ is the bit of duration T_b. $c_{i,t}(t)$ is the unique PN code of the user's traffic signal as $c_i(\cdot)W_k^1$ and $c_{i,p}(t)$ is the unique PN code of user's pilot signal as $c_i(\cdot)W_k^0$. W_k^0 and W_k^1 are the 0th and 1st dimensional Walsh codes, respectively. $\alpha_{l,i}(t)$ is the Nakagami fading signal amplitude, and its probability density function (pdf) is given by

$$P_\alpha(x) = 2\left(\frac{m_l}{\Omega_l}\right)^{m_l} \frac{x^{2m_l-1}}{\Gamma(m_l)} \exp\left(-\frac{m_l}{\Omega_l}x^2\right) \tag{3}$$

where $\Gamma(m_l)$ is the Gamma function, $1 \le l \le L, \Omega_l$ is the average signal power, and $m_l \ge 0.5$ is Nakagami fading parameter.

$a(\theta_{l,i})$ is the array response vector for the lth multipath signal of the ith user as

$$\mathbf{a}(\theta_{l,i}) = \begin{bmatrix} 1 & e^{j\frac{2\pi d_2}{\lambda}\sin(\theta_{l,i})}, & \cdots, & e^{j\frac{2\pi(K-1)d_{K-1}}{\lambda}\sin(\theta_{l,i})} \end{bmatrix}^T \tag{4}$$

where $[\cdot]^T$ denotes the transpose operation, K is the number of antenna elements, and d_l (l=0,...,K-1) is the distance between antenna elements.

We can also easily show that the complex array covariance matrix $\mathbf{R}_{xx,1,i}$ of the pre-correlation signal $\mathbf{x}(t)$ for the 1st multipath signal of the ith user is given by

$$\mathbf{R}_{xx,1,i} = \alpha_{1,i}^2(t)\mathbf{a}(\theta_{1,i})\mathbf{a}^H(\theta_{1,i}) + \mathbf{Q}_{xx,1,i} \tag{5}$$

where $\mathbf{Q}_{xx,1,i}$ is the complex array covariance matrix of the overall undesired signals except for the 1st multipath signal of the ith user. The post-processing complex array covariance matrix for the 1st multipath signal of the ith user's pilot is also obtained by [1]

$$\mathbf{R}_{yy,1,i,p} = 2\alpha_{1,i}(t)\mathbf{a}(\theta_{1,i})\mathbf{a}^H(\theta_{1,i}) + \underbrace{\mathbf{Q}_{yy,1,i,p}}_{SI+MAI+NOISE} \tag{6}$$

3. MVDR Beamforming

In the optimum beamforming-based approach, the goal of an adaptive antenna array is to find a weight vector $\mathbf{w}_{1,i,p}$ for the 1st multipath signal of the ith user's pilot that minimizes the overall interference subject to $\mathbf{a}(\theta_{1,i})^H \mathbf{w}_{1,i,p} = 1$. The weight vector $\mathbf{w}_{1,i,p}$ is chosen to minimize the output power of the adaptive antenna array while maintaining a distortionless response in the direction of the desired signal [2]. The weight vector is therefore found from the solution:

$$\min_\mathbf{w} \mathbf{w}_{1,i,p}^H \mathbf{Q}_{xx,1,i}\mathbf{w}_{1,i,p} \quad \text{subject to} \quad \mathbf{a}(\theta_{1,i})^H \mathbf{w}_{1,i,p} = 1 \tag{7}$$

Thus, the MVDR weight vector is given by

$$\mathbf{w}_{1,i,p} = \varsigma_1 \mathbf{Q}_{xx,1,i}^{-1}\mathbf{a}(\theta_{1,i})$$

$$= \frac{\mathbf{Q}_{xx,1,i}^{-1}\mathbf{a}(\theta_{1,i})}{\mathbf{a}(\theta_{1,i})^H \mathbf{Q}_{xx,1,i}^{-1}\mathbf{a}(\theta_{1,i})} \tag{8}$$

where $\varsigma_1 = \dfrac{1}{\mathbf{a}(\theta_{1,i})^H \mathbf{Q}_{xx,1,i}^{-1}\mathbf{a}(\theta_{1,i})}$. In (8), the weight vector is just a scale version of the weight vector for the minimum mean square error (MMSE) beamforming [3]. The weight vector can also be described the similar form as (8) replaced by $\mathbf{R}_{xx,1,i}$ [4]

$$\mathbf{w}_{1,i,p} = \varsigma_2 \mathbf{R}_{xx,1,i}^{-1} \mathbf{a}\left(\theta_{1,i}\right)$$

$$= \frac{\mathbf{R}_{xx,1,i}^{-1} \mathbf{a}\left(\theta_{1,i}\right)}{\mathbf{a}\left(\theta_{1,i}\right)^H \mathbf{R}_{xx,1,i}^{-1} \mathbf{a}\left(\theta_{1,i}\right)} \tag{9}$$

where $\varsigma_2 = \dfrac{1}{\mathbf{a}\left(\theta_{1,i}\right)^H \mathbf{R}_{xx,1,i}^{-1} \mathbf{a}\left(\theta_{1,i}\right)}$. When $\mathbf{R}_{xx,1,i}$ and $\mathbf{Q}_{xx,1,i}$ are known exactly, the weight vector

in (8) and (9) can be shown to be identical. (9) is usually referred to as MVDR beamforming.

4. Proposed Scheme for performance Improvement

Practically, $\mathbf{a}\left(\theta_{1,i}\right)$ can be estimated by solving for the principal eigenvector corresponding to the

maximum eigenvalue $\hat{\lambda}_{\max}$ of the matrix of $\mathbf{R}_{yy,1,i,p}\mathbf{a}\left(\theta_{1,i}\right) - \hat{\lambda}\mathbf{R}_{xx,1,i}\mathbf{a}\left(\theta_{1,i}\right)$ [5]. From [5], $\mathbf{a}\left(\theta_{1,i}\right)$

is simply obtained as the eigenvector associated with the largest eigenvalue from

$$\Re = \mathbf{R}_{yy,1,i,p} - \mathbf{R}_{xx,1,i} \tag{10}$$

The Toeplitz-plus-Hankel structured covariance estimation is much more accurate than the Toeplitz covariance estimation [6]. Since the subspace of symmetric Toeplitz matrices is a subset of the subspace of symmetric Toeplitz-plus-Hankel matrices, the errors in the Hilbert-Schmidt norm sense will always be smaller than the error using only the Toeplitz approximation [7]. Using this principle, the theoretical array covariance matrix can be estimated by projecting $\Re$ onto the subspace of the Toeplitz-plus-Hankel (TH) matrices. The Real Toeplitz-plus-Hankel Approximation (RTHA) scheme consists of Hermitian persymmetric matrix approximation method and a unitary transformation. Hermitian persymmetric matrix approximation method can be obtained from Toeplitz approximation matrix method or the Forward/Backward Averaging method [8]. A unitary transformation method is used to transform a complex Hermitian persymmetric array covariance matrix into a real Toeplitz-plus-Hankel matrix.

For simplicity, an unitary matrix $\mathbf{U}_K \in C^{K \times K}$ of even order K defined by

$$\mathbf{U}_K = \frac{1}{2} \begin{bmatrix} (1-j)\mathbf{I}_{K/2} & (1+j)\mathbf{J}_{K/2} \\ (1+j)\mathbf{J}_{K/2} & (1-j)\mathbf{I}_{K/2} \end{bmatrix} \tag{11}$$

where $\mathbf{I}_K \in R^{K \times K}$ and $\mathbf{J}_K \in R^{K \times K}$ with all ones on the main antidiagonal and zeros elsewhere,

$\mathbf{J}_K \mathbf{J}_K = \mathbf{I}_K$. Therefore, it can be easily shown that satisfy

$$\mathbf{U}_K^{-1} = \mathbf{U}_K^H, \qquad \mathbf{U}_K^* \mathbf{J}_K = \mathbf{U}_K \tag{12}$$

$\Re_{(T)}$ obtained by Toeplitz approximation matrix method is Hermitian and persymmetric because of

$$\Re_{(T)} = \left(\Re_{(T)}\right)^H \tag{13}$$

Therefore, for any Hermitian persymmetric matrix $\Re_{(T)} \in C^{K \times K}$ and $\mathbf{U}_K \Re_{(T)} \mathbf{U}_K^H$ is a real and persymmetric matrix[7].

Thus, by [6], the real symmetric Toeplitz-plus-skew persymmetric Hankel covariance matrix is obtained by

$$\Re_{(TH)} = \mathbf{U}_K \Re_{(T)} \mathbf{U}_K^H \tag{14}$$

where $\hat{T} = \mathrm{Re}\left[\Re_{(T)}\right]$ and $\hat{H} = \mathrm{Im}\left[\Re_{(T)}\right] \cdot \mathbf{J}_K$. A matrix $\hat{T} = \left\{\hat{t}_{p,q}\right\} \in R^{K \times K}$ is said to Toeplitz if

$$\hat{t}_{p,q} = \begin{cases} c_{p-q+1} & \text{if } p-q \geq 0 \\ r_{q-p+1} & \text{if } q-p > 0 \end{cases} \tag{15}$$

A matrix $\hat{H} = \left\{\hat{h}_{p,q}\right\} \in R^{K \times K}$ is said to Hankel if

$$\hat{h}_{p,q} = \begin{cases} c_{p+q-1} & \text{if} \quad p+q-1 \le K \\ r_{p+q-K} & \text{otherwise} \end{cases} \tag{16}$$

5. Weight Vector Calculation using the Proposed Scheme

In this section, we will introduce the modified weight vector by the proposed scheme $\left[\Re_{(TH)} \right]$ for improving the performance of the MVDR beamforming. The weight vector using the proposed scheme for the 1st multipath signal of the ith user's pilot is given by

$$\hat{\mathbf{w}}_{1,i,p} = \frac{\mathbf{R}_{xx,1,i}^{-1} \hat{\mathbf{a}}(\theta_{1,i})}{\hat{\mathbf{a}}(\theta_{1,i})^H \mathbf{R}_{xx,1,i}^{-1} \hat{\mathbf{a}}(\theta_{1,i})} \tag{17}$$

where $\hat{\mathbf{a}}(\theta_{l,i})$ is obtained by the proposed scheme. (17) is the modified versions of the MVDR beamforming. $\hat{\mathbf{a}}(\theta_{l,i})$ gives an effect of performance enhancement in a weight vector for the MVDR beamforming.

Table 1: Parameters employed in the simulation

Parameter	Specification
Capacity Frequency	2GHz
Link	Reverse Link
Chip Rate	3.84 Mcps
Signal Structure	Pilot & Traffic signal
Geometry of array	Uniform and Nonuniform Linear Array
Radiation Power of antenna	Omni-directional
Processing Gain	32
Symbol Rate	120ksps
Channel model	Frequency-Selective Nakagami fading model
Number of simulation	2000

6. Simulation Results

In the simulation, we consider the reverse link of pilot channel aided coherent DS/CDMA system. Besides Table 1, simulation conditions have $E_{c,t}/E_{c,p} = 4dB$, 4 snapshots, 4 multipaths per a user, and no out-of-cell interference. The power of the pilot transmission would be less than that of the traffic transmission to reduce the effect of the interference due to pilot transmissions from other users. The individual antenna elements of the array are ideally sectorized antennas with a sector of 120^0. It is assumed that each sector provides service to one third of the cell. The DOA value of the desired user's signal is set to $\theta = 0^0$. $\tau_{l,i}$ is independent random variables uniformly distributed over $[0, T_b]$. The angle spread (AS) is defined as 2Δ, i.e., the DOAs of all other user's signals are uniformly distributed over $[-\Delta+\theta, \Delta+\theta]$. Δ maximally can be set until the $\pi/3$ degree owing to 120^0 a sector. Assume that the fading is the fast frequency-selective fading. The channel Doppler spread is approximately 222Hz for a mobile speed of 120km/h and the fading rate for fast fading is given by $0.125/T_b$. Without loss of generality, we used the average signal power with $\Omega = 1.3$ through the simulations. For simplicity, we present the value of the fading parameter used for all channels, e.g., $\bar{m} = 0.5$ implies that all the L channels undergo identical Nakagami fading with $\bar{m} = 0.5$. In the simulation results, we shall here consider two geometry cases of the antenna arrays: 1) Uniform Linear Arrays (ULAs) with $\frac{\lambda}{2}$., 2) Nonuniform Linear Arrays (NLAs) with the antenna elements-spacings of

$$\left\{0,\frac{\lambda}{2},\lambda,\frac{3\lambda}{2},2\lambda,\frac{5\lambda}{2},3\lambda,\frac{7\lambda}{2}\right\}.$$

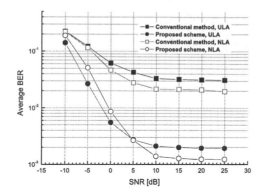

Figure 1: The average Bit Error Rate (BER) versus the received SNR (dB) for comparison between the conventional method and the proposed scheme: AE=Number of antenna elements=8, AS=Angle Spread=120^0, Number of users=20, $\bar{m}=0.5$.

6.1 Effect of the SNR value

Fig.1 shows the simulation results in regard of the BER versus the received SNR (dB) for the proposed scheme according to the SNR values and the geometry structure (ULA, NLA) of the antenna arrays under the angle spread of 120^0, 20 users. Simulation results are shown that the performance of the proposed scheme is much better than that of the conventional method irrespective of the number of antenna elements or the geometry structure of the antenna arrays. Especially, the more the SNR value, the better the performance of the proposed scheme. In the four-branch space diversity and the ULA structure, the proposed scheme needs a SNR of -8dB to obtain a $BER=10^{-1}$, whereas the conventional method needs a SNR of -3dB.

6.2 Effect of the number of users

Next, Fig.2 shows the simulation result compared the proposed scheme to the conventional method according to the number of users and the geometry structure (ULA, NLA) of the antenna arrays in case of the processing gain of 32, the SNR of 10 dB, and $\bar{m}=0.5$. Fig.\ref{main} is shown that the performance of the proposed scheme is much better than that of the conventional method. Generally, the more the number of users, the better the relative performance of the proposed scheme compared to the conventional method. In case of a $BER=10^{-3}$ under the NLA structure in Fig.2, the proposed scheme provides a capacity of 60 users per cell, while the capacity for the conventional method is only 30 users per cell.

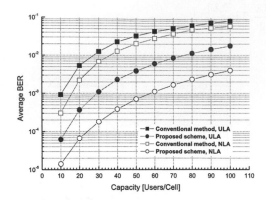

Figure 2: The average Bit Error Rate (BER) versus the number of users for comparison between the conventional method and the proposed scheme: AE=Number of antenna elements=8, AS=Angle Spread=120^0, SNR=10dB, $\bar{m} = 0.5$.

6.3 Effect of the angle spread

Fig.3 shows the simulation results according to the angle spread value $\left(20^0 \sim 120^0\right)$ in case of $\bar{m} = 0.5$, the SNR of 10 dB, and the geometry structure (ULA, NLA) of the antenna arrays. The performance of the proposed algorithm is much better than that of the conventional method irrespective of the angle spread value and the geometry structure of the antenna arrays. Especially, the more the angle spread, the better the relative performance of the proposed scheme compared to the conventional method.

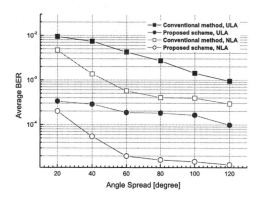

Figure 3: The average Bit Error Rate (BER) versus the Angle Spread for comparison between the conventional method and the proposed scheme: AE=Number of antenna elements=8, SNR=10dB, Number of users=20, $\bar{m} = 0.5$.

6.4 Computational Complexity Comparison

The computational complexity between the conventional method and the proposed scheme is compared using the ``FLOPS" function in MATLAB. Fig.4 summarizes the average results of a run of 300 simulations for the computational complexity of the proposed scheme according to the number of antenna elements. In particular, these values are the averaged value of floating-point operation counts measured by the ``FLOPS" function in MATLAB. The MATLAB measurements of floating-point operations are a rough measure, but is very useful for relative comparison. It can be known through Fig.4 that the proposed scheme has a little higher complexity compared to the conventional method. However, because the proposed scheme give much better performance compared to the conventional method, the proposed scheme can be used enough as the practical technique.

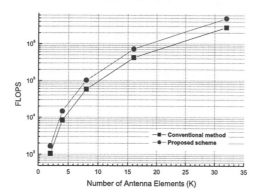

Figure 4: Computational complexity (FLOPS) comparison between the conventional method and the proposed scheme under uniform linear arrays.

7. Conclusions

The efficient scheme to upgrade the performance or the computational complexity of the optimum beamforming for DS/CDMA systems is proposed. The main focus of the proposed scheme is to enhance the practical estimation of an array response vector used at the weight vector of the MVDR beamforming. The proposed scheme for the performance enhancement and the low computational complexity of the MVDR beamforming is the real Toeplitz-plus-Hankel Approximation (RTHA) scheme which has the theoretical property of a overall noise-free signal. Performances of the proposed scheme are confirmed by various simulation parameters such as the number of users, the SNR value, and, the angular spread. Throughout the overall simulations, it can be concluded that the proposed scheme can be an efficient alternative to the conventional MVDR method since they give the better performance and a little higher complexity to implement in real time DS/CDMA systems.

8. References

[1] K. Kim and S. R. Saunders, *Efficient Signal Enhancement Scheme for Adaptive Antenna Arrays in Nakagami Multipath Fading with Power Control Error*, IEICE Trans. Commun., vol.E85-B no.6 (2002) 1105-1114.
[2] M. Torlakand and G. Xu, *Performance of CDMA smart antenna systems*, Proc. The 1995 Twenty-Ninth Asilomar Conference, (1995) 383-387.
[3] N. Y. Wang, P. Agathoklis, and A. Antoniou, *Analysis of MMSE beamformer for DS-CDMA systems*, in Proc. IEEE PACRIM 2001, (2001) 128-31.
[4] B. D. Van Veen and K. M. Buckley, *Beamforming: A versatile approach to spatial filtering*, IEEE Trans. Acoust., Speech Signal processing Mag., (1988) 4-24.
[5] Y. S. Song, H. M. Kwon, and B. J. Min, *Computationally Efficient Smart Antennas for CDMA Wireless Communications*, IEEE Trans. Veh. Technol.}, vol.50, no.6 (2001) 1613-1628.
[6] D. M. Wilkes, S. D. Morgera, F. Noor, and M. H. Hayes, *A Hermitian Toeplitz is Unitarily Similar to a Real Toeplitz-plus-Hankel Matrix*, IEEE Trans. Signal Processing, vol.39, no.9 (1991) 2146-2148.
[7] W. H. Fang and A. E. Yagle, *Two methods for Toeplitz-plus-Hankel approximation to a data covariance matrix*, IEEE Trans. Signal Processing, vol.40, no.6 (1992) 1490-1498.
[8] D. A. Linebarger, R. D. DeGroat, and E. M. Dowling, *Efficient Direction-Finding methods employing Forward/Backward Averaging*, IEEE Trans. Signal Processing, vol.42, no.8 (1994) 2136-2145.

Brill Academic Publishers
P.O. Box 9000, 2300 PA Leiden
The Netherlands

*Lecture Series on Computer
and Computational Sciences*
Volume 4, 2005, pp. 1686-1689

Design of POS System Using the XML-Encryption

Seoksoo Kim[1]

Department of Multimedia Engineering,
Hannam University,
Daejeon, South Korea

Received 7 July, 2005; accepted in revised form 10 August, 2005

Abstract: POS system that become that is supply net administration and computerization fetters of customer management that become point in distribution industry constructed database and use XML-Encryption that is certificate techniques of PKI and standard of security for security that is XML's shortcoming and design distributed processing POS system using XML for data integration by introduction of Ubiquitous concept. This POS system has four advantages. First, because there is no server, need not to attempt authentication and data transmission every time. Second, can integrate data base by XML and improve portability of program itself. Third, XML data in data transmission because transmit data after encryption data safe .Fourth, after encode whenever process data for data breakup anger of POS system client program and elevation of the processing speed, transmit at because gathering data at data transmission

Keywords: POS, XML, Encryption, PKI

1. Introduction

In 21st century, reform of circulation physical distribution is expected. Domestic circulation companies' profitability mend and sale elevation, curtailment of circulation expense and necessity about efficiency raising of circulation system are increasing in actuality that is outside opening of secondary market with factor of increase and pattern change of consumption demand, and internally rise of personnel expenses of purchasing power by elevation of national income level, traffic congestion etc.., and circulation company's customer service improvement and buries of various goods, sale, POS system that do inventory present condition so that can manage all sale information efficiently and so on being the basicest terms desired that permit rationalization of these distribution structure, is going to contribute greatly in distribution structure rationalizations through union with various radio applications.[1]

This treatise studied about method to make use of POS system that the importance is emphasized in whole country as more safe and fast. Existent POS system had begun in basis money receipts and disbursements flag and develop to present web POS system. This web POS system emphasizes safety of data and considers improvement of the processing speed wire or wireless of use device, existent web database constructing distributed processing POS system that is not existing server/client structure using XML the processing speed and portability and integration attribute heighten and this distributed processing POS system using PKI (Public Key Infrastructure) to worm with web each posthumous work and posthumous work connect and supply efficient XML Security because using XML-Encryption. [2]

Composition of this treatise is as following. After explain and design distributed processing POS system after explain structure of existent POS system in 3 chapter about pos system that is connection technology in proverb 2 chapter in 1 chapter introduction, security, XML, XML-Encryption, 4 chapter present conclusion of treatise that see.

[1] Department of Computer & Multimedia Engineering, Hannam University, Daejeon, South Korea . E-mail: sskim@hannam.ac.kr
"This work was supported by a grant No. (R12-2003-004-03003-0) from Ministry of Commerce, Industry and Energy"

2. Authentication and Encryption

Need certification and encryption technology to change environment and connect this breakup system to Internet and use XML database by distributed processing system in existing server/client structure to take advantage of web POS system and advantage of single POS system used in existing to embody distributed processing POS system. Authentication is process that user and computer or Application confirms whether other user, computer, application is going to be anyone and do what. Usually, use credential such as password or PIN (Personal Identification Number), smart card and certification consists Certification in web service consists between applications mainly, and recentralizes ordinary password or certificate.

Because this research produces itself certificate, certificate must offer information (IP, OS, certification number etc.) of hardware (PC, PDA) that program is established and certification server has to be operated with system that see. Encryption acts role that secure integrity doing so that someone may not be able to read this even if get seized data in midway.

Need encryption key and cipher for encryption. Encryption description is 'symmetric' and 'asymmetric' way. Asymmetric says that is 'Public Key algorithm' by the other word, private key method that each manages and distribute public key for decryption to other people be.

3. XML security

XML (Extensible Markup Language) is language of tag form to define and expresses structure of document that has information. 'That can extend' (Extensible) document means thing which can define structure in the document and can change to structure of other document. 'Mark household mascot language' (Markup Language) is expressed and expresses high position information of information by Tag usually. When HTML through internet is used much, in W3C (World Wide Web Consortium) SGML (Standard Generalized Markup Language) thing made curtly by base XML be. XML is situating by standard for expression and exchange of information in industry. [3],[4]

Is expressed in form of tag analogously with HTML, but HTML falls in love though have data of document and information about expression at the same time, XML document separates and expresses these mainly. Data of XML document is hierarchic and structural form and information about expression composes style document (Style Sheet) in form of XML document. Through separation function of these data and expression, XML permits data save and exchange between other programs regardless of Operating System. Also, can define document structure voluntarily and can get into recent publication of Business-to-Business solution because conversion is available. [5]

XML's commission basically beginning tag and end tag matching do. Mark information that arranges data between tag interior and tag. It is known that beginning tag course end tag one pair is Element and Elements are composed hierarchically. Can have Attribute on Element's tag interior. There is capital have other Element that form tag form between Element's tag and Element of Text form can have. Speak that Element, Text, Attribute etc. are all Nodes. [6]

The following is example of simple XML document that display sale recording of bookstore. If use this XML to database, must encrypt and use XML having shortcoming that information of data is revealed just as it is. Application that follow XML encryption standard must embody all elements explained here as long as there is no especial reference. Must include Encrypted Data, Cipher Data, Encryption Properties, ds:KeyInfo necessarily. This XML encryption has following structure Figure 1.

```
<EncryptedData Id? Type?>
<EncryptionMethod/>?
  <ds:KeyInfo>
    <EncryptedKey>?
    <AgreementMethod>?
    <ds:KeyName>?
    <ds:RetrievalMethod>?
    <ds:*>?
  </ds:KeyInfo>?
  <CipherData>
    <CipherValue>?
      (CipherReference URI?)?
  </CipherData>
  <EncryptionProperties/>?
</EncryptedData>
```

Figure 1: Structure of XML encryption

This encryption method is 4 encryption method of encipherment, encipherment of XML element contents, free data encipherment, supermarket encryption and so on of XML elements. By encryption of XML elements encrypts element from beginning tag to end tag on the whole by first, Can hide this element.

By second, encryption of XML element contents is used because element sees but go side by side with element whole encryption that talk in front that enciphers contents of element.

By third, free data encipherment is method to do to treat included whole data by Oktet stream simply, and super encryption must encrypt whole element necessarily in case of use Encrypted Data or super encryption of Encrypted Key element that encrypt again encoded already information finally. This research used encryption of XML element contents.

4. Distributed Processing POS system with XML-Encryption and PKI

As refer in front, POS system is available by real time control of goods in stock and sale analysis at present. Therefore, can make efficient consumer's sale data, and these sale data is essential data to introduce SCM and CRM. This POS system developed by web POS system structure that is server/client structure at single program.

That Figure 2 explains system that present, it is timing diagram that serve. This timing diagram when whole POS system goes state change thing which do diagramming according to flowing of time be .Is decided in visual point that certification and connection work are to flow time if see this reversed character and each program of this timing requires. Can gain four advantages forming this structure.

Because there is no first server, need not to attempt authentication and data transmission every time. Because data that each client program handles by oneself has by oneself, network does impossibility or place that real time network is unnecessary data client voluntarily have and transmit to network for integration of data when need. Can integrate data base by second XML and improve portability of program itself. Because performance of present Mobile improves, if PDA that load OS is most and installs these case client program, transmission of data is available via network gear city certification process as well as itself save of data. Because used XML encryption in third data transmission, transmission time is shortened. Is a present condition that does not recognize Security of data about circulation yet, but can speak that is important information of corporation as well as data that computerization of all circulation data is such if is achieved forward. Fourth, to improve distributed processing and processing speed of POS system client program, transmit at because gathering all datas when do data transmission after encode whenever process data. In the case of existing system, server processes data and server's investment expense was spent much. System that administrator's traffic also was much but present in treatise that see making use of computer that server and administrator are different each client's business processing by itself save of data improve but problem by data amount happens. That quantity of product that a customer a person can happen is average 20, in case medicine 10KB's data is created and sells 1 thatched cottage separated from the main building of house 1 goods in case caught, about 6 persons process data of 864 people for 36 people 24 hours at 1 hour at 1 minute. Data of 864 people is about 8 MB. Client one data uses data of 240 MB in case is grass operation 8 MB two faces month in case is 24 hours. In the case of present POS system hardware, can collect data about 6 months about 2 years because it is general that use hard-disk of 40 GB. Amount of data has shortcoming that must play bulky case backup faithfully by that form these structure.

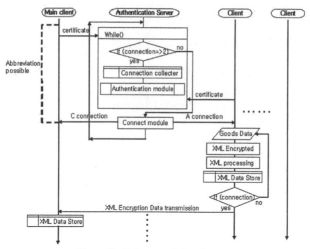

Figure 2: POS system timing diagram

5. Conclusions

So that can load PDA or Mobile etc. to utilize by Ubiquitous, designed way and processing of these data and all-in-one that use itself data base using XML anytime feasibly moment network is linked. Also, because emphasizing the importance of public safety division forward corporation's data that is XML's limitation, focused on certification part for practical use of system that use XML security and see. Even if do not flow certification, itself data processing is available, but must flow certification necessarily to utilize supply net administration and customer management that is last target of POS system. Also, designed certification so that communications between two client may be available in Internet that use Dynamic IP to enhance added value of system. However, need suitable administration of whole equipment lowering because have data while amount of data is bulky case each client yam by that forms these structure However, space about this data is necessarily necessary if computerize data. Can heighten stability of data by that distribute and store doubly data, and need administration by this that been converged to server. About capacity and search of XML data hereafter, must study about method and fast search of this XML data that reduce maximum capacity.

References

[1] Myung-sub Song, Joo-ho Kim : A Study on Development of Intranet based POS System , Press confrence collection of learned papers Vol.- No.1 ,1997.
[2] Ki-rak Son, : XML: New Internet Lingua Franca, Journalism and Mass Communication 3005191 Vol.27 No. ,2001
[3] Kang- hi Kim, Young-il Kwon, Yong Deok Ku : POS system 2003 , Korea Institute of Science and Technology , Ba63, 2003
[4] Jong-sun Leem : Developing an XML Repository for Workflow Management, Society for e-business Studies Vol.8 No.3, 2003
[5] Alexznder Nakhimovsky and Tom Myers : Professional Java XML Programming with Servlets and JSP, Wrox Press, 1999
[6] A Extensible Markup Language(XML), Available at http://www.w3.org/XML/Snag-lae Choi, Young-Min Kim, Jin-su Choi : A Study on the Application Strategies of POS for the Competitiveness Improvement of Retailers , The Korea Logistics Research Association Vol.6 No.2 , 2003

Brill Academic Publishers
P.O. Box 9000, 2300 PA Leiden
The Netherlands

*Lecture Series on Computer
and Computational Sciences*
Volume 4, 2005, pp. 1690-1693

A Study on High Speed Networking and Distributed Security System

Seoksoo Kim[1]

Department of Multimedia Engineering,
Hannam University,
Daejeon, South Korea

Received 7 July, 2005; accepted in revised form 10 August, 2005

Abstract: The study uses two software packages (Packet Capture and Round Robin Test Package) to check packet volume from Virtual Network Structure (data generator, virtual server, Server 1, 2, 3), and finds out traffic distribution toward Server 1, 2, and 3. The functions of implemented Round Robin Load Balancing Monitoring System include round robin testing, system monitoring, and graphical indication of data transmission and packet volume. Graphical output was displayed running on Linux using GTK, which had menu to close a window and icons to tell which protocols are more often used by displaying five different kinds of protocols (Total, TCP, IP, ARP, and Other). As the result of the study shows, Round Robin Algorithm allows servers to ensure definite traffic distribution, unless incoming data loads differ much. Although error levels are high in some cases, they were eventually alleviated by repeated tests for a long period of time.

Keywords: Networking, Security, Session, Load balancing

1. Introduction

In the web environment, numerous information requires security. For example, private information such as social security number and card number needs extra security. To use encryption for the information, Secure Sockets Layer (SSL) is widely used. SSL is a security service that works above TCP connection. It interlinks along with application layer services such as HTTP (Hypertext Transfer Protocol), FTP (File Transfer Protocol), SMTP (Simple Mail Transport Protocol). Thanks to the benefits of varied authentication and encryption techniques, it is recognized one of the most representative security software. Considering this influence, Internet Engineering Task Force (IETF) adopts SSL as the standard of security session service and changes its name to Transport Layer Security (TLS) and works on its standardization. However, TLS protocol does not guarantee high-speed transmission in the web server cluster environment[1,2].

In order to create security session, security keys are preconfigured between communication objects. For this purpose, Handshake Protocol exists. The pre-master secret key that is used in this process needs to interpreted by a server to create master secret key, whose process requires a big calculation, resulting in deteriorating system's transmission performance.

2. Web Server Cluster's Load Balancing

The methods of handling client's requests are implemented in two ways: Relaying front end and TCP handoff. First, one of the examples of relaying front-end methods is an algorithm that is suggested by Rice University. When a request happens from a client, dispatcher finds the corresponding back-end server based on the contents information like URL, distributes loads, and evaluates server's loads in real-time. When overloads happen to a specific server, loads will be distributed to other servers.

---

[1] Department of Computer & Multimedia Engineering, Hannam University, Daejeon, South Korea . E-mail: sskim@hannam.ac.kr
"This work was supported by a grant No. (R12-2003-004-03003-0) from Ministry of Commerce, Industry and Energy"

On the other hand, TCP handoff method has such architecture that the selected back-end server directly sends a reply to a client without going through the dispatcher, although dispatcher manages load distribution. The security policy that this study suggests combines the above two methods. In a specific time frame, sessions will be reused, but the session information that is stored will be regularly initialized. In this way, each server's loads need to be balanced in a long-term basis. To apply this security policy, the above methods of processing client's requests need to be implemented in advance [3,4].

2.1 Purpose (Necessity)

Due to the recent explosive increase of users connected to network and its subsequent services, user's demands cannot be satisfied only with the development of electronic technology and communications equipment. To solve this issue, load balancing that handles recent data as cell unit is suggested. However, scheduling by cell unit requires extra works such as reassembling of data after switching. When switching the packets into a variable length not a fixed length, reassembling process is omitted in the cell unit scheduling method, which results in increasing a process rate of traffic[5].

2.2 Implementation Goal

- To provide a more stable service by traffic distribution across multiple Web servers
- A user's job can be done faster than only using a single computer by evenly distributing loads across computers. Though using tens of computers, if the work is done only in a single computer, desired performance increase is not guaranteed. That's why the load balancing is gaining its impotence.
- To evenly distribute loads, make a virtual server between client and web server. All the requests done by client will be managed by the virtual server that will decide which server will respond. At this time, client does not know which server responds.
- In order to view analyzed traffic, implements software that provides GUI environment using a window box.

2.3 Implementation Features

: Load Balancing using Round Robin Algorithm
- Transfer protocol and application layer protocol needs to be available in order to control network traffic.
- After assuming the packets in a variable length, traffic needs to be controlled using Round Robin Algorithm.
- To visualize the analyzed traffic, it needs to be expressed in graphs.
- With software, users need to print the results in a GUI environment so that they can understand them easily
- Implement a virtual server based on the above theory. The virtual server determines to which server a requesting client can be connected. The client does not know the connecting server, but the server that is connected by the virtual sever handles the request.

3. Round Robin Monitoring System

3.1 Overview of System Architecture

(1) It analyzes header files of random data sent through data generator and shows the results on the generator's monitor.
(2) It sends the analyzed files to the virtual server and distributes across the Web servers using Round Robin algorithm.
(3) It reanalyzes the distributed files and displays the results on the monitor of each server.

3.2 System Architecture (Linux Cluster Implementation)

The Linux Virtual Server is a highly scalable and highly available server built on a cluster of real servers, with the load balancer running on the Linux operating system [6,7].
Shown as the figure below, it depicts general virtual server architecture with one load balancer and several real servers. The real servers are configured to operate the same services and the service contents are copied into a local disk at each server or provide service by sharing a distributed file

system. The load balancer sends client's requests to one real server that is interlinked by adequate scheduling technique (i.e. Round Robin), whenever a client requests.

(Figure 1) Virtual connection architecture, Physical connection architecture

3.3 Scenario

(1) Build a virtual server to analyze incoming packet information from data generator.
(2) Develop software to display analyzed packet information on window screen.
(3) As a final experiment, analyze such packets inputted through data generator as total packet volume, TCP, IP, ARP, and other protocols on the virtual server and find out whether the results can be displayed on the data input computer.
(4) In addition, check what values return when the virtual server sends the analyzed data to server 1, 2, and 3.

3.4 Limitation

(1) Linux operating system is required to test on the real server environment.
(2) A switching hub is required to combine computers into one network environment for Round Robin test.
(3) Data implementation is required to analyze header files of packet data and display the results.

4. Security Session

In the web environment, numerous information requires security. For example, private information such as social security number and card number needs extra security. To use encryption for the information, Secure Sockets Layer (SSL) is widely used. SSL is a security service that works above TCP connection. It interlinks along with application layer services such as HTTP (Hypertext Transfer Protocol), FTP (File Transfer Protocol), SMTP (Simple Mail Transport Protocol). Thanks to the benefits of varied authentication and encryption techniques, it is recognized one of the most representative security software. Considering this influence, Internet Engineering Task Force (IETF) adopts SSL as the standard of security session service and changes its name to Transport Layer Security (TLS) and works on its standardization. However, TLS protocol does not guarantee high-speed transmission in the web server cluster environment.

In the web environment, numerous information requires security. For example, private information such as social security number and card number needs extra security. To use encryption for the information, Secure Sockets Layer (SSL) is widely used [7]. SSL is a security service that works above TCP connection. It interlinks along with application layer services such as HTTP (Hypertext Transfer Protocol), FTP (File Transfer Protocol), SMTP (Simple Mail Transport Protocol). Thanks to the benefits of varied authentication and encryption techniques, it is recognized one of the most representative security software. Considering this influence, Internet Engineering Task Force (IETF) adopts SSL as the standard of security session service and changes its name to Transport Layer Security (TLS) and works on its standardization. However, TLS protocol does not guarantee high-speed transmission in the web server cluster environment.

In order to create security session, security keys are preconfigured between communication objects. For this purpose, Handshake Protocol exists. The pre-master secret key that is used in this process needs to interpreted by a server to create master secret key, whose process requires a big calculation, resulting in deteriorating system's transmission performance. Therefore, it is helpful in increasing transmission speed to reuse secret keys rather than to create them at every connection. However, increasing reuse rates of sessions to increase transmission performance is not applied to the entire network environment. In the cluster environment, excessive reuse of sessions undermines load balancing and overburdens a specific server resulting in explosive network traffics in worst case. As a result, considering a technique that balances between load balancing and security session reuse, a handshake algorithm is required to minimize the transmission speed delay in the overall networks.

5. Research Result and Analysis

As the results show, round robin algorithm ensures traffic distribution if the size of incoming data is not that great difference. In some areas, errors show greatly, but as the times and period of test get longer, the errors are diminished.

The packet volume in the server 1, 2, 3 shows the values between 22000 and 25000. Distribution is evenly made as the values show between 65000 and 70000 sent from data generator. However, when it comes to the packet volume passed through the virtual server, more values are shown from 90000 to 110000 than the volume generated from data generator. It is almost twice as much the volume generated from data generator. It is estimated that the values may be the packet volume that the virtual server sends and receives. In addition, the packet volume in the server 1, 2, 3 shows only the incoming values. Considering the feedback to the virtual server, a much greater volume of traffic is expected to generate.

6. Conclusion

This study analyzes a load balancing technique using Round Robin Algorithm. For this technique, the study used two software packages (Packet Capture Package and Round Robin Test Package). Packet capture module measures packet volume as it outputs text data by protocol, while graphic module creates a window, saves data received from packet capture module, and displays them into visual graphs. Round robin test package includes three modules: data generator module, virtual server module, and data receiver module. Data generator module creates data at random value, which transmits them to virtual server module. Virtual server module distributes the data received from the data generator by round robin algorithm and sends them to data receiver. Data receiver module saves incoming data from virtual server module as files and analyzes them.

The implemented package software measured packet volume that was generated from data generator, virtual server, and server 1, 2, 3, and could find out traffic distribution toward Server 1, 2, 3.

As the result of the study shows, Round Robin Algorithm ensured definite traffic distribution, unless incoming data loads differ much. Although error levels were high in some partial cases, they were eventually alleviated by repeated tests for a longer time.

References

[1] W.R Stevens, TCP/IP Illustarted Volume 1., Addison-Wesley, New york.1996.

[2] V. Pai, M.Aron, G.Banga, M.Svendsen, P.Druschel, W.Zwaenepoel, and E, Nahum Locality Aware Request Distribution in Cluster-based Network Servers. Architectural Support for Programming Languages and Operating systems pp1-12, 1998

[3] E. Levy-Abegnoli, A. Iyengar, J. Song, and D. Dias, "Design and Performance of a Web Server Accelerator," IEEE INFOCOM'99, pp.135-143,1999.

[4] M.Aron, D.Sanders, and P.Druschel, Scalable Content-Aware Distribution in Cluster-based Network Servers. Proceedings of the 2000 USEMIX Technical Conference, 2000

[5] Darwin Streaming Server, http://www. publicsource.apple.com/projects/streaming.

[6]W. Zhang, "Linux Virtual Server for Scalable Network Services," Linux virtural server project,1998.

[7] V.Kumar, A.Grama, and V.N.Rao, Scalable Load Balancing Techniques for Parallel Computers, Journal of Distributed Computing, pp.60-79, 1994

Brill Academic Publishers
P.O. Box 9000, 2300 PA Leiden
The Netherlands

*Lecture Series on Computer
and Computational Sciences*
Volume 4, 2005, pp. 1694-1697

Block Model for Categorizing IT Systems Security

Tai-hoon Kim[1], Haeng-kon Kim[2], Sun-myoung Hwang[3]

[1]SanSungGongSa, Seoul, Korea,
[2]Catholic University of Daegu, Daegu, Korea
[3]Daejeon University, Daejeon, Korea

Received 10 August 2005; accepted in revised form 11 August, 2005

Abstract: Because the networks and systems become more complex, the implementation of the security countermeasures becomes more critical consideration. The designers and developers of the security policy should recognize the importance of building security countermeasures by using both technical and non-technical methods, such as personnel and operational facts. Security countermeasures may be made for formulating an effective overall security solution to address threats at all layers of the information infrastructure. This paper uses the security engineering principles for determining appropriate security countermeasures. This paper proposes a method for building security countermeasures by modeling and dividing IT systems and security components into some blocks.

Keywords: Block Model, Security Engineering

1. Introduction

When we design general or some special IT systems, we may provide a framework for the assessment of quality or security characteristics by considering some approaches and methods. And this framework can be used by organizations involved in planning, monitoring, controlling, and improving the acquisition, supply, development, operation, evolution and support of IT systems.

In the general cases, security countermeasures for IT systems are implemented in buying and installing some security products such as Firewall, IDS and Anti-virus systems. But the scope of IT systems is being extended and the security holes are increased. Most of the threat agents' primary goals may fall into three categories: unauthorized access, unauthorized modification or destruction of important information, and denial of authorized access. Though any cases are occurred, the compromise of IT system may be connected to loss of money or job. Therefore, Security countermeasures must be implemented to prevent threat agents from successfully achieving these goals [1-4].

This paper proposes a method for building security countermeasures by modeling and dividing IT systems and security components into some blocks. In facts, IT systems are very complex and consist of very many components. So we can't help dividing IT systems into some parts. And we categorize security components into some groups.

Security countermeasures should be considered with consideration of applicable threats and security solutions deployed to support appropriate security services and objectives. Our Block model may be used to make security countermeasures in any cases. Because the size of each block expresses parts insufficient in security.

2. Dividing IT Systems

Implementation of any security countermeasure may require economic support. If your security countermeasures are not sufficient to prevent the threats, the existence of the countermeasures is not a

---

[1] Corresponding author. E-mail: taihoonn@empal.com

[2] E-mail: hangkon@cu.ac.kr

[3] E-mail: sunwhang@dju.ac.kr

real countermeasure and just considered as like waste. If your security countermeasures are built over the real risks you have, maybe you are wasting your economic resources.

First step is the division of IT systems into some parts (See Fig.1). In this paper, we divide IT systems into 4 parts. But we think this partition is not perfect one and we are now researching about that.

Each part may have three common components such as Technique, Product, and Operation and Personnel.

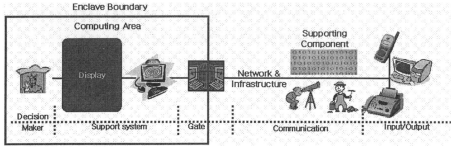

Figure 1: Division of IT systems.

Next step is construction of block matrix by using the parts of IT systems and common components we mentioned above (See the Fig. 2).

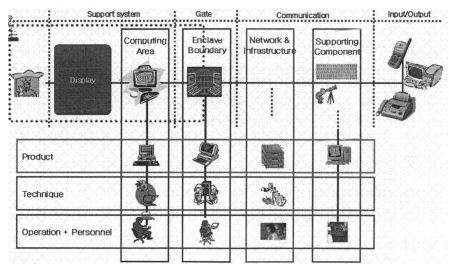

Figure 2: Block matrix.

Each cross point area of Fig.2 may be generalized and reduced into Block and matrix of Fig.3. Each Block may mean the area require security countermeasures or security method.

Next step is determination of security assurance level of IT systems. Security assurance level is related to the robustness. In the concept of our Block model, all cross point area should be protected by security countermeasures.

Robustness is connected to the level or strength of security countermeasures and this idea is expressed like as Fig.4. The last step may be building security countermeasures by using Block Region.

This block matrix can be applied to information engineering and system engineering. Next is the sample applied to design security countermeasures for IT systems.

3. Design Flow of IT Systems Security Countermeasures

We published a design method for IT systems security countermeasures a few months ago. In that model, we identified some components we should consider for building security countermeasures. In fact, the Procedure we proposed is not perfect one yet, and the researches for improving are going on.

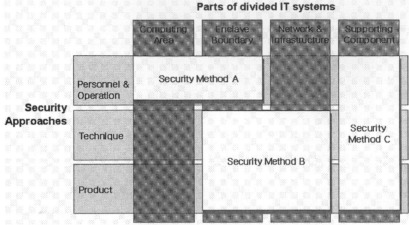

Figure 3: Security Approaches and Methods for divided IT systems.

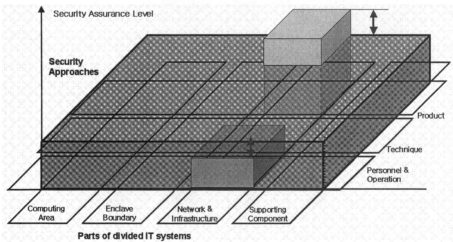

Figure 4: Building security countermeasures by using Block Region.

The discussion of the need to view strength of mechanisms from an overall system security solution perspective is also relevant to level of assurance. While an underlying methodology is offered by a number of ways, a real solution (or security product) can only be deemed effective after a detailed review and analysis that consider the specific operational conditions and threat situations and the system context for the solution.

Assurance is the measure of confidence in the ability of the security features and architecture of an automated information system to appropriately mediate access and enforce the security policy. Evaluation is the traditional method that ensures the confidence. Therefore, there are many evaluation methods and criteria exist. In these days, the ISO/IEC 15408, Common Criteria, replaces many evaluation criteria such as ITSEC and TCSEC.

The Common Criteria provide assurance through active investigation. Such investigation is an evaluation of the actual product or system to determine its actual security properties. The Common

Criteria philosophy assumes that greater assurance results come from greater evaluation efforts in terms of scope, depth, and rigor.

Next figure is the summarized concepts we proposed.

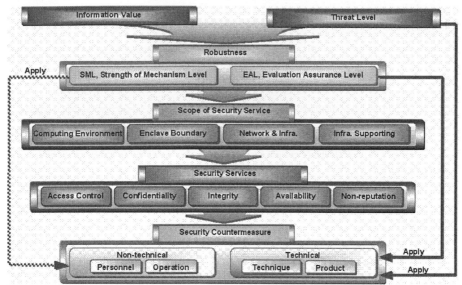

Figure 5: Example of Design Procedure for Security Countermeasures.

4. Conclusion

Our Block model may be used to make security countermeasures in any cases. Because the size of each block expresses parts insufficient in security. For more detail work, we are now researching about the Block Model for Building Security Countermeasures.

References

[1] Eun-ser Lee, Kyung-whan Lee, Tai-hoon Kim and Il-hong Jung: Introduction and Evaluation of Development System Security Process of ISO/IEC TR 15504, ICCSA 2004, LNCS 3043, Part 1, 2004

[2] Tai-hoon Kim , Chang-wha Hong and Sook-hyun Jung: Countermeasure Design Flow for Reducing the Threats of Information Systems, ICCMSE 2004, 2004.

[3] Ho-Jun Shin, Haeng-Kon Kim, Tai-Hoon Kim, Sang-Ho Kim: A study on the Requirement Analysis for Lifecycle based on Common Criteria, Proceedings of The 30th KISS Spring Conference, KISS (2003).

[4] Haeng-Kon Kim, Tai-Hoon Kim, Jae-sung Kim: Reliability Assurance in Development Process for TOE on the Common Criteria, 1st ACIS International Conference on SERA.

[5] Tai-hoon Kim, Yune-gie Sung, Kyu-min Cho, Sang-ho Kim, Byung-gyu No: A Study on The Efficiency Elevation Method of IT Security System Evaluation via Process Improvement, The Journal of The Information Assurance, Vol.3, No.1, KIAS (2003).

Brill Academic Publishers
P.O. Box 9000, 2300 PA Leiden
The Netherlands

*Lecture Series on Computer
and Computational Sciences*
Volume 4, 2005, pp. 1698-1701

Incorporating HVS Parameter into the Transform-based Watermarking

Yoon-ho Kim[1], Byeoung-min Youn[2], Heau-jo Kang[3]

Div. of Computer & Multimedia Content Eng.
Mokwon University, Daejon, 305-729, Korea

Received 13 July, 2005; accepted in revised form 18 August, 2005

Abstract: In this paper, transform-based watermarking scheme based on human visual system is proposed. Human visual parameters are useful to ensure the imperceptibility of watermark image. Not only human visual parameters, such as contrast sensitivity, texture degree but also statistical characteristics are involved to select the optimal coefficients region. The performance of proposed algorithm is evaluated by the experiments of imperceptibility and correctness of watermark. According to some experimental results, contrast sensitivity function is superior in smooth image. On the other hand, statistical characteristics provide good results in rough images. Consequently, how to select the parameters considering image attribute is key problem in effective watermarking.

Keywords: Watermark, HVS, DCT

1. Introduction

The information types of multimedia productions are digital form, which is easy to be stolen by way of transmitting on networks. Therefore, the copyright protection and enforcement of intellectual property rights for digital media has become an important issue. One of the faithful solutions and the most popular method is a digital watermark technology which is the practice of hiding a message into the digital media such as audio, image and video. Since the late 1990s, there has been a rapid increase in the number of digital watermarking algorithms published[1][2]. Applications of digital watermarking include owner identification, authentication, proof of ownership and copy control. Watermarking strategies divided into two major categories: transform-domain and spatial-domain approach. Transform-domain watermarking methods, also called multiplicative watermarks, such as Fourier transform, discrete cosine transform (DCT), discrete wavelet transform (DWT). The watermark is hidden in the middle or lower frequency band is more liable to be suppressed by compression. Therefore, for watermarking, how to select the best frequency band of the image is more important than any other procedures [3] [4].

This paper presents a method of incorporating human visual system (HVS) parameters into the transform-based watermarking. The main goal of this method is to utilize a HVS in order to analyze a performance of watermarking with respect to the image attributes. Further more, statistic parameters are also involved to generate an effective watermarking.

2. Relations between HVS and Transform Domain Coefficients

There are many requirements for a good-designed watermark such as robustness, imperceptibility, unambiguousness etc. Among all types of watermarks, two most common requirements for effective watermarking are required. It should be perceptually invisible, which means the watermark is not visible under typical viewing conditions. It should also be robust to common signal processing and intentional attacks. Spatial domain watermarking which process directly to the location and luminance of the image pixel is one of the fundamental methods in the beginning of digital watermarking researches. This method has the disadvantage that they tend not to be robust congenitally in spite of having simple and easy for implementation. In general, transform domain based approached are

[1] Corresponding author. E-mail: yhkim@mokwon.ac.kr
[2] Corresponding author. E-mail: ybm55@mokwon.ac.kr
[3] Corresponding author. E-mail: hjkang@mokwon.ac.kr

superior to that of spatial domain in preserving contents fidelity and robustness under attacks. In this scheme, three main steps should be specified: image transformation, watermark embedding, and watermark recovery.

The HVS has identified by several phenomena which is related with spatial resolution, intensity resolution, and intensity sensitivity and so on. Let us discuss in more detail the characteristics of these visual properties of HVS which can be described in many terms. Contrast sensitivity means variance or difference of the pixel's brightness. The more contrast property the coefficients region has, the stronger the embedded watermark could be. Contrast sensitivity Cs can be expressed as

$$Cs = \frac{I_{max} - I_{min}}{I_{max} + I_{min}} \tag{1}$$

Where I_{max} and I_{min} denotes the max. and min. brightness for selected region respectively.

The variance of the pixels in sub-image influence the texture, and texture sensitivity means the perceptibility of sine wave. The more complex the background is, the larger the watermark could be. Therefore, texture sensitivity Ts is calculated by DCT coefficients of an image. Namely, quantization results of the DCT coefficients are the same as integer and finally, Ts can be expressed by

$$Ts = \sum \{rnd[X_k(x,y)/Q(x,y)]\} \tag{2}$$

Where $X_k(x, y)$ means the $k'th$ DCT coefficients block and (x, y) refers to the location.

Entropy means the mathematical expectation of the information with respect to the occurrence of the events. When it comes to increase the entropy, watermarking region can be extended. Entropy sensitivity Es can be defined as

$$Es = \sum_{x,y=0}^{7} P_k(x,y) \log \frac{1}{P_k(x,y)} \tag{3}$$

$$p(x,y) = \frac{X_k(x,y)}{\sum_{x,y=0}^{7} X_k(x,y)} \tag{4}$$

Not only these HVS but also statistical parameters such as average, standard deviation are also utilized to generate an effective watermarking. These functions are given by

$$Av_k = \left\{ \sum_{x,y=0}^{7} X_k(x,y) \right\} / size(X_k) \tag{5}$$

$$Std_k = \sqrt{\left\{ \sum_{x,y=0}^{7} (a(x,y) - Av_k)^2 \right\} / size(X_k)} \tag{6}$$

3. The Proposed Method

In the transform-based, embedding the watermark is similar to the multiplicative embedding rule, which can be expressed by

$$O_i^{'} = O_i \cdot (1 + \beta W_i) \tag{7}$$

Where O' and O imply the watermarked material and the original counterpart, respectively, W denotes the watermark, i represents the positions to be embedded, and β is the gain factor. For the watermark with length L, $i \in [0, L-1]$. In order to detect or verify the watermark pattern, a correlate function is often used for full extraction of the watermark. The correlation $(R_{O'',W})$ between the possibly attacked image O'' and the watermark W can be calculated by

$$R_{O^a,W} = \frac{1}{L}\sum_{i=0}^{L-1} O_i^a \cdot W_i \qquad (8)$$

The main issue, in this approach, is to incorporate the HVS with the transform domain watermarking algorithm so as to select the optimal parameters due to the characteristics of original image. Watermarking embedding and extraction process are shown in Fig. 1.

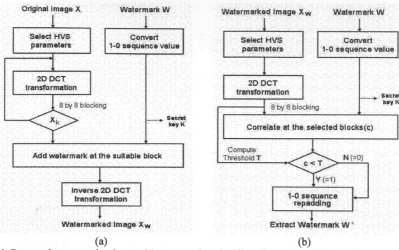

(a)   (b)

Fig. 1. Proposed watermark scheme: (a) watermark embedding, (b) watermark extraction

The main issues of watermarking schemes focus on how to reserve imperceptibility as well as robustness by utilizing HVS [5]. One important feature of our algorithm is it's capability to adaptively calculate the watermark strength and the number of region to be watermarked. We have already reported some results based on the fuzzy inference and the DWT domain [6]. In our approach, two main stages including DCT pre-processing and fuzzy inference can be stated as follows:

Step 1. Definition:
   DCTc: DCT coefficients, GDCTc: Group of DCTc
   MVoDCTc: Maximum variance of DCTc, FIS: Fuzzy inference system
   Sf: Similarity factor for FIS, Sw: Strong watermark, Ww: Weak watermark
/* stage for obtaining DCTc */
Step 2. Preprocessing
      100 : Perform the 2D DCT
Step 3. while image block is not empty repeat
Step 4. Grouping DCTc into 3-level
      Calculating MVoDCTc
Step 5. if MVoDCTc is larger than TH
      goto 100
/* stage for fuzzy inference */
Step 6. Repeat step 7 - step 10.
Step 7. for each value of GDCTc applying FIS
      given fuzzy association map
/* stage for embedding watermark */
Step 8. Find the defuzzification value.
Step 9. Find the Sf
Step 10. if Sf >then embedding a Sw
      else Ww
Step 11. End of algorithm.

4. Experimental Remarks

In order to test the proposed scheme, the watermark image of size 32*32 is generated. Several common numerical processing and geometric distortion were applied with respect the gray scale standard images. To evaluate the imperceptibility of the watermarked image, the peak signal-to-noise ratio (PSNR) is used and, a correlation function is also used for full extraction of the watermark. Experimental results showed that the proposed approach is robust to several signal processing schemes, including JPEG compression, Gaussian noise, histogram equalization and geometric cropping. Fig. 2 shows the response with our method in various attacks. In table 1, meaningful results are illustrated with respect to some HVS parameters, and to some attacks. Finally, regarding the future research, it will be devoted to investigate the trade-off point which is the best case of relations between HVS and various attacks.

Table1. Experimental results with respect to standard images

Attack HVS	Lena Image			Boat Image		
	Loss 1/4 of image	JPEG Q = 50	Gauss. noise mean = 0 Var.= 0.01	Loss 1/4 of image	JPEG Q = 50	Gauss. noise mean = 0 Var.= 0.01
Contrast	0.972	0.890	0.975	0.732	0.894	0.962
Texture	0.963	0876	0.968	0.966	0.872	0.948
Entropy	0.957	0.869	0.917	0.970	0.861	0.908
Std.	0.968	0.884	0.924	0.982	0.911	0.975

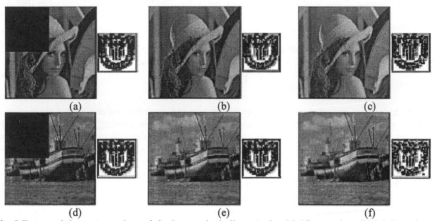

(a)   (b)   (c)

(d)   (e)   (f)

Fig. 2 Extracted the watermarks and the images including attacks: (a),(d) cropping, (b),(e) Gaussian noise, (c),(f) JPEG compression(Q=50)

References

[1] K. Tanaka, Y. Nakamura and K. Matsui, "Embedding Secret Information into a Dithered Multi-level Image," In Proceedings of the 1990, IEEE Military Communications Conference, pp. 216~220, 1990.

[2] R. B. Wolfgang and E. J. Delp, "A Watermark for Digital Images," Proc. of IEEE Int. Conf. on Image Processing, Vol. 3, pp. 219-222, 1999.

[3] I.J. Cox, J. Kilian, T. Leighton, et al., "Secure Spread Spectrum Watermarking for Multimedia," IEEE Trans. On Image Processing, Vol. 6, No. 12, pp. 1673-1687, 1997.

[4] G. C. Langelaar, I. Setyawan, and R.L. Lagendijk, "Watermarking Digital Image and Video Data", IEEE Signal processing magazine, Sep., 2000.

[5] D. C. Lou and T. L. Yin, "Adaptive digital watermarking using fuzzy logic techniques," Optical Engineering, Vol. 41, No. 10, Oct., 2002

[6] Y. H. Kim and H. H. Song, "An Adaptive Digital Watermarking using DWT and FIS," KDCS, Vol. 5, No 2, Jun., 2004.

Brill Academic Publishers
P.O. Box 9000, 2300 PA Leiden
The Netherlands

*Lecture Series on Computer
and Computational Sciences*
Volume 4, 2005, pp. 1702-1705

A Study on Safe Authentication Processing in MIPv6 Multicast

Hoon Ko[1], Uijin Jang[2], Seonho Kim[3]

Department of Computer Engineering,,
Daejin University.
Pocheon-Si, Gyeonngi-do, Korea.
Technical Manager/R&D Strategy Planning Team,
Digicaps Inc..
Bangbae-dong, Seocho-gu, Seoul, Korea
Seoul Metropolitan Fire & Disaster Management Department.
Jung-gu, Seoul, Korea

Received 15 July, 2005; accepted in revised form 12 August, 2005

Abstract: A concern in developing the various devices and service has been increased according to various requests of the Internet applications and development of broadband network. Especially, a structure of high-speed network combining wired with wireless and mobile multicast service based on multimedia-oriented service has been increased. But, a possibility according to flow of private information has been increased owing to impossibility of service according to shortage of IP, hacking attack and damage. A design purpose of mobile IPv6 uses the security, but on the other hand it uses the functions of IPv6. This paper proposes an authentication scheme for secure data transmission to each member in mobile IPv6 multicast environment.

Keywords : Authentication, Mobile, IPv6, Multicast, Security

1. Introduction

The new services are appeared as applying to the finance and securities according to activation of e-commerce using the Internet. Multicast is applied to many Internet services due to minimizing the bandwidth of network. Especially, it is applied to Internet multimedia conference, multimedia education and multimedia instruction. If a new member joins a group and a existing member leaves group, the group key or session key used by members has to update. But, if many groups continually process a key update phase, they have a big problem in normal service and group enlargement. A consideration items for studying and designing the secure multicast is the offer of authentication, access control, security, integrity and non-repudiation[6][7][8]. In situation, a group with frequently joining and leaving such as active multicast group and service using an existing group, if a member leaves a group, a new group key is updated and transmitted because a group key is nominal.

Therefore, this paper proposes a member authentication scheme for receivers, who are participating in a group and joining a new member by focusing on group management and data security[4]. This paper consists of as following. Chapter 2 illustrates the problems of mobile IPv6 multicast and chapter 3 proposes a secure authentication processing of mobile IPv6 for solving the problem. Chapter 4 shows a performance evaluation result of proposed authentication protocol and chapter 5 comes to a conclusion.

2. A Problems of Mobile IPv6 Multicast

Mobile IPv6 multicast has the weak points from the security of binding update. Binding update message is protected by IPSec before. But, the global PKI(Public Key Infrastructure) has to construct for strong authentication of binding update by this method[1][2]. This method is not possible and de-

---

[1] E-mail: simos-editor@uop.gr,

[2] E-mail: neon@digicaps.com

[3] E-mail: shkim2005@fire.seoul.kr

empasized, on the other hand, a feasible authentication method is used such as creating the BSA(Binding Security Association)[2]. For exchanging the binding update between MN and CN, a security problem occurred in binding update is as follow. (1) When MN transmits a binding update message to HA, a attacker can give a different location information with current location of MN. If HA receives the wrong information, MN cannot receive a packet. Therefore, it receives a packet, which is not wanted by other node[6]. (2) When a binding update message is transmitted to CN, a malicious MN can give the wrong information by setting up own HoA as HoA of a subject host to attack. If CN receives the information, the availability and confidentiality of MN are threatened. Because a packet, which is transmitted from CN to a subject host to attack, will be passed time[6].

3. A Secure Authentication Method in Mobile IPv6

(1) RR(Return Routability) Method with Strengthening Security

RR is the confirmation process whether the value of HA and CoA of requested MH is correct value or not. But, the greatest weakness of RR method is that an attacker transfers wrong information between MH and CN. So, it interrupts a secure negotiation. For solving this problem, the ID information and subgroup key of MH, that is, SGK are added to the message transferred between MH and CN. [Figure 1] shows a process step of SRR authentication and the process is as follows. A function for the Hash value uses SHA1.

Figure 1 : Model of SRR(Safe Return Routability)

[STEP 1] Send HoTI, CoTI to CN;

[STEP 2] Send HoT, CoT to MH;

[STEP 3] Kbm=SHA1(home keygen token|care-of keygen token|IDn|SGKn)

[STEP 4] Send BU to CN;

[STEP 5] Send BA to MH;

(2) A Member Authentication Protocol of Group Manager

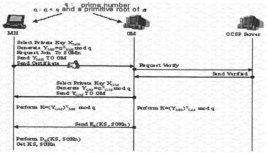

Figure 2 : Member Authentication of GM

Figure 2 illustrates a member authentication process of group manager. First, a member which wants to join a group, provides a group manager with personal information. And then it has to acquire the subgroup key and session key of correspondent subgroup after authentication. This paper proposes the application of DH algorithm.

4. A Result of Performance Evaluation for Authentication Protocol

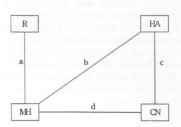

Notation	Content
ρ	λ/μ value
λ	Data Transmission Rate from CN to MH
l_c	Average length of Control Packets
l_d	Average length of Data Packets
l	l_d/l_c Rate
r	Average Processing Time to Control Packet in A Host
C_{auth}	Total Time during move between Domains and Sub networks
C_{nh}	Authentication Time of MH at New Host
C_{dt}	Packets retransmission Time from CN to MH thru HA
C_{of}	Transmission Time during Retransmission Authentication
C_n	Processing Time at each Step
t_r	Processing Time of Control Packets
t_{auth}	Time of Authentication delay

Figure 3 : SRR Model for Cost Analysis

A standard for performance evaluation introduces a cost function and it is calculated by hop counts of node and processing time in each node. But, the unit about distance between nodes and processing time is different, so we calculate a cost function by converting process time to distance. Figure 3 illustrates the SRR model for experiment of cost analysis. A total cost occurred while MH moves from one domain to other domains, illustrates the sum of cost consumed for authentication and retransmission cost, which is occurred by retransmitting the loss packet transmitted for exiting domain in general authentication process.

$$C_{auth}=C_n+C_{of} \qquad (1)$$

An authentication cost of MH from new host is the sum, which is the distance between nodes and cost processed by each node. It is illustrated by expression (2).

$$C_n=(a+b+c+d)+r \qquad (2)$$

Assume that a cost consumed for retransmission of packet from CN to MH through HA is C_{dt}. The retransmission cost according to the loss of data packet, which is transmitted to exiting network for authentication delay, is the cost about packet transmitted for authentication time. It is expressed as follows.

$$C_{of}=\lambda \times t_{auth} \times C_{dt} \qquad (3)$$

C_{dt} in RR of mobile IP is expressed as follows.

$$C_{dt}=l(b+c)+r \qquad (4)$$

Above expression means the cost of single data packet from HA to MH through CN using tunneling. Also, authentication delay time is as follows.

$$t_{auth}=(t_a+t_b+t_c+t_d) \times t_r \qquad (5)$$

Therefore, the total authentication time using expression (2), (3) and (4) is as follows.

$$C_{auth}=(a+b+c+d) \times r+\lambda \times t_{auth} \times C_{dt} \qquad (6)$$

Figure 4 shows the comparison result in case users are 50, 150, 200, and 300 persons using expression (6).

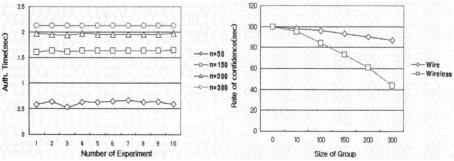

Figure 4 : Time of Authentication   Figure 5 : Rate of Confidence

It shows the result of SRR authentication time and it shows the fast authentication process as the size of group is smaller. Figure 5 is the experimented result of reliability by group size. Finally, as a size of group is bigger as a data loss is bigger because high authentication time is needed.

5. Conclusion

This paper proposes a fast authentication method for adding a group environment based on IPv6 mobility support, which is the mobility support specification of IETF, and guaranteeing a security of

MH. First, we examine a characteristic of authentication about MH based on Mobile IPv6 mobility computing environment closely and applies hash algorithm to authentication process for fast authentication processing. We constitute hierarchical group for quick service of group environment. Also, the fast process of authentication is possible as the subject of authentication process is the manager of sub group. We apply the one-way method with characteristic of hash algorithm to authentication process for fast authentication process rather than existing process. It is possible to secure process than existing process as the ID value of MH and KS, SGK_n are added to input element. We can find the malicious attacker by comparison of hash value using the result of this paper and we can defend against retransmission attack by user authentication. But, as a size of group is bigger as the big difference between authentication rate and data transmission rate as we confirm the result by test. The damage can be occurred due to time difference in case of necessity of realtime processing such as the Internet auction. Therefore, a new alternative plan for overcoming an authentication time difference according to group size is needed.

References

[1] Hoon Ko and Yongtae Shin, A Study on Secure Group Transmission in Group Environment, in *Proceedings of APIS'04*, pp.270-277, Jan. 2004.

[2] Hoon Ko, Uijnin Jang and Yongtae Shin, One-way Authentication Protocol to mobile IPv6 communication Systems, in *Proceedings of PARA'04*, pp.341-348, Jun. 2004.

[3] J. Huang and S. Mishra. Mykil: A highly Scalable and Efficient Key Distribution Protocol for Large group Multicast, *IEEE 2003 Global Communications Conference*, vol.3 pp. 1476-2480, 2003

[4] S. Mishra. Key management in large group multicast. Technical report CU-CS-940-02, Department of Computer Science, University of Colorado, Boulder, CO., 2002

[5] H. Liu and Magada, Zarki, Data and Synchronization Control Middleware to Support Real time Multimedia Services over Wireless PCS Networks, *IEEE journal Communication*, Volume 17, Number 9, pp. 1660-1672, 1999

[6] R. Canetti and B.Pinkas, A taxonomy of Multicast security issues, draft-irtf-smug-taxonomy-01.txt, August., 2000

[7] Pekka Pessi, secure Multicast, *Proc. of Helsinki University of Technology Seminar on Network Security*, 1995.

[8] M. J. Moyer, J. R. Rao and P. Rohotgi, A Survey of Security Issues in Multicast Communications, *IEEE Network, November/December*, 1999.

[9] C. K. Wong, M. Gouda, S.S. Lam, Secure Group Communication Using Key Graphs, *Proceedings of CMSIGCOMM'98*, 1998.

[10] S. Mittra, Iolus : A Framework for Scalable Secure Multicasting, *Proceedings of ACM SIGCOMM'97*, 1997.

[11] G. Caronni, M. Waldvogel, D. Sun and B. Plattner,'Efficient Security for Larget and Dynamic Multicast Groups, *Proceedings of 7th Workshop on Enabling Technologies,(WETICE '98), IEEE Computer Society Press*, 1998.

[12] Jose. L., Munoz Jore and di Forne Juan C. Castro, Evaluation of Certificate Revocation Policies : OCSP Vs Overissaed-CRL, in *Proceedings of IEEE International Symposium on Computer Communication*, pp.710-717, Jun. 2000.

[13] P. McDaniel, A. Prakash and P. Honeyman, Antigone:A Flexible Framework for Secure Group Communication, *Proceedings of th 8th USENIX Security Symposium*, pp.23-36, August, 1999.

[14] M. Handley and V. Jacobson, SDP:Session Description Protocol, IETF RFC 2327, 1998.

[15] M.Handley, C. Perkins and E. Whelan, SAP:Session Announcement Protocol, IETF RFC 2974, 2000.

[16] P. S. Kruus and J. P. Macker, Techniques and Issues in Multicast Security, *Proc. IEEE MILCOM*, 1998.

Brill Academic Publishers
P.O. Box 9000, 2300 PA Leiden
The Netherlands

*Lecture Series on Computer
and Computational Sciences*
Volume 4, 2005, pp. 1706-1707

On Defining Security Metrics for Information Systems

A.H. Koltuksuz[1]

Department of Computer Engineering,
Faculty of Engineering,
Izmir Institute of Technology,
35430 Urla, Izmir, Turkey

Received 30 June, 2005; accepted in revised form 18 July, 2005

1. Introduction

It is quite evident that the need for a security metrics grows rapidly as the security budgets are becoming demanding more than ever. Apart from already known security metrics for cryptosystems such as FIPS 140-1/2 we have SSE-CMM which stands for Systems Security Engineering-Capability Maturity Model. [1] A very good example to the efforts for creating the security metrics there is this one and a half year old security metrics consortium known as the "SECMET" [2]. There are other numerous security metrics projects going on. [3]

According to the International Systems Security Engineering Association the good metrics are those that are specific, measurable, attainable, repeatable, and time dependent. [4]

A security metrics model was defined to consist of three components which are
- the object being measured,
- the security objectives the object is being measured against, and
- The method of measurement. [3]

Although many models to create the security metrics have been proposed the question is whether it can be done or not. In other way of saying is it possible to define a security metric for the information systems?

The rest of this paper will try to answer the above question. In this regard, it is organized as follows; Section 2 will consider the security metrics rationale for cryptography. Section 3 discusses the problems of security metrics for the information systems. Conclusions will delineate the proposed answer.

2. Security Metrics Rationale for Cryptosystems

The Information Theory provides us with a measurement unit for syntactic information which is known as the entropy. The entropy is a measure of uncertainty of a random variable.
Definition. The entropy $H(X)$ of a discrete random variable X is defined by

$$H(X) = -\sum_x p(x) \log p(x)$$

And since the log is to the base 2, the entropy is expressed in bits. [5] This also means that the amount of information in a message is thus measured by the entropy of the message. Shannon defined the perfect security as

$$P_C(M) = P(M) \text{ where;}$$

$P_C(M)$ be the probability that a message M was sent given that C was received, with $C=E(M)$, and $P(M)$ is the probability that message M will occur. [6] [7]

---

[1] Corresponding author. Information Systems Strategy and Security Lab. of IZTECH. E-mail: ahmetkoltuksuz@iyte.edu.tr

While the symmetrical cryptosystems are based on Shannon's Information Theory and thus have an explicit security metric as shown above, the asymmetrical cryptosystems bear the crypto secrecy which actually refers to the concept of intractability.

So far, we have demonstrated that we have a statistically/mathematically proven security metric only for symmetrical cryptosystems.

3. Security Metrics Problem

Defining a security metric for an information system other than the symmetrical cryptosystem is very hard due to fact that there is neither a mathematically proven theory nor any definition for semantic information. This furthermore means that we do not know what we are trying to measure. So now, some questions can be asked such as:

- What exactly is semantic information?
- How can we measure it? In what unit?
- Is it continuous or discrete?
- Is there any proof that it is discrete? Or continuous?
- Is it deterministic or stochastic?
- Would it be possible to process it in a finite state machine if it is continuous and/or stochastic?
- How many dimensions will be needed for the definitions if it is continuous?

4. A Solution Attempt

Although the above asked questions have been circulating around for sometime there seem no clear cut answers as yet. One possible answer as to why not might be due to fact that all of our attempts to define the unit for semantic information and even to define the information itself stems from three dimensional Euclidean geometry.

Trying to define information and/or knowledge in a three dimensional space as a scalar entity is not fruitful. And yet even though the fourth dimension is quite known since Riemann we have yet to include it in our definitions for information and/or knowledge. So the option left is to redefine information in a higher-dimensional space. Here the author's proposal is to conceive the information as a field and thus apply Riemannian tensor analysis to constitute the answers to questions aforementioned. Creating metrics for information security will be an easy task once and if this can be done.

Conclusion

The information and/or knowledge might be a field and if so it should be redefined in a higher-dimensional space by Riemannian tensors. Once it's proven that the information is not a scalar entity then creating security metrics will be possible through tensor analysis.

Acknowledgments

The author wishes to thank the anonymous referees for their careful reading of the manuscript and their fruitful comments and suggestions.

References

[1] SSE-CMM – Metrics; http://www.sse-cmm.org.
[2] SecMet: Security Metrics Consortium; http://www.secmet.org.
[3] S. Katzke, *Security Metrics*, Information Assurance Solutions Group, National Security Agency, USA, 2001.
[4] S. C. Payne, *A Guide to Security Metrics*, SANS Institute, 2002.
[5] T. M. Cover, J. A. Thomas, *Elements of Information Theory*, John Wiley & Sons, USA, 1991.
[6] J. Seberry, J. Pieprzyk, Cryptography: An Introduction to Computer Security, Prentice Hall, USA, 1989.

[7] C. E. Shannon, Communication Theory of Secrecy Systems, *Bell Syst. Tech. J.*, **28**, 656-715(1949).

Brill Academic Publishers
P.O. Box 9000, 2300 PA Leiden
The Netherlands

*Lecture Series on Computer
and Computational Sciences*
Volume 4, 2005, pp. 1708-1711

An Active Key Management Scheme for Conditional Access System on Digital TV Broadcasting System

Han-Seung Koo[1, 2], Il-kyoo Lee[3], Sung-Woong Ra[4], Jae-sang Cha[5], and Kyung-sub Kwak[6]

[2] ETRI, Broadcasting System Research Group, Daejeon, Korea
[3] Division of I&C Engineering Department, Kongju National University, Chung Nam, Korea.
[4] Departments of Electron Engineering, Chung Nam National University, Daejeon, Korea
[5] Seokyeong University, Seoul, Korea
[6] Inha University, Inchun, Korea

Received 27 June, 2005; accepted in revised form 16 July, 2005

Abstract: In this paper, we proposed a novel scheme, which is an active entitlement key management. The proposed scheme not only performs the dynamic entitlement management very efficiently, but also generates and encrypts keys with a small load for CAS, which is just the same as the load of the simple scheme.

Keywords: Conditional Access System, Key Management, Digital TV Broadcasting System

1. Introduction

CAS based on hierarchic key distribution model refreshes keys regularly and irregularly [6]. First of all, CAS refreshes keys regularly because it provides key security and efficient billing. CAS performs efficient billing by synchronizing key refreshment period and service *charging time period* (CTP) [1]. However, since such frequent key refreshment causes a big system load, a trade-off between key security and frequent key refreshment is necessary. This regular key refreshment scheme is called *periodic entitlement management.* Second of all, CAS refreshes keys irregularly when extra key refreshment is necessary. For example, if a user wants to terminate his/her pay service or to change his/her current entitlement to another pay service before the entitlement is originally supposed to be expired, CAS performs irregular key refreshment. In this circumstances, CAS generally refreshes a key related to a channel or service group which a user wants to left, and periodically sends refreshed keys to all users except the one who left his/her entitlement. This irregular key refreshment scheme is called *non-periodic or dynamic entitlement management.* Note that CAS has to send keys periodically because all digital broadcasting standards [7]-[9] specifies one-way system as a mandatory requirement, and two-way system as an optional one. In other words, since CAS can't assure a reception of refreshed keys at a host's side in one-way system, there is no way but to send keys periodically for reliable key transmission. Unfortunately, this mechanism sometimes causes a big system load.

In [1], Tu proposed two kinds of *periodic* and *dynamic entitlement management.* First one is the *simple scheme of four levels hierarchy* [1] (we will call this scheme as the simple scheme in the rest of this paper) based on the *passive entitlement management* consists of *master private key*(MPK), *receiving group key*(RGK), *authorization key*(AK), and *control word*(CW). This scheme is the same as the three levels hierarchy, introduced in section I, except RGK and the fact that head-end CA server refreshes AK and RGK per CTP. Here, RGK is the unique key for each of *receiving group* (RG) [1]-[2], which is a group of subscribers who purchased the same package or channels. The second one is the *complete scheme of four levels hierarchy* [1] (we will call this scheme as the complete scheme in the rest of this paper) is proposed to resolve the problems of the simple scheme when CAS manages the dynamic entitlement. In this scheme, *the receiving group key* matrix, which has the size of $M \times N$, where M and N is the number of charging group [1] and RG, respectively, is used.

An existing solution for *periodic* and *dynamic entitlement management* has a big flaw when it is applied to a big system with tens or hundreds *pay-per-channel* (PPC) and hundreds of thousand or

---

[1] Corresponding author. Member of Engineering Staff in ETRI. E-mail: koohs@etri.re.kr

millions of subscribers. That is a heavy system load for key generation and encryption [1]-[5]. Especially in case of *dynamic entitlement management*, system load problem is getting more serious because a probability of occurring extra entitlement status change events definitely will goes up compared to a small system. This problem is what we resolved with the proposed scheme. With an active entitlement key management proposed in this paper, CAS can handle *periodic* and *dynamic entitlement management* with a small load and securely, even though a system is huge.

2. An active entitlement key management

The proposed active scheme has four levels key hierarchy, such as MPK, RGK, AK, and CW. This key hierarchy model is exactly the same as the complete scheme, but the refreshment period of AK is not charging time unit(CTU), but CTP. Here, the CTU is the refreshing period of AK, and the duration of it could be 24-hour, generally. In the complete scheme, it has to refresh AK per CTU to support dynamic entitlement management because it is based on passive entitlement management scheme. How-ever, our proposed scheme broadcasts ARL to unauthorized subscribers to delete their invalid entitlement, so we don't need to refresh AK when a subscriber lefts his/her entitlement.

ARL is the list of the *record* which includes the identification code of an unauthorized subscriber. We can denote ARL as a group like {*record* 1, *record* 2, ... , *record* M}, where M is the time variant number which varies according to the accumulated number of unauthorized subscribers during a CTP. Each RG has its own ARL, and head-end CA server generates N ARLs, where N is the number of RGs, then broadcasts them to the associated subscribers. We can define a group of ARLs as *ARL set* (ARLS) and denotes it as ARLS={arl_j|j=1,2,...,N}, where the arl_j is the ARL for the receiving group j. And we will use ARL_x notation in the rest of this paper for a arl_j of a certain receiving group. For example, at a subscriber's side which is included in ARL_x, after receiving ARL_x, *security processor* in *access control unit*(ACU) parses the ARL_x which is embedded in the EMM_{AK}, and verifies whether *records* in ARL_x include a matched identification with the ACU. If there is a matched *record* in ARL_x, the *security processor* deletes the entitlement stored in the memory immediately.

One of the reliable transmission schemes of ARLS is periodic transmission of it via EMM_{AK} because there is no way to confirm whether subscribers receive ARLS correctly or not in a one-way digital TV broadcasting system. Note that a two-way digital TV broadcasting system is optional in all type of digital TV broadcasting systems [7]-[10]. The refreshment period of ARLS is like this. Head-end CA server starts to generate new ARLS at every starting point of CTP and discards the old ARLS for previous CTP. And CA server updates ARLS only per a predefined period, i.e., *ARL update period* (AUP) to reduce a system load for ARLS processing because subscribers left their entitlement randomly during a CTP. Therefore, if CA server updates ARLS whenever subscribers left their entitlement, it might cause a system load problem. Thus, in the proposed scheme, CA server temporarily stores incoming subscriber's left of his/her entitlement request and updates ARLS per AUP. CAS takes advantage of EMM_{AK} for periodic transmission of ARLS because CAS originally retransmits EMM_{AK} per *AK retransmission period* (ARP), e.g., 0.1 ~ 15 seconds [2]. As a result, we can transmit ARLS periodically per ARP embedded in EMM_{AK}, and guarantee a reliable ARL transmission with a benefit of such a frequent retransmission period. Beside, we don't broadcast any messages specialized for ARLS, we can reduce the system complexity.

Despite the benefit of ARL update, there is a drawback of periodic ARLS broadcasting for the active scheme. That is an additional bandwidth consumption problem because ARLS is additional information when it is compared to *passive entitlement management*, its periodic transmission surly needs extra transmission bandwidth consumption. Therefore, we also proposed the efficient ARLS transmission scheme based on ARLS transmission table in the followings.

As described in previous section, CAS requires extra transmission bandwidth consumption in the active scheme. Therefore, we propose ARL transmission table to reduce the transmission bandwidth consumption efficiently. There are two different ways of organizing the ARLS transmission table based on the size of ARL_x and the quotient of AUP divided by ARP, i.e., AUP/ARP. First of all, if the size of ARL_x is smaller than the AUP/ARP, head-end CA server organizes the table having a size of AUP/ARP by locating each *record*s of the ARL_x recursively. For example, if the value of AUP/ARP is 5 and the number of *record*s in ARL_x is 2, then the ARLS transmission table for ARL_x will be {$record_1$, $record_2$, $record_1$, $record_2$, $record_1$}. Second of all, if the size of ARL_x is greater than the AUP/ARP, head-end CA server organizes the table in different ways. For example, if the value of AUP/ARP is 5 and the number of *record*s in ARL_x is 7, the ARLS transmission table for ARL_x will be

$$\begin{Bmatrix} record_1 & record_2 & record_3 & record_4 & record_5 \\ record_6 & record_7 & & & \end{Bmatrix}$$

After the completion of organizing ARLS transmission table, CA server broadcasts the tables to subscribers by using two different ways according to the size of ARL_x. First, if the size of ARL_x is smaller than the AUP/ARP, head-end CA server broadcasts each element of the table at every ARP using the EMM_{AK}. For example, if the ARLS transmission table is $\{record_1, record_2, record_1, record_2, record_1\}$, it is broadcasted like case 1 in table 1. Second, if the size of ARL_x is greater than the AUP/ARP, head-end CA server broadcasts each column of the table at every ARP using the EMM_{AK}. For example, if the ARLS transmission table is the same the second example of ARLS transmission organization, it is broadcasted like case 2 in table 1.

Table 1: Example of Broadcasting ARL Transmission Table

ARP timeout order	Case 1	Case 2
1st ARP timeout	Send EMM_{AK} with $\{record_1\}$	Send EMM_{AK} with $\{record_1, record_6\}$
2nd ARP timeout	Send EMM_{AK} with $\{record_2\}$	Send EMM_{AK} with $\{record_2, record_7\}$
3rd ARP timeout	Send EMM_{AK} with $\{record_1\}$	Send EMM_{AK} with $\{record_3\}$
4th ARP timeout	Send EMM_{AK} with $\{record_2\}$	Send EMM_{AK} with $\{record_4\}$
5th ARP timeout	Send EMM_{AK} with $\{record_1\}$	Send EMM_{AK} with $\{record_5\}$

3. Performance and Security Analysis

Table 2 shows the number of times of key generation based on the key refreshment frequency of the complete scheme and the active scheme with T channels, M charging groups, and N receiving groups. Note that, the value of M is the same as the number of days in a month [1], i.e., 30 days. First of all, the complete scheme generates AK T times per CTU because it refreshes AK per CTU, and $T \times M$ times per CTP. In case of RGK, the complete scheme generates RGK $N+f(t)$ times. Here, N is the number of columns in a row of the receiving group key matrix [1], and $f(t)$, $1 \leq t \leq M$, indicates the number of subscribers who left his/her entitlement per CTU. Additionally, the complete scheme generates RGK per CTP $N \times M + \sum_{t=1}^{M} f(t)$ times with the value of M which indicates a total number of days in a month. On the other hand, the active scheme doesn't generate AK and RGK per CTU because it refreshes AK and RGK per CTP, and generates AK and RGK per CTP T and N times, respectively.

Table 2: The number of times of key generation

	The complete scheme [1]	Active scheme proposed
AK per CTU	T	None
AK per CTP	$T \times M$	T
RGK per CTU	$N+f(t)$	None
RGK per CTP	$N \times M + \sum_{t=1}^{M} f(t)$	N

Table 3 shows the number of times of key encryption of AK and RGK with S subscribers, N receiving groups, and M charging groups. First of all, the complete scheme has to encrypt AKs with RGK $N \times M$ times per CTU because there are $N \times M$ packages [1] to be broadcast per CTU. When we consider it for a CTP, the complete scheme has to encrypt AKS with RGK $N \times M \times M$ times because CTP consists of M days. In case of RGK encryption with MPK in the complete scheme, CA system encrypts $S'(t)$ times per CTP, and $S'(t)$ consists of $S + \sum_{t=1}^{M} f(t) + \sum_{t=1}^{M} f'(t)$, $1 \leq t \leq M$, here $f(t)$ indicates the number of subscribers who left his/her entitlement per CTU and $f'(t)$ means the number of subscribers who add his/her entitlement per CTU. On the other hand, the active scheme doesn't need to encrypt AKs with RGK per CTU because it doesn't refresh AKs per CTU, and it encrypts AKs with RGK N times per CTP because there are N receiving groups. In case of RGK encryption with MPK in the active scheme, CA system encrypts $S(t)$ times per CTP, and $S(t)$ consists of $S + \sum_{t=1}^{M} f'(t)$, $1 \leq t \leq M$, here $f'(t)$ means the number of subscribers who add his/her entitlement per CTU.

Table 3: The number of times of key encryption

	The complete scheme [1]	Active scheme proposed
$E_{RGK}\{AKs\}$ per CTU	$N \times M$	N/A
$E_{RGK}\{AKs\}$ per CTP	$N \times M \times M$	N
$E_{MPK}\{RGK\}$ per CTP	$S'(t)$	$S(t)$

One of possible attacks from a hacker is intentional evasion of entitlement message receiving. If a subscriber pulls off ACU from a set-top-box for a while, it is impossible to execute the *active entitlement management*. Therefore, in our scheme, head-end CA server periodically broadcasts ARLS over EMM_{AK} per ARP, and it let the life cycles of ARLS and CTP are the same. As a result, since the ARLS is broadcasted once per AUP at least, a subscriber can't resist receiving the ARLS if he tries to watch pay service during that CTP. Besides, even though a subscriber tries to wait until the CTP is over, he will fail to watch pay service because the RGK is already refreshed at the beginning of the next CTP.

4. Conclusion

In this paper, we proposed an active entitlement key management scheme for CAS on digital TV broadcasting system. We not only introduced a novel concept of ARL for dynamic entitlement management, but also designed key distribution model, including ARL, based on four levels key hierarchy, ARL authentication scheme for secure transmission of ARL, and ARL transmission table for efficient transmission bandwidth consumption. With the proposed scheme, we can reduce key generation and encryption load considerably compared to the complete scheme. Further, we can manage randomly changed users entitlement status securely and efficiently with the proposed scheme. We simulated this remarkable performance improvement by comparing the active scheme and the complete scheme with assumptions of one million subscribers, and one hundred PPC and receiving groups.

Acknowledgments

This research was supported by University IT Research Center Project (INHA UWB-ITRC), Korea

References

[1] F. K. Tu, C. S. Laih, and H. H. Tung, "On key distribution management for conditional access system on Pay-TV system," *IEEE Trans. on Consumer Electronics*, Vol. 45, No. 1, Feb. 1999, pp. 151-158.

[2] H. S. Cho, and S. H. Lee, "A new key management mechanism and performance improvement for conditional access system," *Korea Information Processing Society Journal C*, Vol. 8, No. 1, Feb. 2001, pp. 75-87.

[3] W. Lee, "Key distribution and management for conditional access system on DBS," *Proc. of International Conference on Cryptology and Information Security*, 1996, pp. 82-86.

[4] T. Jiang, S. Zeng, and B. Lin, "Key distribution based on hierarchical access control for conditional access system in DTV broadcast," *IEEE Trans. on Consumer Electronics*, Vol. 50, No. 1, Feb. 2004, pp. 225-230.

[5] ITU Rec. 810, *Conditional Access Broadcasting Systems*, ITU-R, Geneva, Switzerland, 1992.

[6] B. M. Macq, J-J and Quisquater, "Cryptology for digital TV broadcasting," *Proceeding of the IEEE*, 1995, pp. 944-957.

[7] ATSC Std. A/70A, *Conditional Access System for Terrestrial Broadcast, Revision A*, ATSC, Washington, D. C., 2004.

[8] ETSI TS 103 197, *Head-end implementation of DVB SimulCrypt*, Sophia Antiplis Cedex, France, 2003.

[9] ANSI/SCTE 40, *Digital Cable Network Interface Standard*, SCTE, Exton, PA, 2004.

[10] H. S. Cho, and C. S. Lim, "DigiPass: Conditional access system for KoreaSat DBS, *The Institute of Electronics Engineers of Korea Society Papers*, Vol. 22, No. 7, July 1995, pp. 768-775.

Brill Academic Publishers
P.O. Box 9000, 2300 PA Leiden
The Netherlands

*Lecture Series on Computer
and Computational Sciences*
Volume 4, 2005, pp. 1712-1715

An Efficient and Secure Traitor Tracing Method for T-Commerce

Deok-Gyu Lee[1], Im-Yeong Lee[2]

Division of Information Technology Engineering,
Soonchunhyang University,
#646 Eup-nae ri, Shin-chang myun, A-san si, Choongchungnam-do, Korea

Abstract: Broadcast encryption has been applied to transmit digital information such as multimedia, software and paid TV programs on the open networks. One of key factors in the broadcast encryption is that only previously authorized users can access the digital information. If the broadcast message is sent, first of all, the privileged users will decode the session key by using his or her personal key, which the user got previously. The user will get the digital information through this session key. As shown above, the user will obtain messages or session keys using the keys transmitted from a broadcaster, which process requires effective ways for the broadcaster to generate and distribute keys. In addition, when a user wants to withdraw or sign up, an effective process to renew a key is required. It is also necessary to chase and check users' malicious activities or attacking others. This paper presents a method called Traitor Tracing to solve all these problems. Traitor Tracing can check attackers and trace them. It also utilizes a proactive way for each user to have effective renewal cycle to generate keys.

Keywords: Traitor Tracing, T-Commerce, Broadcast Encryption

1   Introduction

With the development of digital technology, various kinds of digital content have been developed. The digital information can be easily copied which has created a lot of damages so far. For example, such digital information stored in CDs or diskettes can be easily copied using CD writers or diskette drivers. The digital information can also be easily downloaded and shared on the Internet. Under these circumstances, broadcast encryption has been applied to transmit digital information such as multimedia, software and paid TV on the open networks. The public key method, one of ways to provide a key, has two kinds of keys; an encoding key of a group to encode a session key and a lot of decoding keys to decode the session key. Therefore, a server can encode the session key and each user can decode the session key using different keys.

One important thing for the broadcast encryption is that only previously privileged user can get the digital information. When a broadcast message is delivered, a privileged user can get the digital information using his or her personal key which the user got previously. The most important one for the broadcast encryption is the process to generate, distribute and renew keys. However, even for the broadcast encryption, illegal users can not be traced if the users use keys maliciously. The method to trace illegal users is to trace them with an hidden simple digital information. In this paper, users who are involved in or related to piracy will be called as traitors, and a technique to trace such traitors as traitor tracing. In this paper, when to trace traitors, proactive techniques are applied to a key which is provided to the users to protect illegal activities beforehand. When a user is involved in illegal activities, an effective way to trace the traitors is presented. This means a new way to trace traitors. In the suggestion method, the broadcast message is composed into block right and block cipher. When a personal key is provided after registration. the key can be transformed effectively from the attackers. In addition, it is designed to sort traitors from privileged users more effectively. After summarizing the broadcast encryption and traitor tracing, this paper will explain each step of suggestion method. Through the suggestion method, we will review the suggestion process and make an conclusion.

[1] Corresponding author. Division of Information Technology Engineering. Soonchunhyang University. E-mail: hbrhcdbr@sch.ac.kr

[2] Division of Information Technology Engineering. Soonchunhyang University. E-mail: imylee@sch.ac.kr

This research was supported by MIC(Ministry of Information and Communication), Korea, under the ITRC(Information Technology Research Center) support program supervised by the IITA(Institute of Information Technology Assessment)

2   Proposed Scheme

The suggestion method is composed of a content provider and n users. Each user will receive one's personal keys. The personal key is required to decode the session key included in the block right. Three personal keys will be transmitted to each user. Through the synchronizing process between the users and the content provider, the broadcasted message will be decoded by each personal keys.

Let assume that all of three users are A, B, and C. The keys transmitted to A are i-3, i-2 , and i-1. For B, they are i-1, i, and i+1 and for C, i+1, i+2, and i+3. There are two purposes to distribute triple keys for each user. First, the key of each user will be used interchangeably so that the attacker cannot user the same key with the users. Secondly, when the user uses the keys illegally, the triple keys will be shifted overall.

When the keys are used illegally, the keys will form an intersection set which will enable to trace the traitors correctly. The steps and assumptions of the suggestion method are summarized in the next part.

- Initial Stage: This is a step to assume the system variables. A content provider will assume the number of participants, and generate public keys and personal keys. The number of keys will be three times more than the number of users.
- Registration Stage: This is a protocol between a content provider and the users who want to receive digital information. The user will get a couple of keys from the content provider.
- Broadcast message encoding stage: This is a stage to generate encoded broadcast message. The user will synchronize a key to use with the content providers. If the transmitted broadcast message is used illegally, the decoded key included in the illegal decoder will be one of the personal keys distributed to each user from the content provider.
- Decoding Stage: Each user will get the digital information from the broadcast message through the decoding process using his or her personal key. At this point, the user can not distinguish his or her own keys. The user can simply decode using his or her keys.
- Key renewal Stage: The content provider will broadcast the related information with the key change using the value of new polynomial. Each user will change his or her personal key using the transmitted value. At this point, the user will perform activities with his or her triple keys, which will be processed automatically.
- Tracing traitors and withdrawal Stage: When a content provider finds pirate decoder, the provider will continue the key renewal process by entering the broadcast message into the decoder to find traitors. The pirate decoder cannot distinguish whether it is to renew the key or to trace traitors. Also, among the triple keys of illegal users, that user who has two keys will be definitely traitors.

2.1   Characteristics of proposed scheme

The suggested method has the following characteristics. First of all, the content provider will insert renewal factors to renew easily the user's keys to set up the system algorithm and related variables, and to keep safely the key of users from illegal users. The users will register to the content providers later. At this point, the allocated personal keys will be transmitted.

When the content provider wants to broadcast a message, only the users who has the encoded encryption for broadcasting data and privileged personal keys can acquire the session keys to constitute a block right. The user can decode using his or her personal keys to acquire messages.

At the stage to transmit and to decode the encoded message, it is required to synchronize messages to use keys between users and servers. When a content provider discovers a traitor and wants to find out a privileged user, and the privileged users decode the broadcasted data after combining his or her personal keys into un-privileged keys, the content provider can find out the used personal keys by checking the relationship between the input and output of the messages.

We assume that the maximum illegal usage by users is k, extracted critical value from traitors is z (because the number of personal keys of traitors is three, the user can be found out exactly.) and the group with the largest decimal order q is G_q .

2.1.1   The initial stage and the process of encoding and decoding the broadcasted

message

Step 1. The content provider selects random number (β) after predicting users $(i = 1,\dots,2n+2)$. The random number at this point will be used as a renewal factors to renew users' keys.

Step 2. The content provider selects the polynomial $f(x) = \sum_{t=0}^{z} \beta_t \alpha_t x^t$ having coefficient z on the

Z_q space. The content provider will publish the public keys after making the polynomials into the secrete keys just as the following.

$\left\langle g, g^{\beta_0 \alpha_0}, g^{f(1)}, \ldots, g^{f(z)} \right\rangle$ Published.

Step 3. When the content provider registers users, he pr she will transmit personal keys $\left((i, f(i-1)), (i, f(i)), (i, f(i+1)) \right)$ to the users. The user will try to verify the correctness of his or her keys. At this point, the content provider will induce to verify with the first key.

$$g^{\beta_0 \alpha_0} = \prod_{t=0}^{z} g^{f(x_t) \lambda_t}, x_0 = 1, x_2 = 2, \ldots, x_{z-1} = z, x_z = i$$

When this equation is verified, the user will acquire personal keys.

Step 4. The content provider calculates the block right after selecting the unused information and the random number $r \in Z_q$.

$$\left(j_1, f(j_1) \right), \left(j_2, f(j_2) \right), \ldots, \left(j_z, f(j_z) \right), C = \left\langle sg^{r\beta_0 \alpha_0}, g^{r\left(j_1, g^{rf(j_1)} \right)}, \ldots, g^{r\left(j_z, g^{rf(j_z)} \right)} \right\rangle$$

After encoding the message into session keys, C and encoded message are transmitted.

Step 5. The process to acquire session keys from the C and encoded message from content provider follows below.

$$s = sg^{r\beta_0 \alpha_0} \Big/ \left[\left(g^r \right)^{f(i)\lambda_z} \cdot \prod_{t=0}^{z-1} \left(g^{rf(x_t)} \right) \right], x_0 = j_1, x_2 = j_2, \ldots, x_{z-1} = j_z, x_z = j_i$$

2.1.2 The Key Renewal Stage

Step 1. A user j requests his or her withdrawal to the content provider.

Step 2. The content provider deletes the renewal factor of a user j from the renewal factor β to renew the existing users' personal key.

Step 3. The content provider will transmit to the user the renewed personal keys after deleting the renewal factors of the withdrawn users. Because three pairs of keys are transmitted for each user, a user j can not use the enabling block to utilize the unused sharing information after fixing the sharing information of the three $B = (\beta_{i-1}, \beta_i, \beta_{i+1})$ related to the renewal.

$$\left(\beta_{i-1}, g^{rf(\beta_{i-1})} \right), \ldots, \left(\beta_{i+1}, g^{rf(\beta_{i+1})} \right) \Rightarrow \left(j_1, g^{rf(j_1)} \right), \ldots, \left(j_{z-3}, g^{rf(j_{z-3})} \right)$$

2.1.3 The conspirator tracing stage

The conspirator tracing will apply the suggested method by WGT in the key renewal process. WGT suggested two kinds of conspirator tracing method. We will explain this method with the suggested method.

Step 1. The content provider will constitute the user's set $\{c_1, c_2, \ldots, c_m\}, (m \le k)$ assuming as a traitor.

Step 2. To trace a traitor, the information will be added like the following block right.

$$\left\langle sg^{r\beta_0 \alpha_0}, g^{r\left(c_1, g^{rf(c_1)} \right)}, \ldots, g^{r\left(c_m, g^{rf(c_m)} \right)} \right\rangle$$

The other method is a way only for the pirate users to decode the block right. The content provider will select a new polynomial $h(x)$ which doe not match with $f(x)$ with other solutions, while $\{(c_1, f(c_1)), \ldots, (c_m, f(c_m))\}$ are part of solutions.

After a traitor is discovered, the content provider can make illegal personal keys not to decode the broadcasted data with the keys of traitors. If we assume $\{c_1, c_2, \ldots, c_m\}, (m \le k)$ as a discovered conspiracy by the content provider, the sharing information can be extracted while not changing the personal keys of other users. The content provider will fix the first m of sharing information as block right with $\left(c_1, g^{rf(c_1)} \right), \ldots, \left(c_m, g^{rf(c_m)} \right)$. The other $z - m$ unused sharing information of $\left(j_1, g^{rf(j_1)} \right), \ldots, \left(j_m, g^{rf(j_m)} \right)$ will constitute the block right.

3 Conclusion

Broadcast encryption has been applied for various kinds of digital information such as multimedia, software and paid TV programs on the open internet. The important one in the broadcast encryption is that only the privileged users can get the digital information. When the broadcast message is transmitted, the privileged users can get the digital information using the previously received personal keys. As stated above, the user can utilize the transmitted key from the broadcaster to acquire message and session keys. In this stage, we need a process to generate and share keys.

In addition, effective key renewal will be required to withdraw or sign up. Only the privileged users can receive the information for the broadcasted message. The privileged users can obtain the session keys using previously transmitted personal keys. This paper applied a proactive method to provide keys to users to protect users' illegal activities beforehand when tracing traitors. Based on this scheme, when a user commit an illegal activity, the traitor can be traced effectively.

This means a new way to trace traitor. The broadcasted message is composed of block right, block renewal, and block code. In addition, because the user will receive a personal key during registration, it is designed more effectively to detect traitors from the original users to trace traitors after transforming the keys effectively from the attackers.

References

[1] Amos Fiat and Moni Naor, "Broadcast Encryption", Crypto'93, pp. 480-491, 1993
[2] A. Narayana, "Practical Pay TV Schemes", to appear in the Proceedings of ACISP03, July, 2003
[3] C. Blundo, Luiz A. Frota Mattos and D.R. Stinson, "Generalized Beimel-Chor schemes for Broadcast Enryption and Interactive Key Distribution", Theoretical Computer Science, vol. 200, pp. 313-334, 1998.
[4] Carlo Blundo, Luiz A. Frota Mattos and Douglas R. Stinson, " Trade-offs Between Communication and Storage in Unconditionally Secure Schemes for Broadcast Encryption and Interactive Key Distribution", In Advances in Cryptology - Crypro '96, Lecture Notes in Computer Science 1109, pp. 387-400.
[5] Carlo Blundo and A. Cresti, "Space Requirements for Broadcast Encryption", EUROCRYPT 94, LNCS 950, pp. 287-298, 1994
[6] Donald Beaver and Nicol So, "Global, Unpredictable Bit Generation Without Broadcast", EUROCRYPT 93, volume 765 of Lecture Notes in Computer Science, pp. 424-434. Springer-Verlag, 1994, 23-27 May 1993.
[7] Dong Hun Lee, Hyun Jung Kim and Jong In Lim, "Efficient Public-Key Traitor Tracing in Provably Secure Broadcast Encryption with Unlimited Revocation Capability", KoreaCrypto 02', 2003
[8] D. Boneh and M. Franklin, "AN Efficient Public Key Traitor Tracing Scheme", CRYPTO 99, LNCS 1666, pp. 338-353, 1999
[9] Dani Halevy and Adi Shamir, "The LSD Broadcast Encryption Scheme", Crypto '02, Lecture Notes in Computer Science, vol. 2442, pp. 47-60, 2002.
[10] Ignacio Gracia, Sebastia Martin and Carles Padro, "Improving the Trade-off Between Storage and Communication in Broadcast Encryption Schemes", 2001
[11] Juan A. Garay, Jessica Staddon and Avishai Wool, "Long-Lived Broadcast Encryption", In Crypto 2000, volume 1880 of Springer Lecture Notes in Computer Science, pages 333--352, 2000.
[12] Michel Abdalla, Yucal Shavitt, and Avishai Wool, "Towards Marking Broadcast Encryption Practical", IEEE/ACM Transactions on Networking, 8(4):443--454, August 2000.
[13] Yevgeniy Dodis and Nelly Fazio, "Public Key Broadcast Encryption for Stateless Receivers", ACM Workshop on Digital Rights Management, 2002
[14] R. Ostrovsky and M. youg, "How to withstand mobile virus attacks", in Proc. 10th ACM symp. on principles of Distributed Computation. pp. 51-61, 1991

Brill Academic Publishers
P.O. Box 9000, 2300 PA Leiden
The Netherlands

*Lecture Series on Computer
and Computational Sciences*
Volume 4, 2005, pp. 1716-1719

An Efficient Scheme of Extra Refreshment Management for CA System in DTV Broadcasting

Il-kyoo Lee[1,2], Han-Seung Koo[3], Sung-Woong Ra[4], Jae-sang Cha[5], Jae-Myung Kim[6], and Bub-joo Kang[7]

[2] Division of I&C Engineering Department, Kongju National University, Chung Nam, Korea.
[3] ETRI, Broadcasting System Research Group, Daejeon, Korea
[4] Departments of Electron Engineering, Chung Nam National University, Daejeon, Korea
[5] Seokyeong University, Seoul, Korea
[6] Inha University, Inchun, Korea
[7] Dongguk University, Kyongju, Kyong Buk, Korea

Received 10 July, 2005; accepted in revised form 12 August, 2005

Abstract: This paper introduces a novel scheme for extra key refreshment management for conditional access system in DTV broadcasting. The proposed scheme sends authorization revocation list (ARL) to subscribers, and it deletes an entitlement only for unauthorized subscribers. With this scheme we can reduce a system load remarkably along with supporting extra key refreshment without any performance deterioration.

Keywords: Conditional Access System, Key Management, Digital TV Broadcasting System

1. Introduction

In conditional access system (CAS), key management is very important and challenging work. One of the important functions in key managing is key refreshment management, and it can be classified into regular scheme and extra one [2]-[5]. In case of regular key refreshment management, CAS refreshes keys once per pre-determined duration. With this scheme, CAS can reduce the possibility of key compromising because frequent key changing guarantees strong key security. In addition to the security reason of keys, regular receiving group key (RGK) refreshment in four levels key hierarchy helps CAS to control billing more efficiently because the duration of RGK refreshment is generally set to the unit of charging period of billing which is called charging time period (CTP) [3]-[4].

CAS needs extra key refreshment management because subscribers could left their entitlement randomly during a CTP, or a hacker might compromise keys unexpectedly. A typical solution for the first case is to simply generate new key and send to all subscribers except the one who lefts his/her entitlement. And a solution for the second case is to generate new key and broadcast it to all authorized subscribers. In this paper, we will introduce a novel extra key refreshment scheme for the case of entitlement left which is occurs randomly during a CTP. Note that we will call this extra key refreshment scheme as the dynamic key refreshment scheme in the rest of this paper.

CAS comprises several blocks which are multiplexer (MUX), scrambler, subscriber authorization system (SAS), subscriber management system (SMS), In-Band modulator, and out-of-band (OOB) modem. Among them scrambler, SAS, and SMS are the components of CAS. First of all, SAS has the responsibility for organizing, sequencing, and delivering entitlement management message (EMM) and entitlement control message (ECM) data streams under direction from the SMS [1]. Second of all, SMS is the business center which issues the access control unit (ACU) [5], sends out bills and receives payments from subscribers. An important resource of the SMS is a database of information about the subscribers, the serial numbers of the decoders and information about the services to which they have subscribed. In commercial terms, this information is highly sensitive [1]. Finally, scrambler has a role to scramble the outputs of MUX. Here, In-Band channel means the forward channel which broadcasts audio, video MPEG-2 TS [9] of all broadcasting channels, and OOB channels is the interactive data

---

[1] Corresponding author. Professor. E-mail: leeik@kongju.ac.kr

channel which transmits data, including EMM, service information (SI) [5]-[6], [10], etc. Note that our proposed dynamic key refreshment scheme utilizes MUX, and the detail idea will be described in section 3.

2. Previous Works

Tu's key distribution model has four levels key hierarchy with control word (CW), authorization key (AK), receiving group key (RGK), and master private key (MPK). This scheme classifies subscribers into groups, and the method of classifying group is basis of subscribing the same channels and the same charging day. In Tu's scheme, subscribers are classified into $M \times N$ classes of disjoint subscribing class u_{ij}, where M is the number of charging group, N is the number of receiving group, and $1 \leq i \leq M$; $1 \leq j \leq N$. The subscribers in the same group have the same subscribed channels and charged time.

Tu's scheme utilizes the RGK matrix which has the size of $M \times N$, and each elements of this matrix are rgk_{ij} which is corresponding to each class u_{ij}. With this condition, head-end CAS refreshes one row of the RGK matrix daily and each row is refreshed once per CTP. For example, if subscribing class u_{11} has three subscribers with mpk_1, mpk_2, and mpk_3, then the system updates rgk_{11} and encrypts it with mpk_1, mpk_2, and mpk_3 at the first day of every month, respectively. Any subscriber who is out of authorization date will not receive the renewed keys in RGK. Since AK is refreshed one per 24 hours, those subscribers who cannot get access to the renewed RGK would lose the ability of viewing program [4]. Note that the advantage of using the RGK matrix is to distribute the system load of regular RGK refreshment and broadcasting to every day's work. Since the RGK of each group is refreshed once per a month, there is no advantage in key security aspect which could be obtained from frequent key update.

3. New Dynamic Key Refreshment Management

Our proposed scheme adopts a novel concept which is authorization revocation list (ARL) including identification of unauthorized subscribers. In the proposed scheme, head-end CAS repeatedly broadcasts the ARL to all subscribers through the In-Band channel by taking advantage of MUX. In addition that, if a subscriber who receives this ARL, and has the matched identification with the one in ARL, then the security processor in ACU deletes corresponding entitlement immediately. Therefore, unlike the previous works, the proposed solution doesn't need to generate and encrypt keys whenever the dynamic key refreshment management required. The details are explained in the followings.

ARL is the list of the record which includes the identification code of an unauthorized subscriber. We can denote ARL as a group like {record 1, record 2, ... , record M}, where M is the time variant number which varies according to the accumulated number of unauthorized subscribers during a CTP. And each record has a set of IDs of unauthorized subscribers, i.e., record #m = {ID #1, ID #2, ..., ID #k}. Each receiving group has its own ARL, and head-end CAS generates N ARLs, where N is the number of receiving groups, and then broadcasts them to the corresponding subscribers. We can define a group of ARLs as ARL set (ARLS) and denotes it as ARLS={arl_j|j=1,2,...,N}, where the arl_j is the ARL for the receiving group j.

The overall processing flow at a subscriber's side which is included in ARL_x is as follows. Firstly, security processor in ACU parses the ARL_x which is embedded in the ECM after receiving ARL_x. And ACU verifies whether records in ARL_x include a matched identification with the ACU. And then, if there is a matched record in ARL_x, the security processor deletes the entitlement stored in the memory immediately.

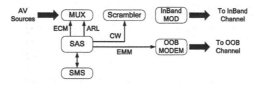

Figure 1: Head-end Architecture, including CAS with the Proposed Scheme

The secure and reliable transmission of ARL is very important in our proposed scheme. Therefore, we take advantage of ECM for transmitting ARL because it is periodically transmitted with a very short time interval, e.g., 0.1 second. And our scheme utilizes MUX for inserting ARL in ECM as shown in figure 1. Originally, CAS utilizes MUX for inserting ECM into MPEG-2 TS audio and video sources with the following steps. Firstly, SAS generates ECM whenever CW is refreshed, and sends ECM to MUX. Here, CW refreshment period would be 15 seconds [8]. Secondly, MUX recursively multiplexes the ECM from SAS with the MPEG-2 TS audio and video sources at every broadcasting period, e.g., 0.1 seconds. The inserting method of ARL into MPEG-2 TS audio and video sources is exactly the same the ECM inserting procedures. In the proposed scheme, MUX receives not only ECM, but also ARL from SAS. And MUX recursively multiplexes those ECM and ARL with audio and video sources from encoders at every broadcasting period, e.g., 0.1 seconds.

As explained before, MUX inserts records into each ECM which are transmitted periodically, e.g., 0.1 seconds. The overall processing scheme is depicted in figure 2. As shown in figure 2, MUX inserts each records using carousel method. MUX recursively inserts record into ECM one by one. For example, if MUX has record #1, record #2, and record #3, it inserts records #1 into the first ECM #1 which is periodically transmitted to In-Band channel, record #2 into the second ECM #1, and so on. Note that, if SAS sends ECM per 15 seconds, and MUX transmits the ECM per 0.1 seconds to In-Band channel, then there will be 150 ECMs will be transmitted before the next ECM is coming from SAS. In this condition, MUX can insert three records 50 times recursively.

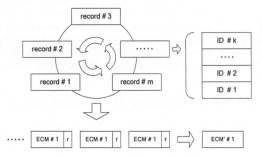

Figure 2: ARL Insertion in ECM on MUX

One of the most important functions in the proposed scheme is the ARL processing in ACU at subscriber's side, and the detail ARL processes are as follows. First of all, ACU receives incoming MPEG-2 TS and parsing them. Second of all, ACU derives ECM from MPEG-2 TS. Third of all, ACU parses ECM and extracts ARL from it. Fourth of all, security processor in ACU checks whether the ID in ARL is the same as the one of mine or not. If there is an ID which is matched with mine, security processor deletes corresponding entitlement in ACU immediately.

4. Evaluations

In our proposed scheme, we used ARL for the dynamic entitlement management. As introduced in the previous section, our scheme doesn't generate any key when a subscriber left his/her entitlement to provide dynamic key refreshment. We just send ARL to subscribers, and make ACU delete its own entitlement if ARL contains the matched identification. Since our scheme utilizes ECM which is periodically transmitted with a very short time interval, we not only guarantee the secure and reliable delivery of ARL, but also reduce the system load of key generation and encryption considerably.

Table 1: The Comparison of Key Refreshment Frequency

Case	Tu's Scheme (AK, RGK)	Proposed Scheme
Regular AK(DEK) Ref.	Per 24 hr	Per 24 hr
Regular RGK(GK) Ref.	Per one month	Per one month
Dynamic AK(DEK) Ref.	No	No
Dynamic RGK(GK) Ref.	Whenever a subscriber left	No

Table 1 shows the comparison of key refreshment frequency between Tu's scheme and the proposed scheme. As it shows, Tu's scheme refreshes each row of the RGK matrix daily, so head-end CAS should generate N+α RGKs and encrypt T AKs with the newly generated RGK per 24 hours, where N is the number of receiving groups, T is the number of PPV channel, and α is additional number of key generation due to the dynamic key refreshment. Note that the number of key generation of RGK in Tu's scheme highly depends on the value of α. In other words, in large system, it could cause considerable system load because the value of α would go high in that system. On the other hand, the proposed scheme doesn't do any dynamic key refreshment when entitlement left event takes place. It just performs regular key management, so CAS based on the proposed scheme can reduce the key generation and encryption load remarkably. Table 2 shows the comparison of the number of times of key generation.

Table 2: The Comparison of the Number of Times of Key Generation

Case	Tu's Scheme (AK, RGK)	Proposed Scheme
AK(DEK) per 24 hr	T	No
AK(DEK) per a month	30T	T
RGK(GK) per 24 hr	$N + \alpha$	No
RGK(GK) per a month	$30N + \alpha$	N

T is the number of PPV channels, N is the number of receiving groups, and α is additional number of key generation due to the dynamic key refreshment. And we consider one month has 30 days

5. Conclusions

We proposed the scheme which performs dynamic key refreshment without new key generation. What makes this possible is the ARL concept. In proposed scheme, instead of generating new key and broadcasting it to subscribers except unauthorized one who left his/her entitlement, head-end CAS periodically broadcasts ARL including the identification of unauthorized subscribers to all subscribers. And the security processor in ACU at a subscriber's side immediately deletes corresponding entitlement which is listed in ARL, if it has the matched identification in ARL with its own.

To broadcast ARL to subscribers securely and reliably, we utilized ECM. Since ECM periodically transmitted with the very short time interval through the In-Band, CAS can guarantee the reliable transmission of it. Besides, since MUX in head-end recursively inserts records of ARL to ECM which is periodically transmitted, the reliability of ARL transmission is very high.

Acknowledgments

This research was supported by University IT Research Center Project (INHA UWB-ITRC), Korea

References

[1] EBU Project Group B/CA, *Functional Model of a Conditional System*, EBU Technical Review, 1995.

[2] ITU Rec. 810, *Conditional Access Broadcasting Systems*, ITU-R, Geneva, Switzerland, 1992.

[3] W. Lee, "Key distribution and management for conditional access system on DBS," *Proc. of International Conference on Cryptology and Information Security*, 1996, pp. 82-86.

[4] F. K. Tu, C. S. Laih, and H. H. Tung, "On key distribution management for conditional access system on Pay-TV system," *IEEE Trans. on Consumer Electronics*, Vol. 45, No. 1, Feb. 1999, pp. 151-158.

[5] B. M. Macq, J-J and Quisquater, "Cryptology for digital TV broadcasting," *Proc. of the IEEE*, 1995, pp. 944-957.

[6] OpenCable™ project, http://www.opencable.com.

[7] ANSI/SCTE 28, *Host-POD Interface Standard*, SCTE, Exton, PA, 2004

[8] H.S.Cho, and S.H.Lee, "A new key management mechanism and performance improvement for conditional access system," *Korea Information Processing Society Journal C*, Vol. 8, No. 1, Feb. 2001, pp. 75-87.

[9] ISO/IEC 13818-1, *MPEG-2 Systems.*

[10] ANSI/SCTE 28, *Service Information delivered Out-of-Band for digital cable television*, SCTE, Exton, PA, 2004.

Brill Academic Publishers
P.O. Box 9000, 2300 PA Leiden
The Netherlands

*Lecture Series on Computer
and Computational Sciences*
Volume 4, 2005, pp. 1720-1726

The Echo Channel Modeling & Estimation Method for Power Line Communication Networks

Jong-Joo Lee[1], ©Myong-Chul Shin[1], Seung-youn Lee[1], Sang-yule Choi[2], Hak-man Kim[3], Hee-seok Suh[4], Jae-Sang Cha[5]

[1]Dept. of Electronic and Electrical Eng., SungKyunKwan Univ., Suwon, Korea
[2]Dept. of Electronic Engineering, Induk Institute of Technology, Seoul, Korea
[3]KERI, 28-1 Sengju-dong, Changwon, Kyung-Nam 641-120, Korea
[4]Dept. of Automation System, Doowon Technical College, Korea
[5]Dept. of Information and Communication Eng., SeoKyeong Univ.,
16-1 Jung-nung dong Sungbuk-ku, Seoul, 136-704, Korea

©mcshin@yurim.skku.ac.kr

Received 24 June, 2005; accepted in revised form 26 July, 2005

Abstract: The power line channel is very complex network topology consist of multi path and variable distance of branches such as distributed nodes, for example, indoor power line network is a branched distribution network and numerous branches. So, the path loss and attenuations, multi paths delay, interference and fading are significant factors of power line channel models. In this paper we presents echo channel modeling technique using pseudo-noise (PN) 63chip sequence for analysis and estimation of power line channels. Using the proposed echo channel modeling and estimation methods, we could build up a more precise power line echo channel models

Keywords: Power line communication, Echo Channel, Channel sounding, Multi-path

1. Introduction

Recently Power Line Communication(PLC) Systems used for communication are highlighted as a communication technique in the Communication Network infra(especially, indoor home network system). The characteristics of the transmission channel of Power Line Communication are as follows. To begin with, power line of network has the various noise characteristics such as background noise and additional impulse noises, etc. On the other hand, various delayed waves generated by multi-path from the transmitter are arrived in the PLC receiver.

Especially, shape of PLC is composed of diverging or expanding distribution network, and these need to consider specific multi-path, piling signals and length of lines causing reduction [1].

Nevertheless, there are only papers researched on alteration and affection of communication-channel showing channel environment of electric-line communication, and there are not enough researches of multi-path caused by distributed power line and branch power line [2][3]. And therefore, on this paper we will study on communication-channels affected by variety of multi-path on electric-line channels caused by distributed power line or branch power line. On this paper we propose new echo channel modeling and estimation method for Power Line Communication Networks. And we confirmed characteristics of impulse responses using defined proposed channel-sound-modeling technique by performance of user defined signal source generator.

©Corresponding author, Dept. of Electronic and Electrical Eng., SungKyunKwan Univ., Suwon, Korea, E-mail: mcshin@skku.edu
[2] Corresponding author, Dept. of Electronic and Electrical Eng., SungKyunKwan Univ., Suwon, Korea, E-mail: mcshin@skku.edu

2. Characteristics Of PLC Channel Model

Channel of electric-line communication have poor communication- and echo-channels by variety of loads operating condition and complex distribution network.

Especially, electric-lines are designed for supplying electrical power and they have characteristics of high attenuation and several of noises (background noise, impulse noise, etc)[4]. Reason is loads that connected on electric-lines always change by time, and diverged electric-line always should consider multi-path affection of reflection wave. Figure 1 shows channel-environment characteristics of electric-line communication listed above [5].

Where At_i is an attenuation factor representing the product of the reflection and transmission factor along the path. The variable τ_i, representing the delay introduced by the path i, is calculated by dividing the path length by the phase velocity.

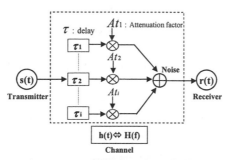

Figure 1. Communication Channel of PLC

3. Proposed Echo Channel Modeling & Estimation Method For Multipath Power Line Characteristics

In this paper, we made simulated test-line to analysis channel characteristics of power-line communication that considers channel characteristics shown above, and used channel-sounding technique(echo channel modeling and estimation method) by simulated test-line to extract Delay Profile which shows delay characteristics of multi-path and echo component and signals caused by branch power line.

Method of extracting Delay Profile is done such as, sample PN63 chip sequence [6] which is easy to catch synchronization, then at transmitting section it makes transmit-signals this sampled PN63 chip, and at receiving section it acquires and collects transmit-signals which went and roam through multi-path channels, after that finally carry out Peak Value by transmit-signals and auto correlation function (ACF).

ACF is expressed as equation shown below [6].

$$R_x(\tau) = \frac{1}{T_0} \int_{-T_0/2}^{T_0/2} pn(t) \cdot pn(t+\tau)dt \tag{1}$$

$$(for \ -\infty < \tau < \infty)$$

Here ACF $R_x(t)$ shows added value of $pn(t+\tau)$ which have time delay value of τ, and power signal $pn(t)$ which have period time value T_0. Figure 2 showed whole flowchart of channel analysis method that is proposed in this paper, and details simulation specification of method to get Delay Profile are as below.

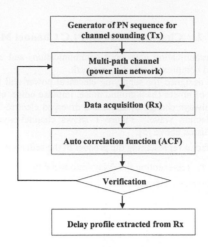

Figure 2. Flowchart of channel analysis method

4. Simulation

4.1 Channel Model Simulation Parameters

Table 1 shows simulation parameters of transmit signals to apply channel estimation technique that proposed in this paper.

Table 1. Conditions for simulation

Items	Contents
Data rate of transmission signal	3MHz
Time delay of transmission signal	3.33×10^{-7} sec per 0.1km
PN code	M-sequence
Spreading factor	63 chip
Modulation	PSK (Phase Shift Keying)
Sampling frequency	150MHz
Attenuation factor	0.97 per 0.1km

4.2 Topology of simulated power line network

To simulate network power-line communication environment, with conditions for simulation above, organized a simulation for multi-path line. Before organizing line, attenuation factor of power-line communication must be considered.

But attenuation factor varies with cable capacity, line factor, values of RLC (the impedance value of actual line) and the frequency of transmit-signals [7], on this simulation we applied 0.1km per 0.97 attenuation factor and simulated it. Path of simulation for Delay Profile extracted from the Figure 2 and we divided blocks □ to □ of simulation for multi-path line. Each section's lengths are □ 0.2 km, □ 0.5 km, □ 1.0 km and □ 1.5 km, and the path which goes through this route is just like below.

- Path A → Length calculation of □: 0.2 km
- Path B → Length calculation of □+□: 1.2 km
- Path C → Length calculation of □+□: 2.2 km

- Path D → Length calculation of □+□+□: 3.2 km
- Path E → Length calculation of □+□+□: 5.2 km
- Path F → Length calculation of □+□+□+□ : 6.2 km

Despite of using the routes shown above, there could be more possible routes and we just putting on representative set up routes of line composition.

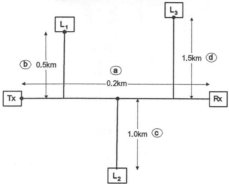

Figure 2. Simulated multi-path power line network

And route □, □, □ represents for line length of returning when setting the distance, because the affection of reflection wave is lower than direction wave, we didn't apply this to the calculation. If there are a lot of diverged-circuits than number of reflection-waves suddenly increase, because of this signal component of reflection-wave's deduction increases and it doesn't affect much. Environmental simulation that is shown above is summarized on table 2.

Table 2. Simulation scenario for delay profile

Path	Length of power line	Time delay	Attenuation factor
Tx	-	0	1.000
Path A	0.2 km	6.66×10^{-7}s (Period delay: 2)	0.941
Path B	1.2 km	3.99×10^{-6}s (Period delay: 12)	0.694
Path C	2.2 km	7.33×10^{-6}s (Period delay: 22)	0.512
Path D	3.2 km	1.06×10^{-5}s (Period delay: 32)	0.377
Path E	5.2 km	1.733×10^{-5}s (Period delay: 52)	0.205
Path F	6.2 km	2.066×10^{-5}s (Period delay: 62)	0.151

4.3 Calculation of Delay Profile using channel-sounding technique

In this simulation, to carry out ACF wave, we made data that refluxed by delay time and attenuation factor based on simulation conditions written above, then correlated-processing changed PN 63 chip 3 sequence cycles at the same time. As shown in Figure 3, we can confirm there are 3 cycles of ACF Peak-value caused by PN 63 chip sequence clearly. Looking at the ACF wave closely, there exist ACF wave components, meaning Delay Profile, made of delay wave constant delay time.

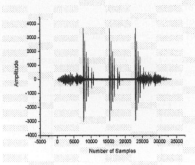

Figure 3. ACF of Simulation result

Figure 4 is a character of Delay Profile carried out from ACF wave of Figure 3. At Figure 4, to get correct timing of ACF wave that of Figure 3, we extracted center of ACF wave then squared the wave and normalized amplitude of Y axis value to 1, and changed X axis into time area. A result, Figure4 means Delay-profile that is fitted above simulation conditions.

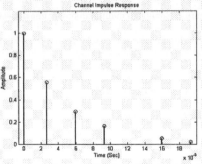

Figure 4. Detected Delay profile from Simulated PLC Channel

Table 3 shows, extracted characteristics of channel-models after organizing simulation for multi-path line using those simulations covered above.

Table 3. Delay properties of each channel

Model Items	Ch 1	Ch 2	Ch 3	Ch 4
Total line length	0.8km	1.4km	4.2km	3.2km
Num. Of Path	2	3	4	6
Num. of Load	3	4	4	5
Delay time	3.33×10^{-7} sec per 0.1km			
Attenuation	0.95	0.95	0.97	0.97
Max. amplitude	Normalization: 1			
Max. delay time	4 μsec	8 μsec	26μsec	20μsec

5.   Implementations

In this paper, we carried on experiment by designing coupler circuit that connects modem devices and power-lines, like Figure 5, using channel-sounding technique to look at channel environment of real power-line communication in details. Table 4 is parameters for experiments.

Table 4. Parameters for simulations

Items	Contents
PN code	M-sequence 63chip
Center frequency	125 KHz
Modulation	BPSK (binary phase shift keying)
Coupler	Toko 707VX-T1002N (resonance freq : 120KHz~140KHz)
Filter	RRC filter (root raised cosine filter)
Detection method	Matched filter
Signal Generator	NI PXI-5421 ADC (16bit, 100MS/s)
Data Actuation (DAQ)	NI PCI-6115 DAC (14bit, 100MS/s)

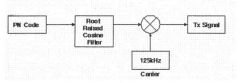

Figure 5. System blocks of channel sounding PN signal using RRC filter block

First step to experiment, based on simulation condition mentioned above, we organized transmit-signals as Figure 5. In Figure 5, we organized PN code that mentioned above, and to get rid of high frequency of PN code there applied an RRC filter (root raised cosine filter)[8]. And I multiplied mostly considerable coupler response frequency of 125 KHz carrier signal, and made transmit-signals that can be generated on transmitter.

With these transmit-signals, it is transmitted through the coupler that expressed on Figure 5's transmitter, and through the power-line transmit-signal is received by coupler of receiver. Figure 6 tells 2 periods of receive signal and transmit signals of ACF that went through power-line channel.

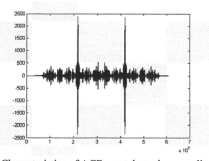

Figure 6. Characteristics of ACF went through power-line channel

Therefore we can obtain both the path diversity by means of the RAKE combining method with delayed antennas and the frequency diversity from multi-carrier [12].

We propose an effective distribution antenna diversity system according to adaptive correlation method for OFDM-DS/CDMA in a frequency selective fading channel environment. The proposed system transmits the different data using several sub-carriers which are correlated, and then, the proposed system transmits the same data using several sub-carriers which are de-correlated. It can achieve combined path and frequency diversity in a variable frequency selective fading channel. It provides high data rate services by transmission the different data using each correlated carrier, and achieves high quality by transmitting the same date with multiple antennas. The proposed system is an effective system to achieve multimedia service of high quality according to channel condition.

2. Distribution Antenna Diversity System according to Adaptive Correlation method

2.1 Base Station Model

We propose a new transmitter diversity systems different data using several correlated sub-carriers, while the proposed system transmits the same data using several sub-carriers which are de-correlated. It can achieve combined path and frequency diversity in a variable frequency selective fading channel. It provides high data rate services by transmitting different data using each carrier, and obtains high quality by transmitting the same data using multiple antennas.

A multi-carrier transmits different data using correlated sub-carriers which provide high-rate data transmission. The proposed system transmits different data on each carrier, which provides an effective system for performing high data rate transmission with efficient frequency utilization such as that required for video, and image. This is because the data rate of the proposed OFDM systems is larger than that of a single carrier system due to the orthogonal overlapping of carriers expressed by equation (1).

On the other hand, the proposed system transmits the same data on a de-correlated carrier. This is an effective scheme for providing a high quality data service in such areas as Internet service which requires low error probability.

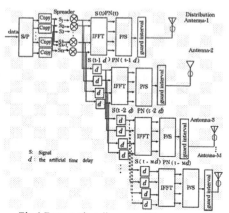

Fig.1 Base station diagram of the proposed

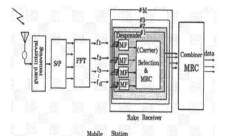

Fig.2 Mobile station diagram of the proposed system

Binary data bits are assigned to different distribution antennas respectively. The k th user's information signal at the m th distributed antenna can be expressed as

$$b_m^k(t) = \sum_{i=-\infty}^{\infty} b_{i,m}^k P_T(t - iT),$$

where P_T is a rectangular pulse with duration T , and $b_{i,m}^k$ is the binary data of duration T at the i th data interval.

The spreading sequence of the k th user is expressed as

$$a^k(t) = \sum_{j=-\infty}^{\infty} a_j^k \psi(t - jT_c),$$

where a_j^k is the j th chip (± 1) of the k th users, and T_c is the period of the spreading sequence, and $\psi(\cdot)$ is a chip waveform, which in this paper is a rectangular pulse with duration T_c .

A transmitted signal from the m th distributed antenna of the k th user's transmitter is given by

$$s_m^k(T) = \sum_{s=1}^{S} \sqrt{2Pb_m^k(t - d_m^k)} a^k(t - d_m^k) e^{j(w_{s,m}t)},$$

where w_s is the sth carrier frequency, P is the power of a bit, and b_m^k / T_c is an integer. T_c is the chip period of a spreading sequence.

The orthogonal frequencies $w_{s,m}(s = 1,2,3...,S)$

are the s th carrier requencies of the mth antenna and have the relation of

$$w_{s,m} = w_1 + (s-1)2\pi/T, where s = 1,2,...S$$

where w_1 is the absolute carrier number.

Fig.1 shows the concept of the proposed system whose base station has multiple antennas with independent fading patterns, and its mobile stations with independent fading patterns, and its mobile stations with single antennas. Several sub-carriers are assigned to each antenna. A set of antennas are fed by a common signal with time delay processing to distinguish signals. Thus, this can makes fading more randomized between the OFDM tones of each antenna than in the conventional system. We can obtain uncorrelated signals from uncorrelated carrier of each antenna; therefore, multipath provides path diversity. The diversity effect of RAKE combining signals delayed by several chips can be obtained by coherent detection at a mobile station.

Fig.1 shows a base station model of the proposed system. The base station transmits different data on near sub-carriers which are correlated such as S_1, S_2, S_3 or S_4, S_5, S_6. This is why multi-carriers are commonly located in successive frequencies and have high correlations among sub-carriers with one antenna. The same data on the de-correlated carriers are at a great distance from one another, such as in the case of S_1, S_{k+1} or S_2, S_{k+2} . In other words, the S_1 carrier transmits the same data with S_{k+1}, and S_2 does the same data with S_{k+2} because they are de-correlated.

Figure 1 shows that input data are converted from serial to parallel, and are spread in the time domain by PN codes in each carrier. Then the signals of each antenna are delayed by several chips, and Inverse Fast Fourier Transform (IFFT) is utilized for multi-carrier modulation. The output of IFFT is parallel to serial converted. A guard interval is inserted between the output signals to prevent ISI.

The complex low path impulse response of a channel of the i th user is assumed to be

$$h_k(t) = \sum_{l=0}^{L-1} \beta_l^k \exp(j\tau_{k,l})\delta(t - lT_c),$$

where l is the number of channel paths, and the path gains $\beta_{k,l}$ are independent identically distributed(i.i.d) Rayleigh random variable (r.v.s) for all k and l , and the angles $\tau_{k,l}$ are i.i.d distributed in $[0,2\pi)$.

system have same structure. The following is a detailed example of domain 2, see figure 4. In domain2, environment agent, safety kernel agent and safety management database agent are realized respectively by task SC, SK and Proc. The security & safety management database is SDL2. To simplify the design of SC, whenever safety kernel has deal with 5 mistaken visiting orders of traffic light in succession, it will send abnormal information to SC. Thus SC will no longer collect abnormal information again. SK provides safety kernel service. WD imitates SK's monitor, once it finds that SK behaves abnormally, it will stop the SK immediately and start another safety kernel SK', which is a back-up of SK. Main-info is auditing document, which records information relating to security.

4.4 The result of the experiment

To prove the validity of the experiment, I set three arrays for COMM: Data1, Data2 and Data3.

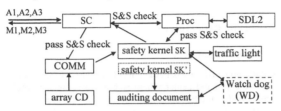

Figure 4. Security & safety system structure of domain 2

Data1 consists of correct orders changing traffic lights' color. Data2 adds incorrect orders on the basis of Data1 in order to check the security assurance function of domain2 if without interference from domain1. Data3 adds more than 5 incorrect orders on the basis of Data2 to check the correctness and security of function of domain 2 or domain 3 if domain1 has interfered them. Results of the experiment prove that the security & safety system are feasible.

5 Conclusion

This paper put forwards an assurance system which combines the security and safety together. The author researches its structure formwork and concrete function. Moreover, the implementation of safety kernel by Metaobject is discussed. Results of simulation show that proposed formwork is reasonable. In fact, safety kernel can be inserted in RTOS and will be automatically provided by RTOS, this work will certainly come down to modification on the kernel of RTOS, which will be our further research.

Acknowledgments

The author wishes to thank the anonymous referees for their careful reading of the manuscript and their fruitful comments and suggestions.

References

[1] Stergiou T, Leeson M S, Green R J. An alternative architectural framework to the OSI security model, computers and security, 2004, 23:137~153
[2] Anon. Open systems interconnection-basic reference model part 2: security architecture, International standards institute (ISO) 7498-2 Standard, 1989
[3] Brucker A D, Wolff B. A case study of a formalized security architecture, Electronic notes in theoretical computer science 2003, http://www.elsevier.nl/locate/entcs/vlume80.html
[4] Lin A, Brown R. The application of security policy to role-based access control and the common data security architecture, Computer communications 2000, 23: 1584~1592
[5] Qin Z G, Lin J D. Design and implementation of global security system in open system environment, Journal of applied sciences, 1999, 17(3):27~32
[6] Butler R W, Finelli G B. The infeasibility of quantifying the reliability of life-critical real-time softwar. IEEE Tran. On software engineering, 1993, 19(1):3~12
[7] Kevin R. Safety kernel enforcement of software safety policies. Ph.D. Thesis, University of Virginia, 1995
[8] Sahraoui A E, Anderson E, Katwijk V, et al. Formal specification of safety shell in real-time control practice. Proceedings of the WRTP'S 2000, 25th IFAC workshop on real-time programming. Oxford: Elsevier. 2000: 117~123
[9] Fabre J C, Perennou T. A metaobject architecture for fault-tolerant distributed systems: the FRIENDS approach, IEEE Trans. On Computers, 1998, 47(1):78~95

Brill Academic Publishers
P.O. Box 9000, 2300 PA Leiden
The Netherlands

*Lecture Series on Computer
and Computational Sciences*
Volume 4, 2005, pp. 1749-1752

An Evolutionist Algorithm to Cryptography

F. Omary[1], A. Tragha[2], A. Bellaachia[3], A. Lbekkouri, A. Mouloudi

[1]Department of mathematics and computer sciences
The science faculty –Rabat University of MohammedV-Morocco
[3]Department of Computer Science
George Washington University Washington, DC 20052

Received 29 June, 2005; accepted in revised form 4 August, 2005

Abstract: In this paper, we present an application of evolutionist algorithms to cryptography and especially to symmetrical ciphering. Our approach reduces the ciphering problem to an optimization problem. Then, we developed an encoding schema similar to the one used for the solution of the Travel Salesman Problem (TSP) using evolutionist algorithm. The paper addresses the coding of the chromosomes, the identification of a fitness function, and the choice of the adapted genetic operators. It also describes the processes of ciphering and decoding a message previously scrambled using the symmetrical key generated by our algorithm. Examples of applications are also presented. Experimental results show that a value between 24 and 32 for the population size yields better performance.

1. Introduction

Encryption is the process of converting a plaintext (original message) into ciphertext (the coded message) which can be decoded back into the original message. An encryption (or ciphering) algorithm along with keys is used in the encryption and decryption of data. It can be either symmetric or asymmetric. The degree of security offered by the algorithm depends on the type and length of the keys utilized.

True ciphering algorithms are counted on the tips of the fingers. Important symmetrical algorithms: DES (1973)[3], IDEA(1990)[4], AES (1997) [3]. And asymmetric algorithm: RSA (1977) [5].

PGP (1991) is a combination of the best functionalities of the asymmetric and symmetric cryptography. It is a hybrid system; it is currently, the dominant crypto-system [9].

Genetic algorithms are used to mimic natural processes of evolution. They have also emerged as practical, robust optimization and search methods [6].

In this paper, we introduce a new symmetric encryption algorithm using genetic algorithm approach. In our approach, we reduced the problem of ciphering a message M to an optimization problem. Then we present a genetic solution based on the method used to solve the traveler salesman problem (TPS) [2].

The paper is organized as follows. The next section describes our encryption algorithm using genetic approach. Section 3 introduces the decoding process of our approach. Experimental results and discussions are presented in Section 4 and section 5 respectively. Finally, Section 6 concludes the paper and gives future directions.

2. Description of our ciphering algorithm

Let M_0 be the message to encode. M_0 consists of n characters where each letter is one of the 256 characters of the ASCII code. We first apply a set of transformations to M_0. Which are a combination of several simple methods such as substitution, permutations, affined ciphering, etc. The transformed message will be denoted as M.

---

[1] E-mail:: omaryfouzia@yahoo.fr
[2] E-mail: a.tragha@univh2m.ac.ma
[3] E-mail: bell@gwu.edu

3. Problem Definition

Let $c_1, c_2, ..., c_m$ be the different characters of M ($m \leq 256$). Denote by L_i ($1 \leq i \leq m$) the list of the different positions of the character c_i in M before the ciphering and by card(L_i) the number of the occurrences of c_i in M. We notice that $L_i \cap L_j$ is empty $\forall i, j \in [1, m]$.

The message M can be represented below by the vector:

(c_1, L_1)	(c_2, L_2)	....	(c_m, L_m)

The objective of our algorithm is to introduce a maximum number of changes in the positions of each character in M to produce a new list L'_i such that the difference of card(L_i) and card(L'_i) is maximal. This becomes an optimization problem that can be solved using evolutionist algorithms.

4. Our Ciphering System

The body of our algorithm is the same as general evolutionist algorithm [1]. Thus we define our ciphering algorithm as follows:

- Step 0: coding schema

An individual (or chromosome) is a vector of size m. In our case, the genes are represented by the L_{p_i} lists ($1 \leq i \leq m$). The ith gene, L_{p_i}, contains the positions of character c_i.

- Step 1: Creation of the initial populations P_0 made of q individuals: $X_1, X_2, ..., X_q$.

Let Original-Ch be the chromosome which genes are $L_1, L_2, ...,$ and L_m lists (placed in this order). These lists represent the message before the application of the algorithm. We apply permutations on Original-Ch in order to get an initial population formed by q potential solutions.

Set i:=0;
- Step 2: Fitness function,

Let X_j be an individual of P_i whose genes are: $L_{j_1}, L_{j_2}, ..., L_{j_m}$. We define the fitness function by:

$$F(X_j) = \sum_{i=1}^{m} | card\ (L_{j_i}) - card\ (L_i) | \qquad (1)$$

- Step 3: Selection of the best individuals

We use the classical method of the roulette that allows keeping the strongest individuals [6].

Since the problem is reduced to a problem of permutations, we then apply the genetic operators adapted to this type of problems as explained below.

- Step 4: Genetic operators

- MPX Crossover : This crossover has been proposed by Mülhenbein [8] .It is applied to the selected individuals with a very precise rate. According to [7], the order of the best rate is from 60% to 100%.

- Mutation of transposition:

We choose a mutation that randomly permutes two genes of a chromosome. This operator is applied to the individuals derived from crossover with an adapted rate preferably from 0.1% to 5% [7].

We place the new offspring in a new population P_{i+1}.

Repeat the steps 2,3 and until a stop criteria.

Define the stop condition:

$$\sum_{i=1}^{m} | card\ (L_{k_i}) - card\ (L_i) | \leq \sum_{i=1}^{m} (card\ (L_{k_i}) + card\ (L_i)) \leq 2*l \qquad (2)$$

The function F is bounded because $0 \leq F(X) \leq 2*m$ for each individual X. In fact:

Theoretically, the function F admits a maximum since it is bounded. According to some results [1], the convergence of fitness function is guaranteed, but perhaps towards a value close to max. This can be determined through experimental results.

Final step of our algorithm: Let Final-Ch be the final solution given by our evolutionist algorithm. The symmetrical key, also called genetic key, is generated using both the Original-Ch and Final-Ch.

5. Decoding

The decoding of the encoded message M' is done in two steps:

Similar to the encoding process, we represent the encoded text, by a vector of lists. Let $c'_1, c'_2, ..., c'_m$ be the different characters of M' and by $L'_1, L'_2, ..., L'_m$ the lists of positions of these characters. Using the genetic key, the positions of the characters in the initial message are recovered. As a matter of fact, the key is a permutation of $\{1, 2, ..., m\}$ that we represent by a vector Key of size m a follows: Key(1)=p_1, Key(2)=p_2, ...Key(i)=1, ... Key(m)=p_m.

Where: The character c'_{p_i} is going to be associated to the list L'_i $(1 \leq i \leq m)$.

Thus we get the message M.

In the second step, we proceed to the decoding of M to get M_0. This is easily achieved the reverse functions used in the encoding process since the genetic key is symmetrical.

6. Experimental Results

We have conducted several experiments using messages of different sizes to illustrate the performance of our algorithm. Table 1 presents a sample of our experiments. This table gives the value of the convergence of the fitness function and the number of the generations reached at the time of this convergence. Figure 1 shows message 3, its encrypted version, and its genetic key. In this example, we have used an affined ciphering defined by: $F(x)=ax+b$ where $a = -1$, $b = 255$ and x is the ASCII code of the character of the initial message.

The experimental results show that the best values of the optimum (value of convergence) are reached for a population of size 24, see Table 1. In some cases, the sizes 30 and 32 also gave good results, but the number of generations in these cases doubled which is expensive.

Abraham Lincoln, the 16th president of the United States, guided his country through the most devastating experience in its national history--the CIVIL WAR. He is considered by many historians to have been the greatest American president. Abraham Lincoln was born Sunday, February 12, 1809, in a log cabin near Hodgenville, Kentucky. He was the son of Thomas and Nancy Hanks Lincoln, Thomas Lincoln was a. Abraham had gone to school briefly in Kentucky and did so again in Indiana. He attended school with his older sister... Both of Abraham's parents were members of a Baptist congregation which had separated from another church due to opposition to slavery. When Abraham was 7, the family moved to southern Indiana As Abraham grew up, he loved to read and preferred learning to working in the fields. This led to a difficult relationship with his father who was just the opposite. Abraham was constantly borrowing books from the neighbors. Lincoln's declining interest in politics was renewed by the passage of the Kansas-Nebraska Act in 1854. In 1860 he furthered his national reputation with a successful speech at the Cooper Institute in New York.

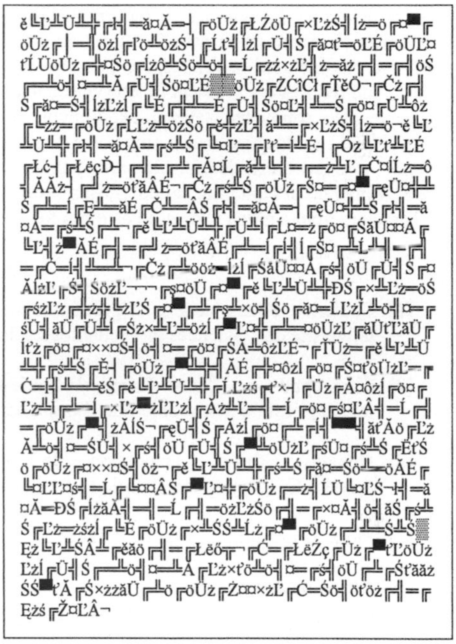

Figure 1 : Encoding of Message 3 (size=1149).

The genetic key is : 9 10 11 12 13 14 15 16 17 18 19 24 25 26 27 28 29 30 31 32 33 0 1 2 3 22 4 5 7 8 23 20 21 6

Table 1: Summary Statement of results.

	Size of population	16	24	30	32
Message1 1026 char.	Number of generations	35	58	71	64
	Value of convergence	1288	1394	1638	1354
Message2 684 char.	Number of generations	26	58	46	53
	Value of convergence	1068	1100	982	1002
Message3 1149 char.	Number of generations	44	58	111	53
	Value of convergence	1458	1622	1512	1444
Message4 803 char.	Number of generations	34	51	38	130
	Value of convergence	886	1096	1088	1160

7. Discussion

In this section, we compare our algorithm with DES, one of the most known symmetrical algorithms. In general, most well known symmetrical algorithms (DES, IDEA, AES) split the message to be encoded into blocks and separately cipher each block. This will allow cryptanalysis to work in parallel on different blocks. Our algorithm ciphers the entire message in only one hold. Therefore, our approach makes an encoded message safer from this type of attacks. The DES has a key of 64 bits out of which only 56 bits are used. This short key encourages attackers to decode a message by an exhaustive search using powerful computers. In our algorithm, if the message contains more than 30 different characters, the key size is at least equal to 240 bits. This makes our approach resistant to attacks. In case the message contains less than thirty different characters, we can complete it by others characters in order to reach a size of 240 bits. Besides, our key is a session key and generated by our own algorithm, while the DES key is not.

8. Conclusion

Most of cryptologists even think that all algorithms of ciphering are long time breakable but the essential is to conceive its ciphering system in such a way that the wished life span for the encoded text must be lower to the one set by the cryptanalyst to break this system. This paper presents a novel encryption algorithm using genetic algorithm approach. The genetic approach used in our algorithm is similar to the one used in solving the Traveler Salesman problem. Our algorithm uses a new encoding schema and reduces the problem of encryption to an optimization problem.

We have also conducted several experiments to analyze the performance of our algorithm. The experimental results show that a range of a population size between 24 and 32 gives better results and therefore it is recommended for the corresponding message sizes. Future work include the introduction of a new method of ciphering called "fusion" which allows the alteration of the appearance frequencies of characters from a given text, used as input for our evolutionist algorithm.

References

[1] Catherine Khanphang : Algorithmes heuristiques et évolutionnistes. Thèse de doctorat université des Sciences et Technologies de Lille, Octobre 1998.

[2] Christophe Caux- Henri Pierreval- Marie-Claude Portmann : Les algorithmes génétiques et leur application aux problèmes d'ordonnancement. APII. Volume 29-N° 4-5/1995 pp 409-443

[3] Douglas Stinson Traduction de Serge Vaudenay, Gildas Avoine et Pascal Junod : Cryptographie Théorie et pratique 2ème edition.

[4] Florin.G et Natkin.S : les techniques de la cryptographie. CNAM 2002.

[5] Alfred J.Menzes, Paul C. Van Oorschot and Scott A. Vanstone : Handbook of applied cryptography. CRC Press October 1996

[6] Goldberg D.E. : Genetic algorithms in search optimization and Machine Learning. Addison-Wesley. Publishing Company, Inc 1989.

[7] Grenfenslette,j.j : optimization of control parameters for genetic algorithms. IEEE translation on systems Man, and cybernitics Vol 16 N°1 pp122-128.1986.

[8] Mühlenbein H., "Evolutionary Algorithms: Theory and applications" (Wiley) 1993.

[9] PGP. "Personal Privacy pour windows 95, 98 et NT", NAI (traduction française:news.misc-cryptologie,1998).

Brill Academic Publishers
P.O. Box 9000, 2300 PA Leiden
The Netherlands

Lecture Series on Computer
and Computational Sciences
Volume 4, 2005, pp. 1753-1756

Spreading SFBC-OFDM Transmission Scheme
for Wireless Mobile Environment

Sang-Soon Park[1], Ho-Seon Hwang[1] and Juphil Cho[2]

[1] Department of Electronic Engineering, Chonbuk National University,
Jeonju, Korea, spwman@chonbuk.ac.kr
[2] School of Electronics and Information Engineering Kunsan National University,
Kunsan, Korea,

Received 28 June, 2005; accepted in revised form 15 July, 2005

Abstract: In this paper, we propose a simple SFBC-OFDM with time and frequency domain spreading for wireless mobile environment. The transmission scheme that we propose in this paper is a simple method with low complexity for improving the performance of conventional SFBC-OFDM. Using Hadamard transformation in time and frequency domain, we can obtain space, time and frequency diversity gain and improve the BER performance. Our spreading SFBC-OFDM is an attractive system in mobile broadband wireless access for ubiquitous

Keywords: OFDM, MIMO, SFBC, spreading

1. Introduction

The explosive growth of wireless communications is creating the demand for high speed, reliable, and spectrally-efficient communications over the mobile wireless medium. There are several challenges in attempts to provide high-quality service in this environment. Space-time block coding (STBC) is a attractive technique as a simple transmit diversity scheme for providing highly spectrally efficient wireless communications[1].

The STBC was applied to the OFDM (Orthogonal Frequency Division Multiplexing) as an attractive solution for a high bit rate data transmission in a multipath fading environment. This system was referred to as the space-time block coded OFDM (STBC-OFDM)[2]. And the space-frequency block coded OFDM (SFBC-OFDM) has been proposed where the block codes are formed over the space and frequency domain[3].

In this paper, we propose a simple SFBC-OFDM with time and frequency diversity gain. This transmission scheme is a simple method for improving a performance of conventional SFBC-OFDM[3] by means of Hadamard transformation in time and frequency domain. Also our spreading SFBC-OFDM can be encoded and decoded with low-complexity, because Hadamard transformation can be performed without multiplication.

2. Spreading SFBC-OFDM transmission scheme

In this section, we will propose a simple SFBC-OFDM transmission scheme to achieve time diversity gain. We can achieve time diversity gain by means of Hadamard transformation in time and frequency domain. The encoding of our SFBC codes is carried out in two successive stages: Spreading in frequency and time domain and SF component coding as Fig. 1.

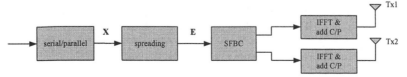

Fig. 1 Spreading SFBC-OFDM transmission scheme

2.1 Spreading in time and frequency domain

At the transmitter, the input data bits are modulated to $x(m)$ where $x(m)$ denotes the input serial data symbols with symbol duration T_S. The modulated symbols are serial to parallel converted and the resulting data vector $\mathbf{x}(n)$ and data block $\mathbf{X}$ are given by

$$\mathbf{x}(n) = \begin{bmatrix} x_0(n) & x_1(n) & \cdots & x_{N-1}(n) \end{bmatrix}^T$$
$$\mathbf{X} = \begin{bmatrix} \mathbf{x}(BK) & \mathbf{x}(BK+1) & \cdots & \mathbf{x}(BK+K-1) \end{bmatrix} \tag{1}$$

Let $\mathbf{E}$ be the Hadamard transform of $\mathbf{X}$ in time and frequency domain, and we can express $\mathbf{E}$ as follows.

$$\mathbf{E} = \left(\mathbf{W}_M \otimes \mathbf{I}_{N/M} \right) \mathbf{X} \mathbf{W}_K \tag{2}$$

where $\mathbf{W}_M$ represents Hadamard transform matrix of order M

$$\mathbf{W}_2 = \frac{1}{\sqrt{2}} \begin{bmatrix} 1 & 1 \\ 1 & -1 \end{bmatrix}, \quad \mathbf{W}_{2^a} = \mathbf{W}_2 \otimes \mathbf{W}_{2^{a-1}} \tag{3}$$

where $\otimes$ represents Kronecker product. Hadamard trans-form spreads each symbol to M subcarriers and K successive OFDM symbols. Therefore matrix $\mathbf{E}$ is a symbol matrix spreaded in time domain.

2.2 Space-frequency component coding

After spreading the data matrix in time and frequency domain, perform the SFBC. During the block instant n, the spreaded symbol vector e(n) can be expressed as follows.

$$\mathbf{e}(n) = \begin{bmatrix} e_0(n) & e_1(n) & \cdots & e_{N-1}(n) \end{bmatrix}^T \tag{4}$$

The spread symbol vector $\mathbf{e}(n)$ is coded into two vectors $\mathbf{e}^{(1)}(n)$ and $\mathbf{e}^{(2)}(n)$ by the SFBC as follows.

$$\mathbf{e}^{(1)}(n) = \begin{bmatrix} e_0(n) & -e_1^*(n) & \cdots & e_{N-2}(n) & -e_{N-1}^*(n) \end{bmatrix}^T$$
$$\mathbf{e}^{(2)}(n) = \begin{bmatrix} e_1(n) & e_0^*(n) & \cdots & e_{N-1}(n) & e_{N-2}^*(n) \end{bmatrix}^T \tag{5}$$

During the block instant n, $\mathbf{e}^{(1)}(n)$ is transmitted from the first transmit antenna Tx1 while $\mathbf{e}^{(2)}(n)$ is transmitted simultaneously from the second transmit antenna Tx2. Let $\mathbf{e}_e(n)$ and $\mathbf{e}_o(n)$ be two length $N/2$ vectors denoting the even and odd component of $\mathbf{e}(n)$

$$\mathbf{e}_e(n) = \begin{bmatrix} e_0(n) & e_2(n) & \cdots & e_{N-2}(n) \end{bmatrix}^T$$
$$\mathbf{e}_o(n) = \begin{bmatrix} e_1(n) & e_3(n) & \cdots & e_{N-1}(n) \end{bmatrix}^T \tag{6}$$

Table 1. shows the SF component coding for two-branch transmit diversity scheme[3]. The SF coded symbols are modulated by IFFT into SFBC-OFDM symbols. After adding the guard interval, the proposed SFBC-OFDM symbols are transmitted.

Table 1 Space-Frequency component coding

Subcarrier index	Tx1	Tx2
Even	$\mathbf{e}_e(n)$	$\mathbf{e}_o(n)$
Odd	$-\mathbf{e}_o^*(n)$	$\mathbf{e}_e^*$

2.3 Decoding of spreading SFBC-OFDM

Decoding processes are the reverse order of the encoding and carried out following two stages.

• First stage: perform the decoding with SF component decoding to obtain $\hat{\mathbf{E}}$, the estimation block of $\mathbf{E}$

• Second stage: perform the reverse of Hadamard transformation to obtain decision matrix $\hat{\mathbf{X}}$ from $\hat{\mathbf{E}}$. And $\mathbf{X}$ is recovered by ML decision.

Let $\mathbf{y}(n)$ be a demodulated signal vector at the receiver and $\mathbf{y}_e(n)$, $\mathbf{y}_o(n)$ be two length $N/2$

vectors denoting the even and odd component of $\mathbf{y}(n)$. Let $\Lambda(n)$ be a diagonal matrix whose diagonal elements are the DFT of the channel impulse response.

$$\Lambda(n) = \text{diag}(\lambda_0(n) \quad \lambda_1(n) \quad \cdots \quad \lambda_{N-1}(n)) \tag{7}$$

Let $\Lambda^{(1)}(n)$, $\Lambda^{(2)}(n)$ be channel responses of transmit antenna1, 2, and $\mathbf{z}(n)$ be the channel noise. The demodulated signal vector at the receiver is given by

$$\begin{aligned}
\mathbf{y}_e(n) &= \Lambda_e^{(1)}(n)\mathbf{e}_e^{(1)}(n) + \Lambda_e^{(2)}(n)\mathbf{e}_e^{(2)}(n) + \mathbf{z}_e(n) \\
\mathbf{y}_o(n) &= \Lambda_o^{(1)}(n)\mathbf{e}_o^{(1)}(n) + \Lambda_o^{(2)}(n)\mathbf{e}_o^{(2)}(n) + \mathbf{z}_o(n)
\end{aligned} \tag{8}$$

Assuming the complex channel gains between adjacent subcarriers are constant, i.e., $\Lambda_e^\alpha(n) \approx \Lambda_o^\alpha(n)$, and the estimate vectors $\hat{\mathbf{e}}(n)$ as follows.

$$\begin{aligned}
\hat{\mathbf{e}}_e(n) &= \Gamma^{-1}(\Lambda_e^{(1)*}(n)\mathbf{y}_e(n) + \Lambda_o^{(2)}(n)\mathbf{y}_o^*(n)) \\
\hat{\mathbf{e}}_o(n) &= \Gamma^{-1}(\Lambda_e^{(2)*}(n)\mathbf{y}_e(n) - \Lambda_o^{(1)}(n)\mathbf{y}_o^*(n))
\end{aligned} \tag{9}$$

where

$$\Gamma = |\Lambda_e^{(1)}(n)|^2 + |\Lambda_e^{(2)}(n)|^2 \approx |\Lambda_o^{(1)}(n)|^2 + |\Lambda_o^{(2)}(n)|^2 \tag{10}$$

The estimate vector $\hat{\mathbf{e}}(n)$ is vectors to estimate the magnitude of $\mathbf{e}(n)$ as well as phase. Let estimate matrix $\hat{\mathbf{E}}$ be the matrix composed of K estimate vectors, and we can express $\hat{\mathbf{E}}$ as follows.

$$\hat{\mathbf{E}} = [\hat{\mathbf{e}}(0) \quad \hat{\mathbf{e}}(1) \quad \cdots \quad \hat{\mathbf{e}}(K-1)] \tag{11}$$

Perform the reverse of Hadamard transformation to obtain decision block $\hat{\mathbf{X}}$ from $\hat{\mathbf{E}}$, and

$$\hat{\mathbf{X}} = (\mathbf{W}_M \otimes \mathbf{I}_{N/M})^{-1}\hat{\mathbf{E}}\mathbf{W}_K^{-1} = (\mathbf{W}_M \otimes \mathbf{I}_{N/M})\hat{\mathbf{E}}\mathbf{W}_K \tag{12}$$

$\hat{\mathbf{X}}$ are vectors to estimate the magnitude of $\mathbf{X}$ as well as phase. Therefore we can recover $\mathbf{X}$ by means of equation (9) in paper [1], even if $\mathbf{X}$ has the unequal energy constellations.

3. Simulation results

In this section, we present the results of computer simulation. In this paper, carrier frequency is 2GHz. The spreading SFBC-OFDM systems using two transmit antennas and one receive antenna are used in this section. And each channel is uncorrelated. And it was assumed that perfect channel estimation was available at the receiver.

Fig. 2(a)shows the simulation results for conventional SFBC-OFDM and the spreading SFBC-OFDM at the velocity = 120km/h. The COST207 six-ray typical urban (TU) channel power delay profile was used throughout the Fig. 2[4]. The OFDM systems employed 512 subcarriers with QPSK modulation at a channel bandwidth = 5×10^6 MHz, and sampling frequency = 5×10^6 MHz.

Fig. 2(b) shows the simulation results for conventional SFBC-OFDM and the spreading SFBC-OFDM at the velocity = 250km/h. The COST207 six-ray hilly terrain (HT) channel power delay profile was used throughout the Fig. 2(b)[4]. The OFDM systems employed 2048 subcarriers with QPSK modulation at a channel bandwidth = 10^7 MHz, and sampling frequency = 10^7 MHz.

4. Conclusions

We have proposed a spreading SFBC-OFDM transmission scheme that is a method for improving the performance of conventional SFBC-OFDM, by means of spreading in time and frequency domain. And we compared the performance of spreading SFBC-OFDM transmission schemes with conventional SFBC-OFDM. In frequency selective channels, we can obtain frequency diversity gain. Also the Doppler spread is large, complex channel gains over K successive OFDM symbols vary rapidly. In this case, the spreading SFBC-OFDM can obtain time diversity gain and improve the BER performance.

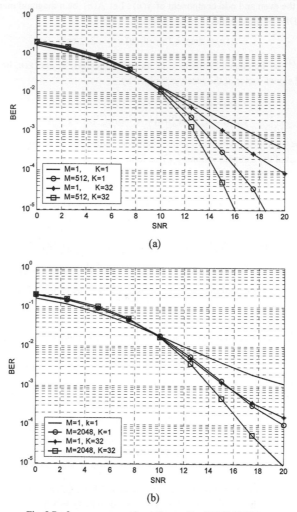

(a)

(b)

Fig. 2 Performance comparison of spreading SFBC-OFDM

References

[1] S. M. Alamouti, "A simple transmitter diversity scheme for wireless communications," *IEEE J. Select. Areas Commun.*, vol. 16, pp. 1451-1458, Oct. 1998.

[2] K. F. Lee and D. B. Williams, "A space-time coded transmitter diversity technique for frequency selective fading channels," *IEEE Sensor Array and Multichannel Signal Processing Workshop*, pp. 149-152, Cambridge, MA, Mar. 2000.

[3] K. F. Lee and D. B. Williams, "A space-frequency transmitter diversity technique for OFDM system," *IEEE GLOBECOM 2000*, vol. 3, pp. 1473-1477, San Francisco, USA, Nov. 2000.

[4] COST207 TD(86)51-REV 3 (WG1), "Proposal on channel transfer functions to be used in GSM tests late 1986," Sept. 1986.

Brill Academic Publishers
P.O. Box 9000, 2300 PA Leiden
The Netherlands

*Lecture Series on Computer
and Computational Sciences*
Volume 4, 2005, pp. 1757-1761

Crypto-Cert Digital Signature Mechanism

Hee-Un Park[1]

Korea Information Security Agency, Garak, Songpa, Seoul, 138-803, Korea

Received 23 June, 2005; accepted in revised form 18 July, 2005

Abstract: In the digital information society, various cryptographic researches have been studied to get the safety and trusty in computer network from illegal act such as hacking, trapping and forgery, and the most common authentication scheme in cryptology is 'Digital Signature.' But because digital signature can be opened by signature confirmers who will be anyone, it has some problems when it is applied into anonymity and confidentiality services such as E-voting, secure E-conference, mobile multicast roaming and E-bidding service etc.

To solve the problems undeniable signature and nominative signature scheme was proposed. But in these schemes, a signer's identity can be opened by confirmer who only signature receiver, so signer's anonymity depends on a confirmer. Because they are also based on digital signature with appendix type, message on transmitted signature information can be opened the third party.

In this paper, we propose 'Crypto-Cert' digital signature scheme which supplies the signer's anonymity and allows that only receiver can confirm the signer's membership. The proposed scheme is applied into diverse branches, because signer's identity can be opened through legal policy or rule set by trust center and confidentiality of message is supported using advanced signcryption.

Keywords: Crypto-Cert, Digital Signature, Anonymity, Confidentiality,

1. Introduction

Based on the development of digital media and network, the more and more we will be dependant on information as time goes by, and most information will be transmitted using digitalized materials instead of using paper based information. And also various applications connected with computer and network have been studied. For example, in IT business top lank areas are multimedia contents, E-commerce, mobile communication and so on.

But, in this applications, message transmission is done through open network. So it is vulnerable to attacks from hacking, illegal tapping and copy etc. Now various crypto methods have been studied to provide safety and trust against some threats. Most of all, 'digital signature' scheme is used for authentication service on a computer network[1][2].

Digital signature can ensure the integrity of messages and settle a dispute between them based on the user's authentication. But, a negative factor is that anyone can check a general digital signature. As a result, it became an issue in applied fields like E-voting, E-conference, mobile multicast roaming and E-bidding because the anonymity and confidentiality should be ensured in this areas. In a case of E-bidding, the signer's identity should be hidden by the time when the bidding is completed.

So, to solve the problems, the special digital signature, that undeniable signature and digital nominative signature, have been proposed up to now[4]~[10]. Yet because these schemes counted on a receiver when checking the user's identity, his identity might be able to be open by the receiver. Besides, the messages of signature information could also be known to the third party because these schemes were based on the digital signature with appendix.

So in this paper, we'll contemplate the security requirements for an anonymous authentication service and investigate the 'Crypto-Cert digital signature' mechanism to ensure the anonymity of a signer and to check his membership by himself. Also this scheme supports confidentiality of message using advanced signcryption. Above all, this scheme can be applied to diverse branches

---

[1] Corresponding author. Korea Information Security Agency, Seoul, Korea. E-mail: hupark@kisa.or.kr

3.3 Analysis of proposed schemes

This scheme satisfies all requirements for applied services above if we consider the former digital signature schemes.

1) Anonymity
A signer can prove which group he belongs to based on the group signature key provided by Crypto-Cert scheme. This can ensure the anonymity because a verifier can't know his any information.

2) Membership confirming
With an public key, a verifier can't know who is the signer, but he can prove which group he belongs to.

3) Confidentiality
With Signcryption mechanism, this sends signature messages in a session key, k_1 encrypted. Therefore, we can keep it confidential from the third part who doesn't know the session key.

4) A receiver confirming
When a session key, k is created, an only verifier can check the signature because it's calculated based on his public key. As a result, this can ensure the recipient nominative.

5) Identity recovery
When a signer commits an illegal action or his identity is open due the feature of protocol, the identity is recovered by only TC after the right process.

These are the results of comparison and analysis on the conventional schemes and new ones if the requirements above are considered[4][5][6][7].

Table 2: The analysis of each schemes

Item schemes	Anonymity	Membership authentication	Confidentiality	Receiver's Confirming	Identity recovery
Undeniable Signature	X	X	X	X	O
Entrusted Undeniable Signature	X	X	X	X	O
Nominative Signature	X	X	O	O	O
Group Signature	O	O	X	X	X
Proposed Scheme	O	O	O	O	O

4. Conclusion

As the Internet becomes a more pervasive part of daily life, various researches are has been done for application. Especially, digital signature methodology granting authentication for information transferring becomes even more critical. However electronic vote and bid application must additionally support confidentiality and anonymity of voter.

Therefore, in this paper, requirements for identification control and authentication service are surveyed. Also, it is concerned how these requirements are applied to the existing scheme.

The existing special digital signature scheme has vulnerability because it is possible for the identification to be exposed when this signature is verified. There is no special digital signature scheme which meets the condition that a signer sends his identification bound by digital signature to assigned person by the signer

The scheme in this paper proposes and applies Crypto-cert methodology and renewed Signcryption methodology. This scheme can provide anonymity control as well as can guarantee the anonymity. The scheme can be applied to various areas.

References

[1] "Specification for a Digital Signature Standard," NIST, FIpS XX. Draft, August 1991[1].
[2] C. Schnorr, "Method for identifying subscribers and for generating and verifying electronic signatures in a data exchange system," US patent #4,995,082, Feb. 1991.
[3] Y. Zheng, "Signcryption and Its Applications in Efficient Public Key Solutions," Proc. ISW'97, LNCS 1397, pp.291-312, 1998.
[4] Chaum, "Undeniable Signature Systems," U.S. Patent #4,914,689, 3 Apr 1990.

[5] S. J. Park, K. H. Lee and D. H. Won, "An Entrusted Undeniable Signature," Proceedings of the 1995 Japan-Korea Workshop on Information Security and Cryptography, Inuyama, Japan, 24-27 Jan 1995, pp. 120-126.

[6] S. J. Kim, S. J. Park and D. H. Won, "Nominative Signatures," Proc. ICEIC'95, pp.II-68 ~ II-71, 1995.

[7] D. Chaum, "Group Signature," Advances in Cryptology -EUROCRYPT 91 Proceedings, Springer-Verlag, 1991, pp.257-265.

[8] D. Chaum, "Blind Signature Systems," US. Patent #4,759,063, 19 Jul 1988.

[9] M. Mambo, K. Usuda, and E. Okamoto, "Proxy Signatures," Proceedings of the 1995 Symposium on Cryptography and Information Security (SCIS 95), Inuyama, Japan, 24-27 Jan 1995, pp. B1.1.1-17.

[10] C. Boyd, "Digital Multisignatures," cryptography and Coding, H.J. Beker and F.C. Piper, eds, Oxford:Clarendon Press, 1989, pp.241-246.

Brill Academic Publishers
P.O. Box 9000, 2300 PA Leiden
The Netherlands

*Lecture Series on Computer
and Computational Sciences*
Volume 4, 2005, pp. 1762-1765

Multi-Criteria Synthetic Evaluation of Software Security[1]

W. J. Wang, S. M. Wang, Z. Chen[2], Z.L.Liu

College of Computer Science and Technology, Jilin University, Changchun 130012, P. R. China

Received 19 June, 2005; accepted in revised form 12 August, 2005

Abstract: A new solution to evaluate software security was proposed. It was a combination of analytic hierarchy process (AHP), principal component analysis (PCA) and cluster analysis (CA). Typical indices were extracted through practical study and analysis of software security. AHP was adopted to determine index weight by means of quantitative analysis. PCA was conducted based on the correlation of the indices. Through the linear transformation, the weighted indices were combined into an independent set which represented the main initial information. CA explored the results of PCA further to group them into specified grades according to user's requirements and make them more direct. The practical evaluation results showed that the proposed solution could evaluate and rank security of softwares with multiple indices correctly and effectively.

Keywords: software security; index extraction; principal component analysis; analytic hierarchy process; cluster analysis

1. Introduction

Security of software in network environment has attracted much concern [1]. Correct evaluation of the security helps to prevent risks and improve software quality [2]. The evaluation relates to many aspects, which consists of complicated and variable elements, which makes it difficult to extract index and analyze directly [3]. Principal component analysis (PCA) can simplify the analytic process through data reduction but tends to reduce the objectivity of evaluation due to no consideration of relative importance between the indices [4], so analytic hierarchy process (AHP) is adopted to determine index weight and then combined with PCA to ascertain the evaluation correctness. Cluster analysis (CA) is employed to group the evaluation results further into specified grades in a direct way.

2. Index Extraction of Evaluation

2.1 Principles of index extraction

Indices are extracted according to following principles: typicalness, feasibility, relative independence, directivity and objectivity.

2.2 Security-related factors

2.2.1 Software secure engineering

The security engineering activities for secure software involve each stage of lifecycle [5]. The indices extracted are: ①the feasibility of secure design and consistency between design and realization; ②the number of potential threats in secure testing; ③the compatibility between software and deployment environment; ④the number of secure threats due to improper modification.

2.2.2 Operational environment

The change or stability of operating context may also result in software-related failures [2]. The indices extracted are: ①the number of authentication or access control holes; ②the feasibility of database secure mechanism; ③the fitness between database and operational infrastructure; ④the number of IIS and Web service holes; ⑤the number of RPC holes;⑥the number of Denial of Service holes; ⑦the number of buffer overflow holes.

---

[1] Supported by National Natural Science Foundation of China (Grant No. 50378042)

2.2.3 Man-made factor

Suffered from insecure man-made factor, software with perfect security will still fail to operate. Two indices are extracted here: ①the secure consciousness of developing team; ②the secure consciousness of users.

3. Algorithms of Evaluation

3.1 The fundamental of AHP

AHP is used to carry out quantitative analysis of non-quantitative things. It can avoid subjective impact to determine index weight [6]. Here follows the calculation process:

1) Analyze the relationship between indices to construct hierarchical index system;

2) Compare indices of level k in 1-9 scale to construct pairwise comparison judgment matrix $B^k = (b_{ij}^k)_{p \times p}$, p is the index number of level k, $k=1,2,...,q$, and q is the number of levels in index system. In 1-9 scale, the value increases with the increase of importance of one index relative to its counterpart;

3) Calculate the weight of index i in level k by the formula:

$$W_i^k = \overline{W_i^k} \bigg/ \sum_{i=1}^{p} \overline{W_i^k} \tag{1}$$

where $\overline{W_i^k} = \sqrt[p]{M_i^k}$, and $M_i^k = \prod_{j=1}^{p} b_{ij}^k$, $i,j=1,2,...,p$;

4) Calculate coherence rate $C_R^k = (C^k / R)$, where $C^k = (\lambda_{max}^k - p) / (p-1)$, and λ_{max}^k is the max eigenvalue of B^k, the value of R is listed in Table 1. If $C_R^k = 0$, B^k is of complete coherence; if $0 < C_R^k < 0.1$, B^k is of satisfactory coherence; in any other situation, B^k should be adjusted or abandoned;

Table 1: Average random coherence index

p	1	2	3	4	5	6	7	8	9	10	11	12	13
R	0	0	0.58	0.90	1.12	1.24	1.32	1.41	1.45	1.49	1.52	1.54	1.56

5) Calculate overall weights of level k according to $B^{k,*} = B^{k-1,*} \times W^{k,k-1}$, where $B^{k-1,*}$ are overall weights of level $k-1$, and $W^{k,k-1}$ are weights of level k relative to level $k-1$;

6) Calculate overall coherence rate and judge whether the result can be accepted.

3.2 The fundamental of PCA

The objective of PCA is to find out some principal components representing the majority information of initial ones through the linear transformation [4]. Suppose that there are n samples and m indices in each sample represented as $X = (x_{ij})_{n \times m}$. Here follows the calculation process:

1) Normalize matrix X to get $X^* = (x_{ij}^*)_{n \times m}$ $i=1,2,...,n, j=1,2,...,m$;

2) Calculate to get correlation coefficient matrix R;

3) Get eigenvalue λ_i and eigenvector α_i of matrix R and rank λ_i in descending order;

4) Calculate the information contribution rate β_k by $\lambda_k \bigg/ \left(\sum_{i=1}^{n} \lambda_i \right)$, then $\left(\sum_{i=1}^{k} \lambda_i \right) \bigg/ \left(\sum_{i=1}^{n} \lambda_i \right)$, which is the accumulative total information contribution rate of the first k components. The value of k is identified ascertaining that the accumulative total information contribution rate is not less than 85%;

5) Calculate the score of each principal component F_i according to the formula:

$$F_i = a_i X^* \quad i=1,2,...,k \tag{2}$$

and then synthetic score

$$F = \beta_1 F_1 + \beta_2 F_2 + \cdots + \beta_k F_k \tag{3}$$

can be calculated.

3.3 The fundamental of CA

The objective of CA is to classify a set of indices according to the degree of similitude. Hierarchical cluster analysis is used here. Firstly, group each sample into an individual cluster, then calculate the distance between clusters and combine two clusters with the nearest distance into one cluster until only one cluster remains. Then clustering result can be got according to one clustering method [7]. The cluster number is determined according to the actual requirements of evaluation.

3.4 The combination of AHP, PCA and CA

Indices may have different effect degrees on overall appraisal; however, in PCA this situation is not regarded. AHP is used to identify index weights and then integrated with PCA. Then CA is used to group the results further. Suppose that there are n samples and m indices in each sample.

1) Firstly the initial data are calculated by the formula:

$$Z= (b_{1j}^{k,*} \times x_{ij}^{*})_{n\times m} \quad i=1,2,\ldots,n, j=1,2,\ldots,m \tag{4}$$

where x_{ij}^{*} is the normalized result of initial data, $b_{1j}^{k,*}$ is the overall weight of index j in level k;

2) Calculate matrix Z according to the steps of PCA;

3) Classify the samples according to the steps of CA, here using the scores of principal components got from step 2) as input data instead of data initially collected.

4. Calculation Results

Table 2: The initial data collected from six network-based softwares

	1	2	3	4	5	6	7	8	9	10	11	12	13
1	-0.80	10	-0.66	3	8	-0.90	-0.85	15	0	0	2	-0.60	-0.50
2	-0.72	12	-0.60	2	6	-0.70	-0.75	20	1	2	14	-0.40	-0.40
3	-0.68	12	-0.43	2	12	-0.80	-0.75	16	0	1	12	-0.30	-0.68
4	-0.60	14	-0.33	3	9	-0.70	-0.70	10	2	1	8	-0.30	-0.60
5	-0.90	9	-0.76	1	0	-0.90	-0.90	0	1	0	1	-0.80	-0.75
6	-0.70	10	-0.66	0	4	-0.80	-0.75	12	0	1	2	-0.40	-0.70

Data source: development document, testing tools, net scanner, judgment of experts.
Indices should have consistent direction of effects on the result. Here negative value will be used if the index helps to increase the security of software, otherwise positive value will be adopted.

Table 3: Overall weight of index got from AHP

index	1	2	3	4	5	6	7	8	9	10	11	12	13
weight	0.097	0.294	0.049	0.033	0.035	0.070	0.035	0.070	0.070	0.070	0.105	0.048	0.024

Through calculation, the overall coherence rate is 0.002, which is between 0 and 0.1, so data in Table 3 can be accepted. And in the obtained component matrix, the accumulative total information contribution rate of four extracted principal components is 96.579% which is much greater than the required rate 85%. And F_1 emphasizes index 6, 7, 1, 12; F_2 emphasizes index 3, 10; F_3 emphasizes index 13, 4, 8, 11; F_4 emphasizes index 9.

Table 4: The comparison of evaluation results in two methods

	PCA						Combination of AHP and PCA					
	F_1	F_2	F_3	F_4	F	rank	F_1	F_2	F_3	F_4	F	rank
1	-1.629	-1.402	-0.150	-1.219	-1.321	2	-0.170	0.078	-0.047	-0.070	-0.097	2
2	1.056	-2.028	1.217	1.392	0.526	4	0.115	0.097	0.071	0.095	0.101	5
3	2.304	0.062	-1.678	-1.254	0.998	5	0.174	0.011	-0.054	-0.055	0.090	4
4	3.383	1.619	1.161	-0.022	2.353	6	0.332	-0.142	0.136	0.039	0.186	6

| 5 | -4.048 | 1.557 | 0.989 | -0.127 | -1.927 | 1 | -0.334 | -0.060 | 0.019 | -0.052 | -0.205 | 1 |
| 6 | -1.362 | 0.498 | -1.170 | 1.432 | -0.679 | 3 | -0.130 | -0.004 | -0.110 | 0.048 | -0.083 | 3 |

Software security increases with the decrease of F, so the secure ranks of six softwares can be got. From Table 4, it can be found that ranks of software 2 and 3 in the two methods invert because of score variance of F_2, F_3, F_4, which emphasize initial index 11, 10, 9, 8; from Table 3, it can be found that the four initial indices have moderately higher weights and according more important effect on results. Software 3 has better behaviors on these indices than its counterpart, and this can be the main reply to the rank change.

From Figure 1, it can be found that the number of clusters is 4, which satisfies the actual requirements of evaluation. With reference on Table 4, it can be analyzed that cluster 2 is secure; cluster 1 is basically secure; cluster 3 is relatively secure; cluster 4 is insecure.

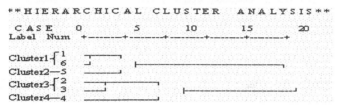

Figure 1: The clustering result of six softwares from CA

5. Conclusion

In the solution, the accumulative total information contribution rate of four extracted principal components is much greater than the required rate; therefore the result can fully reveal the key elements affecting software security. From the comparison of secure ranks in the two methods shown in Table 4, it can be safely concluded that owing to designating index weight in direct ratio to its importance, the combination of AHP and PCA can ensure objectivity of evaluation, and so has evident advantage over sole use of PCA. Besides, the results and secure grades given by CA can respond to user's requirement, which contributes to find out secure situation of software. The proposed method has been applied in practical evaluation of software security and the results confirmatively help to improve secure quality of software.

References

[1] John Viega and Gary McGraw, *Building secure software.* Addison-Wesley Press, Massachusetts, 2001.

[2] J. G. McDanie, Improving system quality through software evaluation, *Computers in Biology and Medicine* 32 127-140(2002).

[3] Fang Liu, Kui Dai and Zhiying Wang, Improving security architecture development based on multiple criteria decision making, *Lecture Notes in Computer Science* (Editor: Chi-Hung Chi, Kwok-Yan Lam), Springer-Verlag GmbH 214~218(2004).

[4] Taiying Zhu, Zhicai Juan and Hongwei Yao, Application of weighted principal component analysis in comprehensive evaluation for variable speed limits, *Theory and Practice of Systematic Engineering* 9 131-136(2004) (in Chinese).

[5] Tai-hoon Kim, Myong-chul Shin and Sang-ho Kim, Security requirements for software development, *Lecture Notes in Computer Science* (Editor: Mircea Gh. Negoita, Robert J. Howlett, Lakhmi C. Jain), Springer-Verlag GmbH 116~122(2004).

[6] Jianquan Zeng, Application of analytic hierarchy process to determining entrepreneur evaluation index weight, *Journal of Nanjing University of Science and Technology* 28(1) 99-104(2004)(in Chinese).

[7] R. A. Johnson and D. W. Wichern, *Applied multivariate statistical analysis*, 4th edn., trans. by Lu Xuan, Tsinghua University Press, Beijing, 2001.

Brill Academic Publishers
P.O. Box 9000, 2300 PA Leiden
The Netherlands

Lecture Series on Computer
and Computational Sciences
Volume 4, 2005, pp. 1766-1769

An Eavesdropping-Proof Identification Method For RFID Tags

J. Zhai[1], T. Jiang, and G. N. Wang

Industrial & Information Systems Engineering Department,
Ajou University,
443-749 Suwon, Korea

Received 27 June, 2005; accepted in revised form 25 July, 2005

Abstract: Radio Frequency Identification (RFID) is considered to be a promising identification approach in ubiquitous sensing technology. The operation of RFID systems in advanced applications may pose security and privacy risks to both organizations and individuals. In this paper, we propose an eavesdropping-proof identification algorithm based on the pseudo-random functions (PRFs) for passive RFID tags. Compared with the existing methods, our algorithm not only shows a distinct security improvement but also gives an adaptive approach to a specific application. While the existing methods might be designed for special situations, our proposed procedure shows a great adaptation capability, which could be well deployed in different realistic applications.

Keywords: RFID (Radio Frequency Identification), Eavesdropping-Proof Algorithm.

1. Introduction

The Radio Frequency Identification (RFID) technology is an emerging technology that uses radio waves to automatically identify individual items [1] [2]. Through automatic and real time data acquisition, this technology can give a great benefit to various industries by improving the efficiency of their operations [3] [4]. RFID applications could be found in the areas such as auto-distribution production line, warehouse security moving in and out check, and smart shelves in the shop. Compared with the existing bar code system, RFID can have special advantages; it does not require physical communication line, it can be reprogrammed easily, it can be used in harsh environment, it can store more data, and it can read many tags simultaneously. We might anticipate RFID will be the next generation solution in object identification and tracking material status under a ubiquitous computing environment.

RFID system basically consists of three components: transceiver (reader), transponder (tag), and data management infrastructure [5]. The reader can read data from and also write data to a transponder. The transponder, or tag, is usually attached to an object to be identified, and it stores data of the item. There are two types of RFID tags: the active one and the passive one. The passive tag seems to be much more attractive than active one because of the price factors, which the preferable price for pervasive deployment of RFID tags would be about $0.05 [5]. The remaining part would be the subsystem usually a data management infrastructure. It may contain both the enterprise application layer and the middle layer software like Savant [6]. These three parts cooperate efficiently to implement all kinds of applications using RFID technology.

The private security becomes extremely crucial along with the drastically increasing deployment of RFID tags. Due to the trend that RFID tags gradually become unified, it is quite easy to detect what other people carries just by a common reader. Except leaking the goods information of customers, one might also track people's location by continuously monitoring the tags people carries. Furthermore, if items in the whole supply chain of a company are equipped with RFID devices, it will be extremely dangerous to employ RFID technology in all wide supply chain process without any data encryption algorithm. The adversary of a company can easily obtain the complete manufacturing process

---

[1] Corresponding author. E-mail: zhaijiaws@vip.sina.com

information of its corresponding manufacture plant by just reading the unprotected tags. However, the inborn properties of RFID devices, such as low-power, low computing ability, storage resource-deficiency, and low-price for pervasive deployment make it quite difficult to apply a perfect data protection method in large scale applications. All the reasonable data encryption solutions for RFID tags try to find a good balance between the cost of hardware and the ability of data protection.

In this paper, we present a data encryption algorithm to prevent the eavesdroppers from monitoring the communication data between RFID devices. In the following sections, we look over some related research works firstly. Then, we propose our eavesdropping-proof identification method. Finally, the conclusion and further studies are also discussed.

2. Related Works

The previous works could be classified into two different aspects: one is concerning mainly with the encryption ability of tags and the other is considering both the protection ability and the corresponding hardware costs. The former has been discussed in previous Stephen A. Weis's work [7]. He gave two algorithms for solving the encryption problem. One is silent tree walking and backward channel key negotiation method, and the other is randomized access control method. However, all these methods just emphasize on improving encryption capability without considering the hardware implementation cost of RFID tags. Due to the price factor, this kind of algorithm might not be widely applied in real applications.

Based on the above discussion, it seems to be more attractive for real application to consider the hardware implementation cost as well as also consider the protection capability simultaneously, which is related to our works. In Sanjay E. Sarma's research work [5], a hash-based access control procedure is presented, which made a balance between the hardware implementation and the data protection task. Nevertheless, their algorithm was not efficient enough and was very easy to be disabled by the middle-attack. An attempt is given to improve the previous Sarma's work which possibly the previous improper points could be corrected or improved by our approach.

3. Problem Definition

To address our proposed procedure, we would like give some assumptions firstly. We assume the communication layer between readers and database infrastructure is not exposed to the eavesdroppers. It means that the eavesdroppers can only listen to the communication signals between tags and readers and can not get any information concerning the reader-database communication layer. Moreover, we assume each tag contains two reading status: locked and unlocked. In the locked status, the tags only send back the fake-key for any enquiries, whereas in the unlocked status, the tags offer all functionalities to the reader. Because the signal sent from tags to reader is relatively much weak, and it may only be monitored by eavesdroppers within the tag's shorter operating range, we also generally assume that in the unlocked status, eavesdroppers can not listen to the functional signals directly without any detection. On contrary, during the locked status, the eavesdroppers may monitor the fake-keys sent by the tags. Thus the key point of our eavesdropping-proof algorithm is to safely change the status of locked tags without being detected by eavesdroppers and also prevent the eavesdroppers from unlocking the tags illegally. The illustration of our assumptions is shown as Fig. 1.

4. Eavesdropping-Proof Identification Algorithm

For describing our algorithm clearly, we look over briefly the existing hash-based method at first. To lock a tag, the database computes a hash value of a random key and sends this hashed value to the tag as a fake-key saving the original keys in the database. In turn, the tag stores the fake-key and enters the locked status. While in the locked status, a tag responds to all queries only by the fake-key value. After the data-base received the fake-key value, it computes all the keys stored and finds which matches the fake-key then sends the matched key back to the tag. The tag hashes this value and compares it to the fake-key. If the value matches, the tag unlocks itself.

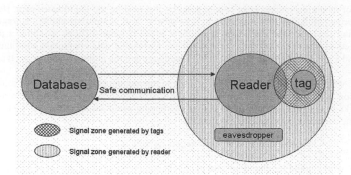

Figure 1: An illustration of eavesdropping scene in the unlocked status. Generally without detection, the eavesdropper can not monitor the functional communication data from tags to reader.

This algorithm is easy to be middle-attacked since one can query tags for getting the fake-key value and send it to the reader, then later unlock the tags with the reader's responding matched key. The crucial weak point of this algorithm is the interrogation key between readers and tags keeping static all the time. Hence, we propose our eavesdropping-proof identification algorithm dealing with this problem successfully next.

The main idea of our proposed algorithm is to keep the interrogation key between readers and tags changing randomly during the runtime of RFID system. We give two functions as mathematical basis below.

$$Y = Rand(X) \tag{1}$$

This is a pseudo-random function (PRFs) which can generate a random value of Y with the maximum value X.

$$Y = f(X) \tag{2}$$

This is a one way hash function.

To lock the tags at beginning, the reader uses (1) to generate two random keys as a pair (m, n) and uses m as input value computed by (2). Then the reader sends the computing result, n value to the tag and stores the computing result as a fake-key value storing n as a flag-key value. All those keys are also saved in the database system: m, n as a pair value (key, flag-key), and the computing result as fake-key value. After tags received the fake-keys and flag-keys, they enter the locked status and respond any enquiries only by the fake-key.

When we need the functionalities of a tag, the reader broadcasts the query signal firstly. After receiving the returned fake-key value from a tag, the reader uses (2) to compute all the keys stored in the database and finds which matches the fake-key value so that we can get a pair value (key, flag-key) related to the received fake-key. Then we generate a random value A by (1) and send (key, flag-key) and value A back to the tag. The tag uses key as input to compute result by (2) and compares the result with stored fake-key value as well as compares the received flag-key value with the one it stored. If the two values all match, the tag unlocks itself. Finally, both the database system and the tag set the flag-key as a new value A. The completed state diagram is shown in Fig. 2.

Before the unlocking action, all the tags check carefully if the flag-key values matched, and the flag-key value stored both in the database system. Since the flag-key value is a stochastic value that keeps changing at each interrogation cycle, our algorithm could resolve the middle-attack efficiently by just employing a PRFs (pseudo-random functions) in the database system. Using the implementation of database system effectively, our approach could be well deployed in a wide application.

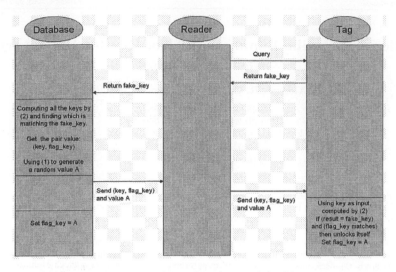

Figure 2: The state diagram of our proposed eavesdropping-proof algorithm.

5. Conclusion

We have demonstrated an eavesdropping-proof algorithm for identifying passive RFID tags safely. Illustrative examples are given to show that our algorithm could be utilized efficiently to stop eavesdroppers from monitoring the data communication in RFID systems. Compared with the existing methods, our method is much more robust as it has successfully solved the middle-attack problem without adding too much hardware burden in the RFID systems. Furthermore, the proposed method could be effectively utilized by giving suitable trade-off between hardware costs and data encryption ability. Finally, the adaptation ability might be also another advantage of our algorithm which paves the way of its universal employment. As a further study, application specific algorithms can be developed based on our research work, which might facilitate the pervasive deployment of RFID system.

References

[1] J. Zhai and G. N. Wang: An Anti-collision Algorithm Using Two-Functioned Estimation for RFID Tags. ICCSA 2005, LNCS 3483. 702-711. (2005)

[2] Klaus Finkenzeller: RFID Handbook: Fundamentals and Applications in Contactless Smart Card and Identification. Second Edition. John Wiley & Sons. 221-228 (2002)

[3] Bornhovd C., Tao Lin, Haller S.; Schaper J.: Integrating Smart Items with Business Processes An Experience Report. System Sciences, 2005. HICSS '05. Proceedings of the 38th Annual Hawaii International Conference on , 03-06 Jan. 227c-227c (2005)

[4] Geng Yang, Jarvenpaa S.L.: Trust and Radio Frequency Identification (RFID) Adoption within an Alliance. System Sciences, 2005. HICSS '05. Proceedings of the 38th Annual Hawaii International Conference on , 03-06 Jan. 208a-208a (2005)

[5] Sanjay E. Sarma, Stephen A. Weis, and Daniel W. Engels: RFID Systems and Security and Privacy Implications. CHES 2002, LNCS 2523 454-469 (2003)

[6] Clark S., Traub K., Anarkat D., Osinski T.: Auto-ID Savant Specification 1.0. Auto-ID Center, White Paper MIT-AUTOID-TM-003, Sep. (2003)

[7] Stephen A.Weis, Sanjay E. Sarma, Ronald L. Rivest and DanielW. Engels: Security and Privacy Aspects of Low-Cost Radio Frequency Identification Systems. Security in Pervasive Computing 2003, LNCS 2802 201-212 (2004)

Brill Academic Publishers
P.O. Box 9000, 2300 PA Leiden,
The Netherlands

Lecture Series on Computer
and Computational Sciences
Volume 4, 2005, pp 1770-1773

Blocking Neville Elimination Algorithm for Exploiting Cache Memories

P. Alonso[1], R. Cortina[2], I. Díaz[2], J. Ranilla[2]

[1]Department of Mathematics.
[2]Department of Computer Science.
University of Oviedo, Campus de Viesques.
E-33271 Gijón, Spain.
E-mail: hpc@aic.uniovi.es

Received 10 July, 2005; accepted in revised form 27 July, 2005

Abstract: Neville elimination is a method for solving a linear system of equations that introduces zeros in a matrix column by adding to each row an adequate multiple of the previous one. In this paper we explore block algorithms for Neville elimination which take into account the memory hierarchies of a computer. These algorithms try to manage the memory movements to optimize them. Thus, the matrix of the system is divided following three different strategies, blocking by rows, columns or submatrices. In each case we study the performance of the algorithm according to the parameter q which evaluates the ratio of flops to memory references. Theoretical estimations show that q depends on data partitioning, being submatrix blocks the best choice.

Keywords: Neville method, performance, blocking matrices.

Mathematics Subject Classification: 65F05, 65Y20, 68W40.

1 Introduction

The recent availability of advanced-architecture computers has had a significant impact on all spheres of scientific computation, including algorithm research and software development in numerical linear algebra. In particular, the solution of linear systems of equations lies at the heart of most calculation in scientific computing.

Regarding linear algebra, Neville elimination is considered as a starting point the interpolation formula of Aitken-Neville. This process appears in a natural way when Neville strategy of interpolation is used for resolving linear systems; this fact also occurs with the Gaussian method and the strategy of Aitken. Neville elimination lies in making zeros in a matrix column by adding to each row an adequate multiple of the previous one (see [10]). This process is an alternative to Gaussian elimination which has been proved to be very useful with totally positive matrices, sign-regular matrices or other related types of matrices (see [1] and [9]), without increasing the error bounds (see [12]).

Several ways of improving the performance of an algorithm can be taken into account. Three remarkable issues are the use of parallel computing (see [3] and [4]), the profitable usage of memory hierarchies and the work with matrix operations (see [2], [6] and [11]).

In this work we analyze the performance of this method working with the memory hierarchy of a computer (see [5] and [13]). Therefore we will analyze the relationship between the number of flops and memory movements (which is noted as q ratio). The performance of Neville method will be weighted in terms of this ratio. Thus, we will try to improve it by studying different matrix blocking schemes and combining them with a suitable computing strategy.

2 Partitioning Algorithms for Exploiting Cache Memories

Whatever algorithm can be implemented in several ways in order to take advantage of this memory structure. For example the order of the loops (if they exist) can be swapped. Then, it is possible to obtain different implementations with different performances.

Let m be the number of memory references and f the floating point operations performed by a subroutine. Then the total running time (if the arithmetic and memory references are not performed in parallel) is as large as

$$ft_{arith} + mt_{mem} = ft_{arith}(1 + \frac{m}{f}\frac{t_{mem}}{t_{arith}}) = ft_{arith}(1 + \frac{1}{q}\frac{t_{mem}}{t_{arith}}), \tag{1}$$

where t_{arith} and t_{mem} is respectively the time needed to make a floating point operation and a memory reference in seconds and

$$q = \frac{f}{m} \tag{2}$$

is the ratio of flops to memory references. Note that algorithms with the larger q values are better building blocks for other algorithms. The parameter q tells us how many flops we can perform on average per memory reference or how much useful work we can do compared to the time moving data.

Our goal is to analyze q when different data partitioning as well as different computing strategies are taking into account in Neville method. Our reference is the Basic Linear Algebra Subroutines (BLAS) (see [5] and [7]) which are standardized and optimized subroutines to perform matrix operations whose efficiency is examined in terms of the memory references and floating point operations.

Therefore, we will examine in detail how to implement Neville elimination algorithm to optimize the number of memory moves and so optimize its performance. To accomplish this task we assume that there are two levels of memory hierarchy, fast and slow, where the slow is large enough to contain the $n \times n$ matrix but the fast memory contains only M words where $2n < M \ll n^2$; this means that the fast memory is large enough to hold two matrix columns or rows but not a whole matrix.

2.1 Partitioning Neville Elimination Algorithm

In this section we block the matrix used to solve a linear equation problem by Neville elimination algorithm and whether it is useful to reduce the number of memory moves. Thus, three possibilities are studied: The matrix elements are moved to fast memory grouped by rows, by columns or by submatrices. Let us study these three possibilities.

2.1.1 Row Blocks

In this case the matrix is divided into p row blocks of r rows ($pr = n$). Once a row block is moved to fast memory, we make zeros below the diagonal in all the columns where we can do it.

As each block has nr elements, the number of moved elements is

$$nrp^2 = pn^2, \tag{3}$$

and then, q is

$$q = \frac{f}{m} = \frac{\frac{2n^3}{3}}{pn^2} = \frac{2r}{3}. \tag{4}$$

On the one hand, the lowest performance (with regard to parameter q) is got when $r = 1$, namely, when the data are moved from slow memory to fast memory row by row. On the other hand, the highest q is reached when $r = n$ (and $p = 1$). Therefore, fast memory must contain in this case n^2 elements, which is a too expensive memory requirement. Therefore there exists a trade-off between the size of the fast memory and q.

2.1.2   Column Blocks

In this case the matrix is divided into p column blocks of r columns ($pr = n$). Following the same strategy as in 2.1.1 the elements moved are pn^2, and q is

$$q = \frac{f}{m} = \frac{\frac{2n^3}{3}}{pn^2} = \frac{2r}{3}. \tag{5}$$

The lowest q is got when $r = 1$ and the highest $r = n$. Once again, when $r = 1$ this case is reduced to the unblocked one. In addition, for certain values of r, the value of q is higher than the obtained for Level 2 BLAS operations.

2.1.3   Blocking by submatrices

In this case the matrix is divided into p^2 blocks of order $r = \frac{n}{p}$. Again, once a block is moved to fast memory, we make as many zeros as possible.

As each block has $r^2 = \frac{n^2}{p^2}$ elements the total movements are pn^2. Then, q is

$$q = \frac{f}{m} = \frac{\frac{2n^3}{3}}{pn^2} = \frac{2r}{3}. \tag{6}$$

As the size of the fast memory M has to be at least $2r^2$ because we need two blocks in fast memory to perform the algorithm, we have that $r \simeq \sqrt{\frac{M}{2}}$ and then

$$q = \frac{2}{3}\sqrt{\frac{M}{2}} \tag{7}$$

which means that q only depends on the size of fast memory. In particular q grows independently of n as M grows, which means that we expect the algorithm to be fast for any matrix size n and go faster if the fast memory size M is increased. There are both attractive properties.

3   Final Remarks

In this work we have studied how to implement Neville elimination algorithm to take advantage of the memory hierarchy of a computer. In this sense we have examined the floating point operations with regard to the memory references (q ratio) when different partitioning and computing strategies are applied. In addition, we have obtained that the best strategy is to partition the coefficient matrix by submatrices.

With regard to the computation strategy the study followed here assures that the best perfor-mance is got when we make all possible zeros over a block stored in fast memory. The performance obtained, weighted in terms of q ranges from $\frac{2}{3}$ to $\frac{2}{3}\sqrt{\frac{M}{2}}$ where M is the size of fast memory. Note that the upper bound depends on M. Then, an increase of M produces an increase of q so that it is possible to overtake the value of $n/2$ got when matrix operations are considered. Nevertheless, it is easy to obtain a performance between Level-2 and Level-3 BLAS if the fast memory has a reasonable size.

Acknowledgment

The research reported in this paper has been supported in part under MEC and FEDER grant TIN2004-05920.

References

[1] T. Ando, Totally positive matrices, *Linear Algebra Appl.* 90: 165-219 (1987).

[2] P. Alonso and J.M. Peña, Development of block and partitioned Neville elimination, *Cr. Acad. Sci. I-Math.* 329: 1091-1096 (1999).

[3] P. Alonso, R. Cortina, V. Hernández and J. Ranilla, Study the performance of Neville elimination using two kinds of partitioning techniques, *Linear Algebra Appl.* 332-334: 111-117 (2001).

[4] P. Alonso, R. Cortina, I. Díaz, V. Hernández and J. Ranilla, A Columnwise Block Striping in Neville Elimination, *Lecture Notes in Comput. Sci.* 2328: 379-386 (2002).

[5] J. Demmel, *Applied Numerical Linear Algebra*, SIAM, Philadelphia (1997).

[6] J. Dongarra, J. Du Croz, I.S. Duff and S. Hammarling, A set of Level-3 basic linear algebra subroutines. *ACM Trans. Math. Softw.* 16: 1-17 (1990).

[7] J. Dongarra and V. Eijkhout, Numerical Linear Algebra Algorithms and Software, *Journal of Computational and Applied Mathematics* 123(1-2): 489-514 (2000).

[8] M. Gasca and C.A. Michelli, eds., *Total Positivity and its Applications, Mathematics and its Applications 359*, Kluwer Academic, Dordrecht (1996).

[9] M. Gasca and G. Mühlbach, Elimination techniques: from extrapolation to totally positive matrices and CAGD, *J. Comput. Appl. Math.* 122: 37-50 (2000).

[10] M. Gasca and J.M. Peña, A matricial description of Neville elimination with applications to total positivity, *Linear Algebra Appl.* 202: 33-45 (1994).

[11] G.H. Golub and C.F. Van Loan, *Matrix Computations*, Johns Hopkins, University Press, Baltimore, Maryland (1989).

[12] N.J. Highham, *Accuracy and Stability of Numerical Algorithms*, SIAM, Philadelphia (1996).

[13] S. Toledo, A survey of out-of-core algorithms in numerical linear algebra, *External Memory Algorithms*, Dimacs Series in Discrete Mathematics and Theoretical Computer Science 50: 161-179, American Mathematical Society, Boston, MA, USA (1999).

Brill Academic Publishers
P.O. Box 9000, 2300 PA Leiden,
The Netherlands

*Lecture Series on Computer
and Computational Sciences*
Volume 4, 2005, pp. 1774-1777

Contact between Solids with Geometric and Material Non-linearities by FETI Domain Decomposition Method

J. Dobias[1], S. Ptak[2]

Institute of Thermomechanics,
Academy of Sciences of the Czech Republic,
Dolejskova 5, 182 00, Prague 8, Czech Republic

V. Vondrak[3], Z. Dostal[4]

Department of Applied Mathematics,
VSB-Technical University Ostrava,
17. listopadu 15, 708 33, Ostrava-Poruba, Czech Republic

Received 7 May, 2005; accepted in revised form 7 May, 2005

Abstract: The mechanical contact between solid bodies belongs to strong non-linear problems and its solution usually poses signi cant computational di culties in non-trivial cases. Trying to solve contact between two or more bodies, we can meet another non-linearities, namely the geometric and material ones, which often occur along with the contact problems. This paper is concerned with the application of the FETI (Finite Element Tearing and Interconnecting) method to solution to contact problems while in addition we also consider another non-linearities, because we assume that the bodies in contact can experience large displacements and rotations and also exhibit inelastic material behaviour. The basic ideas of the FETI methods and its application to the solution to contact problems along with the algorithm considering all the mentioned non-linearities will be presented. The precision and robustness of our method will be documented by the results of numerical experiments in terms of the stress distributions and convergence rate.

Keywords: Contact, Domain decomposition, Geometric non-linearity, Material non-linearity, Finite element method

Mathematics Subject Classi cation: 65N30, 65N55, 37J05

PACS: 02.60.Lj, 02.70.Dh, 46.15.Cc, 46.55.+d

1 Introduction

Contact modelling is still one of the most challenging aspects of non-linear computational mechanics. In spite of the fact that the tractions generated by contact are formally of the same form as the boundary conditions introduced by externally applied loadings, we do not in general know either the distributions of the contact tractions throughout the areas currently in contact or shapes and

---

[1] Corresponding author. E-mail: jdobias@it.cas.cz
[2] E-mail:svatopluk.ptak@it.cas.cz
[3] E-mail:vit.vondrak@vsb.cz
[4] E-mail:zdenek.dostal@vsb.cz

magnitudes of these areas until we have run the problem. Their evaluation have to be part of the solution.

There exist two basic methods to simulate the contact boundary conditions, and some others, which stem from these two methods, see for example [1]. The rst one is the Penalty method which is based on penalization of possible interpenetration of bodies in contact by some suitable penalty parameter. The second one is the Lagrange multiplier method. Consider the displacement based nite element method. Then making use of the latter method results in introduction of new variables in addition to the displacements, which are called the Lagrangian multipliers. These multiplier have, in this context, physical meaning of contact forces.

In 1991 Ch. Farhat and F.-X. Roux introduced a novel domain decomposition method called FETI [2] which is an acronym standing for Finite Element Tearing and Interconnecting method. This method belongs to the class of non-overlapping totally disconnected decompositions. The interface between sub-domains can pass through the bodies, which generates ctitious border, or along the contacting surfaces in the case we consider a system of bodies in mutual contact. The FETI method is based on an idea that the compatibility between the sub-domains can be enforced in terms of forces or the Lagrangian multipliers, which we call in this context the dual variables, while the primal variables, or displacements, are eliminated. It is obvious that the conception that every individual sub-domain, into which the body is partitioned, interacts with its neighbours in terms of the Lagrangian multipliers can naturally be applied to the solution to contact problems.

We developed algorithm for solution to the contact problems within the framework of the FETI method. It is called MPRGP (Modi ed Proportioning with Reduced Gradient Projection). It is directly applicable to the solution to the contact problems, but with other conditions linear, i.e. for linear elasticity with small displacements and rotations, and frictionless contact problems. Any other non-linearity necessitates introduction of additional outer iteration loop.

2 FETI Method as Contact Solver

Let us brie y outline some fundamental ideas and formulae which lay foundation of the FETI method and its application to the solution of the contact problems. Consider a system of solid deformable bodies in contact, discretized in terms of the nite element method and in addition decomposed into sub-domains. The numerical approximation to this problem can be expressed as reads

$$\min \quad \frac{1}{2}u^T A u \quad f^T u \quad \text{subject to} \quad B^I u \quad 0 \quad \text{and} \quad B^E u = 0 \qquad (1)$$

where A stands for a positive semide nite sti ness matrix, B^I and B^E denote the full rank matrices which enforce the discretized inequality constraints describing conditions of non-interpenetration of bodies and inter-subdomain equality constraints, respectively, and f accounts for the external forces. u denotes the vector of unknown nodal displacements.

Making use of the FETI techniques according to [5], the above problem can be transformed by elimination of the primal variable u into the quadratic programming problem with equality constraints and non-negativity bounds as follows

$$\min \quad \frac{1}{2} \quad {}^T B A^\dagger B^T \quad {}^T B A^\dagger f \quad \text{subject to} \quad {}_I \quad 0 \quad \text{and} \quad R^T (f \quad B^T \quad) = 0 \qquad (2)$$

where $A^\dagger$ denotes generalized inverse of A and satis es one of the Penrose axioms, namely $AA^\dagger A = A$. R stands for the full rank matrix with columns spanning the kernel of A. We assemble this matrix in such a way that its entries belong to $\{0, 1\}$ and each column corresponds to a oating sub-domain with the non-zero entries in the positions corresponding to the indices of the nodes belonging to this sub-domain.

Even though problem (2) is much more suitable for computations than (1), further improvement may be achieved by adopting some results by Farhat, Mandel, Roux and Tezaur from [6, 7]. Let us denote

$$F = BK^\dagger B^\mathsf{T}, \quad \widetilde{G} = R^\mathsf{T} B^\mathsf{T}, \quad \widetilde{e} = R^\mathsf{T} f, \quad \widetilde{d} = BK^\dagger f,$$

and let $\widetilde{\ }$ solve $\widetilde{G}\,\widetilde{\ } = \widetilde{e}$, so that we can transform the problem (2) to minimization on the subset of the vector space by looking for the solution in the form $\ = \ + \widetilde{\ }$. Using this notation, the problem (2) is equivalent to

$$\min \; \frac{1}{2}\,{}^\mathsf{T} F\, \ - \ {}^\mathsf{T} d \quad \text{subject to} \quad G\, = 0 \quad \text{and} \quad {}_I \ \geq \ \widetilde{\ }_I \tag{3}$$

where $d = \widetilde{d} - F\widetilde{\ }$ and G denotes a matrix arising from the orthonormalization of the rows of $\widetilde{G}$.

This formulation of the problem turns out to be a convenient starting point for development of e cient algorithms for variational inequalities. It is called MPRGP. Its description and theoretic background can be found in [3, 4].

3   Contact Problem Solution with Other Non-Linearities

Consider a problem where in addition to the contact non-linearity we have to treat another non-linearities, namely the geometric and material ones. To this end we developed algorithm that stems from the MPRGP algorithm, but we add outer iteration loop for treatment of the geometric/material non-linearities. Its simpli ed version is shown in the following owchart

Initial step: Assemble sti ness matrix $A = diag\{A_1, ..., A_p\}$, vector of external load f, and matrix of continuity conditions between sub-domains B_E ;
Set $i = 0$; $\mathbf{u}^0 = 0$, ${}^0 = 0$, $f^0_{int} = 0$;

Step 1: Evaluate contact conditions B^i_I ;

Step 2: Solve contact problem by MPRGP for $\rightarrow$ $\mathbf{u}$;

Step 3: ${}^i = {}^{i-1} + \ $, $\mathbf{u}^i = \mathbf{u}^{i-1} + \ \mathbf{u}$;
$f^i_{int} = \sum\limits_{nelem} \int\limits_{V_e} B_s{}^\mathsf{T}({}^i)\ ({}^i)\, dV$
Assemble residual load vector $res^i = f - B^\mathsf{T}\,{}^i - f^i_{int}$;
Check on convergence criteria $\frac{\|\ \mathbf{u}\|}{\|\mathbf{u}^i\|}$, $\frac{\|res^i\|}{\|f^i_{ext}\|}$;
if met STOP,
otherwise set $i\ \ i+1$ and go to **Step 1**

4   Numerical Experiments

Numerical experiments were carried out with our in-house general purpose nite element package PMD (Package for Machine Design)[8]. Our algorithms were tested on problem of two cylinders in contact with parallel axes (the radius of the upper cylinder is 1 m, and of the lower cylinder is in nite, Young's modulus of both bodies, $E = 2.0 \times 10^{11}$ Pa, and Poisson's ratio, $\ = 0.3$). This problem belongs to the class of Hertzian ones in the case it is geometrically and materially linear. Then its analytic solution is known and can be used for comparison with numerical solutions. Moreover, we also carried out experiments with both geometric and material non-linearities.

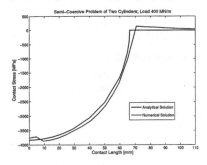

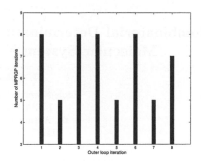

Figure 1: Distribution of Contact Stress along the Surface of the Lower Cylinder.

Figure 2: Number of MPRGP Iterations.

Herein, two gures will su ce for demonstration. Figure 1 shows distribution of the contact stress along the surface of the lower cylinder. The case is geometrically non-linear, but the g-ure also shows analytical solution for the same but geometrically linearly evaluated problem for comparison. Figure 2 depicts numbers of necessary MPRGP iterations in the above algorithm.

Acknowledgment

The authors would like to acknowledge the support of the Czech Science Foundation through grant # 101/02/0423.

References

[1] J. Dobias, Contact Boundary Conditions, *Engineering Mechanics*, **4** 169–182 (1997).

[2] Ch. Farhat, F.-X. Roux, A Method of nite element tearing and interconnecting and its parallel solution algorithm, *Int. J. Numer. Methods Engng.*, **32** 1205–1227 (1991).

[3] Z. Dostal, D. Horak, R. Kucera, V. Vondrak, J. Haslinger, J. Dobias, S. Ptak, FETI Based Algorithms for Contact Problems: Scalability, Large Displacements and 3D Coulomb Friction, *Comput. Methods Appl. Mech. Engrg.*, **194** 395–409 (2005).

[4] Z. Dostal Z., J. Schoberl, Minimizing Quadratic Functions over Non-Negative Cone with the Rate of Convergence and Finite Termination, *Comput. Optim. and Appl.*, **30** 23–43 (2005).

[5] Z. Dostal, et al., Solution of Contact Problems by FETI Domain Decomposition with Natural Coarse Space Projection, *Comput. Methods Appl. Mech. Engrg.*, **190** 1611–1627 (2000).

[6] Ch. Farhat, J. Mandel, F.-X. Roux, Optimal Convergence Properties of the FETI Domain Decomposition Method, *Comp. Meth. in Appl. Mech. and Eng.*, **115** 365-385(1994).

[7] J. Mandel, R. Tezaur, Convergence of Substructuring Method with Lagrange Multipliers *Numerische Mathematik* **73** 473–487 (1996).

[8] Anon., PMD Manuals *www.it.cas.cz/manual/pmd*.

Brill Academic Publishers
P.O. Box 9000, 2300 PA Leiden,
The Netherlands

Lecture Series on Computer
and Computational Sciences
Volume 4, 2005, pp. 1778-1781

Combinatorial Determination of the Volume Spanned by a Molecular System in Conformational Space.

J. Gabarro-Arpa[1]

Ecole Normale Supérieure de Cachan,
LBPA, CNRS UMR 8113,
61, Avenue du Président Wilson,
94235 Cachan cedex, France

Received 10 July, 2005; accepted in revised form 27 July, 2005

Abstract: In a previous paper [1] it was developped a discrete model where the conformational space of a molecule was divided into a finite set of cells by means of a central hyperplane arrangement. This partition is encoded by the face lattice of the arrangement. In a second paper [2] we described a procedure for embbeding $3D$ structures in conformational space. Here we describe an algorithm for solving the following problem: for a molecular system immersed in a heat bath we want to enumerate the subset of cells in conformational space that are visited by the molecule in its thermal wandering. If each cell is a vertex on a graph with edges to the adjacent cells, it is explained how such graph can be built.

Keywords: Molecular Dynamics, Molecular Conformational Space, Hyperplane Arrangement, Face Lattice

Mathematics Subject Classification: 52B11, 52B40, 65Z05

PACS: 02.70.Ns

1 Introduction

Molecular dynamics simulations (MDS) are an essential tool for the modeling of large and very large molecules, it gives us a precise and detailed view of a molecule's behaviour [3]. However, it has two limitations that hamper many practical applications: it is a random algorithm, as such it does not perform a systematic exploration of molecular conformational space (CS); and that currently, the output from an MDS represents only a very small fraction of the volume spanned by the system in CS.

Here it is presented a complementary approach that locally is less precise but that can encompass a broader view of CS. It consists in dividing CS into a finite set of cells, so that the only knowledge we seek about the system is whether it can be located in a given cell or not.

As was extensively discussed in ref. [1] the partition is a variant of the Coxeter $\mathcal{A}_n$ matroid [4-5]: an arrangement of hyperplanes passing through the origin[2] that divides CS into a set of cells shaped as polyhedral cones, such that for a molecule with N atoms we have $(N!)^3$ cells. The set of

---

[1] E-mail: jga@infobiogen.fr
[2] i.e. **central.**

hyperplanes is also a Coxeter reflection arrangement: the arrangement is invariant upon reflection on any of the hyperplanes.

This structure has three important properties [1]:

1. Associated with a Coxeter arrangement there is a polytope [5] whose symmetry group is the reflection group of the arrangement. The **face lattice poset**[3] of the polytope is a hierarchichal combinatorial structure that enables us to manage the sheer complexity of CS, since with simple codes we can describe from huge regions down to single cells.

2. The information needed to encode any face in the polytope is a sequence of integers $3N$ called **non-crossing partition sequence** [1-6].

3. The construction is **modular**: if we consider the CS of two subsets of atoms from a system, the CS of the union set has an associated polytope that is the cartesian product of the polytopes[4] of the two subspaces, and its partition sequence is the union of the partition sequences [1].

The last is particularly important since the CS of the whole system can be built from that of the parts, and the CS of a small number of atoms is very much smaller than that of the molecule and we can reasonably assume that it can be thoroughly explored by an MDS. Moreover, in merging the CSs corresponding to subsets of atoms the number of cells grows exponentially while the length of coding sequences grows only linearly.

2 The orientation problem

As discussed in [2] the discrete model of CS takes into account two symmetries of the Euclidian $3D$ space: translation and scaling, but not **rotation** because it is a **non-linear**[5] transformation.

The consequence is that in CS we cannot distinguish between any two points that correspond to rotated structures. This is a serious drawback, but the solution to this problem also happens to be essential for solving the cell enumeration problem.

As explained above the discrete model of CS is modular and its complexity (i.e. the number of cells) rises exponentially with dimensions; but for the smallest $3D$ structure: the **simplex**[6], the CS has a mere 13824 cells. Simplexes come in a multitude of shapes and morphologies, the approach that was developed in [2] aimed at establishing a morphological classification of simplexes, because a given morphological class can only reach a fraction of the cells in CS (typically about one third) and a procedure can be found to enumerate them [2].

Moreover, there are two ways of going from one cell to another in CS: either through a geometrical deformation or a rotation. The second reason for establishing a morphological classification of simplexes is that constraining a structure into a given morphology greatly restricts the amount of available geometrical deformation.

Thus we sought a classification fulfilling the following condition: **for any structure from a class every cell in the class CS can be reached through a rotation.** That ensures that any accesible cell correponds to an **orientation** of the simplex with respect to an arbitrary reference

---

[3]The faces in the induced decomposition of the polytope ordered by inclusion.

[4]If $P \subset \mathbb{R}^p$ and $Q \subset \mathbb{R}^q$ are polytopes the product polytope $P \times Q$ has the set of vertices $(x, y) \in \mathbb{R}^{p+q}$ where x and y are vertices of P and Q respectively.

[5]That cannot be performed solely by adding vectors and multiplying a vector by a scalar.

[6]A three-dimensional polytope with four vertices.

frame, in what follows the term orientation is also used to designate the cells of adjacent simplexes whose partition sequence is compatible[7] with that of the reference cell.

If (v_1, v_2, v_3, v_4) is the set of vertices of a simplex then

$$e_{ij} = v_i - v_j \qquad 1 \leq i < j \leq 4 \tag{1}$$

$$f_{ijk} = e_{ij} \wedge e_{jk} \qquad 1 \leq i < j < k \leq 4 \tag{2}$$

$$f_{12} = e_{12} \wedge e_{34} \ , \ f_{13} = e_{13} \wedge e_{24} \ , \ f_{14} = e_{14} \wedge e_{23} \tag{3}$$

e_{ij}, f_{ijk} and f_{ij} are the vectors along the edges, perpendicular to the faces and perpendicular to pairs of non-contiguous edges respectively. The 36 quantities

$$SIGN(e_{ij}.e_{kl}) \quad i \leq k \ , \ j < l \quad , \quad SIGN(f_\alpha.f_\beta) \quad \alpha < \beta \tag{4}$$

refer mostly to the angles between adjacent edges and dihedral angles between contiguous faces: 1, 0 and −1 are for acute, right and obtuse angles respectively. There are 3936 realizable 36-digit sequences each representing a morphological class, these classes form a connected graph (excluding sequences containig 0's wich give sets of null measure).

The restrictions imposed on the morphology of simplexes by (4) are still not enough to fulfill the condition defined above. To satisfy this condition each class was further decomposed into a set of derived orientation subclasses that were described in [2], these were determined empirically by Monte Carlo simulation, and a total of 125712 subclasses were found.

The classes (4) offer the possibility of analysing an MDS in terms of discrete entities. The range of morphological variation for simplexes within the molecule on a typical MDS can be summarized as follows

- 90% of simplexes in a structure evolve within less than 20 classes (4).

- The maximum variation observed is somewhat less than 200 classes, about 5% of the total.

A simplex constrained to evolve within a small number of connected geometrical morphologies also occupies a small volume in CS, a reasonable assumption is that it can be thoroughly scanned by a MDS, what can not be completely scanned by a simulation are the combinations of local movements. The CS volumes of individual simplexes obtained from MDSs can be merged progressively to form the CS volume of the system, as explained above the morphological subclasses are such that any cell corresponds to an orientation of the simplex, thus the merging can be done excluding the redundant rotated structures.

3   The construction of the graph

The cells from the volume in CS occupied by a molecular system form a connected graph where each cell is a node with edges towards the adjacent cells, with the formalism described above we can define a procedure for building the graph.

We begin by picking an arbitrary simplex, preferably one with low dynamical activity, and by choosing an orientation among those available to it, this will be the simplex on level 1, the simplexes adjacent to this one form the level 2, and so on. Since adjacent simplexes in a $3D$ structure share

---

[7]Two adjacent cells are said to be compatible if the partition sequences of the three shared vertices are the same for each cell [1].

three vertices the shortest adjacency path between any two simplexes has at most length 4, so we end up with simplexes in 5 levels.

Starting at level 1:

1. From any simplex in level n we select the compatible orientations in the adjoining simplexes in level $n + 1$.

2. From any simplex in the level $n + 1$ we select compatible orientations on the adjoining simplexes at the same level.

3. If $n < 5$ we go to step 1 and continue with level $n + 1$.

A link is created between any two compatible orientations in adjacent simplexes. This is done in two steps:

1. If the simplex in the lower level has not yet been visited any orientation compatible with those from the simplex in the upper level is selected.

2. Otherwise the orientations not already present are discarded. And likewise an orientation that fails to form a link with an adjacent simplex is discarded because of geometrical inconsistency.

The implementation of this procedure as an efficient computer algoritm requires that the CS of a class of simplexes can be quickly searched. Compatible orientations from an adjacent simplex can be retrieved from a hash table [7] whose entries are: the class of the simplexes and the partition sequence of the shared vertices.

References

[1] J. Gabarro-Arpa, A central partition of molecular conformational space. I. Basic structures, *Comp. Biol. and Chem.* **27** 153-159 (2003).

[2] J. Gabarro-Arpa, A central partition of molecular conformational space. II. Embedding 3D-structures, *Proceedings of the 26th Annual International Conference of the IEEE EMBS*, San Francisco, 3007-3010 (2004).

[3] M. Karplus and J.A. McCammon, Molecular dynamics simulations of biomolecules, *Nature Struct. Biol.* **9** 949-852 (2002).

[4] I.M. Gelfand and V.V. Serganova, Combinatorial geometries and torus strata on homogeneous compact manifolds, *Russian Math. Surveys* **42** 133168 (1987).

[5] A.V. Borovik, I.M. Gelfand and N. White, *Coxeter Matroids*, Progress in Mathematics **216**, Birkhäuser, Boston (2003).
http://www.ma.umist.ac.uk/avb/M-title.html

[6] G. Kreweras, Sur les partitions non croisées d'un cycle, *Disc. Math.* **1** 333-350 (1972).

[7] J. Gabarro-Arpa, A central partition of molecular conformational space. III. Combinatorial determination of the volume spanned by a molecular system, in preparation.

Brill Academic Publishers
P.O. Box 9000, 2300 PA Leiden,
The Netherlands

*Lecture Series on Computer
and Computational Sciences*
Volume 4, 2005, pp. 1782-1785

Computed lifetimes of metastable states of CO^{2+}

T. Šedivcová[1], J. Fišer[2], V. Špirko[1]

[1] Department of Molecular Modeling Center for Biomolecules and Complex Molecular Systems
Institute of Organic Chemistry and Biochemistry, Czech Academy of Sciences,
Flemingovo nám. 2, 166 10 Prague 6, Czech Republic
[2] Charles University, Faculty of Science, Department of Physical and Macromolecular Chemistry
Hlavova 2030, Prague 2, Czech Republic

Received 10 July, 2005; accepted in revised form 27 July, 2005

Abstract: Highly correlated icMRCI wavefunctions are used to calculate the potential
energy curves (PEC's) and spin-orbit coupling (SOC) matrix elements for the lowest electronic states of the CO^{2+} dication. The results are employed in calculations of the spectroscopic constants and the spin-orbit induced predissociation lifetimes using the stabilization
method. The computed lifetimes are in good agreement with experimental lifetimes reported previously.

Keywords: Metastable state, lifetimes, Stabilization method, complex scaling

PACS: ab initio calc's 31.15.A, lifetimes 33.70.C, spin orbit coupling 33.15.P, potential
energy surfaces 31.50.G

1 Introduction

Diatomic dications exhibit a wide range of lifetimes depending on the shape of PEC's, SOC's, decay
mechanism and position of a vibronic level with respect to the curve-crossing point or the barrier
height. Though the lowest vibrational levels of CO^{2+} ($X^3\Pi$) can be considered to be relatively
stable against unimolecular dissociation, lifetimes of the order of submicroseconds have been also
reported [1, 2, 3]. The previous theoretical estimates of the mean lifetimes of CO^{2+} [4] differ from
the measured values by some orders of magnitude. Therefore, we find it worthwhile to calculate the
latter characteristics once again using a more accurate theoretical procedure; stabilization method
[5, 6]. These calculations are based on the high-level PEC's and SOC functions. The tunneling
lifetimes are also determined using log-phase-amplitude method [7].

2 Electronic structure calculation

2.1 Potential energy curves and spin-orbit couplings

PEC's were calculated using the internally contracted multireference configuration interaction
(icMRCI). The orbitals for the CI treatment were obtained in state-averaged complete-active-
space SCF (CASSCF) calculations with equal weights for the participating states of the same spin
multiplicity. The CAS consisted of 1σ-6σ, 1π and 2π orbitals in C_{2v} symmetry with all electrons
correlated. The correlation-consistent cc-pV6Z basis set was used. PEC's of eight lowest electronic
states of CO^{2+} are displayed in Fig. 1.

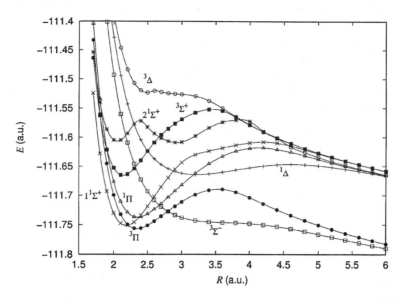

Figure 1: The potential energy functions of the lowest electronic state of CO^{2+}.

The spin-orbit integrals were evaluated with the icMRCI/spdf-cc-pV5Z wavefunctions using Breit-Pauli operator and are shown as a function of internuclear distance in Fig. 2. The vibrational calculations were performed for the six lowest electronic states of CO^{2+} using the Cooley-Numerov integration technique.

3 Energy positions and widths of the metastable vibronic states

Positions and lifetimes of the studied vibronic states were evaluated using different approaches: **Log-amplitude-phase** method based on certain modification of Milne approach [7], As results are tunneling lifetimes of each states alone, without SO interaction. These value are the upper limit for lifetimes of vibronic states.

The key idea of the **Stabilization method** is repeated diagonalization of H in a series of enclosing boxes [6] with the varying size. The result is average density of states $\langle \rho_R(E) \rangle$ from which the resonance energies and widths can be determined by Lorentzian fitting. The six lowest electronic state and S0 interaction among them were included. Vibronic energies and lifetimes obtain in this study using stabilization method are collected and compared with the recent experimental and theoretical data in table 1.

4 Conclusions

The potential energy and spin-orbit coupling functions of the lowest eight electronic states are calculated for the first time on this high level. Calculated lifetimes of the lowest vibrational levels of the X $^3\Pi$, $^1\Pi$ and $^1\Sigma$ states depend profoundly on the spin-orbit interaction and are in very good agreement with experimental data. They are mainly determined by their coupling

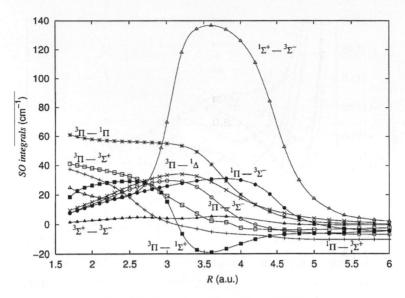

Figure 2: The spin-orbit couplings of CO^{2+}.

Table 1: Positions and lifetimes of the lowest vibronic resonances of CO^{2+}.

$^3\Pi$	E_v cm^{-1}			$\tau(\mu s)$				
v	Ref. [8]	Ref. [9]	this study	Ref. [2]	Ref. [4]	Ref. [1]	Ref. [3]	this study
0	685.5	806.5	732.7	$\gg 3$	7200	$\gg 10$	$> 3.8 \times 10^{+6}$	$13 \times 10^{+6}$
1	2008.3	2258.3	2138.8	> 3	17	> 10	$8 \times 10^{+2}$	$10 \times 10^{+2}$
2	3306.8	3484.3	3550.2	< 0.1	0.00029	0.2		0.61
3	4621.5	4831.2	4885.1	< 0.1		< 0.1		0.0012
4	5807.1	6097.5	6168.9	< 0.1				3.2×10^{-4}
5	—	—	7401.5					8.9×10^{-6}
$^1\Sigma$								
0	2298.6	4081.1	2175.0		0.78	> 10	$6 \times 10^{+3}$	$9 \times 10^{+3}$
1	3911.7	6000.7	4062.5	0.6	0.88×10^{-3}	0.7		0.67
2	5524.9	7670.3	5941.6	< 0.1	—	—		2.8×10^{-2}
3	—	—	7784.6					8.7×10^{-4}
$^1\Pi$								
0	4887.7	5000.6	5030.6	< 0.1		—	0.2	0.833
1	6291.1	6533.1	6512.6	< 0.1		—	< 0.1	2.18×10^{-4}
2	7621.9	8009.1	7952.9	< 0.1				1.01×10^{-5}

with $^3\Sigma^-$ repulsive state. All three metastable electronic states dissociate to the ground state $C^+(^2P) + O^+(^4S)$ ions as was observed.

5 Acknowledgment

The authors wish to thank The Czech Academic Supercomputer Center for time allocation. This work was supported by GAASCR (grant No. A400550511) and Ministry of Education (grant No. LC512); it was also a part of the research project Z4 055 0506.

References

[1] F. Pennet, R.I. Hall, R.Panajotovic, J.H.D. Eland, G. Chaplier, P. Lablanquie, *Phys. Rev. Lett.* **81** 3619 (1998).

[2] M. Lundqvist, P. Baltzer, D. Edvardsson, L. Karlsson, B. Wannberg, *Phys. Rev. Lett.* **75** 1058 (1995).

[3] L.H. Andersen, J.H. Posthumus, O. Vahtras, H. Ågren, N. Elander, A. Nunez, A. Scrinzi, M. Natiello, M. Larsson, *Phys. Rev. Lett.* **71** 1812 (1993).

[4] J.P. Bouhnik,I. Gertner, B. Rosner, Z. Amitay, O. Heber, D. Zajfman, E.Y. Sidky, I. Ben-Itzhak, *Phys. Rev. A* **63** 032509 (2001).

[5] V. Špirko, M. Rozložník, J. Čížek, *Phys. Rev. A* **61** 014102 (2000).

[6] V.A. Mandelshtam, H.S. Taylor, V. Ryaboy, N. Moiseyev, *Phys. Rev. A* **50** 2764 (1994).

[7] E.Y. Sidky, I. Ben-Itzhak, *Phys. Rev. A* **60** 3586 (1999).

[8] M. Hochlaf, R.I. Hall, F. Penent, H. Kjeldsen, P. Lablanquie, M. Lavollee, J.H.Eland *Chem. Phys.* **207** 159 (1996).

[9] G. Dawber, A.G. Mc Konkey, L. Avaldi, M.A. MacDonald, G.C. King, R.I. Hall, *J. Phys. B* **27** 2191 (1994).

Brill Academic Publishers
P.O. Box 9000, 2300 PA Leiden,
The Netherlands

Lecture Series on Computer
and Computational Sciences
Volume 4, 2005, pp. 1786-1789

An Algebraic Analysis of Integrability of Second Order Ordinary Differential Equations

J. Avellar[1], L.G.S. Duarte[2], S.E.S. Duarte[3], L.A.C.P. da Mota[4]

Universidade do Estado do Rio de Janeiro,
Instituto de Física, Departamento de Física Teórica,
R. São Francisco Xavier, 524, Maracanã, CEP 20550–013,
Rio de Janeiro, RJ, Brazil.

Received 29 July, 2005; accepted in revised form 14 August, 2005

Abstract: In [1], we have developed an algorithm to find first order elementary[a] differential invariants (elementary first integrals) for second order ordinary differential equations (SOODEs). Here, we propose a method to, using the machinery presented on [1], algebraically search for regions (in the parameter space) of integrability of the SOODE.

---

[a]For a formal definition of elementary function, see [2].

Keywords: Integrability, elementary first integrals, Darboux procedure, symbolic computation

PACS: 02.30.Hq

1 Introduction

The importance of solving differential equations can not be overstated. In [1], we have presented an algorithm to deal with rational SOODEs. It is a non-classificatory procedure based on a Darboux-type approach and, in essence, converts the solving of the SOODE into solving a system of algebraic equations. Consider the SOODE:

$$y'' = \frac{M(x,y,y')}{N(x,y,y')} = \phi(x,y,y'), \tag{1}$$

where M and N are polynomials in (x,y,y')[5]

Suppose we have a SOODE of the form (1). We will then define the integrating factor as:

$$\mathcal{R}(M - Ny'') = \frac{dI(x,y,y')}{dx} = I_x \, dx + I_y \, dy + I_{y'} \, dy' = 0. \tag{2}$$

By adding the null term $\mathcal{S} y' \, dx - \mathcal{S} \, dy$, where $\mathcal{S}$ is a function of (x,y,y'), we get[3]:

$$\mathcal{R}\left[(M + \mathcal{S} y') \, dx - \mathcal{S} \, dy - N \, dy'\right] = dI = 0, \tag{3}$$

---

[1]E-mail: javellar@dft.if.uerj.br - Permanent address: Fundação de Apoio à Escola Técnica,*E.T.E. Juscelino Kubitschek, 21311-280 Rio de Janeiro – RJ, Brazil*

[2]E-mail: lduarte@dft.if.uerj.br

[3]E-mail: sduarte@dft.if.uerj.br

[4]E-mail: damota@dft.if.uerj.br

[5]From now on, f' denotes df/dx.

Applying the compatibility conditions to (3), we get:

$$\mathcal{R}_y N - \mathcal{R} N_y = N^2 (\mathcal{R}_{y'} S + S_{y'} \mathcal{R}) \tag{4}$$

$$\frac{\mathcal{D}[\mathcal{R}]}{\mathcal{R}} = -(S + N_x + N_y y' + M_{y'}), \tag{5}$$

$$S \frac{\mathcal{D}[\mathcal{R}]}{\mathcal{R}} + \mathcal{D}[S] = -(N M_y - M N_y) \tag{6}$$

where $\mathcal{D}$ is defined as $\mathcal{D} = N \partial_x + y' N \partial_y + M \partial_{y'}$.

From the above and using results presented on [4], we have produced the following results [1]:

- **Theorem 1:** *Consider a SOODE of the form (1), that presents an elementary first integral I. If S is a rational function of (x, y, y'), then the integrating factor $\mathcal{R}$ for this SOODE can be written as:*

$$\mathcal{R} = \prod_i p_i^{n_i} \tag{7}$$

  where p_i are irreducible polynomials in (x, y, y') and n_i are non-zero rational numbers.

- **Theorem 2:** *Consider a SOODE of the form (1), that presents an elementary first integral I. If S (defined above) is a rational function of (x, y, y') ($S = P/Q$, where P and Q are polynomials in (x, y, y')), then*

$$(\mathcal{R}/Q) \mid \mathcal{D}[\mathcal{R}/Q], \tag{8}$$

  i.e. $\mathcal{D}[R/Q]/(R/Q)$ is a polynomial.

  Corollary to Theorem 2: *Consider a SOODE of the form (1), that presents an elementary first integral I. If S is a rational function of (x, y, y') ($S = P/Q$, where P and Q are polynomials in (x, y, y')), then $\mathcal{R}/Q$ can be written as:*

$$R/Q = \prod_i v_i^{m_i} \tag{9}$$

  where v_i are irreducible eigenpolynomials (in (x, y, y')) of the $\mathcal{D}$ operator and m_i are non-zero rational numbers.

In [1], we used these results into equations (4,5,6) to convert (via a Darboux-type procedure) the solving of the SOODE in solving a systems of algebraic equations for coefficients defining the polynomials that construct the S and the integrating factor $\mathcal{R}$. In the next section, we will point out how the strength of the above theorems can be used to search for the integrability regions of the parameter space.

2   Method and Examples

The beauty of such methods as the Prelle-Singer approach (and extensions) to solving first order ordinary differential equations (FOODEs) [4, 5] and the one for SOODEs presented in [1] is that they assure that, if the conditions of applicability of the appropriate method are met, the procedure will eventually (even if it could take a large amount of time to do it) find the solution. This feature combined with the above mentioned conversion of the solving of the SOODE to dealing with an algebraic system opens the door to our method of analyzing the regions of integrability on the parameter space.

The main point is that, according to Corollary to theorem 2, the determination of the first integral for the SOODE requires the determination of the eigenpolynomials (and associated cofactors) for the differential operator $\mathcal{D}$. Since, as mentioned above, the theorems presented on the last section assure that the first integral will be found (for some degree of the eigenpolynomials) we see that we are left with, basically, two possible scenarios: Either the integration factor is trivial or it will be build with the eigenpolynomials of the $\mathcal{D}$ operator. For SOODEs that have limited regions of integrability on the parameter space, the latter case will surely apply.

So, since the whole machinery has to eventually work and, if the SOODE satisfy the conditions for the theorems (is of the form (1) and presents an elementary first integral), one may conclude that, if we include the parameters present on the SOODE as variables, the parameter space regions where the SOODE is integrable has to be present on the solutions to the Darboux equation defining the eigenpolynomials. That will be better explained through an example:

Consider the equation for the force free Duffing-van der Pol oscillator:

$$y'' + \left(\alpha + \beta y^2\right) y' - \gamma y + y^3 = 0 \tag{10}$$

For that equation, the associated differential operator $\mathcal{D}$ becomes:

$$\mathcal{D} = \partial_x + y'\partial_y + \left(-y'\alpha - \beta y^2 y' + \gamma y - y^3\right)\partial_{y'} \tag{11}$$

Regarding the parameters (α, β, γ) as variables in the Darboux equation:

$$\mathcal{D}[p] = \lambda p, \tag{12}$$

where p is an eigenpolynomial and λ is the associated cofactor for the $\mathcal{D}$ operator, we find (up to eigenpolynomials of degree 3) that we have two solutions:

- $\left(p = \dfrac{y + z\beta}{\beta}, \ \lambda = -\dfrac{-1 + \alpha\beta + \beta^2 y^2}{\beta}\right)$, with $\gamma = -\dfrac{-1 + \alpha\beta}{\beta^2}$.

- $\left(p = \dfrac{3y\alpha\beta - 9y + \beta^2 y^3 + 3z\beta}{\beta}, \ \lambda = -3\beta^{-1}\right)$, with $\gamma = -3\dfrac{\alpha\beta - 3}{\beta^2}$

In order to integrate the SOODE (10), we would have to continue the procedure introduced on [1] and described above. By doing that, one can see that neither of the solutions by themselves can fully integrate the SOODE. Actually, for one of the solutions (the second one) one can reduce the SOODE to a FOODE. One is led then to consider both solutions together to achieve full integration. To do that, one is faced with the following obstacle: for each of the solutions, one have different values for γ. So, to make the solutions compatible, one has to find for which values of α (leaving β as a free parameter) both values for γ converge. This leads to:

$$\text{Value for the parameters: } \left(\beta = \beta, \ \alpha = \frac{4}{\beta}, \ \gamma = -\frac{3}{\beta^2}\right) \tag{13}$$

and the following set of eigenpolynomials and cofactors:

$$\left(p_1 = \frac{y + z\beta}{\beta}, \ \lambda_1 = -\frac{3 + \beta^2 y^2}{\beta}\right) \tag{14}$$

$$\left(p_2 = \frac{3y + \beta^2 y^3 + 3z\beta}{\beta}, \ \lambda_2 = -3\beta^{-1}\right) \tag{15}$$

So, after all these calculations, we conclude that to fully integrate the SOODE (10), we have to apply the parameters as in (13). A closer look reveals that this choice for the parameters is the well known result for the integrability of the force free Duffing-van der Pol oscillator [6, 7].

3 Conclusion

The purpose of the method outlined above is the introduction of an algorithmic approach to the integrability analysis. It is based on solving a Darboux eigenvalue equation. Using the theorems derived on [1], we managed to reduce the problem at hand to solving algebraic systems of equations and, furthermore, these are mostly second degree equations on the desired variables. This makes the method introduced here easy to implement in any symbolic computation basin. We are working on a Maple implementation of it to be presented later.

References

[1] J.Avellar, L.G.S. Duarte, S.E.S.Duarte and L.A.C.P. da Mota, *Finding Elementary First Integrals for Rational Second Order Ordinary Differential Equations*, submitted to *J. Phys. A: Math. Gen.*

[2] Davenport J.H., Siret Y. and Tournier E. *Computer Algebra: Systems and Algorithms for Algebraic Computation.* Academic Press, Great Britain (1993).

[3] The function S was first introduced on: L G S Duarte, S E S Duarte, L A C P da Mota and J E F Skea, Solving second order ordinary differential equations by extending the Prelle-Singer method, *J. Phys. A: Math.Gen.*, **34** 3015-3024 (2001).

[4] M Prelle and M Singer, Elementary first integral of differential equations. *Trans. Amer. Math. Soc.*, **279** 215 (1983).

[5] L.G.S. Duarte, S.E.S.Duarte and L.A.C.P. da Mota, *A method to tackle first order ordinary differential equations with Liouvillian functions in the solution*, in *J. Phys. A: Math. Gen.* **35** 3899-3910 (2002); L.G.S. Duarte, S.E.S.Duarte and L.A.C.P. da Mota, *Analyzing the Structure of the Integrating Factors for First Order Ordinary Differential Equations with Liouvillian Functions in the Solution*, *J. Phys. A: Math. Gen.* **35** 1001-1006 (2002); L.G.S. Duarte, S.E.S.Duarte, L.A.C.P. da Mota and J.F.E. Skea, *Extension of the Prelle-Singer Method and a MAPLE implementation*, *Computer Physics Communications*, Holanda, v. 144, n. 1, p. 46-62, 2002; J.Avellar, L.G.S. Duarte, S.E.S.Duarte and L.A.C.P. da Mota, *Integrating first-order differential equations with Liouvillian solutions via quadratures: a semi-algorithmic method*, in *J. Comp. Appl. Math.* **182** 327-332 (2005)

[6] Lakshmanan, M. and Rajasekar, S., *Nonlinear Dynamics: Integrability, Chaos and Pat- terns..* New York: Springer-Verlag (2003).

[7] Chandrasekar, V K ; Senthilvelan, M ; Lakshmanan, M., On the complete integrability and linearization of certain second order nonlinear ordinary differential equations. *http://arxiv.org/abs/nlin/0408053* - accepted for publication on *Proceedings of the Royal Society London Series A*

Brill Academic Publishers
P.O. Box 9000, 2300 PA Leiden,
The Netherlands

Lecture Series on Computer
and Computational Sciences
Volume 4, 2005, pp. 1790-1795

The Integrated Bayesian Framework based on Graphics for Behavior Profiling of Anomaly Intrusion Detection

ByungRae Cha[1]

Dept. of Computer Engineering,
University of Honam,
KOREA

Received 13 June, 2005; accepted in revised form 30 June, 2005

Abstract: Most intrusion detection systems detect only known attack type as IDS is doing based on misuse detection, and active correspondence is difficult in new attack. In this paper, we propose an behavior profiling method using Integrated Bayesian Framework based on graphics from audit data and visualize behavior profile to detect/analyze anomaly behavior. We achieve simulation to t ranslate integrated audit data of host and network into IBF-XML which is behavior profile of semi-s tructured data type for anomaly detection and to visualize IBF-XML as SVG.

Keywords: Anomaly Intrusin Detection, Bayesian Network

1 Introduction

Intrusion detection model is classified into misuse intrusion detection and ano maly intrusion detection. Misuse intrusion detection detects only known intrusion. To overcome the disadvantage of misuse intrusion detection which can not detect new and modified intrusions, research on anomaly intrusion detection is undertaking, which defines unusual behavior patterns to be noticed as intrusion and detects them. For unusual intrusion detection, first, profiles ab out behaviors should be constructed and distinction between normal and anomaly is made throug h comparison between established profiles and new behaviors. Before anomaly detection, it is very important how behavior profiling is constructed. Intrusion detection area is classified into host-based and network-based accord ing to sources of data. Host-based detection system detects anomaly intrusion based on inspection data generated and collected from host. This system obtains inspection data from processor. It col lects original owner and group of the processor, current user and group of processor which is now wo rking, usage of CPU, I/O quota, files used by the processor, system call of the processor and constr ucts normal pattern and intrusion pattern to detect intrusion. The network-based detection system d etects intrusion by collecting packet data on network. It uses IP Address, ports, TCP which are hea der information of network packet. Research of Bayesian technique in intrusion detection field is in the beginning stage. Bayesian technique to detect intruders which pretend to be normal users was presented by Mehdi Nassehi[1] as a technical report, and Steven L. Scott[2] applied this technique to perceiv e patterns of network intrusion detection and to detect intrusion and presented the values obtained i nto probability values, and described relations between data and intrusion model. This study applies Bayesian Framework for production of profile to represent be haviors in integrated intrusion detection area and suggests transformations of IBF-XML

---

[1]Professor of Honam University. E-mail: chabr@hon am.ac.kr

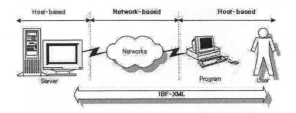

Figure 1: Integrated intrusion detection area of Host-based and Network-based be havior

and SVG for vi sualization. This study overcomes uncertainty of intrusion detection by applying the Bayesian Framework and integrating intrusion detection area. And it provides graphic representation of new intrusion pattern s and modified intrusion patterns for detection/classification information and analysis.

2 Intrusion Detection Model & Bayesian Network

Studies on intrusion detection were first introduced by John Anderson in 1980, and in 1987, a general intrusion detection model was presented. Intrusion detection model in cludes Denning model, Shieh model and Kumar model. Denning model[3] detects intrusion by monit oring abnormal use by the system in log record of system, and uses variables that com pute behaviors of system using statistic method that was previously defined. Shieh model[4] sh ows indirect relations between system state and state transition, and subjects and objects. The system state is captured at audit trace and represented as protection graphs. Kumar model[5] is based on Colored Petri Net by Jensen, intrusion behavior is expressed as Colored Petri N et, and more than one initial status and a single final status are used to define matching i n model. The most urgent matter to be taken care of in intrusion detection is how normal behaviors are constructed. For intrusion detection, profiles should have behaviors described to distinguish intrusion from normal behaviors. Describing systems or behaviors of users is ca lled behavior profiling. Behavior profiling describes characteristics of behaviors of subject s relative to objects or provides description of normal behavior or anomaly symptoms. Behavio r profiling methods include matrix and statistic models.[3] Matrix method provides informat ion for anomaly intrusion detection by presenting estimated values which have been periodically accumulated. Statistic model obtains information by observing behaviors of all the informati on, and does not need an assumption of basic distribution about behaviors.

For a study on construction of intrusion detection profiling, this study is t o apply Bayesian theory for intrusion detection profiling. The Bayesian theory can easily obtain $P(A|B)$, inverse probability of $P(B|A)$, conditional probability, which is a probability that e vent B follows after event A. Inverse probability $P(A|B)$ means probability of A, previous ev ent relative to B, that is, it has been commonly used in medical care and equipment diagnosis in w hich a symptom appears first and reasons of the symptom are identified then. Bayesian network expresses conditional independence of Bayesian theory as a graph of network type. That is, it express es actual knowledge as given directed acyclic graphs. The Byaesian network is called a causal netwo rk or a belief network. Interference is possible through knowledge expressed by Bayesian network.

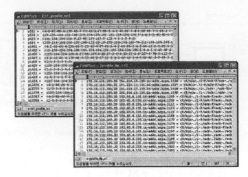

Figure 2: Behavior pattern notation of Host-based Process ID session and Network -based FTP service session

3 Integrated Bayesian Framework for Behavior Profiling

We need methods to express behaviors for intrusion detection. Intrusion detecti on models include Shieh model, Denning model and Kumar model. The Shieh model is designed for a u ser behavior area, and the Denning model presents a model on overall profiling for intrusion detec tion. The Kumar model is a state transition model to prevent intrusion to program behavior area . The Bayesian model presented by this study can express all the three areas and provide graph ic information for easy analysis without installation of new tools. This study proposes how to not ate using string to express behavior of behavior profile and Bayesian framework using Bayesian n etwork of graphic technique. For intrusion detection, audit data have to be processed from log data of syste m or packet data of network. To express behaviors, we have to define a concept called session. S ession indicates that many events are collected to express a behavior. Host-based user behavior defines session as a collection of commands and objects used by users from system log-in to log -out. Program behavior defines session as a collection of system calls by process ID. Network -based behavior defines session as a collection of packet data through IP address, port number, and TCP communication procedure. Figure 1 presents integrated intrusion detection area of user, program and network behaviors by Integrated Bayesian Framework based on XML(IBF -XML).

3.1 Notation using String

For construction of profile of host-based and network-based behaviors, we need how to express to describe a behavior. To express a behavior, a session is used as a basic uni t. To express session, we need notation of behavior pattern to be defined. As signs to describe a beha vior used in this study, "<" and ">" mean the beginning and end of each session. And "-" means division between events(user commands or scripts, system calls and packet information). The method to describe behaviors used in this study is presented in Figure 2. In respect to constructi on of normal behavior profile of network-based packet behavior and host-based program behavior, Figur e 2 presents behavior pattern notation of Host-based Process ID using Sendmail Data[6] of New Maxico University and Network-based FTP service session using DARPA 2000 NT Data[7] of MIT Linclon University for c onstruction of host and network behavior profiles.

3.2 IBF-XML & SVG transformation

XML language can be automatically treated into a semi-structured data type by h aving character-istics of general web language, and taking well-formed and valid document forms and does not need specific database engine. IBF-XML(Integrated Bayesian Framework based on XML) is Bayesian Framework of In tegrated intrusion detection area(user, program and network) defined into XML document. First, IBF_DTD is d efined as in Table 1. IBF-DTD defines elements and property of Bayesian network, and consists of NODE, ARC, P ROBABILITY and TITLE. IBF-DTD is divided two parts into DATA part and STRUCTURE part. DATA part compr ises NODE and OBJECT. NODE is an active object, shows state and uses event information of each session for indication. OBJECT is a passive object and is approached by NODE. STRUCTURE part is com-posed of ARC. AR C indicates relations between NODEs. Relevant nodes are connected by ARC, which shows contexts of eve nts presented by NODEs. Also, for indication of ARC, direct relation is separated from indirect relatio n. User behaviors, program behaviors, and network behaviors are profiled by NODE and ARC.

For vi sualization of XML, it should be transformed into SVG(Scalable Vector Graphics)[8]. In this paper, integrated network-based and host-based behaviors are profiled as Bayesian framework with a use of IBF-XML. When Bayesi an framework is described as IBF-XML, it is automati-cally transformed into SVG file by Perl program. SVG file presents behavior profile described by Bayesian network as graphic, which provides a visual effec t unlike XML. They are presented as SVG, graphic information of behavior profile.

Table 1. IBF-DTD definition for IBF-XML

```
<!- - DTD for Integrated Bayesian Framework - ->
<!ELEMENT IBF-XML ( DATA | STRUCTURE )+>
<!ATTLIST IBF-XML ID CDATA #REQUIRED
NAME CDATA #IMPLIED TITLE CDATA #IMPLIED>
<!- - Node Data declaration section - ->
<!ELEMENT DATA (NODE | OBJECT)*>
<!ELEMENT NODE EMPTY>
<!ATTLIST NODE
NAME NMTOKEN #REQUIRED TYPE (S | S1 | G) "G"
TITLE CDATA #IMPLIED PROBABILITY CDATA #IMPLIED
XPOS CDATA #IMPLIED YPOS CDATA #IMPLIED>
<!ELEMENT OBJECT EMPTY>
<!ATTLIST OBJ
NAME NMTOKEN #REQUIRED TITLE CDATA #IMPLIED
XPOS CDATA #IMPLIED YPOS CDATA #IMPLIED>
<!- - topological dependency structure information - ->
<!ELEMENT STRUCTURE (ARC)*>
<!- - specify dependency arc - ->
<!ELEMENT ARC EMPTY>
<!ATTLIST ARC TYPE (D | I) "D"
PARENT NMTOKEN #REQUIRED CHILD NMTOKEN #REQUIRED
EVENT_NAME CDATA #IMPLIED>
```

4 Behavior Profiling by XML-based Integrated Bayesian Framework

Intrusion detection is classified into host-based and network-based according t o behavior areas. Host-based anomaly behavior is divided into user behavior and program behavior. In this pa per,

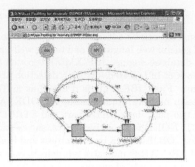

Figure 3: Shieh model description by IBF-XML

Figure 4: SVG transformation of Shieh model

normal and abnormal behaviors from host and network based audit data are described by IBF-XML, and they are transformed into SVG. Transformation into SVG provides graphic information for visualization and easy analysis of normal and abnormal behaviors.

4.1 Behavior Profiling using IBF-XML

Profiling of intrusion behaviors of Shieh model using IBF-XML is presented in F igure 3. Profiles of normal behaviors are described with a use of Sendmail data from New Mexico University for program behavior base and DARPA 2000 NT data from MIT Lincoln University for network base. Using event in formation consisting of sessions in Figures 2, this study created prototyping close to IBF-XML.

4.2 SVG Transformation of IBF-XML

For visualization of XML, it should be transformed into SVG file. Behaviors are profiled as Bayesian framework with a use of IBF-XML. When Bayesian framework is described as IBF-XM L, it is transformed into SVG file by Perl program. SVG file presents behavior profile described by Bayes ian network as graphic, which provides a visual effect unlike XML. This study describes user behavior, progra m behavior and network behaviors using IBF-XML. Behavior profiles described as IBF-XML are automatically transfo rmed into SVG by a program. They are presented as SVG, graphic information of behavior profile in Figure 4, 5 an d 6. Figure 4 shows SVG transformation of IBF-XML of Shieh model on browser. Figure 5 shows SVG transf ormation of IBF-XML for PID 551, normal behavior profile of Sendmail data of New Mexico University. Figure 6 sho ws graphic profiling by SVG transformation of CASESEN Intrusion, an intrusion behavior in FTP network servi ce. CASESEN intrusion is a U2R (User to Root) intrusion for which directory, a sensitive object of NT system, is made misused.

5 Conclusion & Future Work

For anomaly detection, behavior profile should be constructed. By comparing con structed behavior profiles with new behaviors, normal profiles are separated from abnormal ones. Prior to intrusion detection, it is very important to identify how behavior profiling is construct ed. This paper suggests a graphic-based Integrated Bayesian Framework to describe a nd analyse behaviors in host-based and network-based anomaly detection. Bayesian Framework is graphically construc ted

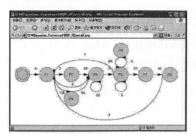

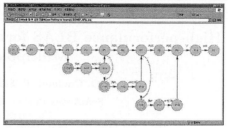

Figure 5: SVG transformation of PID 551

Figure 6: SVG transformation of CASESEN attack

using Bayesian network, and described as semi-structured data using XML. And, for visualization, graphi c representation is realized through SVG transformation. In simulation, host-based user behavior, program be havior, and network behavior are constructed as profiles by Bayesian Framework, and realized into graphic us ing IBF-XML and SVG transformation. In further studies, an effective clustering algorithm using normal behavior pro filing constructed by IBF-XML should be developed. Also we need studies on classification and analysis of mod ified or new behaviors and on automated detection pattern description technology to provide an intrusion dete ction technique for misuse detection.

References

[1] Mehdi Nassehi, "Characterizing Masqueraders for Intrusion Detection ", Computer Science/Mathematics, 1998.

[2] Steven L. Scott, "A Bayesian Paradigm for Designing Intrusion Detec tion Systems", Computational Statistics and Data Analysis, June 20, 2002.

[3] Dorothy E. Denning, "An Intrusion-Detection Model", IEEE Transactio n on Software Engineering, Vol. SE-13, No.2, p222-232, February 1987.

[4] Shiuh-Pyng Shieh and Virgil D. Gligor, "On a Pattern-Oriented Model for Intrusion Detection", IEEE Transaction on knowledge and Data Engineering, Vol. 9, No. 4, July/August, 1997.

[5] Sandeep Kumar and Eugene H. Spafford, "An Application of Pattern Ma tching in Intrusion Detection", Technical Report CSD-TR-94-013, June 17, 1994.

[6] http://cs.unm.edu/ immsec/data/synth-sm.html

[7] http://www.ll.mit.edu/IST/ideval/data/data_index.html

[8] J. David Eisenberg, "SVG essential", O'Reilly, 2002.

[9] Steven A. Hofmeyr, Stephanie Forrest and Anil Somayaji, "Intrusion Detection using Sequences of System Calls", August 18, 1998.

[10] Matthew V. Mahoney and Philip K. Chan, "PHAD : Packet Header Anoma ly Detection for Identifying Hostile Network Traffic", Florida Institute of Tec hnology Technical Report CS-2001-04, 2001.

Brill Academic Publishers
P.O. Box 9000, 2300 PA Leiden,
The Netherlands

*Lecture Series on Computer
and Computational Sciences*
Volume 4, 2005, pp. 1796-1799

High-performance BLAS formulation of the Adaptive Fast Multipole Method

O. Coulaud, P. Fortin[1] and J. Roman

ScAlApplix Project, INRIA Futurs - LaBRI,
Université Bordeaux 1,
351, cours de la Libération,
33405 Talence Cedex

Received 19 July, 2005; accepted in revised form 3 August, 2005

Abstract: Recently, we have presented a new formulation of the uniform version of the FMM (Fast Multipole Method) with matrix products that can be highly efficiently computed thanks to the BLAS (Basic Linear Algebra Subprograms) routines. We propose here to extend this formulation to the adaptive version of the FMM for "structured non uniform distributions": this requires the conception of a new data structure for the octree as well as the detection of the available uniform areas in the non uniform distribution.

Keywords: Fast Multipole Methods, non uniform distributions, BLAS routines, octree data structure.

Mathematics Subject Classification: 65Y20, 70F10, 70-08, 68P05

PACS: 02.70.Ns, 02.30.Mv, 02.30.Em

1   Introduction

The N-body problem in numerical simulations describes the computation of all pairwise interactions among N bodies. The Fast Multipole Method (FMM), developed by Greengard & Rokhlin in [2] for gravitational potentials in astrodynamics and for electrostatic (coulombic) potentials in molecular simulations, solves this N-body problem for any given precision with $\mathcal{O}(N)$ runtime complexity against $\mathcal{O}(N^2)$ for the direct computation. The $3D$ FMM however has the drawback to present a linear complexity with an underlying factor in $\mathcal{O}(P^4)$ where P is the maximum degree in the expansions. As this limits the use of FMM in $3D$ simulations, some schemes have been introduced in order to reduce this cost such as: Fast Fourier Transform (FFT) in [3] and rotations in [4]. All these schemes reduce the $\mathcal{O}(P^4)$ factor to a $\mathcal{O}(P^3)$ or $\mathcal{O}(P^2)$.

In [1] and [5], we have proposed to rewrite the most time-consuming operator in the FMM, namely the multipole-to-local operator $(M2L)$ that converts a multipole expansion into a local expansion, as a matrix product. Thanks to the BLAS (Basic Linear Algebra Subprograms) [6], which are highly efficient routines performing matrix operations, we have thus gained substantial runtime speedup on modern superscalar architectures. Even if we keep the $\mathcal{O}(P^4)$ underlying factor in the operation count, P usually ranges from 3 to 15 in molecular dynamic simulations, which is quite low in terms of operation count, and we obtain therefore comparable, and even better, runtimes than the other schemes (FFT and rotations). In fact, when considering the memory

---

[1]E-mail: fortin@labri.fr

needs of each scheme, their numerical stability and the runtimes they require for some practical accuracies, it appears that the BLAS formulation is the best one.

These comparisons were realized for uniform distributions sequentially computed, and the aim of the current work is to extend this BLAS formulation to the adaptive version of the FMM that can handle non uniform distributions.

2   BLAS formulation of the Adaptive Fast Multipole Method

We will first detail the adaptive version of the Fast Multipole Method we have choosen. Then we will present a new data structure for the octree, the *octree with indirections*, that is efficient in both uniform and non uniform cases. Finally, we will focus on the detection of the uniform areas in non uniform distributions: these uniform areas enable indeed faster computations in the BLAS version.

2.1   The adaptive FMM

The first version of the adaptive FMM has been designed in [7] and improved in [8]. They proceed as follows: each leaf that has more than a fixed s_{max} number of bodies is divided in its eight children which results in an octree with leafs at different levels. The main drawback of this version is the four different lists, instead of 2 in the uniform case, that are now required for each leaf cell: the size of these lists is unbounded which does not ensure anymore the linear operation count of the FMM. Their efficient computation and/or storage is also tricky and hardly predictable when considering the pattern of communications in distributed parallel computations.

A different approach has been used in [9] and [10]: instead of fixing s_{max}, we just fix the height of the octree, and the possibly numerous empty cells will then be omitted. Some improvements, presented in [10], enable to efficiently treat cells containing few bodies and to avoid some computations along the "chains" in the octree (i.e. whenever a cell has only one child). Thanks to these improvements, the linear operation count is guaranteed, as proved in [10], and we keep the two size-bounded and easily computable lists of the uniform case. We have therefore prefered this latter version.

2.2   Octree with indirections

The first issue lies in the conception of a data structure for the octree which can efficiently handle both uniform and highly non uniform distributions. We start by outlining the necessary fast access to any cell's content according to the spatial coordinates of the cell's center: this is accomplished thanks to the Morton ordering of all the cells in the octree (see [9] for exemple).

In uniform distributions, or in almost uniform distributions where there is large uniform areas, the Morton ordering is achieved thanks to linear arrays: at each level l, an array with 8^l pointers is used to store the address of each non empty cell (an empty cell corresponding to the *NULL* address). For uniform distributions the height of the complete octree ranges between 4 and 8, whereas non uniform distributions can lead to unbalanced octrees with height greater than 10. The memory requirements of these arrays becomes then quickly prohibitive while most of the cells are empty.

The *hashed-octree* proposed by Warren & Salmon in [11] allows the storage of such unbalanced octrees with heights up to (theoretically) 20 while maintaining Morton ordering. However in the uniform case, the efficiency of such hashed-octree will decline with growning heights of the octree: each additional level brings new cells whose keys will conflict in the hash table, thus slowing down their access time. We therefore propose here a new data structure, the *octree with indirections*, which is efficient for uniform distributions and able to represent non uniform octrees with heights greater than 10.

The unbalanced octree is divided in several sub-octrees of constant height h. The first sub-octree is rooted at the top of the global octree. If the total height is greater than h, there is other sub-octrees rooted at level h: we will then store only the non empty ones. If the height is also greater than $2h$ some other sub-octrees are used at level $2l$ and so on: see figure 1.

At each of these *indirection levels* $h, 2h, 3h, \ldots$, we use linear arrays to store the addresses of the non-empty sub-octrees: the sub-octrees are indexed in the array according to the Morton index of their root. The interest here is that the linear arrays for the cells are now local to the sub-octrees, and their size will therefore not exceed 8^h.

The additional cost when retrieving the address of a cell in memory, due to the possible indirections required, is expected to be small in practice.

It has to be noted that an efficient implementation of the downward pass of the FMM browses the cells of the octree level by level: in the case of highly unbalanced octrees, we will face the useless browsing of numerous empty cells at the highest levels of the octree. In order to limit this waste of time, we first traverse once the octree in order to build, for each level, a list of intervals that contain indexes of

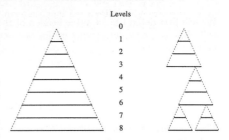

Figure 1: Classic octree (on the left) and octree with indirections (on the right): at each level are represented the linear arrays that allow the access to any cell's content.

non empty cells. Only the indexes of these intervals will then be browsed.

2.3   Uniform areas detection

As shown in [1], a simple level 3 BLAS formulation, that recopies the multipole and local expansions in order to build the matrices, already offers the lowest runtimes when considering the practical accuracy (i.e. the results of the simulations). Nevertheless, the additional cost of the recopies can be avoided when the octree is uniform (or complete) : when all the cells of a given cubical area have multipole and local expansions, we can manage, as detailed in [1], to store the corresponding expansions in memory so that the matrices are already built. These expansions can be stored either "row by row", or "slice by slice".

Since the best BLAS efficiencies are obtained for matrices with sufficient number of rows and columns, this requires long enough rows or big enough slices, which means an "high enough" uniform octree. But with non uniform distributions, the available "uniform areas", which correspond to local uniform octrees, have to be detected, and marked, depending on their size, so that the recopies may be avoided there.

This issue is easily handled when relying on the spatial decomposition as given by the octree. We thus perform an upward pass of the octree in order to detect the cells whose none of the 8 children is empty. If the 8 children were already marked we have an uniform octree of height 2: that's why the mark of each cell c is rather a label containing the height of the uniform sub-octree rooted in c. In [10], cells containing few bodies do not have expansions: this could prevent the use of BLAS without recopies in uniform areas. That is why, when marking the cells, we allow a certain ratio of the children to have few bodies: if it is worth using our BLAS in the surrounding area, we will still use expansions for these cells.

Once this upward pass has been done,we have now detect the biggest uniform areas that fit in a cell of the octree. However there might exist bigger areas: for example at the center of the octree, among 8 cells that do not share any common ancestor but the root. It is now possible to traverse the octree down to the leafs, looking for neighbors among the already detected uniform areas in order to regroup them in bigger ones. After this downward pass, the BLAS formulation of the FMM can proceed using the uniform areas to avoid recopies.

3 Conclusion

We have presented in this extended abstract a new data structure for the octree used in the FMM: this structure allows efficient uniform computations while being able to represent highly unbalanced octrees with heights higher than 10. We have also presented a scheme in order to detect the available uniform areas in a non uniform distribution: these uniforms areas enable a faster BLAS formulation of the FMM. In our talk, we will present simulation exemples, with their execution runtimes, that validate the efficiency of such data structure as well as the gain in exploiting these uniform areas.

Acknowledgment

The authors wish to thank the anonymous referees for their careful reading of the manuscript.

References

[1] O. Coulaud and P. Fortin and J. Roman, High performance BLAS formulation of the multipole-to-local operator in the Fast Multipole Method, submitted to the *SIAM Journal on Scientific Computing* (2005).

[2] L. Greengard and V. Rokhlin, A Fast Algorithm for Particle Simulations, *Journal of Computational Physics*, 73 (1987), 325-348.

[3] W. D. Elliott and J. A. Board, Jr., Fast Fourier Transform Accelerated Fast Multipole Algorithm, *SIAM Journal on Scientific Computing*, 17 (1996), 398-415.

[4] C. A. White and M. Head-Gordon, Rotating around the quartic angular momentum barrier in fast multipole method calculations, *Journal of Chemical Physics*, 105 (1996), 5061-5067.

[5] P. Fortin, Multipole-to-local operator in the Fast Multipole Method: comparison of FFT, rotations and BLAS improvements, *INRIA Research Report* (2005).

[6] J. J. Dongarra and J. Du Croz and S. Hammarling and I. Duff, A Set of Level 3 Basic Linear Algebra Subprograms, *ACM Transactions on Mathematical Software*, 16 (1990), 1-17.

[7] J. Carrier and L. Greengard and V. Rokhlin, A Fast Adaptive Multipole Algorithm for Particle Simulations, *SIAM Journal on Scientific and Statistical Computing*, 9 (1988), 669-686.

[8] H. Cheng and L. Greengard and V. Rokhlin, A Fast Adaptive Multipole Algorithm in Three Dimensions, *Journal of Computational Physics*, 155 (1999), 468-498.

[9] W.T. Rankin, Efficient Parallel Implementations of Multipole Based N-Body Algorithms, *Duke University, Department of Electrical Engineering, Technical Report* (1999).

[10] K. Nabors and F. T. Korsmeyer and F. T. Leighton and J. White, Preconditioned, Adaptive, Multipole-Accelerated Iterative Methods for Three-Dimensional First-Kind Integral Equations of Potential Theory, *SIAM Journal on Scientific Computing*, 15 (1994), 713-735.

[11] M. S. Warren and J. K. Salmon, A Parallel Hashed Oct-Tree N-Body Algorithm, *Supercomputing '93*, 12-21.

Brill Academic Publishers
P.O. Box 9000, 2300 PA Leiden,
The Netherlands

Lecture Series on Computer
and Computational Sciences
Volume 4, 2005, pp. 1800-1803

Bubble Dynamics near Biomaterials in an Ultrasound Field - A Numerical Analysis using Boundary Element Method

S.W. Fong*, E. Klaseboer, C.K. Turangan, K.C. Hung

Institute of High Performance Computing, Singapore.

B.C. Khoo†

Singapore-MIT Alliance, and Department of Mechanical Engineering,
National University of Singapore.

Received 10 July, 2005; accepted in revised form 22 August, 2005

Abstract: Ultrasonic cavitation phenomena play a key role in numerous medical procedures such as ultrasound-assisted lipoplasty, phacoemulsification, lithotripsy, brain tumor surgery, muscle and bone therapies, and intraocular or transdermal drug delivery. This study investigates numerically the interaction of a bubble near a biomaterial involved in the treatments mentioned (fat, skin, cornea, brain, muscle, cartilage, or bone) when subjected to an ultrasound field. A range of frequencies is used to study the bubble behavior in terms of its growth and collapse shapes, and the maximum jet velocity obtained. Simulation results show complex dynamic behaviors of the bubble. For instance, in some cases a jet is formed directed away from the biomaterial, and in others, towards it. In certain cases, the bubble eventually breaks into two, with or without the formation of opposite penetrating jets. Very high maximum jet velocities (700-900 ms^{-1}) are observed in some cases.

Keywords: Ultrasonic Cavitation, Numerical Simulation, Biomaterials, Bubble Collapse, Jet Impact, Jet Velocity.

1   Introduction

The phenomena of ultrasonic cavitation is important in a number of medical therapeutic procedures. The cavitation bubble formed often collapses with a high speed jet which is sometimes utilized to create some desired effects or is to be minimized as an undesirable byproduct. In this study, we focus on several common treatments such as ultrasound-assisted lipoplasty (UAL)[1], phacoemulsification[2], lithotripsy[3], brain tumor surgery, muscle and bone therapies, and intraocular (into the eye via the cornea) or transdermal (through the skin) drug delivery. The biomaterials involved in the therapies mentioned are fat, skin, cornea, brain, muscle, cartilage and bone respectively. To observe the bubble dynamics, we place a bubble near each biomaterial and apply an ultrasound field. A range of frequencies of the sound field is used to study the bubble behavior in terms of its growth and collapse shapes, and the maximum jet velocity obtained. Selected results are briefly described.

---

*Presenting author. E-mail:fongsw@ihpc.a-star.edu.sg

†Corresponding author. E-mail:mpekbc@nus.edu.sg

[1]Ultrasound-assisted lipoplasty is the permanent removal of undesirable or excessive fat deposits located beneath the surface of the skin via an ultrasonically vibrating probe.

[2]Phacoemulsification is a procedure used to emulsify and remove the natural optical lens during cataract surgery.

[3]Lithotripsy is a procedure to remove kidney stones using shockwaves.

2   Numerical Simulations and Results

Our model, using the Boundary Element Method, is largely based on Klaseboer & Khoo [1, 2]. Two fluids (Fluid 1 and Fluid 2) are present, and they are separated by an interface (Fig. 1). Fluid 1 has no elasticity, but Fluid 2 incorporates some elastic properties as it represents the various biomaterials used. An initially spherical bubble with radius R_0, is located in Fluid 1 (distance to biomaterial (H' in Fig. 1) is 1.5). We apply a sound wave as an oscillatory pressure field with pressure $p_\infty(t) = p_0[1 - Asin(2\pi ft)]$, where p_0 represents the static pressure of both liquids; t represents time; p_0A, A, and f are the amplitude, dimensionless amplitude, and the frequency of the pressure perturbation, respectively. Other parameters used in our program include α, κ^*, and f/f_0. The value α is the density ratio of Fluid 1 and 2 ($\alpha = \rho_1/\rho_2$), κ^* depends on the elasticity of Fluid 2 ($\kappa^* = E/2(1 - v^2)p_0$, where E and v are Young's modulus and Poisson ratio of Fluid 2), and f/f_0 is the ratio of the applied frequency with respect to the resonance frequency[4] of the bubble. Dimensionless time (t') and distances (e.g. H') used are scaled by t_0 ($t_0 = R_0\sqrt{\rho_1/p_0}$) and R_0 respectively. Inputs to our program are based on mechanical properties of the biomaterials obtained from various literature (Table 1).

Table 1: Mechanical properties of biomaterials used in simulations. The values are obtained from various studies and references.

No	Material	Density (kg/m^3)	Young's Modulus (kPa)	Poisson Ratio
1	Adipose tissue	950	5.6	0.45
2	Skin	1100	22.6	0.45
3	Cornea	1400	47	0.49
4	Brain	1000	240	0.495
5	Muscle (across fibres)	1060	790	0.45
6	Cartilage (costal)	1300	5000	0.4
7	Cartilage	1230	10000	0.4
8	Bone	2000	14000000	0.43

The high Young's modulus of the bone causes numerical difficulties in our simulation. Since bone is considered a hard material, we have replaced the parameters with that of a solid wall. As shown in Fig. 2, the simulation starts at $t' = 0$ with the bubble denoted by a grey line ($t' = 0$). As the ambient pressure drops, the bubble expands to its maximum size (dashed line), and then, it collapses producing an impinging jet (solid line and the 3D visualization).

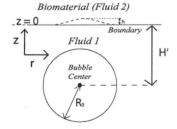

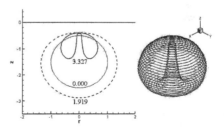

Figure 1: Schematic a bubble immersed in Fluid 1 that is in contact with a biomaterial (Fluid 2). 'h' is the elevation of the interface with respect to its initial equilibrium position.

Figure 2: History profiles of a bubble near a bone (solid) boundary. The dimensionless t' is as indicated near each history profiles. On its right, is the 3D visualization of the collapsed bubble ($t' = 3.327$).

[4] $f_0 = \frac{1}{2\pi R_0}\sqrt{\frac{3\gamma p_0}{\rho_1}}$, where γ is the ratio of specific heats of the bubble's contents

Due to space constraint, the profiles of bubble evolution and biomaterial boundary behavior are discussed only for $A = 0.8$ and $f/f_0 = 1.0$. It is noted that due to the close similarity of the results for coastal cartilage and cartilage, only the former is shown. We observe high sensitivity of bubble behaviors to the different types of biomaterials, as seen in Fig. 3.

For the case with adipose tissue (Fig. 3(a)), a jet directed away from the interface is formed after the bubble expands to its maximum size ($t' = 5.096$, dashed line) and pushing back the fat layer. Similar jet formation is often observed in cavitation bubbles collapse near a free surface. For the cases involving the skin and the cornea (Fig. 3(b) and (c)), the jet formed is directed towards the interface (similar to a bubble collapses in an ultrasound field near a solid wall [3]). As for the brain and muscle tissues (Fig. 3(d) and (e)), the bubble does not split into two in its first two periods of oscillation for the former, and for the latter, it splits in to two with jets forming at the bottom and top of the upper and lower bubbles respectively. The behavior of the bubble near the cartilage is shown in Fig. 3(f). Since the biomaterial is of high κ^*, it is expected not to move much when the bubble expands to its maximum size ($t' = 1.826$, dashed profile). The bubble splits into two, and subsequently forms opposite penetrating jets in top and bottom bubbles.

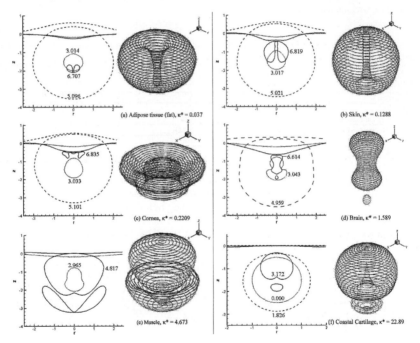

Figure 3: A bubble near various biomaterial boundaries (κ_* values) for $A = 0.8$, and $f/f_0 = 1.0$. The profiles are for its second period of oscillations, with t' as indicated near each profile. The 3D visualization corresponds to the solid line profile.

When the bubble collapses, a high velocity jet is observed. The maximum jet velocity is obtained from the locality on the bubble boundary that is on or nearest to the tip of the jet which eventually penetrates the bubble. For the various materials and frequencies, the results are compiled and presented in Fig. 2. For the case of a boundary consisting of fat, a high speed jet of about 1000 ms^{-1} is obtained. It is observed that only for materials with low κ^*, jet formation in the direction away from the biomaterial boundary is sometimes obtained. The jet speed of a collapsing bubble near skin or cornea is relatively much lower. For the cases involving brain and muscle, the jet speed is low (15 to 37.2 ms^{-1}) for all frequency settings tested. However, as the κ^* of the biomaterial increases, so are the values of jet speed obtained. The jet speed is in the region of hundreds of meter per second for a range of frequencies for cases with cartilage. Another significant observation is that these jet velocities are higher than that obtained when the bubble is near a solid wall (last column). This suggests that severe damages could be sustained by a 'hard' biomaterial (with high κ^*) when a bubble is collapsing close to it.

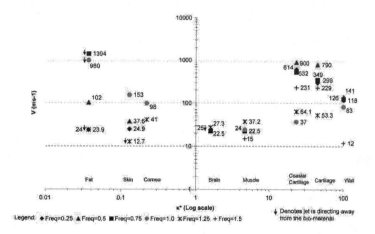

Figure 4: Maximum jet velocity (V) of the bubble near a biomaterial ($H' = 1.5$) in a sound field of frequencies from $f/f_0 = 0.25$ to $f/f_0 = 2.0$. The amplitude is set to $A = 0.8$.

3   Conclusion

Interaction of bubbles with ultrasound in the environment involving biomaterials is very complex. Due to the important role of ultrasound in numerous biomedical applications, we attempt to shed some light on these processes using simulations. Realistic experimental values from the literature are used in the modeling of the biomaterials. We vary the frequency of the ultrasound (f/f_0) for each biomaterials (fat, skin, cornea, brain, muscle, cartilage, and bone) while fixing the distance ($H = 1.5$) between the material and the bubble, and the amplitude of the ultrasound ($A = 0.8$). Interesting bubble profiles and the formation of high speed jets are observed in some of these simulations.

References

[1] E. Klaseboer and B. Khoo. Boundary integral equations as applied to an oscillating bubble near a fluid-fluid interface. *Computational Mechanics*, 33(2):129–138, 2004.

[2] E. Klaseboer and B. Khoo. An oscillating bubble near an elastic material. *Journal of Applied Physics*, 96(10):5808–5818, 2004.

[3] K. Sato, Y. Tomita, and A. Shima. Numerical analysis of a gas bubble near a rigid boundary in an oscillatory pressure field. *Journal of the Acoustical Society of America*, 95(5):2416–2424, 1994.

Brill Academic Publishers
P.O. Box 9000, 2300 PA Leiden,
The Netherlands

Computer letter
Volume 4, 2004, pp. 1804-1807

Application of Rational Approximate Method in Numerical Simulation of Temperature Field of 9SiCr Alloy Steel with Non-linear Surface Heat-Transfer Coefficients During Gas Quenching

Cheng Heming[1], Xie Jianbin, Li Jianyun

Department of Engineering Mechanics,
Faculty of Civil Engineering and Architectrue,
Kunming University of Science and Technology ,
650093 Kunming, China

Received 16 August 2005; accepted in revised form 16 August, 2005

Abstract: The gas quenching is a modern, effective processing technology. On the basis of non-linear surface heat-transfer coefficient obtained by Ref. [1] during the gas quenching, the temperature field of 9SiCr alloy steel with non-linear surface heat-transfer during gas quenching was simulated by means of finite element method. In the numerical calculation, the thermal physical properties were treated as the non-linear functions of temperature. The obtained results show that the non-linear effect should be considered in numerical simulation during gas quenching. In order to avoid effectual the "oscillation"in the numerical solutions, the Norsette rational approximate method was used.
Keywords: gas quenching, surface heat-transfer coefficient, temperature, finite element method, 9SiCr alloy steel, rational approximations.
Mathematics Subject Classification: Finite element methods, quench, temperature
PACS: 78M10, 33.50.Hv, 94.10.Dy

1. Introduction

The gas quenching is a modern, effective processing technology. Because of the smaller temperature difference of between the surface and middle of specimen in gas quenching, the residual stresses in specimen after gas quenching are smaller than that after water or oil quenching. This environment is polluted less with this technique, and it is prevailing in the industries[1,2,3]. But the researches on the mechanism during gas quenching are behind its applications. In order to obtain the distribution of residual stresses and perfect mechanical properties, it is necessary to control the phase transformation and limit distortion. Up to now, thermal strains and thermal stresses can't be measured, the numerical simulation technology is an effective approach to understand the distribution and variety of thermal strains, thermal stresses and microstructure. In recent years, the numerical simulation of quenching processing in various quenching media is prevailing in the world. Although there are now some special soft-wares, which can simulate quenching processing, namely DEFORM-2, DEFORM-3, the important problem in numerical simulation is the boundary condition of stress and temperature. The calculating accuracy of thermal stresses and strains is closely related with the calculating precision of temperature field. For gas quenching processing, the key parameter of the calculation of temperature is the surface heat-transfer coefficient (SHTC).

2. Surface heat-transfer coefficients

During quenching, the surface heat-transfer coefficients (SHTC) have a great influence upon the microstructure and residual stresses in steel specimen. For this reason, many researches into this property have been studied. The variation of this property with temperature has been the subject of investigations. The results obtained are very sensitive to small variation in the experimental conditions, which may lead to considerable discrepancies in the value obtained. Therefore, it must be found necessary to determine the effect of temperature on the surface heat transfer coefficient while using the actual experimental conditions that were to use during the subsequent determination of thermal stress

---

[1] Corresponding author. E-mail: chenghm@public.km.yn.cn

and strain. R.F.Prince and A.J.Fletcher (1980)[4] have an effectual method, which can determine the relationship between temperature and surface heat transfer coefficient during quenching of steel plate. In the Ref.[1], the explicit finite difference method, nonlinear estimate method and the experimental relation between temperature and time during nitrogen gas quenching have been used to solve the inverse problem of heat conduction. The non-linear relationships between surface temperature and surface heat transfer coefficient of cylinder have been given (as shown in Fig.1).

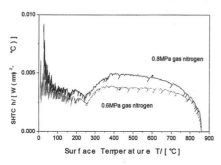

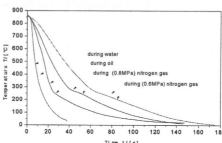

Figure 1   Surface heat-transfer coefficient during gas quenching

Figure. 2 Cooling curves of temperature during various quenching media

Figure 1 shows that (1). The surface heat-transfer coefficients appear as the stronger non-linear property for surface temperature. Therefore, it is inaccurate to simulate the processing of gas quenching by means of linear surface heat-transfer coefficients. (2). In the initial stage of gas quenching, there are stronger heat exchanges between specimen and gas. (3). By increasing the velocity of nitrogen gas, the surface heat-transfer coefficients of the steel go up and the cooling rate also goes up(see figure 2).

3.   The Functional and Finite Element Formula

A $\phi 20 \times 60mm$ cylinder of 9SiCr alloy steel was quenched by 0.6 MPa nitrogen gas from a temperature of 850 □. The non-linear heat conduct equation is[4,5,6]

$$\lambda \frac{\partial^2 T}{\partial z^2} + \lambda \frac{\partial^2 T}{\partial r^2} + \lambda \frac{1}{r} \frac{\partial T}{\partial r} = \rho C_p \frac{\partial T}{\partial t} \tag{1}$$

where C_v and λ denote the specific heat capacity and the thermal conductivity, which are the function of temperature. Therefore, Eq. (1) is a nonlinear equation. The boundary condition of heat transfer is

$$\frac{\partial T}{\partial n}\bigg|_r = h(T)(T_a - T_\infty) \tag{2}$$

The initial condition is

$$T|_{t=0} = T_0(x_k) \qquad k = 1,2,3 \tag{3}$$

here $h(T)$ is the surface heat-transfer coefficients, which are the functions of temperature, and was determined by finite difference method , non-linear estimate method and the experimental relationships between time and temperature[1]。 T_a and T_∞ denote the surface temperature of a cylinder of 45 steel and the temperature of quenching media. The functional[5] above problem is

$$K_n = -\int_{t_{n-1}}^{t_n} \int_\Omega \{\frac{\lambda}{2}[(\frac{\partial T}{\partial r})^2 + (\frac{\partial T}{\partial z})^2] + C_\rho \dot{T} T\} d\Omega dt +$$

$$+ \int_{t_{n-1}}^{t_n} \int_\gamma h(T_{n-1}) \left(\frac{T^2}{2} - T_\infty T\right) ds dt \tag{4}$$

where λ and $C_v \rho$ denote the volume fraction of constituent, thermal conductivity and specific heat capacity, respectively. $h(T_{n-1})$ was the surface heat-transfer coefficients at t_{n-1}, $\{\dot{T}\}$ is fix in variational operation [5]. lThe 8 nodes iso-parameter element was adopted, the finite element formula is:

$$[K]\{T\} + [M]\{\dot{T}\} = \{F\} \tag{5}$$

Here *[K]* is a conductive matrix, *[M]* is a heat capacity matrix, *{F}* is heat supply duo to heat convection on boundary. Under small time step, "oscillation" occurred in numeral solutions. In order to overcome this problem, the rational approximate method was adopted.

4. The Application of Rational Approximate Method

In the functional (4), though $\{\dot{T}\}$ is fix in the variational operation; the various interpolator scheme of $\{\dot{T}\}$ in time field can be found, such as various finite difference methods. The instability of finite difference method leads to the "oscillation"in numeral solutions. In rational approximate method, we don't find out certain interpolator scheme of $\{\dot{T}\}$, then treat approximately the analytic solution of non-linear heat conduct equation(5). It is convential to write Eq. (5) as follows[8]:

$$M\frac{dT}{dt} + KT = P \tag{6}$$

Under initial condition (3), analytic solution of Eq. (6) is

$$T(x,t) = \exp(-tM^{-1}K)(T_0 - K^{-1}P) + K^{-1}P \tag{7}$$

Its incremental form is:

$$T(x,t_i + \Delta t) = \exp(-\Delta t M^{-1}K)(T(x,t_i) - K^{-1}P) + K^{-1}P \tag{8}$$

In order to calculate the $\exp(-\Delta t M^{-1}K)$, let $\overline{C} = \Delta t M^{-1}K$, the power series expansion of $\exp(x)$ is

$$\exp(-\overline{C}) = I - \overline{C} + \frac{1}{2}\overline{C}^2 - \cdots\cdots \tag{9}$$

here I is unit matrix. Provided that the numbers of the terms of right side in Eq. (9) are more enough, we can obtain sufficiently exact results. In the numerical calculation, the $\exp(-C)$ can be replaced by its approximate formula. When the approximate formula is identical with the right side of Eq. (9) in C^2, the error range will be in C^3. In this paper, Norsette approximation was used in finite element method. The Norsette approximate formula can be taken as:

$$\exp(-\overline{C}) \approx \frac{C_n(\overline{C})}{D_N(\overline{C})} = (-1)^n \sum_{j=0}^{n}(-1)^j L_n^{n-j}(\frac{1}{\alpha})\frac{(\alpha\overline{C})^j}{(1+\alpha\overline{C})^{j-1}} \tag{10}$$

Here α is parameter, $L_n^{n-j}(\frac{1}{\alpha})$ is the Laguerre polynomial multinomial of $1/\alpha$. Note that $\overline{C} = \Delta t M^{-1}K$, substituting Eq. (10) into Eq. (8). T_{i+1} And T_i denote $T(x,t_i + \Delta t)$ and $T(x,t_i)$, respectively. We have

$$(M + \alpha\Delta t K)T_{i+1} = M[(-1)^j \sum_{j=1}^{n}(\frac{1}{\alpha})L_n^{n-j}\frac{(\alpha\Delta t K)^j}{(M + \Delta t K)^j}](T_i - K^{-1}P) + K^{-1}P \tag{11}$$

Let $W_i = \alpha\Delta t K/(M + \Delta t K)$, Eq. (5) becomes

$$(M + \alpha\Delta t K)T_{i+1} = M[T_i + \sum_{j=1}^{n-1}(\frac{1}{\alpha})L_n^{n-j}W_i^{j}] + \alpha\Delta t P \tag{12}$$

Solving Eq. (12) T_{i+1} was obtained. In general, n can be taken as 2. As far to the value of α, it is shown that $\alpha = 1 - \sqrt{2}/2$ is better from theoretical analysis and practices[6].

5. Calculated results

For the 9SiCr alloy steel, the thermal physical properties with temperature are shown in figure 3 and figure 4。

Table 1、 Non-linear relationship of thermal physical properties with temperature[1]

Temperature(□)	20	100	200	300	400	500	600	700	800	900
ρC_v $(10^{-3}Ws/mm^3K)$	2.9	2.75	2.6	3.8	3.45	3.6	3.2	3.0	2.64	2.5
Temperature(□)	20	100	200	400	500	600	690	740	800	900
λ $(10^{-2}W/mmK)$	4.7	5.5	5.5	5.6	5.85	6.0	6.1	8.8	6.2	6.3

Figure 2 denotes the cooling curves of temperature during various quenching media. Fig.3 denotes the comparison of temperature field between the calculated values and experimental values during 0.6Mpa nitrogen gas.

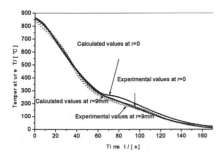

Figure3 the comparison of temperature field between the calculated values and experimental values during 0.6MPa nitrogen gas

6. Conclusions

(1)、 The heat conduction during quenching is a non-linear problem, the calculation of temperature is closely related with surface heat-transfer coefficient. For this reason, the effects of non-linear surface heat-transfer coefficient must be taken into account in simulation of temperature field during gas quenching.

(2) 、 From figure 3, the calculated values coincide with experimental values. It show that the surface heat-transfer coefficient obtained by Ref. [1] during the gas quenching can be used in simulation of temperature field during gas quenching.

(3) 、 Testing of cooling curves of work-pieces during nitrogen gas quenching process shows that the temperature difference of interior work-piece is small. It can be predicted that cooling of the interior work-piece is almost homogeneous and corresponding thermal strains and thermal stresses are rather small.

(4) 、 Different from water and spindle oil quenching, during gas quenching type of gas can be chosen and gas pressure can be adjusted, and the control of quenching parameters can be achieved by this way.

(5) 、 In the numerical simulation of temperature during rapid cooling, the rational approximation can avoid effectual the "oscillation"of the numerical solutions under smaller time step.

Acknowledgments

This work has been supported by the National Natural Science Foundation of China (10162002) and the Key Project of Chinese Ministry Education (No.204138)

References

[1] Xie Jianbin, *The numerical simulation and application of metal and alloy quenched in various quenching media.* Doctor's Dissertation, Kunming, Kunming University of Science and Technology, 2003.

[2] Holm T., Segerberg S., *Gas Quenching Branches out*, Advanced Materials and Processes, 149(6), 64W-64Z, 1996

[3] Segerberg S., *High-Pressure Gas Quenching in Cold Chamber for Increased Cooling Capacity*, 2M Int. Conf. on Quenching and Control of Distortion, Cleveland, Ohio, USA, 4-7 Nov.,1996.

[4] R.F.Prince and A.J. Fletcher, *Determination of surface heat-transfer coefficients during quenching of steel plates.* J. Met. Tech., **2**: 203-215.1980

[5] Wang Honggang, Theory of Thermo-elasticity, Press Qinghua University,1989 (in Chinese)

[6] Cheng Heming, Wang Honggang, Huang Xieqing, Application of Rational Approximate method in Temperature Field and Non-linear Surface Heat-Transfer Coefficient during Quenching, *Metals & Materials*, **5(5)**:345-348, 1999

Brill Academic Publishers
P.O. Box 9000, 2300 PA Leiden,
The Netherlands

*Lecture Series on Computer
and Computational Sciences*
Volume 4, 2005, pp. 1808-1811

Comparison of Surface Heat-transfer Coefficient of T10 Steel Quenched by Different Quenching Media

Lijun HOU [1]; Heming CHENG, Jianyun LI

Department of Engineering Mechanics,
Kunming University of Science and Technology,
Kunming 650093, Yunnan, P.R. China

Received 31 July, 2005; accepted in revised form 10 August, 2005

Abstract: High pressure gas quenching is a new heat treatment technique. In this paper, the surface heat-transfer coefficients during water quenching and nitrogen quenching with 6×10^5Pa pressure are compared. The surface heat-transfer coefficient during water quenching is bigger, but the temperature difference between surface and middle of the cylinder during water quenching is also bigger. The results of calculation coincide with the results of experiment.

Keywords: finite difference method; surface heat-transfer coefficient; gas quenching; water quenching

1. Introduction

The numerical simulation is an access to the understanding of the distribution and variation of thermal stresses, thermal strains and microstructure during quenching. The accuracy of numerical simulation depends on thermal physical and mechanical properties at high temperature and surface heat-transfer coefficient. Many experimental and computational investigations indicate that the surface heat-transfer coefficient during the quenching processes has a significant influence on the resulting component properties such as the hardness distribution, residual stresses and distortion. Therefore, the calculation of surface heat-transfer coefficient has been the subject of many investigations.

In this paper, a T10 steel cylinder was taken as an investigating example. The explicit finite difference method, nonlinear estimate method and the experimental relation between temperature and time during quenching have been used to solve the inverse problem of heat conduction. The relations between surface heat-transfer coefficient in different quenching media and surface temperature of steel cylinders are given.

2. Finite Difference Method

The specimen is a ϕ 20mm×60mm cylinder of T10 steel. The heat conduction equation in middle of specimen is

$$C_V \rho \frac{\partial T}{\partial t} = \lambda (\frac{\partial^2 T}{\partial r^2} + \frac{1}{r} \frac{\partial T}{\partial r})$$

(1)

the boundary condition is

$$\lambda \frac{\partial T}{\partial r}\Big|_\Gamma = h(T_S - T_\infty) + C_V \rho \frac{\partial T}{\partial t}$$

(2)

where λ, C_V and ρ denote thermal conductivity, specific heat capacity and density of material respectively. C_V and λ are the functions of temperature. T, r and t denote temperature, radius and

---

[1] Corresponding author. E-mail: houlijun978@yahoo.com.cn

time. h, T_S and T_∞ denote surface heat-transfer coefficient, specimen temperature of surface boundary and temperature of quenching media, respectively.

The finite difference equations of Eq. (1) and Eq. (2) are

$$T_{i+1}^j = [1 - \frac{\Delta t}{(C_V \rho)_i^j \Delta r}(\frac{\lambda_i^{j+1} + \lambda_i^j}{2\Delta r} + \frac{\lambda_i^j + \lambda_i^{j-1}}{2r_j})] \times T_i^j + [\frac{\Delta t(\lambda_i^{j+1} + \lambda_i^j)}{2(C_V \rho)_i^j \Delta r}(\frac{1}{\Delta r} - \frac{1}{r_j})] \times T_i^{j+1}$$

$$+ [\frac{\Delta t(\lambda_i^j + \lambda_i^{j-1})}{2(C_V \rho)_i^j \Delta r}(\frac{1}{\Delta r} + \frac{1}{r_j})] \times T_i^{j-1}$$

(3)

$$\lambda \frac{T_i^2 - T_i^1}{\Delta r}\bigg|_\Gamma = h(T_i^1 - T_\infty) + C_V \rho \frac{T_i^1 - T_{i-1}^1}{2\Delta t}$$

(4)

where λ_i^j and $(C_V \rho)_i^j$ denote thermal conductivity and specific heat capacity at j node in i time step. Δr, r_j and Δt denote node interval, node radius and time step, respectively.

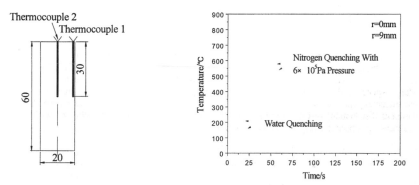

Figure 1: Specimen and position of thermocouple.　　Figure 2: Experimental values of temperature.

3. Experiment

The change of temperature during quenching was measured by thermocouples. Experimental data were recorded by a computer. The thermocouple 1 is placed 1mm distance from the cylinder surface of specimen, thermocouple 2 is on the axis of the cylinder, as shown in Fig. 1. The temperature values measured by thermocouple 1 are key parameters for solving the nonlinear heat conduction equation. The temperature values measured by thermocouple 2 were used to examine the calculated results. The specimen was heated up to 830□, and austenized for 20 min. Then, it was quenched by quenching media. The measured values of temperature are shown in Fig. 2.

4. Calculation Method of Surface Heat-transfer Coefficient during Quenching

4.1 Nonlinear Estimate Method

In order to solve the inverse problem of this nonlinear heat conduction equation, the nonlinear estimate method is adopted. Let a parameter

$$\varphi = \sum (T^{\exp} - T_{l+1}^{cal})^2$$

(5)

to be minimized. We have

$$\frac{\partial \varphi}{\partial h} = (T^{\text{exp}} - T_{l+1}^{cal})\frac{\partial T_{l+1}^{cal}}{\partial h} = 0 \tag{6}$$

where T^{exp} is the measured temperature, T_{l+1}^{cal} is $(l+1)$th iterative computations value of temperature. T_{l+1}^{cal} can be written as Taylor series

$$T_{l+1}^{cal} = T_{l}^{cal} + \frac{\partial T}{\partial h}\Delta h \tag{7}$$

With the help of Eq. (6), Eq. (7) and the convection condition on boundary, we obtain an incremental surface heat-transfer coefficient Δh, that is

$$\Delta h = \frac{\lambda}{T_S - T_\infty}\frac{T_l^{cal} - T^{\text{exp}}}{\Delta r} \tag{8}$$

Table 1: The thermal properties of T10 steel in different temperatures.

Temperature/□	20	200	300	400	500	700	900
$\lambda/10^{-2}(\text{W/(mm·K)})$	4.154	4.320	4.238	4.082	3.890	3.549	3.524
$C_V\rho/10^{-3}(\text{W·s/(mm}^3\text{·K)})$	3.284	4.031	4.146	4.273	4.435	4.791	6.663

4.2 Calculating Procedure

The thermal properties ($\lambda, C_V\rho$) at various node were calculated by using the linear interpolation method according to Table 1. After an estimate value h_0 of surface heat-transfer coefficient was given, the heat conduction equation (3) can be solved and the computation temperature values at node are obtained. If $\left|T_l^{cal} - T^{\text{exp}}\right| > \delta$, a new value of h ($h = h_0 + \Delta h$) will be chosen. T_{l+1}^{cal} is repeated until $\left|T_l^{cal} - T^{\text{exp}}\right| \leq \delta$, in which δ is an inherent error. In this way, the changes of surface heat-transfer coefficient in temperature are obtained. In the calculation, time interval is automatically obtained. According to the stability of explicit finite difference scheme, the maximum Δt is chosen as the time interval during calculation.

$$\Delta t \leq \frac{2\Delta r C_V\rho}{\dfrac{\lambda_{j+1} + 2\lambda_j + \lambda_{j-1}}{\Delta r} - \dfrac{\lambda_{j+1} - \lambda_{j-1}}{r_j}} \tag{9}$$

5. Conclusions

The average cooling velocity of T10 steel during water quenching is bigger than during nitrogen quenching with 6×10^5Pa pressure, but the temperature difference between surface and middle of the cylinder is also bigger, as shown in figure 2. There will be big thermal stress and thermal strain in cylinder during water quenching. Residual stress and distortion of the cylinder is small after nitrogen quenching with 6×10^5Pa pressure.

The relations between surface heat-transfer coefficient and surface temperature are non-linear during water quenching and nitrogen quenching, as shown in figure 3. The surface heat-transfer coefficient during water quenching is bigger than during nitrogen quenching with 6×10^5Pa pressure, but the surface heat-transfer coefficient during nitrogen quenching is equabler than during water quenching. In the initial period during quenching, the increase of the surface heat-transfer coefficient is rapid. This shows that the exchange of the heat flux between the workpiece and quenching media is large. It is noted that at temperature of 700□ during nitrogen quenching, there is a sudden change in the surface heat-transfer coefficient, because of pearlite phase transformation in this temperature. After the end of pearlite phase transformation, the change in the surface heat-transfer coefficient is small.

The results of calculation coincide well with the results of experiment, as shown in figure 4. This method can determine the surface heat-transfer coefficient during quenching effectively.

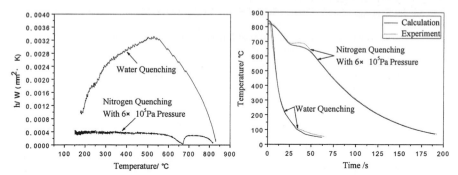

Figure 3: Relations between surface heat-transfer coefficient and surface temperature.

Figure 4: Comparison between calculated values and experimental values of temperature.

Acknowledgements

This project has been supported by the National Natural Science Foundation of China (10162002) and the Key Project of Chinese Ministry of Education (No. 204138).

References

[1] CHENG Heming, WANG Honggang and CHEN Tieli, Solution of Heat Conduction Inverse Problem for Steel 45 during Quenching, *ACTA METALLURGI SINICA*, 33(5), 467-472(1997) (in Chinese).

[2] Z. Li, R.V. Grandhi and R. Shivpuri, Optimum Design of the Heat-transfer Coefficient during Gas Quenching Using the Response Surface Method, *International Journal of Machine Tools & Manufacture*, 42(5), 549–558(2002).

[3] Beichen Liu, *Engineering Computation Mechanics--Theories and Application*. China Machine Press, Beijing, 1994 (in Chinese).

[4] Hu Mingjuan, Pan Jiansheng, Li Bing and Tian Dong, Computer Simulation of Three Dimensional Temperature Field of Quenching Process with Suddenly Changed Surface Conditions, *TRANSACTIONS OF METAL HEAT TREATMENT*, 17(Suppl.), 90-97(1996) (in Chinese).

[5] Heming Cheng, Jianbin Xie and Jianyun Li, Determination of Surface Heat-transfer Coefficient of Steel Cylinder with Phase Transformation during Gas Quenching with High Pressures, *Computational Materials Science*, 29(4), 453-458(2004).

[6] XIE Jianbin, CHENG Heming, HE Tianchun and WEN Hongguang, Numerical Simulation of Temperature Field in Steel 1045 during Its Quenching, *Journal of Gansu University of Technology*, 29(4), 33-37(2003) (in Chinese).

[7] Heming CHENG, Tianchun HE and Jianbin XIE, Solution of an Inverse Problem of Heat Conduction of 45 Steel with Martensite Phase Transformation in High Pressure during Gas Quenching, *J. Mater. Sci. Technol.*, 18(4), 372-374(2002).

Brill Academic Publishers
P.O. Box 9000, 2300 PA Leiden
The Netherlands

*Lecture Series on Computer
and Computational Sciences*
Volume 4, 2005, pp. 1812-1815

A Dynamic Microscopic Model for the Formation of Excess Arsenic in GaAs Layers During Growth at Low Temperature

Sándor Kunsági-Máté,[a,1] Carsten Schür,[b] Eszter Végh,[a] Tamas Marek,[b] Horst P. Strunk [b]

[a] Department of General and Physical Chemistry, University of Pécs, Hungary
Ifjúság 6, H-7624 Pécs, Hungary
[b] Institute of Microcharacterization, Friedrich-Alexander-University Erlangen-Nuremberg,
Cauerstrasse 6, D-91058 Erlangen, Germany

Received 9 July, 2005; accepted 31 July, 2005

Abstract: We examine the formation of interstitial arsenic defects in Low-Temperature grown (LT) GaAs layers by temperature-dependent, direct trajectory molecular dynamics calculations at semiempirical level. A metastable interstitial position of an As_2 molecule is found just below the As-rich c(4x4) reconstructed GaAs(001) surface. This configuration represents a precursor for excess As formation in the LT-GaAs layers. Furthermore, the simulations find a migration layer to exist above the surface in the temperature range of 400 K ... 800 K, where As_2 molecules can freely diffuse. The hopping of As_2 molecules from the interstitial position into the migration layer is identified as plausible microscopic process to determine the experimentally observed dependence of the excess arsenic content on the substrate temperature and arsenic overpressure during growth of LT-GaAs layers.

Keywords: Growth models, Point defects, Molecular beam epitaxy, Semiconducting gallium arsenide
PACS: 68.47.Fg, 61.82.Fk

1. Introduction

GaAs layers grown by molecular beam epitaxy (MBE) under As overpressure at a substrate temperature below 400°C (LT-GaAs) are crystalline, but non-stoichiometric due to an appreciable content of excess As, As_{ex} [1]. It is this excess arsenic that gives rise to the very interesting electrical and optical properties of LT-GaAs layers, such as short carrier lifetime and, after annealing, high resistivity [1]. The As_{ex} content of LT-GaAs shows an inverse dependence on the growth temperature and increases with the As/Ga flux ratio applied during growth. Under very high As/Ga flux ratio the excess arsenic content reaches 1.5% at a substrate temperature of about 200 °C on exactly oriented (001) substrate. This As_{ex} concentration in the LT-GaAs layers is much higher than that derived from the equilibrium phase diagram (0.1% at most), therefore it cannot simply be considered as thermodynamic behavior.

The atomistic processes leading to the As_{ex} formation in the growing layer are not completely understood yet. The observed inverse relation between growth temperature and As_{ex} content in LT-GaAs layers indicates that there must be a surface atomic process which reduces the chance of As_{ex} formation with increasing growth temperature. Two plausible candidates discussed already for such a process are the *direct desorption* of As atoms from the adsorption layer into the vacuum [2] and the *escape* of As_2 molecules from specific interstitial positions in the very substrate surface into the adsorption layer [3,4]. The former process has an activation energy of 1eV, which is too high for the process to occur at appreciable rates at the growth temperature of around 200°C. The latter process, has a low activation energy (64meV), which is much more comparable with the experimental value that results from an Arrhenius evaluation to be ~1meV. Therefore two open problems are obvious. Firstly, a discrepancy between the experimental and theoretical activation energies still exists, which could arise from the difference between the *static* theoretical models (i.e. calculation of energy terms at zero temperature) and the experimental situation, which is highly affected by the *dynamics* of the surface processes. Secondly, the experimentally evaluated activation energy of ~1meV is extremely low and cannot be due to a simple thermodynamic process, since the kinetic energy of particles is at least fifty times higher in the given growth temperature range. Because of this complexity, we conduct

---

[1] Corresponding author. E-mail: kunsagi@ttk.pte.hu

temperature-dependent, direct trajectory molecular dynamics calculations to simulate the kinetic and thermodynamic processes at the same time. Our approach identifies an atomic process to be responsible for the excess arsenic formation in LT-GaAs layers.

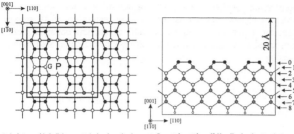

Figure 1: Top and side views of the cluster for representation of the As-rich c(4x4) reconstructed GaAs(001) surface. 'P' denotes the site where the metastable interstitial position of an As_2 molecule is formed.

⊙ As in top and in bulk layers   ● As in adsorption layer   ○ Ga   ○ lowest layer H (dangling bonds saturation)

2. Methods

2.1 Modelling approach

We represent the c(4x4) surface reconstruction by a 9 atomic layers thick slab (including reconstruction layer, Fig.1). For construction of this slab, periodic boundary conditions around a small central box were applied. The central box consists of 4x4x2 bulk unit cells, plus 2 unit cells of the c(4x4) surface reconstruction layer. The periodic boundary was applied to the [-1,1,0] and [1,1,0] directions, forming a 9 atomic layer thick infinite slab. The atoms of the lowest atomic layer (number 8 in Fig.1) were kept in the positions derived from bulk calculations and all atoms within layers above this layer were allowed to move. The atoms at the bottom of this slab are saturated with hydrogen atoms in order to maintain the bulk-like configuration [5].

For explicit simulation of the effect of the presence of a vapor of As_2 molecules just above the surface, a 20 Å wide region is chosen above the central box, and this region was replicated in the same way as described above by the periodic boundary conditions. The density of the As_2 molecule vapor was adjusted by introducing 5, 10, 15, 20, or 25 As_2 molecules into that volume. These densities are indicated in this work with the symbols P5, P10, P15, P20 and P25. The initial positions of the As_2 molecules were randomized by the respective extension of HyperChem package.

The basic reaction studied here is the escape of an As_2 molecule from an energetically metastable interstitial position that has been predicted by our earlier works [3,4]. Figures 1 and 2 show the geometry and the escape reaction. During molecular dynamics calculations the height of the lower As atom of the As_2 molecule above the GaAs surface was recorded as a function of the simulation time. The occurrence of an escape reaction was defined at the point when the lower atom of this As_2 molecule crosses the plane of the top arsenic layer of this surface (Fig. 2).

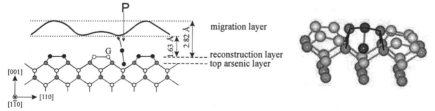

Figure 2: Schematic view of the geometrical shape of the surface of minimum interaction energy (migration layer) and the escape reaction of an As_2 molecule from the interstitial position in the As-rich c(4x4) reconstructed GaAs(001) surface (left). Stick-and-ball presentation of the As-interstitial formation on the As-rich c(4x4) reconstructed GaAs(001) surface (right) due to the formation of a cage by two Ga atoms burying the lower As atom of the As_2 molecule (see text).

2.2 Calculation procedure

The interaction energy between the c(4x4) reconstructed surface and the As_2 molecules was minimized first by variation of the height of molecule above the surface. During this calculation, the mass centre of an As_2 molecule was fixed to axes oriented perpendicular to the surface, while the free rotation of the molecule around its centre was allowed. The intersections of these axes and the surface form a

20x20 mesh (400 points) above one surface unit cell. Then, the dynamic process between As_2 molecules and the cluster-represented GaAs (001) surface in dependence of the temperature was investigated with Langevin molecular dynamic calculations at semiempirical PM3 level. The molecular dynamic calculations were done at five different temperatures between 400 K … and 800 K in steps of 100 K, according to the common temperatures applied in the MBE growth of non-stoichiometric and stoichiometric GaAs layers. Ensembles of N_0=40 trajectories with a time resolution of 10 fs were calculated at each given temperature, for a period of 100 ps or until a reaction occurred during propagation of the trajectory. Reaction rates, k, were determined by fitting k to the calculated fraction of non-reactive trajectories, N/N_0, given by

$$\frac{N}{N_0} = e^{kt}$$ (1)

where t is the reaction time. Accordingly, N_0 is the total number of trajectories and N is the trajectories wich didn't react until the time t. Trajectories which do not end with a reaction within the maximum allowed time (100 ps) contribute to the determination of the rate constant by their contribution to N_0. Static and molecular dynamic calculations were carried out with GAUSSIAN and HyperChem Professional 7 program packages, respectively.

3. Results

The calculated rate constants, k, of the escape reaction (i.e. hopping from the interstitial site to the migration layer) derived from the direct trajectory calculations are shown in the Arrhenius plot in Fig. 3. At each of the five given "density" levels, the $\ln(k)$ values follow the linear classical Arrhenius equation:

$$\ln(k) = \ln A - \frac{E_a}{k_B T}$$ (2)

where A is the preexponential (frequency) factor, E_a is the activation energy, k_B the Boltzmann constant and T is the temperature. However, the activation barrier of the escape reaction, i.e. the slope of $\ln(k)$, decreases with increasing arsenic "density" above the surface. This property can be interpreted by an increased extraction of As_2 molecules to the As_2 bath above the surface at higher As_2 densities. This extraction could arise from an attractive interaction between As_2 molecules.

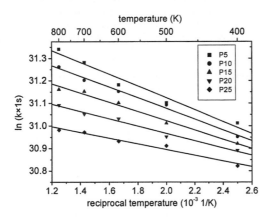

Figure 3: Arrhenius plot of the rate constants of the reaction associated with the escape of As_2 molecules from the metastable interstitial position observed on As-rich c(4x4) reconstructed GaAs(001) surface.

Another interesting property is that the rate of the escape reaction decreases at each individual temperature with increase of the arsenic density above the surface. To interpret this property we consider the following: the escape reaction proceeds from the interstitial position to the migration layer. This reaction requires that there are empty sites in the migration layer to where the As_2 molecules are able to jump. Therefore, higher population of the migration layer at higher pressure inhibits the escape reaction.

The reaction rate values show that this escape reaction is very fast compared to other surface processes. Especially, the growth rate commonly applied in experiments is 1 ML/s. In comparison, the frequency factor of the calculated rates is a few tens of 1/ps. This difference is important for the interstitial formation process. The energy barrier for As_2 molecules to enter into the interstitial position was found to be negligible. Therefore, Boltzmann statistics for the population of the interstitial positions and surface diffusion can be assumed. As a result of the high As flux, an interstitial position will be filled again just after an escape has happened. Accordingly, in a time-averaged view, the interstitial positions are highly populated and the As_2 molecules are continuously hopping between this position and the migration layer. These properties show that formation of excess arsenic in LT-GaAs layers is highly affected by kinetic processes resulting in considerable non-stoichiometry far from thermal equilibrium. We can now apply these aspects to our earlier model of excess As formation during growth that is based on *ab-initio* calculations [5]. Growth means that now, under conditions of excess As gas in the atmosphere also a few Ga atoms impinge onto the migration layer and onto the growth surface. The *ab-initio* calculations indicated that the middle As-dimer in the As_2 dimer triplets of the $c(4x4)$ reconstruction is the most reactive center towards incorporating a newly arriving Ga atom by an exchange reaction with the As there [5]. The present results show that the metastable interstitial As_2 site is directly adjacent to this dimer (see P in Figs. 1 and 2). As a next step in the interstitial formation process, further Ga atoms fill the missing dimer sites. When a second Ga atom arrives nearest to the interstitial As_2 molecule (site G on Fig. 1 and 2, see also in ref. 5), a cage [3] will form by bonding Ga atoms to the upper As atom of the As_2 molecule and to the lattice atoms. Accordingly, the lower As atom now is unable to leave the crystal, it forms an arsenic interstitial defect in the lattice.

4. Conclusion

Our simulation of the As-rich growth on the reconstructed $c(4x4)$ GaAs(001) surface at rather low temperature indicates enhanced surface diffusion of the As_2 molecules in a migration layer located approximately one atomic distance above the reconstruction layer. From this migration layer, As_2 molecules will enter into the crystal onto a metastable position. In this position the lower As atom of an As_2 molecule fills a tetrahedral interstitial site. The rate constant of the escape of an As_2 molecule from this position into the migration layer is smaller at lower temperature and higher arsenic pressure. Accordingly, the lifetime of such an interstitial conformation increases when the growth temperature is lowered and the arsenic overpressure is increased. From this conformation, interstitial excess arsenic can easily be formed by a cage-like conformation, induced by arriving Ga atoms [3,4]. We propose this sequence as a very probable way for excess As formation during the low-temperature MBE growth of GaAs layers [6].

Acknowledgements

Single point calculations were performed on SunFire 15000 supercomputer located in the Supercomputer Center of the Hungarian National Infrastructure Development Program Office. This work was financially supported by MÖB (Magyar Ösztöndíj Bizottság) and DAAD (Deutscher Akademischer Austauschdienst).

References

[1] Low Temperature GaAs and Related Materials, ed.: G.L.Witt et al., MRS Symposia Proceedings Vol.241 (Material Research Society, Pittsburg, 1992).

[2] A. Suda, N. Otsuka, Surface atomic process of incorporation of excess arsenic in molecular beam epitaxy of GaAs, *Surface Science* 458, 162-172 (2000).

[3] T. Marek, S. Kunsági-Máté, H.P. Strunk: Model for the incorporation of excess arsenic into interstitial positions during the low-temperature growth of GaAs(001) layers, *Journal of Applied Physics* 89, 6519-6522 (2001).

[4] S. Kunsági-Máté, T. Marek, C. Schür, H.P. Strunk: Theoretical and experimental energy barriers associated with the incorporation of excess As into GaAs(001), *Surface Science* 515, 219-225 (2002).

[5] S. Kunsági-Máté, C. Schür, T. Marek, H.P. Strunk: Energetics of growth on the c(4x4) reconstructed GaAs(001) surface and antisite formation: An ab initio approach, *Physical Review B* 69, 193301-4 (2004).

[6] S. Kunsági-Máté, C. Schür, E. Végh, T. Marek, H.P. Strunk: Molecular-dynamics-based model for the formation of arsenic interstitials during low–temperature growth of GaAs, *Physical Review B* 72, 75315-9 (2005).

Brill Academic Publishers
P.O. Box 9000, 2300 PA Leiden
The Netherlands

*Lecture Series on Computer
and Computational Sciences*
Volume 4, 2005, pp. 1816-1819

Dynamics of As Atoms in the Reconstruction Layer of the As-rich GaAs(001) c(4x4) Surface

Sándor Kunsági-Máté,[a,1] Carsten Schür,[b] Nikolett Szalay,[a] Tamas Marek,[b] Horst P. Strunk [b]

[a] Department of General and Physical Chemistry, University of Pécs, Hungary
Ifjúság 6, H-7624 Pécs, Hungary
[b] Institute of Microcharacterization, Friedrich-Alexander-University Erlangen-Nuremberg,
Cauerstrasse 6, D-91058 Erlangen, Germany

Received 9 July, 2005; accepted 31 July, 2005

Abstract: The dynamics of As atoms in the reconstruction layer of the As-rich GaAs(001) c(4x4) surface is studied by molecular dynamics calculations with the semi-empirical PM3 formalism. The surface As atoms have, as expected, increased vibrational amplitudes compared to the bulk. In addition, large anisotropies exist between vibrational amplitudes normal *vs.* parallel to the surface and also between the [-1,1,0] *vs.* [1,1,0] surface directions. The signs of these anisotropies are opposite at high (~900K) and low (~500K) temperatures, where the MBE growth of stoichiometric (High-Temperature-) and non-stoichiometric (Low-Temperature-) GaAs is usually performed, respectively. As a consequence, these anisotropies cause, at low temperature growth, a rather high As surface concentration (desorption reduced) and, due to the increased in-plane vibrations an enhanced surface diffusion, particularly in the [1,1,0] direction. The anisotropy of the surface diffusion offers a model to explain the experimentally observed dependence of excess As formation during growth on substrate surfaces tilted away from the (001) orientation.

Keywords: Growth models, Point defects, Molecular beam epitaxy, Semiconducting gallium arsenide
PACS: 68.47.Fg, 61.82.Fk

1. Introduction

The growth of GaAs layers by molecular beam epitaxy (MBE) at low temperature (LT-GaAs) using high As/Ga flux ratio is of special technological interest. The resulting epitaxial layers, in an annealed state, represent highly resistive buffer material for semiconductor devices and – due to their extremely short carrier lifetimes – are ideal for high-speed photoconductive switches. These very interesting properties of the LT-GaAs layers are related to their excess arsenic (As_{ex}) content. On an exactly oriented (001) substrate, the content increases up to 1.5% when the the substrate temperature decreases to about 200 ^{0}C [1]. The excess As content of LT-GaAs layers can be controlled by the MBE growth parameters, particularly by substrate temperature and As/Ga flux ratio. These parameters affect the rates of adsorption, desorption and surface diffusion of Ga and As species, and therefore influence the properties of the resulting crystalline layer.

A very interesting aspect is that the formation of As_{ex} was recently found to be anisotropic on the tilted surface [2]. This observation is rather surprising and points to a specific growth-induced formation process of the excess arsenic defects. Plausible processes on atomic level have already been suggested and discussed [3,4,5]. However, all the activation energy barriers derived in the models are far from the experimentally determined values. This discrepancy may be due to the fact that all models are based on a *static* approach (*i.e.* calculation of energy terms at zero temperature), but the experimental situation at finite temperature is expected to be affected by the *dynamics* of the surface processes.

In this work we extend our earlier model [3,4] from the 'static' to the 'dynamic' view in coupling it with molecular dynamics simulations. With such calculations we examine the thermal vibrations of the surface atoms parallel and perpendicular to the GaAs(001) surface. According to the known growth processes, the atomic vibrations normal to the surface support the desorption reactions, while the vibrations in the surface plane mainly result in surface diffusion. Therefore, we focus on the determination of the ratio of the normal *vs.* in-plane vibrations, as well as their variation with the

---

[1] Corresponding author. E-mail: kunsagi@ttk.pte.hu

temperature. We find an anisotropy of the in-plane vibrations. In consequence our approach supports a model for the source of increased interstitial As concentration observed experimentally in LT-GaAs layers grown on <111>A misoriented GaAs(001) substrate [2].

2. Methods

2.1 Modeling of the process

Two different modeling procedures were applied to calculate the bulk and surface atomic vibrations (Fig.1). On the one hand, the bulk and on the other hand the As-rich surface of the crystal is modeled. In the first case the bulk crystal was represented by a finite three-dimensional cluster under periodic boundary conditions (Fig.1, a and b). This model simulates the bulk crystal by continuous replication of a small central box in all directions. The box consists of 4x4x3 unit cells, with edges parallel to the [-1,1,0], [1,1,0] and [0,0,1] directions, respectively.

In the second case, the (0,0,1) surface of the crystal is modeled (Fig.1, c and d). For these calculations periodic boundary conditions were applied again, however only to the [-1,1,0] and [1,1,0] directions, forming a 12 atomic layer thick infinite slab. The center of mass of the lowest atoms of the layer was kept in the positions derived from the bulk calculations and the atoms within layers above this layer were allowed to move. The atoms at the bottom of this slab were saturated with hydrogen atoms in order to maintain the bulk-like configuration. The As-rich $c(4x4)$ reconstructed GaAs surface was modeled by adding a further arsenic layer with 0.75 ML on the top of central box defined above, representing the $c(4x4)$ reconstruction. Note, that the surface layer is identified here as the bulk termination arsenic layer, and the adsorption layer is identified as a further (0th) layer on the surface layer, where the reconstruction mainly happens.

Simulations were done at temperatures within the range of 400 K to 1000 K (in steps of 100 K) where the MBE growth of such layers usually is performed. The coordinates describing the atomic movements were collected, then the mean-squared amplitudes of vibration $<\mu_x^2>$, $<\mu_y^2>$ and $<\mu_z^2>$ (with: x≡[-1,1,0], y≡[1,1,0], z≡[0,0,1], see Fig. 1c) were calculated from which follows the anisotropy of the vibration amplitudes as described by their ratios.

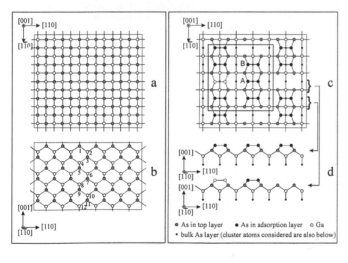

Figure 1. a, b.: The central box used for bulk calculations (left). c, d.: The $c(4x4)$ reconstructed surface GaAs(001) (right). Top (a, c) and side (b, d) view of the cluster used in the molecular dynamic calculations.

2.2 Computational details

The molecular dynamics calculation was performed within the Langevin model by the HyperChem program package. The Hamiltonian was constructed in the frame of the PM3 semi-empirical quantum-chemical approximation.

To find an appropriate initial condition for molecular dynamics, the 'heating' algorithm implemented in the HyperChem program package was used. This procedure smoothly heats up the system from 0 K to the temperature T at which the molecular dynamics simulation is to be conducted. The starting geometry for this heating phase is the static initial structure as represented by the energy-optimized PM3 geometry of the GaAs cluster. Temperature and time steps in the heating phase were set to 2K and 0.1 fs, respectively. Following this heating phase, after an additional 100 fs equilibration period, ten thousand data points during a 1 ps time interval and with 0.1 fs resolution were recorded and the mean-squared amplitudes of the vibrations were calculated.

3. Results

3.1 Atomic vibrations in the bulk GaAs crystal and their dependence on the temperature

Bulk GaAs equilibrium lattice parameters were calculated first. In order to obtain reliable results care has to be taken to ensure a strain-free model system. Therefore the system geometry was optimized first at a given temperature by collecting for each atom the vibrational amplitudes during a 1ps time interval (after the heating phase and the 100fs equilibration period). This procedure was repeated by varying the size of the central box until all amplitudes of same type atoms (*i.e.* As or Ga) became homogeneous. Since the size of the central box reflects the lattice parameter, the state where the amplitudes are equal, indicates the equilibrium lattice parameter at the selected temperature. Figure 2 shows the calculated crystal unit cell dimensions as a function of the temperature. Although the theoretical values follow quite well the lattice parameters calculated using room temperature experimental parameter and linear expansion coefficient (straight line), the PM3 method seems to underestimate the real lattice constants, especially at higher temperature (Fig. 2). For example, at 298 K, our theoretically determined lattice constant is 5.6539Å, which is a little bit larger than the appropriate experimental value (5.6536 Å). The linear thermal expansion coefficient by our calculations was found to be $5.02*10^{-6}$ 1/K, while the respective experimental value is $5.97 \cdot 10^{-6}$ 1/K.

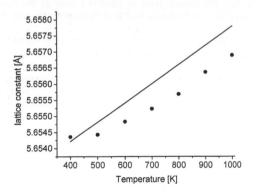

Figure 2: Temperature dependence of the lattice constant of bulk GaAs crystal calculated theoretically in this work. The straight line shows the experimental lattice parameters calculated using room temperature experimental parameter and the linear expansion coefficient, for comparison.

3.2 Anisotropic vibrations of As atoms in the bulk and in the reconstructed surface

Figure 3 shows the temperature dependence of the three orthogonal mean-squared vibrational amplitudes in x, y, and z, $<\mu_x^2>$, $<\mu_y^2>$ and $<\mu_z^2>$, of As atoms in the bulk (Fig. 3, dotted lines) and also in the reconstruction layer of the GaAs(001) surface (Fig. 3). The Figs. 3A and 3B refer to the As atoms in the surface at the respective positions A (at the dimer located at the middle of the triplets) and B (at a dimer at the end of the triplets). Note, that in the bulk $<\mu_x^2>=<\mu_y^2>=<\mu_z^2>$ since the vibration is by definition isotropic in the crystal.

In the bulk and in both A and B cases on the surface, mean-squared amplitudes in the [-1,1,0] direction follow the classical high-temperature linear relation:

$$<\mu^2_{x,K}>=\beta_{x,K}T; \ K=A \ or \ B \ or \ bulk \quad (1)$$

The $\beta_{x,A}$, $\beta_{x,B}$ and β_{bulk} constants are $3.70x10^{-5}$ Å²/K, $3.27x10^{-5}$Å²/K and $3.18x10^{-5}$Å²/K, respectively. The corresponding $\beta_{x,A}/\beta_{bulk}$ and $\beta_{x,B}/\beta_{bulk}$ proportionality constants were found to be 1.16 and 1.03, respectively. In contrast, the $<\mu_y^2>$ and $<\mu_z^2>$ values show a non-linear dependence on the temperature. This unexpected result reflects the considerable coupling between the [1,1,0] and [0,0,1] vibrations. Furthermore, it indicates a considerable anisotropy of the vibrations between [-1,1,0] and [1,1,0] directions at the surface. This anisotropy moreover exhibits temperature dependence and it has opposite character at low and high temperatures. In particular, at low temperature, enhanced vibrations in [1,1,0] direction are observed, therefore enhanced diffusion in this direction can be assumed. A significant difference is obtained for the characteristic temperature where the in-plane vibrations are isotropic, *i.e.* where the curves on Fig.3 for [1,1,0] and [-1,1,0] directions cross: this temperature is at about $T_{c,A}$=920K in case A, and about $T_{c,B}$=770K in case B.

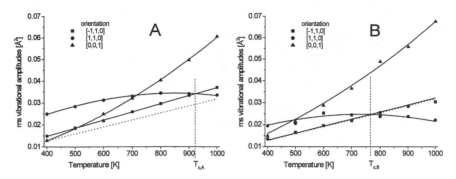

Figure 3: Mean-squared vibrational amplitudes of atoms located in the reconstruction layer #0 on the As-rich GaAs(001) surface. Inclined dotted lines show the corresponding bulk values. Perpendicular dotted lines indicate the characteristic temperature where the anoisotropy in the surface vibrations vanishes. Figures A and B refer to the As atom position A or B of Fig.1c, respectively.

4. Conclusion

Surface atomic vibrations on the As-rich GaAs(001) c(4x4) reconstructed surface at a temperature range where the MBE growth of GaAs is usually performed, were studied by molecular dynamic simulations. A large anisotropy of vibrational amplitudes exists between vibrations normal and parallel to the surface and between the vibration along both<110> directions within the surface. The normal/parallel anisotropy exhibits an unexpected temperature dependence which results in an opposite distribution of kinetic energy between normal and in-plane surface vibrations at high and low temperature. At low temperature, the significant decrease of vibrations normal to the surface hinders the desorption of the As atoms from the surface, while the enhanced in-plane vibrations support the surface diffusion of As atoms. The parallel/parallel in-plane anisotropy supports anisotropic surface diffusion, particularly along the [1,1,0] direction, *i.e.* diffusion perpendicular to the [1,-1,0] steps on the <111>A surface. This behavior correlates with the increased interstitial As concentration observed experimentally in LT-GaAs layers grown on <1,1,1>A misoriented GaAs(001) substrates.

Acknowledgements

Part of the calculations were performed on SunFire 15000 supercomputer (Supercomputer Center of the Hungarian National Infrastructure Development Program) This work was financially supported by MÖB (Magyar Ösztöndíj Bizottság) and DAAD (Deutscher Akademischer Austauschdienst).

References

[1] Low Temperature GaAs and Related Materials, ed.: G.L.Witt et al., MRS Symposia Proceedings Vol.241 (Material Research Society, Pittsburg, 1992).

[2] C. Schür, T. Riedl, T. Marek, H.P. Strunk, and S. Kunsági-Máté, in: P. Specht *et al.* (eds.), 4th Symposium on Non-Stoichiometric III-V Compounds, Series 'Physik Mikrostrukturierter Halbleiter' (University of Erlangen, Institut for Microcharacterization) 27, 19 (2002).

[3] T. Marek, S. Kunsági-Máté, H.P. Strunk: Model for the incorporation of excess arsenic into interstitial positions during the low-temperature growth of GaAs(001) layers, *Journal of Applied Physics* 89, 6519-6522 (2001).

[4] S. Kunsági-Máté, T. Marek, C. Schür, H.P. Strunk: Theoretical and experimental energy barriers associated with the incorporation of excess As into GaAs(001), *Surface Science* 515, 219-225 (2002).

[5] S. Kunsági-Máté, C. Schür, T. Marek, and H.P. Strunk: Energetics of growth on the c(4x4) reconstructed GaAs(001) surface and antisite formation: An ab initio approach, *Physical Review B* 69, 193301-4 (2004).

Brill Academic Publishers
P.O. Box 9000, 2300 PA Leiden
The Netherlands

*Lecture Series on Computer
and Computational Sciences*
Volume 4, 2005, pp. 1820-1823

Calculation of the Surface Heat-transfer Coefficients of W18Cr4V Steel during High Velocity Gas Quenching

Jianyun LI[1]; Heming CHENG; Lijun HOU

Department of Engineering Mechanics,
Kunming University of Science and Technology,
Kunming 650093, Yunnan Province, P.R. China

Received 3 August, 2005; accepted in revised form 14 August, 2005

Abstract: The surface heat-transfer coefficients during high velocity gas quenching is calculated by explicit finite difference method, nonlinear estimate method and the experimental relation between temperature and time. The relations between surface heat-transfer coefficient in 60m/s and surface temperature of steel cylinders are given. Based on the solved surface heat-transfer coefficients, the temperature field is obtained by solving heat conduction. In calculation, the thermal physical properties of material are treated as the functions of temperature. The results of calculation coincide with the results of experiment.

Keywords: surface heat-transfer coefficient; high velocity gas quenching; finite difference method; inverse problem of heat conduction

PACS: 78M10, 33.50.Hv, 94.10.Dy

1. Introduction

Heat treatment is a process of heating and cooling a solid metal to obtain the desired properties. After undergoing the quenching process, the properties of metal and alloy such as surface hardness, strength, microstructure, service life, and mechanical properties etc., can be greatly improved. Treating metal and alloy with high velocity gas quenching is a kind of heat treatment technique that can improve the material properties by cooling the specimen down quickly. In order to find out the best quenching way to quench steel researches should be made on the technique of the high velocity gas quenching.

The surface heat-transfer coefficient is a key parameter in numerical simulation of quenching process. The explicit finite difference method, nonlinear estimate method and the experimental relation between temperature and time during quenching have been used to solve the inverse problem of heat conduction. The relationships between surface temperature and surface heat transfer coefficient of cylinder have been given. In calculation, physical properties were treated as the function of temperature and volume fraction of constituent. The results obtained have been shown that this technique can determine effectual the surface heat transfer coefficients during gas quenching.

2. Experiment

The experimental investigations were carried out using cylinder specimen made out of the Steel W18Cr4V containing 0.75 wt.-% of carbon, 18.0 wt.-% of tungsten, 4.10 wt.-% of chromium and 1.2 wt.-% of vanadium. The specimen is a cylinder of W18Cr4V steel which dimension is Φ 20mm×60mm. After austenized for 30 minutes, the specimens were quenched by high speed Nitrogen gas which velocity is 60m/s. The change of temperature during quenching was measured by thermocouples. A computer recorded experimental data. The thermocouple 1 is placed 1mm distance from the cylinder surface of specimen, thermocouple 2 is on the axis of the cylinder. The temperature values measured by thermocouple 1 are key parameters for solving the nonlinear heat conduction equation. The temperature values measured by thermocouple 2 were used to examine the calculated results. The measured values of temperature are shown in Fig. 1.

---

[1] Corresponding author. E-mail: jean_lee_km@yahoo.com.cn

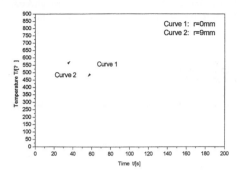

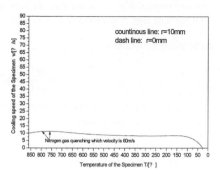

Figure 1: Experimental values of temperature.(60m/s)   Figure 2: Average cooling speed of the specimen

3. Calculation Method of Non-linear Heat Transfer Coefficients

During gas quenching, the surface heat transfer coefficients have a great influence upon the microstructure and residual stresses in steel specimen. The results obtained are very sensitive to small variation in the experimental conditions, which may lead to considerable discrepancies in the value obtained. Therefore, it must be found necessary to determine the effect of temperature on the surface heat transfer coefficient while using the actual experimental conditions that were to use during the subsequent determination of thermal stress and strain. Prince & Fletcher (1980) have an effectively method, which can determine the relationship between temperature and surface heat transfer coefficients during quenching of steel plate.

3. 1 The Non-linear Heat Conduct Equation and Its Finite Difference Scheme
The specimen is a ϕ 20mm×60mm cylinder of W18Cr4V steel. For the axial symmetry specimen, the heat conduction equation in middle of specimen is

$$C_V\rho\frac{\partial T}{\partial t} = \lambda(\frac{\partial^2 T}{\partial r^2} + \frac{1}{r}\frac{\partial T}{\partial r})$$

(1)

where λ, C_V and ρ denote thermal conductivity, specific heat capacity and density of material respectively. C_V and λ are the functions of temperature. T, r and t denote temperature, radius and time. According to the finite difference discrete method of nonlinear heat conduct equation, the finite difference scheme of the nonlinear heat conduct equation can be given by

$$T_j^{i+1} = F\times T_j^i + G\times T_{j+1}^i + D\times T_{j-1}^i$$

(2)

in Eq. (2)

$$F = 1 - \frac{\Delta t}{[(C_v\rho)_j^i + \Delta H]\Delta r}(\frac{\lambda_{j+1}^i + \lambda_j^i}{2\Delta r} + \frac{\lambda_j^i + \lambda_{j-1}^i}{2r_j})$$

(3)

$$G = \frac{\Delta t(\lambda_{j+1}^i + \lambda_j^i)}{2[(C_v\rho)_j^i + \Delta H]\Delta r}(\frac{1}{\Delta r} - \frac{1}{r_j})$$

(4)

$$D = \frac{\Delta t(\lambda_j^i + \lambda_{j-1}^i)}{2[(C_v\rho)_j^i + \Delta H]\Delta r}(\frac{1}{\Delta r} + \frac{1}{r_j})$$

(5)

$$\lambda_j^i = \sum_{k=1}^4 \phi_k\lambda_k \quad (C_v\rho)_j^i = \sum_{k=1}^4 \phi_k(C_v\rho)_k$$

(6)

where k=1 austenite, k=2 ferrite/pearlite, k=3 bainite, k=4 martensite, ϕ_k is the volume fraction of constituent, λ_j^i and $C_v\rho_j^i$ denote the thermal conductivity and specific heat capacity at j nodal in i time step respectively, Δr is the distance between the two inner nodal of the finite difference girding, Δt is the time step, rj is the radius of the j nodal, and ΔH is the latent heat of phase transformation.

the boundary condition is

$$\lambda \frac{\partial T}{\partial r}\bigg|_\Gamma = h(T_S - T_\infty) + C_V \rho \frac{\partial T}{\partial t}$$

(7)

h, T_S and T_∞ denote surface heat-transfer coefficient, specimen temperature of surface boundary and temperature of quenching media.

In order to solve the heat conduct equation with nonlinear surface heat transfer coefficient, the nonlinear estimate method is adopted, i.e. let a parameter

$$\varphi = \sum (T^{exp} - T_{l+1}^{cal})^2$$

(8)

to be minimized, we have

$$\frac{\partial \varphi}{\partial h} = (T^{exp} - T_{l+1}^{cal}) \frac{\partial T_{l+1}^{cal}}{\partial h} = 0$$

(9)

where T_{l+1}^{cal} is $l+1$ the iterative value of temperature, T^{exp} is the measured temperature. T_{l+1}^{cal} can be written as Taylor polynomial

$$T_{l+1}^{cal} = T_l^{cal} + \frac{\partial T}{\partial h} \Delta h$$

(10)

With the help of Equation 16, Equation 17 and the convection condition on boundary, an incremental surface heat transfer coefficient Δh can be expressed in the following form:

$$\Delta h = \xi \frac{\lambda}{T_s - T_\infty} \frac{T^{cal} - T^{exp}}{\Delta r}$$

(11)

If the nodal of finite difference don't pace on the position of thermocouple, it is necessary to obtain a predicted value of temperature at the thermocouple position by interpolation between the appropriate nodal temperatures.

3.2 Calculating Procedure

The thermal properties ($\lambda, C_V \rho$) at various node were calculated by using the linear interpolation method according to Table 1. After an estimate value h_0 of surface heat-transfer coefficient was given, the heat conduction equation (4) can be solved and the computation temperature values at node are obtained. If $\left| T_l^{cal} - T^{exp} \right| > \delta$, a new value of h ($h = h_0 + \Delta h$) will be chosen. T_{l+1}^{cal} is repeated until $\left| T_l^{cal} - T^{exp} \right| \leq \delta$, in which δ is an inherent error. In this way, the changes of surface heat-transfer coefficient in temperature and the temperature field are obtained. In the calculation, time interval is automatically obtained. According to the stability of explicit finite difference scheme, the maximum Δt is chosen as the time interval during calculation.

$$\Delta t \leq \frac{2\Delta r C_V \rho}{\dfrac{\lambda_{j+1} + 2\lambda_j + \lambda_{j-1}}{\Delta r} - \dfrac{\lambda_{j+1} - \lambda_{j-1}}{r_j}}$$

(12)

Table 1: The thermal properties of W18Cr4V steel in different temperatures.

Temperature/□	20	200	300	400	500	700	900
$\lambda /10^{-2}$(W/(mm·K))	4.402	4.248	4.097	3.904	2.079	2.288	2.477
$C_V \rho /10^{-3}$(W·s/(mm³·K))	3.538	3.850	4.003	4.207	4.565	4.754	4.951

4. The Calculated results and Conclusions

The relations between surface heat-transfer coefficient and surface temperature are non-linear during high pressure gas quenching, as shown in figure 3. In the initial period during quenching, the increase of the surface heat-transfer coefficient is rapid. This shows that the exchange of the heat flux between the workpiece and quenching media is large. It is noted that at temperature of 350□, there is a sudden change in the surface heat-transfer coefficient, because of martensite phase transformation in this temperature. After the end of martensite phase transformation, the change in the surface heat-transfer coefficient is small.

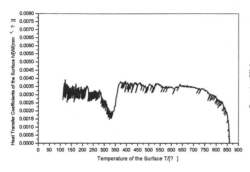

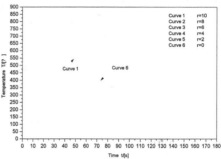

Figure3: The relationship between surface heat transfer coefficients and surface temperature

Figure4: The calculated temperature field.

The calculated result of temperature field is shown in figure 4. These five curves denote the change of temperature with time at the position of $r = 0,2,4,6,8$mm of the cylinder, respectively. In the calculation of temperature field, the non-linear relations between surface heat-transfer coefficient and surface temperature were taken into account.

From figure 1 and 4, the calculated values coincide with experimental values. It show that the surface heat-transfer coefficient obtained during the gas quenching can be used in simulation of temperature field during gas quenching.

Testing of cooling curves of work-pieces during nitrogen gas quenching process shows that the temperature difference of interior work-piece is small. It can be predicted that cooling of the interior work-piece is almost homogeneous and corresponding thermal strains and thermal stresses are rather small.

Acknowledgments

This project has been supported by the National Natural Science Foundation of China (10162002) and the Key Project of Chinese Ministry of Education (No. 204138).

References

[1] R.F.Prince and A.J. Fletcher, *Determination of surface heat-transfer coefficients during quenching of steel plates.* J. Met. Tech., **2:** 203-215.1980

[2] S.Denis, D.Farias and A.Simon:*Mathematical model coupling phase transformations and temperature evolutions in steels.* ISIJ International, 32, 316-325(1992).

[3] Cheng Heming, Wang Honggang and Chen Teili: *Solution of heat conduction inverse problem for steel 45 during quenching.* ACTA Metallurgica Sinica, 33,465-472(1997).

[4] Heming Cheng, Tianchun He and Jianbin Xie: *Solution of an Inverse Problem of Heat Conduction of 45 Steel with Martensite Phase Transformation in High Pressure.* J. Mater. Sci. Technol., 4,372-374 (2002).

[5] Xie Jianbin, Cheng Heming and He Tianchun: *The research of non-linear problems for steel quenching.* Journal of Guilin Institute of Technology, 3, 265-270 (2001).

[6] Xie Jianbin, *The numerical simulation and application of metal and alloy quenched in various quenching media.* Doctor's Dissertation, Kunming, Kunming University of Science and Technology, 2003

Brill Academic Publishers
P.O. Box 9000, 2300 PA Leiden,
The Netherlands

*Lecture Series on Computer
and Computational Sciences*
Volume 4, 2005, pp. 1824-1827

Large-Scale Simulation of the Visual Cortex: Classical and Extraclassical Phenomena

Jim Wielaard[1] and Paul Sajda

Laboratory of Intelligent Imaging and Neural Computing
Department of Biomedical Engineering
Columbia University
New York, USA

Received 11 July, 2005; accepted in revised form 5 August, 2005

Abstract: We have developed a large-scale computational model of a 4x4 mm^2 patch of a primary visual cortex (V1) input layer. The model is constructed from basic established anatomical and physiological data. Based on numerical simulations with this model we are able to suggest neural mechanisms for a wide variety of classical response properties of V1, as well as for a number of extraclassical receptive field phenomena. The nature of our model is such that we are able to address stationary as well as dynamical behaviour of V1, both on the single cell level and on a population level of up to about 10^4 cells.

Keywords: large-scale simulation, brain, visual cortex

1   The Model

Our model consists of 8 ocular dominance columns and 64 orientation hypercolumns (i.e. pinwheels), representing a 16 mm^2 area of a macaque V1 input layer $4C\alpha$ or $4C\beta$. The model contains approximately 65,000 cortical cells and the corresponding appropriate number of LGN cells. Our cortical cells are modeled as conductance based integrate-and-fire point neurons, 75% are excitatory cells and 25% are inhibitory cells. Our LGN cells are rectified spatio-temporal linear filters. The model is constructed with isotropic short-range cortical connections ($< 500\mu m$), realistic LGN receptive field sizes and densities, and realistic sizes of LGN axons in V1. Details of the model can be found in [1]. Some background information can also be found in previous work [2,3] by one of the authors (JW).

Dynamic variables of a cortical model-cell i are its membrane potential $v_i(t)$ and its spike train $S_i(t) = \sum_k \delta(t - t_{i,k})$, where t is time and $t_{i,k}$ is its kth spike time. Membrane potential and spike train of each cell obey a set of N equations of the form

$$C_i \frac{dv_i}{dt} = -g_{L,i}(v_i - v_L) - g_{E,i}(t, [S]_E, \eta_E)(v_i - v_E) - g_{I,i}(t, [S]_I, \eta_I)(v_i - v_I) \, , \; i = 1, \ldots, N \, . \quad (1)$$

These equations are integrated numerically using a second order Runge-Kutta method with time step 0.1 ms. Whenever the membrane potential reaches a fixed threshold level v_T it is reset to a fixed reset level v_R and a spike is registered. The equation can be rescaled so that $v_i(t)$ is dimensionless and $C_i = 1$, $v_L = 0$, $v_E = 14/3$, $v_I = -2/3$, $v_T = 1$, $v_R = 0$, and conductances (and currents) have dimension of inverse time.

---

[1]Corresponding author. Active Member of the American Physical Society. E-mail: djw21@columbia.edu

The quantities $g_{E,i}(t, [\mathcal{S}], \eta_E)$ and $g_{I,i}(t, [\mathcal{S}], \eta_I)$ are the excitatory and inhibitory conductances of neuron i. They are defined by interactions with the other cells in the network, external noise $\eta_{E(I)}$, and, in the case of $g_{E,i}$ possibly by LGN input. The notation $[\mathcal{S}]_{E(I)}$ stands for the spike trains of all excitatory (inhibitory) cells connected to cell i. Both, the excitatory and inhibitory populations consist of two subpopulations $\mathcal{P}_k(E)$ and $\mathcal{P}_k(I)$, $k = 0, 1$, a population that receives LGN input ($k = 1$) and one that does not ($k = 0$). In the model presented here 30% of both the excitatory and inhibitory cell populations receive LGN input. We assume noise, cortical interactions and LGN input act additively in contributing to the total conductance of a cell,

$$g_{E,i}(t, [\mathcal{S}]_E, \eta_E) = \eta_{E,i}(t) + g_{E,i}^{cor}(t, [\mathcal{S}]_E) + \delta_i g_i^{LGN}(t)$$

$$g_{I,i}(t, [\mathcal{S}]_I, \eta_I) = \eta_{I,i}(t) + g_{I,i}^{cor}(t, [\mathcal{S}]_I) , \tag{2}$$

where $\delta_i = \ell$ for $i \in \{\mathcal{P}_\ell(E), \mathcal{P}_\ell(I)\}$, $\ell = 0, 1$. The terms $g_{\mu,i}^{cor}(t, [\mathcal{S}]_\mu)$ are the contributions from the cortical excitatory ($\mu = E$) and inhibitory ($\mu = I$) neurons and include only isotropic connections,

$$g_{\mu,i}^{cor}(t, [\mathcal{S}]_\mu) = \int_{-\infty}^{+\infty} ds \sum_{k=0}^{1} \sum_{j \in \mathcal{P}_k(\mu)} \mathcal{C}_{\mu',\mu}^{k',k}(||\vec{x}_i - \vec{x}_j||) G_{\mu,j}(t - s) \mathcal{S}_j(s) , \tag{3}$$

where $i \in \mathcal{P}_{k'}(\mu')$ Here $\vec{x}_i$ is the spatial position (in cortex) of neuron i, the functions $G_{\mu,j}(\tau)$ describe the synaptic dynamics of cortical synapses and the functions $\mathcal{C}_{\mu',\mu}^{k',k}(r)$ describe the cortical spatial couplings (cortical connections). The length scale or excitatory and inhibitory connections is about $200\mu m$ and $100\mu m$ respectively.

In agreement with experimental findings, the LGN neurons are modeled as rectified center-surround linear spatiotemporal filters. A cortical cell, $j \in \mathcal{P}_1(\mu)$ is connected to a set $N_{L,j}^{LGN}$ of left eye LGN cells, or to a set $N_{R,j}^{LGN}$ of right eye LGN cells,

$$g_j^{LGN}(t) = \sum_{\ell \in N_{Q,j}^{LGN}} [g_\ell^0 + g_\ell^V \int_{-\infty}^{\infty} ds \int d^2y \, G_\ell^{LGN}(t - s) \, \mathcal{L}_\ell(||\vec{y}_\ell - \vec{y}||) \, I(\vec{y}, s)]_+ , \tag{4}$$

where $Q = L$ or R. Here $[x]_+ = x$ if $x \geq 0$ and $[x]_+ = 0$ if $x \leq 0$, $\mathcal{L}_\ell(r)$ and $G_\ell^{LGN}(\tau)$ are the spatial and temporal LGN kernels respectively, $\vec{y}_\ell$ is the receptive field center of the ℓth left or right eye LGN cell, which is connected to the jth cortical cell, $I(\vec{y}, s)$ is the visual stimulus. The parameters g_ℓ^0 represent the maintained activity of LGN cells and the parameters g_ℓ^V measure their responsiveness to visual stimuli. The LGN kernels are of the form

$$G_\ell^{LGN}(\tau) = \begin{cases} 0 & \text{if } \tau \leq \tau_\ell^0 \\ k \, \tau^5 \left(e^{-\tau/\tau_1} - c \, e^{-\tau/\tau_2}\right) & \text{if } \tau > \tau_\ell^0 \end{cases} \tag{5}$$

and

$$\mathcal{L}_\ell(r) = \pm(1 - K_\ell)^{-1} \left\{ \frac{1}{\pi \sigma_{c,\ell}^2} e^{-(r/\sigma_{c,\ell})^2} - \frac{K_\ell}{\pi \sigma_{s,\ell}^2} e^{-(r/\sigma_{s,\ell})^2} \right\} , \tag{6}$$

where k is a normalization constant, $\sigma_{c,\ell}$ and $\sigma_{s,\ell}$ are the center and surround sizes respectively, and K_ℓ is the integrated surround-center sensitivity.

The connection structure between LGN cells and cortical cells, given by the sets $N_{Q,j}^{LGN}$, is made so as to establish ocular dominance bands and a slight orientation preference which is organized in pinwheels [4]. It is further constructed under the constraint that the LGN axonal arbor sizes in V1 do not exceed the anatomically established values. A sketch of the model geometry is given in Figure 1.

Brill Academic Publishers
P.O. Box 9000, 2300 PA Leiden
The Netherlands

*Lecture Series on Computer
and Computational Sciences*
Volume 4, 2005, pp. 1828-1831

Multiscale Simulation of the Mushroom-Shaped Nanoscale Supramolecular System: From Self-assembly Process to Functional Properties

Hsiao-Ching Yang*[1], Tien-Yau Luh[1], Cheng-Lung Chen[2]

Department of Chemistry[1], National Taiwan University, Taipei, Taiwan
Department of Chemistry[2], National Sun Yat-sen University, Kaohsiung, Taiwan

Received 10 July, 2005; accepted in revised form 4 August, 2005

Abstract: The miniaturized rod-coil triblock copolymers BPEIS ((biphenyl ester)^3-(isoprene)^9-(styrene)^9) had been found to self-assemble into well-ordered nanostructures and unusual head to tail multilayer structure[1]. The purpose of our study is to obtain fundamental understanding the connection of the inherent morphological characterization of single molecule and the mechanism of phase behavior of this polar self-assembly system, which result in significant optical properties. Dissipative particle dynamics simulation was carried out to study the mechanism of phase behavior of the solvent-copolymers system. We found that the solvent-induced polar effect under different temperature is important in the process of self-assembly of block copolymers. In different temperature the solvent induces hybrid structure aggregation. Our results are consistent with experimental observations and give more understanding of the mechanism governing the self-assembly process and phase behavior. Furthermore, the sizes and stabilization energies of mushroom-shaped supramolecular clusters were predicted by molecular modeling method. Clusters of sizes from 16 to 90 molecules were found to be stable. In combination of classical and simple quantum mechanical calculations, the band gaps of BPEIS clusters with various sizes were estimated. The band gap was converged at 2.45 eV for cluster contains 90 molecules. Nonlinear optical properties of the material were investigated by the semi-empirical quantum mechanical calculations of molecular dipole moment and hyperpolarizabilities. Significant second-order nonlinear optical properties were shown from these calculated properties.

Keywords: self-assembly, block copolymer, phase behavior, dissipative particle dynamics, cluster size, π– electron
PACS: 61.43.Bn and 78.20.-e

1. From Molecular Character to Phase Behavior

We adopted mesoscopic simulation dissipative particle dynamics (DPD) method [2-3] to study the dynamics process and phase behaviors of the self-assembly functional block copolymers. DPD method allows the simulation of hydrodynamic behavior of larger systems up to the microsecond (10^{-6} s) range. In DPD simulation, the molecule is divided into beads based on some physical and chemical rules. To reproduce the experimental result qualitatively interaction parameters between beads are obtained by the combination of quantum mechanical and Monte Carlo methods[2-3]. Then the beads simply follow Newton's equations of motion. All bead masses are set equal to unity for simplicity, and the total force on bead i contain three parts, each of which is a pairwise additive.

$$F_i = \sum_{i \neq j} (F_{ij}^C + F_{ij}^D + F_{ij}^R)$$

(1)
where the sum is over index j with i = j excluded. In which Fij^C is conservative force; Fij ^{D} and Fij ^{R} are, respectively, the dissipative and random forces, which effectively act as a thermostat and result in fast equilibration to the Gibbs–Boltzmann canonical ensemble. The significance of these terms is investigated in detail elsewhere [1-2].

The particular architecture of this miniaturized rod-coil triblock copolymer ((biphenyl ester)$_3$-(isoprene)$_9$-(styrene)$_9$) short named BPEIS as shown in Figure 1, hence, special interactions between solvent and beads lead to the very unusual phase behavior of the system. We are interesting on the self-

---

1 Author to whom correspondence should be addressed. E-mail: hcyang@chem.sinica.edu.tw, yanghc@gmail.com

assembly process and phase behaviors of these BPEIS block copolymers in solvents at different concentrations and temperature. We found that the solvent-induced effect and the control of temperature are important in the process of self-assembly of block copolymers. In different temperature the solvent induces hybrid structure aggregation. The simulation result shows that only chloroform solvent at critical concentration can induce forming mushroom shaped self-assembling aggregates as shown in Figure 2. The origin of forming stem-to-cap ordered layer was also investigated. Our results are consistent with experimental observations[1] and give more understanding of the mechanism governing the self-assembly process and phase behavior.

2. On the Cluster Size and Optical-electronic Functional Properties

In mushroom BPEIS cluster, the balance of attractive and repulsive forces among the different blocks of molecules leads to the formation of stable finite sized mushroom-shaped supramolecules. Both energy minimization and Molecular dynamics simulation show that the cluster is in the mushroom shape and has considerable stability. Our simulations predicted the sizes and stabilization energies of the mushroom-shaped supramolecular clusters. The possible sizes of clusters are ranged from 16 to 90 chains. These clusters may be formed depending on the concentration of copolymer in solvent. The largest size of the predicted mushroom cluster contains about 90-chain and in the nanoscale dimension. In the BPEIS cluster, the polar array is related to the dipole moment of BPEIS molecule. Given in Figure 3a is the calculated dipole vector respect of a single molecule obtained from quantum mechanical AM1 calculation. In the mushroom aggregates the total dipole vector $\mu_{cluster}$ can be treated as the vector sum of individual molecular dipole vectors μ_{chain}. These molecular dipoles were obtained from their geometries at each MD trajectory point. The total dipole vectors of mushroom aggregates are the sum of individual dipole vectors of molecules as shown in Figure 3c.

Some special optical-electronic functional properties had also been found for this mushroom shaped BPEIS system[1,4-5]. The optical-electronic functional properties are strongly correlated to the special structure of the cluster. Alternatively, simplified approximation may be adopted to estimate band gap and other optical-electronic properties[6]. We interest to investigate the size effect to the band gap of the BPEIS cluster. In the band gap calculation, only the rod blocks of BPEIS molecules were considered. Figure 4 shows the calculated π-orbital energies of clusters with different number of chains. The figure shows that as the size of cluster increases the energy states become closer and formed bands. Figure 4 also shows the correlation between HOMO-LUMO energy gap and the size of the BPEIS cluster. The decrease of the energy gap is obviously associated to the delocalization of the π-electron in the rod blocks of the cluster. This is the evidence that the delocalization of π-electrons not only happened on the intrachain frame but also happened over interchain rod blocks. It has been well established that the π-electron delocalization effect plays an important role in reproducing the electronic and optical properties of molecules[6-8]. Our calculation confirms the inter-chain delocalization of π-electrons on the BPEIS clusters. This leads to unusual electronic-optical properties of the BPEIS materials.

3. Conclusion

The present work combined classical molecular dynamics and quantum mechanical methods to investigate the mushroom-shaped nanoscale supramolecular system from mesoscopic phase behavior, atomic structure aggregate to electronic optical property. We successfully used DPD simulation to investigate the self-assembly process and the unusual phase behavior of this system. From the information obtained from mesoscopic simulation, we had basic realization that the morphology characteristics of BPEIS system, which make this copolymer attractive for optical-electronic materials. Then molecular dynamics applied to further study microstructure and related properties of mushroom aggregates of BPEIS. Our simulations predicted the sizes and stabilization energies of the mushroom-shaped supramolecular clusters. These clusters may be formed depending on the concentration of copolymer in solvent. The largest size of the predicted mushroom cluster contains about 90-chain and in the nanoscale dimension.

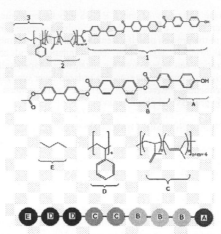

Figure 1: The miniaturized triblock polymer BPEIS and each BPEIS molecule is then divided into five types of DPD beads.

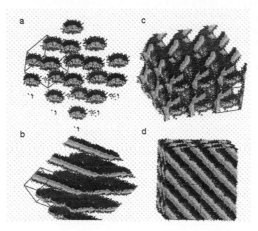

Figure 2: Phase behavior of the miniaturized triblock polymer BPEIS (a) The DPD results of system at molar ratio of BPEIS to chloroform 1: 9 and temperature at 400K; mushroom shaped aggregates in chloroform solvents; (b) at molar ratio of BPEIS to chloroform 3: 7 and temperature at 500K. (c) at molar ratio of BPEIS to chloroform 5: 5 and temperature at 500K. (d) The result of particle density from our DPD simulation at high solute concentration (BPEIS : chloroform = 8 : 2) at 600K.

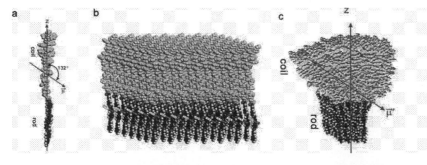

Figure 3: (a) The orientation of the dipole vector respect to a single geometry optimized BPEIS molecule. (b) Illustration of the dipole array of the ideal molecules. (c) The total dipole vectors of mushroom aggregates are the sum of individual dipole vectors of molecules.

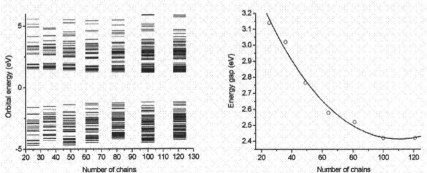

Figure 4: The calculated energy band of different number of chain flowed as 16, 25, 36, 49, 64, 81, 100, 121 of chains and. The fit curve of the correlation between HOMO-LUMO gap and BPEIS cluster size.

The electronic properties of BPEIS clusters were investigated by the simplified quantum mechanical calculations. From the result of our simplify quantum mechanical approach, it suggested there had strong correlation between the size and band gap energy. The electronic calculation also indicates that inter-chain delocalization of π-electrons in the cluster. The calculated HOMO-LUMO energy gap converged in number of chain about 90. The calculated energy gap of 90-chain-cluster is about 2.45 eV compared to the 2.33 eV of Stupp's experiment.

References

[1] S. I. Stupp, V. LeBonheur, K.Walker, L. S. Li, K. Huggins, M. Keser, A. Amstutz, *Science* **1997**, *276*, 321.

[2] R. D. Groot, P. B. Warren, *J. Chem. Phys.* **1997**, *107*, 4423.

[3] M. Y. Kuo, H. C. Yang, C. Y. Hua, C. L. Chen, *ChemPhysChem* **2004**, *5*, 575.

[4] M. Sayar, F. J. Solis, M. Olvera de la Cruz, S. I. Stupp, *Macromolecules*, **2000**, *33*, 7226

[5] Leiming Li ; Samuel I. Stupp. *Appl. Phys. Lett.*, **2001**, *78*, 4127

[6] H. C. Yang, C. Y. Hua, Ming-Yu Kuo, Q. Huang, C. L. Chen, *ChemPhysChem* **2004**, *5*, 373

[7] D. Hu, J. Yu, K. Wong, B. Bagchi, P. J. Rossky, P. F. Barbara, *Nature* **2000**, *405*, 1030.

[8] S. N. Yaliraki, R. J. Silbey, *J. Chem. Phys.* **1996**, *104*, 1245.

Brill Academic Publishers
P.O. Box 9000, 2300 PA Leiden,
The Netherlands

Lecture Series on Computer
and Computational Sciences
Volume 4, 2005, pp. 1832-1835

Cutting Segments Configuration in Square Cutting

Y. Teshima[1], K.Kase, and A. Makinouchi

Integrated VCAD System Research Program,
RIKEN (The Institute of Physical and Chemical Research),
Wako, Saitama 351-0198, Japan

Received 5 August, 2005; accepted in revised form 20 August, 2005

Abstract: The number of equivalence classes of configured points/segments for square cutting is investigated in this paper. In Volume Computer-Aided Design (VCAD), the shape of three-dimensional objects is approximated in discrete cubic lattice. With our unique shape approximation method called Kitta cube, three-dimensional shape is approximated by triangles(cutting triangles) which are held in the cubes of a cubic lattice. The intersection between the edge of a cube of a cubic lattice and three-dimensional shape is called cutting point. Combinatorial analysis of Kitta cube is essential for better shape approximation. In Kitta square, which is a two-dimensional analogue of Kitta cube, the boundary of two-dimensional shape is approximated by segments(cutting segments). Two endpoints of the segment are cutting points located on the edge of a square of a square lattice. The purpose of this paper is to provide mathematical foundations to Kitta square by enumerating the number of configured cutting points/segments using Pólya-Redfield's theory of counting. We assume that the number of cutting points on one edge of the square is at most one.

Keywords: VCAD, Kitta cube, Kitta square, Cutting segment, Cutting point, Equivalence class, Pólya-Redfield's Theory of Counting

Mathematics Subject Classification: 68R05

1 Introduction

Volume CAD (VCAD), a next generation Computer-Aided Design (CAD), can retain not only the shape but also the physical attributes of three-dimensional objects[3]. In VCAD, the shape of three-dimensional objects is approximated in discrete cubic lattice. The intersection between the edge of a cube of a cubic lattice and three-dimensional shape is called cutting point. With our unique shape approximation method called Kitta cube(KC), three-dimensional shape is approximated by triangles which are held in the cubes of a cubic lattice. The triangle is called cutting triangle and its vertices are cutting points. Combinatorial analysis of KC is essential for better shape approximation. In previous papers, enumerations of cutting points/triangles configuration for KC were investigated [5] [6]. A problem to be solved in future is to classify the present triangle patterns into detailed category. This paper describes a study on the enumeration of cutting points/segments configuration for two-dimensional Kitta cube or Kitta square(KS). It is very important to investigate KS in order to classify the triangle patterns of KC into detailed category because KS appears on six faces of a cube.

---

[1]Corresponding author. E-mail: kippoh@riken.jp, Member of Interdisciplinary Institute of Science, Technology and Art.

Other papers also provide mathematical foundations for KC and VCAD but from other points of view, for example, information theory[1][2] and topology[4].

2 Definition of square cutting: Kitta square

In accordance with the definition of KC[5][6], the definition of two-dimensional KC is given in this section.

In each square, the boundary line of the shape is not retained but shape approximation that satisfies the following three conditions are made: (i) Approximation with segments, (ii) Endpoints of the segment lie on the square edges, and not on the vertices of square, (iii) Number of the endpoints of the segment on each edge of the square is at most one (the single cutting point condition). This approximate segment is called *cutting segment*, and the intersection between the boundary and edge of the square is called *cutting point*. We call squares with such approximate segments *Kitta squares* or *two-dimensional Kitta cube* (In Japanese, Kitta means to cut). A cutting point is an endpoint of the cutting segment. The location of a cutting point on an edge is not identified during enumeration.

In KS, we assume that there is only one single cutting point, but generally speaking, two or more cutting points can appear on one edge. They can be unified into a single cutting point by unification disposal[4]. After this, we assume that the disposal has already been completed for the cutting point on each edge.

The above defines the original KS. If we place cutting points not only on each of the four edges but also on each of the four vertices of the square, in which cutting points on the vertices are independent of cutting points on the edges, other cutting segments will be obtained. This type of KS is called additional KS in this paper. The number of patterns of cutting points/segments configuration for the additional KS will also be enumerated in Section 4.3 and Section 4.4 .

3 Symmetry group on square

Three-dimensional rotation group of a square is considered as the permutation group for the square. There are eight permutations in the three-dimensional rotation of a square. The permutation group G of a square, can be classified into the following five kinds: $\pi_1, \cdots, \pi_5$.

π_1. An identity permutation.

π_2. A permutation arising from a rotation 180° around a four-fold rotational axis of a square.

π_3. Permutations arising from a rotation 90° around a four-fold rotational axis of a square. There are two such permutations according to two directions of the rotation 90° (clockwise and unclockwise).

π_4. Permutations arising from a flipping around an axis through the midpoints of two parallel edges of a square. There are two such permutations according to two choices for the axis.

π_5. Permutations arising from a flipping around an axis through the diagonal of a square. There are two such permutations according to two choices for the axis.

4 Enumeration

Under the group action denoted in Sec. 3, we will apply Pólya-Redfield's theory of counting to our enumeration.

Brill Academic Publishers
P.O. Box 9000, 2300 PA Leiden
The Netherlands

*Lecture Series on Computer
and Computational Sciences*
Volume 4, 2005, pp. 1836-1841

A Molecular Dynamics Study of the Interaction-Induced Effects on the Far-Infrared and Infrared Spectrum of HCl Dissolved in CCl$_4$

Georgios Chatzis and Jannis Samios[*]

*Laboratory of Physical Chemistry, Department of Chemistry,
University of Athens,Panepistimiopolis 15771, Athens, Greece*

Received 6 August, 2005; accepted in revised form 16 August, 2005

Abstract: The molecular dynamics simulation technique was used to study the interaction-induced dipole contributions in terms of "dipole-induced dipole" and "back-induced dipole" mechanisms between solute and solvents,. The simulated dipole time correlation functions and spectral line shapes for far-infrared and infrared were compared with corresponding available experimental results and reasonable agreement was observed. It is found that the well-known interaction-induced dipole effects contribute insignificantly to the infrared, whereas they are not negligible in the case of the far-infrared absorption profile.

1. Introduction

Solutions of small polar molecules in polar or neutral solvents are among the systems for which the relative importance of the intermolecular interactions of different natures governing the solute dynamics still remains debatable. Note that solutions of hydrohalogens, and, in particular, those of hydrogen chloride (HCl) in nonpolar solvents, are representative solutions of such systems which have also received much attention from the experimentalists. To the best of our knowledge, far-infrared (FIR) and infrared (IR) spectroscopic techniques have been widely used by several authors[1-7] who studied the rotational dynamics of HCl dissolved in CCl$_4$. Still, some questions remain open as one attempts to elucidate spectral features of the system at different thermodynamic conditions. In general, these spectral features reflect the influence of specific molecular interactions upon the HCl/solute dynamics. We have employed the molecular dynamics (MD) simulation technique in order to investigate the structural, thermodynamical, and single molecule dynamical properties of the system. This computational approach is central in the present study. One aims to investigate the well-known interaction-induced effects on the total dipole moment time correlation functions (CFs), which are related to the FIR and IR absorption spectra of HCl in CCl$_4$.

The calculations were carried out at thermodynamic conditions for which the experimental spectral profiles of the solution available in the literature do not show isolated rotational quantum absorption lines. In that particular case, a typical atomistic simulation of the system, such as the classical MD simulation technique, is an alternative approach to obtain directly the total dipole moment time CF of the solution as well as the corresponding absorption spectral line shape (via Fourier transform).
Concretely, in the case of Far-Infrared the absorption coefficient $\alpha(\omega)$ is connected with the Fourier transform (FT) of the system's total dipole moment CF, $C_M(t)$, according to the following relation:

$$\alpha(\omega) = \frac{4\pi n\omega}{3\hbar\eta(\omega)D(\omega)c}\tanh\left(\frac{\hbar\omega}{2k_BT}\right)\text{Re}\left\{\int_0^\infty e^{i\omega t}C_M(t)dt\right\}. \tag{1}$$

The evaluation of the total dipole moment at time t, may be expressed as a sum of four different CFs:

$$C_M(t) = \left(\left(\vec{M}_P(0)+\vec{M}_I(0)\right)\cdot\left(\vec{M}_P(t)+\vec{M}_I(t)\right)\right) = \left(\vec{M}_P(0)\cdot\vec{M}_P(t)\right)+\left(\vec{M}_I(0)\cdot\vec{M}_I(t)\right)+\left(\vec{M}_P(0)\cdot\vec{M}_I(t)\right)+\left(\vec{M}_I(0)\cdot\vec{M}_P(t)\right) = \tag{2}$$

---

[*] Corresponding Author e-mail: isamios@cc.uoa.gr

$$C_M^{PP}(t) + C_M^{II}(t) + C_M^{PI}(t) + C_M^{IP}(t) = C_M^{PP}(t) + C_M^{II}(t) + C_M^{+}(t)$$

where, $C_M^{PP}(t)$ and $C_M^{II}(t)$ represent the autocorellation function of the total permanent and induced dipole moment in the sample, respectively. $C_M^{PI}(t)$ and $C_M^{IP}(t)$, or the sum of them, $C_M^{+}(t)$, describe the interference between the total permanent and the total induced dipole moments of the sample. Additionally, the calculation of the dipole induced dipoles and back induced dipoles of the molecules was based upon an isotropic polarizability of the solvent ($\alpha^{CCl_4} = 11.2 \, \overset{o}{A}{}^3$ [22]) and an anisotropic polarizability of the axially symmetric solute molecule. The data used in our calculation for the isotropic, $\alpha^{HCl} = 2.60 \, \overset{o}{A}{}^3$, and the anisotropic, $\gamma^{HCl} = 0.311 \, \overset{o}{A}{}^3$, component of the polarizability of HCl were borrowed from the literature[12, 13].

Thus, according to the first order DID mechanism, the electric field originated by the permanent dipole moment of the solute molecule j, $\vec{\mu}_j^P$, at the center of mass of another solute or solvent molecule i can be expressed as

$$E_{ij,\alpha}^{DID} = \frac{\mu_{HCl}^P}{R_{ij}^5}\left[3\left(\hat{u}_j \cdot \vec{R}_{ij}\right)R_{ij,\alpha} - R_{ij}^2 u_{j,\alpha}\right],\tag{3}$$

and the induced dipole moment on each molecule is obtained by means of the following equation

$$\vec{\mu}_{ij}^{DID}(t) = \hat{\alpha}_i(t) \cdot \vec{E}_{ij}^{DID}(t),\tag{4}$$

where $\hat{\alpha}_i$ is the polarizability tensor of the ith molecule. Note that in the case of the CCl$_4$ molecule, the first order in the polarizability tensor was used, since for the molecular symmetry T$_d$ the polarizability is isotropic (i.e. scalar). Additionally, we calculate the two total electric field components, $\vec{E}_i^{DID,//}$, and the normal, $\vec{E}_i^{DID,\perp}$, to the molecular axis of the HCl:

$$\left[\left(\vec{E}_i^{DID} \cdot \hat{u}_i\right)\hat{u}_i\right] = \vec{E}_i^{DID,//}, \quad \left(\vec{E}_i^{DID} - \vec{E}_i^{DID,//}\right) = \vec{E}_i^{DID,\perp}.\tag{5}$$

In the second step of our procedure, the expected total DID moment, $\vec{\mu}_{I,i}^{DID}$, of the molecule is obtained via the relations:

$$\vec{\mu}_{I,i}^{DID,//} = \alpha_i^{//} \cdot \vec{E}_i^{DID,//}, \quad \vec{\mu}_{I,i}^{DID,\perp} = \alpha_i^{\perp} \cdot \vec{E}_i^{DID,\perp}\tag{6}$$

$$\vec{\mu}_{I,i}^{DID} = \vec{\mu}_{I,i}^{DID,//} + \vec{\mu}_{I,i}^{DID,\perp}.\tag{7}$$

$\alpha_i^{//}$, $\alpha_i^{\perp}$ denote the parallel and the normal to the molecular axis of HCl components of the polarizability, respectively.

$$\alpha_i^{HCl} = \frac{1}{3}\left(\alpha_i^{//} + 2\alpha^{\perp}\right), \quad \gamma_i^{HCl} = \alpha_i^{//} - \alpha_i^{\perp}.\tag{8}$$

Finally, according to the BID mechanism the total induced dipole moment on each solvent molecule, $\vec{\mu}_{I,i}^{DID}$ (i=1(1)N$_A$), induces back a dipole moment on each of the other molecules in the sample. Thus, the resulting induced dipole moment on each molecule, $\vec{\mu}_{I,i} \equiv \vec{\mu}_{I,i}^{CCl_4}$ or $\vec{\mu}_{I,i}^{HCl}$, may be simply calculated as the sum of two terms, the contributions due to the DID and the BID induction mechanism.

According to the above theoretical considerations, the related to the Infrared absorption total dipole moment of a probe solute molecule, $\vec{\mu}_{tot,i}^{HCl}$ is constituted by its permanent dipole, $\vec{\mu}_i^P$, and the parallel component to the molecular axis, $\vec{\mu}_{I,i}^{//}$, of its total induced dipole moment $\vec{\mu}_{I,i}$

$$\vec{\mu}_{tot,i}^{HCl}(t) = \vec{\mu}_i^P(t) + \vec{\mu}_{I,i}^{//}(t).\tag{9}$$

Following the well-known procedure outlined in ref. 14, the IR dipole CF may be written as follows

$$C_{IR}(t) = \left(\frac{\partial \vec{\mu}_{tot,i}^{HCl}}{\partial q_l} \right) \left(\hat{v}_i(0) \cdot \hat{v}_i(t) \right). \tag{10}$$

Here, $\hat{v}_i(t)$ denotes the unit vector of the total dipole moment, $\vec{\mu}_{tot,i}^{HCl}(t)$, of the solute molecule i. In general, the direction of the vector $\vec{\mu}_{tot,i}^{HCl}(t)$ may be the same or opposite to that of the permanent dipole $\vec{\mu}_i^p(t)$ of the solute. Specifically, from the pertinent results presented herein it can be seen that the average magnitude of $\vec{\mu}_{I,i}^{''}(t)$ of HCl is quite small when compared to its permanent dipole moment value. Thus, the resulting total dipole moment, $\vec{\mu}_{tot,i}^{HCl}(t)$, and the permanent dipole, $\vec{\mu}_i^p(t)$, of a solute exhibit permanently the same direction. As we can see from the calculated CFs $C_{IR}(t)$ presented below, this result leads to the conclusion that the lineshape of the IR absorption spectra of this solution is mainly due to the permanent dipole moment contribution of the solute molecules in the sample.

2. Results and Discussion

The calculated dipole moment CFs and the corresponding spectral line shapes presented here are based on NVE-MD simulations of two solutes in 254 solvents with the usual cubic periodic boundary conditions. The analytical description of the employed potential model has been determined in our previous simulation study[8] along with the fundamental technical details applied in the present simulations[8,9]. We notice that, for a given configuration of the system, the interaction-induced dipoles among the molecules have been calculated for center of mass intermolecular separations, R_{ij}, fulfilling the condition $R_{ij}<L/2$, where L denotes the length of the simulation box. Finally, it should be mentioned that, after equilibration, each MD run has been extended up to 150ps and the pertinent dipole time CFs were obtained using the usual method of the direct calculation (DC).

The three dipole moment time CFs of interest $C_A(t) = \left(\vec{A}(0) \cdot \vec{A}(t) \right)$, $\vec{A} \equiv \vec{M}_{tot}, \vec{\mu}_i^p, \vec{\mu}_{tot,i}^{HCl}$ were obtained at three different temperatures (273, 290 and 343K) and at solvent densities corresponding to the orthobaric pressures.In all cases, the time dependent quantities $\vec{M}_{tot}$ and $\vec{\mu}_{tot,i}^{HCl}$ were calculated at each time step of the simulation Moreover, as mentioned in the previous section, $\vec{\mu}_i^p(t)$ denotes the orientation of the time dependent permanent dipole vector of the ith HCl molecule, obtained at each time step in the simulation run. Note that each CF, $C_A(t)$, was obtained in time intervals of five time steps ($5 \, \delta t = 0.01ps$) over a large total time interval and for a large number of time origins.

In what follows we will first discuss the results obtained for the above-mentioned dipole CFs in conjunction with data from prior FIR and IR experimental studies of the mixture under investigation. Thus, the computed CFs, $C_A(t)$, at 290K are shown in Fig. 1. Note that the three dipole CFs are normalized to their own amplitudes at t=0. The insert figure illustrates the experimental IR[4,8] and FIR[7] corresponding dipole CFs of this highly diluted solution at room temperature. The time evaluation of both CFs confirms that the qualitative trends observed in the IR dipole CF are also observed in the corresponding correlation obtained using the FIR technique. Specifically, the two curves practically overlap each other at very short times and up to t $\cong$ 0.08ps. Afterwards, each curve exhibits a positive minimum followed by a submaximum and converges to zero after relatively long time. Note also that the IR CF goes through a shallower positive minimum than the FIR function. In addition, the positive hump in the IR CF is located at shorter time and shifted to a slight higher ordinate value compared to that in the FIR correlation. However, the direct comparison among the FIR[7] and the most recent available experimental IR[4] data CFs, reveals that the feature of the FIR CF is in semiquantitative agreement with that from the IR band of HCl in CCl_4 at comparable thermodynamical conditions.

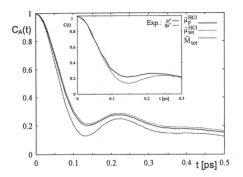

Figure 1. Comparison of the simulated dipole moment autocorellation functions (ACFs) of the diluted mixture HCl/CCl$_4$ at 290K. The solid line (———) corresponds to the HCl molecule permanent dipole $\vec{\mu}_p^{HCl}$ ACF; the short dash (- - - -) to the HCl total dipole moment $\vec{\mu}_{tot}^{HCl}$ ACF ; and the long dash (— —) corresponds to the total dipole $\vec{M}_{tot}$ ACF of the sample

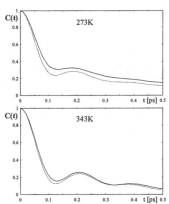

Figure 2. Permanent and induced dipole contributions to the total dipole moment ACFs with temperature. In any case, solid line (———) denotes the $\vec{\mu}_{tot}^{HCl}$ and the dashed line (- - - -) the $\vec{M}_{tot}$ ACFs, respectively

It is now convenient to inspect the simulated CFs, $C_A(t)$, in conjunction to the experimental curves. The MD results for the three dipole CFs are depicted in Fig. 1 up to 0.5ps. As we can see, the two first-order Legendre reorientational CFs of the solute molecules ($C_A(t)$, $A \equiv \vec{\mu}_i^p, \vec{\mu}_{tot,i}^{HCl}$) exhibit quantitatively a quite similar behavior in going from one curve to another. Note, however, that the average strength of $\vec{\mu}_{I,i}''(t)$ has been found to be quite small in comparison to the corresponding permanent dipole moment. This result leads to the conclusion that both $\vec{\mu}_{tot,i}^{HCl}(t)$ and $\vec{\mu}_i^p(t)$ reveal the same direction at any time. Moreover, the overall behavior of each of these two correlations is found to be in quite good accordance with the corresponding IR experimental CF. Thus, our MD results support the suggestion made in previous experimental studies[1,4,5], that the IR absorption band shape of HCl in CCl$_4$ is not affected by the well-known interaction induced dipole effects among the species in the solution.

The time evolution of the experimental FIR CF is fairly well reproduced by the MD collective total dipole moment correlation $C_M(t)$. At very short times, the $C_M(t)$ decreases faster than the aforementioned $C_{IR}(t)$ correlations. Also $C_M(t)$ exhibits a deeper positive minimum than the $C_{IR}(t)$ curves. It is of course due to different orientation of the total dipole moment of the sample $\vec{M}(t)$ compared to that of $\vec{\mu}_i^p(t)$ and $\vec{\mu}_{tot,i}^{HCl}(t)$. Summing up, it appears that the interaction-induced dipole effects are not negligible in the case of the FIR absorption lineshape of this solution. In this particular case, it is very interesting to compare our FIR results for the HCl/CCl$_4$ system with previous published theoretical ones for similar mixtures.

Another point of particular interest here is to examine the temperature dependence of the simulated $C_A(t)$ ($A \equiv \vec{\mu}_i^p, \vec{M}_{tot}$) CFs. The results obtained for these correlations at 273and 343 K are displayed in Fig. 2. As we can see from Fig 2, the two $C_A(t)$ curves become nearly super-imposed with increasing temperature. This MD result appears to be in agreement to what is expected in this particular case. Again, this illustrates the fact that at higher temperatures or lower densities, the strength of the interaction-induced effects is weak compared to lower temperatures.

Figure 3 *Temperature dependence of the IR corresponding ACFs and its reflection on the HCl spectrum*

We present in Fig. 3a the behavior of the calculated $C_A(t)$ $(A \equiv \vec{\mu}_{tot}^{HCl})$ CF at 273 and 343 K. At very short times both curves follow the well known behavior of the free rotor. Also, the initial decrease of $C_A(t)$ is more rapid at higher temperature, as expected. Generally, the $C_A(t)$ curves show a slower decay with decreasing temperature. It means that the reorientational motion of the solute is somewhat slowed with increasing density. The IR absorption lineshapes CFs of HCl in CC_4 corresponding to the simulated $C_A(t)$ $(A \equiv \vec{\mu}_{tot}^{HCl})$ are shown in Fig. 3b. The aforementioned IR absorption spectra were estimated by Fourier transformation of these CFs at 273 and 343 K. Note also that these spectra are normalized to the same amplitude. The spectra show three absorption maxima, which will be designated as the usual P, Q and R branches of the rotation-vibration absorption band of linear molecules, observed also in experiment. Eventually, we found that the line shapes of the simulated IR spectra are in quite good agreement with corresponding experimental[1, 2, 4] data.

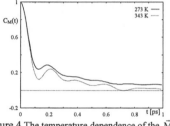

Figure 4 The temperature dependence of the $\bar{M}_{tot}$ ACFs

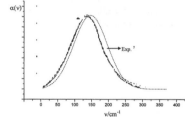

Figure 5 Comparison of the simulated with the corresponding experimental[7] fir absorption line shapes, $\alpha(v)$, at room temperature, normalized to the same amplitude

The simulated total dipole moment CFs, $C_M(t)$, at 273 and 343 K are compared in Fig. 4. These functions show a relatively similar temperature behavior to that obtained for the IR CFs discussed above. Unfortunately, temperature dependent experimental FIR absorption spectra and corresponding CFs for HCl in CCl_4 are not available in the literature. Therefore, the reliability of the temperature dependence of the simulated FIR CFs obtained in the frame work of the present study must be judged experimentally.

Finally, Fig. 5 displays the simulated FIR absorption spectrum of the solution at 290 K in comparison with corresponding experiment[7]. The calculated spectrum has been obtained by numerical Fourier transformation of the simulated total dipole moment CF $C_M(t)$. In order to compare the calculated with the experimental spectrum profile, the absorption maximum of the calculated curve is set equal to that of the experimental one. By inspecting these curves, we see clearly that the simulated FIR spectrum profile is almost identical to the experimental one. This fact denotes the significance of the interaction induced contributions to the FIR absorption of the system.

3. Concluding Remarks

The main purpose of this work was to investigate systematically the permanent and electrostatical dipole-induced contributions to the IR and FIR absorption spectra and corresponding correlation functions

(CFs). In order to realize this, the collective total dipoles of the system were calculated by assuming two distinct induction mechanisms among the species in the sample, namely the "dipole-induced dipole" as well as the "back-induced dipole". Thus the most interesting dipole CFs, corresponding to the IR and FIR absorption spectra, were calculated and found to be in good agreement with available experimental data. In this way, it was possible to estimate the influence of the interaction induced effects on the formation of the IR and FIR spectra and CFs. Finally, we found that the above mentioned effects are not negligible in the case of FIR absorption profile, whereas they do not contribute to the IR absorption of this mixture, as expected.

References

(1) Keller, B.; Knenbuehl, F. Helv. Phys. Acta 1972, 45, 1127.

(2) Lascombe, J.; Besnard, M.; Caloine, P. B.; Devaure, J.; Perrot, M. Molecular Motion in Liquids, Lascombe, J. (ed.); D. Reidel Publ. Company: Dordrecht Holland, 1974.

(3) Turrell, G. J. Mol. Spectr. 1978, 19, 383.

(4) Mushayakarara, E. C.; Turrell, G. Mol. Phys. 1982, 46, 991. Idrissi, A.; Arroume, M.; Turrell, G. J. of Mol. Struc. 1993, 294, 103.

(5) Ohikubo, Y.; Ikawa, S.; Kimura, M. Chem. Phys. Lett. 1976, 43, 138.

(6) Ikawa, S.; Sato, K.; Kimura, M. Chem. Phys. 1980, 47, 65.

(7) Ikawa, S.; Yamazaki, S.; Kimura, M. Chem. Phys. 1980, 51, 151.(8) Chatzis, G.; Chalaris, M.; Samios, J. Chem. Phys. 1998, 228, 241.

(9) Chatzis, G.; Samios, J. Chem. Phys. 2000, 257, 51. Chatzis, G.; Samios, J.,J. Phys. Chem. A, 2001,105, 9527

(10) Tildesley, D. J.; Madden, P. A. Mol. Phys. 1983, 48, 129.

(11) Dorfmüller, Th.; Samios, J.; Mittag, U. Chem. Phys. 1986, 107, 397.

(12) Gray, C. G.; Gubbins, K. E. Theory of Molecular Fluids, Oxford University Press: Oxford 1984.

(13) Maroulis, G. J. Chem. Phys. 1998, 108, 5432.

(14) Dellis, D.; Samios, J. Chem. Phys. 1995, 192, 28

Brill Academic Publishers
P.O. Box 9000, 2300 PA Leiden
The Netherlands

*Lecture Series on Computer
and Computational Sciences*
Volume 4, 2005, pp. 1842-1848

Axisymmetric Stagnation – Point Flow of a Viscous Fluid On a Rotating Cylinder with Time-dependent Angular Velocity and Uniform Transpiration

A. Baradaran Rahimi[*], R. Saleh[**][†]

*Professor, Faculty of Engineering, Islamic Azad University of Mashhad,
P.O. Box 91735-413,
Mashhad, Iran,
**Ph. D. student and lecturer, Azad Univ. of Mashhad

Received 13 June 2005; accepted in revised form 31 July, 2005

Abstract: The unsteady viscous flow in the vicinity of an axisymmetric stagnation point of an infinite rotating circular cylinder with transpiration is investigated when the angular velocity varies arbitrarily with time. The free stream is steady and with a strain rate of . An exact solution of the Navier-Stokes equations is derived in this problem. A reduction of these equations is obtained by use of appropriate transformations for the most general case when the transpiration rate is also time-dependent but results are presented for uniform values of this quantity. The general self-similar solution is obtained when the angular velocity of the cylinder varies as specified time-dependent functions.

Keywords: stagnation flow, time-dependent rotation, viscous flow, exact solution

1. Introduction

The problem of finding exact solutions of the Navier-Stokes equations is a very difficult task. This is primarily due to the fact that these equations are nonlinear. An exact solution of these equations governing the problem of two-dimensional stagnation flow against a flat plate has been given by Hiemenz [1]. The first exact solution of the problem of axisymmetric stagnation flow on an infinite circular cylinder was obtained by Wang [2]. Gorla [3-6], in a series of papers, studied the steady and unsteady flows over a circular cylinder in the vicinity of the stagnation-point for the cases of constant axial movement. In more recent years, Cunning, Davis, and Weidman [7] have considered the stagnation flow problem on a rotating circular cylinder with constant angular velocity. They have also included the effects of suction and blowing in their study. Takhar, Chamkha, and Nath [8] have investigated the unsteady viscous flow in the vicinity of an axisymmetric stagnation point of an infinite circular cylinder for the particular case when both the cylinder and the free stream velocity vary inversely as a linear function of time.
The effects of cylinder rotation with time-dependent angular velocity, perhaps of interest in centrifugal processes in industry, have not yet been investigated.
In the present analysis, the unsteady viscous flow in the vicinity of an axisymmetric stagnation point of an infinite rotating cylinder with uniform transpiration is considered when the angular velocity varies arbitrarily with time, though the reduction of the Navier-Stokes equations is obtained for the more general case of time-dependent transpiration rate. An exact solution of the Navier-Stokes equations is obtained. The general self-similar solution is obtained when the angular velocity of the cylinder varies in a prescribed manner. The cylinder may perform different types of motion: it may rotate with constant speed, with exponentially increasing/decreasing angular velocity, with harmonically varying rotation speed, or with accelerating/decelerating oscillatory angular speed.
For different forms of azimuthal component of velocity, sample distribution of shear stresses are presented for Reynolds numbers ranging from 0.1 to 1000 and selected values of uniform suction and blowing rates. Particular cases of these results are compared with existing results

---

[†] Corresponding Author. E-mail: rahimiab@yahoo.com

of Wang [2], and Cunning, Davis, and Weidman [7], correspondingly.

2. Problem Formulation

We consider the laminar unsteady incompressible flow of a viscous fluid in the neighborhood of an axisymmetric stagnation-point of an infinite rotating circular cylinder with uniform normal transpiration U_0 at its surface, where $U_0 \succ 0$ corresponds to suction into the cylinder, though the formulation of the problem is for the more general case of time-dependent transpiration rate. The flow configuration is shown in Figure 1 in cylindrical coordinates (r, θ, z) with corresponding velocity components (u, v, w).

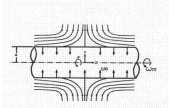

Fig. 1 Schematic diagram of a rotating cylinder under radial stagnation flow in the fixed cylindrical coordinate system (r, θ, z)

The cylinder rotates with time-dependent angular velocity. A radial external flow of strain rate k impinges on the cylinder with radius a and centered on r = 0. The unsteady Navier-Stokes equations in cylindrical polar coordinates governing the axisymmetric flow by applying the transformations

$$u = -k \frac{a}{\sqrt{\eta}} f(\eta, \tau) \quad v = \frac{a}{\sqrt{\eta}} G(\eta, \tau) \quad , \quad w = 2k f'(\eta, \tau) z , \quad p = \rho k^2 a^2 P \tag{1}$$

are as follows:

$$\eta f''' + f'' + \text{Re}[1 - (f')^2 + f f'' - \frac{\partial f'}{\partial \tau}] = 0 \tag{2}$$

$$\eta G'' + \text{Re}[f G' - \frac{\partial G}{\partial \tau}] = 0 \tag{3}$$

$$P - P_o = -[\frac{f^2}{2\eta} + \frac{1}{\text{Re}} f' + 2(\frac{z}{a})^2 - \frac{1}{2\bar{k}^2} \int_1^\eta \frac{G^2(\xi)}{\xi^2} d\xi] \tag{4}$$

where $\tau = 2kt$ and $\eta = (r/a)^2$ are the dimensionless time and radial variable. In these equations, primes indicate differentiation with respect to η and $\text{Re} = ka^2 / 2\upsilon$ is the Reynolds number. The boundary conditions for (2) and (3) are as following:

$$\eta = 1: \quad f = s(\tau), \ f' = 0, \ G = \omega(\tau) \qquad \qquad \eta \to \infty : f' = 1, \ G = 0 \tag{5}$$

in which, $s(\tau) = \dfrac{U_0(\tau)}{k a}$ is the dimensionless wall-transpiration rate, taken as a constant here.

3.   Self-Similar Equations

Equation (3) can be reduced to an ordinary differential equation if we assume that the function $G(\eta,\tau)$ in (3) is separable as:

$$G(\eta,\tau) = g(\eta).\phi(\tau)$$ (6)

Substituting these separation of variables into (3), gives:

$$\eta\frac{g''}{g} + \text{Re}.f\frac{g'}{g} = \text{Re}\frac{d\phi(\tau)/d\tau}{\phi(\tau)}$$ (7)

The general solution to the differential equation in (7) with τ as an independent variable is as the following:

$$\phi(\tau) = b.Exp[(\alpha + i\beta)\tau]$$ (8)

Here, $i = \sqrt{-1}$ and b , α and β are constants. The boundary conditions are:

$$G(1,\tau) = \omega(\tau) = \phi(\tau).g(1) \quad \rightarrow \quad \phi(\tau) = \omega(\tau)$$ (9)

which for g(1)=1, gives

$$G(\infty,\tau) = 0 = \phi(\tau).g(\infty) \rightarrow g(\infty) = 0$$ (9a)

Substituting the solution (8) into the differential equation (7) with η as independent variable results in:

$$\eta g'' + \text{Re}[\,fg' - \alpha g - i\beta g] = 0$$ (10)

Note, in (9) that $b = 0$ corresponds to the case of non-rotating cylinder, as of Wang [2]. If $b \neq 0$ and $\alpha = \beta = 0$, (9) gives the case of a uniformly rotating cylinder with constant angular velocity, as in Cunning et al. [7].   If $b \neq 0$, $\alpha \neq 0$ and $\beta = 0$, (9) gives the case of rotating cylinder with an exponential angular velocity. $b \neq 0$, $\beta \neq 0$ and $\alpha = 0$, corresponds to the case of pure harmonic rotation of the cylinder.   For non-zero b , α , and β , (9)   gives the cases of accelerating and decelerating oscillatory motions of the cylinder.

To obtain solutions of equation (10), it is assumed that the functions $g(\eta)$ is a complex function as:

$$g(\eta) = g_1(\eta) + ig_2(\eta)$$ (11)

Substituting (11) into (10), the following coupled systems of differential equations are obtained:

$$\begin{cases} \eta g_1'' + \text{Re}(fg_1' - \alpha g_1 + \beta g_2) = 0 \\ \eta g_2'' + \text{Re}(fg_2' - \alpha g_2 - \beta g_1) = 0 \end{cases}$$ (12)

Considering the boundary conditions (5), the boundary conditions for functions f and g become:

$$\eta = 1: \quad f = 0, \quad f' = 0, \quad g = 1$$
$$\eta \rightarrow \infty : f' = 1, \quad g = 0$$ (13)

Hence, the boundary conditions on functions g_1 and g_2 are:

$$\eta = 1: \quad g_1 = 1, \quad g_2 = 0,$$ (14)

$$\eta \rightarrow \infty : g_1 = 0, \quad g_2 = 0,$$ (15)

The coupled system of equations (12) along with boundary conditions (14) and (15) have been solved by using the fourth-order Runge-Kutta method of numerical integration along with a shooting method.

4. Shear-Stress

The shear-stress at the cylinder surface is calculated from, [7]:

$$\sigma = \mu [r \frac{\partial}{\partial r}(\frac{v}{r})\hat{e}_\theta + \frac{\partial w}{\partial r}\hat{e}_z]$$
(16)

where μ is the fluid viscosity. Using definition (1), shear-stress at the cylinder surface for semi-similar solutions becomes

$$\sigma = 2\mu [G'(1,\tau) - \omega(\tau)]\hat{e}_\theta + 4\mu \frac{kz}{a}f''(1)\hat{e}_z .$$
(17)

Thus the axial and azimuthal shear stress components are proportional to $f''(1)$, which has been presented in Ref. [7], and $[G'(1,\tau) - \omega(\tau)]$, respectively. Azimuthal surface shear-stress for self-similar solutions is presented by the following relation:

$$\sigma_\theta = \sigma_{\theta_1} + i\sigma_{\theta_2} = 2\mu b Exp(\alpha\tau)[\{\cos(\beta\tau)(g_1'(1)-1)-\sin(\beta\tau)g_2'(1)\}+i\{\sin(\beta\tau)(g_1'(1)-1)+$$

$$\cos(\beta\tau)g_2'(1)\}].$$
(18)

Some numerical values of σ_{θ_1} will be presented later for few examples of angular velocities.

5. Presentation of Results

In this section, the solution results of the self-similar equation (10) and the semi- similar equation (3) along with surface shear stresses for different functions of angular velocity are presented. Also, the azimuthal component of velocity, $v(\eta,\tau)$, for self-similar case is given. Figure (2) presents the sample profiles of $g(\eta)$ for $\omega(\tau)$ in exponential form for accelerating and decelerating case, at Re=1 and transpiration rate of S=0. It is interesting to note that for $\alpha \prec 0$, as the absolute value of α increases, fluid velocity in the vicinity of the cylinder cannot decrease with the same rate as the cylinder rotation velocity and therefore in this region , as the figure shows, the fluid velocity is greater than the cylinder velocity. Also, $\alpha = 0$ indicates the case of rotating cylinder with constant angular velocity, Ref. [7].

Figure (3) presents the sample profiles of $g(\eta)$ for $\omega(\tau)$ in exponential form for decelerating case At Re=1.0 and transpiration rate of S=0. It is interesting to note that as α decreases, the depth of the diffusion of the fluid velocity field decreases.

Figure (4) presents the sample profiles of $g(\eta)$ solution when $\omega(\tau)$ represents the accelerating and decelerating oscillatory moti on of the cylinder, at $Re = 1000$ and for transpiration rate of $S = 0$. Again, like the exponential angular velocity case, the depth of the diffusion of the fluid velocity field for $\alpha > 0$ and $\alpha < 0$ is less and more than that for the case of $\alpha = 0$, respectively.
Figure (5) presents the sample profiles of the real part of azimuthal component of the velocity when $\omega(\tau)$ represents pure harmonic motion of the cylinder at Re=1000. and transpiration rate S=0.
Figure (6) presents the real part of azimuthal shear-stress on the surface of the cylinder with harmonic rotation at $Re = 1000$ and transpiration rate of $S = 0$.

This shear-stress is for a complete perid between 0 and 2π. It can be seen from this figure that as frequency of the oscillation increases, the maximum of the absolute value of the shear-stress increases, and $\beta = 0$ corresponds to the case of constant angular velocity in which the imaginary part of the azimuthal shear-sress is zero and its real part is a constant, as in Ref. [7].

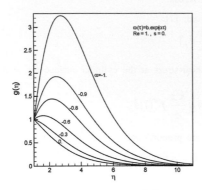 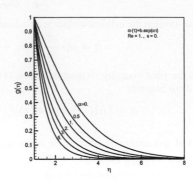

Fig.2. Sample profiles of $g(\eta)$ for cylinder with exponential angular velocity.   Fig. 3. Sample profiles of $g(\eta)$ for cylinder with exponential angular velocity.

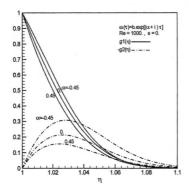 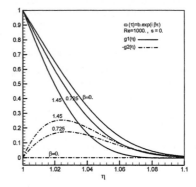

Fig. 4. Sample profiles of $g(\eta)$ for cylinder with accelerating and decelerating oscillatory motion.   Fig. 5. Sample profiles of $g(\eta)$ for cylinder with harmonic rotation.

Figures (7) and (8) show the azimuthal shear-stress on the surface of the cylinder for exponential angular velocity in terms of acceleration rate for different transpiration rates and $Re = 1000$ and $Re = 0.1$. It is concluded from these figures that the slope of the curves decreases as α increases, meaning that the sensitivity of shear-stress with respect to variation of α, decreases as α increases. The case $\alpha = 0$ corresponds to the same shear-stress value as the case in Ref. [7]. Figure (8) shows that as suction rate increases, the absolute value of the azimuthal shear-stress increases. The

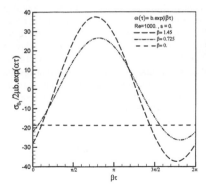

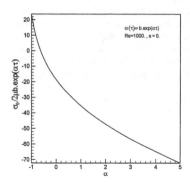

Fig. 6. Real part of azimuthal shear-stress component for cylinder with harmonic rotation.

Fig. 7. Azimuthal shear-stress component for cylinder with exponential angular velocity.

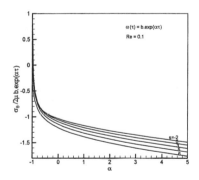

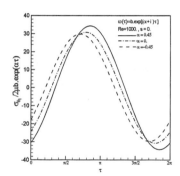

Fig. 8. Azimuthal shear-stress component for cylinder with exponential angular velocity.

Fig. 9. Real part of azimuthal shear-stress component for cylinder with acccelerating and decelerating oscillatory motion.

practical application of this interesting result is that by providing blowing on the surface of a cylinder, reduction of resistance against its rotation inside a fluid can be achieved. It is also interesting to note that for a particular value of negative α, the value of shear-stress is zero. This interesting result opens the way for an analysis into flows for which a cylinder spun up from rest in an exponential manner is azimuthally stress-free for certain combinations of Reynolds number and α.

Figure (9) presents the real part of azimuthal shear-stress on the surface of the cylinder with accelerating and decelerating oscillatory motion at $Re = 1000$ and transpiration rate of $S = 0$.

6. Conclusions

An exact solution of the Navier-Stokes equations is obtained for the problem of stagnation-point flow on a circular cylinder with uniform transpiration rate. A general self-similar solution is obtained when the cylinder has different forms of rotational motions including: constant angular velocity rotation, exponential angular velocity rotation, pure harmonic rotation, both accelerating and decelerating oscillatory rotations. The azimuthal component of fluid velocity and surface azimuthal shear-stress on the cylinder are obtained in all the above situations, and for different values of Reynolds numbers and transpiration rates. The absolute value of the azimuthal shear stress corresponding to all the cases increases with increasing Reynolds number and suction rate. Also, the maximum value of shear-stress increases with increasing oscillation frequency and accelerating and decelerating parameter α in the exponential amplitude function. It is shown that by providing blowing on the surface of a cylinder, a reduction of resistance against its rotation inside a fluid can be achieved. It is also shown that a cylinder spun up from rest in an exponential manner is azimuthaly stress-free for certain combinations of Reynolds number and rate of this exponential function.

References

[1] K.Hiemenz ,Die Grenzchicht an einem in den gleichformingen Flussigkeitsstrom eingetauchten geraden Kreiszylinder. Dinglers Polytech. J. 326(1911)321-410.

[2]C. Wang, Axisymmetric stagnation flow on a cylinder, Quarterly of Applied Mathematics, Vol. 32, 1974, pp. 207-213.

[3] R.S.R. Gorla, Unsteady laminar axisymmetric stagnation flow over a circular cylinder, Dev. Mech. 9(1977) 286-288.

[4] R.S.R. Gorla, Nonsimilar axisymmetric stagnation flow on a moving cylinder, Int. J. Engineering Science, 16(1978) 397-400.

[5] R.S.R. Gorla, Transient response behaviour of an axisymmetric stagnation flow on a circular cylinder due to time dependent free stream velocity, Int. J. Engineering science, 16(1978) 493-502.

[6] R.S.R. Gorla, Unsteady viscous flow in the vicinity of an axisymmetric stagnation-point on a cylinder, Int. J. Engineering Science, 17(1979) 87-93.

[7] G.M. Cunning, A.M.J. Davis, P.D. Weidman, Radial stagnation flow on a rotating cylinder with uniform transpiration, Journal of Engineering mathematics 33:113-128, 1998.

[8] H.S. Takhar, A.J. Chamkha, G. Nath, Unsteady axisymmetric stagnation-point flow of a viscous fluid on a cylinder, Int. Journal of Engineering Science, 37(1999) 1943-1957.

Brill Academic Publishers
P.O. Box 9000, 2300 PA Leiden
The Netherlands

*Lecture Series on Computer
and Computational Sciences*
Volume 4, 2005, pp. 1849-1855

Unsteady Free Convection From a Sphere in A Porous Medium with Variable Temperature

Jalal Talebi[1]*, Asghar Baradaran Rahimi**

* Graduate student
** Professor, Department of Mechanical Engineering , Ferdowsi University of Mashhad,
P.O. Box 91775-1111
Mashhad, Iran.

Received 13 June 2005; accepted in revised form 31 July, 2005

Abstract: In this paper a transient free convection flow from a sphere with variable temperature on surface which embedded in a porous medium has been considered . The temperature of the sphere is suddenly raised and subsequently maintained at values that varies with position on surface . A finite-difference scheme is used to solve the problem, numerically, for finite values of Rayleigh numbers. Transient and steady-state flow and temperature patterns around the sphere are discussed in details and a comparison between numerical and analytical results has been done .

Keywords: unsteady, free convection, sphere, time-dependent surface

1. Introduction

Studies on natural convection over a sphere in fluid-saturated porous media , are of interest in many engineering processes , such as thermal insulation systems , nuclear waste management , the storage of grain , in petroleum reservoirs and catalytic reactors. Yamamoto [1] was the first to consider the problem of steady natural convection around an isothermal sphere in porous medium . He obtained asymptotic solutions for small Rayleigh numbers . Subsequently, Merkin [2] , Cheng [3] , Nakayama and Koyama [4] , and Pop and Ingham [5] considered high Rayleigh number steady natural convection around a sphere with both an isothermal and non-isothermal surface . Sano and Okihara [6] have studied the transient natural convection from a sphere in a porous medium using asymptotic solutions in terms of small Ra , Nguyen and Paik [7] have investigated numerically the unsteady mixed convection from a sphere in a porous medium saturated with water using a Chebyshev-Legeure spectral method . Yan et.al. [8] performed a numerical study of unsteady free convection from a sphere which embedded in a fluid-saturated porous medium and whose surface is impulsively changed to a constant temperature or constant heat flux .

All works on natural convection over a sphere in porous media have been conducted only for the constant temperature and constant heat flux on surface of sphere . In this paper , we consider the problem of unsteady convection around a sphere in a porous medium when the temperature of surface is changing with position. Initially the temperature of surface is at a certain value and then it suddenly changes with location on the surface . By using a finite-difference method, the results are obtained, numerically, for finite values of Rayligh numbers. For higher Rayleigh numbers, we compared the results obtained by numerical method with those obtained by Yan et.al.[8] for constant temperature of surface .

2. Governing Equations

Consider a sphere of radius r_0 , which is immersed in a fluid-saturated porous medium which is at a constant temperature . Suppose that initially the sphere has the same temperature as the porous medium and at time τ' , it is suddenly heated and subsequently the temperature of surface changes with position .

---

[1] Corresponding Author . E-mail:rahimiab@yahoo.com

A spherical polar coordinate system (r',θ,ϕ), with the origin at the center of the sphere , is chosen with $\theta = 0$, vertically upwards , as shown in Fig.1 .

Both the flow and temperature are assumed to be axially symmetric and hence independent of the azimuthal coordinate ϕ. The fluid motion is described by radial and transversal velocity component (u',v') in a plane

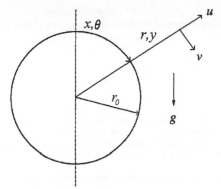

Fig. 1. Polar coordinate system

through the axis of symmetry . The velocity component are expressed in terms of a dimensionless stream function $\psi(r,\theta)$ by the equations,

$$u = \frac{1}{r^2 \sin\theta} \frac{\partial \psi}{\partial \theta} \qquad v = -\frac{1}{r \sin\theta} \frac{\partial \psi}{\partial r} \tag{1}$$

If the physical properties of the fluid are assumed constant and the Darcy-Boussinesq approximation holds , then the non-dimensional governing equations in terms of the stream function ψ and temperature T can be written. The dimensionless variables are defined as,

$$T = \frac{T'-T_\infty}{T_w - T_\infty}, r = \frac{r'}{r_0}, \tau = \frac{\alpha_m \tau'}{r_0^2}, u = \frac{u'}{U_r}, v = \frac{v'}{U_r}$$

$$U_r = \frac{Kg\beta(T_w - T_\infty)}{\upsilon} \;,\; Ra = \frac{Kg\beta(T_w - T_\infty)r_0}{\upsilon\alpha_m} \;,\; \psi = \frac{\psi'}{U_r r_0^2}$$

(2)

here, K is the permeability of the thermal porous medium , β the coefficient of thermal expansion , υ the kinematic viscosity of the fluid , g the acceleration due to gravity , U_r the characteristic velocity , Ra the Rayleigh number and α_m the effective thermal diffusivity of the fluid-saturated porous medium .

Since the flow experiences larger gradients near the surface of the sphere we introduce the following transformation for use in numerical method,

$$r = \frac{\alpha}{1-x} - \alpha + 1$$

(3)

where α is a constant which , to some extent , can be used to control the mesh density when we set up the finite-difference scheme . The governing equations in terms of the new variables (x,θ) become,

$$\nabla^2 \psi = \frac{1+(\alpha-1)x}{(1-x)} \sin\theta \left\{ \begin{array}{l} \sin\theta \dfrac{(1-x)^2}{\alpha} \dfrac{\partial T}{\partial x} + \\[3mm] +\dfrac{(1-x)\cos\theta}{[1+(\alpha-1)x]} \dfrac{\partial T}{\partial \theta} \end{array} \right\}$$

(4)

$$\frac{\partial T}{\partial \tau} + \frac{Ra}{\sin\theta} \frac{(1-x)^4}{\alpha[1+(\alpha-1)x]^2} \left(\frac{\partial T}{\partial x}\frac{\partial \psi}{\partial \theta} - \frac{\partial \psi}{\partial x}\frac{\partial T}{\partial \theta} \right) = +\frac{2(1-x)}{[1+(\alpha-1)x]} \left\{ \frac{(1-x)^2}{\alpha}\frac{\partial T}{\partial x} + \right.$$
$$\left. +\frac{(1-x)}{[1+(\alpha-1)x]}\cos\theta \frac{\partial T}{\partial \theta} \right\}$$

(5)

$$\nabla^2 = \frac{(1-x)^4}{\alpha^2}\frac{\partial^2}{\partial x^2} - \frac{2(1-x)^3}{\alpha^2}\frac{\partial}{\partial x} + \frac{(1-x)^2}{[1+(\alpha-1)x]^2}\left(\frac{\partial^2}{\partial\theta^2} - \cot\theta\frac{\partial}{\partial\theta} \right)$$

(6)

At initial instant, $t \le 0$, we have

$$\psi = 0, T = 0 \quad , \quad 0 \le x \le 1, \quad 0 \le \theta \le \pi$$ (7a)

For t > 0 , we have

$$\psi = 0, T = f(\theta) \quad , \quad x = 0 \ (r = 1)$$ (7b)

$$\psi \to \frac{1}{2}r^2 \sin^2\theta \ , \ T \to 0 \quad , \ x = 1$$
$$(r \to \infty)$$ (7c)

or for numerical method,

$$T = \frac{\partial \psi}{\partial x} = 0 \qquad , \ x = 1 \ (r \to \infty)$$ (7d)

Finally , the symmetrical boundary conditions are :

$$\psi = \frac{\partial T}{\partial x} = 0 \ , \quad \theta = 0, \pi \ , \quad 0 \le x \le 1$$ (7e)

3. Numerical Solution

Equations (4) and (5) are now solved numerically using a finite-difference scheme , subject to boundary conditions (7a)-(7e). We use central-difference and fully explicit schemes for equations (4) and (5), respectively. The discretized stream equation and energy equation are solved by using a line by line TDMA (Tri Diagonal-Matrix Algorithm) . In each iteration , first , equation (4) is solved by using a point relaxation and in the iteration process for the ψ the most updated values for ψ and T on the adjacent lines are used . Then , equation (5) is solved and the same procedure is done . The convergence criterion for iterations is chosen as follows,

$$\sum \left\{ \left| \psi^{(m)} - \psi^{(m-1)} \right| + \left| T^{(m)} - T^{(m-1)} \right| \right\} < \varepsilon$$

(8) where ε is prescribed tolerance and the summation takes place over all the mesh points . For the steady state solution, the above procedure is carried out until τ is sufficiently large that the solutions for two successive time steps are almost identical.

4. Results and Discussions

Numerical results were obtained for variable temperature on surface of sphere for $0.1 \leq Ra \leq 50$ and the mesh sizes and the constant α for the calculations presented in this paper vary with Rayleigh number. It is found that numerical results are more sensitive to the mesh sizes in x-direction that to that in the θ-direction. For small Rayleigh numbers, we compared results obtained by numerical method with those obtained by analytical method. Since, there were no results for variable temperature in large Rayleigh numbers, we compared numerical results obtained in this work, with those that presented in [8] for constant temperature case. We evaluated many function for variation of temperature on surface of sphere and in this paper we show the main results for function T(s) = 1+ .1 Cos θ.

Figure (2) shows the instantaneous streamlines for Ra=50 at time τ = 1, 15 and 50 for constant temperature case. It is seen that the results obtained in the present investigation are in very good agreement with those obtained by Yan et.al. [8].

Figures (3)   & (4)show the instantaneous streamlines for Ra = .4 and 10 and time τ = 1, 10 and 50 for T(s) =1.0+ 0.1Cos θ. Fluid is entrained towards the hot sphere and an upward flow is generated along the sphere surface. In the early stages, the fluid motion is mainly confined to the vicinity of the sphere whilst at later times, due to convection from sphere , the flow motion spreads outwards and upwards. It is seen, that there appears a vortex ring which surrounds the sphere and it is surrounded by the main flow. For small Ra numbers, the streamlines are symmetrical with respect to plane $\theta = 90$ and, as Ra number increases the core of vortex ring moves up due to effect of convection. From about $\tau = 10$, the flow pattern very close to sphere does not change very much but there is a large difference far away from the sphere. The steady state may never reached, but after a long times, the flow near the sphere will approach a steady-state situation.

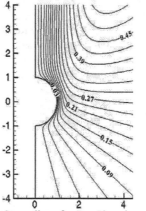

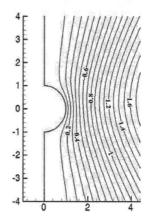

Fig. 2. Streamlines for Ra=50 and constant temp. at different times.

Fig. 3.Streamlines for Ra=0.4 at different times.

Figure (5) show the temperature contours for Ra = .4 and 10 and $\tau = 1,10$ and 50. For Ra = .4, the isothermal lines looks circular and as Ra increases, convection starts to become more important and for Ra =10 and $\tau = 50$, we observe a very clear cap of the plum which moves upwards as the convection process continues . As Ra increases, the thermal boundary layer thickness decreases in bottom of sphere and increases in top of sphere .

If we consider equation (20) and differentiate it with respect to θ, we have,

$$\frac{\partial Nu}{\partial \theta} = (-2a_1 + \frac{a_0}{4}Ra)\sin\theta \qquad (9)$$

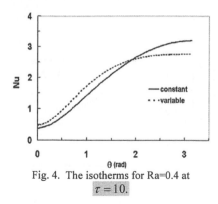

Fig. 4. The isotherms for Ra=0.4 at $\tau = 10$.

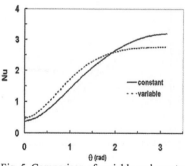

Fig. 5. Comparison of variable and constant surface temperature

setting $\dfrac{\partial Nu}{\partial \theta} = 0$, for steady-state condition yields,

$$Ra\big)_{crit} = \frac{8a_1}{a_0} \quad , \quad Nu\big)_{crit} = a_0 + .4776 a_1 Ra\big)_{crit}$$

(10)

For example , if $a_0 = 1, a_1 = .1$, we have , $Ra\big)_{crit} = .8, Nu\big)_{crit} = 1.0382$. This means that the local Nusselt number for this value of Ra number and in steady-state condition is independent of θ and this suggest that the buoyancy effects on the local Nusselt number at the top and bottom of the sphere cancel each other.

Figure (2) shows the instantaneous streamlines for Ra= 50 at time $\tau = 15$ for special case of constant surface temperature. These results compare very well with those obtained by Ref. [8]. The instantaneous streamlines for Ra=0.4 and at time $\tau = 10$ is shown in Figure (3).for surface temperature function T(s) =1.0 + 0.1Cos θ. As it is seen from this figure, fluid is entrained towards the hot sphere and upwards flow is generated along the sphere surface. In the early stages, the fluid motion is mainly confined to the vicinity of the sphere whilst at later times due to convection from sphere, the flow motion spreads outwards and upwards. It is also seen that a vortex ring surrounds by the main flow appears which surrounds the sphere. The flow pattern very close to sphere does not change very much but variation of streamlines with respect to time increases in far away distances from it. Reaching steady state conditions may not become possible but the flow near the sphere will approach to steady state situation after a relatively long time.

The temperature contours for Ra=0.4 and $\tau = 10$ is depicted in Figure (4) . The isothermal lines are nearly circular and as Ra increases convection effects become stronger.

In Figure (5), the results obtained for variable temperature case is compared with those for constant temperature case, for Ra=10 and at $\tau = 1$. It is observed that the magnitudes of radial velocities for variable temperature case and at $\theta = 0$ are higher and at $\theta = 180°$ are smaller than those for the constant temperature on the surface of the sphere.

Figure (6) shows the variation of steady-state local Nusselt number with respect to θ for Ra= 0.4 and for sphere surface temperature $T(s) = 1.0 + 0.1 Cos \theta$. As from this figure, there is a good agreement between analytical and numerical results. In Figure (7), the distribution of transient local Nusselt number in terms of θ for sphere surface temperature $T(s) = 1.0 + 0.1 Cos \theta$ and for Ra=0.1 at $\tau = 3$ is shown. Again we see that there is a very good agreement between numerical and analytical results. Figure (8) shows the value of steady-state local Nusselt number obtained analytically for Ra= 0.1, 0.4, 0.8 and 1.0. It is clear that for lower Ra numbers like 0.1 and 0.4 variation of local Nusselt number is different than for greater Ra numbers like 0.8 and 1.0. For the former Ra numbers, the local Nusselt number decreases from its maximum value at $\theta = 0$ and reaches to its minimum value at $\theta = 180$.

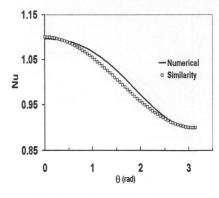
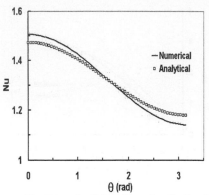

Fig. 6. Steady-state Nusselt number for Ra=0.1

Fig. 7. transient Nusselt number for Ra=0.1

For Ra=0.8 variation of local Nusselt number is independent of θ and for Ra=1.0 its variation changes completely so that at $\theta = 0$ its value is a minimum and at $\theta = 180$ is a maximum. Also it is evident from the figure that around $\theta = 0$ this quantity decreases with increasing Ra number. This is because both the convection and the surface temperature function affect the local Nusselt number. The surface temperature function is such that temperature is a maximum at the top of the sphere $(\theta = 0)$ and minimum value at the bottom $(\theta = 180)$. Therefore, at initial moments and relative to constant temperature function $T(s) = 1.0$ the temperature difference between top surface and its near flow and its local Nusselt number is greater than that at the bottom of the sphere. But as time passes and as the effect of convection gets stronger, the fluid temperature around the sphere warms up and heat moves upwards. This is why at the same time that local Nusselt number decreases with respect to time the difference between Nusselt number at the top and bottom of the sphere decreases and at Ra=0.8 becomes equal and that is why the local Nusselt number is independent of θ. As Ra increases, the convection effect is so large that at the initial stages the local Nusselt number at the top of the sphere is smaller than the bottom. In this state this quantity at $\theta = 0$ is a minimum and at $\theta = 180$ is a maximum. Also the thickness of thermal boundary layer at the top of the sphere and for large Ra numbers increases considerably. For small ra numbers like 0.4, since the convection effects are weak and thermal conduction effects are stronger and therefore the effects of surface temperature variation causes a greater local Nusselt number at the top relative to the bottom of the sphere.

In Figure (9), the distribution of transient local Nusselt number in terms of θ for sphere surface temperature $T(s) = 1.0 + 0.1Cos\theta$ and for Ra=0.1 at $\tau = 10$ is shown. Again the good agreement between numerical and analytical results is noted from the figure.

5. Conclusions

Transient free convection heat transfer from a sphere with variable temperature on surface and in a porous medium has been studied for finite values of Ra number numerically. The results obtained are in excellent agreement with those obtained by analytical method for small Ra numbers and with those presented in [8] for finite values of Ra numbers and constant temperature on surface of sphere. For high Ra numbers, such as Ra = 10 and 50, a buoyancy plume with a mushroom-shaped front is formed above the sphere. A vortex ring is formed that moves up by increasing Ra number. This causes higher values of radial velocities in above of sphere and smaller values in bottom of sphere for higher value of Ra number.

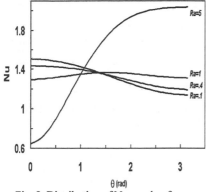

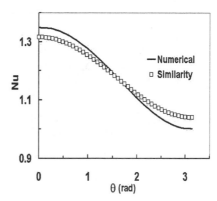

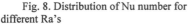

Fig. 8. Distribution of Nu number for different Ra's

Fig. 9. Transient Nu number for Ra=0.1 at $\tau = 10$

For $T(s) = 1.0+0.1 \cos \theta$, the magnitudes of steady-state local Nusselt number for Ra =.1 and .4 decrease from maximum value at $\theta = 0$ to minimum value at $\theta = 180$. For Ra = .8 , the local Nusselt Number is independent of θ and for higher Ra numbers, this variation of local Nusselt number is changed. It is seen that if the function of temperature on surface of sphere defines as $T(s) = 1.0+0.1 \cos \theta$, the values of local Nusselt number are higher than those for constant temperature case, up to approximately $\theta = 110$ and after this, these values become smaller than those for constant temperature case. In this case the vortex ring moves upper and the radial velocities at $\theta = 0,180$ become higher and smaller for variable temperature case, than those obtained for constant temperature case, respectively.

References

[1]   K. Yamamoto, Natural convection about a heated sphere in a porous medium. Journal of the Physics Society of Japan, 1974, **37**, 1164-1166.

[2]   J. H. Merkin, Free convection boundary layers on axi-symmetric and tow-dimensional bodies of arbitrary shape in saturated porous medium . International Journal of Heat and Mass Transfer, 1979, **22**, 1461-1462.

[3]   P. Cheng, Natural convection: Fundamentals and Applications. Hemisphere, 1985, Washington, DC, p.475.

[4]   A. Nakayama and H. Koyama, Free convection heat transfer over a nonisothermal body of arbitrary shape embedded in a fluid-saturated porous medium, Journal of Heat Transfer, 1987, **109**, 125-130.

[5]   I. Pop and D. B. Ingham, Natural convection about a heated sphere in a porous medium, Proceeding of the Ninth International Heat Transfer Conference, 1990, vol. 2, pp. 567- 572.

[6]   T. Sano and R. Okihara, Natural convection around a sphere immersed in a porous medium at small Rayleigh numbers, Fluid Dynamics Research, 1994, **13**, 39-44.

[7]
 H. D. Nguyen and S. Paik, Unsteady mixed convection from a sphere in water-saturated porous medium with variable surface temperature/heat flux, International Journal of Heat and mass transfer, 1994, **37**, 1783-1793.

[8] B. Yan and I. Pop and D.B. Ingham, A numerical study of unsteady free convection from a sphere in a porous medium, International Journal of Heat and Mass Transfer, 1997 , Vol. 40, No. 4, pp. 893-903.

[9] T. Sano and K. Makizono, Unsteady mixed convection around a sphere in a porous medium at low peclet numbers, Fluid Dynamics Research, 1998, **23**, 45-61.

Brill Academic Publishers
P.O. Box 9000, 2300 PA Leiden
The Netherlands

*Lecture Series on Computer
and Computational Sciences*
Volume 4, 2005, pp. 1856-1858

The Comparisons of Heart Rate Variability and Perceived Exertion during Simulated Cycling with Various Viewing Devices

Tien-Yow Chuang [1]

Department of Physical Medicine and Rehabilitation, Taipei Veterans General Hospital and National Yang-Ming University, School of Medicine, Taipei Taiwan

Received 23 June 2005; accepted in revised form 18 July, 2005

Abstract: The sympathovagal modulation during immersion in virtual environment might be one of the important issues influencing the complete task performance of the human. The aim of this study was to investigate the frequency domain of heart rate variability and perceived exertion during virtual reality (VR) biking with different display devices. The study was a randomized 4x4 Latin square cross-over design balanced for carryover effects, involving 16 healthy, young volunteers provided with 3 viewing devices (i.e. desktop monitor, HMD, projector) and no-VR display served as control. The subjects strived to achieve their maximal physical capacity before experiment. After a week, the quick ramp exercise test was conducted and then maintained at anaerobic threshold intensity for each session to evaluate power spectral density and the rating of perceived exertion (RPE). Post hoc LSD analysis showed a significant difference between no-display and projector groups for total power (TP) and very low frequency (VLF) components. There was a significant difference when comparing HMD and no-display exercise RPE curves within 6 minutes of cycling (Breslow test, p=0.0023) and at the termination of the exercise (log rank test, p=0.0026). However, the significant difference was also achieved in projector vs no-VR display comparison at the termination of the exercise (log rank test, p=0.008). In conclusion, our results indicated that using HMD and projected VR cycling system may assist in carrying out anaerobic exercise in longer duration. The fact that projected VR could reduce the TP and VLF power spectral density leads to possible implications with regard to both humeral and sympathetic activities.

Keywords: Heart rate variability; Power spectral analysis; Autonomic function; Virtual reality;

Cardiopulmonary.

Mathematics SubjectClassification: 87.80.-y Biological techniques and instrumentation; biomedical engineering

1. Research Protocol

After reviewing their medical records, participants were scheduled for four assessment sessions (desktop monitor, projector, HMD, and no simulation display) with a 1-week interval between the tests. To incorporate balance so as to avoid any carryover effects, the order of the tests was subjected to a 4 × 4 Latin square design in a randomized manner. One week before the experiment began, all of the volunteers performed a progressive cardiorespiratory test on a friction-braked cycle ergometer until they reached subjective exhaustion, plateauing of oxygen intake or clinical contraindication and were unable to continue. Following the conclusion of physical capacity in each person, each participant was then asked to perform a symptom-limited bicycle exercise on each sequential protocol in an upright position with different displays. The exercise was started at 30W and continued at this rate for about 1 minute, and was then increased to the work rate at the subject's anaerobic threshold. The examiner monitored each subject's Electrocardiogram (ECG), blood pressure (BP), and the rating of their perceived exertion (RPE) according to the Borg's 6-20 point scale. Readings from the scale (RPEs) were taken from the subjects every 1 minute until the conclusion of the investigation.

---

[1] Corresponding author: Tien-Yow Chuang, MD, Department of Physical Medicine and Rehabilitation, Veterans General Hospital Taipei, No.201, Shih-Pai Rd, Sec. 2, Taipei, 11217 Taiwan. telephone: +886 2 28757296, e-mail: tychuang@vghtpe.gov.tw

Table 1: The comparisons of mean (standard error) of power spectral density in subjects receiving different viewing devices by use of a 4x4 Latin square balanced for order.

Ms^2	No Display	LCD	HMD	Projector	P Value
TP	2151(307)*	1547(307)	1769(295)	1070(307)*	0.017*
ULF	851(130)	732(130)	832(125)	524(130)	NS
VLF	629(102)*	489(102)	571(98)	336(102)*	0.049*
LF	226(43)	150(43)	215(41)	103(43)	NS
HF	93(35)	37(35)	87(33)	27(35)	NS

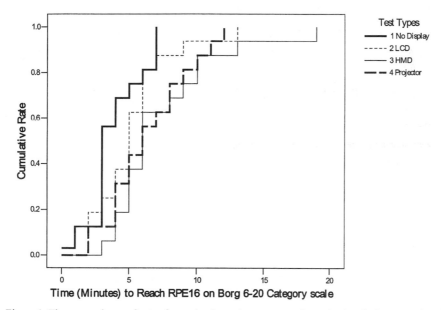

Figure 1: The comparisons of rate of perceived exertion among various viewing devices were based on the Kaplan-Meier method.

Acknowledgments

The author wished to thank the anonymous referees for their careful reading of the manuscript and their fruitful comments and suggestions.

References

[1] S.R. Ellis, What are virtual environments? IEEE Computer Graphics and Applications, 14(1),17-22 (1994).

[2] D.R. Pratt, M. Zyda, and K. Kelleher, Virtual reality: in the mind of the beholder, IEEE Computer, 28(7),17-19(1995).

[3] C.Y. Shing, C.P. Fung, T.Y. Chuang, I.W. Penn, and J.L. Doong, The study of auditory and haptic signals in a virtual reality-based hand rehabilitation system, Robotica, 21, 211-218 (2003).

[4] K. Hayashibe, M. Hara, K. Tsuji, and T. Matsuzawa, Japanese monkeys' cardiac responses to visual depth, Perceptual & Motor Skill, 67(1), 303-10(1988).

[5] B. Gilmartin, E.A. Mallen, and J.S. Wolffsohn, Sympathetic control of accommodation: evidence for inter-subject variation, Ophthalmic & Physiological Optics, 22(5), 366-71(2002).

[6] B. Xue, K. Skala, T.A. Jones, and M. Hay, Diminished baroreflex control of heart rate responses in otoconia-deficient C57BL/6JEi head tilt mice, American Journal of Physiology - Heart & Circulatory Physiology, 287(2),H741-7(2004).

[7] T. Brandt, J. Dichgans, and W. Wagner, Drug effectiveness on experimental optokinetic and vestibular motion sickness. Aerospace Medicine, 5(11),1291-7(1974).

[8] M.D. Delp, R.B. Armstrong, D.A. Godfrey, M.H. Laughlin, C.D. Ross, and M.K.Wilkerson, Exercise increases blood flow to locomotor, vestibular, cardiorespiratory and visual regions of the brain in miniature swin, Journal of Physiology, 533(Pt 3),849-59(2001).

[9] B. Sayers, Analysis of heart rate variability, Ergonomics, 16,17-32(1973).

[10] R.L. Kohl, Autonomic function and plasma catecholamines following stressful sensory stimuli, Aviation Space & Environmental Medicine, 64(10),921-7(1993).

[11] D. Mouginot and G. Drolet, Brain angiotensin receptors and adaptation to stress, Medecine Sciences,19(10),972-5(2003).

[12] R.G. Victor, D.R. Seals, A.L. Mark, Differential control of heart rate and sympathetic nerve activity during dynamic exercise, J Clin Invest, 79,508-516(1987).

Brill Academic Publishers
P.O. Box 9000, 2300 PA Leiden,
The Netherlands

Lecture Series on Computer
and Computational Sciences
Volume 4, 2005, pp. 1859-1863

Prediagonalized Block-Davidson Scheme : an efficient tool to selectively calculate highly excited vibrational states

Christophe Iung[1] and Fabienne Ribeiro[2]

LSDSMS (UMR-CNRS 5636) CC 014 , Université de Montpellier II
34095 Montpellier Cedex, France

This paper is a part of the Symposium: "Trends and perspectives in Computational Chemistry"

Received 25 August, 2005; accepted in revised form 29 August, 2005

Abstract: Most of the ro-vibrational methods used to compute excited vibrational states of polyatomic systems rely on some iterative scheme. A very popular and efficient one consists of the straight Lanczos algorithm which first converges the low density part of the spectrum. Consequently, such a basic approach is not the ideal method to use to selectively obtain highly excited vibrational states which are located in a dense part of the spectrum. We will show that a Davidson scheme following a specific prediagonalization step is more efficient to focus selectively on one part on the spectrum. Application of this new scheme have been performed to calculate highly excited overtones in HFCO and H_2CO. Such studies provide some ideas on the numerical methods which should be developed in the future to tackle higher energy excited states in polyatomics.

Keywords: Overtones, Calculation, Davidson, Spectroscopy

Mathematics Subject Classification: Here must be added the AMS-MOS or PACS Numbers

PACS: Here must be added the AMS-MOS or PACS Numbers

1 Introduction

Intramolecular Vibrational-energy Redistribution (IVR) is a fundamental energy transfer mechanism that occurs in polyatomic molecules. In spite of the remarkable interest of the RRKM theory which assumes that complete redistribution of an initial vibrational excitation over all vibrational modes occurs prior to reaction, several examples of "non statistical" behavior have been reported and raise questions about the completeness of IVR[1]. ¿From the experimental side, different groups have developed very sophisticated experiments in order to obtain fully resolved spectra of highly excited polyatomic systems such as HFCO[2], CF_3H [3] and benzene [4]. However, it should be emphasized that such very accurate data can not be fully understood using simple models. *To reproduce and analyze the mine of information present in fully resolved spectra, it is thus vital to perform a quantum mechanical study which takes into account all the couplings between modes.* Numerical simulations demonstrate that non-resonant states which are weakly coupled to the bright state often play a minor role individually but a significant one collectively[5]. *Very accurate quantum mechanical simulations providing eigenvalues and eigenvectors are thus required*

---

[1] E-mail: iung@univ-montp2.fr

[2] New Address: IRSN-Major Accident Prevention Division, Cadarache, Bat 702 Saint-Paul-lez-Durance 13108 (France)

for the analysis of sophisticated experimental data. The expression of the eigenvectors in the primitive basis set allows the labelling of the different lines of the spectrum. It also yields information on the localization of the energy in a given state. Futhermore, whatever the quantum method of vibrational spectrum computation, it requires a potential energy surface (PES) which correctly describes the states studied. However, the calculation of an accurate PES of high vibrationally excited system constitutes a real challenge. A comparison between experimental data and simulated spectrum is required to establish the quality of the PES used to describe a system and to predict its dynamical behavior. It also allows refinement of a PES obtained by *ab initio* quantum calculations. The present study does not focus on PES calculations, but rather on numerical methods providing highly excited states when the PES is known.

Different numerical strategies yielding the excitation spectrum of a polyatomic system have recently been developed. What are the features of an ideal method that provides highly excited vibrational states?

- *(i) the method should be general*, i.e. adapted to calculate the energy spectrum of a large variety of systems.

- *(ii) the method has to provide the eigenvalues, but also the eigenvectors or, at least, their projections on some given state.* This point is crucial because calculations must be able to provide the intensities of the spectral lines, and some physical assignment for the different bands in the spectrum.

- *(iii) the method has to be robust*, i.e. able to provide the eigenvalues when the state density becomes significant and/or when Fermi resonances are important and couple several modes together.

- *(iv) the accuracy of the calculations should be set by the user.* It is useless and time-consuming to converge eigenvalues and eigenstates with a great accuracy (e.g. 10^{-2} cm^{-1} for an energy level) if the PES used is not accurate enough, or if the basis set used is not sufficiently large to allow such an accuracy.

- *(v) the memory requirement should be as small as possible* in order to generalize the use of the method to larger or/and more excited polyatomics.

- *(vi) the calculation should be fast.* This point can be decisive if the calculation is used to refine a PES : in such a case, the energy spectrum has to be computed several times.

- *(vii) the method should be able to focus on a selected part of the spectrum,* for instance a highly excited overtone.

- *(viii) finally, the method should be easy to use.* The number of parameters in the method should be small and their use must be transferable to a large variety of problem. A method using a myriad of unphysical parameters will never be adopted by a large number of groups.

In the present study, the efficiency of the Davidson scheme[6] coupled to a specific prediagonalization step to focus selectively on a few highly excited states is presented. In Sec.2, the Davidson scheme which provides eigenstates and eigenvalues is presented. Previous studies[9, 10, 8] adapted the Block-Davidson scheme to provide all the spectral lines up to a given energy. This study is devoted to the use of the Davidson (or Block-Davidson) scheme for calculating some specific, highly excited states. Consequently, a new way to define the $\mathcal{P}$ subspace in which the Hamiltonian matrix is prediagonalized is presented. This step is crucial to be able to easily focus on highly excited states located in a dense part of a spectrum. This new scheme has been applied to calculate highly excited out of plane overtones, $|n\nu_6 >$ (n=6,8,10), in HFCO[7]. In Sec.3, discussion on the interest of such a method based on the Davidson scheme is presented.

2   Specific Prediagonalized Davidson Algorithm Scheme

We adopt in this study the modified Davidson Algorithm scheme developed by Leforestier *et al*[9] to calculate H_2CO vibrational excitation spectra up to 9,500 cm^{-1}. More recently, Ribeiro *et al*[10, 8] have implemented a Block-Davidson scheme to provide spectrum which presents Fermi resonances. The modified regular Davidson[9] and/or Block-Davidson[10, 8] schemes developed previously exhibit the following advantages :

- *No energy window has to be defined a priori.* The method requires the knowledge of a satisfactory zero-order description of the studied state. Then, the method itself iteratively optimizes the energies and the eigenstates studied.

- *Convergence of the levels is assessed during the iterations by looking at the residual* $\|(H - E_M)\psi_M\|$. We have demonstrated[10] that the error in an eigenvalue associated with a residual smaller than $\epsilon = 20$ cm^{-1} is less that 0.1 cm^{-1} for moderately excited states in HFCO and H_2CO. More recently[8], we showed that the residual also constitutes a good tool to estimate the accuracy of the eigenstates provided by the Davidson scheme for moderated excited states.

The limitations of the Davidson schemes previously developed[9, 10, 8] come from the core memory required to store the $N_\mathcal{P}$ first lines of the $N_B x N_B$ Hamiltonian matrix expressed in the primitive basis set and the M Davidson vectors u_i (c.f Table 1) known from their expansions onto the primitive basis set ($N_\mathcal{P}$ denotes the dimension of the $\mathcal{P}$ subspace in which the Hamiltonian is prediagonalized while N_B is the dimension of the primitive basis set). This imposes some constraints on the size of the $\mathcal{P}$ subspace and on the maximum number of Davidson iterations, denoted M_{max}, performed. In order to preclude such core memory problems, we propose first to precise the definition of $\mathcal{P}$ subspace, and second, to add a restart option in the Davidson scheme previously developed. The new scheme used in this study is given in Table 1. If the convergence of the studied eigenstate is not achieved after M_{max} Davidson iterations (step 4 in Table 1), the procedure is restarted in step (1). In order to profit from the initial Davidson scheme, $N_{restart}$ vectors $|u_j >$ ($j = 1, \ldots, N_{restart}$), associated with the largest projections onto the initial subspace $\mathcal{H}$ defined in step (0), are retained in step (1) of the general procedure. In the 9th out-of-plane overtone calculation in HFCO[7], M_{max} and $N_{restart}$ are set to 550 and 100, respectively.

We focus in this study on several highly excited overtone states located in a dense part of the spectrum. Calculation of such highly excited states requires a specific prediagonalization in a subspace $\mathcal{P}$ which has to be as small as possible. $\mathcal{P}$ is called *active space* because it should contain the zero-order states which plays an active role during the computation of the studied state. *How should we build such an active space $\mathcal{P}$?* Different strategies have been explored. As our aim is to develop a general method which is easy to adopt and to use, the following scheme is proposed :

- (i) First, a regular Davidson calculation is performed *in a small basis*, denoted B_{small} whose dimension is limited by using some drastic (but realistic) energy criterion and some constraints on the allowed v_i quantum numbers. The Davidson calculation applied directly to the guess vector $|u_0 >$ provides an estimation of the studied eigenstate.

- (ii) Then, the zero-order states associated with the main contributions of this approximated eigenstate are retained in the $\mathcal{P}$-subspace.

- (iii) Finally, the Hamiltonian is diagonalized in $\mathcal{P}$. The estimation of the studied state obtained after the first Davidson scheme in B_{small} is used to initialize the second Davidson scheme performed in the large working basis set.

0)	Initialization (M = b)	Define the initial subspace $\mathcal{H} = \{u_0 \ldots u_{b-1}\}$
1)	Diagonalization	Diagonalize $\mathbf{H}$ in the $\{u_0, \ldots, u_{M-1}\}$ basis set
		Select the b eigenvectors $\{\Psi_m^{(M)}, m = 1 \ldots b\}$ with the largest projections onto $\mathcal{H}$
2)	Convergence	Form residual $\mathbf{q}_m = (\mathbf{H} - E_m^{(M)})\Psi_m^{(M)}$
3)	Propagation	Only the eigenvectors $\Psi_m^{(M)}$ for which $\|\mathbf{q}_m\| > \epsilon$
	Preconditioning	with zero order Hamiltonian $\mathbf{H}^o : \bar{\mathbf{q}}_m = \left(E_m^{(M)} - \mathbf{H}^o\right)^{-1} \mathbf{q}_m$
	Orthonormalization	$\xi = \left\{1 - \displaystyle\sum_{m=0}^{M-1} |u_m\rangle\langle u_m|\right\} \bar{\mathbf{q}}_m; u_M = \xi/\|\xi\|; M := M + 1$
4)	if $M < M_{max}$	back to 1)
	if $M = M_{max}$	$N_{restart}$ vectors u_j vectors associated to the largest projections onto the initial subspace $\mathcal{H}$ are retained in step 1, back to 1)

Table 1 : Block-Davidson Scheme with a restart option

This approach has been successfully used to calculate highly excited out-of-plane overtones in HFCO[7]. We emphasize that this numerical approach to determine the active space uses only one parameter, c_{min}, whose physical meaning is obvious. Consequently, such an approach is very easy to implement.

3   Conclusion

In conclusion, we have reviewed an efficient, robust, reliable and general method which is able to extract a very highly excited state located in a dense part of the spectrum. This approach can be coupled to any other numerical method which provides the action of the Hamiltonian on a vector known by its expansion onto the primitive basis set. One has also to be able to define a zero-order Hamiltonian, H^o used in the preconditioning step (step 3 in Table 1). We propose a general way to determine the active space $\mathcal{P}$ which is easy to implement. However, it would be worthwhile to explore other strategies for building $\mathcal{P}$ based on, for instance, the notion of a polyad defined with different quantum numbers or based on the wave operator theory used previously by Iung *et al.*[5].

This new scheme presents the following characteristics concerning the general requirements stated in the Introduction section :

- *(i)* The approach is general : it only requires the action of the Hamiltonian operator on a vector expressed in some contracted basis set.

- *(ii-iii)* The method can accurately compute eigenvalues/eigenstates even in the quasi-degenerate case.

- *(iv)* The accuracy of the calculations, with respect to both the eigenvalues and eigenvectors, can be controlled by the residual criterion $\epsilon = \|(\mathbf{H} - E_\alpha)\Psi_\alpha\|$.

- *(v)* One possible limitation in the use of the Davidson scheme in its present implementation could arise from the core memory required to store either the Hamiltonian matrix expressed in the $\mathcal{P}$-subspace, or the M Davidson vectors successively generated during the iteration process. A restart option (Table 1) allows to limit the number of Davidson vectors stored in core memory.

- *(vi)* The block-Davidson approach has been shown to be significantly faster than the regular Davidson scheme, especially in the case of quasi-degeneracy which becomes the rule at higher energies. Both schemes are much faster than the Lanczos algorithm if eigenvectors are also of interest.

- *(vii)* In this study, we used a prediagonalized version of the block-Davidson scheme that provide selectively highly excited states.

- *(viii)* Finally, the method is very easy to use, as it is controlled by two parameters only : ϵ, the residual value $\|(\mathbf{H} - E_\alpha)\Psi_\alpha\|$ which defines the accuracy of the calculations, and N_{max} the size of the blocks considered.

Acknowledgment

Prof. Claude Leforestier is warmly thanked for many fruitful discussions.

References

[1] A.H. Zewail, *Femtochemistry*. Wiley-VCH, Edited by F. C. De Schryver and S. De Feyter and G.Schweitzer, 2001.

[2] Y. Choi and C.B. Moore, *J. Chem. Phys.* **94** (1991), 5414.

[3] O. Boyarkin, M. Kowalszyk, and T. Rizzo. *J. Chem. Phys.* **118** (2003), 93.

[4] R. Page, Y. Shen, and Y. Lee. *J. Chem. Phys.* **88** (1988), 4621.

[5] C.Iung and R.E. Wyatt. *J. Chem. Phys.* **99** (1993), 2261–2264.

[6] E. Davidson. *J. Comp. Phys.* **17** (1975), 87.

[7] C.Iung and F.Ribeiro. *J. Chem. Phys.,to be published* (2005).

[8] F.Ribeiro, C.Iung, and C.Leforestier. *J. Chem. Phys.,in press* (2005).

[9] F.Ribeiro, C.Iung and C.Leforestier, *Chem. Phys. Lett.* **362** (2002), 199.

[10] F. Ribeiro, C. Iung, and C. Leforestier. *J. Theo. Comp. Chem.* **2** (2003), 609.

Brill Academic Publishers
P.O. Box 9000, 2300 PA Leiden
The Netherlands

Lecture Series on Computer
and Computational Sciences
Volume 4, 2005, pp. 1864-1867

The Examples of the Artificial Intelligence Applications in Materials Engineering

L.A. Dobrzański[1], J. Trzaska, W. Sitek

Institute of Engineering Materials and Biomaterials,
Faculty of Mechanical Engineering,
Silesian University of Technology,
44-100 Gliwice, Poland

Received 27 July, 2005; accepted in revised form 8 August, 2005

Abstract: The paper presents the examples of application of the artificial neural networks for modelling and simulation of steels' properties. It presents the method of modelling of hardenability curves according to the chemical composition of steel. Using this method, the other one was developed - the method of designing of the chemical composition of the steel showing the required properties, including the required hardenability. The third method, presents the neural network model for modelling of Continuous Cooling Transformation (CCT) diagrams for constructional and engineering steels.

Keywords: Neural network, chemical composition, steel, hardenability, CCT diagram, modelling

PACS: 07.05.Mh; 07.05.Tp; 81.30.Bx; 81.30.Kf; 81.40.Ef, 87.15.La

1. Introduction

Simultaneously, rapid development of computer science and technology as well as of modern computer tools among them artificial intelligence prompts their increasingly common use in different fields of science and technology, also in materials science. Especially the modern computer tools, artificial intelligence among them, make it possible to extend the investigations of the constructional and engineering steels, for example, for the reduction of production costs by the replacement of the labour-consuming and expensive metallurgical experiments with the computer simulation. In many cases its correct results are the basis for the analyses carried on later.

Over the last few years one can observe a significant increase of interest in the modern computer techniques in the area of material and production engineering. The authors' own papers [1-8] present the examples of application of the artificial neural networks for modelling and simulation of constructional and engineering steels' properties.

2. Modelling of the hardenability curves of constructional alloy steels basing on

its chemical composition

Basing on the experimental results of the hardenability investigations, which employed Jominy method, the model of the neural networks was developed and fully verified experimentally. The model makes it possible to obtain Jominy hardenability curves basing on the steel chemical composition. Hardenability assessment, being one of the main criteria for the selection of steel for constructional elements, makes it possible to accomplish the expected properties' distribution in the element transverse section. The

---

[1] Corresponding author. E-mail: ldobrzan@zmn.mt.polsl.gliwice.pl

authors' own works performed so far indicate to the necessity of developing a new method of Jominy curves calculation, more adequate to the curves obtained experimentally than the models published so far [2, 3]. Quantitative evaluation of the hardenability calculation methods requires the employing the numerical coefficient, which limiting value, which would make it possible to approve the hardenability calculation methods investigated and would meet the expectations concerning accuracy, is determined as the result of the analysis of repeatability of the experimental results obtained in Jominy end-quench test. Therefore, the numerical coefficient evaluating the adequacy of hardenability curves calculation methods is suggested in the paper. It is defined by the relationship:

$$s = \sqrt{\frac{(h_{mi} - h_{ci})^2}{n}} \qquad (1)$$

where: hmi - hardness measured at a distance i from the face, hci - hardness calculated for a distance i from the face, n - number of measurements.

Investigations' results referred to indicate that the value of 2.5 HRC may be assumed as the limiting value of the coefficient s that may classify positively the hardenability calculation method. Multi-layer feedforward neural networks with learning rule based on the error backpropagation algorithm were employed for modelling of the hardenability curves. As the result of the learning error analysis the 6-30-15 two layer neural network with the learning rate $\eta = 0.15$ and momentum parameter $\alpha = 0.3$ were assumed. Six nodes of the neural network input layer correspond to the concentration of the alloying elements (C, Mn, Si, Cr, Ni, Mo). Hardness on the hardenability curve is calculated at fifteen points. The calculations verifying the neural network models were made for each class of the steels analysed. Average values of coefficient s calculated for each model of self-developed neural network and method known from literature are included in table 1. Figure 1 presents graphical comparison of the hardenability curves calculated using different methods and the experimental ones for some heats of the alloy constructional steels'. Calculation results obtained using neural network indicate good conformity of the hardenability calculations with the experimental data. The average values of the adequacy evaluation coefficient s are lower than the value 2.5 HRC determining the required accuracy of the calculations.

a)   b)

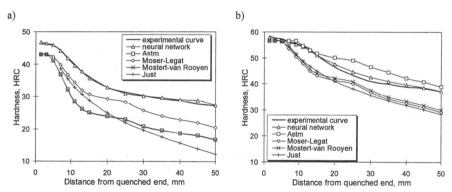

Figure 1. Comparison of the calculations results with experimental data for the a) 15HGN and b) 42CrMo4 exemplary steel heats.

Table 1. Values of the coefficient *s* obtained from the calculations

Model	Carburizing steels			Heat-treatable steels		
	Class I	Class II	Class III	Class I	Class II	Class III
neural network	2.1	1.6	1.9	1.9	2.1	2.4
ASTM	10.6	5.9	3.2	4.6	2.6	3.6
Moser - Legat	2.7	2.1	2.2	5.7	5.3	4.1
Mostert - van Rooyen	10.5	5.8	2.8	5.6	5.3	3.6
Just	6.0	4.5	4.9	10.2	6.3	4.4

3. Designing of the chemical composition of the alloy constructional steels basing on the assumed Jominy hardenability curves' shapes

Multi-layer feedforward neural networks with backpropagation learning rule were employed for the designing of the chemical composition of the steel having the required hardenability [4]. After the learning error analysis, the 15-30-6 network model was accepted. 15 input nodes of the neural network correspond to fifteen values of hardness at successive points in fixed distances from the specimen face and 6 nodes of the output layer correspond to the concentration of the alloying elements (C, Mn, Si, Cr, Ni, Mo). The neural networks developed were experimentally verified. Verification of the model consists in designing of such a chemical composition of a heat so that after its casting and carrying out the relevant hardenability investigations it would yield the required hardenability curve shape. The hardenability curves, using the hardenability model developed, were calculated for the chemical compositions obtained as a result of the calculations mentioned. These curves were compared with the ones with the required shape, calculating the value of the coefficient of the calculation methods adequacy s for each case. The average value of the coefficient of the methods' adequacy was calculated for each class. The obtained values are specified in table 2.

As the example of the calculations made, the results for two hardenability shapes assumed (No.1 for the carburizing steel and No.2 for the heat-treatable one) are presented. Figure 2 present graphical comparison of the assumed hardenability curves and the experimental ones of the steel heats of the designed chemical composition.

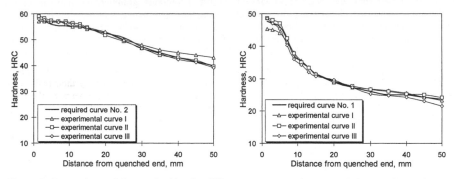

Figure. 2. Comparison of the required hardenability curve No. 1 and No. 2 and the experimental ones of the steels' heats of the designed chemical composition.

Table 2. The average value of the coefficient of the methods' adequacy *s* obtained for each steels' class

Steels' group					
Carburizing steels			Heat-treatable steels		
Class I	Class II	Class III	Class I	Class II	Class III
1.4	2.2	1.6	1.8	1.6	2.2

4. Application of neural networks for the prediction of continuous cooling transformation diagrams

The relationship between the chemical composition and austenitizing temperature, and the Continuous Cooling Transformation diagram for steels were developed using neural networks.. Input data are chemical composition and austenitising temperature. Results of calculation of neural networks consist of temperature of the beginning and the end of transformation in the function of cooling rate, the participation of the structural components and the hardness of steel cooled from austenitising temperature with a fixed rate. Presented quantities enable to draw the CCT diagram. The model presented in the paper enables the analysis of the influence of the chemical composition and austenitising temperature on CCT diagrams. In order to work out the methods the set of experimental data worked out on the basis of information available in the literature consisting of 400 CCT diagrams made for constructional and engineering steels were used. The numerical verification of the developed method was carried out making 50 CCT diagrams for data that was not used for model development. Examples of the CCT diagrams, worked out on the basis of the calculations carried out, along with the experimental plots, are presented in Figure 3. The detailed description of the developed method of determining the complete CCT diagrams is presented in [5-8].

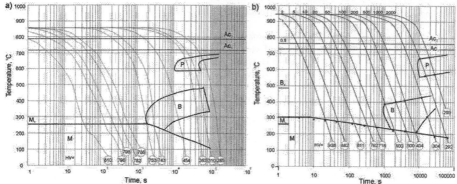

Figure 3. CCT diagram for steel with concentrations: 0.52% C. 0,52% Si, 1.09%Cr, 1.72% Ni, 0.43% Mo, 0.14% V, austenitised at a temperature of 950°C; a) experimental, b) calculated.

5. Final remarks

The synergetic effect of the alloying elements on constructional and engineering steels' properties is the reason for which analysing the influence of single elements does not reveal fully their real effect. The influence of the alloying elements should be analysed for the particular elements at the fixed concentration of the remaining constituents in the analysed steel. The artificial intelligence tools, including the neural networks, make it possible to substitute partially the costly and time consuming experimental investigations with the computer simulation and using the obtained results as data for further analyses.

References

[1] L.A. Dobrzański: Material selection principles for machine design, Proceedings of the International Seminar on Principles and Methods of Engineering Design, vol. I, Naples, Italy, 267-293 (1997).

[2] L.A. Dobrzański, W. Sitek: Comparison of hardenability calculation methods of the heattreatable constructional steels, *Journal of Materials Processing Technology*, 1997, Vol. 64 (1 3), 117-126 (1997).

[3] L.A. Dobrzański, W. Sitek: Application of neural network in modelling of hardenability of constructional steels, Proceedings of the International Conference on Advanced Materials and Processing Technologies, AMPT'97, vol. I, Guimaraes, Portugal, 24-31 (1997).

[4] L.A. Dobrzański, W. Sitek, Designing of the chemical composition of constructional alloy steels, *Journal of Materials Processing Technology*, Vol. 89-90, 467-472 (1999).

[5] L.A. Dobrzański, J. Trzaska: Application of neural network for the prediction of continuous cooling transformation diagrams, *Computational Materials Science*, Vol. 30, 3-4, 251-259 (2004).

[6] L.A. Dobrzański, J. Trzaska: Application of neural networks for prediction of critical values of temperatures and time of the supercooled austenite transformations, *Journal of Materials Processing Technology*, Vol. 155-156, 1950-1957(2004).

[7] L.A. Dobrzański, J. Trzaska: Application of neural networks to forecasting the CCT diagram, *Journal of Materials Processing Technology*, Vol. 157-158, 107-113 (2004).

[8] L.A. Dobrzański, J. Trzaska: Application of neural networks for prediction of hardness and volume fractions of structural components in constructional steels cooled from the austenitizing temperature, *Material Science Forum*, Vol. 437–438, 359–362 (2003).

Brill Academic Publishers
P.O. Box 9000, 2300 PA Leiden
The Netherlands

*Lecture Series on Computer
and Computational Sciences*
Volume 4, 2005, pp. 1868-1870

Modeling of Magnetic and Mechanical Properties of Hard Magnetic Composite Materials Nd-Fe-B

L.A. Dobrzański[1], M. Drak, J. Trzaska

Institute of Engineering Materials and Biomaterials,
Faculty of Mechanical Engineering,
Silesian University of Technology,
Gliwice, Poland

Received 27 July, 2005; accepted in revised form 8 August, 2005

Abstract: The paper presents a neural network model for evaluation of the magnetic and mechanical properties of the polymer matrix hard magnetic composite materials with particles of Nd-Fe-B strip with addition of metallic powders. A neural network was established based on the results from the investigations carried out.

Keywords: neural network, magnetic properties, mechanical properties, Nd-Fe-B magnets

PACS: 07.05.Mh; 75.75.+a; 87.15.La; 07.55.Db;

1. Introduction

Hard magnetic materials based on the rare earth and transition metal compounds, in particular on the $SmCo_5$, Sm_2Co_{15} and $Nd_2Fe_{14}B$ phases, are materials with big application possibilities. They are more and more commonly used in the instrument and devices for telecommunication, computer industry, control and measurement technology and lastly their use in electric machines is growing. Their use has revolutionized the zone of permanent magnet application. These materials, with high magnetic properties, make able to miniaturizing and to make simplification of device shapes. [1-3]

The most dynamic growth is observed in the sintered, hot compacted, upset and composite materials. Depending on the way of obtaining they show different magnetic, mechanical properties and corrosion resistance. Magnets fabricated as polymer matrix composite with the hard magnetic particles from the milled rapid quenched Nd-Fe-B strip have many advantages. One can make magnets with complicated shapes and with very accurate dimensions and minimize material losses in their manufacturing process; therefore manufacturing of such magnets is cheaper and easier compared to the sintered magnets.

Permanent magnets made from powders have low mechanical properties. During assembling and in normal work in magnets occur some mechanical stresses which can damage the magnet. It is necessary to develop magnets with better mechanical properties. The mechanical properties of magnets depend mainly on their composition: amount of hard magnetic powder, resin amount and manufacturing technology. One of the ways of improving mechanical properties of Nd-Fe-B composite materials is strengthening them by mixing the hard magnetic powder with metal powder [4-7].

Neural networks are used as a general-purpose tool for numerical modelling that is suitable to map complex functions. No need to develop an algorithm, no computer program is required to adapt neural networks to carry out a specific task. Instead, they are capable of learning using a series of standard stimuli and corresponding emulated responses. The strongest reason for using neural networks is their ability to generalise when confronted with new situations. Neural networks do not require a priori knowledge about problems, which they are intended to solve, they have ability to tolerate disruptions or discontinuities, accidental gaps or loss in the learned data set. Over the last several years, neural networks have gained increasing interest in the field of materials engineering. The growing popularity

---

[1] Corresponding author.E-mail: ldobrzan@zmn.mt.polsl.gliwice.pl

of neural networks is due their ability to model relations between investigated variables with no need to know the physical model of the phenomena. The results provided by neural networks very often exhibit better correlation with experimental data than those obtained from empirical explorations or mathematical models of the processes under investigation. [8-10]

The goal of the paper is modelling of magnetic and mechanical properties of composite materials Nd-Fe-B with addition of metallic powders by the use of neural networks.

2. Material and experimental method

The investigation were made on samples of the polymer matrix hard magnetic composite materials with particles of the powdered rapid quenched Nd-Fe-B strip with addition of metallic powder: iron, aluminum, CuSn10 type casting cooper-tin alloy and X2CrNiMo17-12-2 high alloy steel with the range 0 - 15 wt. %. The samples were subjected magnetic (remanence, coercive force, maximum energy product) and mechanical (hardness, compressive strength) investigations.

Based on the results of these investigations the neural network was used to computer simulation of these properties. There was established the relationship between the mass contribution of the powdered addition and the magnetic or mechanical properties. In order to evaluate the quality of the model, the following methodology was applied: the error-mean square, the standard error deviation, the standard deviation ratio for errors and data (that for an „ideal" prognosis assumes the value of 0) and Pearson's correlation coefficient. The task of the development of a neural network required to determine the following quantities: type of the neural network, the size of hidden layers and the number of neurons in individual layers, the type and form of the activation function, variable scaling procedure, function of error and neural network training technique and parameters. All the parameters mentioned above were selected after the analysis of their influence on the assumed quality coefficients. The number of neurons in both the input and output layers were established. The neural activation level in the input layer was made dependent upon the percentage of the metallic powder. A single neuron within the output layer meant the percentage of the magnetic/mechanical properties..The neural networks parameters are presented in Table 1.

Table 1. Neural networks parameters

	Properties				
	Remanence	Coercive force	Maximum energy product	Hardness	Compressive strength
Network type[*]	MLP	MLP	MLP	MLP	MLP
Network structure	5-2-1	5-1-1	5-3-1	5-3-1	5-2-1
Training method[**]	CG	CG	CG	BP	BP, CG
No of epochs	51	77	67	23	50,11

[*]MLP – multi layer perceptron,
[**]CG – conjugate gradients method, BP – backpropagation method

3. Discussion

Based on investigated data models of neural network was developed. Table 2 present the result of the quality evaluation for the optimised neural network models.

Table 2. Quality assessment coefficients for neural networks

Quality assessment	Set	Properties				
		Remanence	Coercive force	Maximum energy product	Hardness	Compressive strength
Mean error	Training	0,01	4,08	2,04	1,36	3,33
	Validating	0,01	4,52	2,54	1,02	2,29
	Testing	0,01	4,72	2,68	1,16	3,60
Standard deviations ratio	Training	0,14	0,15	0,34	0,42	0,29
	Validating	0,12	0,20	0,30	0,30	0,22
	Testing	0,15	0,13	0,33	0,31	0,31
Pearson's correlation coefficient R	Training	0,99	0,99	0,94	0,94	0,96
	Validating	0,99	0,98	0,95	0,95	0,98
	Testing	0,99	0,99	0,97	0,95	0,96

These results, namely the low value of the mean error, close to zero standard error deviation ratio for errors and data and Pearson's correlation coefficient close to 1 confirm the validity of the neural network-based mapping approaches. Merely the same coefficients for the sets: training, validating and testing ones indicate that the develop neural networks are able to generalize. Figure 1 illustrate examples of comparison between real and calculates values of magnetic/mechanical properties of composite materials.

a)   b)

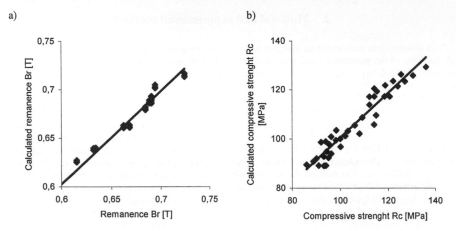

Fig. 1. Real and calculated: a) remanence, b) compressive strength of composite materials

4. Conclussion

There was proposed a simulation approach to model the magnetic/mechanical properties in metallic powder added polymer-matrix composite materials filled with magnetically hard Nd-Fe-B particles in function of the chemical composition, employing a neural network technique. The obtained results indicate that the developed neural networks are able to generalize that justify using these models to the simulation of magnetic/mechanical properties of hard magnetic composite materials.

References

[1] D.C. Jiles: Recent advances and future directions in magnetic materials, *Acta Materialia*, 51(2003) 5907
[2] M. Leonowicz, *Nanocrystalline magnetic materials*, WNT, Warszawa, 1998 (in Polish)
[3] B. Ślusarek, *Dielectromagnets NdFeB*, Wrocław, 2001 (in Polish)
[4] Dobrzański L., Drak M.: Structure and properties of composite materials with polymer matrix reinforced Nd-Fe-B hard magnetic nanostructured particles, Journal of Materials Processing and Technology 157-158 (2004) 650
[5] Dobrzański L. A., Drak M.: Polymer matrix composite materials reinforced with hard magnetic particles Nd-Fe-B, *Inżynieria Materiałowa*, Vol. 137 No. 6 (2003), pp. 593-597 (in Polish)
[6] Dobrzański L.A.: *Metallic engineering materials*, WNT, (Warszawa 2004) (in Polish)
[7] Drak M.: Structure and application properties of polymer matrix hard magnetic composite materials reinforced with Nd-Fe-B nanocrystalline particles with metal addition, PhD thesis - unpublished, Main Library of the Silesian University of Technology, Gliwice 2004 (in Polish)
[8] Bhadeshia H.K.D.H.: Neural Networks in Materials Science. *ISIJ International*. Vol. 39, 10 (1999) 966
[9] Dobrzański L.A., Trzaska J.: Application of neural networks to forecasting the CCT diagram, *Journal of Materials Processing Technology*, 157-158, (2005) 107
[10] Dobrzański L.A., Trzaska J.: Application of neural networks for prediction of critical values of temperatures and time of the supercooled austenite transformations, J*ournal of Materials Processing Technology*, 155-156 (2004) 1950

Brill Academic Publishers
P.O. Box 9000, 2300 PA Leiden
The Netherlands

Lecture Series on Computer
and Computational Sciences
Volume 4, 2005, pp. 1871-1875

Active Linear Modeling of Cochlear Biomechanics Using Hspice

You Jung Kwon[1], Soon Suck Jarng[1]
Department of Information, Control & Instrumentation,
Chosun University, South Korea

Received 10 June, 2005; accepted in revised form 15 June, 2005

Abstract: This paper shows one and two dimensional active linear modeling of cochlear biomechanics using Hspice. The advantage of the Hspice modeling is that the cochlear biomechanics may be implemented into an analog IC chip. This paper explains in detail how to transform the physical cochlear biomechanics to the electrical circuit model and how to represent the circuit in Hspice code. There are some circuit design rules to make the Hspice code to be executed properly. The comparison between one and two dimensional models is figured.

Keywords: Active, Linear, Cochlea, Biomechanics, Hspice, Modeling, Electric Circuit Model

Mathematics SubjectClassification: 07.05.Tp

PACS: 07.05.Tp

1. Introduction

Since *von Be'ke'sy* first time experimentally analyzed the electrical functioning of the cochlea in 1938, not only physiological but also morphological researches have been done about the cochlea [1]. The measurement of the otoacoustic emission(OAE) by Kemp in 1970 promoted new approaches for the cochlear research because he proved that the cochlea actively generated biological energy from inside the cochlea [2]. Last decades were tediously spent for searching the exact biomechanics of the cochlea. And one of the most significant results is that the outer hair cell (OHC) of the cochlea has the core role of the active biomechanics of the cochlea [3-5]. And the other important result is that the sharp frequency selectivity of the ear is mainly processed at the organ of corti rather than at the matrix of auditory nerves [6-8]. There are two important issues to be considered in the cochlear study; Frequency Selectivity, Intensity Sensitivity.

This paper shows one and two dimensional active linear modeling of cochlear biomechanics using Hspice [9]. The present cochlear biomechanical modeling is based on the deductive approach as others but is processed using Hspice which is the most popular analog electric circuit simulator among electrical and electronic engineers. Because Hspice is so generally used for active filter design, the shortage of the cochlear model may be supplemented with secondary analog active filters. This extra filtering may be called as second filters. The advantage of the Hspice modeling is that the cochlear biomechanics may be implemented into an analog IC chip. This paper explains in detail how to transform the physical cochlear biomechanics to the electrical circuit model and how to represent the circuit in Hspice code.

2. Methods

The one-dimensional linear and active model of the cochlea suggested by Neely and Kim [10] was previously transformed to an electrical transmission line model by the author [11]. The electrical transmission line model was previously solved by the finite difference method (FDM). The same model is now extended to two dimensions and solved by Hspice in this paper.

The basic idea of Neely and Kim [10] for the cochlear biomechanics may be described as Fig. 1. Two masses coupled with four springs describe a simplified model of the segmented cochlear partition. Both the BM and the tectorial membrane (TM) are attached at the spiral ligament and the spiral limbus respectively. This can be described as two masses attached at a reference ground by two damped springs. The OHC and the stereocilia connect the BM with the TM. This can be also described as the two masses coupled by two damped springs in series. This mechanical system of two masses with four springs results in the fourth order dynamic mechanics. The active force generated by the OHC is transferred directly to the BM as well as indirectly to the TM through the stereocilia. The forth order passive resonating system is actively tuned by the extra force. Neely and Kim suggested that the OHC generates the active force in proportion to the velocity of the OHC stereocilia [10].

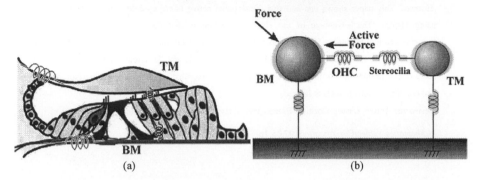

Figure 1 : (a) The cross section of the segmented cochlear partition. (b) Two masses coupled with four springs describe a simplified model of the cochlear biomechanics. BM: Basilar Membrane, TM: Tectorial membrane. The OHC and the stereocilia connect the BM with the TM.

The mechanical resonant system can be transformed to an electrical filter circuit by through/across analogy (Table 1). Mechanical variables such as force and velocity are analogies with current and voltage. Therefore the independent applied force and the dependent active force of the two masses - four springs system may be transformed to the independent current source, I_S, and the dependent current source, I_{OHC}, respectively (Fig. 2). Since the mechanical active force is in proportion to the velocity of the OHC stereocilia, I_{OHC} is depending on the voltage across the stereocilia, V_{ST}. The actual value of I_{OHC} is the current passing through the OHC impedance, L'_4 and G'_4 where L'_4 and G'_4 are an electrical inductor and a conductor in analogy with the OHC mechanical compliance and damping. C'_1, L'_1 and G'_1 are a capacitor, an inductor and a conductor in analogy with the BM mechanical mass, compliance and damping respectively. In the same way, C'_2, L'_2 and G'_2 are electric analogies of the TM elements, and L'_3 and G'_3 are electric analogies of the stereocilia elements

Table 1 : Mechanical / Electrical Analogies

Analogy	Variables	Elements
Through / Across	Force ↔ Current Velocity ↔ Voltage	Mass (M) ↔ Capacitance (C) Compliance (Cm) ↔ Inductance (L) Damping (Rm) ↔ Conductance (G)

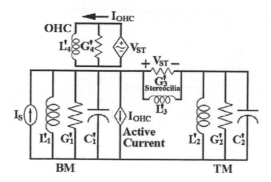

Figure 2 : The electrical through/across analogue of the two mass and four spring system.

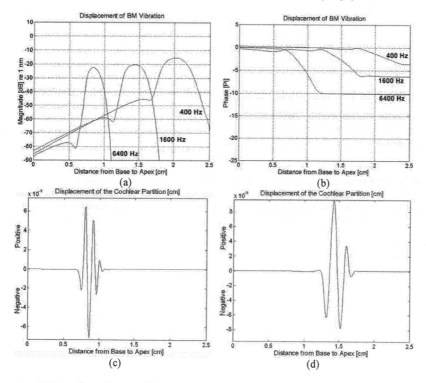

Figure 3 : (a) Two dimensional cochlear model displacement amplitude responses. (b) Displacement Phase responses. (c) Displacement time response (6400 Hz). (d) Displacement time response (1600 Hz). (e) Displacement time response (400 Hz). Amplification Gain (γ) = 1. LF=3E-3 [H].

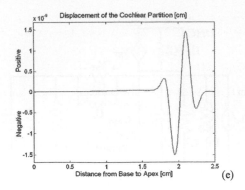

(e)

Figure 3 : Continued

3. Result and Discussion

Fig. 3 shows the simulation results of the two dimensional cochlear model with different frequencies; 6400 [Hz], 1600 [Hz], 400 [Hz]. The amplification gain (γ) is 1.0. The cochlear fluid mass, LF, is increased as 5E-3 [H]. Fig. 3 (a) and Fig. 3 (b) are the displacement amplitude and phase responses respectively. As the input frequency decreases, the peak moves toward the apex. The increase of the cochlear fluid mass not only increases the sharpness of the bandpass filter shape but also increases the amount of the phase shift. The increase of the sharpness is more significant than the increase of the phase shift. Fig. 3 (c), Fig. 3 (d) and Fig. 3 (e) show the displacement temporal responses of the basilar membrane vibration at each different input frequency respectively.

4. Conclusion

This paper shows one and two dimensional active linear modeling of cochlear biomechanics using Hspice. The advantage of the Hspice modeling is that the cochlear biomechanics may be implemented into an analog IC chip. This paper explains in detail how to transform the physical cochlear biomechanics to the electrical circuit model and how to represent the circuit in Hspice code. There are some circuit design rules to make the Hspice code to be executed properly. The comparison between one and two dimensional models is figured.

References

[1] *von Be'ke'sy G.*, "Experiments in hearing", Wiley, New York, 1960.

[2] Kemp D.T., "Stimulated acoustic emissions from within the human auditory system", J. Neurophysiology, vol. 34, pp:802-816, 1978.

[3] Ashmore J.F., "A fast motile response in guinea-pig outer hair cells: the molecular basis of the cochlear amplifier", J. of Physiology (London), vol. 388, pp:323–347, 1987.

[4] Iwasa K.H. and Chadwick R.S., "Elasticity and active force generation of cochlear outer hair cells", J. Acoust. Soc. Am., vol. 92(6), pp:3169–3173, 1992.

[5] Dallos P., "Cochlear neurobiology", The cochlea, Edited by P. Dallos, A.N. Popper, and R.R. Fay, New York, Springer, pp:186–257, 1996.

[6] Sellick P.M., Patuzzi R., Johnstone B.M., "Measurement of basilar membrane motion in the guinea pig using the Mössbauer technique", J. Acoust. Soc. Am., vol. 72, pp:131-141, 1982.

[7] Khanna S.M., Leonard D.G.B., "Basilar membrane tuning in the cat cochlea", Science, vol. 215, pp:305-306, 1982.

[8] Narayan S.S., Temchin A.N., Recio A., and Ruggero M.A., "Frequency tuning of basilar membrane and auditory nerve fibers in the same cochleae", Science, vol. 282, pp:1882–1884, 1998.

[9] Hspice User's Guide , 2002.

[10] Neely S.T., Kim D.O., "A model for active elements in cochlear biomechanics", J. Acoust. Soc. Am., vol. 79, pp:1472-1480, 1986.

[11] Jarng S.S., "Electrical transmission line modeling of the cochlear basilar membrane" J. of Korea Soc. of Med. and Bio. Eng., vol. 14, no. 2, pp:125-136, 1993.

Brill Academic Publishers
P.O. Box 9000, 2300 PA Leiden
The Netherlands

Lecture Series on Computer and Computational Sciences
Volume 4, 2005, pp. 1876-1879

Electronic and magnetic properties of La$_{1-x}$Ca$_x$MnO$_3$

M. -H. Tsai[1] and Y. -H. Tang

Department of Physics, National Sun Yat-Sen University, Kaohsiung, 804 Taiwan

Received 8 July, 2005; accepted 29 July, 2005

Abstract: The electronic and magnetic properties of manganites have usually been studied by phenomenological approaches such as the double-exchange model, because Mn 3d electrons are strongly correlated. The first-principles calculation methods have been questioned to be suitable for these systems. However, the electronic structures of La$_{1-x}$Ca$_x$MnO$_3$ obtained by the spin-polarized first-principles pseudofunction method, which uses a muffin-tin-orbital basis set, explain reasonably well the experimentally observed dopant-induced insulator-like to metal-like and anti-ferromagnetic to ferromagnetic transitions and reverse transitions at x=~0.5 at low temperatures. Delocalization of majority-spin Mn e_g bands that straddle the Fermi level, E$_F$, and the lowering of the minority Mn t_{2g} bands down to E$_F$ are found to play the major role in these properties.

Keywords: electronic structures, ferromagnetism, anti-ferromagnetism, manganites
PACS: 75.47.Lx; 75.30.-m;71.20.Ps

1. Introduction

Rare-earth manganites doped with alkaline-earth metals exhibit colossal magnetoresistance (CMR) and interesting electronic and magnetic properties such as anti-ferromagnetic to ferromagnetic and insulator-like to metal-like transitions. The understanding of CMR and their electronic and magnetic properties has usually been based on the strongly correlated phenomenological type of approaches such as the double exchange model or its variations. The phenomenological type of approach has been argued to be necessary for treating localized Mn 3d orbitals. The first-principles calculation methods based on linear combination of atom-centered orbitals or plane waves have been argued to be inadequate for strongly correlated electronic systems [1]. However, the argument against the first-principles calculation methods was based on an incorrect analogy with the hydrogen molecule [1]. In principle any localized and singly occupied state can be expanded in spin-polarized Bloch states if the basis set and the unit cell are sufficiently large. A localized state will show as a flat energy band. The repulsive intra-atomic Coulomb energy, U, which characterizes a strongly correlated system, in fact is automatically included in the Hartree part of the Kohn-Sham effective potential.

Based on the decrease of resistivity, ρ, with the increase of temperature, T, the low-temperature pure and highly Sr or Ca doped LaMnO$_3$ are described as anti-ferromagnetic insulator [2], despite that ρ has an order of only 10^0 Ω-cm when T→0 for La$_{0.1}$Ca$_{0.9}$MnO$_3$ [3]. Note that ρ→∞ when T→0 for an intrinsic insulator and a typical insulator has a ρ of 10^{13} Ω-cm at room temperature. On the other hand, La$_{1-x}$Ca$_x$MnO$_3$ with x<~0.50 exhibit ferromagnetic properties and are described as metals, despite that their ρ's are in a range between 10^{-1} and 10^0 Ω-cm [3], while a typical ρ of metals has an order of 10^{-6} Ω-cm at room temperature. The double exchange model [4-6] is based on the co-existence of Mn^{3+} and Mn^{2+} ions or Mn^{3+} and Mn^{4+} ions in electron- and hole-doped manganites, respectively, i.e. the mixed valence state. However, validity of the mixed valence picture of Mn ions is questionable for (1) the Mn-O bond is not 100% ionic, (2) the total electrostatic energy of the mixed valence state is more repulsive than the uniform valence state, and (3) the chemical-shift difference due to the differing on-site electrostatic potentials at Mn^{3+} and Mn^{4+} ions was not observed in Mn $L_{3,2}$-edge x-ray absorption spectroscopy (XAS) spectra [7-9]. Thus, a new interpretation of the electronic and magnetic properties of manganites is needed.

---

[1] Corresponding author. E-mail: tsai@mail.phys.nsysu.edu.tw

2. Calculation method and crystal structures

The spin-polarized first-principles calculation method used in this study is the pseudofunction (PSF) method [10], which uses local-spin-density approximation (LSDA) of von Barth and Hedin [11]. The PSF method calculates spin-polarized full charge densities, which includes the core charge densities, and the corresponding potentials self-consistently through iterations. The basis set of the PSF method is composed of Bloch sums of muffin-tin orbitals with spherical Hankel and Neumann tailing functions. The linear theory of Andersen [12] is used to solve the Schrödinger equations inside the muffin-tin spheres for the radial component of the muffin-tin orbitals. Since the muffin-tin orbitals change rapidly inside the muffin-tin spheres, a large number of plane waves have to be used for calculating the nonspherical and interstitial parts of the Hamiltonian matrix elements. Therefore, the use of pseudofunctions, which are smooth mathematical functions expanded by a minimal number of plane waves, are to improve the efficiency for calculating the nonspherical and interstitial parts of the Hamiltonian matrix elements by the fast Fourier transform (FFT) technique. In the calculation of the charge densities, the special k point scheme of Monkhorst and Pack [13] with q=4 in each direction of the reciprocal vector is used to approximate the integration over the first Brillouin zone. In the current spin-polarized formalism of the PSF method, the sub-matrices of majority-spin ($\uparrow$-spin) and minority-spin ($\downarrow$-spin) states in the total Hamiltonian matrix are diagonalized separately. The off-diagonal sub-matrices, which accounts for the explicit coupling between $\uparrow$-spin and $\downarrow$-spin states, are ignored. However, coupling between $\uparrow$-spin and $\downarrow$-spin states is implicitly included through LSDA. Relative stability of the ferromagnetic and anti-ferromagnetic states is determined by comparison of their total energies. In this study only x=0.00, 0.25, 0.50, 0.75, and 1.00 five cases are considered. The tetragonal structure is used for CaMnO₃, i.e. x=1.00 case, and the orthorhombic Pnma structure with distortion is used for other four cases. Experimental lattice parameters given in references [14-18] are used for x=0.00, 0.25, 0.50, 0.75, and 1.00 cases, respectively

3. Magnetic states

In the ferromagnetic state, LaMnO₃, La₀.₇₅Ca₀.₂₅MnO₃, La₀.₅₀Ca₀.₅₀MnO₃, La₀.₂₅Ca₀.₇₅MnO₃ have total energies per 20-atom unit cell of 0.1375, -0.6625, -0.0025, 0.0175eV, respectively, relative to those in the A-type anti-ferromagnetic state. For CaMnO₃, the ferromagnetic state has a total energy of 0.95eV per 40-atom unit cell relative to that in the G-type anti-ferromagnetic state. The total energy result shows that LaMnO₃, La₀.₂₅Ca₀.₇₅MnO₃ and CaMnO₃ are antiferromagnetic, while La₀.₇₅Ca₀.₂₅MnO₃ and La₀.₅₀Ca₀.₅₀MnO₃ are ferromagnetic in agreement with the phase diagram of La₁₋ₓCaₓMnO₃ [2]. Note the energy difference between ferromagnetic and anti-ferromagnetic states is very small for x=0.5, which suggests that La₀.₅₀Ca₀.₅₀MnO₃ is near the borderline between the two magnetic states.

4. Electronic properties

For the distorted structure, LaMnO₃ has a gap of about 0.7eV between Mn $\uparrow$-spin e_g and $\downarrow$-spin t_{2g} subbands as shown in Fig. 1. The calculated spin polarized partial densities of states (PDOS) of ferromagnetic La₀.₇₅Ca₀.₂₅MnO₃ and La₀.₅₀Ca₀.₅₀MnO₃ are shown in Figs. 2(a) and (b), respectively. These figures show that the leading edge of the $\uparrow$-spin e_g band is broadened and extend above E_F. The O 2p states near E_F are also found to be delocalized and spread above E_F. Another change is that the $\downarrow$-spin t_{2g} band is lowered, so that its leading edge touches E_F. The $\downarrow$-spin t_{2g} PDOS is small at E_F, so that these states do not contribute dominantly to the conductivity. The observed semimetallic behavior is dominantly due to delocalized $\uparrow$-spin e_g and O 2p states.

The calculated spin polarized PDOS's of anti-ferromagnetic La₀.₂₅Ca₀.₇₅MnO₃ and CaMnO₃ are shown in Figs. 3(a) and (b), respectively. CaMnO₃ has an energy gap, while La₀.₂₅Ca₀.₇₅MnO₃ has a deep trough just below E_F. The "insulator" behavior refers to a decrease of ρ with the increase of T. The conductivity, σ, which is the inverse of ρ, depends on both carrier density and mobility. The carrier density in turn depends on PDOS near E_F, while the mobility depends on the effective mass and scattering of carriers. The sharp Mn $\downarrow$-spin t_{2g} peak just above E_F in the PDOS of La₀.₂₅Ca₀.₇₅MnO₃ indicates a sharp increase of the number of states contributing to σ with the increase of T. Thus, the present study can explain the relatively small ρ at T→0 and the insulator/semiconductor-like dependence of ρ on T for La₁₋ₓCaₓMnO₃ with a large x.

5. Magnetic properties

The average magnetic moments of Mn ions for La₁₋ₓCaₓMnO₃ are 1.89, 1.65, 1.59, 1.36, and 1.17μ_B, respectively, for x=0.00, 0.25, 0.50, 0.75 and 1.00 cases. The average magnetic moment of Mn ions decreases with the increase of x. The magnetic moments of the rest of ions are quiet small. In the ferromagnetic state, all Mn ions have the same spin orientations. In the A-type anti-ferromagnetic state, Mn ions in each plane perpendicular to the c-axis have the same spin orientation, while the spin changes orientation alternatively along the c-axis. In the G-type anti-ferromagnetic state, the nearest neighboring Mn ions have opposite spin orientations. The calculated effective charges are 1.29e, 1.86e, 1.55e, 1.62e and 1.44e for Mn ions and -1.00e, -1.02e, -0.98e, -0.92e and −0.84e for O ions, respectively, for x=0.00, 0.25, 0.50, 0.75 and 1.00. Note that the effective charges of Mn and O ions are not whole numbers as depicted in the mixed-valence picture.

There are two competing mechanisms that determine the magnetic properties of manganites. First, the O-mediated super-exchange coupling between adjacent Mn spins favors anti-ferromagnetism [19]. Another is the mediation by itinerant or delocalized ↑-spin empty states in the vicinity of E_F, which favors ferromagnetism. For pure $LaMnO_3$ and $CaMnO_3$, there is a deficiency of ↑-spin e_g states immediately above E_F, so that the O mediated super-exchange coupling dominates and these materials are anti-ferromagnetic. In contrast, ↑-spin e_g band extends above E_F for Ca doped manganites as shown in Figs. 2(a) and (b). The delocalized empty ↑-spin e_g states immediately above E_F enhance delocalized-state mediated Mn-Mn spin couplings, so that these materials become ferromagnetic. For $La_{0.25}Ca_{0.75}MnO_3$, however, the ↑-spin e_g PDOS near E_F as shown in Fig. 3(a) is reduced, which reduces delocalized-state mediated magnetic coupling, so that the O-mediated super-exchange coupling dominates again and the manganite becomes anti-ferromagnetic.

6. Conclusion

The spin-polarized first-principles calculations of the electronic structures of $La_{1-x}Ca_xMnO_3$ show that Ca doping induces delocalization of near-E_F ↑-spin Mn e_g and O $2p$ states, so that these manganites become semi-metallic. The delocalized ↑-spin Mn e_g states mediate Mn-Mn spin coupling, so that they become ferromagnetic. The insulator-like temperature dependence of the resistivity for $La_{1-x}Ca_xMnO_3$ with a large x can be attributed to the existence of the sharp ↓-spin Mn t_{2g} bands immediately above E_F, so that the density of carriers increases sharply with the temperature. The reversal of the magnetic state for $La_{1-x}Ca_xMnO_3$ with a large x can be attributed to the reduction of delocalized ↑-spin Mn e_g states near E_F, which renders O-mediated Mn-Mn super-exchange coupling to become dominant.

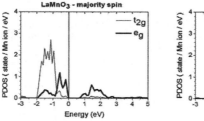
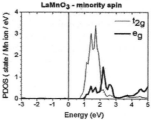

Figure 1: Calculated partial densities of states (PDOS) of Mn majority- and minority- spin t_{2g} and e_g for antiferromagnetic $LaMnO_3$.

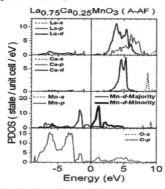
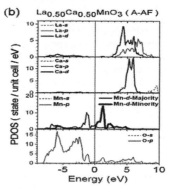

Figure 2: Calculated partial densities of states (PDOS) for ferromagnetic (a) $La_{0.75}Ca_{0.25}MnO_3$ and (b) $La_{0.50}Ca_{0.50}MnO_3$.

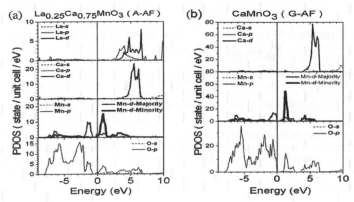

Figure 3: Calculated partial densities of states (PDOS) for anti-ferromagnetic (a) La$_{0.25}$Ca$_{0.75}$MnO$_3$ and (b) CaMnO$_3$.

References

[1] U. Mizutani, " Introduction to the electron theory of metals, Chapter 14 ", Cambridge University Press, New York 2001.

[2] R. Schiffer, A. P. Ramirez, W. Bao, and S. –W. Cheong, Phys. Rev. Lett. **75**, 3336(1995).

[3] R. Mahendiran, S. K. Tiwary, A. K. Raychaudhuri, T. V. Ramakrishnan, R. Mahesh, N. Rangavittal, and C. N. R. Rao, Phys. Rev. **B53**, 3348(1996).

[4] C. Zener, Phys. Rev. **81**, 440(1951).

[5] C. Zener, Phys. Rev. **82**, 403(1951).

[6] P. W. Anderson and H. Hasegawa, Phys. Rev. **100**, 675(1955).

[7] T. Saitoh, A. E. Bocquet, T. Mizokawa, H. Namatame, A. Fujimori, M. Abbate, Y. Takeda, and M. Takano, Phys. Rev. **B51**, 13942 (1995).

[8] M. Croft, D. Sills, M. Greenblatt, C. Lee, S. –W. Cheong, K. V. Ramanujachary, and D. Tran, Phys. Rev. **B48**, 8726 (1997).

[9] K. Asokan, J. C. Jan, K.V. R. Rao, J. W. Chiou, H. M. Tsai, S. Mookerjee, W. F. Pong, M. –H. Tsai, Ravi Kumar, Shahid Husain, and J. P. Srivastava, J. Phys.: Condens. Matter **16**, 3791 (2004).

[10] R. V. Kasowski, M. -H. Tsai, T. N. Rhodin, and D. D. Chambliss, Phys. Rev. **B34**, 2656 (1986).

[11] U. von Barth and L. Hedin, J. Phys. **C5**, 1629 (1972).

[12] O. K. Andersen, Phys. Rev. **B12**, 3060 (1976).

[13] H. J. Monkhorst and J. D. Pack, Phys. Rev. **B13**, 5188 (1976).

[14] J. B. A. A. Elemans, B. V. Laar, K. R. V. D. Veen, and B. O. Looptra, J. Solid State Chem. **3**, 238 (1971).

[15] P. G. Radaelli, G. Iannone, M. Marezio, H. Y. Hwang, S-W. Cheong, J. D. Jorgensen, and D. N. Argyriou, Phys. Rev. **B56**, 8265 (1997).

[16] P. G. Radaelli, D. E. Cox, M. Marezio, S-W. Cheong, Phys. Rev. **B55**, 3015 (1997).

[17] M. Pissas and G. Kallias, Phys. Rev. **B68**, 134414 (2003).

[18] G-Q Gong, C. Canedy, G. Xiao, J. Z. Sun, A. Gupta, and W. J. Gallagher, Appl. Phys. Lett. **67**, 1783 (1995).

[19] P. W. Anderson, Phys. Rev. **79**, 350 (1950).

Brill Academic Publishers
P.O. Box 9000, 2300 PA Leiden,
The Netherlands

Lecture Series on Computer
and Computational Sciences
Volume 4, 2005, pp. 1880-1883

New Enzyme Algorithm, Tikhonov Regularization and Inverse Parabolic Analysis

Xin-She Yang[1]

Received 6 August, 2005; accepted in revised form 21 August, 2005

Abstract:

Many modelling problems require the inverse analysis and optimization of parameters. Inverse problems are usually more complicated than the forward computational modelling such as the finite element analysis. A new enzyme algorithm is developed for the optimization and inverse analysis of parabolic equations based on the effective mechanism of biological enzyme reactions by using the Michaelis-Menton fitness function. The implementation shows how the enzyme algorithm can be applied to solve the function optimization and the inverse parabolic analysis. Simulations also show that the errors of the best estimates for optimization and inverse analysis can be controlled using a given criterion. Furthermore, the nature of the enzyme algorithm suggests that parallel implementation can significantly increase its efficiency.

Keywords: Enzyme algorithm, inverse analysis, parabolic, Tikhonov regularization.

1 Introduction

Biology-derived algorithms form an important part of computational sciences that are essential to many scientific disciplines and engineering applications. Engineering problems with optimization objectives are often difficult and time consuming, and the application of nature or biology inspired algorithms in combination with the conventional optimization methods has been very successful in the last several decades [1-6]. We aim to develop a new enzyme algorithm (EA) for optimization based on the kinetics of biological enzyme actions. Its activity involves the inhibition and cooperativity [7]. We will use the Michaelis-Menton kinetics to derive the fitness function so as to increase the forward reaction rate. The paper will be organized as follows: First, we outline the new enzyme algorithm; then we apply it to the function optimization to test its efficiency. After that, we apply it to the inverse analysis of parabolic equations. Error estimations in the inverse analysis will also be discussed.

2 An Enzyme Algorithm

Enzymes are catalysts that convert proteins and other molecules called substrates into products, while enzymes themselves are not changed by the reaction. Each enzyme contains active site(s) and it works by binding to a given substrate at the active sites. An enzyme (E) combines with a substrate (S) to form an intermediate complex (C), then the complex breaks down into a product releasing the enzyme E in the process. The general formula for a simplified enzyme reaction can be written as

$$S + E \rightleftharpoons C \rightarrow P + E, \tag{1}$$

[1]Corresponding author. Department of Engineering, University of Cambridge, Trumpington Street, Cambridge CB2 1PZ E-mail@ xy227@eng.cam.ac.uk

which usually leads to the Michaelis-Menton equation

$$v = \frac{[S]V_m}{K_m + [S]}, \tag{2}$$

where $[S]$ is the substrate concentration. V_m is the maximum velocity and K_m the Michaelis-Menton constant. This equation suggests that there are a limited number of enzyme sites in a solution of enzymes and substrates. The enzyme algorithm (EA) involves the encoding of an optimization function as arrays of bits or character strings to represent the enzyme reaction rate, the manipulation operations of strings by genetic operators, and the selection according to their fitness in the aim to find a solution to the problem. This is often done by the following procedure: 1) encoding of the objectives or optimization functions; 2) defining a fitness function or selection criterion based on the Michaelis-Menton fitness $v = V_m/(1 + K_m/[S])$; 3) creating a population of individuals including the random generation of enzyme and substrate concentrations; 4) evolution cycle or iterations by evaluating the fitness of all the individuals in the population, creating a new population by selecting the fittest, and iterate using the new population; 5) decoding the results to obtain the solution to the problem.

3 Tikhonov Regularization

In general, inverse problem can be expressed as

$$\mathbf{A}\mathbf{u} = \mathbf{f}, \tag{3}$$

where $\mathbf{u}$ and $\mathbf{f}$ are known, while the aim is to estimate the $n \times n$ stiffness matrix $\mathbf{A}$ with m unknown parameters $\mathbf{k} = (k_1, k_2, \cdots, k_m)$. However, the inverse analysis does not necessarily have a unique solution even if the data are exact and complete. For perturbed data $\mathbf{f}^\epsilon$, $||\mathbf{f} - \mathbf{f}^\epsilon|| \leq \epsilon$, where ϵ is the noise level and $||.||$ is the norm in the Hilbert space. The improved inversion is carried out by the Tikhonov regularization

$$\mathbf{u}_\gamma \rightarrow ||\mathbf{A}\mathbf{u} - \mathbf{f}^\epsilon||^2 + \gamma||\mathbf{u}||^2, \tag{4}$$

which is applicable to a wide range of applications including linear and nonlinear problems. The second term, namely the regularization term, will stabilise the inverse procedure. In any regularization procedure, the choice of regularization parameter γ is very important as it represents a compromise between stability and accuracy. If the parameter γ is too small, errors may be amplified too strongly. If γ is too large, the approximate is not good enough. The choice of γ shall be optimized in such a way that $||\mathbf{A}\mathbf{u}_\gamma - \mathbf{f}^\epsilon|| \sim \epsilon$.

As the matrix $\mathbf{A} = [a_{ij}](i, j = 1, 2, \cdots, n)$ is usually a nonlinear function of the parameters $\mathbf{k} = (k_1, k_2, \cdots, k_m)$. The equation can be rewritten as

$$\mathbf{F}(\mathbf{k}, \mathbf{u}, \mathbf{f}) = \mathbf{A}(\mathbf{k})\mathbf{u}_\gamma - \mathbf{f}^\epsilon = 0. \tag{5}$$

Nonlinear iterations shall be used. We use an adaptive iteration of the Newton-Raphson type

$$\mathbf{R}(\mathbf{k}; \mathbf{k}^n) = \mathbf{F}(\mathbf{k}^n) + \mathbf{J}(\mathbf{k}^n)(\mathbf{k} - \mathbf{k}^n), \quad \mathbf{J}(\mathbf{k}) = \nabla \mathbf{F}, \tag{6}$$

where $\mathbf{J} = [J_{ij}] = \partial F_i/\partial u_j$ is the Jacobian. To find the next approximation $\mathbf{k}^n$ from the $\mathbf{R}(\mathbf{k}^{n+1}; \mathbf{k}^n) = 0$, one has to solve a linear system with $\mathbf{J}$ as the coefficient matrix

$$\mathbf{k}^{n+1} = \mathbf{k}^n - \mathbf{J}^{-1}\mathbf{F}(\mathbf{k}^n). \tag{7}$$

The aim is to obtain a convergent estimation so that $||\mathbf{k}^{n+1} - \mathbf{k}^n|| \leq \epsilon$. The choice of these parameters is essentially an optimization problem and there are various well-established methods for optimizations [1].

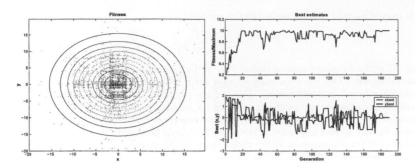

Figure 1: Function optimization for $f(x,y) = 10\exp[-0.01(x^2 + y^2)]$. Michaelis-Menton fitness (left) and best estimates of (x^*, y^*) at different generations (right).

4   Simulation and Results

4.1   Function Optimization

By implementing the enzyme algorithm, we can simulate various optimization problems. In the rest of this section, we will give examples of the new enzyme algorithm and its applications in engineering. As an example, we use the enzyme algorithm to optimize $f(x,y) = 10\exp[-0.01(x^2 + y^2)]$. It has a maximum $f^* = 10$ at $(x^*, y^*) = (0, 0)$. Figure 1 shows the evolution of the Michaelis-Menton fitness (left) and the best estimates of the function maximum at various generations (right). The best estimates are $f^* = 9.986$, $(x^*, y^*) = (0.016, 0.005)$ after 190 generations. The contour corresponds to the function to be optimized, and the scattered diagram corresponds to location of the evolving estimates.

4.2   Parabolic Equations and Inverse Analysis

The inverse initial-value, boundary-value problem (IVBN) is another optimization paradigm where genetic algorithms have been used successfully [5]. We now use the enzyme algorithm to show how the new algorithm can be applied to the parabolic equation and its inverse analysis as an IVBV optimization. On a square plate of unit dimensions, the diffusivity $\kappa(x, y)$ varies with locations (x, y). The equation and its boundary conditions can be written as:

$$\frac{\partial u}{\partial t} = \nabla \cdot [\kappa(x,y)\nabla u], \quad x, y \in \Omega, \quad t > 0 \tag{8}$$

$$u(x, y, 0) = 1, \quad u|_{\partial\Omega} = 0. \tag{9}$$

For a rectangular domain that is discretized as a $N \times M$ grid, and the measurements of values at (x_i, y_j, t_n), $(i = 1, 2, , N; j = 1, 2, ..., M; n = 1, 2, ..., K)$. The data set consists of the measured values at $M \times N$ points at three different times t_1, t_2, t_3. The objective is to invert or estimate the $N \times M$ diffusivity values at the $N \times M$ distinct locations. An error estimate of the Karr's type [5] is defined as

$$E_\kappa = A \frac{\sum_{i=1}^{N} \sum_{j=1}^{M} |\kappa_{i,j}^{known} - \kappa_{i,j}^{predicted}|}{\sum_{i=1}^{N} \sum_{j=1}^{M} |\kappa_{i,j}^{known}|}, \tag{10}$$

where A=100 is just a constant. For the 2-D case on a grid of 16×16 points with a target matrix of diffusivity, after 40,000 random $k(i, j)$ matrices were generated, the best value of the error

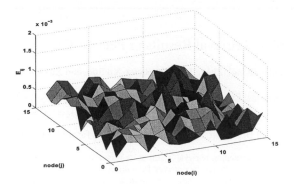

Figure 2: Error estimation in 2-D inverse parabolic analysis.

metrics $E_k = 1.250 \times 10^{-2}$. Figure 2 shows the error metric E_k associated with the best solution determined by the enzyme algorithm. The small values in the error metric imply that the inverse diffusivity matrix is very close to the true diffusivity matrix.

5 Discussion

By using the effective mechanism of enzyme reactions, we have formulated a new enzyme algorithm for optimization and inverse analysis. We used two examples to show how the bio-inspired enzyme algorithm can be applied to solve the optimization problems in function optimization and the inverse analysis of diffusivity matrix as an initial-value, boundary-value problem. Simulations show that the errors of the best estimates are usually within a given accuracy. Furthermore, the present method can be applied to model other optimization and inverse problems.

References

[1] A. Binder , H. W. Engl, C. W. Groestch , A. Neubauer and O. Scherzer, Weakly closed nonlinear operators and parameter identification in parabolic equations by Tikhonov regularization, *Appl. Anal.*, **55**, 215-234 (1994).

[2] K. Deb, *Optimization for Engineering Design: Algorithms and Examples*, Prentice-Hall, New Delhi, (1995).

[3] M. Mitchell, *An Introduction to Genetic Algorithms*, Cambridge, Mass.: MIT Press, (1996).

[4] D. E. Goldberg, *Genetic Algorithms in Search, Optimization and Machine Learning*, Reading, Mass.: Addison Wesley (1989).

[5] C. L. Karr, I. Yakushin, K. Nicolosi, Solving inverse initial-value, boundary-valued problems via genetic algorithms, *Engineering Applications of Artificial Intelligence*,**13**, 625-633 (2000).

[6] A. J. Keane, Genetic algorithm optimization of multi-peak problems: studies in convergence and robustness. Artificial Intelligence in Engineering, **9**,75-83 (1995).

[7] X. S. Yang, Pattern formation in enzyme inhibition and cooperativity with parallel cellular automata, *Parallel Computing*, **30**, 741-751 (2004).

Brill Academic Publishers
P.O. Box 9000, 2300 PA Leiden
The Netherlands

*Lecture Series on Computer
and Computational Sciences*
Volume 4, 2005, pp. 1884-1887

Merging Retrieval Results in Peer-to-Peer Networks

Qian Zhang[1], Zheng Liu, Xia Zhang, Yu Sun, and Xuezhi Wen

National Engineering Research Center for Computer Software
Northeastern University, China.

Received 10 August, 2005; accepted in revised form 15 August, 2005

Abstract: P2P-based information retrieval is still at its infant stage and confronted with many challenges. And how to rank and merge the results retrieved from different peers is one of the most urgent problems. In this paper, we propose a result merging algorithm to address the challenge. Our algorithm has the advantage of being simple, since as input it only uses result lengths and document ranks. First, we rank the results from each neighboring leaf node based on result lengths returned by them. Then we merge results retrieved from directory nodes. Our experiments demonstrate that the new approach is effective.
Keywords: Peer-to-Peer; results merging; information retrieval

Mathematics Subject Classification: Data Mining and Information Retrieval

PACS: 07.05.Kf

1. Introduction

Hybrid P2P architectures include two types of nodes. There are *leaf nodes* that provide information as well as post requests. Leaf nodes can be used to model an information resource. There are also *directory nodes* that do not have content of their own but which provide regionally centralized directory services to the network to improve the routing of information requests.

In hybrid P2P networks, result merging naturally takes place at directory nodes because they already provide directory services to route query requests. The results that need to be merged at a directory node may include not only the results from other directory nodes down the query path, but also the results from the neighboring leaf nodes. Result merging in hybrid P2P networks is not a simple adaptation of existing approaches due to unique characteristics of P2P environments, for example, skewed collection statistics, and higher communication costs. About the related work, in[1], Zhiguo Lu have proposed a deterministic strategy, but it requires full cooperation from participants. [2] also presents a deterministic strategy, but it is just suitable for pure structured P2P systems and need global statistics such as the inverse document frequency, which can only be obtained in completely cooperative environments where each node share its document and corpus statistics. Jie Lu [3] adapted Semi-Supervised Learning (SSL) result merging algorithm to hybrid P2P networks with multiple directory nodes. A directory node that uses SSL learns a query-dependent score normalizing function for each of its neighboring nodes. This function transforms neighbor-specific documents to directory-specific document scores. The training data needed by SSL is pairs of neighbor-specific and directory-specific scores for a set of documents [4]. However, the modified SSL algorithm requires downloading documents from neighboring nodes to server as training data that means higher communication costs. Jie Lu also proposed another result merging algorithm, which is called SESS. SESS is a cooperative algorithm that requires neighboring nodes to provide information for each of their top-ranked documents. We have proposed a novel results merging algorithm (called *RMNC* [5]) for hybrid P2P networks, which does not require cooperation from neighboring nodes.

In short, all algorithms mentioned above require cooperation from participants with the exception of modified SSL and *RMNC*. However, SSL requires downloading documents from neighboring nodes to serve as training data that means higher communication costs [3]. *RMNC* requires directory nodes learn neighboring nodes' content by observing which queries each neighboring node responds to. Given a new query, the directory node computes relevance scores for its neighboring nodes, which

---

[1] Qian Zhang. E-mail: zhangqian@neusoft.com

may be used to merge results from different peers. So the directory nodes are required to record the full query terms of past queries, which is space inefficient.

The remainder of the paper is organized as follows. Section 2 describes our new approach to merge results. Section 3 and Section 4 explain our experimental methodology. Section 5 concludes.

2. Result Merging

Recall that the results that need to be merged at a directory node may include not only the results from the neighboring leaf nodes, but also the results sent back by other directory nodes down the query path. For a given directory node D, we suppose the set of neighboring leaf nodes $LP = \{LP_1, LP_2,...LP_{|LP|}\}$ with goodness scores $LS_{1,q}, LS_{2,q},...LS_{|LP|,q}$ has been selected for query q. We also suppose the set of neighboring directory nodes $SP = \{SP_1, SP_2,...SP_{|SP|}\}$ with goodness scores $SS_{1,q}, SS_{2,q},...SS_{|SP|,q}$ has been selected for query q. Each of the neighboring nodes returns a set of documents, called result set, ranked by their relevance scores with respect to the given query q. $O_{u,v,q}$ denotes the rank of document u in neighboring node v for a given query q. There are two major steps to merge results from neighboring nodes. First, we merge results from neighboring leaf nodes based on result lengths returned by them. The result set after being merged is called $M1$. Then we merge results from neighboring directory nodes together with $M1$.

For the first step, we assume that the number of documents contributed by neighboring leaf node to the final merged result is proportional to its goodness score. The underlying idea is to use weight in order to increase documents scores from peers having scores greater than the average score, and decrease those from any peers having scores less than the average score. The score for the ith neighboring leaf node is determined by:

$$LS_{i,q} = \log\left(1 + \frac{LN_{i,q}}{\sum_{j=1}^{|LP|} LN_{j,q}}\right) \qquad (1)$$

Where: $LN_{i,q}$ is the number of documents retrieved by the ith neighboring leaf node for a given query q. Based on the assumption mentioned above, we define the following local document rank to directory-specific relevance score mapping.

$$LG_{u,i,q} = \frac{LS_{i,q}}{LO_{u,i,q} \times LS_{maxq}} \qquad (2)$$

Where $LO_{u,i,q}$ is the rank of document u in neighboring leaf node i, $LG_{u,i,q}$ is the global relevance score of document u in neighboring leaf node i, and $LS_{max,q}=max\{ LS_{m,q} \mid m=1,2,....|LP|\}$. The new algorithm we propose is inspired in part by results merging algorithm proposed in [6]. The difference is that our algorithm takes into account the distribution of document ranks within each of the neighboring nodes, which could be used to calculate a score for each peer. Directory node D collects results from its neighboring leaf nodes and ranks them according to score $LG_{u,i,q}$. The result set after first being merged is called $M1$.

For the second step, we use the same method discussed at the first step to merge results from different directory nodes including directory node D. The score for the ith directory node is determined by:

$$SS_{i,q} = \log\left(1 + \frac{SN_{i,q}}{\sum_{j=1}^{|SP|} SN_{j,q} + |M1|}\right) \qquad (3)$$

Where: $SN_{i,q}$ is the number of documents retrieved by the ith neighboring directory node for a given query q. Then the score for the uth document from the ith directory node is determined by:

$$SG_{u,i,q} = \frac{SS_{i,q}}{SO_{u,i,q} \times SS_{maxq}} \qquad (4)$$

Where $SO_{u,i,q}$ is the rank of document u in neighboring directory node i, $SG_{u,i,q}$ is the global relevance score of document u in directory node i, and $SS_{max,q}=max\{ SS_{m,q} \mid m=1,2,....|SP|+1\}$. In the final stage, the resulting global relevance scores are sorted in a decreasing order of score $SG_{u,i,q}$, and the final results will be returned to the next directory node down the query path.

3. Tested

We developed one test set, which is a 6 gigabyte, 1.1 million document set downloaded from Internet. The test data was divided into 2,400 collections based on document URLs. The HTML title fields of the documents were used as document names. Each of the 2,400 collections defined a leaf node in a hybrid P2P network.

Prior research shows that 85% of the queries posted at web search engines have 3 or less query terms [7], so for most documents, we only extract a few key terms as queries. Here we extract key terms from document names. Most queries had 2-3 terms and no query had more than 7 terms. 12,000 queries were randomly selected from the automatically-generated queries to be used in our experiments.

We use the Power-Laws method described by [8] to generate the connections between directory nodes. A directory node had on average 5 directory node neighbors and 16 leaf node neighbors. As in the Gnutella protocol [9], each message had a time-to-live (TTL) field that determined the maximum number of times it could be relayed in the network. In our experiment, the initial TTL was set to 5 for query messages routed to directory nodes. When the query messages were routed to leaf nodes by directory nodes, the TTL was set to 1 since leaf nodes were not supposed to further route query messages.

4. Evaluation Methodology

Recall and Precision are two classical metrics to evaluate the information retrieval technology. Here we use 11-point recall-precision to evaluate the merged retrieved results from the hybrid P2P networks.

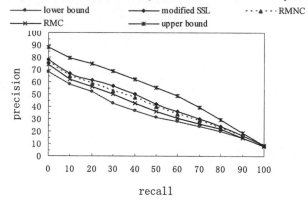

Figure 1: The 11-Point Recall-Precision Curves for Different Result Merging Algorithms

Figure 1 shows the 11-point recall-precision curves for different result merging algorithms when 11,000 queries have been randomly issued. The lower bound was generated by directly merging documents from different neighboring nodes using their local document scores. The upper bound was generated by the documents that were returned by retrieval in the hybrid P2P networks and sorted by their corresponding scores in the retrieval results from the whole single collection. The modified SSL method improved the average precision by 23.7% compared with the lower bound, and the *RMNC* algorithm improved the average precision by 21.5% compared with the lower bound. Our *RMC* algorithm had near optimal performance compared with modified SSL algorithm and *RMNC* algorithm, which improved the average precision by 18.3% compared with the lower bound. However, modified SSL requires downloading large documents from neighboring nodes to serve as training data, which means higher communication costs. Compared with *RMNC* algorithm, *RMC* had little performance loss in average precision, which is the value of recording query terms of past queries that neighboring node has responded to.

5. Summary and Conclusions

In this paper, we propose a novel result merging algorithm used by directory nodes in hybrid P2P networks. Our algorithm does not require cooperation from neighboring nodes, which makes it different from other algorithms except modified SSL and *RMNC* algorithm. However, modified SSL can induce higher communication costs, and *RMNC* algorithm is space inefficient for recording full query terms of past queries. Experimental results demonstrate that our algorithm is effective and has near optimal performance compared with *RMNC* algorithm and modified SSL.

References

[1] Zhiguo Lu1, Bo Ling1, Weining Qian1,etc , A Distributed Ranking Strategy in Peer-to-Peer Based Information Retrieval Systems. *The Sixth Asia Pacific Web Conference*, (2004), 279-284.

[2] M. M. Chunqiang Tang, Zhichen Xu. Peersearch: Efficient information retrieval in peer-to-peer networks. In Proceedings of HotNets-1, *ACM SIGCOMM*,(2002).

[3] J. Lu and J. Callan. Merging retrieval results in hierarchical peer-to-peer networks (poster description). In *Proceedings of the 27th Annual International ACM SIGIR Conference on Research and Development in Information Retrieval*. (2004).

[4] L.Si and J.Callan. A semi-supervised learning approach to merging search engine results. *ACM Transactions on Information Systems*. Vol. 21, No. 4, (2003), 457–491.

[5] Qian Zhang, Xia Zhang, Zheng Liu. A Novel Ranking Strategy in Hybrid Peer-to-Peer Networks, *In: Proceedings of 6th International Conference on Web-Age Information Management*,(2005).

[6] Yves Rasolofo, Faiza Abbaci, Jacques Savoy. Approaches to Collection Selection and Results Merging for Distributed Information Retrieval, *In: Proceedings of CIKM2001*,ACM Press, (2001),191-198.

[7] M. Jansen, A. Spink and T. Saracevic. Real Life, real users, and real needs: A study and analysis of user queries on the web. *Information Processing and Management*, 36(2),(2000),207-227.

[8] C.R. Palmer and J. G. Steffan. Generating Network Topologies That Obey Power Laws. *In Proceedings of Global Internet Symposium*, (2000).

[9] The Gnutella protocol specification v0.6. http://rfcgnutella.sourceforge.net.

Brill Academic Publishers
P.O. Box 9000, 2300 PA Leiden
The Netherlands

*Lecture Series on Computer
and Computational Sciences*
Volume 4, 2005, pp. 1888-1891

Universal Attributes of Evolution of Dissipative Structures Emerging Under the Effect of Thermal Shock and Ultrashort Pulses of Laser Radiation

V.T.Punin, A.Ya. Uchaev[1], A.P.Morovov, A.S.Konkin, N.I.Sel'chenkova, L.A.Platonova, E.V.Kosheleva, N.A.Yukina

Russian Federal Nuclear Center – VNIIEF,
Russia, 607188, Nizhni Novgorod region, Sarov, Mira avenue 37,

Received 5 August, 2005: accepted in revised form 25 August, 2005

Abstract: A new method of studying the structural materials dynamic failure process allowing a significant expansion of the area of nonequilibrium matter states is application of laser radiation ultrashort pulses. Employment of lasers with picosecond and, especially, femtosecond pulse duration opens up unique possibilities for studying matter nonequilibrium states and expands the field under study to the earlier inaccessible ultrashort negative pressure range what is significant for determining common regularities of solid bodies' behavior under the effect of powerful pulses of penetrating radiation in the wide time interval. In an effort to determine the universal attributes of metals behavior in the dynamic failure phenomenon there was determined the mass of percolation clusters of inner surfaces roughness of failure centers on the nanoscale level, arising as a result of effect of ultrashort pulses of laser radiation (UPLR), high-current beams of relativistic electrons.

It is shown that the process of metals dynamic failure under the effect of UPLR and under the effect of high-current beams of relativistic electrons, in the studied metals is similar on different scale levels: beginning from nanolevel and up to the scale of the failing body.

Keywords: high-current beams of relativistic electrons, ultrashort pulses of laser radiation, dynamic failure dissipative structures, mass of percolation clusters.

As a result of earlier performed studies it is shown that at amplitudes of pulsed pressure of units-hundreds kilobar in the longevity range $t \sim 10^{-6} \div 10^{-10}$ s the evolution of micro- and mesoscopic defects, failure center cascades in the dynamic failure phenomenon is determining in common regularities of invariant behavior of solid bodies under the effect of thermal shock caused by powerful penetrating radiation pulses (initial temperature range $T_0 \sim 4K \div 0.8 T_{melt.}$, energy input rate $dT/dt \sim 10^6 \div 10^{12}$ K/s, absorbed energy density $10 \div 10^4$ J/g) [1-5].

A new method of studying the structural materials dynamic failure process allowing a significant expansion of the area of nonequilibrium matter states is application of laser radiation ultrashort pulses. Employment of lasers with picosecond and, especially, femtosecond pulse duration opens up unique possibilities for studying matter nonequilibrium states and expands the field under study to the earlier inaccessible ultrashort negative pressure range what is significant for determining common regularities of solid bodies' behavior under the effect of powerful pulses of penetrating radiation in the wide time interval [4, 5].

Metal response on laser radiation can be conventionally subdivided into three modes:

1. Duration of a single pulse of laser radiation is $\tau_i > \tau_z$, ($\tau_z \sim L/c$ – hydrodynamic times, where c – sound speed, and L – size of the order of the region under laser radiation action). If radiation rate J is sufficient there is formed a pit. No shock-wave phenomena are observed.

2. $\tau_i \sim \tau_z$ ($\tau_i \sim 10^{-10} \div 10^{-11} s$). At sufficient rate J a pit is formed, and destructive processes emerge due to the shock-wave phenomena and the thermal shock phenomenon.

3. $\tau_i \sim 10^{-13} \div 10^{-14} s$. At the particular J laser radiation rate due to Stoletov light pressure the electrons exit the region of laser radiation action what can lead to the ion core Coulomb explosion.

Characteristic electron heat time by laser radiation can be estimated using an expression for the force affecting the electrons from the side of radiation:

---

[1] Corresponding author. Head of Department RFNC-VNIIEF. E-mail: uchaev@expd.vniief.ru

$$F \sim \varepsilon_\gamma \cdot v \cdot \sigma_1/c,$$

where σ_1 is a cross-section of bremsstrahlung electron radiation, ε_γ – quantum energy, v – electron velocity, $dv/dt = -F/m*$, where $m*$ – effective electron mass. Taking that $F/m* \sim v/\tau_1$ (τ_1 – electron heat time), we have

$$\tau_1 = 1/\left(z^2 \cdot \alpha\right) \cdot \left(m*/m_e\right)^2 \cdot \left(kT_e/\left(\varepsilon_\gamma \cdot \sigma_T \cdot c\right)\right),$$

where σ_T – Thomson cross-section, z – order number, and $\alpha = e^2/\hbar c$. Electrons transfer energy to photons and the crystal lattice. The temperature field $T(x,y,z)$ in the lattice can be described by the heat conduction equation.

$$\frac{dT}{dt} = \chi \Delta T.$$

This equation is correct in the case of incompressible medium at times $\tau \leq 10^{-12}s$.

When the irradiation pulse effect ends one can accept that radiation is thermalized with electrons $T_e \leq T_\gamma$ (T_γ – radiation temperature).

In this case the crystal lattice will be subjected to compressive pressure

$$P_\gamma \sim \Gamma \rho c_p \int_{T_{n.con}}^{T_\gamma} dT,$$

which leads to the local phenomenon of thermal shock (G - Gruneisen parameter).

Also this region experiences a pulsed pressure $P \sim \rho c v_m$ due to evaporation of a part of metal, where v_m - mass velocity of matter adjacent to the region under the laser radiation action.

After the laser radiation pulse action ends the negative pressure emerges due to load relief in the non-equilibrium region equaling to

$$P \sim -\Gamma \rho c_p T_v + \rho c v_m.$$

At pulse duration $\tau_i \sim 10^{-10} \div 10^{-11}s$ in metals destructive processes emerge due to shock-wave phenomena and the phenomenon of thermal shock.

To study dynamic failure in the sub-nanosecond ($10^{-9} \div 10^{-11}$ s) longevity range, there were applied ultrashort pulses of laser radiation (UPLR) with laser radiation power density up to $J \sim 10^{14}$ W/cm^2 [4, 5].

To obtain quantitative characteristics of dissipative structures arising in the course of dynamic failure, there were studied metallographic sections of structural material loaded samples exposed to relativistic electron high-current beams and UPLR.

There was studied a structure of the loaded metal samples exposed to high-current relativistic electron beams ($t \sim 10^{-6} \div 10^{-10}$ s) as well as to UPLR ($t \sim 10^{-9} \div 10^{-11}$ s) on nanoscale level with the aid of scanning tunnel microscope [6].

With the aid of specialized mathematical program package of interactive image analysis system (IIAS) [1] there were processed images of roughness of failure centers inner surfaces.

It is shown that roughness of failure centers inner surfaces at two above-listed types of action on the nanoscale level is a percolation cluster.

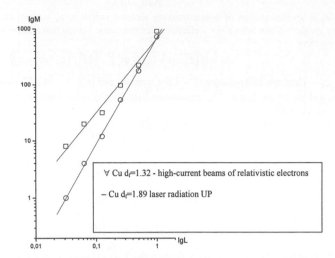

Figure 1: – Dependences of mass of percolation clusters on dimension of lattice L parameter, covering the cluster on the nanoscale level, under the effect of UPLR, high-current beams of relativistic electrons.

In an effort to determine the universal attributes of metals behavior in the dynamic failure phenomenon there was determined the mass of percolation clusters of inner surfaces roughness of failure centers on the nanoscale level, arising as a result of effect of UPLR, high-current beams of relativistic electrons.

Results presented in figure 1 show that roughness of failure center inner surfaces at two above-described types of effect on the nanoscale level is a percolation cluster, since the mass of structural elements grows depending on the size of the lattice parameter covering the cluster (failure surface).

Thus, it is shown that the process of metals dynamic failure under the effect of UPLR and under the effect of high-current beams of relativistic electrons, in the studied metals is similar on nanoscale level.

It is shown that the cascade of dissipative structures at the macro-failure threshold when there takes place a topology phase transition – emergence of connectivity in the system of dissipative structures, is a percolation cluster what determines the possibility for application of the universal apparatus of critical phenomena theory to the quantitative description of the process and determines metal universal behavior in the dynamic failure phenomenon. This allows prediction of the unstudied materials behavior in the mentioned time-temperature ranges.

References

[1] R. I. Il'kaev, A. Я. Uchaev, S. A. Novikov, N.I. Zavada, L. A. Platonova, N. I. Sel'chenkova. Universal metal properties in the dynamic failure phenomenon // DAN, 2002, vol. 384, № 3, P. 328-333.

[2] R. I. Il'kaev, V.T.Punin, A. Ya. Uchaev, S. A. Novikov, E.V. Kosheleva, L. A. Platonova, N. I. Sel'chenkova, N.A. Yukina. Time regularities of metals dynamic failure process conditioned by hierarchic properties of dissipative structures – failure centers cascades //
DAN, 2003, vol. 393, № 3.

[3] E.K.Bonjushkin, S.A.Novikov, A.Ya.Uchaev Kinetics of dynamic metal failure in the mode of pulsed volume heating up. Scientific edition. Sarov, RFNC-VNIIEF, 1998, 275 p.

[4] Uchaev A.Ya., Il`kaev R.I., Punin V.T., Novikov S.A., Sel`chenkova N.I. Metals behavior in the extreme conditions. ABSTRACTS VII Russian-Chinese Symposium New Materials and

Technologies. September 13-18 2003. Agoy, Krasnodar region, Russia. Moscow – Agoy, 2003. P. 16-17.

[5]   Scaling properties of dissipative structures formed on the nanolevel in metals under the action of ultrashort pulses of laser radiation in subnanosecond longevity range A.Ya. Uchaev, R.I. Il'kaev, V.T. Punin, A.P. Morovov, S.A. Novikov, N.I. Sel'chenkova, N.A. Yukina, V.V. Zhmailo Lecture Series on Computer and Computational Sciences. International Conference of Computational Methods in Sciences and Engineering 2004 (ICCMSE 2004), V.1, 2004, pp.523-526.

[6] A.L.Suvorov Modern methods of study of condensed material in the application developments of MinAtom RF. Employment of advances of fundamental studies in nuclear technologies. Collection of reports of the 4-th scientific practical conference of MinAtom RF. M.: MinAtom Russia. P 34-49.

Brill Academic Publishers
P.O. Box 9000, 2300 PA Leiden
The Netherlands

*Lecture Series on Computer
and Computational Sciences*
Volume 4, 2005, pp. 1892-1895

Universal Attributes and Possibility to Predict Metals Behavior Including Uranium Under Extreme Conditions

R.I.Il'kaev, V.T.Punin, S.A.Novikov, A.Ya. Uchaev[1], N.I.Sel'chenkova

Russian Federal Nuclear Center – VNIIEF,
Russia, 607188, Nizhni Novgorod region, Sarov, Mira avenue 37,

Received 5 August, 2005: accepted in revised form 25 August, 2005

Abstract: Application of apparatus of critical phenomena and theory of second-order phase transitions for quantitative description of dynamic failure process at the final stage allowed determination of universal properties of metals behavior in the dynamic failure phenomenon conditioned by self-organization and instability in dissipative structures, for example, failure center cascades.

A dynamic invariant relating the loading parameters with energy parameters of crystal lattice was determined. The invariant has similar values for all studied metals.

With the aid of earlier determined invariants of the process of dynamic metals failure there were predicted failure surfaces for nature uranium, determining the limit of failure region, in coordinates: longevity t ($t \sim 10^{-6} \div 10^{-10}$ s), pressure P ($P \sim 1 \div 10\ GPa$), temperature T (T $\sim 4K \div 0.8T_{melt}$).

Keywords: high-current beams of relativistic electrons, dissipative structures, hierarchy of structural levels, ferromagnetic Ising model, the model of lattice gas, apparatus of critical phenomena and theory of second-order phase transitions

The dynamic failure phenomenon caused by powerful pulse energy matter action is related to the ultimate capabilities of current technology and unique scientific facilities: spaceships - effect of micro-meteorites on the spaceship skin; resistance of the first wall of the thermonuclear reactor; resource of the pulse reactor core; supercollider's inner elements resistance. That is why the knowledge of matter behavior under extreme conditions is important [1-5].

Application of classic research methods such as, for example, classic kinetics apparatus for adequate description of nonequilibrium, non-linear processes is problematic. Priority methods for description of unique phenomenon of dynamic failure of metals involving uranium (natural) are methods of non-linear physics that allow determining of universal attributes of nonequilibrium systems evolution conditioned by collective effects, phenomena of self-organization in the dissipative structures. These methods simplify quantitative description of systems exposed, for example, to powerful pulses of penetrating rays.

Modern literature shows analogy between the ferromagnetic Ising model and the model of lattice gas (similarity of functions of distribution of large canonic ensemble). Magnetic phase transitions refer to second order phase transitions which are characterized by universal attributes of behavior near the critical point. The function of Ising model distribution is the following

$$Z_i = \sum_S exp\left(\frac{J}{kT}\sum_{i,j}s_i s_j + \frac{\mu H}{kT}\sum_{i=1}^{N}s_i\right), \tag{1}$$

where J - exchange energy; μ - Bohr magneton; H – field intensity.
For large canonical Gibbs distribution

$$Z = \sum_{N=0}^{\infty}z^n exp(\frac{J}{kT}\sum_{i,j}e_i e_j) \tag{2}$$

---
[1] Corresponding author. Head of Department RFNC-VNIIEF. E-mail: uchaev@expd.vniief.ru

$z = exp\dfrac{\mu}{kT}$, μ - chemical potential.

After small transformation we have

$$Z = \sum_{e} exp\left(lnz \sum_{i=1}^{n} e_i + \frac{J}{kT} \sum_{i,j} e_i e_j \right) \tag{3}$$

After substitution $e_i e_j = \frac{1}{4} s_i s_j + \frac{1}{4}(s_i + s_j) + \frac{1}{4}$, we receive expression (1).

From this follows an analogy in percolation cluster and ferromagnetic behavior.

Earlier the paper authors [2] demonstrated that in the dynamic longevity range the thermodynamic potential – enthalpy is a parameter which controls the dynamic failure process. Relation between density of absorbed energy and energy parameters of crystal lattice (enthalpy and phase transition heat) is an invariant of metals behavior with respect to the external action. In this case the ultimate density of strain energy of the local volume of failing body (near the failure centers) can be taken to be equal to, see, for instance, [2]

$$Q^{max} = \int_{T'}^{T_m} c_p^{(T)} dT + L_m ,$$

where c_p – heat capacity, T' - temperature, below which one can neglect contribution of atomic temperature oscillation to density of internal energy, T_m – melting temperature, L_m – latent crystal lattice melting heat. Expression $H = \int_{T'}^{T_m} c_p dT$ - is enthalpy.

One can write down the potential energy of n-level system:
$$U_2 = U_2 (\rho_1, U_1)$$
$$U_3 = U_3 (\rho_2, U_2)$$
$$\dots \dotsm \dots$$
$$U_n = U_n (\rho_{n-1}, U_{n-1}),$$
ρ_n – density of structural units on the n-level.
The expression for the total energy is of the form:

$$U_{tot} = \sum_{i} U_i \cdot \rho_i$$

For the maximum density of energy E, which can be accumulated by crystal lattice instability zones [3] (without changing aggregative state), has the value $E = H + L_m$, we obtain the expression:

$$\sum_{i} U_i \cdot \rho_i = \int_{T_0}^{T_m} c_p dT + L_m \tag{4}$$

This expression allows a significant simplification of description of behavior of dynamically loaded solid body (in the dynamic longevity range).

Earlier a similar model was successfully used for numerical simulation of percolation clusters emulating failure centers cascade in structural materials under pulse effect of high-current beams of relativistic electrons [1-5].

Application of apparatus of critical phenomena and theory of second-order phase transitions for quantitative description of dynamic failure process at the final stage allowed determination of universal properties of metals behavior in the dynamic failure phenomenon conditioned by self-organization and instability in dissipative structures, for example, failure center cascades.

A system approach (involving micro-, meso- and macro-levels of failure) to the study of the process of dynamic failure allowed revealing of universal attributes of incipient structures' evolution which can be studied as a form of self-similarity in the behavior of metals under impulsive loading. As an example, fig.1 gives masses of percolation clusters M of failure surfaces as the lattice L scale of a number of metals increases under the effect of high-current beams of relativistic electrons.

Results of processing have shown that the fractal dimensions D obtained at processing of a number of metals failure surfaces (see figure 1) is close to the data given in the modern literature obtained by calculations $D_p = 1,89$. Fractal dimensions (see fig. 1) of mass of failure surface percolation clusters as the lattice scale of a number of metals increases has the value $D \sim 1,8$.

It is shown that a failing body in the dynamic longevity range is a hierarchical, nonergodic system [2].

The foregoing proves that it is possible to obtain quantitative characteristics of metals behavior under extreme conditions on the macro-level on the basis of behavior regularities on the mesolevel as well as grounds the possibility for predicting the unstudied metals behavior under extreme conditions.

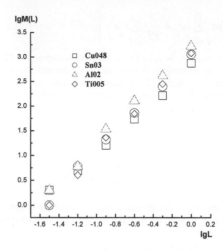

Figure 1: Masses of percolation clusters of failure surfaces as the lattice scale of a number of metals increases under the effect of high-current beams of relativistic electrons

A dynamic invariant relating the loading parameters with energy parameters of crystal lattice was determined. The invariant has similar values for all studied metals [1].

With the aid of earlier determined invariants of the process of dynamic metals failure there were predicted failure surfaces for nature uranium, determining the limit of failure region, in coordinates: longevity t ($t \sim 10^{-6} \div 10^{-10}$ s), pressure P ($P \sim 1 \div 10 \ GPa$), temperature T (T $\sim 4K \div 0.8T_{melt}$).

References

[1] R. I. Il'kaev, A. Я. Uchaev, S. A. Novikov, N.I. Zavada, L. A. Platonova, N. I. Sel'chenkova. *Universal metal properties in the dynamic failure phenomenon* // DAN, 2002, vol. 384, № 3, P. 328-333.

[2] R. I. Il'kaev, V.T.Punin, A. Ya. Uchaev, S. A. Novikov, E.V. Kosheleva, L. A. Platonova, N. I. Sel'chenkova, N.A. Yukina. *Time regularities of metals dynamic failure process conditioned by hierarchic properties of dissipative structures – failure centers cascades //* DAN, 2003, vol. 393, № 3.

[3] E.K.Bonjushkin, S.A.Novikov, A.Ya.Uchaev *Kinetics of dynamic metal failure in the mode of pulsed volume heating up.* Scientific edition. Edited by R.I.Il'kaev, Sarov, RFNC-VNIIEF, 1998, 275 p.

[4] A.Ya.Uchaev, N.I.Zavada, R.I.Il'kaev, E.V.Kosheleva, S.A.Novikov, L.A.Platonova, V.T.Punin, N.I.Sel'chenkova, N.A.Yukina. *Invariants in the phenomenon of metal dynamic failure//* RFNC-VNIIEF transactions: Scientific research edition/ Chief editor R.I.Il'kaev, - Sarov: RFNC-VNIIEF. Issue 3.-2002. Transactions of VNIIEF.

[5] *Universal properties of metals behavior in the dynamic failure phenomenon in a wide longevity range under the action of powerful pulses of penetrating radiation* A.Ya. Uchaev,

R.I. Il'kaev, V.T. Punin, S.A.Novikov, N.I. Zavada, L.A. Platonova, N.I. Sel'chenkova Lecture Series on Computer and Computational Sciences. International Conference of Computational Methods in Sciences and Engineering 2004 (ICCMSE 2004), V.1, 2004, pp.527-529.

Brill Academic Publishers
P.O. Box 9000, 2300 PA Leiden
The Netherlands

Lecture Series on Computer
and Computational Sciences
Volume 4, 2005, pp. 1896-1898

On New Methods of Obtaining Quantitative Characteristics of Metals Behavior Under Extreme Conditions

V.T.Punin, A.Ya. Uchaev[1], N.I.Sel'chenkova, E.V.Kosheleva, L.A.Platonova, N.A.Yukina

Russian Federal Nuclear Center – VNIIEF,
Russia, 607188, Nizhni Novgorod region,
Sarov, Mira avenue 37

Received 5 August, 2005: accepted in revised form 25 August, 2005

Abstract: In the paper there was applied a complex approach to determining of quantitative characteristics of dissipative structures originating in metals exposed to pulse action on different scale levels.

The complex approach is based on application of: methods of fractal theory and percolation theory, methods of quantitative fractography with application of new developed specialized package of mathematical programs – interactive image analysis system (IIAS) allowing identification and obtaining of quantitative characteristics of studied structure objects. Studies of such a kind are automated methods of search of quantitative characteristics of fractal systems.

Keywords: high-current beams of relativistic electrons, ultrashort pulses of laser radiation, shock-wave loading, dissipative structures, hierarchy of structural levels, nanoscale level, mesolevel I, II, macrolevel

In the paper there was applied a complex approach to determining of quantitative characteristics of dissipative structures originating in metals exposed to pulse action on different scale levels.

The complex approach is based on application of: methods of fractal theory and percolation theory, methods of quantitative fractography with application of new developed specialized package of mathematical programs – interactive image analysis system (IIAS) allowing identification and obtaining of quantitative characteristics of studied structure objects. Studies of such a kind are automated methods of search of quantitative characteristics of fractal systems [1-4].

It is shown that failure in the dynamic longevity range on different scale levels at different time-amplitude characteristics of the external action is represented by time-spatial evolution of the hierarchical organized system what allows description of the process of dynamic failure within the limits of the unique concept. Specific features of such time-spatial hierarchies are collective effects in the system behavior – sample failure and self-organization processes.

According to the modern approaches on the concept of material structure under conditions of external pulse action the material properties are determined by structure formation processes. These processes, as a rule, are of irreversible and nonequilibrium character and are described by mathematical apparatus of non-linear physics.

At external action on the system stable regular stochastically self-similar structures on the different scale levels can be formed on different scale levels.

One of promising approaches for solving the problem of quantitative description of structures arising, for example, at high-intense external action on metals is fractal geometry. It should be noted that self-similar natural fractals are observed in the limited scale interval, where power relations of the kind $N(D) \sim D^{-\alpha}$ are true, where $N(D)$ is a number of structures of size D, $\alpha > 1$.

Employment of new approaches of multi-fractal and fractal analysis together with the methods of quantitative fractography using mathematical package of programs of interactive image analysis system allow quantitative characteristics of dissipative structures formed at different kinds of pulse action on different scale levels.

To obtain quantitative characteristics of dissipative structures emerging in the course of dynamic failure process, there were studied metallographic sections of loaded samples of a number of metals

---
[1] Corresponding author. Head of Department RFNC-VNIIEF. E-mail: uchaev@expd.vniief.ru

under shock-wave loading ($t \sim 10^{-6} \div 10^{-8}$ s) and exposed to high-current beams of relativistic electrons ($t \sim 10^{-6} \div 10^{-10}$ s) as well as to ultrashort pulses of laser radiation ($t \sim 10^{-9} \div 10^{-11}$ s).

There were mathematically processed dissipative structures formed under different types of pulse action such as:

- Roughness of inner surfaces of failure centers (nanoscale level – study conducted with the aid of a scanning tunnel microscope [5]);
- Cascade of slipbands of crystal lattice (mesolevel I);
- Clusters – failure centers cascade (mesolevel II);
- Failure surface roughness (macrolevel).

Above-listed quantitative characteristics of dissipative structures were obtained with the aid of the new developed package of mathematical programs - IIAS.

There is shown an analogy of the dynamic failure process under the effect of ultrashort pulses of laser radiation and under shock-wave loading which manifests itself in the similarity of off-orientation of crystal lattice slipbands.

It is shown that roughness of inner surfaces of failure centers at three above-listed types of action on the nanoscale level is a percolation cluster.

As an example, figure 1 gives masses of percolation clusters of roughness of failure center inner surfaces on the nanoscale level, occurring as a result of effect of laser radiation ultrashort pulses (UP), high-current beams of relativistic electrons and under shock-wave loading.

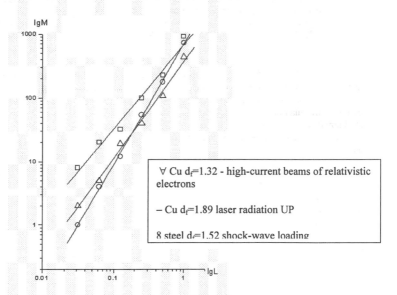

Figure 1: Dependences of mass of percolation clusters on the lattice parameter L size covering the cluster on nanoscale level under the effect of laser radiation UP, high-current beams of relativistic electrons and under shock-wave loading

Results presented in fig. 1 show that roughness of inner surfaces of failure centers at three above–mentioned types of action on nanoscale level is a percolation cluster, for structural elements mass grows depending on the lattice parameter size. It is shown that the dynamic metals failure revealing itself on different scale levels as different structural formations is a nonequilibrium process. Nonequilibrium processes are characterized by fractal structures' formation.

Dissipative structures show as a formation: roughness of inner surfaces of failure centers on the nanolevel, cascade of crystal lattice slipbands and failure center cascade in the metallographic sections which are perpendicular and parallel to the failure surface (mesolevels I and II) and roughness of the failure surface on the microlevel, are fractal clusters.

As a result of conducted studies it was shown that the character of behavior of a number of materials on four levels: nanolevel; mesolevel I, II; macrolevel has common properties when exposed to high-current beams of relativistic electrons, laser radiation UP and shock-wave loading conditioned by collective effects, phenomenon of self-organization in different dissipative structures which possess fractal properties, i.e. they are scale – invariant.

Thus, the new developed mathematical apparatus allowed determination of time-space scalings of dissipative structures emerged in metals under pulse action.

References

[1] R. I. Il'kaev, A. Я. Uchaev, S. A. Novikov, N.I. Zavada, L. A. Platonova, N. I. Sel'chenkova. *Universal metal properties in the dynamic failure phenomenon* // DAN, 2002, vol. 384, № 3, P. 328-333.

[2] R. I. Il'kaev, V.T.Punin, A. Ya. Uchaev, S. A. Novikov, E.V. Kosheleva, L. A. Platonova, N. I. Sel'chenkova, N.A. Yukina. *Time regularities of metals dynamic failure process conditioned by hierarchic properties of dissipative structures – failure centers cascades* // DAN, 2003, vol. 393, № 3.

[3] E.K.Bonjushkin, S.A.Novikov, A.Ya.Uchaev *Kinetics of dynamic metal failure in the mode of pulsed volume heating up.* Scientific edition. Sarov, RFNC-VNIIEF, 1998, 275 p.

[4] *Scaling properties of dissipative structures formed on the nanolevel in metals under the action of ultrashort pulses of laser radiation in subnanosecond longevity range* A.Ya. Uchaev, R.I. Il'kaev, V.T. Punin, A.P. Morovov, S.A. Novikov, N.I. Sel'chenkova, N.A. Yukina, V.V. Zhmailo Lecture Series on Computer and Computational Sciences. International Conference of Computational Methods in Sciences and Engineering 2004 (ICCMSE 2004), V.1, 2004, pp.523-526.

[5] A.L.Suvorov Modern methods of study of condensed material in the application developments of MinAtom RF. *Employment of advances of fundamental studies in nuclear technologies.* Collection of reports of the 4-th scientific practical conference of MinAtom RF. M.: MinAtom Russia. P 34-49.

Brill Academic Publishers
P.O. Box 9000, 2300 PA Leiden,
The Netherlands

Lecture Series on Computer
and Computational Sciences
Volume 1, 2005, pp. 1899-1891

Animation Contents for Mobile Based on the JAVA

Sung-Soo Hong[1], K.Kase[2] and A. Makinouchi[3]

Integrated VCAD System Research Program,
RIKEN (The Institute of Physical and Chemical Research),
Wako, Saitama 351-0198, Japan

Received 7 July, 2005; accepted in revised form 31 August, 2005

Abstract: The Korean animation has enjoyed the brisk formation and establishment of its world-class infra for the last several years without unified titles or concepts, under the name of a national strategic project in the age of digital image. It also enjoys its new evaluation as digital animation that it's one of the greatest money making business in the non-education and frivolous culture and has the closest relations with the modern time, For instance, in Cyber Clam Museum, 1000 gesture contents, the visual processes were used to design a screen with a realistic image and create an animation that makes possible show at 360 and every such transformation as translation, rotation, and scaling can be applied in the image interactively for the convenient and effective viewing. Also they are applied to 1000 of character movement expression, natural phenomena, dinosaur movement expression, 1000 of virtual shellfish museums, and 600 of virtual fossil museums.

1 Introduction

One may naturally conjures up the term "mobile contents" when hearing IT sector of 2005. Mobile Contents refers to as a personal machine, that is mobile and portable, including PDA, PCS, notebook PC, Guam PC, handset, messenger, cellular phone, mobile phone. In particular, the scale of domestic cellular phone market is around 35 million, while overseas market is over 300 million, both are formidable markets. With this, they are turning their eyes to mobile game from online game, which used to be subject of their investment.[3]

Mobile Contents remains the nascent stage of mobile contents phenomenon including "phone ring tone", "character", "download service" and "stock trade", but it is expected to focus on upgrading "mobile game", "entertainment", "tele course", "VOD and "intranet". In light of the potential for mobile market, it is a very promising sector. [2] Mobile Contents has many strong points such as easy access to service, security through personal handset and easy combination of mobile and wireless Internet.[7]

2 Mobile Platform

VM (Virtual Machine), hardware-independent form without influenced by the kind of handset or operating system, is a kind of middle ware that can be equipped only by revising software. It refers to as basic technology that is operated on the wireless handset. The technology began to be practically used by mobile telecommunication providers in Korea, along with Japan, faster than

[1]Corresponding author. E-mail: sshong@office.hoseo.ac.kr
[2]E-mail: scchoseo@paran.com
[3]E-mail: chiww@hanmail.net

any other countries. Today, five standard handset platforms are rampant in Korea alone. VM is essential platform that is mainly equipped in colour mobile phone, which is earning popularity fast. Therefore, each mobile telecommunication provider selects a different platform and provides the service.

It was LGT's Java Station that first provided VM service. Java is most widely used on fixed-line Internet, and has environment for supporting development of various application programs. It was followed by GVM (General Virtual Machine), developed by Sinjisoft, that began to provide the service in SKT. GVM, based on C language, provided environment for faster calculation and image processing than Java.[4] KTF introduced the service using MAP (Mobile Application S/W Plug-in) of Mobiletop, while map, based on C language like GVM, put its base on Microsoft's visual studio. XCE, which supports n-Top wizards environment of SKT, provides environment that is compatible with standard DLDC /MIDP of Son. In addition, it is developing self-based WR-TOP, based on Java, in SKT.

Mobile phone has many constraints such as small memory and slow access. However, it is about to have colour, fixed-line network and speed, similar to PC. Mobile industry is the core business that will lead the culture of the 21th century.

3 Animation Contents Design and Implementation

3.1 Movement Expression Animation Contents Design

The model I intended to suggest in this paper is based on the model[1] suggested by J. Latta. This model consists of human, virtual environment and interface connecting human and virtual environment.

The ultimate aim that virtual world animation system seeks is to make users not feel the difference between virtual objects and real objects by making them feel the virtual object more real in color and shape. Display is that users virtually see real objects through monitor. Sensing module is to sense active human movements and expression of opinions of human beings, referring to various input/output units, digital camera and sensor. Information input through this is processed in virtual perception module, and extracts intension of users. This module connects physical sensor and logical sensor, and environment of virtual world and range form of interaction are determined in accordance with user's intension. This is determined in interaction module.

Simulation module is where determined environment and interaction are practically operated. It, of course, operates basic movements even when the user does not set up interaction.

Rendering module is to draw the changed virtual world. It provides the drawing to monitor by zoom-in and zoom-out, and changing direction of real object.

3.2 Algorithm for Animation

The animation algorithm consists of an Init processor, a Display processor, a Zoom processor, and a Speed processor and all processors run by calling a run processor. The animation is done by creating a file from the image to be animated taken by a camera and by utilizing a Java program that is saved in the memory address of an image. In the run processor, a thread runs first to determine a rotation/stop and left/right rotation situation and then the animation is performed sequentially if the satisfaction is made. The reverse operation is performed in the case that the nix is reducing. Then, the Speed processor and Display processor are called.

3.3 Animation Contents Wireless Internet model

I seek to suggest and realize in this paper is VM based on Java, and used by 600 of world's manufacturers and domestic firms including SK Telecom, LG Telecom and KVM. Movement expression animation contents are divided into 9 as seen in table 4-1. They are applied to 1000 of character movement expression, natural phenomena, dinosaur movement expression, 1000 of virtual shellfish museums, and 600 of virtual fossil museums.

4 Conclusion

Mobile Contents is likely to be extended to almost every sector of our life by adding mobile game, intranet, VOD, telemedicine to existing basic services, such as "phone ring tones', "MP3 download", "character download". Among other things, Mobile Contents will exceed PC game and on-line game, and existing technologies will be spurred to be mobilized. Also, only Mobile Contents that understand characteristics of mobile and intention of users can survive. Since Mobile Contents has risks in commercialization, mobilization will prevail for time being due to revival of existing games and animations, in order to minimize the risks. Recently, animations account for a large share in mobile phone or commercial advertizement. In addition, since these contents are sensitive to the trend, they may disappear unless they attract users' interest.

Although many domestic universities created animation-related majors in order to meet the demand for animation, they confine to hand work or editing animations due to lack of high-quality contents to manufacture animations.

This paper designed and realized more than 1000 animation models such as gestures of objects or men, natural phenomena and animals' movements so that students easily manufacture animations. Also it developed models that students can animate in an exciting manner. The next step will be to develop game manufacturing system by using guidelines of suggested algorithm.

References

[1] J. Latta, D. Orberg, A conceptual Virtual Reality Model, IEEE Computer Graphic and Application; vol4, No.1, pp23-20, Jan, 1994.

[2] JSR-184 "Mobile 3D Graphics API for JAVA TM2 Micro Edition", vol 1.0 Nov 2003.

[3] Ing-Ray Chen, Ngoc Anh Phan, and I-Ling Yen, "Algorithms foe supporting disconnected write operations for wireless web access in mobile client-server environments" IEEE Transactions' on Mobile Computing, Vol. 1, No. 1, pp. 46-58. Jan. 2002.

[4] Kah-Seng Chung, Yia fourg chen, "A versatile Digital Mobile Channel Simulation." Apcc 2002, pp10 14, Sep. 17. 2002.

[5] P.Debvec,"Randering Synthetic Objects into Real Scenes : Bridging Traditional and Image Based Graphics with Global Illumination and High Dynamic Range Photography", SIGGRAPH98, pp.189-198. 1998.

[6] T.Kanade *et al*, "Constructing Virtual Words Using Dense Stereo", ICCV98, Bombay, Indiay, p.3-10, JAN. 1998.

[7] I.EEE STANDARD FOR Imformation technology - Telecommumication and information exchange between systems - LAN/MAN specific Requirements, Jan, 2005

Brill Academic Publishers
P.O. Box 9000, 2300 PA Leiden
The Netherlands

Lecture Series on Computer
and Computational Sciences
Volume 4, 2005, pp. 1902-1905

Multi-layered Student Modeling in Game Environments

YoungMee Choi [1] MoonWon Choo, Jinwoo Choi,

Division of Multimedia, Sungkyul University
147-2, Anyang-8 Dong, Manan-Gu, Anyang-City, Kyunggi Province, Korea

Abstract: Many coaching systems in game environments have been developed in an attempt to achieve a student's discovery discovery learning. The coaching systems usually try to improve localized weak points of students based on their current behavior. This implies that the previous coaching systems cannot exactly capture student's learning process. Therefore, they cannot support the student-oriented interaction that is crucial to the provision of individualized instruction, which intelligent tutoring systems try to achieve. This paper has presented a new student modeling, multi-layered diagnose, which manipulates student's chronological knowledge in three different layers. In the three layers, student's knowledge is diagnosed according to the degree of ability, the degree of certainty, and the time, respectively. Using the three layered diagnoses, the eWEST can provide useful information for articulately instructing student's weak points, such as increasing rate of student's knowledge, degree of student's ability, student's learning deficiencies, and learning needs.

Keywords: multi-layered diagnose, student modeling, Intelligent Tutoring Systems, Student-oriented coaching, chronological records

1. INTRODUCTION

The spread of personal computer has brought students to play computer-based games during much of their free time. Games provide enticing problem-solving environments in which a student explores at will, free to create his own ideas for solving problems, and to invent his own strategies for utilizing the ideas[2]. While educational games are usually successful in increasing student engagement, they often fail in triggering learning. If the student does not have sufficient information to change his behavior as a result of the perceived error, then the student is in plateau[1]. At this place coach's interventions are needed. The interventions must be effective in discovering plateaus without destroying interest in the activity. The coach can provide proactive help tailored to the specific needs of each individual student.

Coach is one of the major paradigms in the existing intelligent tutoring systems which are mixed initiative tutor, diagnostic tutor, coach, microworld[4], and articulate coach. The coach observes student's performance and provides advices for helping the student to perform better by unobtrusively monitoring student's behavior, recognizing misconceptions and inefficient operations, and then offering pertinent advice about concepts and commands. Coach paradigm is best suited to problem-solving types of programs, such as simulations and games. Examples of such programs are WEST[3] and WUSOR. These coaching systems in game environment have been developed in an attempt to achieve student's discovery learning. Especially WEST is a computer game program that was originally designed for the PLATO Elementary Mathematics Project. The purpose of the game is to exercise arithmetic skills. It is the first computer coach system that assists a student to play the game "How the West was Won". The object of the game is to move a player across an electronic gameboard by the number of moves equal to the value of an algebraic expression that the student

---

[1]YoungMee Choi. Professor of the Sungkyul University, E-mail : choiym@sungkyul.edu

formulates. WEST uses "differential modelling" technique to diagnose student's behavior. The differential modelling technique compares student's performance with that of a computerized expert, and constructs a model of the differences. These differences suggest hypotheses about what the student does not know or has not yet mastered.

However, the previous coaching systems usually try to instruct localized weak points of students by explaining the difference between what the student is doing and what an expert would do in his place. This implies that the previous coaching systems cannot exactly capture student's learning process. Therefore, they cannot support the student-oriented interaction that is crucial to the provision of individualized instruction, which intelligent tutoring systems try to achieve. This paper has designed a "multi-layered student modeling" technique in game environment, eWEST, to diagnose the student's behavior by using his chronological records. The eWEST can articulately instruct student's weak points so that it can support student-oriented coaching environments. The content of the paper is arranged as follows. Chapter 2 describes the gaming situation; Chapter 3 describes the multi-layered student modeling; Chapter 4 describes the evaluation of eWEST; Chapter 5 discusses conclusions and future work.

2. THE GAMMING SIUATION

The game situation is as follows. In each turn, each player receives three numbers from the eWEST, which must be used in an arithmetic expression (using the operations addition, subtraction, multiplication, and division as well as parentheses) with the constraint that the same operator cannot be used more than once. The value of the expression is the number of spaces that the student moves along the board. To make the student's task more complicated than just making the biggest number, there are several kinds of special moves(BUMP, TOWN, SHORTCUT). The object of the game is to be the player to reach the goal. Figure 1 shows the gameboard and Figure 2 shows student's input interface.

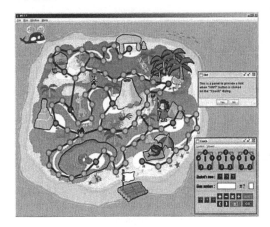 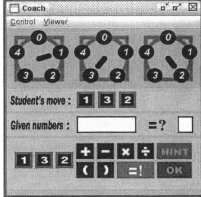

Figure 1: Gameboard   Figure 2: Student input interface

3. MULTI-LAYERED STUDENT MODELING

This section simulates a hypothetical game of eWEST, shows how the eWEST works, and demonstrates that the eWEST globally evaluats the student's behavior using the chronological knowledge.

For an example of the game, Talbe 1 shows two students's expression and expert's optimal expression at the same game situation. From the student's input and the optimal solution, the **eWEST** can abstract issues which indicate skills of the game and arithmetic knowledge, for exmaple, PARENTHESES, SUBTRACTION, EXPRESSION PATTERN, TOWN, SHORTCUT, BUMP, FORWARD, GAME STRATEGY, and etc. In the following, we show abstracted issues for the first three turns in Table 1.

Table 1 : Simulation of the game

Turn	Student	Given 3 number	Student's input	Optimal solution
1	A	1 3 2	(3+2)*1	(3+2)*1
	B	6 2 1	6+2-1	6+2-1
2	A	4 1 0	1-4+0	1-4+0
	B	3 2 7	7+2-3	7+2-3
3	A	1 3 7	(3+7)*1	3+7*1
	B	3 2 5	3*5+2	3*5+2
4	A	1 2 3	3+2-1	3+2-1
	B	4 2 1	(4+1)*2	(4+1)*2
5	A	1 2 1	(1+1)*2	(1+1)*2
	B	0 1 3	3+1-0	3+1-0
6	A	3 1 2	3-2+1	3-2+1
	B	3 2 0	3*2+0	3*2+0

...

turn 3 : Student A's input is abstracted by the **eWEST** which are $(a + b)* c$
EXPRESSION PATTERN, special move TOWN, ADDITION, MULTIPLICATION, and PARENTHESES issues.
Student B's input is abstracted by the **eWEST** which are $a * b + c$
EXPRESSION PATTERN, number order ORIGINAL, ADDITION, and SUBTRACTION issues.

...

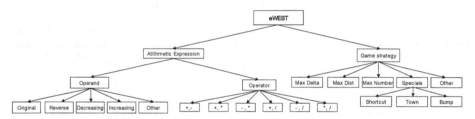

Figure 1. Issue Analysis of the eWEST

For the notational convenience, we use the following notations. Let each student's input data be S_i and let each recognized issues from S_i be $I_{i,j,k,l}$, where the subscript i indicates the temporal order, the subscript j indicates the type of issues, the subscript k indicates the degree of ability for the issue, and the subscript l indicates the degree of certainty for the issue

Whenever the student plays the game with self-training facility, the **eWEST** abstracts the student's response, diagnoses the student's behavior, and gives student-oriented prescriptions. In order to achieve the student-oriented coaching, The **eWEST** performs following steps.

Step 1: First layered student modeling
In the first layered diagnosis, issues are constructed comparing student's input data(S_i) with the optimal solution. The issues from student's behavior are partitioned seven groups by their degrees of ability: BEST, GOOD, FAIR, POOR, BAD, MISSING, MISCONCEPTION. In the recognized issues $I_{i,j,k,l}$, the value of subscript k is 1 when the degree of ablity is BEST, 2 when the degree of ablity is GOOD, and so on.
Step 2: Second layered student modeling
In the second layered diagnosis, the eWEST calculates certainties of the issues using Bayesian Network.
Step 3: Third layered student modeling
In the third layered diagnosis, student's learning processes are analyzed according to the time. In this step, the eWEST analyzes the change of the student's ability for each issue, represents student's global learning process using the record movement of issues, and calculates the increasing rate of student's knowledge according to the time. The eWEST also determine the student's learning preference, learning deficiencies, and learning needs.

step 4: Instruct student-oriented coaching
The three layered diagnoses provide many useful information about student's learning preference, student's learning deficiencies, and student's learning needs. From this information, the eWEST globally observe the progress of student's learning so that the eWEST can instruct effective student-oriented coaching

4. THE EVALUATION OF eWEST

The eWEST diagnoses student's behavior using the student's chronological knowledge. The advantages of using multi-layered student modelling are as follows.

1.Student's learning process could be globally observed. Since the chronological knowledge can be used to represent the whole process of student's behavior, the eWEST can observe the changing process of student's knowledge. This mechanism can represent the global learning process of student.
2.Diagnoses are careful and accurate. Consider the following situation. Suppose that some issues are obtained from the previous poor-behavior and these issues are also obtained from the recent good-behavior. In that case, the eWEST evaluates issues except the commonly obtained issues, as "delay". After that, the same issues are marked again as "delay", then eWEST ask the student some questions for taking more clear information.
3.Various coaching informations are available. Since the eWEST use the three layered diagnoses, the eWEST can provide various information at each diagnosis, such as increasing rate of student's knowledge, degree of student's ability, certainy of student's behavior, student's learning deficiencies, and learning needs.
4.Student-oriented coaching is available. The diagnosed student's learning process provides information for determining adaptable coaching strategies for the student's learning style. This mechanism can introduce the effective student-oriented coaching in the eWEST.

5. CONCLUSION AND FUTURE WORKS

We have implemented an adaptive coaching system, eWEST, in game environments using JAVA. The eWEST is a educational game to enhance student's arithmetic expression formulating skills. To globally diagnose student's learning behavior, our system manipulates student's chronological knowledge by the multi-layered student modeling. Our multi-layered student modeling provide useful information about student's learning preference, student's learning deficiencies, student's learning needs, and student's globally learning style. Using this information, student-oriented coaching environment can be supported.

To design student model more precisely, we need the assessments from a probabilistic student model using Dynamic Bayesian Networks[5], consult with elementary school math teachers in order to reflect proper knowledge context of the targeted domain knowledge of our game, and evaluate with the real student about the usability of the game eWEST.

Reference

[1] Adrian Y. Zissos and Ian H. Witten, "User modelling for a computer coach: a case study", Int. J. Man-Machine Studies, 23, pp. 729_750, 1985.
[2] M. When Does The Use Of Computer Games And Other Interactive Multimedia Software Help Students Learn Mathematics? NCTM Standards 2000 Technology Conference, 1998, Arlington, VA, U.S.A.
[3] Richard R. Burton and John Seely Brown, "An investigation of computer coaching for informal learning activities," Intelligent Tutoring Systems, D. Sleeman et al.(Eds.), New York: Acadenic Press, pp.79_98, 1982.
[4] Seymour Papert, "Microworlds: Transforming Education," Artificial Intelligence and Education Volume One, R. W. Lawler and M. Yazdani(Eds.), New Jersey: Albex Publishing Corporation, pp.79_94, 1987.
[5] Xiaohong Zhao, "Adaptive Support for Student Learning in Educational Games," *Thesis for the master degree*, The Univ. of British Columbia, 2002.